Grzimek's Animal Life Encyclopedia focuses on the diversity of animal life on Earth. Other books and collections of articles have attempted to deal with either ancient extinctions or the modern biodiversity crisis. These volumes unite these two lines of biological research into a single, stylistically and editorially unified work, providing in-depth discussions on the social, political, and economic implications of current extinction research findings.

Grzimek's Animal Life Encyclopedia
Extinction

• • • •

Grzimek's Animal Life Encyclopedia
Extinction

● ● ● ●

Volume 2

Norman MacLeod, Editor in Chief
J. David Archibald and Phillip S. Levin, Advisory Editors

● ● ● ●

GALE
CENGAGE Learning

Detroit • New York • San Francisco • New Haven, Conn • Waterville, Maine • London

Grzimek's Animal Life Encyclopedia

Extinction

Project Editor: Deirdre S. Blanchfield

Editorial: Jason Everett

Rights Acquisition and Management:
Margaret Chamberlain-Gaston

Composition: Evi Abou-El-Seoud

Manufacturing: Wendy Blurton

Imaging: John Watkins

Product Design: Kristine Julien

Product Manager: Douglas Dentino

Publisher: Peter Scott

For product information and technology assistance, contact us at
Gale Customer Support, 1-800-877-4253.
For permission to use material from this text or product,
submit all requests online at **www.cengage.com/permissions.**
Further permissions questions can be emailed to
permissionrequest@cengage.com

Volume 1 cover photograph reproduced by permission of F. Hart/The Bridgeman Art Library/Getty Images (picture of a dodo bird). Volume 2 cover photograph reproduced by permission of Louie Psihoyos/Corbis (picture of Archaeopteryx).

While every effort has been made to ensure the reliability of the information presented in this publication, Gale, a part of Cengage Learning, does not guarantee the accuracy of the data contained herein. Gale accepts no payment for listing; and inclusion in the publication of any organization, agency, institution, publication, service, or individual does not imply endorsement of the editors or publisher. Errors brought to the attention of the publisher and verified to the satisfaction of the publisher will be corrected in future editions.

LIBRARY OF CONGRESS CATALOGING-IN-PUBLICATION DATA

Grzimek, Bernhard. [Tierleben. English] Grzimek's animal life encyclopedia: exintction / Norman MacLeod, editor in chief; J. David Archibald and Phillip S. Levin, advisory editors.
  volumes cm.
  Includes bibliographical references and index.
  ISBN 978-1-4144-9067-0 (set : alk. paper) -- ISBN 978-1-4144-9068-7 (vol. 1 : alk. paper) -- ISBN 978-1-4144-9069-4 (vol. 2 : alk. paper)— ISBN 978-1-4144-9070-0 (ebook : alk. paper)
  1. Zoology—Encyclopedias. I. MacLeod, Norman, 1953- II. Archibald, J. David. III. Levin, Phillip S. IV. Title. V. Title: Animal life encyclopedia.

QL7.G7813 2013 Suppl.
590.3--dc23   2012049977

Gale
27500 Drake Rd.
Farmington Hills, MI, 48331-3535

ISBN-13: 978-1-4144-9067-0 (set)   ISBN-10: 1-4144-9067-4
ISBN-13: 978-1-4144-9068-7 (vol. 1)   ISBN-10: 1-4144-9068-2
ISBN-13: 978-1-4144-9069-4 (vol. 2)   ISBN-10: 1-4144-9069-0

This title is also available as an e-book.
ISBN-13: 978-1-4144-9070-0   ISBN-10: 1-4144-9070-4
Contact your Gale, a part of Cengage Learning sales representative for ordering information.

Printed in China
1 2 3 4 5 6 7 17 16 15 14 13

• • • • •

Editorial board

Contents

Preface

Like its conceptual converse—evolution—extinction was held to be a nonissue by most naturalists, scholars, clergy, and laypeople for the overwhelming majority of human history. As most religious doctrines teach that all species were created by one or more loving and benevolent deities and placed in this perfect world for humanity's instruction and use, the notion that such a power would, or could, allow any part of that creation to disappear was, to most people, (obviously) impossible. But over the course of the 1800s, and in a manner that was, in many ways, similar to the development of evolutionary theory, the notion of the impossibility of extinction fell prey to observations made during the exploration of new continents by Western scientists as a result of various voyages of discovery. In both cases the fact of extinction/evolution was accepted on the basis of the empirical evidence of patterns in nature long before the mechanisms responsible for those patterns were accepted. In both cases acceptance of extinction/evolution challenged the social, cultural, and political fabrics that prevailed in Western thought and sparked a reexamination of long-established societal systems. In both cases the acceptance of extinction/evolution also set off fierce debates within the scientific community over a set of "eternal questions" (see Gould 1977) involving (1) the nature of organic change (directional versus directionless), (2) the causes of organic change (externally forced versus internally channeled), and (3) the tempo of organic change (progressive and constant versus episodic and variable). But unlike evolution, whose primary initial evidence was found in the comparative anatomy of modern organisms, the initial evidence for extinction came from the past, specifically, the discovery of morphologically bizarre fossils whose anatomy corresponded to no known living species. Plainly, these fossils were the remains of once-living creatures, in some cases of gigantic proportions. Just as plainly, there was no proof that species exhibiting such features occurred in the living world. Had science simply not discovered these creatures yet? Did monsters really exist somehow, somewhere, in some remote and, as yet, undiscovered part of the globe? Or were these fossils evidence that extinction had indeed occurred in the past? Most importantly, if species had gone extinct in the past could extinction occur in the future, and what did that mean for the interpretation of the past, present, and future of the human species?

While literary allusions have been made to the idea of extinction since time immemorial, most scholars trace the first scientific identification of an extinct species to the French naturalist Georges Cuvier and his analysis of the so-called American *incognitum*, a fossil collected from the Ohio territory by French soldiers in 1739. After careful comparison of the fossil with modern proboscideans, Cuvier declared it to be a new species previously unknown to science in 1806 and speculated that such a large animal would require a large territory to survive and would likely be noticed both by indigenous peoples and foreign visitors. Since there had been no reports of any living proboscideans occurring in North America in living memory, and no living elephant skeletons were known to exhibit the characteristic morphologies of the American *incognitum*, Cuvier concluded that this elephant species was quite possibly extinct.

Cuvier's extinction-related investigations were not confined only to taxonomic questions. He, along with his colleague and coauthor, Alexandre Brongniart, became famous for advancing the hypothesis of revolutions in life— a concept that later came to be associated with the term *catastrophism*—in which sudden and geographically regional (perhaps global) extinctions were held to have occurred at various times in Earth history. Cuvier and Brongniart argued that these revolutions were driven by unknown processes that had the power to overwhelm creatures of the land and the sea, with their remains intermingled in deposits that preserved these remains as assemblages of fossil species that could not have occurred together in life.

Opposing this doctrine of revolutionary change was the idea that change in the natural world proceeded at a slow and stately pace, driven by the same processes naturalists observe happening everywhere on a day-to-day basis (actualism). An extreme version of this idea postulated that the history of life constituted a cyclic procession of the same, or nearly the same, types of organisms appearing, flourishing, declining, and disappearing in a pattern that had "no vestige of a beginning,—no prospect of an end" (Hutton 1788, p. 304). This is canonical statement of the theory that was known at the time as uniformitarianism (see Gould 1987). These theories were championed by James Hutton in Scotland and later by Charles Lyell and Charles Darwin in England. The resulting actualism/catastrophism debates were intense, acrimonious, and in no small part driven as much by political agendas as scientific data. While Cuvier and Brongniart's catastrophism enjoyed some popularity in the late 1800s and early 1900s, especially in continental Europe, British and American geologists came down firmly on the side of actualism/uniformitarianism, only mentioning catastrophism to deride it as a discredited theory in virtually all the standard historical geology textbooks. But as had always been insisted on by the supporters of catastrophism, the fossils themselves told a different story.

Meanwhile, throughout the 1800s naturalists had begun to become alarmed by a much more immediate form of extinction than the ancient events recorded in the fossil record, for it had begun to be noticed that some modern species were being driven to extinction, largely by the activities of humans. While many histories of extinction list the dodo as the first documented case of extinction, its demise was preceded by that of the aurochs (*Bos primigenius*), the wild ancestor of modern-day cattle. The aurochs was one of the few large mammals that escaped the Late Pleistocene extinction event. Prior to the arrival of *Homo sapiens* it ranged throughout Europe, much of Asia, and parts of the Middle East. The aurochs was domesticated at least twice between 10,000 and 8,000 years ago, once in the Near East and once in India (Rokosz 1995; Vuure 2005). These domestication events eventually resulted in the production of humpless (*Bos taurus*) and humped (*Bos indicus*) cattle, respectively—an interpretation that has been confirmed by independent mitochondrial DNA analyses (Edwards et al. 2010). Nonetheless, the aurochs began disappearing from its ancestral range in Roman times. By the 1500s the only known population was confined to the Polish Royal Forests west of Warsaw. The last aurochs known to science died in 1627 in Poland's Jaktorów Forest. The causes of the aurochs's extinction are held to be hunting, a narrowing of habitat resulting from the development of farming, climate changes, and diseases transmitted by domestic cattle (Rokosz 1995; Vuure 2002, 2005). After its demise, in relatively quick succession, a host of other species extinctions were documented, including the dodo (*Raphus cucullatus*, 1662), Steller's sea cow (*Hydrodamalis gigas*, 1768), the great auk (*Pinguinus impennis*, 1844), the passenger pigeon (*Ectopistes migratorius*, 1914), and the Tasmanian tiger (*Thylacinus cynocephalus*, 1930), to name but a few. Interestingly, many species extinctions have occurred following the introduction of legislation designed by conservationists and enacted specifically to help preserve the species.

While popular, economic, and political concern with extinction was growing through the late 1800s and early 1900s, paleontologists were engaged in a scramble to collect dinosaurs and enormous mammals, especially from the North American West. Led in the popular imagination by the famous professional rivalry between Othniel Charles Marsh (1831–1899) of Yale University and Edward Drinker Cope (1840–1897) of the Academy of Natural Sciences (Philadelphia), the purpose of this research program was to document the ancient natural history of North America and obtain specimens that could be displayed, for the amazement and education of the public, in the new natural history museums, including the American Museum of Natural History (New York), the Smithsonian Institution's National Museum of Natural History (Washington, DC), and the Carnegie Museum (Pittsburgh), in addition to those connected with

Marsh and Cope. Indeed, so impressive were these specimens that they became fetish objects for a form of both personal and national potlatch, such as Andrew Carnegie's gift of casts of the reconstructed skeleton of *Diplodocus carnegii* to the major European natural history museums (see Mitchell 1998). Suddenly, extinction and extinct species were at the center of Western science, education, culture, and politics.

But from a scientific perspective, dinosaurs and large mammals were hopeless subjects to use to study the dynamics of extinction events. There were too few of them preserved in the fossil record, and their remains were very difficult to date accurately. Instead it was to the humbler and far more speciose groups such as trilobites, brachiopods, bivalves, ammonites, and (later) microfossils that scientists turned to establish time relations between bodies of rock located at different places on Earth's surface. Patient taxonomic and stratigraphic research work by several generations of paleontologists had established such a detailed record of the comings and goings of ancient extinct species that an experienced biostratigrapher could often assign rock layers to a specific interval of geological time based on a fragment of a fossil that had been collected from them. Then, in the 1950s and 1960s Otto H. Schindewolf (1955, 1962) in Germany and Norman D. Newell (1963, 1967) in the United States turned this research program on its head and began to compare rates of extinction in the various geological time intervals. Both these researchers, but Newell (1963) in particular, were able to demonstrate that, contrary to the claims of the actualists/uniformitarians, the fossil record did record short time intervals of extremely high extinction rates, which Newell termed "mass extinctions." These were located at the ends of geological stages of the Cambrian, Ordovician, Devonian, Permian, Triassic, and Cretaceous. While Schindewolf regarded these data as vindicating catastrophism, Newell was more cautious, drawing a distinction between catastrophic-looking patterns in his stratigraphic data and the processes responsible for creating those patterns, which he regarded as being consistent with actualistic/uniformitarian theory. Still, the magnitude of the species extinctions implied by Schindewolf's and Newell's data was undeniable. Later, in the 1970s and 1980s, using a data set assembled by J. John Sepkoski Jr., David M. Raup (1979) showed that the end-Permian mass extinction could have involved the elimination of as much as 96 percent of all fossilizable species living in the Late Permian oceans. In a subsequent article published in 1982, Raup and Sepkoski used an advanced statistical analysis of the extinction intensities present in the latter's database to distinguish between the latter five of Newell's mass extinctions and the rest of the extinction record, which they regarded as being characterized by lesser, or "background," levels of extinction intensity.

While Newell's reconciliation of the mass extinction patterns existing in the fossil record with the processes accepted by actualists/uniformitarians was both elegant and logical, it was to receive its most stringent challenge in the 1980s by a discovery that was so unexpected and astounding that it shook the entire scientific community to its foundations. In an effort to develop a new tool to estimate the time interval represented by sediment layers, University of California, Berkeley, geologist Walter Alvarez, along with his father, the Nobel laureate physicist Luis Alvarez, and two other Berkeley colleagues, used a (then) new and highly precise instrument to measure small amounts of the rare earth element iridium that were present in sedimentary rock layers. Most of Earth's native iridium is located in its mantle and core. On occasion this deep iridium is found in igneous or metamorphic rocks as an ore body, having been carried from the deep Earth by various volcanic and/or tectonic mechanisms. But the iridium found in sediments largely comes from outer space in the form of meteoroids that burn up in Earth's atmosphere, causing a minute, but statistically constant, amount to fall in a gentle rain across Earth's surface. When attempting to use this source of iridium to estimate the time interval over which the uppermost Cretaceous sediments in Italy had been deposited, the Alvarez team ran across an unexpected result—a spike or anomaly in the concentration of iridium was found coincident with the boundary between Cretaceous and overlying Tertiary sediments. After confirming this observation was real, the team settled on an exciting interpretation: that a large comet or asteroid had collided with Earth at the time of the transition between the Cretaceous and Tertiary (now Paleogene) time periods—a transition that was marked by one of Newell's mass extinctions. The implications were obvious. Here was an unquestionably catastrophic mechanism that not only resulted in copious and widespread extinctions, but literally changed the course of evolution by eliminating ecologically entrenched groups (e.g., dinosaurs, ammonites) and opening up this ecological space for occupation by groups (e.g., mammals, fish) that, up until that point, had been excluded from entering certain ecological spaces by these incumbents. Perhaps asteroid impacts were responsible for other mass extinctions as well. Perhaps they were responsible for them all!

The 1980 article by the Alvarez team detailing the discovery and interpretation of the Cretaceous–Tertiary iridium anomaly caught the attention of both the scientific community and the public at large. Here was a link between two of the most popular scientific topics around—dinosaurs and outer space—that, in one fell swoop, appeared to overturn more than a century of actualistic/uniformitarian dogma that had underpinned not only geology, but much of science in general. The public loved it. The media loved it. The scientific community loved it. But by and large the paleontological community remained skeptical for a variety of reasons.

Of course, the ensuing controversy raised the public, and the media's, interest in the topic of extinction to even higher levels. Had a rock fallen out of the sky and killed the dinosaurs? Could it happen again, this time to humans? If so, what steps should humanity take to protect itself from this (suddenly) tangible threat from outer space? While few geologists today seriously dispute the Alvarez team's interpretation of the iridium anomaly as resulting from an asteroid impact, debate continues with regard to whether this impact was the sole cause or just a contributing factor to the end-Cretaceous extinction event and with regard to the role asteroid impacts may have played in causing other extinction events.

Drawing support from the paleontological data on mass extinctions, modern-organism ecologists began calibrating the loss of modern biodiversity against the standard set by the fossil record of extinction. Perhaps unsurprisingly, their results showed that, while the rates of documented extinctions fell well within the intensity range of background-level events, the historical interval over which these losses occurred was vanishingly short by geological standards. In other words, if the frequency of modern species extinctions is extrapolated into the future, levels of extinction intensity comparable to those of paleontological mass extinction would be achieved in surprisingly short intervals of time (hundreds of years or so) for many groups. This work has led to the modern biodiversity crisis being labeled a "sixth mass extinction"— one that is being driven not by any natural catastrophe such as an asteroid impact or a volcanic eruption, but by the thoughtless use of natural resources by humankind (e.g., the release of greenhouse gases that alter Earth's climate, poor land-use practices that destroy species' habitats). Proponents of the sixth extinction concept (e.g., Leakey and Lewin 1995; Wilson 2002) argue that continued exploitation of the planet in the manner to which Western societies have become accustomed is not only unethical and morally indefensible, it places human life in jeopardy in the form of driving species that may be useful to humans in the future extinct and by degrading the planet's environment to the point at which even humans may not be able to survive. Both the extreme economic development and extreme biodiversity conservation positions along the opinion spectrum seem untenable to most people. Nonetheless, debate continues as to where the most appropriate balance can be struck that preserves the greatest proportion of natural habitat and species diversity while at the same time causing minimal economic, social, and political disruption.

Grzimek's Animal Life Encyclopedia is a 17-volume set, covering the entire animal kingdom from protozoa to mammals. Edited originally by Bernhard K.M. Grzimek, the noted Silesian-German director of the Frankfurt Zoological Garden (1945–1974), zoologist, author, editor, and conservationist, and published in German in 1968, all 11 original volumes were translated into English in the 1970s and went on to become a standard reference work in the fields whose topics they cover. This two-volume addition to the series is designed to be a companion and complement to the *Evolution* volume, which was published in 2011.

In these volumes, which are the first of their kind on the topic of extinction, J. David Archibald, Philip Levins, and I, along with the authors of the entries, have tried to provide a comprehensive overview of the many disparate topics covered by extinction research. Other books and collections of articles have attempted to deal with either ancient extinctions or the modern biodiversity crisis. But to our knowledge this is the first work that attempts to unite these at once obviously related but also disparate lines of biological research with summaries of the social, political, and economic implications of extinction research findings into a single, stylistically and editorially unified work. As the readers of these entries will learn, there is and will likely continue to be a healthy diversity of opinion regarding the importance of particular data to particular issues in the field of extinction research, the interpretations of those data, and the priorities different researchers assign to the data that remain to be collected and interpreted. Regardless, what comes across most vividly to me after reading through these contributions is the passion that the idea of extinction invokes in scientists, and indeed in people generally. There is something undeniably tragic about the concept, notwithstanding the fact that extinction, along with natural selection, is one of the most common processes in nature. The world today, and the history of life on planet Earth, would be very different if no species had ever become extinct. As will also be appreciated by reading the articles collected together here, extinction is a topic in which the results of scientific research often have a direct bearing on cultural practices, political policy, and legislation. At their heart, however, many extinction-related phenomena remain profound mysteries, which is perhaps the most intriguing, as well as the most frustrating, aspect of the topic. Unlike evolution, ecology, and ethology, which are about understanding what patterns exist in the natural world and what processes brought them into existence, extinction research is about what once was part of the natural world but is now absent from it and the causes of that absence. This fact alone makes the study of extinction especially challenging and eternally captivating for everyone—constituting a fascination that itself has "no vestige of a beginning and no prospect of an end." Long may it continue to be so.

In closing I would like to express my profound appreciation and gratitude to my associate editors, J. David Archibald and Philip Levins; our project editor, Deirdre S. Blanchfield; and our product manager, Douglas Dentino, as well as to the authors of the entries and to all the Gale Cengage staff for all the work they have done for, and the talent they have brought to bear on, this project. I can only hope the result that you, dear reader, now holds in your hands is viewed by all those who made this project happen with as deep a sense of satisfaction and accomplishment as I have for both the project and its product.

London, England, 2013
Norman MacLeod
Editor in Chief

Resources

Books

Gould, Stephen Jay. "Eternal Metaphors of Paleontology." In *Patterns of Evolution as Illustrated by the Fossil Record*, edited by A. Hallam, 1–26. Amsterdam: Elsevier, 1977.

Gould, Stephen Jay. *Time's Arrow, Time's Cycle: Myth and Metaphor in the Discovery of Geological Time*. Cambridge, MA: Harvard University Press, 1987.

Leakey, Richard, and Roger Lewin. *The Sixth Extinction: Patterns of Life and the Future of Humankind*. New York: Doubleday, 1995.

Mitchell, W.J.T. *The Last Dinosaur Book: The Life and Times of a Cultural Icon*. Chicago: University of Chicago Press, 1998.

Newell, Norman D. "Revolutions in the History of Life." *Uniformity and Simplicity: A Symposium on the Principle of the Uniformity of Nature*, edited by Claude C. Albritton Jr., 63–91. Geological Society of America Special Paper 89. Boulder, CO: Geological Society of America, 1967.

Vuure, Cis (T.) van. *Retracing the Aurochs: History, Morphology, and Ecology of an Extinct Wild Ox*. Sofia, Bulgaria: Pensoft, 2005.

Wilson, Edward O. *The Future of Life*. New York: Knopf, 2002.

Periodicals

Alvarez, Luis W., Walter Alvarez, Frank Asaro, and Helen V. Michel. "Extraterrestrial Cause for the Cretaceous–Tertiary Extinction." *Science* 208, no. 4448 (1980): 1095–1108.

Edwards, Ceiridwen J., David A. Magee, Stephen D.E. Park, et al. "A Complete Mitochondrial Genome Sequence from a Mesolithic Wild Aurochs (*Bos primigenius*)." *PLoS ONE* 5, no. 2 (2010): e9225. Accessed January 7, 2013. doi:10.1371/journal.pone.0009255.

Hutton, J. "Theory of the Earth; or an Investigation of the Laws Observable in the Composition, Dissolution, and Restoration of Land Upon the Globe." *Transactions of the Royal Society of Edinburgh* 1 (1788): 209–304.

Newell, Norman D. "Crises in the History of Life." *Scientific American* 208, no. 2 (1963): 76–93.

Raup, David M. "Size of the Permo-Triassic Bottleneck and Its Evolutionary Implications." *Science* 206, no. 4415 (1979): 217–218.

Raup, David M., and J. John Sepkoski Jr. "Mass Extinctions in the Marine Fossil Record." *Science* 215, no. 4539 (1982): 1501–1503.

Rokosz, Mieczyslaw. "History of the Aurochs (*Bos taurus primigenius*) in Poland." *Animal Genetics Resources Information* 16 (1995): 5–12.

Schindewolf, Otto H. "Die Entfaltung des Lebens im Rahmen der geologischen Zeit." *Studium Generale* 8 (1955): 489–497.

Schindewolf, Otto H. "Neokatastrophismus?" *Deutsch Geologische Gesellschaft Zeitschrift Jahrgang* 114 (1962): 430–445.

Vuure, Cis (T.) van. "History, Morphology, and Ecology of the Aurochs (*Bos primigenius*)." *Lutra* 45, no. 1 (2002): 1–16.

Contributing writers

J. David Archibald
Department of Biology
San Diego State University

Nan Crystal Arens
Department of Geoscience
Hobart & William Smith Colleges

Catherine Badgley
Department of Ecology & Evolutionary
    Biology; Museum of Paleontology
University of Michigan

Anthony D. Barnosky
Department of Integrative Biology and
    Museums of Paleontology and
    Vertebrate Zoology
University of California, Berkeley

Paul M. Barrett
Department of Earth Sciences
The Natural History Museum

Roger B. J. Benson
Department of Earth Sciences
University of Oxford

Peter J. Bowler
School of History and Anthropology
Queen's University, Belfast

Charlotte Boyd
School of Aquatic and Fishery Sciences
University of Washington

Evan R. Buechley
PhD Candidate, Department of Biology,
    University of Utah
University of Oxford

David A. Burney
National Tropical Botanical Garden

James T. Carlton
Williams College–Mystic Seaport
    Maritime Studies Program

Clark R. Chapman
Department of Space Sciences
Southwest Research Institute

Amy E. Chew
Department of Anatomy
Western University of Health Sciences

Per Christiansen
Department of Biotechnology,
    Chemistry, and Environmental
    Engineering
University of Aalborg

Euan N.K. Clarkson
Professor Emeritus of Palaeontology,
    School of Geosciences
University of Edinburgh

Margaret Clegg
Department of Earth Sciences
Natural History Museum London

Daryl Codron
Florisbad Quaternary Research
    Department, National Museum,
    Bloemfontein, South Africa
Clinic for Zoo Animals, Exotic Pets
    and Wildlife, University of Zürich,
    Switzerland

Dalia A. Conde
Institute of Biology
University of Southern Denmark

Whitney P. Crittenden
The Nature Conservancy

Samuel A. Cushman
Rocky Mountain Research Station,
    Flagstaff, Arizona
U.S. Forest Service

Brian Czech
Virginia Tech, National Capitol Region,
    Arlington, Virginia
Center for the Advancement of the
    Steady State Economy, Arlington,
    Virginia

José Maria Cardoso da Silva
Conservation International

Mark A. Davis
Department of Biology
Macalester College

John E. Fa
Durrell Wildlife Conservation Trust/
Imperial College London

Marina García-Llorente
Social-Ecological Systems Laboratory,
    Department of Ecology
Universidad Autónoma de Madrid,
    Spain

J. Whitfield Gibbons
Savannah River Ecology Lab
University of Georgia

Felix M. Gradstein
Geology Museum
University of Oslo

Donald K. Grayson
Department of Anthropology and
    Quaternary Research Center
University of Washington

Anne D. Guerry
Lead scientist, Natural Capital Project
Stanford University

Oyvind Hammer
Geology Museum
University of Oslo

David A. T. Harper
Department of Earth Sciences
Durham University

Noel A. Heim
Department of Geological &
 Environmental Sciences
Stanford University

Gene S. Helfman
 School of Ecology
University of Georgia

C. Michael Hogan
President, Lumina Technologies Inc.

Chris Johnson
School of Zoology,
University of Tasmania

Peter Kareiva
Department of Environmental Studies
 and Sciences
Santa Clara University

Stephen L. Katz
Research Department
Channel Islands National Marine
 Sanctuary, National Oceanic and
 Atmospheric Administration

Andrew C. Kitchener
Principal Curator, Vertebrate Biology
National Museums Scotland

Spencer G. Lucas
New Mexico Museum of Natural
 History and Science

Eimear Nic Lughadha
Royal Botanic Gardens, Kew

Marcin Machalski
Institute of Paleobiology
Polish Academy of Sciences, Warszawa,
 Poland

Norman MacLeod
The Natural History Museum

Berta Martín-López
Social-ecological systems laboratory,
 Department Ecology
Universidad Autónoma de Madrid

George R. McGhee
Department of Earth & Planetary
 Sciences
Rutgers University

Alistair J. McGowan
School of Geographical and Earth
 Sciences
University of Glasgow

W. Scott McGraw
Department of Anthropology
The Ohio State University

Michael L. McKinney
Earth & Planetary Sciences
University of Tennessee,
 Knoxville

Darren Naish
University of Southampton

Jeffery C. Nekola
Biology Department
University of New Mexico

Gabi Ogg
West Lafayette, IN

Maria Rita Palombo
 Dipartimento di Scienze della Terra
Università di Roma "La Sapienza"

P. David Polly
Department of Geological Sciences
Indiana University

Donald R. Prothero
Department of Vertebrate
 Paleontology
Natural History Museum of Los
 Angeles County

Amanda P. Rehr
NOAA Northwest Fisheries Science
 Center

Mikhail Rogov
Geological Institute of Russian
 Academy of Sciences

Robert Rozzi
Dipartimento di Scienze della Terra
Università di Roma "La Sapienza"

Peter M. Sadler
Department of Earth Sciences
University of California Riverside

Leah H. Samberg
The Nature Conservancy

Stuart A. Sandin
Center for Marine Biodiversity and
 Conservation, Scripps Institution of
 Oceanography
University of California, San Diego

M. Sanjayan
The Nature Conservancy

R. Paul Scofield
Senior Curator, Department of
 Vertebrate Zoology
Canterbury Museum

Kevin E. See
Integrated Status & Effectiveness
 Monitoring Program
NOAA Fisheries

Cagan H. Sekercioglu
Department of Biology
University of Utah

Stephen Self
Visiting professor, Department of Earth
 and Environmental Sciences
The Open University, UK

Alan H. Simmons
Department of Anthropology
University of Nevada, Las Vegas

Peter Stoett
Loyola Sustainability Research Centre
Concordia University, Montreal

Nigel E. Stork
Environmental Futures Centre, Griffith
 School of Environment
Griffith University, Australia

Anthony John Stuart
Biological and Biomedical Sciences
Durham University, UK

Yadong Sun
China University of Geosciences

Kostas A. Triantis
Department of Ecology and Taxonomy,
Faculty of Biology
National and Kapodistrian University of
Athens, Athens

Samuel T. Turvey
Institute of Zoology
Zoological Society of London

Tommy Tyrberg
Lake Tåkern Field Research Station

Charlie Underwood
Birkbeck College

Eric J. Ward
Northwest Fisheries Science Center,
NMFS, NOAA

Peter D. Ward
Dept of Biology
The University of Washington

Stephen R. Westrop
Sam Noble Oklahoma Museum of
Natural History, and School of
Geology and Geophysics
University of Oklahoma

Quentin D. Wheeler
Virginia M. Ullman Professor of Nat-
ural History and the Environment
School of Life Sciences, Arizona State
University

Steven M. Whitfield
School for Field Studies
Bocas del Toro, Panama

Robert J. Whittaker
School of Geography and the
Environment

Paul B. Wignall
School of Earth and Environment
University of Leeds

Jamie R. Wood
Landcare Research, New Zealand

Bernd Würsig
Marine Biology,
Texas A&M University

Tara J. Zamin
Department of Biology,
Queen's University

• • • • •

The end-Jurassic extinction

Consideration of the end-Jurassic extinction should start with a brief introduction to the Jurassic–Cretaceous (J–K) boundary problem. The single-system J–K boundary is still not defined by a global boundary stratotype section and point (GSSP). As of 2012, the section and key events were under discussion.

In the middle of the nineteenth century, scientists placed the J–K boundary at the boundary between the marine Portlandian and the non-marine Purbeckian stages (or the marine Neocomian in the Alps; see Wimbledon 2008). Later, after the introduction of the Valanginian and Berriasian stages in the lowermost Cretaceous, the J–K boundary was set at the base of the Berriasian or Valanginian stages. In the higher latitudes this boundary was traditionally placed between the Volgian and Ryazanian (the Boreal Berriasian) stages. Over the course of the late twentieth and early twenty-first centuries, the correlation of the Upper Volgian and Lower Berriasian, albeit based on indirect evidence, became accepted widely (Zeiss 1983; Sey and Kalacheva 1999; among others). For example, this correlation was used in databases, such as the *Compendium of Fossil Marine Animal Genera* (2002) by J. John Sepkoski Jr.,and in numerous publications referencing the Paleobiology Database. Nevertheless, paleomagnetic data (Houša et al. 2007) along with new ammonite records from the Ryazanian of the Russian Platform (Mitta 2007; Mitta and Sha 2009) suggest that the Volgian more nearly corresponds to the Tithonian and that only the uppermost zone of the Volgian should be correlated with the lowermost Berriasian. It should be noted that the correlation of the Upper Volgian and Lower Berriasian superimposed Late Tithonian extinction rate likely resulted from a strong turnover in boreal molluscan faunas near to the Middle to Late Volgian transition and a decrease in Lower Berriasian origination rates.

Mass extinction at the J–K boundary was first recognized in a 1984 study by David M. Raup and Sepkoski on the basis of an analysis of ranges of marine invertebrate families. This conclusion was later confirmed at the genus level (Raup and Sepkoski 1986). The nature and significance of this extinction event have been controversial ever since.

When analyzing bivalve diversity through the Jurassic, Anthony Hallam concluded in a 1976 study that increase of the generic extinction rate at the end of the period correlated with a regional marine regression. In studies in 1986 and 1989, Hallam contended that the end-Jurassic extinction in marine invertebrates had occurred on a regional, but not a global, scale. Only bivalves in Western Europe showed a high rate of species turnover within the Middle Volgian, whereas the brachiopods, foraminifera, and coccoliths showed no particularly marked change across the J–K boundary. Also, this extinction was observed only in those regions that showed a shallowing trend, whereas in areas characterized by end-Jurassic transgression the bivalve fauna showed no change upward in the sequence (Hallam 1986). The end-Jurassic extinction was later recognized in non-marine tetrapods (Benton 1995; Barrett and Upchurch 2005; Butler et al. 2011), marine vertebrates (Kriwet and Klug 2008; Benson et al. 2010 and references therein; Young et al. 2010), marine benthic organisms (Kiessling and Aberhan 2007a, 2007b), and radiolarians (Vishnevskaya and Basov 2007; Vishnevskaya and Kozlova 2012). Among boreal ammonites, a so-called Late Volgian crisis has been shown to have occurred (Mitta and Bogomolov 2007; Mitta and Sha 2009; Rogov et al. 2010), but in this case the J–K boundary does not coincide with any significant changes in ammonite faunas.

Some previously suggested Tithonian extinction events (e.g., the ichthyosaur extinction) now should be reappraised, because Tithonian ichthyosaur genera have been found in the Lower Cretaceous (Fischer et al. 2012). Other groups of marine reptiles (e.g., metriorhynchid crocodiles) show strong drops in diversity into the Early Tithonian (Young et al. 2010).

Hallam and Paul B. Wignall, in their 1997 book *Mass Extinctions and Their Aftermath*, along with results published in 1995 by Mike J. Benton showed that the end-Jurassic extinction was much more significant for terrestrial than for marine organisms, the records of which have been biased by the so-called Lagerstätten effect—that is, the presence of many last-appearance datums of taxa in localities characterized by exceptional preservation (e.g., the Kimmeridge Clay Formation in England and the Solnhofen Limestone in southwestern Germany). Nevertheless, the strong fall in diversity in sauropod dinosaurs (Barrett and Upchurch 2005; Butler et al. 2011) could reflect real processes, such as an increasing influence of predators on this relatively small and specialized fossil group.

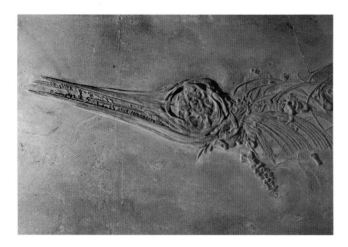

Ichthyosaur fossil. © Natural Visions/Alamy.

Biases in the data

There are different sources of bias inherent in J–K boundary paleodiversity studies. The end-Jurassic extinction exhibits a clear relation between the quantity of preserved sedimentary rock and taxonomic diversity (Smith and McGowan 2007). The strong influence of the European fossil record on global biodiversity curves should also be recognized from the good level of correspondence of the Mesozoic extinction events with fossil ranges derived from the French naturalist Alcide d'Orbigny's publications of the mid-nineteenth century (see Ruban 2005). Additional problems with this extinction are related to the possible influence of the aforementioned Lagerstätten effect and commonly encountered regressive facies in one of the most studied regions in Europe. Bias of database sources and problems with correlation (especially the boreal–Tethyan correlation) also affect the estimates of this extinction's magnitude and scope. Nevertheless, generic richness curves, based on Sepkoski's compendium of fossil marine genera and the Paleobiology

Database, compiled by Alexander V. Markov (2009) show a clear drop in diversity in the Early Berriasian for both the genera of mollusks and of all marine animals (see Figure 1). As is also indicated by the analysis of taxic first and last appearance datums, the Late Tithonian is characterized by peaks in appearance of new taxa and especially in extinction intensity (see Figure 2). In ammonites such high levels in both extinction and the origination rate reflect the existence of many short-lived taxa and a high rate of diversification.

Geological settings around the world

The Jurassic–Cretaceous boundary is characterized by a complex set of global and regional events that could be at least partially responsible for the observed biotic changes. There are three bolide impact events dated as occurring near to the Jurassic-Cretaceous transition. These are the Mjølnir impact, which occurred on the Barents Sea shelf (Dypvik et al. 2010); the Morokweng impact in South Africa (Reimold et al. 2002); and the Gosses Bluff impact in Australia (Haines 2005). The Mjølnir event is well dated by buchiid bivalves and ammonites and occurs very close to the Volgian–Ryazanian boundary. This event could be responsible for the rarity of occurrences of the Volgian–Ryazanian transitional beds in the Barents Sea shelf and Svalbard (Wierzbowski et al. 2011), but it is not

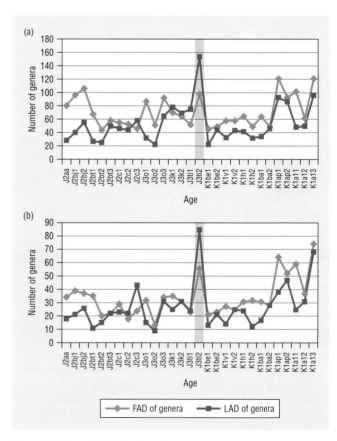

Figure 2. FAD and LAD of all marine genera (a) and molluscan genera (b) through the Middle Jurassic to latest Early Cretaceous, based on Markov's (2009) compilation of Sepkoski's compendium and the Paleobiology Database. Late Tithonian is marked by grey stripe. Reproduced by permission of Gale, a part of Cengage Learning.

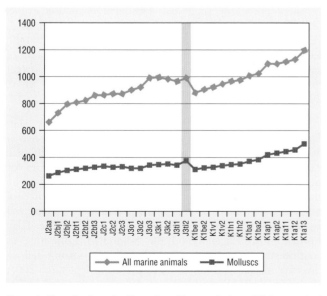

Figure 1. Generic diversity. Reproduced by permission of Gale, a part of Cengage Learning.

associated with any significant faunal changes except a distinct but short-term bloom of the prasinophycean alga *Leiosphaeridia*, which has been recognized in post impact deposits (Dypvik et al. 2010).

Zircons recovered from the quartz norite of the Morokweng structure have yielded U–Pb ages of 145 ± 0.8 million years ago (MYA), and biotites have provided Ar–Ar ages of 144 ± 4 million years ago (Hart et al. 1997). These dates correspond approximately to the base of the Cretaceous (Mahoney et al. 2005). The Australian Gosses Bluff crater has not been dated precisely. Soon after its identification it was dated as Early Cretaceous (133 ± 3 MYA; see Milton et al. 1972), but its age was later redetermined as earliest Cretaceous but close to the J–K boundary (142.5 ± 0.8 MYA; see Haines 2005). The diameters of all these impact structures are relatively small: less than 50 miles (80 km) for the Morokweng structure, about 15 miles (24 km) for the Gosses Bluff crater, and 25 miles (40 km) for Mjølnir.

There are no unusually large global sea-level oscillations near the J–K boundary (Snedden and Liu 2010). The so-called Portland–Purbeck regression of northwestern Europe had a strong impact on the European fossil record (including the Russian Platform), but has been recognized as having been of limited regional extent. In addition to Europe, J–K boundary sea-level fall is recognized in the Kutch region of India (Bardhan et al. 2007). By contrast, there are many regions that show sea-level rises as having occurred during the J–K

transition (e.g., western Siberia, northern Siberia, Barents Sea, northeastern Asia, Far East, Gulf of Mexico, Himalayas).

Late Jurassic and earliest Cretaceous climates have been characterized as warm and equable. The Middle to Late Jurassic transition was marked by regional cooling in Europe (Wierzbowski and Rogov 2011), while the Late Jurassic showed gradual warming (Dera et al. 2011). Tithonian cooling, indicated by changes in bivalve δ^{18}O values, is not recognized in isotope data derived from belemnite rostra (Dera et al. 2011). In boreal areas the whole Late Jurassic is characterized by a gradual increase in temperatures (Price and Rogov 2009; Žák et al. 2011), accompanied by aridization (Abbink et al. 2001). The beginning of high-latitude cooling at this time—recognized by the shift in δ^{18}O values (Price et al. 2000), the appearance of tillites (Alley and Frakes 2003), and glendonite occurrences—might be correlated with the latest Berriasian/latest Ryazanian to earliest Valanginian. Nevertheless, the significance of the oscillations in δ^{18}O and δ^{13}C values around the J–K boundary remains obscure (Föllmi 2012). In high-latitude sites, including many boreal areas, from the Norwegian Sea in the west to the Lena River Basin in the east, as well as in austral regions such as Antarctica and the Argentine Neuquén Basin, laminated organic-rich facies are very typical in the J–K transition (Föllmi 2012).

Paleoceanographic changes in the Tethys occurred in the Late Tithonian. During this time Ammonitico Rosso facies

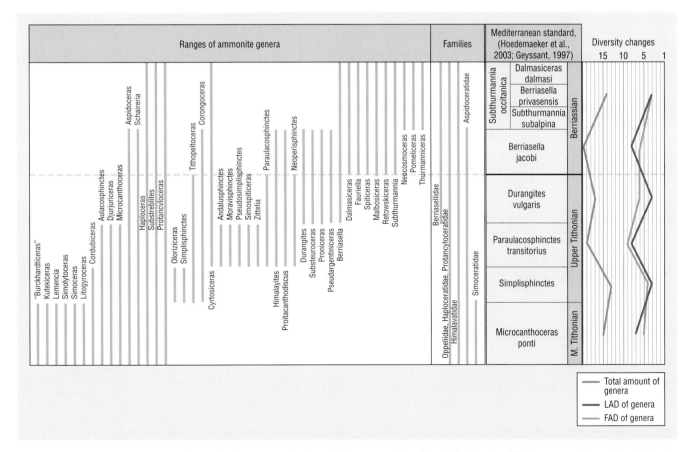

Figure 3. Ammonite ranges and diversity oscillations across the Jurassic–Cretaceous boundary in the Western Mediterranean. Base of the Berriasian (= base of the *Berriasella jacobi* Zone) marked by grey dotted line (see also Figures 4 and 6). Reproduced by permission of Gale, a part of Cengage Learning.

began to disappear, and the onset of pelagic white nannoconid-rich limestones with calpionellids have been dated to the same period (Cecca 1997). These changes coincided with a turnover in nannofossil assemblages: The "*Conusphaera* world" of the Middle Tithonian changed to the "*Nannoconus* world," which began in the Berriasian (Tremolada et al. 2006).

Ammonites case study

Ammonites are among the most intensively studied fossils from the J–K transition. Because this group is characterized by rapid evolution and occurrence in a wide spectrum of environments, it provides the most accurate biostratigraphic dating in comparison with other fossil groups. Even for Middle to Late Volgian transition beds, which show strong faunal provincialism, accurate zonal and infrazonal Panboreal correlations can be made by focusing on ammonites (Rogov and Zakharov 2009).

Latest Jurassic Tethyan ammonites are characterized by relatively high rates of diversification. Nearly half of Late Tithonian ammonite genera in Western Europe exhibit relatively small stratigraphic ranges (one or two zones). Such short-ranged genera were especially diverse within the uppermost two biozones of the Tithonian stage (see Figure 3). Thus,

there are many Late Tithonian ammonite genera (17 or half of the Late Tithonian genera) that never cross the Tithonian–Berriasian boundary; except for three genera, all ammonites disappear before the boundary horizon. Ammonite diversity in southwestern Europe grew constantly during the whole Late Tithonian (see Figure 3). Only one ammonite family disappears during the Late Tithonian. In contrast to the diversity pattern in Mediterranean and sub-Mediterranean ammonites, ammonites from the Panboreal Superrealm show a decrease in species richness during the latest Middle Volgian and the whole Late Volgian (see Figure 4). This decrease coincides with abrupt changes in ammonite morphology. In the Russian Platform, this event has been called the "Late Volgian crisis" (Mitta and Bogomolov 2007; Mitta and Sha 2009) and is believed to have been related to sea-level fall. Sea-level regression has also been invoked as the primary extinction driver for latest Jurassic bivalves (Hallam 1986—but see Zakharov and Yanine 1975, a study showing the proximity of latest Jurassic and earliest Cretaceous bivalves in Siberia and the southern USSR on the generic level), neoselachian fishes (Kriwet and Klug 2008), and radiolarians (Vishnevskaya and Kozlova 2012). Nevertheless, analysis of boreal ammonite assemblages (see below) has revealed that the significant faunal changes encountered in those areas during the J–K transition are also associated with a continuous sea-level rise.

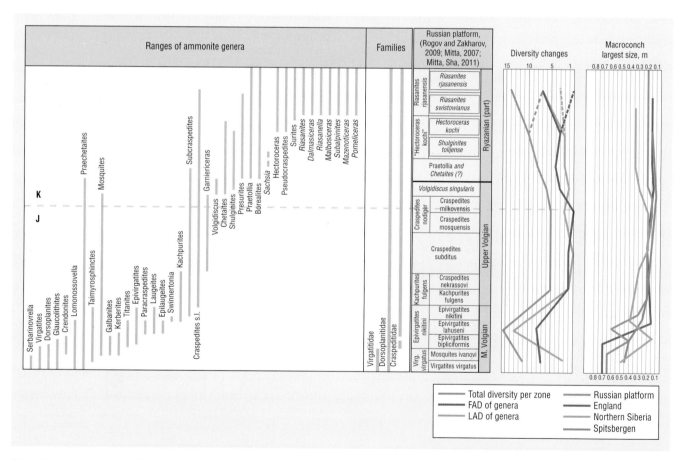

Figure 4. Ammonite ranges, diversity, and shell size oscillations across the Jurassic–Cretaceous boundary in the Panboreal Superrealm (after Rogov et al., 2010, with minor changes). Immigrant taxa of the Tethyan origin in the lowermost Cretaceous are written by italics. Dotted lines in graph showing changes in ammonite diversity belong to biodiversity changes without immigrant taxa of Mediterranean origin. Reproduced by permission of Gale, a part of Cengage Learning.

The uppermost Middle Volgian is characterized by remarkable changes in Boreal ammonite faunas. Large-sized and well-ribbed ammonites abruptly change to small-sized relatively smooth ammonites (see Figure 5) characterized by simplified sutures (see Figure 6). These changes were especially strong in England, where assemblages with "Portlandian giants" are followed by those that include

Figure 5. Typical large-sized and well-ribbed Middle Volgian dorsoplanitid ammonite *Epivirgatites* sp, (left, field photo, Gorodischi section, Volga river, Russia) and small smooth-shelled craspeditid ammonite (this is nearly full-growth *Craspedites okensis* (d'Orb.), one of the biggest (diameter ~4.5 in [~12 cm]) ammonite from the Late Volgian of the Russian Platform. Courtesy of Mikhail A. Rogov.

small-sized *Swinnertonia*. Similar but more gradual changes have been recognized in other boreal areas.

In the Russian Platform the uppermost subzone of the *Epivirgatites nikitini* zone is characterized by the co-occurrence of small-sized dorsoplanitids with early craspeditids belonging to two lineages (*Subcraspedites* and *Kachpurites*; see Rogov and Zakharov 2009). Dorsoplanitid ammonites disappear here at the end of the Middle Volgian. The same changes in ammonite faunas can be recognized in Svalbard (Rogov 2010; Wierzbowski et al. 2011). The same features, that is, a decrease in shell size, as well as a smoothness of ribbing, can be recognized in J–K ammonite faunas of northern Siberia.

The peak in ammonite richness and extinction in the latest Middle Volgian *Epivirgatites nikitini* zone (see Figure 4) corresponds partially to a very quick diversification in dorsoplanitid ammonites and partially to the strong paleo-biogeographical segregation. The maximum ammonite richness from the single bed/biohorizon in the latest Middle Volgian does not exceed six genera. During the Middle to Late Volgian transition, ammonite diversity in the whole pan-boreal superrealm decreased significantly. Except for the uncommon records of phylloceratid and lytoceratid ammonites in northern Siberia (excluded from this analysis because of their persistent records through the whole Late Jurassic and Neocomian), ammonite diversity through the Late Volgian fluctuated from one to three genera.

Such low generic richness was accompanied by extremely low extinction and origination rates (see Figure 4). Only during the beginning portion of the Ryazanian was there marked diversification in craspeditid ammonites. At the same time ammonite diversity in the Russian Platform increased significantly due to mass immigration of Mediterranean berriasellid and himalayitid ammonites. The maximal size of Boreal ammonites also began to increase during the Early Ryazanian. The dorsoplanitid to craspeditid transition was marked by the independent appearance of ammonites with craspeditid morphologies within at least three semi-isolated basins: northwestern Europe, the Russian Platform, and the Arctic. Moreover, some characteristically craspeditid features (such as small size, smooth ribbing, and specific types of the septal suture) appeared mosaically within different dorsoplanitid lineages during the Middle Volgian.

The causes of such complex changes in ammonite morphologies remain unclear. These changes were more significant in those areas characterized by a sea-level fall, but have also been well recognized in other Boreal regions. Strong size reduction in Boreal ammonites in the latest Middle Volgian can be considered an example of the Lilliput effect, reflecting a pronounced ecological crisis. On the Russian Platform, new craspeditid taxa exhibit mass occurrences after the first phase of dorsoplanitid size reduction, forming bedding planes covered by hundreds of dwarf *Kachpurites* shells (1.2–2.0 inches [3–5 cm] in diameter). These changes in ammonite faunas partially coincide with an extinction event in Boreal bivalves, recognized in England (Hallam 1986) and caused by the sea-level fall. But analysis of the published data on bivalve and brachiopod ranges across the Middle and Upper Volgian of the Russian Platform (Gerasimov et al. 1995) shows only a minor turnover at the

Northern Siberia		NW Europe	Russian Platform	
Zone	Subzone	Zone	Zone	Subzone
Chetaites chetae		Volgidiscus lamplughi	Volgidiscus singularis	
Craspedites (Taimyroceras) taimyrensis		Subcraspedites preplicomphalus	Craspedites (Trautsch.) nodiger	Cr. (Trautsch.) milkovensis
				Mosquites mosquensis
	Craspedites (Taimyroceras) originalis		Craspedites (Cr.) subditus	
Craspedites (Cr.) okensis			Kachpurites fulgens	Mosquites nekrassovi
	Craspedites (Craspedites) okensis			Kachpurites fulgens
Praechetaites exoticus			Epivirgatites nikitini	Epivirgatites nikitini
Epivirgatites variabilis		Swinnertonia primitivus		E. lahuseni
		Paracraspedites opressus		
Taimyrosphinctes exoticus		Titanites anguiformis		Epivirgatites bipliciformis
		Kerberites kerberus		

Figure 6. Craspeditid evolution during the latest Middle Volgian and Late Volgian. Zonal succession after Rogov, Zakharov, 2009, 2011. Craspeditid ammonites evolved independently within 3 semi-isolated basins (only cross-sections are shown): Swinnertonia, Subcraspedites, and Volgidiscus lineage mainly inhabited NW European region; Kachpurites and Garniericeras lineage was typical for the Russian Platform; while Craspedites (Craspedites), C. (Trautscholdiceras), and C. (Taimyroceras) lineages were widely distributed in Arctic. Reproduced by permission of Gale, a part of Cengage Learning.

species level in bivalves accompanied by the extinction of only one brachiopod genus. Changes in belemnite faunas (e.g., the regional extinction of cylindroteuthid belemnites and the appearance of short-rostra *Acroteuthis*) are well recognized in northwestern Europe and the Russian Platform (Gustomesov 1964; Mutterlose 1990) but cannot be traced in Siberia (Dzyuba 2004). Perhaps changes in belemnite assemblages are connected directly with sea-level changes because they are observed only in regions showing clear evidence of sea-level fall.

In spite of the abrupt changes in ammonite morphologies, the end–Middle Volgian crisis exhibits some clear features of its own internal, biological nature, suggesting it was not caused directly by abiotic changes (e.g., increasing of speciation and extinction rates before the peak of the crisis, small-sized crisis fixed before the main crisis, the appearance of many short-ranged taxa near to the crisis event, changes in ecological dominants; see Kalandadze and Rautian 1993). The Late Volgian phase of the crisis is also characterized by the small magnitude of both extinction and origination rates and by very low species richnesses. The end of the Late Volgian is characterized by increasing extinction and origination rates in boreal ammonites and increasing generic richness (see Figure 4).

Changes in ammonite faunas clearly correspond to changes in radiolarians. As was shown in a 2012 study by Valentina S. Vishnevskaya and Genrietta E. Kozlova, specimens of

Dinosaur excavation. Protecting a limb bone of a Sauropod dinosaur with plaster prior to transit. Africa. © The Natural History Museum/The Image Works.

Parvicingula from the Late Volgian are very small compared to their Middle Volgian counterparts, possess small pore frames, and have post-abdominal chambers with weakly developed circumferential ridges or almost none. Such simultaneous changes traced through the different areas of the pan-boreal superrealm in these two fossil groups could indicate that the crisis event was connected with changes in planktonic taxa and could reflect turnover in planktonic food webs. Similar size reductions in radiolarians have also been detected during other extinction events, such as the end-Permian extinction (De Wever et al. 2006). Such size decreases could be explained partly by warming because, at high temperatures, plankton may be able to offset some of the effects of reduced buoyancy caused by reduced water density and viscosity by reducing shell size (see Atkinson 1994). During Neogene climate changes, however, different microplankton groups showed different modes of shell size changes (Schmidt et al. 2006). Unfortunately, the diet of ammonites and the possible role of plankton in it are not well understood. Advances in the study of ammonoid paleobiology and feeding habitat (Kruta et al. 2011), based on the investigation of very specific heteromorph taxa, cannot be accepted as a valid model for other ammonites. As Marion Nixon pointed out in a 1996 contribution, the crop/stomach contents from the body chambers of ammonites show a wide spectrum of ammonite prey, ranging from other ammonites to crinoids, ostracods, and foraminifers. Simultaneous species richness falls in Tethyan radiolarians (Vishnevskaya and Basov 2007), corresponding to nannoplankton turnover, partially overlapped with the size reduction of calpionellids at the base of the Berriasian.

Significance of the end-Jurassic extinction event

In spite of its regional nature, the end-Jurassic extinction led to significant changes in at least some marine and terrestrial faunas. In the boreal seas this extinction event was responsible for strong changes in ammonite assemblages. While pre-extinction ammonite diversity occurred during the earliest Cretaceous (Ryazanian), large Boreal ammonites appeared later. Other groups of Boreal animals crossed the J–K boundary without significant changes. The majority of marine groups show a drop in diversity only in those regions influenced by the sea-level fall during the Jurassic–Cretaceous transition. Paleoceanographic changes in the Tethys ocean led to changes in sedimentation, connected directly with bloom of new groups of microfossils with calcareous shells (especially calpionellids and nannoconids). These changes possibly led to a gradual increase in the diversity and abundance of plankton-feeding heteromorph ammonites during the Early Cretaceous. End-Jurassic extinction among the terrestrial faunas exerted an especially strong influence on sauropod dinosaurs. This could reflect an increase in predator activity on these specialized dinosaurs. Among the other groups of terrestrial biota, only minor changes in diversity and/or changes of dominant groups have been verified.

As was pointed out in 1997 by Hallam and Wignall, there is no evidence of a global mass extinction among the marine invertebrate fauna at or near to the Tithonian–Berriasian boundary. But there is good evidence, in Europe, of a regional event, significant only at the species level, among bivalves and a few other groups. In some cases this regional extinction in marine invertebrates (bivalves and belemnites of boreal Europe) is associated with a marine regression. The major turnover in Boreal ammonite faunas in the latest Middle Volgian and Late Volgian was perhaps amplified by sea-level oscillation, but its causes also quite possibly had a strong biological component, reflecting changes in the planktonic assemblages, especially changes in radiolarians. Tethyan ammonite faunas are characterized by relatively high speciation and extinction rates during the Tithonian, but the J–K boundary is not marked by any significant changes in the overall taxonomic makeup of these ammonites.

Resources

Books

Bardhan, Subhendu, Sabyasachi Shome, and Pinaki Roy. "Biogeography of Kutch Ammonites during the Latest Jurassic (Tithonian) and a Global Paleobiogeographic Overview." In *Cephalopods Present and Past: New Insights and Fresh Perspectives*, edited by Neil H. Landman, Richard Arnold Davis, and Royal H. Mapes, 375–395. Dordrecht, Netherlands: Springer, 2007.

Barrett, Paul M., and Paul Upchurch. "Sauropodomorph Diversity through Time: Paleoecological and Macro Evolutionary Implications." In *The Sauropods: Evolution and Paleobiology*, edited by Kristina A. Curry Rogers and Jeffrey A. Wilson, 125–156. Berkeley: University of California Press, 2005.

Dypvik, Henning, Filippos Tsikalas, and Morten Smelror, eds. *The Mjølnir Impact Event and Its Consequences: Geology and Geophysics of a Late Jurassic/Early Cretaceous Marine Impact Event*. Berlin: Springer, 2010.

Dzyuba, Oksana S. *Belemnites (Cylindroteuthidae) and Biostratigraphy of the Middle and Upper Jurassic of Siberia*. [In Russian.] Novosibirsk, Russia: Publishing House of SB RAS, Department "Geo," 2004.

Gerasimov, Petr A., Vasily V. Mitta, and Maria D. Kochanova. *Fossils of the Volgian Stage of Central Russia*. [In Russian.] Moscow: VNIGNI, 1995.

Hallam, Anthony, and Paul B. Wignall. *Mass Extinctions and Their Aftermath*. Oxford: Oxford University Press, 1997.

Nixon, Marion. "Morphology of the Jaws and Radula in Ammonoids." In *Ammonoid Paleobiology*, edited by Neil H. Landman, Kazushige Tanabe, and Richard Arnold Davis, 23–42. New York: Plenum Press, 1996.

Sepkoski, J. John, Jr. A Compendium of Fossil Marine Animal Genera. Bulletins of American Paleontology, no. 363. Ithaca, NY: Paleontological Research Institution, 2002.

Periodicals

Abbink, Oscar, Jordi Targarona, Henk Brinkhuis, and Henk Visscher. "Late Jurassic to Earliest Cretaceous Palaeoclimatic Evolution of the Southern North Sea." *Global and Planetary Change* 30, nos. 3–4 (2001): 231–256.

Alley, N.F., and L.A. Frakes. "First Known Cretaceous Glaciation: Livingston Tillite Member of the Cadna-owie Formation, South Australia." *Australian Journal of Earth Sciences* 50, no. 2 (2003): 139–144.

Atkinson, D. "Temperature and Organism Size—a Biological Law for Ectotherms?" *Advances in Ecological Research* 25 (1994): 1–58.

Benson, Roger B.J., Richard J. Butler, Johan Lindgren, and Adam S. Smith. "Mesozoic Marine Tetrapod Diversity: Mass Extinctions and Temporal Heterogeneity in Geological Megabiases Affecting Vertebrates." *Proceedings of the Royal Society* B 277, no. 1683 (2010): 829–834.

Benton, M.J. "Diversification and Extinction in the History of Life." *Science* 268, no. 5207 (1995): 52–58.

Butler, Richard J., Roger B.J. Benson, Matthew T. Carrano, et al. "Sea Level, Dinosaur Diversity, and Sampling Biases: Investigating the 'Common Cause' Hypothesis in the Terrestrial Realm." *Proceedings of the Royal Society* B 278, no. 1709 (2011): 1165–1170.

Cecca, Fabrizio. "Late Jurassic and Early Cretaceous Uncoiled Ammonites: Trophism-Related Evolutionary Processes." *Comptes Rendus de l'Académie des Sciences*, series IIA, *Earth and Planetary Science* 325, no. 8 (1997): 629–634.

Dera, Guillaume, Benjamin Brigaud, Fabrice Monna, et al. "Climatic Ups and Downs in a Disturbed Jurassic World." *Geology* 39, no. 3 (2011): 215–218.

De Wever, Patrick, Luis O'Dogherty, and Špela Goričan. "The Plankton Turnover at the Permo-Triassic Boundary, Emphasis on Radiolarians." *Eclogae geologicae Helvetiae* 99, supp. 1 (2006): S49–S62.

Fischer, Valentin, Michael W. Maisch, Darren Naish, et al. "New Ophthalmosaurid Ichthyosaurs from the European Lower Cretaceous Demonstrate Extensive Ichthyosaur Survival across the Jurassic–Cretaceous Boundary." *PLoS ONE* 7, no. 1 (2012): e29234.

Föllmi, K.B. "Early Cretaceous Life, Climate, and Anoxia." *Cretaceous Research* 35 (2012): 230–257.

Gustomesov, V.A. "Boreal Late Jurassic Belemnites (Cylindroteuthinae) of the Russian Platform." [In Russian.] *Transactions of the Geological Institute* 107 (1964): 89–216.

Haines, P.W. "Impact Cratering and Distal Ejecta: The Australian Record." *Australian Journal of Earth Sciences* 52, nos. 4–5 (2005): 481–507.

Hallam, Anthony. "Stratigraphic Distribution and Ecology of European Jurassic Bivalves." *Lethaia* 9, no. 3 (1976): 245–259.

Hallam, Anthony. "The Pliensbachian and Tithonian Extinction Events." *Nature* 319, no. 6056 (1986): 765–768.

Hallam, Anthony. "The Case for Sea-Level Change as a Dominant Causal Factor in Mass Extinction of Marine Invertebrates." *Philosophical Transactions of the Royal Society* B 325, no. 1228 (1989): 437–455.

Hart, Rodger J., Marco A.G. Andreoli, Marian Tredoux, et al. "Late Jurassic Age for the Morokweng Impact Structure, Southern Africa." *Earth and Planetary Science Letters* 147, nos. 1–4 (1997): 25–35.

Houša, Václav, Petr Pruner, Victor A. Zakharov, et al. "Boreal–Tethyan Correlation of the Jurassic–Cretaceous Boundary Interval by Magneto- and Biostratigraphy." *Stratigraphy and Geological Correlation* 15, no. 3 (2007): 297–309.

Kalandadze, Nikolai N., and Alexander S. Rautian. "Symptomatology of Ecological Crises." *Stratigraphy and Geological Correlation* 1 (1993): 473–478.

Kiessling, Wolfgang, and Martin Aberhan. "Environmental Determinants of Marine Benthic Biodiversity Dynamics through Triassic–Jurassic Time." *Paleobiology* 33, no. 3 (2007a): 414–434.

Kiessling, Wolfgang, and Martin Aberhan. "Geographical Distribution and Extinction Risk: Lessons from Triassic–Jurassic Marine Benthic Organisms." *Journal of Biogeography* 34, no. 9 (2007b): 1473–1489.

Kriwet, Jürgen, and Stefanie Klug. "Diversity and Biogeography Patterns of Late Jurassic Neoselachians (Chondrichthyes, Elasmobranchii)." *Geological Society, London, Special Publications* 295 (2008): 55–69.

Kruta, Isabelle, Neil H. Landman, Isabelle Rouget, et al. "The Role of Ammonites in the Mesozoic Marine Food Web Revealed by Jaw Preservation." *Science* 331, no. 6013 (2011): 70–72.

Mahoney, J.J., R.A. Duncan, M.L.G. Tejada, et al. "Jurassic–Cretaceous Boundary Age and Mid-Ocean-Ridge–Type Mantle Source for Shatsky Rise." *Geology* 33, no. 3 (2005): 185–188.

Markov, Alexander V. "Alpha Diversity of Phanerozoic Marine Communities Positively Correlates with Longevity of Genera." *Paleobiology* 35, no. 2 (2009): 231–250.

Mikhailov, Nikolai P. "Pavlovia and Allied Groups of Ammonites." [In Russian.] *Bulletin of the Moscow Society of Naturalists, Geology Section* 37, no. 6 (1962): 3–30.

Milton, D.J., B.C. Barlow, Robin Brett, et al. "Gosses Bluff Impact Structure, Australia." *Science* 175, no. 4027 (1972): 1199–1207.

Mitta, Vasily V. "Ammonite Assemblages from Basal Layers of the Ryazanian Stage (Lower Cretaceous) of Central Russia." *Stratigraphy and Geological Correlation* 15, no. 2 (2007): 193–205.

Mitta, Vasily V., and Yuri I. Bogomolov. "Late Jurassic of the Russian Platform: Ammonite Evolution and Paleoenvironments." *Bulletin of the Geological Society of Greece* 40 (2007): 1485–1490.

Mitta, Vasily V., and JinGeng Sha. "Late Jurassic Ammonite Evolution and Paleoenvironment of the Russian Platform." *Science in China*, series D, *Earth Sciences* 52, no. 12 (2009): 2024–2028.

Mutterlose, Jörg. "A Belemnite Scale for the Lower Cretaceous." *Cretaceous Research* 11, no. 1 (1990): 1–15.

Price, Gregory D., and Mikhail A. Rogov. "An Isotopic Appraisal of the Late Jurassic Greenhouse Phase in the Russian

Platform." *Palaeogeography, Palaeoclimatology, Palaeoecology* 273, nos. 1–2 (2009): 41–49.

Price, Gregory D., Alastair H. Ruffell, Charles E. Jones, et al. "Isotopic Evidence for Temperature Variation during the Early Cretaceous (Late Ryazanian–Mid-Hauterivian)." *Journal of the Geological Society* 157, no. 2 (2000): 335–343.

Raup, David M., and J. John Sepkoski Jr. "Periodicity of Extinctions in the Geologic Past." *Proceedings of the National Academy of Sciences of the United States of America* 81, no. 3 (1984): 801–805.

Raup, David M., and J. John Sepkoski Jr. "Periodic Extinction of Families and Genera." *Science* 231, no. 4740 (1986): 833–836.

Reimold, Wolf Uwe, Richard A. Armstrong, and Christian Koeberl. "A Deep Drillcore from the Morokweng Impact Structure, South Africa: Petrography, Geochemistry, and Constraints on the Crater Size." *Earth and Planetary Science Letters* 201, no. 1 (2002): 221–232.

Rogov, Mikhail A. "New Data on Ammonites and Stratigraphy of the Volgian Stage in Spitzbergen." *Stratigraphy and Geological Correlation* 18, no. 5 (2010): 505–531.

Rogov, Mikhail A., and Victor A. Zakharov. "Ammonite- and Bivalve-Based Biostratigraphy and Panboreal Correlation of the Volgian Stage." *Science in China*, series D, *Earth Sciences* 52, no. 12 (2009): 1890–1909.

Rogov, Mikhail A., Victor A. Zakharov, and Boris L. Nikitenko. "The Jurassic–Cretaceous Boundary Problem and the Myth on J/K Boundary Extinction." *Earth Science Frontiers* 17 (2010): 13–14.

Ruban, Dmitry A. "Mesozoic Marine Fossil Diversity and Mass Extinctions: An Experience with the Middle XIX Century Paleontological Data." *Revue de Paléobiologie, Genève* 24, no. 1 (2005): 287–290.

Schmidt, Daniela N., David Lazarus, Jeremy R. Young, and Michal Kucera. "Biogeography and Evolution of Body Size in Marine Plankton." *Earth-Science Reviews* 78, nos. 3–4 (2006): 239–266.

Sey, Irina I., and Elena D. Kalacheva. "Lower Berriasian of Southern Primorye (Far East Russia) and the Problem of Boreal–Tethyan Correlation." *Palaeogeography, Palaeoclimatology, Palaeoecology* 150, nos. 1–2 (1999): 49–63.

Smith, Andrew B., and Alistair J. McGowan. "The Shape of the Phanerozoic Marine Palaeodiversity Curve: How Much Can Be Predicted from the Sedimentary Rock Record of Western Europe?" *Palaeontology* 50, no. 4 (2007): 765–774.

Snedden, John W., and Chengjie Liu. "A Compilation of Phanerozoic Sea-Level Change, Coastal Onlaps, and Recommended Sequence Designations." *Search and Discovery*, article 40594 (2010). Accessed July 22, 2012. http://www.searchand-discovery.com/documents/2010/40594snedden/ndx_snedden.pdf.

Spath, Leonard Frank. "Additional Observations on the Invertebrates (Chiefly Ammonites) of the Jurassic and Cretaceous of East Greenland. I. The Hectoroceras Fauna of the S.W. Jameson Land." *Meddelelser om Grønland* 132, no. 3 (1947): 1–69.

Tremolada, Fabrizio, André Bornemann, Timothy J. Bralower, et al. "Paleoceanographic Changes across the Jurassic/Cretaceous Boundary: The Calcareous Phytoplankton Response." *Earth and Planetary Science Letters* 241, nos. 3–4 (2006): 361–371.

Vishnevskaya, Valentina S., and Ivan A. Basov. "New Data on Biotic Events during the Santonian–Campanian Transition: Evidence from Microplankton Fossils of the Russian Pacific Margin." *Doklady Earth Sciences* 417, no. 2 (2007): 1299–1303.

Vishnevskaya, Valentina S., and Genrietta E. Kozlova. "Volgian and Santonian–Campanian Radiolarian Events of the Russian Arctic and Pacific Rim." *Acta Palaeontologica Polonica.* 57, no. 4 (2012): 773–790.

Wierzbowski, Andrzej, Krzysztof Hryniewicz, Øyvind Hammer, et al. "Ammonites from Hydrocarbon Seep Carbonate Bodies from the Uppermost Jurassic–Lowermost Cretaceous of Spitsbergen and Their Biostratigraphical Importance." *Neues Jahrbuch für Geologie und Paläontologie: Abhandlungen* 262, no. 3 (2011): 267–288.

Wierzbowski, Hubert, and Mikhail A. Rogov. "Reconstructing the Palaeoenvironment of the Middle Russian Sea during the Middle–Late Jurassic Transition Using Stable Isotope Ratios of Cephalopod Shells and Variations in Faunal Assemblages." *Palaeogeography, Palaeoclimatology, Palaeoecology* 299, nos. 1–2 (2011): 250–264.

Wimbledon, William A.P. "The Jurassic–Cretaceous Boundary: An Age-Old Correlative Enigma." *Episodes* 31, no. 4 (2008): 423–428.

Young, Mark T., Stephen L. Brusatte, Marcello Ruta, and Marco Brandalise de Andrade. "The Evolution of Metriorhynchoidea (Mesoeucrocodylia, Thalattosuchia): An Integrated Approach Using Geometric Morphometrics, Analysis of Disparity, and Biomechanics." *Zoological Journal of the Linnean Society* 158, no. 4 (2010): 801–859.

Žák, Karel, Martin Košťák, Otakar Man, et al. "Comparison of Carbonate C and O Stable Isotope Records across the Jurassic/Cretaceous Boundary in the Tethyan and Boreal Realms." *Palaeogeography, Palaeoclimatology, Palaeoecology* 299, nos. 1–2 (2011): 83–96.

Zakharov, Victor A., and Boris T. Yanine. "Les bivalves à la fin du Jurassique et au début du Crétacé." *Mémoires du Bureau des Recherches Géologiques et Minières* 86 (1975): 221–228.

Zeiss, Arnold. "Zur Frage der Äquivalenz der Stufen Tithon/Berrias/Wolga/Portland in Eurasien und Amerika: Ein Bietrag zur Klärung der weltweuten Korrelation der Jura-/Kreide-Grenzschichten im marinen Bereich." *Zitteliana* 10 (1983): 427–438.

Mikhail Rogov

The end-Cretaceous extinction

Earth has undergone at least five episodes of mass extinction during the Phanerozoic Eon, which began approximately 540 million years ago (MYA). In decreasing magnitude of severity, these are the end-Permian extinction (251 MYA), the Late Ordovician extinction (c. 445 MYA), the Late Devonian extinction (c. 367 MYA), the end-Triassic extinction (205 MYA), and the end-Cretaceous extinction (65.5 MYA). For comparison, the end-Permian mass extinction, called the "mother of all mass extinctions" by Douglas H. Erwin (1993, 2006), may have resulted in an incredible 95 percent extinction of all marine species. Relative to this event, the end-Cretaceous extinction was less severe, with about 50 percent of the genera and about 20 percent of the families becoming extinct (Sepkoski 1993). Nonetheless, the end-Cretaceous extinction constituted a major reorganization of much of Earth's biota. It is also well known because all non-avian dinosaurs (Dinosauria excluding birds) disappeared and because this is the only mass extinction that can be closely tied to the impact of an extraterrestrial body.

Patterns of extinction

A major issue in the investigation of any patterns in Earth history is the quality of the rock and fossil record. This becomes especially germane when trying to investigate events that may have occurred in a matter of weeks, months, or years—time frames that usually fall well below the resolution of data collected from the fossil and rock record. In addition, the quality of the end-Cretaceous fossil record varies greatly from environment to environment. Accordingly, a much more geographically and stratigraphically better-sampled record of the end-Cretaceous extinctions has been produced for the marine realm than for the terrestrial and freshwater realms. Not surprisingly, then, scientists' understanding of the patterns of extinction of, for example, planktonic foraminifera and nannoplankton is considerably better than that of the dinosaurs. Relative to foraminifera, nannoplankton, and ammonites, dinosaurs were fewer in number, lived in environments less conducive to fossilization, and had skeletons that tended to fall apart soon after death.

Terrestrial and freshwater realm

Late Cretaceous terrestrial and freshwater faunas are known from almost all continents, but unfortunately the record for these faunas up to, through, and following the Cretaceous–Paleogene (K–Pg) boundary is limited to the Western Interior of North America. More specifically, the K–Pg fauna is best known from the northern parts of the western continent of Laramidia, which bordered the large epicontinental Pierre Seaway, a body of water that separated Laramidia from the eastern continent of Appalachia. Although not always stated, in the following discussion the terrestrial and freshwater vertebrate data come from the eastern coastal regions of Laramidia exclusively. It must be remembered that this region may not be representative of the entire globe at this time.

VERTEBRATES Of the three major clades of living mammals, two are known from Laramidia—Metatheria and Eutheria—which include, respectively, the living marsupials and placental mammals. Another major player is the wholly extinct Multituberculata, a group of rodent-like mammals that became extinct about 35 million years ago. Tallying what is now known for the region, five of ten Cretaceous multituberculates survived the K–Pg boundary, seven of seven eutherians survived, and one of eleven metatherians survived. These differences in survival among mammal groups points to an important pattern for extinctions at the K–Pg boundary; the extinctions were highly selective or differential. The implications that follow from some groups being devastated and others not must be considered when scientists try to find the ultimate and proximate cause(s) of these extinctions.

Crocodilians, pterosaurs, and dinosaurs (including birds) together constitute the great clade known as Archosauria, or ruling reptiles. The most basal branch of this clade is the crocodilians (crocodiles, alligators, and relatives), followed by pterosaurs and dinosaurs as each other's nearest relative.

Starting with crocodilians, four that survived are the well-known species *Borealosuchus* (formerly *Leidyosuchus*) *sternbergi*, thought to be a crocodile or alligator; *Thoracosaursus neocesariensis*, thought to be more closely related to the gavials among living crocodilians; and two unnamed alligator relatives. *Brachychampsa montana*, still regarded as an alligator relative, did not make it through the boundary.

One latest Cretaceous pterosaur is known and has been assigned to *Quetzalcoatlus*, the same genus that is better known from Texas, which with a wingspan of perhaps up to 40 feet

(12 m) was the largest organism to fly. All that can really be said with certainty is that the last occurrence of (admittedly rare) *Quetzalcoatlus* fossils has been found below the local K–Pg boundary.

Over the last 10 million years of the Cretaceous in the northern part of the Western Interior, the number of non-avian dinosaurs dropped from 47 to 26, a 45 percent reduction in species diversity, even though the fossil-bearing exposures preserving the last of the dinosaurs were some 16 percent greater in extent. Even more interesting, the most common kinds of dinosaurs, the hadrosaurids and ceratopsians, have been found to have decreased in taxonomic diversity by precipitous rates—82 percent and 50 percent, respectively—over the last 10 million years of the Cretaceous in northern Laramidia, whereas the rarer theropods do not show a similar trend (Archibald 2011; see Table 1). A somewhat broader study echoed these findings:

> Large-bodied bulk-feeding herbivores (ceratopsids and hadrosauroids) and some North American taxa declined in [morphological] disparity during the final two stages of the Cretaceous, whereas carnivorous dinosaurs, mid-sized herbivores, and some Asian taxa did not. Late Cretaceous dinosaur evolution, therefore, was complex: there was no universal biodiversity trend and the intensively studied North American record may reveal primarily local patterns. At least some dinosaur groups, however, did endure long-term declines in morphological variability before their extinction. (Brusatte et al. 2012, 1)

Unfortunately, the poor quality of the non-avian dinosaur record in last few million years of the Cretaceous has made it difficult to assess how rapidly the last of these animals became extinct.

The record of Cretaceous birds is also fragmentary, but researchers are beginning to understand more about their fate. The crown group of birds—that is, the groups that include the living birds—appears to have originated before the K–Pg boundary. This group is Neornithes, or literally "new birds." All birds that are stem taxa to Neornithes appear to have become extinct before or at the K–Pg boundary. Beyond these two points there is little agreement, especially between molecularly, anatomically, and paleontologically based results.

Another major group of reptiles known in these geological sections are the turtles, or Testudines. In a 2009 study, Patricia A. Holroyd and colleagues recorded a minimum of eight families and 20 genera of turtles from the Cretaceous part of the section, certainly among the taxonomically richest turtle assemblages known anywhere or at any time. Except for very rare species, the number of turtle species remains consistent through the vertical extent of the rocks as the K–Pg boundary is approached. Relative abundances do vary greatly, with chelydrids (snapping turtles) and plastomenine trionychids (soft-shelled turtles) being especially common in the lower part of the section. Higher in the section, no groups are dominant. The greatest numbers of species was found higher in the section, correlating to a warming trend 400,000 to 500,000 years before the K–Pg boundary. One of the few turtles to become extinct was *Basilemys*, a terrestrial, tortoise-like form.

The reptilian group Squamata includes lizards and snakes. Of the two, the lizards are much more common, and thus scientists are much more confident regarding their pattern of extinction and survival across the K–Pg boundary in the region. Among the Cretaceous tetrapods of the region, lizards show the greatest diversity in adult body size, ranging from about 6 inches (15 cm) in the smaller teiids to over 10 feet (3 m) for the possible varanid *Palaeosaniwa*, a distant relative of the Komodo dragon. From what is known, most of these lizards were terrestrial, although a semiaquatic habitus cannot be precluded, such as among relatives of the teiids, which today include semiaquatic species. In regard to diet, again the little that is known suggests insectivory through carnivory, depending on size. Terrestrial squamates were little affected by the end-Cretaceous crisis, with the exception of the boreoteiioids, a clade of large herbivorous lizards from Asia and North and South America (Evans and Jones 2010).

Of all the major groups of reptiles from the region, the only totally extinct lineage and probably the least well known is Choristodera—the champsosaurs. At first blush, these creatures seem to have been a form of narrow-snouted crocodilian such as the extant gavials of Asia. On closer inspection of the skull and skeleton, however, it becomes clear that champsosaurs were clearly not crocodilians. Characters of the skull, notably the two openings behind the eyes, show that champsosaurs belonged to Diapsida, the larger group of reptiles that includes lizards, snakes, crocodiles, pterosaurs, and dinosaurs (including birds). Champsosaurs first appeared in the Late Cretaceous of North America, and except for possible Early Cretaceous relatives, few clues as to their origin have been discovered. They were quite common, surviving the K–Pg boundary and lasting into the Early Paleogene.

It may at first seem odd that elasmobranchs (sharks and their relatives) have been found in the freshwater deposits of the Western Interior. But their presence can be explained by the relative proximity to the east of the marine waters of the Pierre Seaway. Elasmobranchs are predominately marine or brackish-water creatures, but they quite often travel many miles inland into freshwater habitats. Still fewer species were restricted to freshwater. One and possibly two sawfishes, a possible guitarfish relative, and two sharks are known from the Cretaceous of the region. None is found immediately after the K–Pg boundary in the region, likely because of the loss of the marine connection at the end of the Cretaceous, but with a return of this connection in the mid-Paleocene, as many as fifteen new taxa are known from nearby regions.

Almost half of the diversity of vertebrate species today is found within the class Actinopterygii—the ray-finned fishes—most of which are marine. In the freshwater deposits in the Late Cretaceous faunas of the Western Interior, these fish were also important, but they unfortunately have not been thoroughly studied. Interestingly, seven of the fifteen fish species known from the region in the Late Cretaceous—sturgeon, paddlefish, bowfin, and gar—still have living descendants in the Mississippi River drainage. The remaining eight species are teleosts, overwhelmingly the most diverse group of living fishes. This was not the case for the Late Cretaceous bony fish fauna for two reasons. First, teleosts had

Dinosaur Park Fm. (Judithian, Late Campanian)*		Lance Fm. (Lancian, Late Maastrichtian)∞		
Genus	# species	Genus	# species (M & D)	Change in # of species (with J, T, M & D)
Theropoda		**Theropoda**		−9(−7)
Albertosaurus (J)	(1)	Albertosaurus	1	
Bambiraptor (T)	(1)	Caenagnathus (D)	(1)	
Caenagnathus	2	Chirostenotes (M, D) Elmisaurus	(1)	
Chirostenotes	1	Dromaeosaurus	1	
Daspletosaurus	2	Nanotyrannus (M)**	(1)	
Dromaeosaurus	1	Ornithomimus	1	
Dromiceiomimus	1	Ricardoestesia	1	
Elmisaurus	1	Saurornitholestes	1	
Gorgosaurus	1	Struthiomimus (D)	(1)	
Hesperonychus	1	Troodon	1	
Ornithomimus	1	Tyrannosaurus	1	
Richardoestesia	1			
Saurornitholestes	1			
Struthiomimus	1			
Therizinosaurid	1			
Troodon	1			
Ankylosauria		**Ankylosauria**		−1(−2)
Edmontonia (1 is from J)	1(+1)	Ankylosaurus	1	
Euoplocephalus	1	Edmontonia	1	
Panoplosaurus	1			
Euornithopoda		**Euornithopoda**		+1(0)
Hypsilophodontid	1	Bugenasaura	1	
Orodromeus (T)	(1)	Thescelosaurus	1	
Hadrosauridae		**Hadrosauridae**		−6(−9)
Brachylophosaurus	1	Edmontosaurus	2	
Corythosaurus	1			
Gryposaurus	2			
Hypacrosaurus (T)	(1)			
Lambeosaurus	2			
Maiasaura (T)	(1)			
Parasaurolophus (1 is from T)	1(+1)			
Prosaurolophus	1			

Table 1. Generic and species counts of non-avian dinosaurs from Dinosaur Park Formation compared to Lance Formation, with additions from the Judith River, Two Medicine, and Hell Creek Formations (from Archibald 2011). Reproduced by permission of Gale, a part of Cengage Learning.

not yet undergone their tremendous evolutionary radiation. As noted, even today in the Mississippi River drainage, a ghost of the bony fish fauna remains with the presence of paddlefish, sturgeon, gar, and bowfin. Second, the Late Cretaceous record is rather fragmentary compared to that of the Tertiary, and many Late Cretaceous teleostean fossils remain to be examined.

Finally among vertebrates are the amphibians, with about one species of anuran or frog and seven of caudates or salamander in the Late Cretaceous of the Western Interior, with the same eight species in the Early Paleocene. Amphibians had a collective yawn at whatever happened at the K–Pg boundary, but they may have responded to climatic changes before the boundary. In a 2009 study, Grace E. Carter and

Dinosaur Park Fm. (Judithian, Late Campanian)*		Lance Fm. (Lancian, Late Maastrichtian)∞		
Genus	# species	Genus	# species (M & D)	Change in # of species (with J, T, M & D)
Pachycephalosauria††		**Pachycephalosauria††**		−2(−1)
Stegoceras	3	Pachycephalosaurus	1	
		Stegoceras (M, D)	(1)	
Ceratopsia		**Ceratopsia**		−3(−6)
Achelosaurus (T)	(1)	Diceratops (or Diceratus)	1	
Avaceratops (J)	(1)	Leptoceratops	1	
Anchiceratops†	1	Triceratops	2§	
Centrosaurus	1			
Chasmosaurus	3			
Einiosaurus (T)	(1)			
Leptoceratops	1			
Styracosaurus	1			
Total numbers of species	38(48)		18(23)	−20(−25)

*Additional genera from the Judith River(J) and Two Medicine (T) formations of central Montana.

**Although here maintained as a genus, the taxonomic status of *Nanotyrannus* is equivocal, with some advocating generic status, others arguing that it is a species of *Tyrannosaurus*, and others suggesting it is a juvenile *T. rex*. Phil Currie (pers. comm. 2009) noted, "Logic says it is a young *T. rex* (why would there be only juvenile *Nanos*), but there is still some evidence to suggest it is a distinct species. I will probably go with the former interpretation if no new evidence is forthcoming." See also Currie 2003.

†Found south, outside of Dinosaur Park.

††Mark Goodwin (pers. comm. 2009) indicated that the recognition of three species of *Stegoceras* from the Dinosaur Park Formation may be obscured by the possibly juvenile status of the specimens. For the Lancian, Horner and Goodwin (2009) recognized two monotypic species, one belonging to *Stegoceras* (= *Sphaerotholus*, = *Prenocephale*) and the other belonging to *Pachycephalosaurus* (= *Stygimoloch*, = *Dracorex hogwartsia*). They wrote: "*Dracorex hogwartsia* (juvenile) and *Stygimoloch spinifer* (subadult) are reinterpreted as younger growth stages of *Pachycephalosaurus wyomingensis* (adult)" (1).

∞Additional genera from the Hell Creek Formation of eastern Montana (M) and western part of the Dakotas (D).

§The number of species of *Triceratops* is uncertain. I use the higher number of two. Further, according to Scannella (2009), *Torosaurus latus*, from the Lance Formation, represents older individuals of *Triceratops* and is thus not recognized here.

Table 1 continued. Reproduced by permission of Gale, a part of Cengage Learning.

Gregory P. Wilson tracked the structural patterns of amphibian paleocommunities through much of the last 2 million years of the Cretaceous in the region. These researchers found little change in the taxonomic composition, with seven genera of caudates or salamanders persisting through the time studied. There were some fluctuations of relative abundances during the last 500,000 years of the Cretaceous, notably with a high relative abundance of the salamander *Opisthotriton* and low relative abundances of the salamanders *Scapherpeton* and *Habrosaurus*, correlating with changes in the relative abundances of mammal species.

FRESHWATER INVERTEBRATES With the notable exceptions of freshwater snails and bivalves, a rather paltry record of invertebrates exists across the K–Pg boundary in the Western Interior of North America. For a 2007 study, Henning Scholz and Joseph H. Hartman examined the patterns, causes, and ecological significance of extinction of the bivalves near the end of the Cretaceous in the Western Interior of North America. For the Hell Creek Formation they found an exceptionally high taxonomic diversity of about 30 species of bivalves, not unlike the diversity of the modern Mississippi River drainage. This fauna underwent a significant turnover near the end of the Cretaceous and a decrease in taxonomic diversity in the interval spanning the K–Pg boundary associated with a significant shift in morphospace occupation—that is, different forms came to dominate the fossil record. These authors noted that this is congruent with a decrease in habitat stability. Four processes and events are possible causes of this change in habitat stability: the extraterrestrial impact at Chicxulub, global climate changes related to Deccan volcanism, regional environment changes caused by the emerging Rocky Mountains, and global changes in sea level. They concluded that the extraterrestrial impact was not the major event for bivalve extinctions. Scholz and Hartman argued that the effects of the emerging Rocky Mountains and global changes in sea level were much more important for explaining the turnover of bivalve species at the end of the Cretaceous.

PLANTS Ideas on plant extinction at the K–Pg boundary have changed considerably after the early 1990s. In 1992 an almost 80 percent extinction of the megaflora (leaves, fruits, seeds, etc.) was reported across the K–Pg boundary. By 2004 Peter Wilf and Kirk R. Johnson were painting a quite different portrait of megafloral extinction across the K–Pg boundary. These researchers calculated a much lower (57%) megafloral extinction rate, one that was regarded as the maximum percentage of plant extinction. Their figure dropped to as low as 30 percent when pollen was used instead of megaflora fossils, both of which were from sites in southwestern North Dakota. Pollen offers tighter stratigraphic control in part because it is very small and often very numerous. But pollen offers less taxonomic resolution than megafloral remains. Wilf and Johnson concluded that their 57 percent extinction estimate is significantly lower than previous 80 percent megafloral observations, but the latter were based on a thicker sequence of latest Cretaceous strata.

Another paleofloral study (Thompson et al. 2009) in the Western Interior found the extinction percentage of pollen and spore species across the K–Pg boundary could be as high as the 30 percent level found by Wilf and Johnson elsewhere or as low as 15 percent. While not denying the importance of an extraterrestrial impact at the boundary, these authors argued that environmental changes in the millennia before the impact, as well as the role these changes may have played in the differential extinctions, have not been well examined. They documented a decline from 44 to 11 pollen and spore species in the uppermost 11.5 feet (3.5 m) of Cretaceous sediments for taxa of dicotyledonous angiosperms. Importantly, they observed no changes in the depositions of sediments or the way the pollen was preserved that could account for this decline. Similar declines have been reported in this upper interval at other localities throughout the Western Interior, although the species lost at one site may persist elsewhere. This study indicates that the observed declines in species diversity are local events, not a general gradual extinction. This biostratigraphic/geographic pattern suggests that the latest Cretaceous plant communities were becoming more heterogeneous, which in turn altered the habitats for some vertebrates that depended on these plants. Local disappearances of the plant species may have rendered these particular vertebrates more vulnerable to extinction from major environmental disturbances such as an extraterrestrial impact. As noted above, these more complex environmental interactions better explain the differential vertebrate extinctions at the K–Pg boundary (Arens and West 2008).

The preceding discussion concerns vertebrate and invertebrate faunas as well as paleofloras from the Western Interior of North America. Additional comparative paleofloral data are available from elsewhere that shed light on the patterns of extinction and survival across the K–Pg boundary. One such megaflora has been described from southern Argentina in a 2007 study by Ari Iglesias and coauthors. Recall that a megaflora includes larger specimens such as leaves, fruits, flowers, seeds, and cones—hence the name. This particular megaflora included 36 species of angiosperms or flowering plants, ferns, and conifers. The Argentine stratigraphic record in question does not include the K–Pg boundary interval, but at about 61.7 million years ago, some 4 million years of Paleocene sediments are represented. These floras represent low-lying floodplains in a humid, warm temperate climate, and they are paleoenvironmentally quite comparable to Paleocene floras from the Western Interior of North America. Yet this South American flora is more than 50 percent richer in species than that of the Western Interior. The authors suggest that this means a more vibrant terrestrial ecosystem existed in South America than in the comparable North American floodplain environments just 4 million years after the K–Pg boundary. Further, it indicates high floral diversity in South America 10 million years earlier than previously thought. The authors speculate that the considerable differences at this time in the Paleocene between comparable North and South American floras could involve reduced effects of an extraterrestrial impact on the South American flora because of the greater distance from the impact site in the Yucatán, faster recovery or immigration rates after the K–Pg boundary, or initially higher latest Cretaceous diversity.

Whichever of these scenarios is correct, and they could all be correct, the evidence from the plant realm points to the differential effects of extinction and survival at the K–Pg boundary. As noted for mammals earlier, this differential pattern of extinction and survival also holds true in that metatherian mammals fared much more poorly than did the eutherians in terms of survival at the K–Pg boundary in the Western Interior of North America, but by the earlier Paleocene in South America, both had done well, with eutherians mostly herbivores and metatherians omnivores and carnivores.

Marine realm

In order to understand the nature of the end-Cretaceous marine extinctions and the implications that patterns within this data have for the inference of extinction causes, it is necessary to appreciate the physical, ecological, and temporal context within which they occurred as well as to trace the histories of all major organismal groups.

GLOBAL SETTING During the close of the Cretaceous the continents were in different positions than those they occupy today (see Figure 1). This has implications for both marine and atmospheric circulation patterns as well as for the distribution and intensities of climates across Earth's surface. In the Atlantic Basin, rifting along the mid-ocean ridge continued and spread south over the entire Cretaceous period. By the end of the Cretaceous, Africa had separated completely from South America to the west and from the remainder of the southern supercontinent Gondwana to the south. Indeed, Africa existed as an island continent for most of the Cretaceous. Similarly, India and Madagascar had broken away from Africa and existed as smaller continental fragments in the proto-Indian Ocean. Australia and Antarctica were the only remnants of old Gondwana still together, but both had drifted south with Antarctica coming to lie over the South Pole. To the north the formerly deep and wide Tethys Sea had all but disappeared as Africa began to collide with southern Europe and Asia. Most of Western Europe was submerged beneath a shallow sea that had resulted from a sea-level highstand. A shallow seaway had also opened between North America and Greenland, while North America continued its northerly drift. In the vicinity of the present-day Ural Mountains, the Turgai Strait separated present-day Europe from Asia with a seaway oriented north to south. The Late Cretaceous North Pole was covered by the shallow marine waters of a flooded Siberian continental platform.

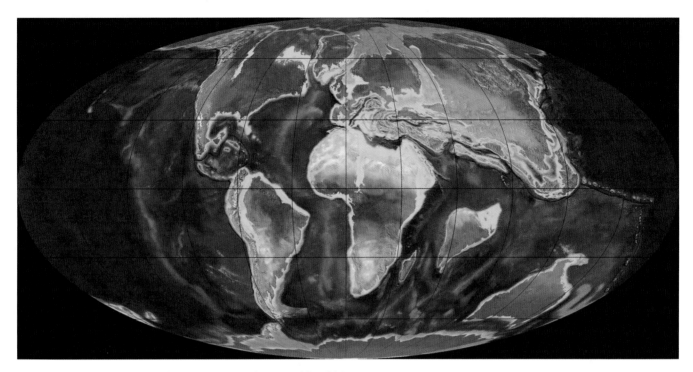

Figure 1. End Cretaceous (65 MYA) paleogeography. Couresty of Ron Blakey.

Over this Cretaceous interval, sea level rose steadily as the climate warmed. Short-term sea-level regressions occurred in the Late Jurassic (Tithonian Age) and the early Late Cretaceous (Cenomanian Age). Interestingly, these short-term reversals in global sea-level trends are associated with minor extinction-intensity peaks, although some have argued that these extinctions are more regional than global in character. At the end of the Cretaceous, however, the sea level underwent quite a large and quite a rapid fall, dropping between 330 and 500 feet (100 and 150 m) in less than 1 million years. The magnitude and rapid onset of this regression remain enigmatic causally insofar as there is little physical evidence of glacial activity sufficiently extensive to account for such a large drop in sea level in the uppermost Cretaceous stage (the Maastrichtian), although this lack of evidence could well reflect the fact that exposures containing the uppermost layers of Maastrichtian sediments are relatively rare and certainly not distributed world-wide.

During the Cretaceous the reorganization of global climates, begun in the Jurassic, developed further, aided by the appearance of circumglobal ocean-surface current circulation in the northern high latitudes. An equatorial zone of tropical climates developed in the Early Cretaceous that promoted the preservation of extensive coal deposits in northern Africa, northern South America, and southern Central America. This tropical warm and wet zone was bounded in the northern and southern hemispheres by an arid zone as evidenced by Cretaceous evaporite deposits in South America, Africa, North America, and southern China. This zone then passed into warm-temperate zones between the 30° and 60° N and S latitudes that promoted coal formation, with cool-temperate environments above and below, respectively, the 60° N and S latitudes. Despite the large degree of climatic

variation, however, the Late Cretaceous high latitudes were quite warm, with both crocodilian and dinosaur faunas occurring above the Arctic Circle.

In terms of the overall magnitude of marine extinctions, the end-Cretaceous extinction is the least intense of the Big Five mass extinction events, as noted above. Nevertheless, although the estimated percentage of species lost appears modest relative to the figures for the end-Permian (95% ± 2%) and end-Ordovician (85% ± 3%) extinction events, it should be remembered that the overall taxonomic biodiversity increased exponentially throughout the Mesozoic. Over 2,400 genera have been found in the Maastrichtian fossil record compared to 1,239 for J. John Sepkoski Jr.'s richest Permian stage, the Guadalupian, and 1,978 for his end-Ordovician Caradocian Stage (see Sepkoski 2002). Accordingly, although the extinction percentage is decidedly lower in the Maastrichtian, the total number of taxa lost is of the same order of magnitude as extinction events that appear much larger. Extinction records for the organismal groups traditionally associated with the end-Cretaceous extinction event are described below.

NANNOPLANKTON Calcareous nannofossils include three principal groups: coccoliths (remains of haptophyte algae), nannoliths (probable remains of haptophyte algae), and calcispheres (remains of dinophytes). Haptophytes are unicellular, planktonic algae. They compose an important phytoplankton group but probably contribute less than 10 percent of modern open-ocean primary productivity. Haptophytes may have been significantly more abundant in the Late Cretaceous because they are exceptionally abundant in the Upper Cretaceous sedimentary rocks, forming the core constituents of the widespread chalks that typify this interval of Earth history. Calcareous nannofossils are the smallest and most abundant fossils routinely studied (with a typical size of 2 to 20 microns [0.000077 to 0.00079 inches). This means they are vulnerable to reworking (that is, the erosion of fossil forms from older sediments and their redeposition within younger sediments), but are available in essentially infinite abundance.

Vanishing Cretaceous nannoplankton species are those regarded by most qualified biostratigraphers as being confined to the Cretaceous during their lives. Some of these species occur in overlying Danian strata occasionally, but this is not the case consistently. Moreover, these species are not thought to have been the direct ancestors of any Paleocene descendants. Members of this fauna typically form more than 90 percent of Maastrichtian assemblages and include numerous long-ranging species. In the nannoplankton data there does not appear to be any foreshadowing of the K–Pg event by uppermost Cretaceous nannofossil species extinctions.

During their decline the relative abundances of these vanishing nannofossil species remained essentially constant, and, despite extensive efforts, consistent last-appearance datums within the Danian have not been identified for any of these taxa. There are various plausible explanations for these Danian occurrences of Maastrichtian nannofossils. Survivorship of these taxa into the Paleocene is an obvious, simple explanation of the record. The main problems with this scenario are as follows: (1) why, if the vanishing group survived the apparently anomalous conditions of the earliest Danian ocean, did they then fail to recover when more normal conditions were resumed? (2) why do members of this large group of species not exhibit widely varying, biogeographically related behavior in the (presumably) highly variable Early Danian ocean (as the surviving and incoming species do)? and (3) why do they not exhibit clear and consistent last-appearance levels?

Persistent (or survivor) species are the Late Cretaceous nannofossil species regarded as occurring in the Danian because they exhibit high relative abundances, exhibit consistent occurrences in Danian strata, and/or represent the inferred ancestors of fully Danian species. These species dominate nannofossil assemblages between the decline of the Cretaceous nannoplankton flora and the evolution of new Paleocene species. Most of these taxa do not belong to the major coccolith or nannolith families but instead belong to evolutionarily conservative taxa of uncertain affinities. These groups are usually rare but become much more abundant during the earliest Danian. By contrast, only a very few of the more common Cretaceous taxa persist. Calcispheres compose another group of nannoplankton, which are exceptionally abundant in the survivor assemblages. Ulrike Kienel, in a 1994 publication, suggested that a major turnover occurred within this group but during the Early Danian rather than at the K–Pg boundary.

DINOFLAGELLATES Dinoflagellates are marine unicellular algae that possess two flagella and construct a rigid test of cellulose plates. Some modern species also construct resting cysts of sporopollenin. The K–Pg dinoflagellate record is difficult to interpret with confidence because only a small proportion of modern dinoflagellate species produce fossilizable organic-walled cysts. Therefore, it may be highly misleading to infer turnover patterns among ancient floras on the basis of data from cyst-forming species alone.

Details of the K–Pg dinoflagellate record are best known from sections at El Kef, Tunisia, and Seymour Island, Antarctica. In both successions the last appearances of dinocyst species occur throughout the uppermost Cretaceous and lowermost Paleocene interval in a broadly progressive pattern. The turnover rate does increase in the uppermost Maastrichtian (close to the K–Pg boundary), but this interval precedes the emplacement of impact debris in all studied sections and is probably best correlated with the uppermost Maastrichtian eustatic (or global) sea-level rise.

DIATOMS Diatoms are microscopic unicellular algae whose valves are made of opaline silica. The presence of diatoms diminishes down the geological column, and in pre-Eocene sediments they are know from only a handful of localities worldwide (see Harwood and Gersonde 1990). Through the mid-1980s, the lack of diatomaceous K–Pg successions led to varying estimates of diatom survival across the K–Pg boundary (from 13% to 100% survivorship). In a 1988 study, David M. Harwood described a well-documented diatom flora from the K–Pg succession on Seymour Island, Antarctica. This high-latitude locality contains the only diatomaceous K–Pg boundary succession known at that time. From that succession, Harwood reported that 84 percent of Late Cretaceous diatom species survived into overlying Paleocene sediments.

High diatom survivorship rates across the K–Pg boundary event horizon were dismissed as evidence for a prolonged extinction event by some supporters of the impact–extinction link who suggested that the ability of some diatom species to form dormant resting spores during times of environmental stress (especially during low nutrient/light conditions) enabled these species to survive a short-term catastrophic environmental event. But the ability of diatoms to (effectively) hibernate through unfavorable environmental events cannot be used to identify the nature of those events, and this adaptation certainly did not arise in response to any single source of environmental selection pressure. Indeed, Harwood, in the same 1988 study, noted a great increase in the relative abundance of resting cysts in the Seymour Island succession both at and beyond the K–Pg boundary. Consequently, the hypothesis that an increase in diatom resting spores in the Seymour Island section might result from local environmental factors extending over a considerable ecological time interval cannot be ruled out.

BENTHIC FORAMINIFERA Foraminifera are single-celled amoeboid protists, most species of which are able to construct internal shells or tests of calcite, aragonite, or agglutinated inorganic sedimentary particles. These tests are usually multichambered, with new chambers being formed as the animal increases in size during growth. The coiling modes adopted, positions of apertures (for the extrusion of cytoplasm), the canal systems of the walls and intercameral septa, and many other structures allow species to be identified readily, reliably, and rapidly. In the early twentieth century fossil foraminifera became the most reliable tool for biostratigraphical inference during petroleum exploration.

Benthic foraminifera do not show a significant drop in global species diversity across the K–Pg boundary. Instead, benthic faunas from El Kef, Tunisia, the South Atlantic, the southern high latitudes, and Caravaca, Spain, all exhibit a change in faunal composition, with infaunal and presumably opportunistic forms dominating in lowermost Danian beds followed by a progressive increase in the epifaunal component and diversity of the assemblages through Lower Danian planktonic foraminiferal zones. These compositional changes were not sudden and coincident with the K–Pg boundary in these sections. Instead, they occupied at least the last 20,000 years of the Maastrichtian, with depressed diversities in the Danian lasting for at least an additional 50,000 years.

PLANKTONIC FORAMINIFERA Much controversy surrounds the planktonic foraminiferal record across the K–Pg boundary. In many sections and cores the difference between the characteristically large-sized Cretaceous species and comparatively minute Danian species of this microfossil group is so striking that it can be seen with the naked eye. Reports of the first detailed studies of this group's K–Pg extinction pattern indicated that all but two or three Maastrichtian species existed at normal diversities right up to the boundary, at which point all but one or two species disappeared simultaneously (see Figure 2a). But after 1988 the picture of the end-Cretaceous planktonic foraminiferal extinction grew steadily more complicated, interesting, and well corroborated. In particular, Gerta Keller and her colleagues (e.g., Canudo et al. 1991) have documented a complex pattern of planktonic foraminiferal species turnover globally that involves extinctions prior to and at the K–Pg boundary as well as a sizable and taxonomically consistent assemblage of "Cretaceous"

species that are found routinely in lowermost Paleogene sediments (see Figure 2b). Some of the occurrences of these latter species may be a result of redeposition from eroded Upper Maastrichtian sediments, a possibility that has always been acknowledged by Keller and her colleagues. However, the consistency with which these Cretaceous species appear in lowermost Paleogene sediments globally, along with their high abundance, excellent state of preservation, and (in some cases) isotopic signatures all point to the majority of these fossils representing remains from living, Paleogene survivor populations. Independent estimates have suggested that as many as 30 percent of the Late Maastrichtian planktonic foraminifer species may have survived the end-Cretaceous extinction event.

Interestingly, these K–Pg survivor species tend to be small forms with simple morphologies that appear superficially similar to the morphologies exhibited by the fully Paleogene species that first appear in lowermost Danian sediments. This occurrence pattern suggests the planktonic foraminiferal victims of the end-Cretaceous extinction event were large, ornate, ecologically specialized, and mostly tropical species, whereas the survivors were small, morphologically simple, ecologically generalized, cosmopolitan species. The patterns of faunal turnover in this group also mirror similar patterns of turnover in the phytoplankton groups that form the basis of primary productivity in the marine realm. Unlike the wholly inferential interpretations of productivity collapse during the end-Permian and end-Triassic extinctions (where there is lack of a fossil record of phytoplankton species that the diversity of marine filter-feeders suggests must have existed), there is direct evidence for phytoplankton extinctions and abundance reductions in the uppermost Maastrichtian interval (see above and MacLeod et al. 1997).

OSTRACODS Ostracods are small crustaceans with an external bivalved carapace composed of calcite. They reproduce both

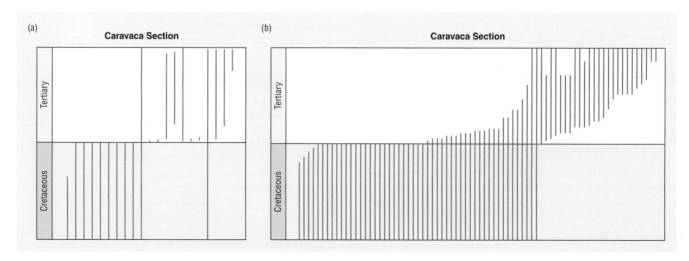

Figure 2. Two views of the end-Cretaceous planktonic foraminifera record at the Caravaca section in southern Spain. An early interpretation that shows most species terminating at the Maastrichtian–Danian boundary horizon with only a single survivor species (from Smit 1982) and a later interpretation documenting extinctions of "Cretaceous" species prior to, at, and after the boundary horizon (From Canudo et al. 1991). Reproduced by permission of Gale, a part of Cengage Learning.

sexually and by parthenogenesis, and they grow intermittently by a process of ecdysis. Thus, each animal can be represented by a series of juvenile and adult shells. Modern ostracod species live in essentially all aquatic habitats as plankton and nekton, as well as infaunal and epifaunal benthos.

The Late Cretaceous appears to have been a time of major turnover for ostracods. In a comprehensive synoptic study published in 1990, Robin Whatley recorded 70 generic originations (representing 678 species originations) and 230 generic extinctions (representing 1,372 species extinctions) during the uppermost Cretaceous interval. For a 1990 study, Graham Coles reviewed Cenozoic North Atlantic deepwater and adjacent-shelf ostracod faunas and found that diversity was lower than at any other time in the Paleocene. These reviews, however, lack the precise stratigraphical control needed to determine the timing, magnitude, and geographical scope of the extinctions occurring close to the K–Pg boundary interval itself. Accordingly, the patterns and rate of ostracod faunal turnover in the vicinity of the K–Pg boundary remain unknown.

CORALS Unlike previous estimates that had reported a coral extinction rate across the K–Pg boundary of as high as 50 percent, a 2004 analysis by Wolfgang Kiessling and Rosemarie C. Baron-Szabo based on a substantially superior data set indicates that the true extinction rate is substantially lower: 30 ± 8 percent for genera, which is one of the lowest rates recorded for any major marine invertebrate group. But because the data used to perform this analysis were, of necessity, parsed over a somewhat vaguely defined "Upper Maastrichtian" interval, the true rate of coral extinction that coincides with the Chicxulub impact event is lower than this, perhaps substantially so. Within this pattern, photosymbiotic corals were significantly more affected than non-photosymbiotic corals, suggesting, but not demonstrating, that attenuation of sunlight may have been a factor—and that wide geographic distribution did confer a measure of extinction resistance, especially in high-latitude biomes. Also, many of the newly identified coral survivor species record their last stratigraphic appearances in the immediately overlying Danian strata. This latter pattern provides evidence that what mechanism(s) were responsible for this turnover pattern operated in the global environment for a considerable amount of time (in the area of hundreds of thousands to as much as a million years).

BRYOZOANS Bryozoans are suspension-feeding, benthic metazoans that are closely related to brachiopods (see below) but form colonies of small, asexually budded zooids. Species belonging to two orders (Cheilostomata and Cyclostomata) having mineralized calcite skeletons are present in Cretaceous and Paleocene marine-shelf sediments. Poor knowledge of their systematics and stratigraphical distributions constrain interpretations of bryozoan extinction and survival across the K–Pg boundary.

Bryozoans are the dominant macrofossil group in the Danish Maastrichtian and Danian. They occur in chalks, including the gray Maastrichtian chalk of Stevns Klint where their skeletons are carbon stained and rich in iridium (Hansen et al. 1987), and in bryozoan limestones that often accumulated as mounds on the seabed. More than 500 species are present in this section, but no detailed data have been published on species ranges across the K–Pg boundary. In 1979 Eckart Håkansson and Erik Thomsen summarized some unpublished data from the Nye Kløv section. These authors found species diversity to be high in the white chalk at the top of the Maastrichtian (where approximately 70 cheilostome species were found), dropping to four species in the basal Danian marl and then recovering to a peak of more than 40 species in a bryozoan limestone approximately 20 feet (6 m) above the K–Pg boundary before declining again in the overlying Danian pelagic chalk. Of 115 cheilostome species, only eleven are found in both the Maastrichtian and Danian parts of the sequence at Nye Kløv. This very low survivorship ratio (9.6%), however, is a local feature because a number of the Maastrichtian species reappear in the Danian elsewhere in the basin. The four basal Danian marl species at Nye Kløv form a specialized community of rooted and free-living colonies very different from the faunas above and below this unit.

Tentative global genus-level and family-level data are available at stratigraphic stage level for the Cretaceous and Paleocene. Ehrhard Voigt recognized the extinction of 38 cheilostome genera and 32 cyclostome genera during the Maastrichtian (reported in Taylor and Larwood 1988). L. A. Viskova compiled data on Late Cretaceous and Paleogene bryozoan genera for a 1994 study, finding that the Maastrichtian contained high diversities of cyclostomes (178 genera) and of cheilostomes (172 genera) but that the generic diversity of both groups declined equally and by more than a half by the Late Paleocene. Therefore, the selective extinction of cheilostomes mentioned above is not apparent in Viskova's analysis based on her global generic database. At a higher taxonomic level still, the diversity of cyclostome and cheilostome families shows no more than a slight fall across the K–Pg boundary (Taylor and Larwood 1988; Figure 6 of Lidgard et al. 1993).

BRACHIOPODS Brachiopods are bivalved marine lophophorates that have hard "valves" (shells) on their upper and lower surfaces (unlike the left and right arrangement in bivalve mollusks, which they resemble superficially). Brachiopod valves are hinged at the rear end, whereas the front can be opened for feeding or closed for protection. A typical brachiopod also exhibits a stalk-like pedicle that projects from an opening in the rear of one of the valves; this pedicle is used for attaching the animal to the seabed.

Two studies—one from 1984 by Finn Surlyk and Marianne Bagge Johansen and the other from 1987 by Johansen—provide evidence suggesting a clear and indisputable, sudden diminution in brachiopod species at the K–Pg boundary in Denmark. Johansen's 1987 analysis in particular was based on a large collection of minute and immature specimens, most of which she concluded had been derived from the underlying Upper Maastrichtian chalk. Thirty-five species were named, of which six were said to be restricted to the basal Danian Fish Clay, 6 were common in the Maastrichtian and not present in Lower Danian strata, and the remainder were held to be new to the Danian. The weakness in this analysis, however, lies in the confident assignment of immature specimens to established species. Many brachiopod genera and species assume different morphological shapes during their development. Such is the case particularly in species of *Terebratulina*, *Gisilina*, and *Rugia*, which form a substantial part of the fauna Johansen examined. It is also not uncommon to find a paucity of brachiopod species within clay facies. An obvious analogy to the basal Danian Fish Clay is that of the Plenus Marls in Britain. This facies supports very few brachiopod species but is followed by a chalky facies of Lower Turonian age that supports a diverse brachiopod fauna related both generically and specifically to Lower and Upper Cenomanian forms. These results and observations suggest that the interpretations of Surlyk and Johansen may need to be revised.

GASTROPODS AND BIVALVE MOLLUSKS The snail (gastropod) and clam (bivalve) molluscan records across the K–Pg boundary are best known from the Brazos River sections in Texas (Hansen et al. 1987, 1993). For the Brazos River Cottonmouth Creek section, Thor A. Hansen and colleagues reported in 1993 that 19 percent of the combined Upper Maastrichtian bivalve and gastropod fauna exhibit their last appearances at least 63 inches (160 cm) below the K–Pg boundary horizon, followed by gradual species loss for the next 24 inches (60 cm) and then an abrupt drop in species richness coincident with a disconformable clastic unit (that is, the so-called tsunami deposit [Bourgeois et al. 1988]). Paleoecological analyses of these data suggest that a Signor– Lipps effect resulting from sampling and/or relative abundance bias cannot account for this apparently progressive turnover pattern. Thus, the Brazos River record appears to contain a genuine example of long-term or progressive faunal turnover among the non-cephalopod mollusks. In these successions there is a definite shift from epifaunal suspension feeders in the Upper Cretaceous to grazers and infaunal detritus feeders in the lowermost Danian—a shift that appears to corroborate a 1986 hypothesis by Peter M. Sheehan and

Hansen that detritus-feeder food chains were able to sustain some species through the environmental catastrophe wrought by the Chicxulub impact. Nevertheless, this ecological shift begins well below the local K–Pg boundary and proceeds in a progressive manner. Both of these observations are contrary to predictions of the Sheehan and Hansen model. In fact, two of the three species interpreted to be molluscan K–Pg survivors in these successions are suspension feeders.

CEPHALOPOD MOLLUSKS Ammonites represent the invertebrate group most associated with the end-Cretaceous extinction. Ammonoid mollusks had almost become extinct on several occasions during their long evolutionary history (House 1993). At each of these occasions there was a progressive decline in diversity that ceased just short of extinction. In this sense the only difference characterizing the K–Pg event is that the ammonoids finally failed to pull through. Regardless, the terminal decline in ammonoid diversity began at least 20 million years before their final extinction. This decline was marked not just by a reduction in numbers of taxa but also by an increase in the patchiness of their areal distribution. Moreover, no new ammonoid families appeared during the last 11 million years of the Cretaceous. At the same time, however, some of the longest-living ammonoid families extended to almost the end of the Maastrichtian.

The Upper Maastrichtian ammonite fossil record has been studied most extensively in the Zumaya section in northern Spain (see Figure 3). At a coarse level of stratigraphic resolution (see Figure 3*a*), it appears as though 12 of the 30 ammonite species present in the Zumaya section disappear from the fossil record coincident with the Upper Maastrichtian stage boundary (the K–Pg boundary), an overall boundary extinction rate for this interval of 40 percent. Focusing on the interval just prior to the K–Pg boundary in more detail (see Figure 3*b*), however, it can be seen that none of these 12 species is actually recorded from the boundary horizon. If

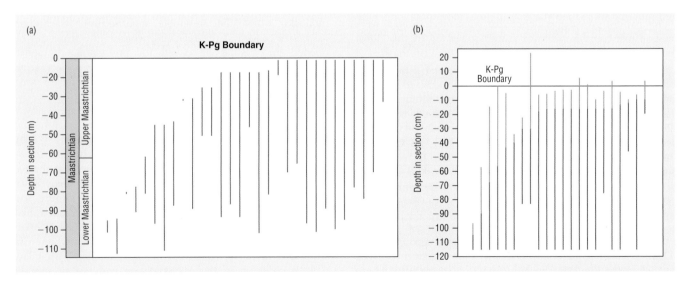

Figure 3. Aspects of the ammonite fossil record. Coarse representation of the end-Cretaceous ammonite record at Zumaya, Spain (a). Detail of the of the uppermost Cretaceous ammonite record at Zumaya, Spain with 50% stratigraphic confidence intervals shown in red (b). Both figure redrawn from Marshall and Ward (1996). Reproduced by permission of Gale, a part of Cengage Learning.

stratigraphic confidence intervals are placed on these ammonite ranges, the extinction limits of most species fall well below the K–Pg boundary, although five species do have fossil occurrence patterns that suggest they may have reached the boundary or indeed may even occur in lowermost Paleogene sediments. The prediction of this stratigraphic confidence interval analysis has been met, at least in part, with the 2005 publication of Marcin Machalski and Claus Heinberg's well-corroborated report of the discovery of a Danian ammonite occurrence.

MARINE FISH For the cartilaginous fishes (sharks, rays, and chimaeras), 35 families passed through the K–Pg boundary, seven became extinct, and one originated in the Danian. Expressed another way, the K–Pg boundary was survived by 80 percent of cartilaginous fish families. The survival rate between the Campanian and Maastrichtian is 95 percent and that between the Danian and Thanetian is 94 percent. Therefore, there is a reduced survival rate at the K–Pg boundary, but this reduction is not significantly acute to implicate a catastrophic, boundary-focused mechanism.

Henri Capetta (1987) has compiled the most comprehensive database for elasmobranchs (sharks and rays) by documenting generic occurrences throughout the Cretaceous and the Tertiary. His counts show a significant disappearance of genera within the Maastrichtian: Twenty-three out of the 53 genera (or 43.4%) become extinct, and this compares with a 26.3 percent disappearance rate within the Campanian and an 8.1 percent rate in the Danian. Although the extinction rate is high, so is the origination rate: In the Maastrichtian 22.6 percent of genera are first occurrences (comparative figures for the Campanian and Danian are 21.4 percent and 18.9 percent, respectively). It therefore appears that the Maastrichtian was a time of high generic turnover and not only a period of high extinction. In addition, these counts were calculated over the entire Maastrichtian interval, and so they do not represent patterns of turnover characteristic of the very latest Maastrichtian.

For ray-finned fishes the record consists of 43 families that passed through the K–Pg boundary, four of which became extinct within the Maastrichtian and nine that originated within the Danian. Therefore, about 10 percent of teleost fish families became extinct and 90 percent survived. Comparing this with stages on either side of the K–Pg boundary, the evidence shows an 81 percent survival rate from Campanian to the Maastrichtian, a 90 percent survival rate from the Maastrichtian to the Danian, and an 85 percent survival rate from the Danian to the Thanetian (see MacLeod et al. 1997). At this crude level of analysis, then, there is no reason to believe that a mass extinction of ray-finned fish took place in uppermost Maastrichtian time.

Proposed causes and scenarios of extinction

Although a relatively small part of the end-Cretaceous mass extinction taxonomically, the fate of the dinosaurs captures the imagination of the public and skews people's perception of this extinction event. Many scenarios have been proposed for the extinction of the dinosaurs, which thus provide a coarse proxy for the number of reasonable hypotheses proposed for the end-Cretaceous extinction overall. The use of the term *scenario* here is intentional as few of the proposed extinction mechanisms rise to the level of an hypothesis let alone a full-fledged theory. Scenarios of the extinction of non-avian dinosaurs started appearing soon after they were first discovered and named in the first half of the nineteenth century. In the twentieth and twenty-first centuries, Glenn L. Jepsen (1963), Alan J. Charig (1983, 1994), M. J. Benton (1990), and J. David Archibald (2012) provided estimates of the numbers of dinosaur, as well as, by linkage, end-Cretaceous extinction scenarios. Depending on how one parses the tabulations, Jepsen listed something like 48 scenarios. The highest number of extinction scenarios claimed to have been complied were the eighty or so reported to Archibald by Charig. Although Charig had discussed various possibilities, he did not publish an extensive list. Benton provided a much more methodically organized and referenced list of possible causes. He also provided opinions as to whether these were "deliberate jokes," "speculative ideas," or "supported by some evidence." Again, the exact numbers are somewhat difficult to compute, but Benton tabulates something like sixty-six scenarios, to which Archibald later added about ten.

Only the three most consistently proposed and examined mechanisms are explored here—massive eruptions of flood basalts, marine regression and habitat fragmentation, and an asteroid (or, more generally, bolide) impact. It must be emphasized that these are ultimate causes that may have a number of overlapping proximate causes or results. For example, if there was a chilling of the global climate, was it caused by one or some combination of these three events? None of these three major ultimate causes are mutually exclusive, and a number of scenarios have been proposed that combine parts of each. In general the three causes, in the order listed above, range from very long term to very short term in both duration and effect.

Massive eruptions of flood basalts

Massive eruptions of flood basalts called the Deccan Traps on the Indian subcontinent occurred over a much longer interval than marine regression, perhaps millions of years surrounding the K–Pg boundary (Courtillot 1990), although as much as 80 percent of the extrusives may have been emplaced in an interval of 50,000 years bracketing the boundary. The volume of material estimated to have been erupted over this 4-million-year interval would cover both Alaska and Texas to a depth of 2,000 feet (610 m). Proximate causes resulting from such massive volcanism have not been as well investigated as those relating to either marine regression or asteroid impact, but some studies have been launched (see, e.g., Keller et al. 2008). Proximate causes, however, have been argued to be similar for both volcanism and impact (see below). Climatic changes caused by massive eruptions would have been longer term, lasting as long as the emplacement sequence itself.

Marine regression and habitat reduction

Marine regression and habitat fragmentation occurred over a shorter interval of time, from tens to hundreds of thousands of years. Marine regression refers to the draining of epicontinental seas (shallow seas that have periodically covered lower-lying areas of continents, possibly caused by increased rates of seafloor spreading). A decrease in rates of seafloor spreading is likely the greatest cause of eustatic (or global) marine regression. At such times, spreading ridges deflate somewhat, allowing water to drain from continental regions into deeper oceanic basins. One of the greatest such regressions is recorded in rocks near the end of the Cretaceous Period, some 65.5 million years ago. Estimates suggest that 11.2 million square miles (29 million sq km) of land were exposed during this interval (Smith et al. 1994), more than twice the next largest such addition of land during the past 250 million years. That amount of land is approximately equivalent to the size of Africa. There were marked proximate effects of this regression—major loss of low coastal plain habitats, fragmentation of the remaining coastal plains, establishment of land bridges, extension of freshwater systems, and climatic change with a general trend toward cooling on the newly emerged landmasses.

Extraterrestrial impact

The asteroid impact theory (Alvarez et al. 1980) contends that a 6-mile-wide (10-km-wide) asteroid struck Earth 65.5 mya, producing ejecta and a plume reaching far enough into the atmosphere to spread around the globe, blocking the sun. A cessation of photosynthesis would have resulted in the death and eventual extinction of many plants, the herbivores that fed on them, and the carnivores that in turn fed on the herbivores (but see Pope 2002). The probable crater, named the Chicxulub crater, has been located near the tip of the Yucatán Peninsula (e.g., Hildebrand 1991). At 110 miles (180 km) across, it is one of the larger such structures on Earth. In addition to the crater, there are two other important pieces of physical evidence supporting the occurrence of an impact at this time: an increased abundance of the element iridium in stratigraphic sections at horizons coincident with the K–Pg boundary and minerals, especially quartz grains showing shocked lamellae (see below) in two directions, at these same levels. A high level of iridium at the surface of Earth and double lamellae are together indicative of an impact. Some of the more proximate effects of an asteroid impact that have been suggested include acid rain, global wildfires, sudden temperature increases and/or decreases, infrared radiation, tsunamis, and super-hurricanes (e.g., Archibald 2011; Archibald and Fastovsky 2004).

Single versus multiple-cause scenarios of extinction

Regarding the single-cause scenario of K–Pg extinctions, there is little doubt within the geologic community that some object from space slammed into Earth about 65.5 million years ago and in effect defined the K–Pg boundary. When Luis W. Alvarez, his son Walter, and their colleagues proposed in 1980 that an asteroid impact was the cause of the mass extinction at the end of the Cretaceous, there was much consternation and doubt that such an event had even occurred, let alone caused extinctions. When the hallmark of the impact hypothesis—the platinum group metal iridium—began to show up in not just marine sections in Italy, Denmark, Spain, and New Zealand but also in terrestrial sections in Montana, more people began to be convinced that an impact had indeed occurred. Another piece

of evidence was the discovery of so-called shocked quartz that shows multiple planes of fracture that leave their mark in the quartz mineral stishovite formed in this process. Although very-high-pressure events on Earth might form it, the real culprit appears to be instantaneous, high-pressure events such as extraterrestrial impacts. In fact, it was first named from occurrences at Meteor Crater in Arizona. Even more convincing was the discovery of the Chicxulub impact crater in the Yucatán in 1991.

Some supporters of the single-cause scenario (e.g., David E. Fastovsky; see Archibald and Fastovsky 2004) argue that the single-cause argument for the K–Pg extinction is fundamentally a parsimony argument. Any hypothesis purporting to explain events at the K–Pg boundary must meet two criteria. First, the hypothesis must be testable; second, the hypothesis must be able to explain as much of what is known about extinction patterns at and in the vicinity of the boundary as is possible. Accordingly, the argument that a variety of causes (asteroid impacts, volcanism, and regression) produced a variety of effects on a variety of different organisms seems unsatisfying to some. If more possible causes were known, would it then be possible to better explain the effects? Simply because these events are known to have occurred is not an a priori reason to consider them causes. Many of these events have occurred many times and even in conjunction, but are they associated with a mass extinction? In effect, those who prefer a single-cause model feel that those who conclude that multiple causes were responsible for the extinctions observed at the K–Pg boundary are confusing correlation and causation.

Fastovsky (in Archibald and Fastovsky 2004) has argued that the asteroid hypothesis predicts that the K–Pg mass extinction was globally both abrupt and synchronous. As humanity has never experienced such a large impact of this magnitude, the kinds of corollary effects such an asteroid impact might bring are unknown. He continues that it is not known whether acid rain, global wildfires, or another "grisly death-and-destruction scenario" could result from an impact occurrence. Although these possible by-products have been hypothesized and researched, they should not be confused for predictions of the asteroid impact hypothesis. Accordingly, the only legitimate prediction for the impact of an extraterrestrial body is that the pattern of extinction must be abrupt and synchronous.

Supporters of the single-cause scenario argue that a general pattern has emerged for synchronous and abrupt extinctions in both the terrestrial and marine realms, with ammonite, foraminiferan, and angiosperm extinctions presented as being examples of this. The single-cause extinction scenario argues that further work will show that the extinction of various groups, notably the dinosaurs, will be shown to have been abrupt and synchronous. Supporters point to the ongoing studies of terrestrial K–Pg boundaries in China, India, Spain, France, and Romania (e.g., Buffetaut 2003), which may provide the key to reconstructing dinosaur population fluxes in the latest Cretaceous.

The argument is that at the boundary primary production dropped synchronously around the globe. It is much more parsimonious to postulate that the synchronicity of the extinctions was caused by a single event. The impact of an asteroid is thought to be potentially big enough to do the job, and thus why add other mechanisms when a sufficiently powerful means of killing exists? Although exact proximate causes are not completely clear, if sunlight were reduced or lost over several months, photosynthesis and thus primary production would shut down. In this sense, then, the apparent loss of primary productivity correlates with a rather specific pattern of fossil occurrences that might be expected from a cessation or reduction of sunlight.

Thus, according to Fastovsky (in Archibald and Fastovsky 2004), two important features of the single-cause scenario are that of parsimony of the argument and that the asteroid impact was large enough to cause the known pattern of extinctions. Because the ramifications of massive impacts are poorly understood, precise mechanistic scenarios are difficult to test. In addition, the vagaries of the fossil record do not permit a highly refined analysis. The chaotic and catastrophic scenario that resulted from the impact made survivorship in part a matter of luck.

Regarding the multiple-cause scenario of K–Pg extinctions, some authors (e.g., Archibald 2011; MacLeod, 2013) argue that there is no single pattern of extinction and survival at the K–Pg boundary; thus, those who support the multiple-cause scenario for the K–Pg mass extinction argue it was an extinction of multiple causes—volcanism, marine regression, and asteroid impact. They argue that the fossil record is not unambiguously consistent with a sudden and globally instantaneous pattern of extinctions and that the great phylogenetic and ecological differences between who survived and who did not, point to multiple causes. Together, volcanism, marine regression, and an extraterrestrial impact are cited to explain the observed length of the taxonomic turnover interval and the differential pattern of survival in a much more parsimonious manner than that of the single-cause impact-driven extinction scenario.

According to this multiple-cause scenario, one must begin at least 10 million years before the boundary to understand what transpired. The record of vertebrate change during this time is largely limited to the Western Interior of North America, thus this scenario at least for vertebrates must by necessity be similarly limited, although the marine invertebrate record is such that a similar constraint is unnecessary. Taking the terrestrial system as a model, approximately 75 million years ago an open plain with scattered trees existed on the eastern shore of Laramidia (Western Interior). Rivers of moderate size meandered through the landscape. In the distance, the shallow Pierre Seaway stretched to the horizon, lapping the shore of the plains, which were dominated by vast herds of several species of duck-billed and horned dinosaurs, much like the large mammals on the Serengeti Plain of Africa today. Other ornithischian dinosaurs and the infrequent meat-eating theropods crossed the landscape. The streams were populated by various species of turtles, amphibians, crocodilians, and fish, including the occasional skate or shark swimming up from the nearby shallow sea. Other inhabitants of the plain were mice and rat-sized mammals.

Fast forward to about 66 million years ago, by which point the shallow seaway had begun to slip rapidly away to the southeast. As the exiting waters reached lower-lying, flatter terrain, the rate of the waters' exodus quickened, with the final stages of withdrawal occurring in, at most, tens of thousands of years. Simultaneously, the great herds of duck-billed and horned dinosaurs diminished. As the dwindling refugia of low coastal plains rapidly decreased, first one, then another of the species dwindled until the great herds were reduced to at most two or three remaining species, much like the herds of bison that once roamed North America. Dinosaurs, like large vertebrates everywhere and at every time, were the first to experience biotic stresses leading to decline and disappearance. The plants changed too, from more homogeneous stands stretching as far as the eye could see to more heterogeneous stands in the dwindling plains. The skies were also darker than they had been, caused by the faraway eruption of the Deccan basalt flows. This brought a cooling climate that may have put a burden on the more warmth-loving species of plant and animal.

It is not known what was happening to vertebrates in more inland areas just before the K–Pg boundary, as few such areas are well preserved and none has been studied. The coastal plains dinosaurs certainly were capable of migrating from one shrinking coastal habitat to another, but finally even this could not stop further declines in population sizes—just like the relentless encroachments of increasing human populations that decimated the American bison and are causing many biotas to shrink today. Other large vertebrates suffered. The Komodo dragon–sized lizards and the single exclusively terrestrial turtle, *Boremys*, also experienced declines. Populations of smaller terrestrial vertebrates were also declining, but because of shorter life spans and quicker turnover rates, they adapted more quickly to the environmental stresses caused by the loss and fragmentation of the coastal plains.

Marsupials had flourished for some 25 million years in North America. Newly emerging land bridges appeared as the seas retreated. Invaders appeared. In North America there were newly arriving diminutive archaic ungulates, possibly from Asia, that joined the few that had reached North America during the Late Cretaceous. In the Western Interior, at least, these species outcompeted the marsupials for dwindling resources. In South America events were different. Both groups of mammals appeared in South America soon after the K–Pg boundary, but in South America they divided the guilds, with marsupials becoming the carnivores and the ungulates the herbivores. This coevolutionary arrangement lasted for almost 50 million years, with only an infusion of rodents and primates from the outside world.

Unlike the terrestrial vertebrates, freshwater species faced far less stress, especially because the size of their habitat at least held its own as the lengthening streams followed the retreating seas. Not all aquatic vertebrates fared so well. With the loss of close ties to the seas, sharks and skates ventured into the rivers in the area less and less frequently, as the distance to the sea expanded from tens to thousands of miles, eventually reaching much farther south.

Plants and nearshore species also showed added stresses as their respective habitats shrank. Certainly, some species must have done fine as new habitats were formed as the seas regressed. As with vertebrates, however, no clear record exists of these environments away from the coastal areas.

The Deccan basalts continued to spew forth, adding further stresses. The additional particulate matter in the atmosphere continued to cool and dry some areas of the globe, probably at an accelerating rate. Then, suddenly, a

literally Earth-shattering event magnified the differences between the "have" and "have-not" species. A very large extraterrestrial object struck what today is called the Yucatán Peninsula. Fine material and aerosols injected into the upper atmosphere formed a cover of darkness, shielding the sun to the point that photosynthesis ceased or diminished for many weeks, depending on the location. A rain of material in at least some parts of the globe created a searing blanket of infrared radiation that microwaved any exposed plants or animals. This alone spelled doom for some species.

The effects of darkness were especially acute at lower latitudes and areas closer to the impact, such as in North America. Plants unaccustomed to lower light regimes caused by seasonal changes in the sun's position were especially hard hit. Higher-latitude plants accustomed to seasonally lower light regimes survived much better, as did the animals that fed on them. The effects on higher-latitude plants and animals were tempered by which season they were experiencing when the impact occurred. Extinction rates for coastal plants in North America soared because of the cumulative effects of continued habitat loss, drought, and loss of sunlight, as well as possibly because of being closer to the impact site.

Except for the sharks and relatives, who had already departed or become extinct as the seas regressed, all ectothermic, aquatic vertebrate species (bony fishes, amphibians, turtles, champsosaurs, and crocodilians) weathered the impact quite well in their still-flourishing freshwater habitats. Their watery cover also did a good job of shielding them from infrared radiation.

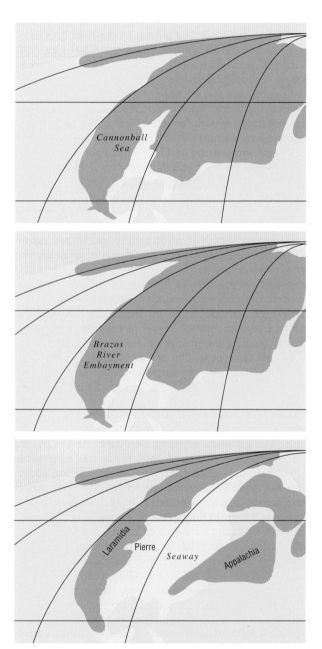

Figure 4. North American biogeography in the latter part of the Cretaceous Period showing the major western continent of Laramidia, the eastern continet of Applachia. Separated by the shallow epicontinental Pierre Seaway (bottom), the same region near the K–Pg boundary showing the Brazos River Embayment (middle) and near the mid Paleocene showing the Canonball Sea (top) (Archibald 2011). Reproduced by permission of Gale, a part of Cengage Learning.

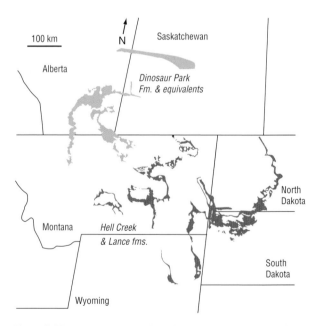

Figure 5. The mapped areas of the Campanian (-75 MYA) Dinosaur Park Formation and its equivalents with 48,000 sq km yielding 47 dinosaur species, compared to the mapped areas of the Maasirichian (-66 MYA) Lance and Hell Creek formations with 57,000 sq km but only 26 dinosaur species. This is a 45 percent drop in species diversity (after Archibald 2011). Reproduced by permission of Gale, a part of Cengage Learning.

With the added loss of more plant species and the reduction of biomass that the impact wrought in the already highly stressed ecosystem on land, other vertebrate species rapidly succumbed. Most notable were the last of the large herbivorous non-avian dinosaurs. The remaining predaceous non-avian dinosaurs followed soon after, with the larger species disappearing first. In some places on the globe the great saurians may have lingered a while longer, but finally, for the first time in more than 150 million years, no large land vertebrates graced Earth. The landscape was open and waiting for evolution's next gambit—the rise of the mammals.

Resources

Books

Archibald, J. David. *Dinosaur Extinction and the End of an Era: What the Fossils Say.* New York: Columbia University Press, 1996.

Archibald, J. David. *Extinction and Radiation: How the Fall of Dinosaurs Led to the Rise of Mammals.* Baltimore: Johns Hopkins University Press, 2011.

Archibald, J. David. "Dinosaur Extinction: Past and Present Perceptions." In *The Complete Dinosaur*, 2nd ed., edited by M.K. Brett-Surman, Thomas R. Holtz Jr., James O. Farlow, and Bob Walter, 1027–1038. Bloomington: Indiana University Press, 2012.

Archibald, J. David, and David E. Fastovsky. "Dinosaur Extinction." In *The Dinosauria*, 2nd ed., edited by David B. Weishampel, Peter Dodson, and Halszka Osmólska, 672–684. Berkeley: University of California Press, 2004.

Buffetaut, Eric. *La fin des dinosaures* [The end of the dinosaurs]. Paris: Fayard, 2003.

Charig, Alan J. *A New Look at the Dinosaurs.* New York: Facts on File, 1983.

Charig, Alan J. "The Cretaceous–Tertiary Boundary and the Last of the Dinosaurs." In *Evolution and Extinction*, edited by William G. Chaloner and Anthony Hallam, 147–158. Cambridge, U.K.: Cambridge University Press, 1994.

Coles, Graham. "A Comparison of the Evolution, Diversity, and Composition of the Cainozoic Ostracoda in the Deep Water North Atlantic and Shallow Water Environments of North America and Europe." In *Ostracoda and Global Events*, edited by Robin Whatley and Caroline Maybury, 71–86. London: Chapman and Hall, 1990.

Erwin, Douglas H. *The Great Paleozoic Crisis: Life and Death in the Permian.* New York: Columbia University Press, 1993.

Erwin, Douglas H. *Extinction: How Life on Earth Nearly Ended 250 Million Years Ago.* Princeton, NJ: Princeton University Press, 2006.

Evans, Susan E., and Marc E.H. Jones. "The Origin, Early History, and Diversification of Lepidosauromorph Reptiles." In *New Aspects of Mesozoic Biodiversity*, edited by Saswati Bandyopadhyay, 27–44. Heidelberg, Germany: Springer, 2010.

Håkansson, Eckart, and Erik Thomsen. "Distribution and Types of Bryozoan Communities at the Boundary in Denmark." In *Cretaceous–Tertiary Boundary Events Symposium.* Vol. 1, *The Maastrichtian and Danian of Denmark*, edited by Tove Birkelund and Richard G. Bromley, 78–91. Copenhagen: University of Copenhagen, 1979.

Harwood, David M. "Upper Cretaceous and Lower Paleocene Diatom and Silicoflagellate Biostratigraphy of Seymour Island, Eastern Antarctic Peninsula." In *Geology and Paleontology of Seymour Island, Antarctic Peninsula*, edited by Rodney M. Feldmann and Michael O. Woodburne, 55–129. Geological Society of America Memoir 169. Boulder, CO: Geological Society of America, 1988.

House, M.R. "Fluctuations in Ammonoid Evolution and Possible Environmental Controls." In *The Ammonoidea: Environment, Ecology, and Evolutionary Change*, edited by M.R. House, 13–34. Oxford: Clarendon Press, 1993. Published for the Systematics Association.

Johansen, Marianne Bagge. *Brachiopods from the Maastrichtian–Danian Boundary Sequence at Nye Kløv, Jylland, Denmark.* Fossils and Strata 20. Oslo: Universitetsforlaget, 1987.

MacLeod, Norman. *The Great Extinctions: What Causes Them and How They Shape Life.* London: Natural History Museum Publications, 2013.

Smit, Jan. "Extinction and Evolution of Planktonic Foraminifera after a Major Impact at the Cretaceous/Tertiary Boundary." In *Geological Implications of Impacts of Large Asteroids and Comets on the Earth*, edited by Leon T. Silver and Peter H. Schultz, 329–352. Geological Society of America Special Paper 190. Boulder, CO: Geological Society of America, 1982.

Smith, Alan G., David G. Smith, and Brian M. Funnell. *Atlas of Mesozoic and Cenozoic Coastlines.* Cambridge, U.K.: Cambridge University Press, 1994.

Taylor, P.D., and G.P. Larwood. "Mass Extinctions and the Pattern of Bryozoan Evolution." In *Extinction and Survival in the Fossil Record*, edited by G.P. Larwood, 99–119. Oxford: Clarendon Press, 1988. Published for the Systematics Association.

Viskova, L.A. "The Dynamics of Diversity of Gymnolaemata around the Cretaceous–Paleogene Crisis." In *Fossil and Living Bryozoa of the Globe*, 61. Perm: All-Russian Paleontological Society, 1994.

Whatley, Robin. "Ostracoda and Global Events." In *Ostracoda and Global Events*, edited by Robin Whatley and Caroline Maybury, 3–24. London: Chapman and Hall, 1990.

Periodicals

Alvarez, Luis W., Walter Alvarez, Frank Asaro, and Helen V. Michel. "Extraterrestrial Cause for the Cretaceous–Tertiary Extinction." *Science* 208, no. 4448 (1980): 1095–1108.

Arens, Nan Crystal, and Ian D. West. "Press-Pulse: A General Theory of Mass Extinction?" *Paleobiology* 34, no. 4 (2008): 456–471.

Benton, M.J. "Scientific Methodologies in Collision: The History of the Study of the Extinction of the Dinosaurs." *Evolutionary Biology* 24 (1990): 371–400.

Bourgeois, Joanne, Thor A. Hansen, Patricia L. Wiberg, and Erle G. Kauffman. "A Tsunami Deposit at the Cretaceous–Tertiary Boundary in Texas." *Science* 241, no. 4865 (1988): 557–570.

Brusatte, Stephen L., Richard J. Butler, Albert Prieto-Márquez, and Mark A. Norell. "Dinosaur Morphological Diversity and the End-Cretaceous Extinction." *Nature Communications* 3 (2012): 804.

Canudo, J.I., G. Keller, and E. Molina. "Cretaceous/Tertiary Boundary Extinction Pattern and Faunal Turnover at Agost and Caravaca, S.E. Spain." *Marine Micropaleontology* 17, nos. 3–4 (1991): 319–341.

Capetta, Henri. "Extinctions et renouvellements fauniques chez les Sélachiens post-jurassiques." *Mémoires de la Société géologique de France*, n.s., no. 150 (1987): 113–131. Translated by Jess Duran in 2002 as "Extinctions and Faunal Renewals among Post-Jurassic Selachians."

Courtillot, Vincent E. "A Volcanic Eruption." *Scientific American* 263, no. 4 (1990): 85–92.

Hansen, Thor A., R.B. Farrand, H.A. Montgomery, et al. "Sedimentology and Extinction Patterns across the Cretaceous–Tertiary Boundary Interval in East Texas." *Cretaceous Research* 8, no. 3 (1987): 229–252.

Hansen, Thor A., B. Upshaw III, E.G. Kauffman, and W. Gose. "Patterns of Molluscan Extinction and Recovery across the Cretaceous-Tertiary Boundary in East Texas: Report on New Outcrops." *Cretaceous Research* 14 (1993): 685–706.

Harwood, David M., and Rainer Gersonde. "Lower Cretaceous Diatoms from ODP Leg 113, Site 693 (Weddell Sea). Part 2: Resting Spores, Chrysophycean Cysts, an Endoskeletal Dinoflagellate, and Notes on the Origin of Diatoms." *Proceedings of the Ocean Drilling Program, Scientific Results* 113 (1990): 403–425.

Hildebrand, Alan R., Glen T. Penfield, David A. Kring, et al. "Chicxulub Crater: A Possible Cretaceous/Tertiary Boundary Impact Crater on the Yucatán Peninsula, Mexico." *Geology* 19, no. 9 (1991): 867–871.

Iglesias, Ari, Peter Wilf, Kirk R. Johnson, et al. "A Paleocene Lowland Macroflora from Patagonia Reveals Significantly Greater Richness than North American Analogs." *Geology* 35, no. 10 (2007): 947–950.

Jepsen, Glenn L. "Terrible Lizards Revisited." *Princeton Alumni Weekly* 64, no. 10 (1963): 6–10, 17–19.

Keller, G., T. Adatte, S. Gardin, et al. "Main Deccan Volcanism Phase Ends Near the K–T Boundary: Evidence from the Krishna-Godavari Basin, SE India." *Earth and Planetary Science Letters* 268, nos. 3–4 (2008): 293–311.

Kienel, Ulrike. "Die Entwicklung der kalkigen Nannofossilien und der kalkigen Dinoflagellaten-Zysten an der Kriede/Tertiär-Grenze in Westbrandenburg im Vergleich mit Profilen in Nordjütland und Seeland (Dänemark)." *Berliner Geowissenschaftliche Abhandlungen*, Reihe E 12 (1994): 1–87.

Kiessling, Wolfgang, and Rosemarie C. Baron-Szabo. "Extinction and Recovery Patterns of Scleractinian Corals at the Cretaceous–Tertiary Boundary." *Palaeogeography, Palaeoclimatology, Palaeoecology* 214, no. 3 (2004): 195–223.

Lidgard, Scott, Frank K. McKinney, and Paul D. Taylor. "Competition, Clade Replacement, and a History of Cyclostome and Cheliostome Bryozoan Diversity." *Paleobiology* 19, no. 3 (1993): 352–371.

Machalski, Marcin, and Claus Heinberg. "Evidence for Ammonite Survival into the Danian (Paleogene) from the Cerithium Limestone at Stevns Klint, Denmark." *Bulletin of the Geological Society of Denmark* 52, no. 2 (2005): 97–111.

MacLeod, Norman, P.F. Rawson, P.L. Forey, et al. "The Cretaceous–Tertiary Biotic Transition." *Journal of the Geological Society* (London) 154, no. 2 (1997): 265–292.

Marshall, Charles R., and Peter D. Ward. "Sudden and Gradual Molluscan Extinctions in the Latest Cretaceous of Western European Tethys." *Science* 274, no. 5291 (1996): 1360–1363.

Pope, Kevin O. "Impact Dust Not the Cause of the Cretaceous–Tertiary Mass Extinction." *Geology* 30, no. 2 (2002): 99–102.

Scholz, Henning, and Joseph H. Hartman. "Fourier Analysis and the Extinction of Unionoid Bivalves Near the Cretaceous–Tertiary Boundary of the Western Interior, USA: Pattern, Causes, and Ecological Significance." *Palaeogeography, Palaeoclimatology, Palaeoecology* 255, nos. 1–2 (2007): 48–63.

Sepkoski, J. John, Jr. "Ten Years in the Library: New Data Confirm Paleontological Patterns." *Paleobiology* 19, no. 1 (1993): 43–51.

Sepkoski, J. John, Jr. "A compendium of fossil marine animals." Bulletins of American Palaeontology 363 (2002): 1–563.

Sheehan, Peter M., and Thor A. Hansen. "Detritus Feeding as a Buffer to Extinction at the End of the Cretaceous." *Geology* 14, no. 10 (1986): 868–870.

Surlyk, Finn, and Marianne Bagge Johansen. "End-Cretaceous Brachiopod Extinctions in the Chalk of Denmark." *Science* 223, no. 4641 (1984): 1174–1177.

Wilf, Peter, and Kirk R. Johnson. "Land Plant Extinction at the End of the Cretaceous: A Quantitative Analysis of the North Dakota Megafloral Record." *Paleobiology* 30, no. 3 (2004): 347–368.

Other

Carter, Grace E., and Gregory P. Wilson. "Amphibian Paleocommunity Dynamics of the Hell Creek Formation in Northeastern Montana and the Cretaceous–Tertiary Extinction Event." *Abstracts for the North American Paleontological Convention (NAPC 2009)*: 331.

Holroyd, Patricia A., Gregory P. Wilson, and J. Howard Hutchison. "Turtle Diversity through the Latest Cretaceous of the Hell Creek Formation, Montana." *Abstracts for the North American Paleontological Convention (NAPC 2009)*: 146.

Thompson, Anna, Nan Crystal Arens, and A. Hope Jahren. "Vegetation Indicators of Environmental Stress Precede the Cretaceous/Tertiary Boundary." *Abstracts for the North American Paleontological Convention (NAPC 2009)*: 151.

J. David Archibald
Norman MacLeod

Paleocene–Eocene turnover

The Paleocene–Eocene (P–E) epochal boundary marks an episode of extremely rapid, planet-wide biotic turnover that occurred approximately 56 million years ago (MYA). Although not a mass extinction event along the lines of the end-Permian or the end-Cretaceous events, some land and marine organisms did suffer this fate whereas many others rapidly evolved, migrated vast distances around the globe, and/or changed their life strategies. For mammals, turnover at the P–E boundary was perhaps the most extensive of any in their history. Only recently has the probable cause of this widespread biotic event been discovered. In the early 1990s, ocean sediment cores revealed a vast perturbation in Earth's carbon cycle at the P–E boundary that raised global temperatures by 9°F to 14°F (5°C to 8°C) in a geological instant. In the years since its discovery, this severe warming has become known as the Paleocene–Eocene Thermal Maximum (PETM) and most now believe it drove the P–E boundary biotic turnover. Life, climate, and the environment during the PETM have been studied extensively, and more than 400 scientific publications in two decades have been devoted to its causes and consequences. The result is that this event is now extraordinarily well documented. Many also consider it to be the best available analogue for modern, anthropogenic climate change.

The evidence

The Paleocene and Eocene (65–56 and 56–34 MYA, respectively) are the first two epochs of the Cenozoic era. The Cenozoic is known as the Age of Mammals and encompasses Earth's history from the demise of non-avian dinosaurs to the present. The fossil record of the early part of the Cenozoic provides a reasonably good, though far from complete, understanding of life on most of the continents and in the oceans during the Late Paleocene and Early Eocene. However, until recently, the P–E transition itself was virtually undocumented. In the mid-1980s, Philip D. Gingerich realized that a few fossils collected from a distinctive sandstone between Clarkforkian- and Wasatchian-aged deposits (Late Paleocene and Early Eocene North American Land Mammal Ages, respectively) in the northern part of Wyoming's Bighorn Basin represented a unique, short-lived assemblage of mammals, fish, and reptiles from the earliest part of the Eocene. Although fossil collecting in this area was arduous,

Gingerich and his field crews from the University of Michigan persisted, eventually amassing a sample of nearly 450 specimens. This enabled Gingerich to publish a description of this unique assemblage in 1989. The assemblage was found to be transitional between Clarkforkian and Wasatchian faunas in some respects and surprisingly unique in others. Gingerich assigned it to a new biozone, Wa0, which preceded the previously recognized Wasatchian biozones Wa1 to Wa7. The Wa0 sample was correlated with the only other earliest Eocene mammals known at the time, fragmentary assemblages from the Sparnacian (now Ypresian stage) in Europe.

Two years later, the Wa0 fauna assumed much greater significance following an analysis of the chemical composition of seafloor sediments from the Weddell Sea (Antarctica). Nearly complete, finely layered sedimentary sequences can be found on all ocean floors, the oldest stretching back to the mid-Jurassic. Beginning in the early 1970s, international scientific ocean drilling programs began recovering high-quality sediment cores from the floors of the world's oceans. Many of these cores span the P–E boundary. Their sediments consist primarily of the fossilized hard parts of several groups of unicellular microorganisms loosely referred to as plankton. The hard parts of these tiny organisms are formed from elements absorbed from the surrounding seawater. Foraminiferans (forams) is the most prominent of these groups, including both benthic (bottom-dwelling) and planktic (freely floating) ameboid forms. Another well-represented group is Ostracoda, which consists of small, flattened, mostly benthic crustaceans sometimes called seed shrimp. Surface-water dwelling plankton with hard parts include the coccolithophores, algae with tiny plates that accumulate in vast quantities to form chalk, and the dinoflagellates, which have whip-like protuberances and protective cysts. When these various microorganisms die, their hard parts rain down through the ocean water and form continuous layers on the ocean floor.

In 1991, James P. Kennett and Lowell D. Stott sought to shed light on the cause of the mass extinction of benthic forams that occurred at the P–E boundary by examining the ratios of stable oxygen and carbon isotopes preserved in forams in the Ocean Drilling Project 690B core from the Weddell Sea. They identified a brief, abrupt decrease in the ratios of heavy (^{18}O) to light (^{16}O) oxygen and heavy (^{13}C) to

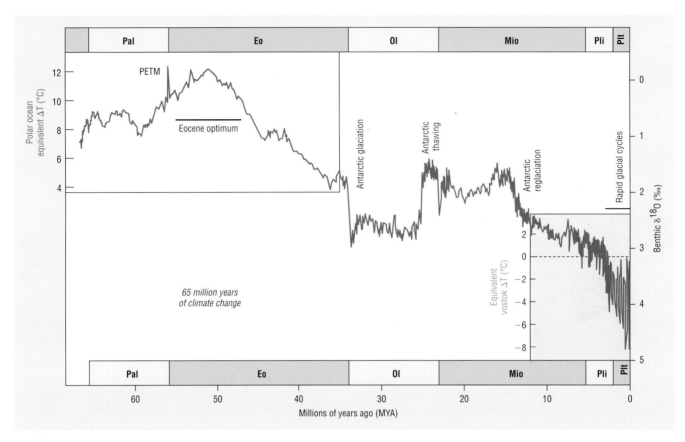

Climate change during the last 65 million years. Reproduced by permission of Gale, a part of Cengage Learning.

light (^{12}C) carbon isotopes at the P–E boundary. The ratio of heavy to light oxygen in water is strongly related to the temperature of the water when other factors are held constant (e.g., salinity, latitude). Because of slight differences in the reactive properties of the isotopes, cooler water tends to be enriched in heavy oxygen, whereas warmer water is enriched in light oxygen. The shift in the foram oxygen isotopes at the P–E boundary suggested that, for a few hundred thousand years, ocean waters were much warmer than previously. The coincident decrease in carbon isotope ratios held the key to

Badlands in the Bighorn Basin of Wyoming. © franzfoto.com/Alamy.

the cause of the ocean warming. The ratio of heavy to light carbon is indicative of the major sources of carbon in the ocean-atmosphere system. Plants preferentially fix light carbon during respiration, so plants and plant-based carbon reservoirs (e.g., peat, coal, natural gas) are described as sources of light carbon. The decrease in the foram carbon isotope ratios suggested that there was a sudden injection of an enormous amount of light, organically derived carbon into the ocean-atmosphere system that resulted in global warming through greenhouse effects.

Within a few months of Kennett and Stott's publication, the shift in carbon isotope ratios, which is now known as the carbon isotope excursion (CIE), had been identified in the equatorial Pacific, North and South Atlantic, and Indian oceans. The following year (1992), Paul L. Koch and colleagues identified the CIE in the carbon isotopes of Wa0 mammal tooth enamel and co-occurring fossil soil nodules from the northern Bighorn Basin in Wyoming, thereby firmly linking this marine event to terrestrial biotic changes. At that time, it was generally agreed that the abrupt warming must have been global in nature, and after several iterations (Initial Eocene Thermal Maximum, IETM, and Late Paleocene Thermal Maximum, LPTM), it became known as the PETM.

The importance of the PETM was apparent to the scientific community immediately, and an extensive, collaborative effort began to search for more records of this important interval and to better study those already known.

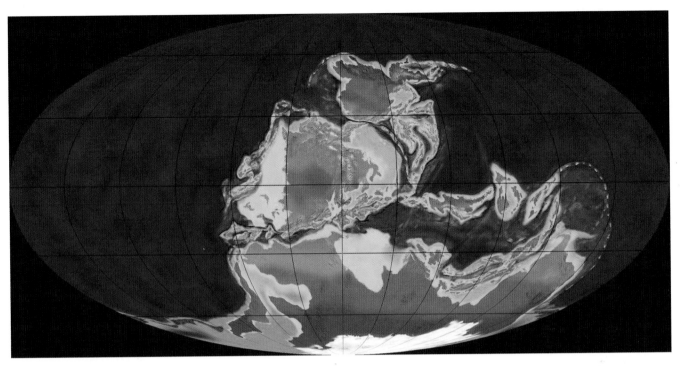

Configuration of the continents during the Eocene. Courtesy of Ron Blakey.

Two decades later, Francesca A. McInerney and Scott L. Wing (2011) counted more than 30 marine cores and nearly 50 continental marine outcrops documenting the PETM described in the scientific literature. These reveal oceanic conditions and planktonic response to the PETM, but they include no record of ocean life larger than plankton.

Terrestrial records of the PETM remain rare in spite of extensive exploration. By far the most complete and informative is Wyoming's Bighorn Basin. During the Early Cenozoic, enormous amounts of sediment accumulated in northwestern Wyoming and southern Montana as the Rocky Mountains rose. These are now exposed in widespread, deeply eroded badlands. Following Gingerich's discovery of the Wa0 fauna, the Polecat Bench and other parts of the Bighorn Basin in which the Clarkforkian/Wasatchian boundary was preserved have been the focus of intensive paleontological work. Suzanne G. Strait from Marshall University described a collection of approximately 300 isolated teeth of PETM mammals found by sifting sediments through finely meshed screens in the Honeycombs area in the southern Bighorn Basin. The most extensive PETM faunal assemblage now known from anywhere in the world is from the Sand Creek Divide near the middle of the basin. Kenneth D. Rose and crews from Johns Hopkins University collected in this area intensively from 2001 to 2009, eventually describing a sample of more than 1,000 mammal specimens. After targeted searching, Scott L. Wing and crews from the Smithsonian Institution found the first fossilized leaves from the PETM in the Cabin Fork area of the Bighorn Basin. Others have studied fossil pollen, fossil soils, and trace fossils of soil fauna and herbivorous insects around the basin. Further stable isotope work has documented the CIE and coincident fluctuation in oxygen isotopes in the teeth of other groups of fossil mammals, freshwater fish scales, bivalve shells, and fossil soils.

A handful of other terrestrial sites in North America (Mississippi, Texas, Utah), South America (Colombia, Venezuela), Europe (Belgium, France, Spain, United Kingdom), Asia (China, India), and New Zealand document PETM faunas, floras, pollen, and soils. PETM faunal records outside of the Bighorn Basin are fragmentary and restricted to a few sites on the northern continents (Belgium, France, China). PETM floras, pollens, and soils are more widely distributed and these are extremely useful for reconstructing continental climate and environment, which cannot be extrapolated directly from oceanic patterns. Terrestrial mean annual temperature (MAT) can be estimated both from oxygen isotope ratios and by leaf margin analysis, which is based on the strong positive correlation between MAT in modern settings and the proportion of plant species with untoothed leaf margins. Moisture and broad vegetation patterns can be estimated from fossil pollen and soil characteristics. At present, knowledge of regional climatic and environmental conditions and short-term biotic response to the PETM warming is limited to the smallest ocean dwellers and the northern continents, particularly the western interior of North America. Fossil pollen has been found in New Zealand, but terrestrial biotic response on the southern continents is otherwise unknown during the PETM.

The setting

During the Paleocene, the continents were arranged in a nearly modern configuration, except that the Atlantic Ocean

A microscopic view of ocean sediment cores laid down before (top) and after (bottom) sudden releases of carbon dioxide during the Paleocene–Eocene Thermal Maximum 55.8 million years ago. The "sudden" release of CO_2 caused acidification of the oceans and the dissolution of the cocolithophores that are a major component of the ocean food web. It also caused major extinctions in the ocean. Courtesy of Brian Huber.

was narrower and the Indian subcontinent was separate from Asia until the end of the epoch. Global temperatures were cooler than they had been during the Cretaceous, but still much warmer than today with higher concentrations of greenhouse gases and no polar ice. Ocean temperatures were equitable, while on land, tropical and paratropical conditions extended to the middle latitudes and subtropical conditions extended to the poles. There was little seasonality and rainfall was much higher and probably more continuous throughout the year than at present. This time is generally referred to as the "Cenozoic Greenhouse." In the Late Paleocene (60–58 MYA), global temperatures entered into a 10-million-year-long trend of concerted and continuous warming that culminated in the Early Eocene Climatic Optimum (EECO) at the end of the Early Eocene (53–50 MYA). The EECO was the warmest period of the Cenozoic. This gradual warming is generally attributed to mantle-related processes such as continental drift, seafloor spreading, sea-level changes, and increased atmospheric carbon dioxide from mantle or volcanic sources. Volcanic activity was much more frequent in the

Early Cenozoic than it is today. The PETM is the first, best-studied, and most severe of several "hyperthermal" events of brief and intense warming now known to have punctuated this gradual trend.

The Paleocene began immediately after the well-known Cretaceous–Paleogene (K–Pg) mass extinction 65 million years ago that eradicated non-avian dinosaurs and many other groups of organisms. Many families of plants disappeared at the K–Pg boundary, and initially Paleocene forests were dominated by ferns. However, flowering plants and conifers soon rebounded. Various kinds of dense tropical and paratropical forests, swamps, and jungles covered much of Earth. With no large animals to thin them, the forests were probably denser than at other times. Grassland habitats had not yet evolved. By the Late Paleocene, dawn redwood forests were found in the Arctic Circle. These high-latitude forests were unique in that they endured polar light regimes. Trees in these forests had broad leaves to capture the summer sun and likely shed these leaves during the winter darkness. Most Late Paleocene plant species are attributed to modern groups. Mid-latitude swamp forests were dominated by conifers related to the bald cypress. Drier forests consisted of sycamore, elm, walnut, tea, laurel, and birch relatives. Modern palms and cacti made their first appearances at this time. Plant-eating insects were depressed throughout the Paleocene, recovering from the K–Pg event only by the end of the epoch. Although insects rarely fossilize, herbivorous insects leave traces of their activity in feeding damage on fossil leaves.

The K–Pg mass extinction eliminated all land animals weighing more than approximately 50 pounds (25 kg). This included the pterosaurs, some families of birds, and many metatherian mammals in North America, but not South America. The surviving birds and eutherian mammals began to radiate into the ecological niches vacated by the larger animals and eventually became the dominant land animals of the era. Because their bones are fragile and they lack teeth, which readily fossilize, very little is known about the earliest Cenozoic radiation of birds. From the early part of the Paleocene, a few forms of shorebirds (including the flamingo-like *Scaniornis*) have been discovered, as well as the humeri (upper arm bone) of apparently goose-like and petrel-like forms. By the Late Paleocene, more fossil birds are known, including large, flightless meat-eating birds in Europe (gastornithids) and "terror birds" in South America (phorusrhacids), penguin and owl-like birds, and possibly the earliest cuckoo. This is undoubtedly a great underrepresentation of the variety and abundance of birds that existed at this time.

In contrast, the sturdier bones and teeth of mammals are more readily converted into rock, and Early Cenozoic mammals have left a fairly extensive fossil record. Earliest Paleocene mammals were initially small, insectivorous, or generalized-feeder forest-dwellers, but diversified rapidly. Before the end of the Paleocene, mammals included larger, stumpier forms (the graceful, long-legged runners of the open grasslands had not yet evolved) and more varied diets. Many Paleocene mammal groups are described as archaic, meaning they have no living close relatives. The archaic group that typifies and epitomizes the Paleocene is the condylarths, an informal assemblage of hoofed and clawed predecessors of most of the modern and extinct groups of ungulates. In the Late Paleocene, the major condylarth families included the hyopsodontids, small, squirrel-like creatures; arctocyonids, raccoon or bear-like omnivores; phenacodontids, common, small to sheep-sized omnivores and herbivores that often comprise half or more of the specimens recovered from North American Late Paleocene fossil localities; and didolodontids, a group similar to phenacodontids but limited to South America.

Other archaic Late Paleocene mammal groups include the multituberculates, mouse to beaver-sized rodent-like animals; plesiadapiformes, small, primitive animals closely related to primates that had a variety of diets and lifestyles; palaeodonts, burrowing anteater-like forms; oxyaenid creodonts, flat-footed cat-like ambush predators; mesonychids, relatively rare, hyena-like scavengers; and pantodonts and uintatheres, large, semi-aquatic river dwellers. In North America there were also taeniodonts, a group of tusked and clawed diggers of roots and tubers. In South America, the insectivorous and carnivorous mammals were marsupials (didelphids and borhyaenids, respectively); eutherian mammals evolved into myriad omnivores and herbivores; and the first armadillo relative appeared. In Asia, there were arctostylopids, hyrax-like animals with unusual, cresty teeth. Some modern groups had also made their appearances by the Late Paleocene, including the first rodents, which originated in Asia and spread rapidly throughout the northern continents, likely outcompeting the plesiadapiformes and multituberculates as they migrated. In Europe and North America, there were some true insectivorans distantly related to modern shrews and hedgehogs, and the first true carnivorans, small marten or civet-like forms that may have been distant ancestors of modern cats.

Additional important components of Late Paleocene terrestrial faunas include insects, snails, lizards, snakes, amphibians, and aquatic reptiles, such as turtles, crocodilians, and champsosaurs, a freshwater, gharial-like fish-eater. These reptiles were important members of the ecosystem. Crocodilians were among the apex predators in North America and were found as far north as the Arctic Circle. Some large champsosaurs reached 10 feet (3 m) in length. In the oceans, many groups had disappeared at the K–Pg boundary, including the large marine reptiles, ammonites, belemnites, and families of mollusks, plankton, and teleost (bony) fishes. Marine diversity was initially depressed, but by the Late Paleocene ocean faunas began to resemble modern ones except that they lacked marine mammals and charcharinid sharks. Mackerel sharks, sand tiger sharks, and the first small-toothed white sharks had replaced the lost marine reptiles as the dominant ocean predators, and new groups of teleost fishes had become quite common, along with modern forms of sea urchins, forams, gastropods, and bivalves.

The PETM

Experts estimate that between 4,740 and 16,976 gigatons (4,300 and 15,400 petagrams) of light, organic carbon were released into the ocean-atmosphere system at the beginning of the PETM (1 gigaton equals about 2 trillion pounds). All of the carbon in Earth's fossil fuel reservoir today is estimated to weigh approximately 4,950 gigatons (4,500 petagrams). The most likely form of this carbon was methane, which is a greenhouse gas that readily oxidizes to form carbon dioxide, the most notorious of the greenhouse gases. Where did this vast amount of methane come from and what triggered its release? In 1995, Gerald R. Dickens and colleagues proposed that it came from oceanic methane hydrates. Methane is produced by anaerobic bacteria on the ocean floor and normally trapped there in lattices of frozen water, stable only at the near-freezing temperatures and high pressure of the deep ocean. Methane hydrates are relatively unstable and can be induced to release their trapped methane gas through the warming of deep ocean water, which may have occurred during the long-term Late Paleocene warming trend or through the reversal of deep ocean water circulation triggered by volcanism around the Caribbean. Methane hydrates might also release their methane through physical disruption, such as from submarine seismicity, volcanism, or deep-sea sediment slumping and slope failure.

Although most scientists favor the methane hydrate hypothesis, there are many alternatives, some of which have recently begun to attract attention. In 2004 Henrik Svensen and colleagues described an extensive submarine network of volcanic gas and fluid seeps in the northeast Atlantic. The injection of magma into these organic-rich sediments is likely to have released vast amounts of methane. Others (see McInerney and Wing 2011) have suggested that enormous quantities of light carbon may have accumulated as permafrost and peat in Antarctica during the Paleocene and melted at the time of the PETM. The discovery of other, smaller hyperthermals following the PETM has raised the possibility that orbital forcing played a role as well. Orbital forcing is the variation in the intensity of the solar radiation warming Earth through quasi-periodic fluctuations in Earth's orbital shape, axial tilt, and precession. The PETM and subsequent hyperthermals appear to correspond to maxima in long-term (100,000 year) precessional cycles, which may have periodically boosted long-term, Late Paleocene warming or resulted in climatic instability. More than one source of light carbon might also have been tapped sequentially as an initial release of methane might have started a feedback loop in which rising global temperatures precipitated the release of further methane from similar or other stores. Evidence is mounting for multiple releases and therefore possibly multiple sources of PETM methane.

Whatever the source and trigger, the PETM event is usually considered to have occurred in three distinct phases: initiation, body, and recovery. Most estimates of the length of the initiation phase are between 10,000 and 20,000 years. During this time, enormous amounts of methane were released into the ocean-atmosphere system, causing one to three doublings of the concentration of atmospheric carbon dioxide and dramatic environmental change. Global temperatures began to warm, and this warming was amplified at high latitudes and in the deep oceans to the extent that oceanic latitudinal and depth temperature gradients virtually disappeared. The loss of latitudinal oceanic temperature gradients probably led to a decrease in the intensity of atmospheric heat transport and circulation. The loss of surface-to-deep ocean water temperature gradients probably led to a profound change in oceanic circulation, resulting in warm, salty bottom waters with low levels of dissolved oxygen. Also during this phase, atmospheric carbon dioxide would have begun dissolving back into the oceans, causing acidification and lowering of the carbonate ion content of seawater. Near-shore marine waters indicate freshening and nutrient influx from increased continental runoff.

During the body of the PETM, the Earth settled into an alternative state for approximately 100,000 years. Carbon isotope ratios were low and stable indicating that methane release had ceased, although global temperatures apparently continued to rise during the first 30,000 years of the body of the PETM. Peak temperatures were eventually reached and maintained during this phase. In total, sea surface temperatures increased by up to 9°F (5°C) at low latitudes and up to 16°F (9°C) at high latitudes. Deep ocean temperatures rose by 7°F to 9°F (4°C to 5°C). Bighorn Basin temperature proxies suggest that MAT rose between 9°F and 13°F (5°C and 7°C) in the mid-latitude lowlands of the western interior of North America. This translates to a MAT of around 75°F (24°C), suggesting that temperature in the northern Rocky Mountain region was similar to that of south Florida today. Temperature increase in nearby highland regions was closer to 4°F (2°C). Increased continental runoff in marginal marine sections has been interpreted as evidence of high-intensity precipitation or of highly seasonal, episodic precipitation and/or more intense storms on the continents. PETM floras, pollen, and soil traits in North America and parts of Europe indicate that rainfall was probably more seasonal than it had been in the Paleocene in the mid-latitudes at least. Oceanic and terrestrial biotic assemblages stabilized and existed throughout the body of the PETM.

The mechanisms that maintained the high temperatures and negative carbon isotope ratios during the body of the PETM are not clear. Following the addition of vast amounts of methane and carbon dioxide into the atmosphere, natural carbon cycle processes (e.g., weathering, formation of organic matter) would be expected to immediately begin to sequester, or return this carbon from the atmosphere to available carbon reservoirs, such as soils, forests, or marine sediments. Instead, this did not begin until the recovery phase of the PETM, which is estimated to have lasted 40,000 to 80,000 years following the body of the PETM. Once begun, the recovery phase proceeded rapidly at first and then steadied. To some, this suggests that the sequestration of carbon primarily occurred through the burial of organic matter, such as soil or peat, which would preferentially use up the plentiful light carbon. If so, this would favor the hypothesis of melting Antarctic permafrost as the source of the PETM methane, as this reservoir was then relatively rapidly recharged during the recovery phase. Also during the recovery phase, global temperatures gradually dropped and biotic assemblages adjusted to conditions typical of the Early Eocene. Biotic changes may have been the only permanent effects of the PETM.

Biotic consequences

Biotic response to the PETM included (1) limited extinction; (2) sudden appearance of new forms; (3) rapid turnover; (4) range extensions; and (5) transient dwarfing. Of these responses, only the first two resulted in permanent biotic alteration. The PETM is best known for the sudden appearance of major groups of land animals. These new groups appeared through rapid evolution and swift migration, dramatically and permanently changing the composition, community dynamics, and evolutionary trajectories of their newly adopted communities. Many groups of organisms also experienced dramatic turnover, which is the brisk replacement of individual species with new ones. A fourth hallmark of biotic response to the PETM was the sudden global distribution of a few notable species that had been previously limited to tropical regions. Finally, some mammals and insects were smaller during the PETM than either before or after the event.

The biggest casualties of the PETM were the benthic forams whose tests originally led to the discovery of the event. About 30 to 50 percent of benthic foram species on the floors of all of the oceans went extinct. Late Paleocene benthic forams were diverse and relatively large with heavily calcified tests. Many of these larger forms disappeared at the beginning of the PETM, leaving a few small survivors with thinner-walled or non-calcified tests, including the buliminids and *Nuttallides truempyi*. These extinctions are generally believed to have been caused by warming, low levels of oxygen and carbonate, and increased acidity of deep ocean water. On land, the only mammal family to experience major extinctions at the PETM was the Plesiadapidae, a family of ground squirrel or marmot-like plesiadapiformes that had been common in Europe and North America. The demise of the abundant Late Paleocene plesiadapid genera (particularly *Plesiadapis*) may have been related to competition with rodents that first appeared in the Late Paleocene coupled with a change in food resources or availability at the PETM.

One of the hallmarks of biotic response at the PETM is the rapid appearance and dispersal of major groups of land animals within and across the northern continents. The most striking example of this is the nearly instantaneous appearance of the modern Primates (monkeys, apes, and humans), Perissodactyla (horses, rhinos, and tapirs), Artiodactyla (deer, camels, and pigs), and hyaenodontid creodonts (archaic carnivores) across the northern hemisphere. All four groups flourished during the Eocene, becoming important members of their communities. The primates and hyaenodontids were particularly diverse, whereas the perissodactyls and artiodactyls came to typify later epochs. The primates, perissodactyls, and artiodactyls remain principal components of modern mammal faunas. Although the hyaenodontids went extinct by the end of the Miocene (c. 5 MYA), they were extremely successful during the Early Cenozoic and were much more widespread, diverse, and common than the oxyaenids, the original creodont family that had first appeared during the Paleocene. Later hyaenodontids became some of the largest carnivorous mammals that have ever existed (early Miocene *Megistotherium* may have weighed more than 1,763 pounds [800 kg]).

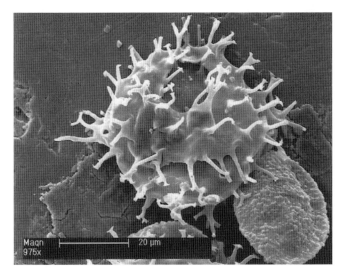

A dynamic climate and ecosystem state during the Paleocene–Eocene Thermal Maximum: inferences from dinoflagellate cyst assemblages on the New Jersey Shelf. Courtesy of Appy Sluijs.

The PETM perissodactyls consisted of the earliest horse, *Sifrihippus* (formerly *Hyracotherium*) *sandrae*, a browsing species the size of a small dog with three to four hoofed toes. The PETM artiodactyls consisted of the rabbit-sized *Diacodexis ilicus*, another browser similar to the Asian mouse deer. The earliest true primates included the omomyid genus *Teilhardina*, tiny animals that weighed about 1 ounce (30 grams) and resembled modern bush babies, and the adapoid genus *Cantius*, similar to modern sportive lemurs. There were several genera of PETM hyaenodontids, a group of dog-sized, slender carnivores that walked on their toes. Where these species originated is unclear, although some evidence exists that hyaenodontids lived in the latest Paleocene in Asia. The others likely arose within a few hundreds or thousands of generations in response to climate change at the PETM. High-resolution studies of *Teilhardina* have suggested a sequence of rapid emergence in Asia and dispersal to Europe and thence to North America. Migration routes were probably across northern land bridges opened up by global warming. Dispersal occurred within the first 10,000 to 25,000 years of the PETM, although *Cantius* may have appeared slightly later than the other species within the body of the PETM.

The sudden appearance of these species drove up diversity during the PETM and coincided with an enormous turnover that is best documented in the Wa0 mammal fauna from the Sand Creek Divide in the Bighorn Basin. Comparison of this sample with preceding late Clarkforkian and succeeding early Wasatchian Bighorn Basin faunas (Rose et al. 2012) demonstrates that diversity was conspicuously higher in Wa0. Moreover, 80 percent of species and nearly 50 percent of genera in the Wa0 sample were not present in the late Clarkforkian, which represents a dramatic turnover (both proportions were probably less than 25% at other times). Nearly one quarter of the new species was limited to Wa0 (this proportion was 5–10% in the other biozones). There were also relatively large proportions of new genera and

species at the beginning of the subsequent Wa1 biozone. This implies that the Sand Creek Divide Wa0 fauna was unique compositionally. Most of the species at the beginning of the PETM were new and many of them disappeared at the end of the event. Such high levels of diversity and turnover have not been documented at any other time in the Bighorn Basin except for the onset of the EECO more than one million years later, and even then changes in diversity and turnover were much less rapid.

Some evidence exists that turnover also occurred in European and Asian PETM mammal faunas, but the samples from these continents lack the resolution of the Bighorn Basin. Evidence also exists of rapid turnover and increased diversity in other groups of organisms. At least seven lizard species appeared and spread across North America. Several turtle species also appeared during or at the end of the PETM in North America and Asia. Pollen records from Colombia demonstrate that equatorial forests flourished during the PETM with few disappearances and enhanced speciation that

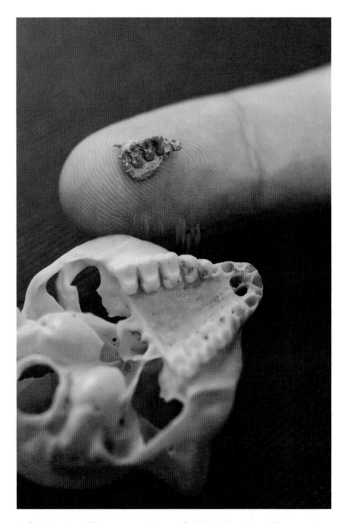

A Smithsonian Museum specimen of *Teilhardina brandti*'s upper jaw (top) compared to a Florida Museum of Natural History specimen of the skull of a tarsier. Notice how small the primate with the first known fingernails was, given that it had such a tiny jaw. Courtesy of Kristen Grace.

must have led to increased richness. In the warm surface waters of the oceans, new forms of planktic forams rapidly appeared and existed only during the PETM, while coccolithophores experienced their highest rates of origination and extinction in the Cenozoic.

Some groups of organisms reacted to changing temperatures and environmental conditions with sudden, dramatic, temporary shifts in geographic range. In the western interior of North America, the Late Paleocene forests of dawn redwood, alders, sycamores, walnuts, sassafras, and dogwoods were temporarily and entirely replaced by a unique plant community adapted to dry tropical to subtropical conditions. The newcomers were mostly legumes from the bean family, and warm-adapted relatives of the poinsettias, sumacs, and paw-paws. Many of these species must have moved rapidly northward more than 621 miles (1,000 km) from around the Mississippi Embayment, precursor of the Gulf of Mexico. As temperatures decreased in the recovery phase of the PETM, relatives of the lindens and wingnuts arrived from Europe and remained to become part of Eocene plant communities in North America. In the oceans, the tropical surface-dwelling foram genera *Morozovella* and *Acarinina* spread to polar waters during the body of the PETM. The tropical dinoflagellate genus *Apectodinium* became so abundant and globally distributed that its record during the PETM is known as the "*Apectodinium* acme." Warm-water coccolithophores also spread to high latitudes replacing cool-adapted forms.

One final, notable biotic response to the PETM was a temporary decrease in the body size of many different herbivorous mammals and soil insects. One hypothesis relating these changes to the elevated concentrations of atmospheric carbon dioxide is attractive because it accounts for dwarfing in mammals and soil insects as well as another observation of PETM biotic response: a distinct increase in the frequency of insect feeding damage on fossilized leaves.

Elevated concentrations of carbon dioxide have been shown to reduce leaf digestibility and protein content. Herbivorous mammals and soil-dwelling insects that feed on plants might, therefore, be expected to be smaller, whereas leaf-eating insects would have had to feed more voraciously to extract the proper nutrition. However, it must be noted that simpler and more direct explanations exist for each of these biotic changes. The simplest explanation of dwarfing in mammals is that the smaller forms came from elsewhere, likely moving from lower to higher latitudes. In the face of increasing temperatures, herbivorous insects respond with elevated metabolic rates and year-round breeding, which might lead to smaller forms and more intense feeding. Dwarfing in soil insects could also be related to decreased soil moisture.

Implications for the future

Study of the PETM has allowed scientists to determine the connections between different components of Earth's biotic and abiotic systems; to determine how these systems recover from perturbation; and to predict the transient versus irrevocable consequences of dramatic and severe global warming. However, much about the PETM is still unknown, including the causes of the carbon cycle perturbation and the impacts of the PETM in the southern hemisphere and in marine organisms larger than plankton. It is estimated that combustion of the entire fossil fuel reservoir (around 4,950 gigatons [4,500 petagrams) will occur in the next 300 years. This amount may equal or exceed that expelled during the PETM, but the rate of expulsion will be much faster, suggesting that the effects of anthropogenic activity will be more severe. It is important to know whether human activity might initiate a cascade of carbon release, leading to uncontrollable warming that could be devastating for the modern world. At the very least, melting polar ice sheets would flood many densely populated coastal areas and impact agriculture, leading to widespread human displacement and starvation.

Resources

Books

Aubry, Marie-Pierre, Spencer G. Lucas, and William A. Berggren, eds. *Late Paleocene-Early Eocene Climatic and Biotic Events in the Marine and Terrestrial Records*. New York: Columbia University Press, 1998.

Janis, Christine. "Victors by Default." In *The Book of Life: An Illustrated History of the Evolution of Life on Earth*, edited by Stephen Jay Gould, 169–218. New York: Norton, 2001.

Rose, Kenneth D. *The Beginning of the Age of Mammals*. Baltimore: Johns Hopkins University Press, 2006.

Periodicals

Bowen, Gabriel J., David J. Beerling, Paul L. Koch, et al. "A Humid Climate State during the Paleocene/Eocene Thermal Maximum." *Nature* 432, no. 7016 (2004): 495–499.

Bowen, Gabriel J., Timothy J. Bralower, Margaret L. Delaney, et al. "Eocene Hyperthermal Event Offers Insight into Greenhouse Warming." *Eos* 87, no. 17 (2006): 165–169.

Bowen, Gabriel J., William C. Clyde, Paul L. Koch, et al. 2002. "Mammalian Dispersal at the Paleocene/Eocene Boundary." *Science* 295, no. 5562 (2002): 2062–2065.

Dickens, Gerald R., James R. Oneil, David K. Rea and Robert M. Owen. "Dissociation of Oceanic Methane Hydrate as a Cause of the Carbon-Isotope Excursion at the End of the Paleocene." *Paleoceanography* 10 (2005): 965–971.

Fricke, Henry C., and Scott L. Wing. "Oxygen Isotope and Paleobotanical Estimates of Temperature and δ^{18}O-latitude Gradients over North America during the Early Eocene." *American Journal of Science* 304, no. 7 (2004): 612–635.

Gingerich, Philip D. "Environment and Evolution through the Paleocene-Eocene Thermal Maximum." *Trends in Ecology and Evolution* 21, no. 5 (2006): 246–253.

Jardine, Phil. "The Paleocene-Eocene Thermal Maximum." *Palaeontology* [online]. January 10, 2011. Accessed April 25, 2012. http://www.palaeontologyonline.com/articles/2011/the-paleocene-eocene-thermal-maximum.

Kennett, James P., and Lowell D. Stott. "Abrupt Deep-Sea Warming, Palaeoceanographic Changes and Benthic Extinctions at the End of the Paleocene." *Nature* 353, no. 3641 (1991): 225–229.

Koch, Paul L., James C. Zachos, and Philip D. Gingerich. "Correlation Between Isotope Records in Marine and Continental Carbon Reservoirs near the Paleocene/Eocene Boundary." *Nature* 358, no. 6384 (1992): 319–322.

Kraus, Mary J., and Susan Riggins. "Transient Drying during the Paleocene-Eocene Thermal Maximum (PETM): Analysis of Paleosols in the Bighorn Basin, Wyoming." *Palaeogeography, Palaeoclimatology, Palaeoecology* 245, no. 3–4 (2007): 444–461.

McInerney, Francesca A., and Scott L. Wing. "The Paleocene-Eocene Thermal Maximum: A Perturbation of Carbon Cycle, Climate, and Biosphere with Implications for the Future." *Annual Review of Earth and Planetary Sciences* 39 (2011) 489–516.

Meng, Jin, and Malcolm C. McKenna. "Faunal Turnovers of Palaeogene Mammals from the Mongolian Plateau." *Nature* 394, no.6691 (1998): 364–367.

Nel, Andre, Gael de Ploeg, Jean Dejax, et al. "An Exceptional Sparnacian Locality with Plants, Arthropods and Vertebrates (Earliest Eocene, MP7): Le Quesnoy (Oise France)." *Comptes Rendus de l'Academie des Sciences Serie II Fascicule A-Sciences de la Terre et des Planetes* 329, no. 1 (1999): 65–72.

Rohl, Ursula, Thomas Westerhold, Timothy J. Bralower, and James C. Zachos. "On the Duration of the Paleocene-Eocene Thermal Maximum (PETM)." *Geochemistry, Geophysics, Geosystems: An Electronic Journal of the Earth Sciences* 8 (2007): 1–13. Accessed April 25, 2012. doi:10.1029/2007GC001784.

Rose, Kenneth D., Stephen G.B. Chester, Rachel H. Dunn, et al. "New Fossils of the Oldest North American Euprimate *Teilhardina Brandti* (Omomyidae) from the Paleocene-Eocene Thermal Maximum." *American Journal of Physical Anthropology* 146, no. 2 (2011): 281–305.

Smith, Jon J., Stephen T. Hasiotis, Mary J. Kraus, and Daniel T. Woody. 2009. "Transient Dwarfism of Soil Fauna during the Paleocene–Eocene Thermal Maximum." *Proceedings of the National Academy of Sciences* 106, no. 42 (2009): 17655–17660.

Strait, Suzanne G. "New Wa-0 mammalian fauna from Castle Gardens in the southeastern Bighorn Basin. *Paleocene-Eocene Stratigraphy and Biotic Change in the Bighorn and Clarks Fork Basins, Wyoming.*" University of Michigan Papers on Paleontology No. 33, (2001): 127–143.

Suess, Erwin, Gerhard Bohrmann, Jens Greinert, and Erwin Lausch. "Flammable Ice." *Scientific American* 281, no. 5 (1999): 76–83.

Svensen, Henrik, Sverre Planke, Anders Malthe-Sorenssen, et al. "Release of Methane from a Volcanic Basin as a Mechanism for Initial Eocene Global Warming." *Nature* 429, no. 6991 (2004): 542–545.

Wing, Scott L., Guy J. Harrington, Francesca A. Smith, et al. "Transient Floral Change and Rapid Global Warming at the Paleocene-Eocene Boundary." *Science* 310, no. 5750 (2005): 993–996.

Yans, Johan, Suzanne G. Strait, Thierry Smith, et al. "High-Resolution Carbon Isotope Stratigraphy and Mammalian Faunal Change at the Paleocene-Eocene Boundary in the Honeycombs Area of the Southern Bighorn Basin, Wyoming." *American Journal of Science* 306, no. 9 (2006): 712–735.

Zachos, James C., Gerald R. Dickens, and Richard E. Zeebe. "An Early Cenozoic Perspective on Greenhouse Warming and Carbon-Cycle Dynamics." *Nature* 451, no. 7176 (2008): 279–283.

Other

Gingerich, Philip D. "New Earliest Wasatchian Mammalian Fauna from the Eocene of Northwestern Wyoming: Composition and Diversity in a Rarely Sampled High-Floodplain Assemblage," University of Michigan Papers on Paleontology No. 28. Ann Arbor: University of Michigan, 1989.

Gingerich, Philip D., ed. "Paleocene-Eocene Stratigraphy and Biotic Change in the Bighorn and Clarks Fork Basins, Wyoming," University of Michigan Papers on Paleontology No. 33. Ann Arbor: University of Michigan, 2001.

Gingerich, Philip D. "Mammalian Responses to Climate Change at the Paleocene-Eocene Boundary: Polecat Bench Record in the Northern Bighorn Basin, Wyoming," Geological Society of America Special Paper 369 (2003): 463–478.

Rose, Kenneth D., Amy E. Chew, Rachel H. Dunn, et al. "Earliest Eocene Mammalian Fauna from the Paleocene-Eocene Thermal Maximum at Sand Creek Divide, Southern Bighorn Basin, Wyoming," University of Michigan Papers on Paleontology No. 36. Ann Arbor: University of Michigan, 2012.

Amy E. Chew

The Eocene–Oligocene extinction

The transition between the Eocene and Oligocene epochs (from about 37 to 30 million years ago, MYA) represents a crucial event in earth history, when the greenhouse conditions of the Early Eocene were transformed into the icehouse conditions of the Early Oligocene to the present (Prothero 1994, 2006, 2009; Prothero and Berggren 1992; Prothero et al. 2003). This global change can be documented in marine and terrestrial faunas and floras, in carbon and oxygen isotopes from both marine and terrestrial rocks and fossils, and in many other indicators such as paleosols (soils formed long time ago) on land and many other climatic proxies. Consequently, there has been intense interest over the past few decades in documenting these changes in detail and dating them as precisely as possible.

Although not among the *Big Five* events in terms of extinction of families and genera, the extinctions at the end of the Eocene have received attention because they are recent enough to have left a detailed record and may be associated with impacts. At one time, these events were treated as a single *Terminal Eocene Event* (TEE), and iridium anomalies were reported near the Eocene–Oligocene boundary (Asaro et al. 1982; Alvarez et al. 1982). However, as the interval was studied in greater detail and the dating method improved, a completely different picture emerged.

The oxygen isotope record of the deep sea, as mentioned in the 2001 James Zachos paper, shows the detailed global temperature story. There is a peak of warmth (lighter, more negative, more oxygen-16 rich values; oxygen-16 is the isotope of oxygen with eight protons and eight neutrons) in the Early Eocene, followed by a steady drop in global temperature (heavier, more positive, more oxygen-18 rich values; oxygen-18 is the isotope with eight protons and 10 neutrons) through the Middle and Late Eocene. Occasionally these data show some steps and reversals caused by short-term warming events and more rapid cooling pulses. But the overall trend is clear. From values indicating average sea bottom temperatures around 55°F (13°C) in the Early Eocene, average sea-water temperatures dropped steadily so that, by the Early Oligocene, the deep oceans were close to freezing on a global basis and ice caps formed in the Southern Ocean. To place this in context, each Pleistocene glacial-interglacial cycle changed the global temperature by only 34–35°F (1–2°C), yet in the course of the 15 million years within the Middle and Late Eocene, the temperature of the oceans dropped by almost 54°F (12°C).

These oxygen isotope data show there was a dramatic drop in sea-water temperatures just after the Eocene/Oligocene boundary. Current estimates contend that mean deep-sea temperatures cooled from 9–11°F (5–6°C), so that the global mean deep-sea temperature in the Early Oligocene was only 41°F (5°C). This was by far the most dramatic temperature change in the entire Cenozoic until a similar cooling event in the Middle Miocene signaled the development of the modern East Antarctic ice sheet.

However, oxygen isotopes are not the only evidence of cooling. Sand and pebbles shed by icebergs were retrieved from both lower and upper Oligocene sediments in numerous deep-sea cores from the Southern Ocean. Drill holes in the Antarctic margin at the Ross Sea and at Prydz Bay encountered thick lower and Middle Oligocene glacial deposits, suggesting that, by the Early Oligocene, there was a major ice sheet over much of Antarctica that lasted several hundred thousand years. That ice sheet retreated in the late Early Oligocene, but returned again by the Middle Oligocene, as indicated by the oxygen isotope record and by the pulse of Middle Oligocene glacial deposition.

So why did Earth move from a greenhouse to an icehouse? Possible explanations are debated intensely even now. One school of thought (e.g., Kennett 1977; Berggren and Hollister 1977) points to evidence that major changes in oceanic circulation occurred in the Eocene and Oligocene, which rearranged the distribution of heat over the surface of the Earth and triggered the development of Antarctic ice caps. According to this model, the opening of the circum-Antarctic current as Australia pulled away from Antarctica thermally isolated the southern oceans and led to their freezing. Critics of this model argue that it only explains short-term changes in local climate and not the long-term cooling of the Earth over 55 million years. In addition, this model does not explain how the greenhouse gases (methane and possibly excess carbon dioxide) were removed from the atmosphere and locked in Earth's sediments.

Another model (Ruddiman and Kutzbach 1991; Raymo and Ruddiman 1992) contends that the rapid uplift of new mountain ranges (especially the Himalayas, which are north of the Indian subcontinent) increased the rate of crustal weathering. This, in turn, removed carbon dioxide from the atmosphere as this gas combined with the newly formed minerals in soils. Although this hypothesis may work well for the last 15 million years, it does not explain the Eocene cooling, when the Himalayan uplift had just begun.

As of 2012, the most popular explanation is the greenhouse gas hypothesis (e.g., Berner et al. 1983, Pagani et al. 2005; DeConto and Pollard 2003), in which greenhouse conditions occurred when there were high rates of seafloor spreading as a result of the release of carbon dioxide from the mantle (e.g., during the Cretaceous). By contrast, slow rates of seafloor spreading produce less carbon dioxide, so this gas is pulled out of the atmosphere by weathering in soils. Indeed, there is evidence that seafloor spreading rates were at their highest in the Early and earliest Middle Eocene, followed by plate rearrangement events at 54 million years ago and again at 44 million years ago, after which spreading slowed down in parallel with the long-term cooling trend. However, for the past 15 million years, spreading rates have been increasing, while climates have continued to cool. Consequently, this model does not work well for the Late Cenozoic. In addition, many scientists doubt that the carbon dioxide and methane taken up by weathering in either the crustal uplift model or the seafloor spreading model is sufficient to account for the loss of these greenhouse gases in the atmosphere. These scientists argue that other reservoirs, for example, of limestones or coals, which are relatively rare in the Middle and Late Eocene, are needed or possibly more methane hydrates (methane molecules frozen into a cage of water molecules) on the seafloor, though there is no way to test this hypothesis in ancient sediments.

Whatever the ultimate cause(s) of the cooling trend seen in the oxygen isotopes, it is a reality. Evidence of cooling can be seen everywhere, including the formerly ice-free Antarctic continent. By 49 MYA, there were small glaciers on the Antarctic Peninsula (Birkenmajer et al. 2005). By the late Middle Eocene, icebergs were melting in the Pacific sector of the Southern Ocean and dropping ice-rafted sediment, which has been retrieved from deep-sea cores (Wei 1989). However, Antarctica still had not frozen over, because cool-temperate forests of southern beeches, araucarias, podocarps, and abundant ferns were still cloaking the continent through most of the Middle and Late Eocene.

Marine biotic response

How did marine life respond to the long-term Eocene–Oligocene cooling trend? The story is complicated, with a mixture of signals. According to a 1987 research study by Anne Boersma et al., planktonic foraminifera show that the beginning of the Middle Eocene (49 MYA) was marked by several degrees of cooling in high latitudes and in bottom waters that flowed from high latitudes all the way to the mid-Atlantic and the Gulf of Mexico.

The most dramatic event, however, occurred 37 to 38 million years ago, at the end of the Middle Eocene. Bottom waters became thermally decoupled from tropical surface waters, and warm tropical waters were prevented from mixing with polar waters. Polar waters became trapped in the Antarctic region, increasing the temperature difference between the equator and the South Pole. Although the tropics did not yet cool significantly, the temperate and polar regions did. As a result, there was a significant extinction of warm-water planktonic foraminiferans, with over 18 species going extinct (Boersma et al. 1987).

The end of the Middle Eocene was a terrible time for other tropical forms as well. Large coin-shaped nummulitid foraminiferans or common marine plankton species disappeared completely from the shrinking Tethys Seaway (now cut in half by the collision of India), and only smaller nummulitids (large lenticular marine protozoa, a type of foraminiferan) managed to struggle through the end of the Eocene and into the Oligocene. As referred to in the 2003 study by Prothero et al., major extinctions occurred in the benthic foraminifera of the Caribbean and the Gulf Coastal Plain of the United States. Further, in 1992, Marie-Pierre Aubry found that the tropical planktonic coccolithophorid algae also underwent a major extinction at this time as the oceans became cooler and surface-to-deep-water thermal gradients more stratified. American paleontologist Jack Baldauf wrote in 1992 that diatoms (one-celled algae), by contrast, showed a large increase in diversity and abundance at the end of the Middle Eocene, reflecting the more vigorous circulation and the increase in nutrients they require.

According to research by David T. Dockery (1986), Dockery and Pierre Lozouet (2003), and Thor A. Hansen (1988 and 1992), marine mollusks from the rich fossil beds of the US Gulf Coastal Plain were hit hard at the beginning of the Middle Eocene. Mollusk richness values then recovered in the later Middle Eocene and diversified to even higher levels, reaching their peak at the end of the Middle Eocene. However, the end of the Middle Eocene marked a severe decline in mollusks, with 89 percent of the snail species and

The large coin-sized foraminiferans known as nummulitids were the most common fossils of the shallow tropical seaways that ran from the Mediterranean to Indonesia. Known as the Tethys Seaway, it was eventually fragmented in the late Eocene when India collided with Asia. The nummulitids were so abundant in the middle and late Eocene that most tropical limestones are full of them, and the Great Pyramids of Gizeh in Egypt are built of them. Courtesy of Don Prothero.

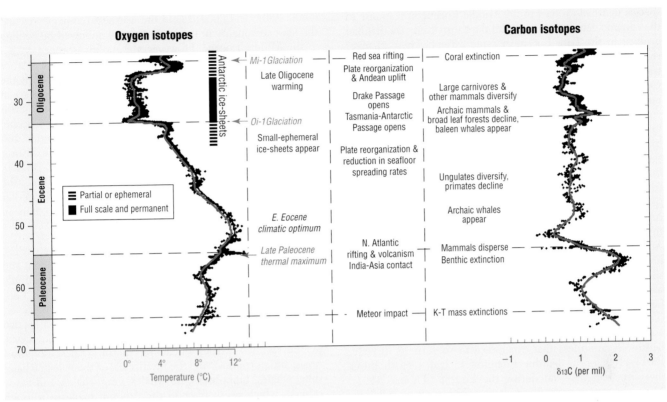

Oxygen and carbon isotope fluctuations in the world's deep oceans in the early Cenozoic, with major events of the Paleogene labeled. The oxygen isotopes are largely a proxy for global oceanic temperature, while the carbon isotopes show big changes when there are re-arrangements of oceanic circulation and recycling of deep-water nutrients (Modified from Zachos et al. 2001). Reproduced by permission of Gale, a part of Cengage Learning.

84 percent of bivalve species disappearing from the fossil record. Another lesser extinction occurred in the early Late Eocene, which eliminated 72 percent of the species of snail and 63 percent of bivalve species. Molluscan diversity then stayed at a low level through the rest of the Late Eocene. A similar pattern is demonstrated in the Atlantic Coastal Plain of the Carolinas (Campbell and Campbell 2003). In the Pacific Coast of North America, 48 percent of the snail species went extinct at the end of the Middle Eocene, according to Squires (2003). Nesbitt (2003) and Hickman (2003) noted a tremendous loss in molluscan species richness, and especially of tropical taxa at the end of the Middle Eocene in the Pacific Northwest. Likewise, in Europe the late Middle Eocene Bartonian Stage of the Paris Basin contains some 1300–1400 species of mollusks, but the Late Eocene retains only 800 species (Lozouet 1997; Cavelier 1979).

The other well-studied shelled marine invertebrate group is the echinoids. McKinney et al. (1992) showed that the Early and early Middle Eocene were peaks in echinoid diversity, after which their numbers declined. This decline is particularly apparent at the Middle to Late Eocene boundary in the Gulf Coastal Plain of North America, especially in Florida (Carter 2003).

The most severe extinction event of the Cenozoic, which wiped out a huge number of tropical and warm-water taxa and fundamentally rearranged the marine faunas, occurred at the cooling event at the end of the Middle Eocene, not a TEE as was suggested in the 1980s. The end of the Middle Eocene

was one of the most profound faunal turnovers in the Cenozoic history. However, it is usually neglected or misunderstood in the misguided focus on the Eocene–Oligocene boundary, which was a relatively minor event.

Response to bottom water currents

How did marine life respond to the sudden cold bottom water currents and the rapid change in temperature in the earliest Oligocene? According to Boersma et al. (1987), planktonic foraminiferans of the Early Oligocene were a homogeneous fauna that was small in size, low in diversity, and found almost worldwide. Only a handful of species (mainly the hantkeninids) went extinct at the Eocene–Oligocene boundary or in the earliest Oligocene. These patterns suggest that the cold oceans were well circulated, not stratified by depth or temperature. The abundance of deep-water cold-tolerant foraminiferans is particularly diagnostic of the upwelling and mixing of cold bottom waters with the surface. In the tropics, however, some depth stratification remained, as the carbon isotope differences between shallow and deep-water foraminifers remained high.

The cooling trend definitely affected temperature-sensitive marine algae. Among the coccolithophorid algae, as Aubry reported in 1992, there was extinction among temperate-zone taxa. This was the final blow for the tropical Eocene forms in this group. All but one of the long-ranging species that had evolved in the Early Eocene, several of the long-ranging species that had evolved in the Middle Eocene in response to the climatic deterioration, and all the short-lived taxa that had evolved in the Late Eocene became extinct. These changes reduced diversity to 30 percent of the original number of species since the extinctions began in the Middle Eocene. Cold-tolerant coccolithophorid algae bloomed in huge numbers due to the release of nutrients provided by the vigorous deep-water circulation. By the end of the Early Oligocene, the division of the oceans into latitudinal water masses led to increased coccolith provincialism as different species became segregated into different temperature water masses. The diatoms also responded dramatically. According to Baldauf (1992), about 45 percent of the Eocene diatom species became extinct by the end of the Early Oligocene, replaced by almost the same number of new species. This turnover occurred in all latitudes, but especially in the Antarctic, where there was a great increase in siliceous productivity due to the cold deep-water currents that brought silica to the surface waters (Baldauf and Barron 1990). There was also significant extinction in the dinoflagellates at the beginning of the Oligocene (Brinkhuis 1992).

The cold bottom waters naturally had a great effect on the benthic foraminifera that experience these currents directly. Earliest Oligocene benthic foraminifera had a much lower diversity than they did during the Eocene (Thomas 1992). Most species retreated to living within the sediment, suggesting that the deep cold waters were corrosive to their calcite shells. The US Gulf Coast saw a high rate of extinction of benthic foraminiferans as well (Gaskell 1991; Fluegeman 2003). In the tropical Tethys, the last of the warm-water taxa, such as the coin-shaped nummulitids, disappeared. Agglutinated foraminiferans, which build their shells from cemented sand grains, were also severely affected, again because of the cold corrosive bottom water.

Moreover, larger marine species were not spared. In the US Gulf and along the Atlantic coasts there was another major extinction of mollusks, with about 95 percent of the snail species and 89 percent of bivalve species that survived the Middle Eocene extinction dying out in the Early Oligocene (Hansen 1992; Dockery 2003; Campbell and Campbell 2003; Hansen et al. 2004). Similar trends occurred on the Pacific Coast, where there was mass extinction in the snails (Squires 2003), with most of the new species coming down from the cold North Pacific and Arctic regions (Oleinik and Marincovich 2003), and clams were also strongly affected (Nesbitt 2003; Hickman 2003). There was also a dramatic drop from 43 Late Eocene echinoid species to only 15 species found in lower Oligocene deposits of the US Gulf Coastal Plain (Carter 2003), and about 50 percent of echinoid species became extinct worldwide (McKinney et al. 1992). Tropical echinoids were the most severely affected, and the groups that flourished were the cold-adapted marsupiate echinoids, which protect their eggs in a pouch; these managed to survive and spread during the chilly Oligocene.

Another indicator of conditions in shallow marine habitats is the oxygen isotopes recorded not only in mollusk and foraminiferan shells but also in the tiny ear bones of fish, known as otoliths. Three studies by Linda C. Ivany et al.

(2000, 2003, and 2004) found the warmest temperatures and the least seasonality in the Early Eocene, followed by cooling in the Middle Eocene. However, the otolith isotopes do not show a dramatic temperature drop in the Gulf Coast of Alabama and Mississippi at the end of the Eocene or at the Eocene/Oligocene boundary. Instead, there was an increase in seasonality, with ranges of only 35°F (2°C) (77–80°F [25–27°C] between summer and winter in the Early Eocene) to 41°F (5°C) (59–68°F [15–20°C] between summer and winter in the Late Eocene). In the Early Oligocene, winter temperatures dropped another 39°F (4°C) (to 52°F [11°C]), but summer temperatures were not affected. Thus, the increased seasonality was enhanced by dramatically cooler winters in the Gulf Coast in the Oligocene, even though mean temperatures had not changed appreciably.

At the top of the Eocene marine food web are the whales. They had evolved into huge archaeocetes by the Middle and Late Eocene, which apparently flourished in low-latitude areas such as the warm Tethyan seaway and the US Gulf Coast (Fordyce 2003). By the latest Eocene, however, archaeocetes had vanished and were rapidly replaced by two new groups that dominate the oceans today: the odontocetes, or the toothed whales (including sperm whales, dolphins, orcas, and porpoises), and the mysticetes, or the baleen whales (blue whales, humpbacks, fin whales, gray whales, and many others), which use a filter in their mouth made of a tough horny protein, known as baleen, to screen plankton from seawater. Early members of both groups are found mostly at high latitudes in the Oligocene (especially New Zealand and the North Pacific). According to Fordyce (1980, 1989, 1992, and 2003), their evolution and diversification in the Early Oligocene was triggered by the explosive blooms of plankton when rising deep bottom waters (especially in the Southern Ocean) brought up nutrients. This population explosion in diatoms and coccolithophorids fed the crustaceans, which are the main source of nourishment for baleen whales.

In summary, the Early Oligocene extinctions of marine life were not as severe as those occurring at the end of the Middle Eocene, but still there was tremendous turnover, especially with tropical and warm-adapted taxa being replaced by cold-adapted taxa. By the late Early Oligocene, diversity in the oceans had stabilized at a new, much lower level than that during the Eocene, and held on at these lower numbers of species throughout the cold harsh conditions of the rest of the Oligocene.

Terrestrial climate and biotic response

What about the terrestrial record at the Eocene–Oligocene transition? The White River Group in the Big Badlands of South Dakota preserves details regarding the history of many kinds of organisms. For example, the 1983 research by Gregory J. Retallack studied the color bands visible in the Badlands sections and found that they were paleosols, or ancient soil horizons. Those from the upper Eocene Chadron Formation were formed under forests with closed canopies of large trees (the huge root casts are particularly conspicuous) with between 20 and 35 inches (500 and 900 mm) of rainfall per year. In the overlying lower Oligocene (Orellan) Brule

Formation, the paleosols indicate more open, dry woodland with only 20 inches (500 mm) of rainfall per year. In eastern Wyoming, in 1992, Emmett Evanoff and his colleagues studied the sediments and found that the moist Chadronian floodplain deposits abruptly shifted to drier, wind-blown deposits by the Orellan. In the same beds are climate-sensitive land snails. According to Evanoff et al. (1992), Chadronian land snails are large-shelled taxa similar to those found in wet subtropical regions such as modern Central America. Based on modern analogues, these snail species indicate a mean annual temperature of 63°F (16.5°C) and a mean annual precipitation of about 18 inches (450 mm); very similar to the results obtained by Retallack (1983) for neighboring South Dakota. By contrast, Orellan land snails are drought-tolerant small-shelled taxa indicative of warm-temperate open wood-lands with a pronounced dry season. Their living analogues are found today in Baja California.

The amphibians and reptiles suggest similar trends of cooling and drying in the Early Oligocene (Hutchison 1982, 1992). The Eocene was dominated by aquatic species (especially salamanders, pond turtles, and crocodilians) that had been steadily declining in numbers in the Middle and Late Eocene. Crocodiles were regionally extinct by the Chadronian, but a few fossil alligators have been recovered from the Chadron Formation. By the Oligocene, however, only land tortoises were common, indicating a pronounced drying trend. In fact, these tortoises (*Stylemys nebraskensis*) were so common in the Orellan that these beds were originally called the turtle-oreodon beds after their two most common vertebrate fossils.

The North American floras show a clear trend. Based on leaf-margin analysis, Wolfe (1971 1978, 1985, 1992) suggested that mean annual temperatures in North America

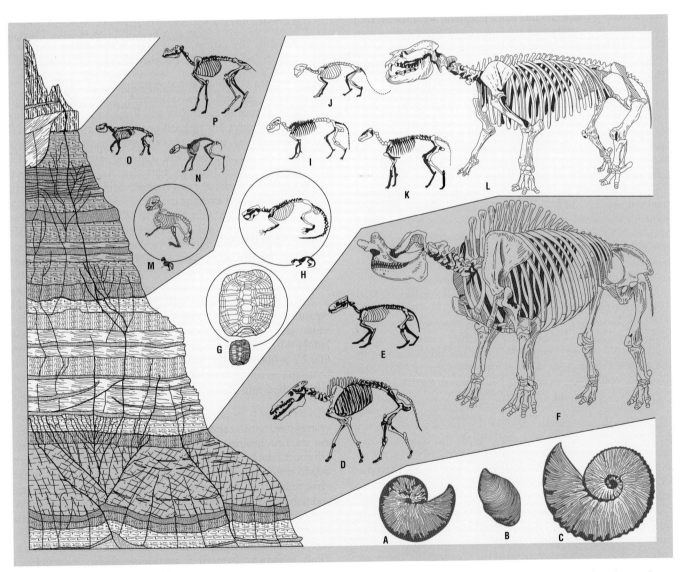

Sketch of the Big Badlands of South Dakota, showing typical mammals and reptiles. The huge brontotheres (F) vanished at the end of the Eocene (lower orange band). D) *Archaeotherium*; E) *Hyaenodon*. In the early Oligocene (Orellan) Scenic Member of the Brule Formation (middle white layer), there are (G) tortoises; (H) *Ischyromys*; (I) *Merycoidodon*; (J) *Hoplophoneus*; (K) *Mesohippus*; (L) *Metamynodon*. The late Oligocene (upper orange band) yields (M) *Palaeolagus*; (N) *Leptomeryx*; (O) *Leptauchenia*; (P) *Protoceras*. Reproduced by permission of Gale, a part of Cengage Learning.

cooled about 13–23°F (8–12°C) in less than a million years. This would be by far the most dramatic cooling event of the entire North American floral record and, as noted above, was the original basis for the phrase TEE (even though revised dating now places it in the Early Oligocene). Perhaps Wolfe's 1971 earlier phrase *Oligocene deterioration* would be a better term.

However, recent work re-dating many of these floras and re-interpreting their climatic signature has changed this story somewhat. The Rocky Mountains of central Colorado yield several important floras that span the Eocene–Oligocene transition. The floras of the famous Late Eocene Florissant Formation, dated at 34.07 ± 0.10 million years ago, records the final phase of Late Eocene warmth before the Early Oligocene deterioration. Even though it was at 6,560 to 9,840 feet (2,000 to 3,000 m) of elevation in the Eocene, the Florissant flora is believed to represent warm-temperate climatic conditions of moderate rainfall and a mean annual temperature of 55.4–57.2°F (13°14°), compared to modern mean annual temperatures of 39.2°F (4°C). As Prothero (2008) demonstrated, there was then a gradual cooling in the high-altitude Rocky Mountain floras, from the latest Eocene Antero flora (contemporaneous or slightly younger than Florissant), the Early Oligocene Pitch-Pinnacle flora, to the Late Oligocene Creede flora.

The longest and most complete sequence of floras spanning the Eocene–Oligocene transition occurred in the Eugene and Fisher formations near Eugene, Oregon. Retallack et al. (2004) showed that the paratropical Middle Eocene Comstock flora (39.7 MYA) suggests a mean annual temperature of 72.3°F (22.4°C), reminiscent of warm subtropical conditions. The Late Eocene Goshen flora, dated at 33.4 MYA, yields a mean annual temperature of 67.5°F (19.7°C). The Rujada flora, dated at 31.3 MYA, shows temperature estimates consistent with that of a post-deterioration flora, with a mean annual temperature of 55.4°F (13.0°C). The Early Oligocene Willamette flora, dated at 30.9 MYA, yields a mean annual temperature of 55.8°F (13.2°C), while the much less well-known Coburg flora is very similar. Thus, the Eugene-Fisher floral sequence shows a gradual sequence of climatic change through the Middle Eocene to the earliest Oligocene (consistent with the global record), with no catastrophic Oligocene deterioration as argued by Wolfe (1971, 1978, 1985, and 1992).

Despite these dramatic changes in the soils, land plants, land snails, and reptiles and amphibians, the change in the mammalian fauna is not impressive (Prothero, 1994; Prothero and Heaton 1996; Prothero 1999). Most of the archaic Eocene taxa (especially the forest dwellers and arboreal forms) were already gone by the last Eocene, with only a few multituberculates straggling on to the Middle Chadronian. A few groups, such as the brontotheres, the camel-like oromerycids, the mole-like epoicotheres, and two groups of rodents, did die out near the end of the Chadronian, but none was around to witness the big Early Oligocene climatic deterioration. Most of the taxa that were present before the climatic crash showed no change whatsoever, except for a dwarfing event in one lineage of the oreodont *Miniochoerus*.

Apparently, the groups that were present in the Late Chadronian were already adapted to the drier more open woodlands habitats, so the vegetational change did not make that much difference—or else the responsiveness of mammals to short-term changes in climate has been oversold, and they are not as sensitive as we have long believed (Prothero and Heaton 1996; Prothero 1999).

The terrestrial records on some other continents are well studied and yield interesting contrasts. In China and Mongolia, there is a major extinction of most of the Late Eocene lineages, leaving a depauperate fauna with only a few huge rhinos and entelodonts, no medium-sized mammals, and a huge diversity of burrowing rodents and rabbits. Known as the *Mongolian remodeling* (Meng and McKenna 1998), it presents an interesting contrast to the surprising non-response of North American mammals to climate change.

Europe, on the other hand, is harder to interpret (Prothero 2006, 2009). There was a huge turnover event in the earliest Oligocene, called *La Grande Coupure* (the great cutoff), which is recognized by the extinction of most of the archaic mammal groups that dominated the Eocene. However, it was mostly due to a major immigration event of Asian mammals (rhinoceroses, anthracotheres, entelodonts, and most rodents), so it is impossible to tell how much of this turnover was due to climate change and how much was due to competition from more advanced immigrant groups. Although the terrestrial fossil record is improving from South America, Africa, and Australia, as of 2012 there was not enough evidence to clearly show how their land faunas responded to the climate changes at the end of the Eocene.

Impacts without impact

When the idea occurred in 1980 that a bolide impact killed off the dinosaurs at the end of the Cretaceous Period (66 MYA), it quickly spurred scientists to look at other extinction events and see if they too could have been caused by impact. An iridium anomaly was discovered near the Eocene–Oligocene boundary (Alvarez et al. 1982; Asaro et al. 1982; Ganapathy 1982; Glass et al. 1982). The impact advocates quickly jumped on the bandwagon and declared the Eocene extinctions to be caused by an asteroid or cometary impact (Hut et al. 1987).

The impacts of the Late Eocene are a particularly good example of some scientists getting ahead of their data. When the iridium anomaly was first discovered near the Eocene–Oligocene boundary, the impact advocates assumed there must be a connection and bragged that they had found the cause. However, those with better knowledge of the Eocene and Oligocene rocks and fossils performed the hard detective work that the impact advocates had neglected. (Studies from Prothero and Berggren [1992] and Prothero [2003] discuss this scenario.) The impact layers were dated at 35.5 to 36.0 MYA, in the middle of the Late Eocene, when no extinctions had ever been proposed. They occurred at least one to two million years before the Eocene–Oligocene boundary and even earlier than the early Oligocene extinctions; they were at least one to two million years after the terminal Middle

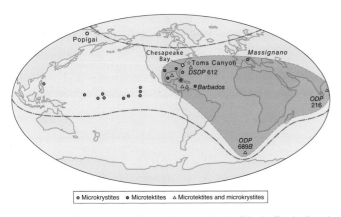

Map showing the major late Eocene impact sites in Siberia (Popigai) and in Chesapeake Bay and Toms Canyon in the Atlantic, and the distribution of debris from those impacts. Reproduced by permission of Gale, a part of Cengage Learning.

Eocene extinctions at 37 to 38 MYA. In a few Caribbean sections that did contain the impact layer, there were five radiolarian species (out of dozens) that disappeared at the impact (Maurrasse and Glass 1979; Glass and Swart 1977), but nothing else suffered—not the rest of the microfossils (especially the sensitive foraminifera—Hut et al. 1987), nor the marine invertebrates (mollusks, echinoids, crustaceans), nor the land plants or land mammals (Prothero 1994, 2009; Prothero and Heaton 1996).

Later, the sources of the impact debris and the iridium were identified. Deep drilling in the Chesapeake Bay region (Poag et al. 1992; Poag 1999) uncovered evidence of a crater buried deeply beneath Maryland and Virginia, which was dated at 35.5 million years ago (Poag et al. 2003). Eventually, scientists found evidence of a buried crater about 60 miles (100 km) in diameter (about two-thirds the size of the K–PG impact crater), and a second crater on the continental shelf 60 miles (100 km) east of New Jersey in Toms Canyon, which was the same age and about 9 miles (15 km) in diameter. Presumably, both craters were from the same Late Eocene shower of rocks, and they left a debris field that was detected as far as the Indian Ocean and South Atlantic. About the same time, a third crater was identified at Popigai in eastern Siberia (Masaitis et al. 1975; Bottomley et al. 1997), which was dated at 35.7 MYA and left a crater about 56 miles (90 km) in diameter. Thus, there were three well-dated large impacts in the Late Eocene, yet the evidence was still clear: These huge impacts had no effect on life.

The observation that the 60- and 56-mile (100- and 90-km) craters at Chesapeake Bay, Toms Canyon, and Popigai had no effect on life has profound implications for the idea that impacts cause extinctions. David M. Raup (1991) had originally fit a *kill curve* of impact size versus number of species disappearing from the fossil record, using the end-Cretaceous crater as a model. Then, Poag (1997) replotted the kill curve with the Eocene crater data, and its shape changed dramatically. If craters over 60 miles (100 km) in diameter have no effect, then only the biggest impacts have any chance of causing extinction, and the kill curve must shoot up rapidly between diameters of 60 and 112 miles (100 and 180 km).

This was further confirmed when other Cenozoic impact sites were reexamined. The Montagnais impact structure, off the coast of Nova Scotia, was formed by an impact 50 MYA, yet had no effect on life, as determined by Bottomley and Derek York (1988), Aubry et al. (1990), and L. F. Jansa et al. (1990). Likewise, the Miocene Ries crater in Germany had no effect on life, according to K. Heissig (1985). Prothero (2005) surveyed all of the Cenozoic impacts in the impact database on the Web and found that none corresponded with peaks of extinction on any diversity curve; John Alroy (2003) reached a similar conclusion.

Because of the overwhelming evidence that the Late Eocene impacts had little or no biotic effect, impact advocates have instead argued that impacts might have caused some of the Late Eocene climatic perturbations that preceded the extinctions in the earliest Oligocene (Poag 1999; Vonhof et al. 2000; Coccioni et al. 2000; Poag et al 2003; Fawcett and Boslough 2002). What is peculiar about these explanations is that they predict opposite effects. The direct effect of an impact should produce a debris ring and global cooling (Vonhof et al. 2000; Fawcett and Boslough 2002), yet the isotopic and paleoclimatic records of the Late Eocene shows that the exact opposite, a short-term warming event, actually occurred (Poag 1999; Poag et al. 2003). Clearly, impacts cannot cause global warming and cooling simultaneously. Even if they could do so, there is no clear explanation for how either climatic change might have caused extinctions in the Early Oligocene, almost two million years later, given that the

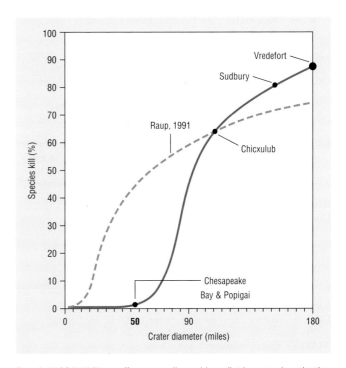

Raup's (1991) "kill curve" supposedly could predict how much extinction would be caused by an impact of a given size. But the existence of huge late Eocene impacts like Chesapeake Bay and Popigai (just slightly smaller than Chicxulub) which caused almost no extinctions completely changes the shape of the curve, and shows that only the largest impacts have the capability of causing mass extinction. (After Poag 1997). Reproduced by permission of Gale, a part of Cengage Learning.

effects of such impact events diminish in years or decades, not millions of years.

From the last quarter of the twentieth century into the early 2000s, scientists were intensively studying the record of the Eocene–Oligocene transition, and the verdict was clear: The huge impacts in the Late Eocene had no significant effect. More importantly, the theme of certain presentations at several scientific meetings held in the early 2000s was that most mass extinctions were not caused by impacts. The Eocene-–Oligocene extinctions are the great exception to the idea that impacts cause mass extinctions and force a reassessment of the entire impact-extinction hypothesis.

Resources

Books

Aubry, Marie-Pierre. "Late Paleogene Calcareous Nannoplankton Evolution: A Tale of Climatic Deterioration." In *Eocene–Oligocene Climatic and Biotic Evolution*, edited by Donald R. Prothero and William A. Berggren. Princeton, NJ: Princeton University Press, 1992.

Baldauf, Jack G. "Middle Eocene through Early Miocene Diatom Floral Turnover." In *Eocene–Oligocene Climatic and Biotic Evolution*, edited by Donald R. Prothero and William A. Berggren. Princeton, NJ: Princeton University Press, 1992.

Baldauf, Jack G., and John A. Barron. "Evolution of Biosiliceous Sedimentation Patterns—Eocene through Quaternary: Paleoceanographic Response to Polar Cooling." In *Geological History of the Polar Oceans: Arctic Versus Antarctic*, edited by U. Bleil and Jörn Thiede. Amsterdam: Kluwer Academic, 1990.

Brinkhuis, Henk. "Late Paleogene Dinoflagellate Cysts with Special Reference to the Eocene/Oligocene Boundary." In *Eocene–Oligocene Climatic and Biotic Evolution*, edited by Donald R. Prothero and William A. Berggren. Princeton, NJ: Princeton University Press, 1992.

Campbell, David C., and Matthew R. Campbell. "Biotic Patterns in Eocene–Oligocene Mollusks of the Atlantic Coastal Plain, U.S.A." In *From Greenhouse to Icehouse: The Marine Eocene–Oligocene Transition*, edited by Donald R. Prothero, Linda C. Ivany, and Elizabeth A. Nesbitt. New York: Columbia University Press, 2003.

Carter, Burchard D. "Diversity Patterns in Eocene and Oligocene Echinoids of the Southeastern United States." In *From Greenhouse to Icehouse: The Marine Eocene–Oligocene Transition*, edited by Donald R. Prothero, Linda C. Ivany, and Elizabeth A. Nesbitt. New York: Columbia University Press, 2003.

Dockery, David T., III, and Pierre Lozouet. "Molluscan Faunas across the Eocene/Oligocene Boundary in the North American Gulf Coastal Plain, with Comparison to Those of the Eocene and Oligocene of France." In *From Greenhouse to Icehouse: The Marine Eocene–Oligocene Transition*, edited by Donald R. Prothero, Linda C. Ivany, and Elizabeth A. Nesbitt. New York: Columbia University Press, 2003.

Evanoff, Emmett, Donald R. Prothero, and R. H. Lander. "Eocene–Oligocene Climatic Change in North America: The White River Formation near Douglas, East-central Wyoming." In *Eocene–Oligocene Climatic and Biotic Evolution*, edited by Donald R. Prothero and William A. Berggren. Princeton, NJ: Princeton University Press, 1992.

Fluegeman, Richard H. "Late Eocene-early Oligocene Benthic Foraminifera in the Gulf Coastal Plain: Regional vs. Global Influences." In *From Greenhouse to Icehouse: The Marine Eocene–Oligocene Transition*, edited by Donald R. Prothero, Linda C.

Ivany, and Elizabeth A. Nesbitt. New York: Columbia University Press, 2003.

Fordyce, R. Ewan. "Cetacean Evolution and Eocene–Oligocene Environments." In *Eocene–Oligocene Climatic and Biotic Evolution*, edited by Donald R. Prothero and William A. Berggren. Princeton, NJ: Princeton University Press, 1992.

Fordyce, R. Ewan. "Cetacean Evolution and Eocene–Oligocene Oceans Revisited." In *From Greenhouse to Icehouse: The Marine Eocene–Oligocene Transition*, edited by Donald R. Prothero, Linda C. Ivany, and Elizabeth A. Nesbitt. New York: Columbia University Press, 2003.

Glass, B. P., and M. J. Zwart. "North American Microtektites, Radiolarian Extinctions and the Age of the Eocene/Oligocene Boundary." In *Stratigraphic Micropaleontology of the Atlantic Basin and Borderlands*, edited by Frederick M. Swain. Amsterdam: Elsevier, 1977.

Hansen, Thor A. "The Patterns and Causes of Molluscan Extinction across the Eocene/Oligocene Boundary." In *Eocene–Oligocene Climatic and Biotic Evolution*, edited by Donald R. Prothero and William A. Berggren. Princeton, NJ: Princeton University Press, 1992.

Hickman, Carole S. "Evidence for Abrupt Eocene–Oligocene Molluscan Faunal Change in the Pacific Northwest." In *From Greenhouse to Icehouse: The Marine Eocene–Oligocene Transition*, edited by Donald R. Prothero, Linda C. Ivany, and Elizabeth A. Nesbitt. New York: Columbia University Press, 2003.

Hutchison, J. H. "Western North American Reptile and Amphibian Record across the Eocene/Oligocene Boundary and Its Climatic Implications." In *Eocene–Oligocene Climatic and Biotic Evolution*, edited by Donald R. Prothero and William A. Berggren. Princeton, NJ: Princeton University Press, 1992.

Ivany, Linda C., Kyger C. Lohmann, and William P. Patterson. "Paleogene Temperature History of the U.S. Gulf Coastal Plain Inferred from Fossil Otoliths." In *From Greenhouse to Icehouse: The Marine Eocene–Oligocene Transition*, edited by Donald R. Prothero, Linda C. Ivany, and Elizabeth A. Nesbitt. New York: Columbia University Press, 2003.

Masaitis, V., M. V. Mikhailov, and T.V. Selivanovskaya. *Popigai Meteorite Crater* [in Russian]. Moscow: Nauka Press, 1975.

McKinney, M. L., B. D. Carter, K. J. McNamara, and S. K. Donovan. "Evolution of Paleogene Echinoids: A Global and Regional View." In *Eocene–Oligocene Climatic and Biotic Evolution*, edited by Donald R. Prothero and William A. Berggren. Princeton, NJ: Princeton University Press, 1992.

Nesbitt, Elizabeth A. "Changes in Shallow-marine Faunas from the Northeastern Pacific Margin across the Eocene/Oligocene Boundary." In *From Greenhouse to Icehouse: The Marine Eocene–Oligocene Transition*, edited by Donald R. Prothero,

Linda C. Ivany, and Elizabeth A. Nesbitt. New York: Columbia University Press, 2003.

Oleinik, Anton E., and Louie Marincovich Jr. "Biotic Response to the Eocene–Oligocene Transition: Gastropod Assemblages in the High-latitude North Pacific." In *From Greenhouse to Icehouse: The Marine Eocene–Oligocene Transition*, edited by Donald R. Prothero, Linda C. Ivany, and Elizabeth A. Nesbitt. New York: Columbia University Press, 2003.

Poag, C. Wylie. *Chesapeake Invader*. Princeton, NJ: Princeton University Press, 1999.

Poag, C. Wylie, Edward Mankinen, and Richard D. Norris. "Late Eocene Impacts: Geologic Record, Correlation, and Paleoenvironmental Consequences." In *From Greenhouse to Icehouse: The Marine Eocene–Oligocene Transition*, edited by Donald R. Prothero, Linda C. Ivany, and Elizabeth A. Nesbitt. New York: Columbia University Press, 2003.

Prothero, Donald R. *The Eocene–Oligocene Transition: Paradise Lost*. New York: Columbia University Press, 1994.

Prothero, Donald R. *After the Dinosaurs: The Age of Mammals*. Bloomington: Indiana University Press, 2006.

Prothero, Donald R. *Greenhouse of the Dinosaurs*. New York: Columbia University Press, 2009.

Prothero, Donald R., and William A. Berggren, eds. *Eocene–Oligocene Climatic and Biotic Evolution*. Princeton, NJ: Princeton University Press, 1992.

Prothero, Donald R., Linda C. Ivany, and Elizabeth A. Nesbitt, eds. *From Greenhouse to Icehouse: The Marine Eocene–Oligocene Transition*. New York: Columbia University Press, 2003.

Raup, David M. *Extinction: Bad Genes or Bad Luck?* New York: Norton, 1991.

Squires, Richard L. "Turnovers in Marine Gastropod Faunas during the Eocene–Oligocene Transition, West Coast of the United States." In *From Greenhouse to Icehouse: The Marine Eocene–Oligocene Transition*, edited by Donald R. Prothero, Linda C. Ivany, and Elizabeth A. Nesbitt. New York: Columbia University Press, 2003.

Thomas, Ellen. "Middle Eocene–late Oligocene Bathyal Benthic Foraminifera (Weddell Sea): Faunal Changes and Implications for Oceanic Circulation." In *Eocene–Oligocene Climatic and Biotic Evolution*, edited by Donald R. Prothero and William A. Berggren. Princeton, NJ: Princeton University Press, 1992.

Wolfe, Jack A. "Climatic, Floristic, and Vegetational Changes Near the Eocene/Oligocene Boundary in North America." In *Eocene–Oligocene Climatic and Biotic Evolution*, edited by Donald R. Prothero and William A. Berggren. Princeton, NJ: Princeton University Press, 1992.

Periodicals

Adams, C. Geoffrey, Daphne E. Lee, and Brian R. Rosen. "Conflicting Isotopic and Biotic Evidence for Tropical Sea-surface Temperatures during the Tertiary." *Palaeogeography, Palaeoclimatology, Palaeoecology* 77 (2009): 289–313.

Alroy, John. "Cenozoic Bolide Impacts and Biotic Change in North American Mammals." *Astrobiology* 3 (2003): 119–132.

Alvarez, Walter, Frank Asaro, Helen V. Michel, et al. "Iridium Anomaly Approximately Synchronous with Terminal Eocene Extinctions." *Science* 216 (1982): 886–888.

Asaro, Frank, Luis W. Alvarez, Walter Alvarez, et al. "Geochemical Anomalies near the Eocene/Oligocene and

Permian/Triassic Boundaries." *Geological Society of America Special Paper* 190 (1982): 517–528.

Aubry, Marie-Pierre, Felix M. Gradstein, and Lubomir F. Jansa. "The Late Early Eocene Montagnais Meteorite: No impact on Biotic Diversity." *Micropaleontology* 36, no. 2 (1990): 164–172.

Berggren, William A., and C. D. Hollister. "Paleogeography, Paleobiogeography, and the History of Circulation in the Atlantic Ocean." *SEPM Special Publication* 20 (1974): 126–176.

Berner, Robert A., Antonio C. Lasaga, and Robert M. Garrels. "The Carbonate-silicate Geochemical Cycle and Its Effect on Atmospheric Carbon Dioxide over the Past 100 Million Years." *American Journal of Science* 283 (1983): 641–683.

Birkenmajer, Krzysztof, Andrzej Gaździcki, Krzysztof P. Krajewski, et al. "First Cenozoic Glaciers in West Antarctica." *Polish Polar Research* 26 (2005): 3–12.

Boersma, Anne, Isabella Premoli-Silva, and N. J. Shackleton. "Atlantic Eocene Planktonic Foraminiferal Paleohydrographic Indicators and Stable Isotope Paleoceanography." *Paleoceanography* 2, no. 3 (1987): 287–331.

Bottomley, Richard, Richard A. F. Grieve, Derek York, et al. "The Age of the Popigai Impact Event and Its Relation to Events at the Eocene/Oligocene Boundary." *Nature* 388 (1997): 365–368.

Bottomley, Richard, and Derek York. "Age Measurements of the Submarine Montagnais Impact Crater and the Periodicity Question." Geophysical Research Letters 14, no. 12 (1988): 1409–1412.

Cavelier, Claude. "La limite Eocène-Oligocène en Europe occidentale." *Sci. Géol. Inst. Géol. Strasbourg (Mém.)* 54 (1979): 1–280.

Coccioni R., D. Basso, H. Brinkhuis, et al. "Marine Biotic Signals across a Late Eocene Impact Layer at Massignano, Italy: Evidence for Long-term Environmental Perturbations?" *Terra Nova* 12 (2000): 258–263.

DeConto, Robert M., and David Pollard. "Rapid Cenozoic Glaciation of Antarctica Induced by Declining Atmospheric CO_2." *Nature* 421 (2003): 245–249.

Dockery, David T., III. "Punctuated Succession of Paleogene Mollusks in the Northern Gulf Coastal Plain." *Palaios* 1 (1986): 582–589.

Fawcett, Peter J., and Mark B. E. Boslough. "Climatic Effects of an Impact-induced Equatorial Debris Ring." *Journal of Geophysical Research* 107, no. D15 (2002): 10129–10146.

Fordyce, R. Ewan. "Whale Evolution and Oligocene Southern Ocean Environments." *Palaeogeography, Palaeoclimatology, Palaeoecology* 31 (1980): 319–336.

Fordyce, R. Ewan. "Origins and Evolution of Antarctic Marine Mammals." *Special Publications of the Geological Society of London* 47 (1989): 269–281.

Ganapathy, R. "Evidence for a Major Meteorite Impact on the Earth 34 Million Years Ago: Implications for Eocene Extinctions." *Science* 216 (1982): 885–886.

Gaskell, Barbara A. "Extinction Patterns in Paleogene Benthic Foraminiferal Faunas: Relationship to Climate and Sea Level." *Palaios* 6 (1991): 2–16.

Glass, B. P., David L. DuBois, and R. Ganapathy. "Relationship between an Iridium Anomaly and the North American

Micro-Tektite Layer in Core RC9-58 from the Caribbean Sea." *Journal of Geophysical Research* 87 (1982): 425–428.

Hansen, Thor A. "Early Tertiary Radiation of Marine Molluscs and the Long-term Effects of the Cretaceous-Tertiary Extinction." Paleobiology 14 (1988): 37–51.

Hansen, Thor A., Patricia H. Kelley, and David M. Haasl. "Paleoecological Patterns in Molluscan Extinctions and Recoveries: Comparison of the Cretaceous-Paleogene and Eocene–Oligocene Extinctions in North America." *Palaeogeography, Palaeoclimatology, Palaeoecology* 214 (2004): 233–242.

Heissig, K. "No Effect of the Ries Impact Event on the Local Mammal Fauna." *Modern Geology* 10 (1985): 171–179.

Hut, Piet, Walter Alvarez, William P. Elder, et al. "Comet Showers as a Cause of Mass Extinctions." *Nature* 329 (1987): 118–126.

Hutchison, J. H. "Turtle, Crocodilian, and Champsosaur Diversity Changes in the Cenozoic of the North-Central Region of the Western United States." *Palaeogeography, Palaeoclimatology, Palaeoecology* 37 (1982): 149–164.

Ivany, Linda C., William P. Patterson, and Kyger C. Lohmann. "Cooler Winters as a Possible Cause of Mass Extinctions at the Eocene/Oligocene Boundary." *Nature* 407 (2000): 887–890.

Ivany, Linda C., Bruce W. Wilkinson, Kyger C. Lohmann, et al. "Intra-annual Isotopic Variation in *Venericardia* Bivalves: Implications for Early Eocene Temperatures, Seasonality, and Salinity on the U.S. Gulf Coast." *Journal of Sedimentary Research* 74 (2004): 7–19.

Jansa, Lubomir F., Marie-Pierre Aubry, and Felix M. Gradstein. "Comets and Extinctions: Cause and Effect?" *Geological Society of America Special Paper* 247 (1990): 223–232.

Kennett, James P. "Cenozoic Evolution of Antarctic Glaciation, The Circum-Antarctic Ocean, and Their Impact on Global Paleoceanography." Journal of Geophysical Research 82 (1977): 3843–3860.

Lozouet, Pierre. "Nouvelles espèces de gasteropodes (Mollusca: Gastropoda) de l'Oligocène et du Miocene d'Aquitaine (Sud-Ouest de la France)." Partie 2: *Cossmanniana* 6 (1997): 1–68.

Maurrasse, Florentin, and B. P. Glass. "Radiolarian Stratigraphy and North American Microtektites in Caribbean Core RC9-58: Implications Concerning Late Eocene Radiolarian Chronology and the Age of the Eocene–Oligocene Boundary." *Seventh Caribbean Geological Conference Proceedings* (1976): 205–212.

Meng, Jin, and Malcolm C. McKenna. "Faunal Turnovers of Palaeogene Mammals from the Mongolia Plateau." *Nature* 394 (1998): 364–367.

Pagani, Mark, James C. Zachos, Katherine H. Freeman, et al. "Marked Decline in Atmospheric Carbon Dioxide Concentrations during the Paleogene." *Science* 309 (2005): 600–603.

Poag, C. Wylie. "Roadblocks on the Kill Curve: Testing the Raup Hypothesis." *Palaios* 12 (1997): 582–590.

Poag, C. Wylie, David S. Powars, Larry J. Poppe, et al. "Deep Sea Drilling Project Site 612 Bolide Event: New Evidence of Late Eocene Impacts-wave Deposits and a Possible Impact Site, U.S. East Coast." *Geology* 20 (1992): 771–774.

Prothero, Donald R. "Does Climatic Change Drive Mammalian Evolution?" *GSA Today* 9, no. 9 (1999): 1–5.

Prothero, Donald R. "Did Impacts, Volcanic Eruptions, or Climatic Change Affect Mammalian Evolution?" *Palaeogeography, Palaeoclimatology, Palaeoecology* 214 (2005): 283–294.

Prothero, Donald R. "Magnetic Stratigraphy of the Eocene–Oligocene Floral Transition in Western North America." *Geological Society of America Special Paper* 435 (2008): 71–87.

Prothero, Donald R., and Timothy H. Heaton. "Faunal Stability during the Early Oligocene Climatic Crash." *Palaeogeography, Palaeoclimatology, Palaeoecology* 127 (1996): 239–256.

Raymo, M. E., and William F. Ruddiman. "Tectonic Forcing of Late Cenozoic Climate." *Nature* 359 (1992): 117–122.

Retallack, Gregory J. "Late Eocene and Oligocene Paleosols from Badlands National Park, South Dakota." *Geological Society of America Special Paper* 193 (1983).

Retallack, Gregory J., William N. Orr, Donald R. Prothero, et al. "Eocene–Oligocene Extinctions and Paleoclimatic Change near Eugene, Oregon." *Geological Society of America Bulletin* 116 (2004): 817–839.

Ruddiman, William F., and John E. Kutzbach. "Plateau Uplift and Climatic Change." *Scientific American* (March 1991): 66–75.

Vonhof, Hubert B., Jan Smit, Henk Brinkhuis, et al. "Global Cooling Accelerated by Early-Late Eocene Impacts?" *Geology* 28 (2000): 687–690.

Wei, Wucijang. "Reevaluation of the Eocene Ice-rafting Record from Subantarctic Cores." *Antarctic Journal of the United States* (1989): 108–109.

Wolfe, Jack A. "Tertiary Climatic Fluctuations and Methods of Analysis of Tertiary Floras." *Palaeogeography, Palaeoclimatology, Palaeoecology* 9 (1971): 27–57.

Wolfe, Jack A. "A Paleobotanical Interpretation of Tertiary Climates in the Northern Hemisphere." *American Scientist* 66 (1978): 694–703.

Wolfe, Jack A. "Distributions of Major Vegetational Types during the Tertiary." *American Geophysical Union Geophysical Monographs* 32 (1985): 357–376.

Zachos, James C., Mark Pagani, Lisa C. Sloan, et al. "Trends, Rhythms, and Aberrations in Global Climate 65 Ma to Present." *Science* 292 (2001): 686–693.

Donald R. Prothero

Late Quaternary extinctions

The Quaternary system (including the Pleistocene and the Holocene), covers the last 2.6 million years of geological time. This was predominantly a cold interval, but featured numerous and complex climatic fluctuations at different time scales, which on the one hand included interglacial periods as warm or warmer than today, and on the other episodes of major glaciation. These fluctuations resulted in major shifts in plant and animal distributions. At times within the cold phases ice sheets covered much of North America and northwestern Eurasia with the accompanying reduction in sea levels of more than 330 feet (100 m) exposing several land bridges that are now submerged—such as the Bering Straits—enabling animals and humans to cross in both directions. During the last 50,000 years (the Late Quaternary) many, often rapid, temperature changes occurred on a scale of thousands and hundreds of years. Major climatic events include the last glacial maximum (LGM; c. 27,500–14,700 years ago), the late glacial warming (c. 14,7000–12,900 years ago), the renewed cold of the Younger Dryas (c. 12,900–11,700 years ago), and the Holocene (postglacial) warming (c. 11,700 to the present day).

The last 50,000 years featured unprecedented losses of large terrestrial vertebrates (megafauna). Earlier extinctions in the Pleistocene affected small as well as large species, and megafaunal losses were mostly replaced by the evolution or immigration of ecologically similar forms. In contrast, Late Quaternary extinctions (excepting those within the last few hundred years) primarily affected large mammals, together with a few large birds and reptiles, leaving marine biotas unscathed.

The term *megafauna* is used here (following Paul S. Martin) for species with mean adult body weights of 97 pounds (44 kg) or over. In general, the larger the animal the more it is at risk of extinction because large size usually correlates with slow breeding and smaller number of individuals in the population. Megafaunal extinctions were highly variable in their severity between different zoogeographical regions, with the greatest impact in North America, South America, Australasia and northern Eurasia. In contrast, sub-Saharan Africa and southern Asia were relatively little affected, so that elephants and rhinos survive there, albeit precariously, to the present day. From a geological perspective the Late Quaternary extinctions can be seen as either a relatively minor episode or just the beginning of a major mass extinction—the so-called sixth extinction—that is in progress now.

That there had been a major extinction of megafauna in the recent geological past was recognized by nineteenth-century researchers, including Alfred Russel Wallace and Charles Darwin. Wallace's much-quoted observation bears repeating for its remarkable insight: "we live in a zoologically impoverished world from which all the hugest and fiercest and strangest forms have recently disappeared. It is surely a marvelous fact and one that has hardly been sufficiently dwelt upon, this dying out of so many large Mammalia, not in one place only, but over half the land surface of the globe" (1876, p. 150).

Not surprisingly, since Wallace's day there have been substantial advances in understanding the geographical and temporal patterns of megafaunal extinctions. Most significantly, following the invention of radiocarbon dating in the 1950s by the American chemist Willard Libby, reliable chronologies have been emerging for North and South America and northern Eurasia. The technique can be used to date megafaunal material: bones, teeth, antlers, and, more rarely, soft tissues and dung directly. As there is relatively little published information on megafaunal extinctions in sub-Saharan Africa and southern Asia, these regions are not discussed here.

Herein, all radiocarbon dates are given in calibrated years before present (closely approximating to calendar years)—that is, they are corrected from the measured radiocarbon dates. Other methods of absolute dating, such as uranium–thorium, optically stimulated luminescence, and amino acid racemization, are used for material beyond the radiocarbon limit of about 50,000 years. Major advances in ancient DNA and stable isotope analyses are also contributing significantly to scientists' understanding of the genetic population history and ancient diets of Quaternary megafaunal species.

Northern Eurasia

Northern Eurasia (the Palearctic ecozone) is an especially fruitful region for the study of Late Quaternary extinctions

Examples of global Late Quaternary megafauna. Species shown in brown survive today; the rest are extinct. Illustration by Joseph E. Trumpey.

not only because of the wealth of available archaeological, paleontological, and environmental data, but also because it has by far the largest number of radiocarbon dates made directly on megafaunal remains. These have made it possible to construct detailed chronologies for most of the extinct species and also many of the survivors. Modern humans arrived in Europe from Africa via the Middle East around 40,000 years ago, eventually replacing the Neanderthals who preceded them. Most megafaunal extinctions in the region occurred many millennia after this event.

The importance of megafaunal extinctions in this region has been played down by some authors as far fewer species disappeared compared with North America. It should be recognized, however, that losses in northern Eurasia were also significant: About 14 species out of a total of 49 with body weights greater than 97 pounds (44 kg) disappeared, or about 29 percent. As elsewhere, the largest animals were most affected, so that three species of elephant, two rhinos (each exceeding 4,400 pounds [2 metric tons]), and most other large mammals weighing over 1,100 pounds (500 kg) disappeared. A striking feature of the extinctions in northern Eurasia is that they were staggered over many millennia; moreover, there were important differences in the timing of extinctions in different geographical areas. In addition to those with good radiocarbon coverage (discussed below), there are a few species for which available information is

inadequate to infer the time of extinction. These include the spiral-horned antelope (*Spiroceros kiakhtensis*) and *Camelus knoblochi* (a camel) (both from southern Siberia) and a giant deer (*Sinomegaceros yabei*) from China and Japan. The well-dated extinctions fall into four broad chronological phases: (1) early last cold stage, (2) close to the onset of the LGM, (3) late glacial and Early Holocene, and (4) Late Holocene.

Early last cold stage (110,000–40,000 years ago)

This first wave of extinctions occurred mostly beyond the range of reliable radiocarbon dating, so that the chronology is very approximate. During the last interglacial (c. 130,000–117,000 years ago), the range of the hippopotamus (*Hippopotamus amphibius*) extended from Africa to Mediterranean Europe and Iberia northward to Britain. Intolerant of cold, it was probably extirpated from Europe at the beginning of the last glacial. The straight-tusked elephant (*Palaeoloxodon antiquus*) was widespread in Europe in the last interglacial period in association with regional temperate and Mediterranean forests but apparently retreated south of the Pyrenees and Alps in response to cooler temperatures and the spread of open vegetation in the early part of the last glacial. It may have survived in Iberia as late as 50,000 years ago. The narrow-nosed rhino (*Stephanorhinus hemitoechus*) had a similar history but probably survived rather later, again in southern Europe, to perhaps 42,000 years ago.

North America

1. Woolly mammoth (*Mammuthus primigenius*)	4. Mastodon (*Mammut americanum*)	7. Tapir (*Tapirus veroensis*)
2. Columbian mammoth (*Mammuthus columbi*)	5. Western horse (*Equus occidentalis*)	8. Sabretooth cat (*Smilodon fatalis*)
3. Western camel (*Camelops hesternus*)	6. Giant armadillo (*Glyptotherium floridanum*)	9. Harlan's ground sloth (*Paramylodon harlani*)

South America

10. Litoptern (*Macrauchenia*)	13. A notoungulate (*Toxodon*)	16. A ground sloth (*Mylodon*)
11. Giant ground sloth (*Megatherium americanum*)	14. A horse (*Equus* sp.)	
12. Giant armadillo (*Glyptodon*)	15. Sabretooth cat (*Smilodon populator*)	

Northern Eurasia

17. Giant deer (Megaloceros *giganteus*)	21. Cave lion (*Panthera spelaea*)	25. Narrow-nosed rhino (*Stephanorhinus hemitoechus*)
18. Woolly rhino (*Coelodonta antiquitatis*)	22. Steppe bison (*Bison priscus*)	26. Straight-tusked elephant (*Palaeoloxodon antiques*)
19. Woolly mammoth (*Mammuthus primigenius*)	23. Spotted hyaena (*Crocuta crocuta*)	
20. Cave bear (*Ursus spelaeus*)	24. Hippopotamus (*Hippopotamus amphibius*)	

Africa

27. African elephant (*Loxodonta africana*)	29. Black rhino (*Diceros bicornis*)	31. Spotted hyaena (*Crocuta crocuta*)
28. Hippopotamus (*Hippopotamus amphibius*)	30. White rhino (*Ceratotherium simum*)	

Southern Asia

32. Indian rhino (*Rhinoceros unicornis*)	33. Asian elephant (*Elephas maximus*)

Australasia

34. Giant wombat (*Diprotodon optatum*)	35. Giant short-faced kangaroo (*Proceoptodon goliah*)	36. Marupial lion (*Thylacoleo carnifex*)

Key to Figure 1. Examples of global Late Quaternary megafauna. Reproduced by permission of Gale, a part of Cengage Learning.

Extinctions close to the onset of the LGM (30,000–27,000 years ago)

The extinction of the largely vegetarian cave bear (*Ursus spelaeus*), which at around 27,500 years ago correlates well with the onset of the LGM, can be attributed to decreased temperatures and concomitant deterioration in vegetational quality and productivity. The spotted hyena (*Crocuta crocuta*), which still occurs in sub-Saharan Africa today, also disappeared at about the same time, perhaps because of a decrease in available prey and scavenging opportunities. Other species, such as the cave lion and the giant deer, withdrew from Europe during the LGM but survived farther east. A series of direct dates on Naumann's elephant (*Palaeoloxodon naumanni*) from Japan terminates at around 28,000 years ago, so this species may also have disappeared at the onset of the LGM. A sensational date of about 32,000 years ago on a single find of *Homotherium latidens* (a sabertooth cat) dredged from a North Sea bed needs to be corroborated by other dated finds. Otherwise, all the latest sabertooth records from Europe are Middle Pleistocene. The chronology of *Homo neanderthalensis* extinction is controversial, but broadly Neanderthals seem to have survived in many areas until around 35,000 years ago and to 28,000 years ago or later in Iberia. Clearly, Neanderthals were in decline well before the onset of the LGM, so climatic deterioration was probably not the main cause of their extinction. Possibly they succumbed to competition from modern humans, although the available evidence indicates that there was a substantial overlap between the two species of at least five millennia.

Late glacial and Early Holocene (15,000–4,000)

Both the woolly rhino (*Coelodonta antiquitatis*) and cave lion (*Panthera spelaea*) went extinct during the late glacial interval at about 14,000 years ago, probably in response to increased temperatures and the accompanying replacement of vast areas of open herb-dominated vegetation (the so-called mammoth steppe) by shrubs and trees, a process that started approximately 14,700 years ago. In the case of the cave lion there was possibly a concomitant reduction in the abundance of available prey. The range of the woolly mammoth (*Mammuthus primigenius*) was also drastically reduced at this time, but it survived until about 10,800 years ago (Early Holocene) in a restricted area of northern Siberia (Taimyr and New Siberian Islands). In 1993 the Russian paleontologist Sergey L. Vartanyan and his colleagues published their sensational finding, substantiated by many radiocarbon dates, that woolly mammoths had survived many millennia later than previously thought on Wrangel Island off northeastern Siberia. Radiocarbon dating has shown that woolly mammoths continued to at least around 4,600 years ago (contemporary with early Egyptian civilization), although attempts to find late records from mainland Siberia have been unsuccessful so far. There are rather few available dates for the northern Eurasian steppe bison (*Bison priscus*), but it evidently survived into the late glacial

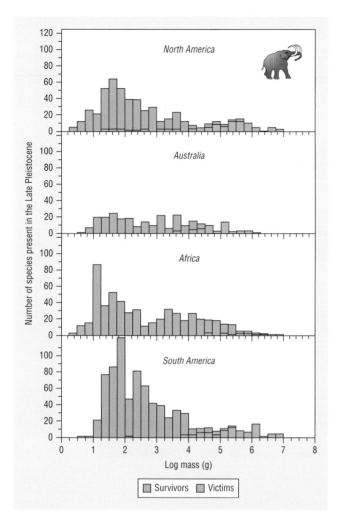

Late Pleistocene body size frequency distributions compared for four continents. Africa (12–15 KYR), Australia (ca.50 KYR), North America (12–15 KYR) and South America (12–15 KYR). The relationship between large size and likelihood of extinction is clearly shown. Reproduced by permission of Gale, a part of Cengage Learning.

interval in Siberia and European Russia and to the earliest Holocene in Taimyr. Another Holocene survivor, the Irish elk (*Megaloceros giganteus*), disappeared from western Europe with the onset of the Younger Dryas cold phase around 12,900 years ago and previously was thought to have gone extinct at that time. More recent research has revealed, however, that it persisted until at least 7,700 years ago in western Siberia and European Russia (Anthony J. Stuart et al. 2004).

Interestingly, one small mammal also became extinct. The vole *Pliomys lenki* had been widespread in Europe earlier in the Pleistocene, but by the late glacial interval it had retreated to a restricted area of northern Spain, where it finally disappeared by the beginning of the Holocene. The likely reasons for its demise are climatic change and competition from other small rodents.

Later extinctions (4,000 years ago to present)

Musk ox (*Ovibos moschatus*), now restricted to Arctic North America and Greenland, survived until at least 3,000 years ago

in the tundra of northern Siberia (Taimyr). In the Late Pleistocene the European ass (*Equus hydruntinus*) ranged across southern Europe to southwestern Asia. A compilation by Jennifer Crees shows that its area of distribution progressively shrank during the Holocene, with the latest records dating from approximately 3,500 years ago from the Caucasus.

Other species have disappeared within the last few hundred years, almost certainly because of hunting by humans. The last recorded individual of aurochs (*Bos primigenius*), the wild ancestor of domestic cattle, died in 1627 in Poland's Jaktorów Forest. The lion (*Panthera leo*) had probably disappeared from its foothold in southeastern Europe by 2,000 years ago, although it was still present within the last 200 years in the Middle East and North Africa.

Survivors

The larger surviving mammals of the region include the tiger (*Panthera tigris*), polar bear (*Ursus maritimus*), brown bear (*Ursus arctos*), horse (*Equus ferus*), onager (*Equus hemionus*), wild boar (*Sus scrofa*), Eurasian elk or moose (*Alces alces*), red deer (*Cervus elaphus*), wapiti (*Cervus canadensis*), sika deer (*Cervus nippon*), fallow deern (*Dama dama*), reindeer or caribou (*Rangifer tarandus*), saiga antelope (*Saiga tatarica*), bighorn sheep (*Ovis canadensis*), Spanish ibex (*Capra pyrenaica*), European bison (*Bison bonasus*), Bactrian camel (*Camelus ferus*) and yak (*Bos mutus*). The 2011 Red List of Threatened Species of the International Union for Conservation of Nature (IUCN) classifies the saiga and Bactrian camel as "critically endangered," while designating the tiger, wild horse, and ass "endangered."

North America

During the last glacial period North America (the Nearctic ecozone) had an extraordinarily rich and diverse large mammal fauna. In part this was a legacy of the so-called Great American Biotic Interchange in the Late Pliocene about 3 million years ago when, after millions of years of isolation, North and South America became connected via the Isthmus of Panama. This event allowed endemic South American animals such as ground sloths, glyptodonts, and capybaras to colonize North America. Another factor was the lower rate of extinction throughout the earlier Pleistocene, compared with northern Eurasia, so that by the Late Quaternary there were many more megafaunal species in total. The extent of the Late Quaternary extinctions was correspondingly great. Out of the 54 species with a body weight greater than 97 pounds (44 kg), around 37 went extinct, or about 69 percent. Curiously, all the horses in the Americas died out in the Late Pleistocene, but they survived in northern Eurasia. They thrived when reintroduced to North America by the Spanish in the sixteenth century.

North America experienced very similar climatic changes to those of northern Eurasia, although during the LGM a vast ice sheet covered most of the northern half of the continent, but the interior of Alaska and part of the Yukon was ice free. For many millennia this ice sheet prevented most or all biotic interchange between Alaska/Yukon and what is now the

Representative extinctions in northern Eurasia 1) woolly rhino (*Coelodonta antiquitatis*), 2) cave bear (*Ursus spelaeus*), 3) giant deer or "Irish elk" (*Megaloceros giganteus*), 4) spotted hyaena (*Crocuta crocuta*), 5) steppe bison (*Bison priscus*) 6) cave lion (*Panthera spelaea*), 7) woolly mammoth *Mammuthus primigenius*). Illustration by Joseph E. Trumpey.

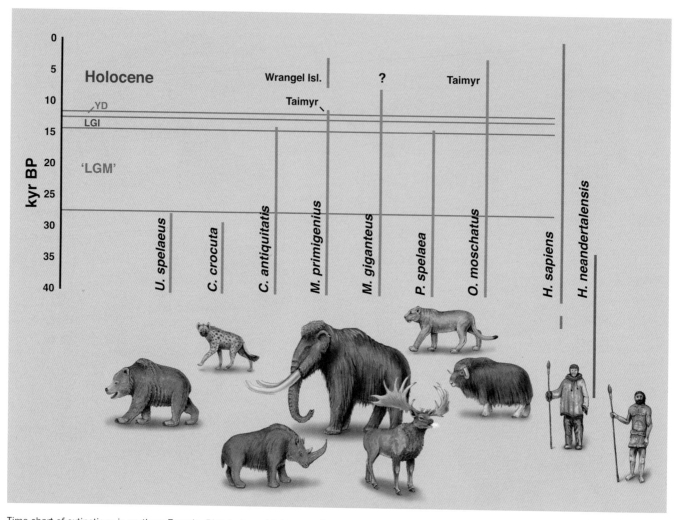

Time chart of extinctions in northern Eurasia. Blue text = cold phase, red = warm. Last Glacial Maximum (LGM), Late Glacial Interstadial (LGI), Younger Dryas (YD). Green bars show the known time-span (from 40 KYR) of a range of large mammals. Illustration by Joseph E. Trumpey.

contiguous United States. During the late glacial interval, however, a corridor opened between the western Cordilleran and eastern Laurentide ice sheets, allowing animals and humans to migrate in both directions.

Humans reached the Americas only late in the Pleistocene, entering Alaska from northeastern Siberia, but the timing of the colonization is highly controversial. There is good archaeological evidence for humans in Alaska by about 14,000 years ago, although until the opening of the ice-free corridor, they were probably unable to migrate southward to the present contiguous United States. There is abundant evidence in the form of distinctive Clovis projectile points from across most of the United States that have been dated (via, for example, associated charcoal) to a maximum "window" extending from 13,200 to 12,800 years ago, as well as some evidence for human occupation up to a few thousand years earlier.

Alaska/Yukon

There are many available radiocarbon dates made directly on megafauna from Alaska/Yukon (eastern Beringia). These

show that *Equus* species (horses) survived there until around 14,600 years ago, saiga antelope (*Saiga tatarica*) until 14,500 years ago, woolly mammoth (*Mammuthus primigenius*) until 13,400 years ago, and cave lion (*Panthera spelaea*) until 13,300 years ago. The saiga survives to the present day only in a limited area of northern Eurasia (mainly Kazakhstan). Available dates on *Homotherium serum* (a sabertooth cat) and the short-faced bear (*Arctodus simus*) are all pre-LGM, suggesting a pattern of staggered extinctions, as in northern Eurasia. *M. primigenius* survived on St. Paul Island in the Bering Sea until about 6,500 years ago—an interesting parallel to the Holocene survival of the same species on Wrangel Island.

Contiguous United States

Unfortunately, the available dates from the contiguous United States are as of 2012 inadequate for constructing a reliable chronology. A compilation by Gary Haynes (2009) shows latest dates pre-LGM for five or six species, including *Glyptotherium floridanum* (a giant relative of the armadillo) and Harlan's ground sloth (*Paramylodon* [*Glossotherium*] *harlani*). Twenty have latest dates in the range about 12,000 to 15,000

Representative extinctions in North America: Harlan's ground sloth (*Paramylodon [Glossotherium] harlani*), 2) Western camel (*Camelus hesternus*), 3) short-faced bear (*Arctodus simus*), 4) sabretooth cat (*Smilodon fatalis*), 5) western horse (*Equus occidentalis*), 6) Columbian mammoth (*Mammuthus columbi*), 7) American mastodon (*Mammut americanum*). Illustration by Joseph E. Trumpey.

years ago, including the American lion (*Panthera atrox*), *Smilodon fatalis* (a sabertooth cat), short-faced bear (*Arctodus simus*), Shasta ground sloth (*Nothrotheriops shastensis*), Jefferson's ground sloth (*Megalonyx jeffersonii*), Columbian mammoth (*Mammuthus columbi*), woolly mammoth (*Mammuthus primigenius*), mastodon (*Mammut americanum*), *Equus* spp. (horses), *Platygonus compressus* and *Mylohyus nasutus* (peccaries), western camel (*Camelops hesternus*), *Palaeolama mirifica* (a llama), and *Cervalces scotti* (a moose).

Because it is incomplete, the available evidence could be interpreted in two radically different ways: (1) most or all megafauna went extinct within a very short time—around 12,500 to 13,500 years ago, or perhaps a much shorter period within that time span—and that more dating work would show that apparent earlier extinctions are an artifact of inadequate sampling, or (2) the apparent pattern of extinctions in two or more phases, pre-LGM and late glacial, is real—much as in northern Eurasia. The second interpretation would imply that the earlier extinctions most likely occurred before the arrival of humans and therefore resulted from environmental changes. Clearly many more dates are needed to clarify this picture.

Additional evidence is provided by studies of the changing abundance of the dung fungus *Sporormiella* in lake sediments. Marked declines in the abundance of *Sporormiella* at several localities have been interpreted as reflecting a decline in megafaunal population size a few millennia before their accepted time of extinction. Research aiming to improve and refine this promising method is ongoing in the early 2000s.

Survivors

The larger surviving mammals of the region include: grizzly or brown bear (*Ursus arctos*), black bear (*Ursus americanus*), polar bear (*Ursus maritimus*), mountain lion (*Puma concolor*), jaguar (*Panthera onca*), pronghorn (*Antilocapra americana*), wapiti or North American elk (*Cervus canadensis*), caribou (*Rangifer tarandus*), mule deer (Odocoileus hemionus), white-tailed deer (Odocoileus *virginianus*), moose (*Alces alces*), musk ox (*Ovibos moschatus*), bighorn sheep (*Ovis canadensis*), Dall sheep (*Ovis dalli*), mountain goat (*Oreamnos americanus*), and American bison (*Bison bison*). None of these are listed as endangered in the IUCN Red List. The America bison was brought close to extinction during the nineteenth century as a result of commercial hunting and a deliberate policy of extermination (to deprive the Plains Indians of their food source), but the species' numbers were substantially increased following active conservation efforts in the late twentieth and early twenty-first centuries.

South America

In the Late Pleistocene, South America (the Neotropical ecozone) exhibited a rich and varied fauna, resulting from the admixture of often bizarre endemic animals that had evolved during millions of years when it was an isolated island continent and the immigration of North American forms that began in the Late Pliocene. Late Quaternary megafaunal extinction in South America appears to have been more

extensive than on any other continent. Taxonomic uncertainties for the extinct species preclude an accurate estimate of losses, but at a generic level approximately 80 percent became extinct. Interestingly, as in North America, all horse species went extinct. The severity of the extinctions can also be judged from the fact that the largest modern South American mammal is Baird's tapir with a maximum weight of about 660 pounds (300 kg).

Unlike the northern continents, South America was not extensively glaciated in the Pleistocene. The last glacial period, however, saw a contraction of the tropical forests at the expense of open vegetation well suited to grazing herbivores.

Although much good research has been done and is in progress as of 2012, at present the number of available dates is inadequate for constructing a reliable chronology for extinctions. Undoubtedly, however, there is a very exciting and interesting story to be uncovered here. Fortunately, as most of the South American extinctions seem to have occurred within the range of radiocarbon dating, the prospects for future research are excellent. The brief account given here is based largely on a 2010 review by Anthony D. Barnosky and Emily L. Lindsey, in which the researchers evaluated the available radiocarbon dates on the megafauna and early human settlement of South America.

Excluding the fanciful suggestion that Paleolithic people somehow crossed the Atlantic from Africa or Europe, the only possible route for human entry is via North and Central America. Yet the claimed earliest archaeology in the Americas, from Monte Verde in Chile, is dated to 14,500 years ago, significantly older than Clovis or even the oldest-known archaeology from Alaska, and some archaeologists remain unconvinced by the evidence. Other South American sites date from around 13,000 years ago or later.

Barnosky and Lindsey listed 93 dates for 15 megafaunal genera. Their list, however, does not include dates for one of the most extraordinary endemic mammals, the litoptern *Macrauchenia*, described as resembling a humpless camel with a short trunk; probably it had disappeared earlier. At face value, the latest radiocarbon-dated occurrences fall into the following three groups:

1. those dating from earlier than 18,000 years ago—*Holmesina* (giant armadillo), *Glyptodon* (a glyptodont), and *Haplomastodon* (a mastodon)

2. those dating from between 18,000 and 11,000 year ago—*Cuvieronius* (a gomphothere mastodon); *Mylodon*, *Glossotherium*, *Nothrotherium*, and *Eremotherium* (ground sloths); *Toxodon* (a notoungulate); and *Hippidion* and *Equus* (horses)

3. those dating from between 11,500 and 8,000 years ago—*Smilodon* (a sabertooth cat), *Megatherium* (giant ground sloth), *Catonyx* (a ground sloth), and *Doedicurus* (a glyptodont)

Much further research is required, however, to see whether there is any reality to this apparent chronological pattern.

Radiocarbon dates indicating that *Doedicurus*, *Catonyx*, *Megatherium*, and *Smilodon* survived into the Holocene—that is, later than around 11,700 years ago—also need corroboration.

It is evident that a substantial number of extinct megafauna survived until close to the end of the Pleistocene, and some perhaps into the Early Holocene. Beyond this, in view of the unsatisfactory dating evidence, it is premature to attempt correlations with either environmental changes or human arrival in South America.

Survivors

The larger surviving mammals of the region include lowland or Brazilian tapir (*Tapirus terrestris*), mountain tapir (*Tapirus pinchaque*), Baird's tapir (*Tapirus bairdii*), giant anteater (*Myrmecophaga tridactyla*), jaguar (*Panthera onca*), mountain lion (*Puma concolor*), spectacled bear (*Tremarctos ornatus*), capybara (*Hydrochoerus hydrochaeris*), vicuña (*Vicugna vicugna*), guanaco (*Lama guanicoe*), marsh deer (*Blastocerus dichotomus*), taruca or North Andean deer (*Hippocamelus antisensis*), and huemul or South Andean deer (*Hippocamelus bisulcus*). The 2011 IUCN Red List classifies mountain and Baird's tapirs as "Endangered."

Australasia

Australasia comprises Australia, Tasmania, and New Guinea, all of which were connected at times of lowered sea levels during the Pleistocene. Unlike the northern continents, Australasia was not glaciated, but there were changes between more arid and wetter periods. The first humans (*Homo sapiens*—the ancestors of the modern aborigines) arrived via Southeast Asia around 62,000 to 43,000 years ago. Although sea levels were lower at this time, these humans would have needed boats to cross several intervening water gaps.

In the Late Quaternary, as today, the terrestrial mammalian faunas were dominated by marsupials. There were also a few monotremes (egg-laying mammals unique to this continent in the Quaternary, although also present in South America in the Palaeocene), some endemic placental rodents and bats, and the dingo, which originated in eastern Asia and was introduced by human agency perhaps around 3,000 to 4,000 years ago. Because of stratigraphic uncertainties and problems of direct dating, it is difficult to be sure that all the species that are supposed to have occurred in the Late Pleistocene actually did so. Uniquely, in Australasia most extinctions seem to have taken place close to or beyond the range of radiocarbon dating, so that other, less precise methods of absolute dating have to be used. Establishing a robust chronology for extinctions is hampered at present because of these difficulties, although of course future improvements in dating techniques might radically change this situation. Nevertheless it is clear that Australasia suffered spectacular extinction in the Late Pleistocene, with the loss of many bizarre megafaunal species. An estimated 94 percent of megafaunal genera (15 out of 16) became extinct. The extinctions included 19 species of marsupials exceeding 220 pounds (100 kg) and 22 out of 38 species weighing between 22 and 220 pounds (10 and 100 kg).

Another curious and unique characteristic of Australian Pleistocene mammal faunas, in comparison with faunas from other continents, is that as a whole the size spectrum was shifted toward the smaller end. This phenomenon probably relates to the general environmental depletion in nutrients and the restricted area habitable by large mammals because of the arid continental interior. The largest known Australian mammal of all time was the giant wombat-like *Diprotodon optatum*, which weighed about 4,400 pounds (2 metric tons; roughly comparable to a white rhinoceros or a hippopotamus), but there was no marsupial equivalent in size to an elephant or mastodon. Other large mammals related to *Diprotodon* were the cow-sized marsupial tapir (*Palorchestes azael*), which had long claws and probably a moderately long trunk, and *Zygomaturus trilobus*, which had a smaller trunk. The extinct megafauna also included giant wombat (*Phascolonus gigas*), several species of *Sthenurus* and *Simosthenurus* (short-faced kangaroos), giant short-faced kangaroo (*Procoptodon goliah*), *Protemnodon* species (large wallaby-like kangaroos), marsupial lion (*Thylacoleo carnifex*), and the monotreme giant echidna (*Zaglossus hacketti*).

Extinctions also occurred in some nonmammalian megafauna. The giant goanna or monitor lizard (*Megalania prisca*), which at perhaps 1,550 pounds (700 kg) was much larger than its living relative, the Komodo dragon (*Varanus komodoensis*), would have been a formidable predator able to tackle even the largest marsupials. There is no certainty, however, that it survived into the last glacial period.

Most authorities believe that all, or nearly all, of the extinct Australasian megafauna had disappeared by 46,000 years ago, which is much earlier than in other regions. As elsewhere, however, the possible reasons for the extinctions are hotly disputed. Proponents of the overkill hypothesis point to the rather close correspondence in time between the arrival of humans and the demise of the megafauna and that the key factors were predation and habitat modification (especially burning of vegetation) by humans. Others argue, however, that the extinctions resulted from increasing aridity during the Late Pleistocene.

Very few sites provide any evidence of extinct megafauna in association with human artifacts, and Cuddie Springs in New South Wales has provided the best evidence for the coexistence of extinct megafauna and humans from Australia. At this site, sediments dated to 36,000 to 30,000 years ago contain the remains of extinct megafauna and stone tools. The evidence has been much disputed, however, as it appears that material of various ages is mixed together. Moreover, a 2010 study by Rainer Grün and colleagues involving the dating of tooth enamel from Cuddie Springs indicated a minimum age for the megafaunal remains of around 45,000 years ago, consistent with other Australian sites.

One extinct species, the large flightless goose *Genyornis newtoni* (a mihirung), was the subject of an extensive 1999 study by Gifford H. Miller and colleagues, who dated 700 eggshell fragments by amino acid racemization and optically stimulated luminescence dating of the enclosing sediment. The youngest dates obtained were about 50,000 years ago, the

Examples of extinct Australian megafauna. Clockwise from top left: *Genyornis newtoni* (a giant flightless goose), *Diprotodon optatum* (a giant wombat-like marsupial), *Procoptodon goliah* (giant short-faced kangaroo), *Thylacinus cynocephalus* (Tasmanian wolf or tiger, survived in Tasmania until 1936); and *Thylacoleo carnifex* (marsupial lion) disputing a kill with *Megalania prisca* (a giant goanna or monitor lizard). Courtesy of Peter Trusler.

inferred time of extinction. In marked contrast, a series of dates for emu (*Dromaius novaehollandiae*) shell fragments from the same sites continued through to the present day. Miller and colleagues suggest that the extinction of *Genyornis* and other Australian megafauna resulted from habitat modification resulting from the burning of vegetation by humans. As Steve Webb has observed, "Without an extensive dating programme of individual megafauna species in collections and field assemblages, however, many of the above arguments will remain as propositions only" (2008, p. 15).

Survivors

The only surviving mammal species that qualify as megafauna (i.e., greater than 97 pounds [44 kg] in weight) are the red kangaroo (*Macropus rufus*) at 190 pounds (85 kg) and the rather smaller western and eastern gray kangaroos (*Macropus fuliginosus* and *Macropus giganteus*). The flightless ratite emu (*Dromaius novaehollandiae*) averages only about 80 pounds (36 kg), although larger individuals reach 110 pounds (50 kg), whereas females of the somewhat larger southern cassowary (*Casuarius casuarius*) can reach 128 pounds (58 kg). There have been several recent extinctions among the smaller marsupials, and the inexcusable fate of Tasmanian wolf or Tasmanian tiger (*Thylacinus cynocephalus*) is well known. Bounty hunters were actually paid by the Tasmanian government for dead "tigers." The last wild specimen is recorded as having been shot in 1932, and the last captive animal died in 1936.

Island extinctions

This section briefly considers megafaunal extinctions on islands that are more or less isolated from continents. Island extinctions are especially interesting in showing the impact of human arrival on established endemic faunas in the Holocene. The extinction around 1680 of the endemic flightless dodo (*Raphus cucullatus*) on the island of Mauritius is both iconic and infamous. It is virtually certain that the demises of the dodo and many other Holocene island extinctions were caused by human activities: hunting, habitat destruction, and the introduction of alien species such as rats, pigs, and goats. Weighing only about 44 pounds (20 kg), however, the dodo does not qualify as megafauna, and neither do most of the other species that compose oceanic island extinctions, such as the 50 percent loss of 140 species of native birds recorded historically from the Hawaiian Islands.

New Zealand is separated from Australia by about 930 miles (1,500 km) of sea and by similar distances from major Pacific islands. Because of its remoteness, it was not colonized by humans in boats (from Pacific islands) until about 720 years ago. Prior to the arrival of the Maoris, there were in total eleven species of moa—flightless birds ranging from 44 to 440 pounds (20 to 200 kg)—on North and South Islands. Within 100 years they were all extinct, well before the first contact with Europeans in the seventeenth century. The abundance of moa remains in early archaeological sites provides compelling corroborative evidence that they were exterminated by Maori hunters. The extinction of the giant Haast's eagle (*Harpagornis moorei*), which weighed up to 30 pounds (14 kg), is likely to have resulted from the loss of its moa prey.

The rich endemic fauna of Madagascar, off the east coast of Africa, included a range of giant lemurs, pigmy hippopotamus, and the elephant birds *Aepyornis* and *Mullerornis*. Humans arrived by boat from Africa or Indonesia about 2,500 years ago. As described by Brooke E. Crowley in 2010, numerous direct radiocarbon dates record a decline in extinct large animals about 500 years after human colonization but also indicate that all of these survived until at least around 1,000 years ago. At most, however, there were only rare survivors when Europeans first arrived about 500 years ago. So unlike the case of New Zealand, for Madagascar there was a considerable lag of about 1,500 years between human arrival and extinctions.

All species of ground sloth went extinct in North and South America before 12,000 years ago. Remarkably, however, endemic ground sloths survived as late as 4,800 years ago in Cuba and Haiti. Humans had arrived on both by about 6,000 years ago, so again there was a significant lag before the extinctions occurred.

A dwarf mammoth (*Mammuthus exilis*) descended from the full-sized mainland *M. columbi*, is known from the Channel Islands off California. Radiocarbon dates on a skeleton from Santa Rosa indicate survival to about 13,000 years ago. The earliest evidence of humans from the island is dated to around 12,000 years ago. The Holocene survivals of woolly mammoths on Wrangel Island, off northeastern Siberia, and St. Paul Island, off Alaska, have already been mentioned.

Brief mention should be made here of the numerous extinctions of endemic species, both large and small, on the various Mediterranean islands. At present much of the complex story of repeated immigrations and extinctions throughout the Pleistocene is poorly understood. Dwarfed hippos, elephants, and deer evolved independently on different islands from full-sized ancestors that had arrived by chance from mainland Europe. Reliable dating is currently available only in a few instances, but work in progress by Victoria Herridge and Adrian M. Lister (Natural History Museum 2012) promises to improve the situation. Radiocarbon dates on charcoal associated with remains of pygmy hippo (*Phanourios minor*) and pigmy elephants on Cyprus suggest that both were present until around 11,000 years ago, about the same time as the earliest evidence of humans on the island. The Balearic Islands cave goat (*Myotragus balearicus*), which was endemic to Minorca and Majorca, went extinct about 5,000 years ago. The timing of human arrival on the Balearic Islands is disputed but possibly occurred as late as 4,000 to 5,000 years ago.

The search for a cause

Nearly half a century after Paul S. Martin reignited interest in this fascinating topic, the cause or causes of the Late Quaternary extinctions remain controversial and continue to generate a large amount of literature and debate. The major alternative hypotheses are overkill by human hunters and climatic/environmental change. Additional contenders include hyperdisease, extraterrestrial impact, and solar flare.

A number of questions about the Late Quaternary extinctions need to be answered, including the following:

1. Why were the extinctions mostly confined to larger terrestrial mammals, together with a few large birds and reptiles?

2. Why were they not synchronous on different continents?

3. Why were they most severe in North and South America and Australasia, moderate in northern Eurasia, and slight in sub-Saharan Africa?

4. Why did different species go extinct at different times in northern Eurasia and probably also elsewhere?

5. Why did similar extinctions not occur earlier in the Pleistocene?

6. Why did West Indian ground sloths and Siberian and Alaskan woolly mammoths survive thousands of years longer on islands than on the mainland?

Overkill

It is generally accepted that the numerous extinctions of terrestrial vertebrates (large and small) on oceanic islands were a direct consequence of human colonization in the postglacial period, mostly in the last thousand years or so. Ecosystems that had evolved in the absence of humans were rapidly devastated by habitat destruction, by the introduction of alien species such as rats and goats, and by hunting. However, the role of humans in the Late Pleistocene extinctions on the continents is less clear. With his prehistoric overkill hypothesis, Martin (1967) postulated that unsustainable levels of hunting of megafauna by humans with Upper Paleolithic technologies resulted in extinctions. The larger species are thought to have been especially vulnerable because they were slow-breeding and existed in relatively low populations. Computer modeling by John Alroy (2001) and others suggests that hunting by humans could result in extinctions, especially if the model allows the hunters to switch to other more abundant prey but still continue opportunistic hunting of the increasingly rare slow-breeding species.

It is very important to distinguish between the overkill hypothesis and its extreme variant—the Blitzkrieg hypothesis. Martin proposed that when modern humans (*Homo sapiens*) first colonized North America (and by extension also South America and Australia) in the Late Pleistocene they encountered naive prey highly vulnerable to the novel and aggressive human predator. As a result the populations of many species underwent catastrophic collapse as humans swept through each continent in just a few hundred years. Martin envisaged a "dreadful syncopation" in which human colonization was swiftly followed by extinctions. In response to criticisms that there were far fewer "kill sites"—that is, sites with direct association of projectile weapons with megafaunal remains—in North America than would have been predicted by the overkill hypothesis, Martin proposed that the slaughter had occurred so rapidly as to leave little trace in the fossil record.

Whereas today the overkill hypothesis enjoys plenty of support, the more extreme Blitzkrieg hypothesis boasts very few adherents. Subsequent researchers have pointed out that there is no mystery about the paucity of kill sites in North America. On the contrary, given the incompleteness of the fossil record, it is remarkable that several skeletons of Columbian mammoth have been recovered with associated Clovis spear points from a time window of only a few centuries. The Blitzkrieg hypothesis predicts that all of the North American megafauna would have disappeared at almost the same time, correlating with the advent of modern humans with Clovis technology. As described in the section on North America, this was evidently not the case, as there was probably more than one wave of extinction and mastodons and mammoths possibly outlived the other extinct megafauna.

The hypothesis predicts that major extinctions occurred soon after the arrival of modern humans (with stone- or bone-tipped projectile weapons) in a given region. In northern Eurasia, however, most megafaunal extinctions occurred many millennia after the arrival of modern humans. Moreover, it is difficult to see how hunting alone could have led to the extinction of so many species with wide geographical distributions. For example, woolly mammoth ranged from Spain and the British Isles across northern Asia to Alaska and Canada.

Climatic/environmental change

Proponents of the climatic/environmental change hypothesis point to the marked climatic changes and resulting major disruption of biota that occurred toward the end of the Pleistocene. Major changes in North America and northern Eurasia included the replacement of vast areas of open vegetation with grasses and herbs—the mammoth steppe—by forests, mainly boreal conifer forest and temperate deciduous forest. Why were the animals that went extinct not able to find suitable habitat somewhere? The geological record shows a complex pattern of climatic/environmental changes throughout the Pleistocene, in particular during the epoch's last 780,000 years, which featured repeated glacial–interglacial transitions resembling the most recent last glacial–Holocene transition. If these changes were responsible for major extinctions, then clearly the last cold stage must have been in some way unique, because no such catastrophic extinction occurred previously. A final point here is that the Late Quaternary extinctions were demonstrably not synchronous on a global scale, which argues against a single global climatic cause.

Combined hypothesis

A number of authors have suggested a combination of overkill mechanisms and environmental change, in which extinctions resulted from human hunting of megafaunal populations subject to habitat fragmentation and the stress of climatic and resulting vegetational changes. Hunting pressure became critical only when populations were already reduced by climatic/vegetational changes.

Hyperdisease

In 1997 Ross D.E. MacPhee and Preston A. Marx postulated that extinctions could have been driven by a lethal pathogen introduced by humans or their dogs as they spread around the globe. However, there is no known disease at the present day that would preferentially select larger species or be infectious across the ecological spectrum. Hyperdisease is subject to the same objections as Martin's Blitzkrieg model, as both predict that extinctions closely followed first contact with humans. Similarly, the staggered pattern of extinctions in northern Eurasia and Alaska/Yukon does not fit with this hypothesis.

Extraterrestrial impact

In 2007 a sensational new proposal was made by Richard B. Firestone and colleagues. They described a range of features in sediment profiles across North America dating from the onset of the Younger Dryas stadial (c. 12,900 years ago) that they attributed to the impact of a comet or other extraterrestrial body. The event is supposed to have been a major contributor to megafaunal extinction in North America and also caused, among other effects, the release of vast quantities of meltwater from the destabilized North American (Laurentide) ice sheet—which in turn triggered the Younger Dryas cold episode, very extensive wildfires, and a postulated devastation of the Paleo-Indian population.

Although it has attracted a great deal of media attention, this idea has not been well received by many in the scientific community. Other researchers have been unable to find corroborative evidence, such as "impact markers" (nanodiamonds or high concentrations of magnetic spherules) from the originally studied localities, or charcoal in lake sediments resulting from "massive burning." Moreover, most North American megafaunal extinctions seem to have occurred before or after the Younger Dryas, and there is no evidence from the archaeological record of a marked decline in human populations. Clearly the hypothesis could not apply to other continents, such as northern Eurasia, where extinctions were staggered and none corresponded with the Younger Dryas event.

Solar flare hypothesis

Another dramatic hypothesis postulates that about 12,800 years ago a massive increase in radiation from a solar flare caused global mass extinctions. This idea is easily dismissed because the extinctions were demonstrably not synchronous on a global scale, as would have resulted from such a catastrophic event. Additionally, if the radiation dose was sufficiently high to wipe out much of the megafauna, then most other terrestrial species and humans would also have been seriously affected, which clearly did not happen.

Conclusions—The story so far

Without a detailed chronology of extinctions for each region, to use in comparison with the climatic and archaeological records, it is not possible to pin down the cause or causes of the Late Quaternary extinctions. Researchers need to know first and foremost when each species became extinct both in time and geographically. To this end, the most important tool is radiocarbon dating applied directly to megafaunal remains. So far, as detailed above, northern Eurasia has by far the best coverage in terms of radiocarbon dates. There are substantial numbers of radiocarbon dates for North America, but many more are needed to assess the overall picture, especially as there are many more extinct North American taxa than ones from northern Eurasia. The number of available dates for South America is insufficient as of 2012 but growing steadily. There is a particular problem with Australasia as most of the extinctions there evidently occurred beyond the range of radiocarbon dating, so that other, less precise dating methods have to be used. Future refinements in these dating methods are likely to provide the key to understanding the causes of the extinctions on that continent.

Analyses of ancient DNA in megafaunal remains, in combination with radiocarbon dating, are producing some exciting results, including the estimation of changing past population sizes. During the last glacial period in northern Eurasian and North America, several extinct megafaunal species (e.g., *Panthera spelaea*, *Mammuthus primigenius*) declined in genetic diversity but subsequently recovered. To what extent this contributed to their eventual extinction is less clear, however, as others such as *Ovibos moschatus* also experienced similar bottlenecks but nevertheless have survived to the present day. Another significant development is the application of niche modeling, in which geographical ranges of megafauna at different times in the past are related to climate and vegetation. These niche models are used to simulate the changing potential range of each species and thus determine whether species became extinct as a result of habitat loss or whether other factors such as human hunting might have been involved.

Claims by some authors that the problem has been solved, one way or another, are definitely premature. Although there is a vast body of available data, it is apparent that researchers still have some way to go in tracking down the cause or causes of the Late Quaternary megafaunal extinctions.

Resources

Books

Haynes, Gary, ed. *American Megafaunal Extinctions at the End of the Pleistocene*. Dordrecht, Netherlands: Springer, 2009.

Lister, Adrian M., and Paul G. Bahn. *Mammoths: Giants of the Ice Age*. Rev. ed. Berkeley: University of California Press, 2007.

MacPhee, Ross D.E., ed. *Extinctions in Near Time: Causes, Contexts, and Consequences*. New York: Kluwer Academic/Plenum, 1999.

MacPhee, Ross D.E., and Preston A. Marx. "Humans, Hyperdisease, and First-Contact Extinctions." In *Natural Change and Human Impact in Madagascar*, edited by Steven M. Goodman and Bruce D. Patterson, 169–217. Washington, DC: Smithsonian Institution Press, 1997.

Martin, Paul S. "Prehistoric Overkill." In *Pleistocene Extinctions: The Search for a Cause*, edited by Paul S. Martin and Herbert E. Wright Jr. New Haven, CT: Yale University Press, 1967.

Martin, Paul S., and Richard G. Klein. *Quaternary Extinctions: A Prehistoric Revolution*. Tucson: University of Arizona Press, 1984.

Martin, Paul S., and David W. Steadman. "Prehistoric Extinctions on Islands and Continents." In *Extinctions in Near Time: Causes, Contexts, and Consequences*, edited by Ross D.E. MacPhee, 17–55.

Simmons, Alan H., and associates. *Faunal Extinction in an Island Society: Pygmy Hippopotamus Hunters of Cyprus*. New York: Kluwer Academic/Plenum, 1999.

Wallace, Alfred R. *The Geographical Distribution of Animals: With a Study of the Relations of Living and Extinct Faunas as Elucidating the Past Chances of the Earth's Surface*, Vol. 1, 150. London: Macmillan, 1876.

Periodicals

Alroy, John. "A Multispecies Overkill Simulation of the End-Pleistocene Megafaunal Mass Extinction." *Science* 292, no. 5523 (2001): 1893–1896.

Barnosky, Anthony D., and Emily L. Lindsey. "Timing of Quaternary Megafaunal Extinction in South America in Relation to Human Arrival and Climate Change." *Quaternary International* 217, nos. 1–2 (2010): 10–29.

Crowley, Brooke E. "A Refined Chronology of Prehistoric Madagascar and the Demise of the Megafauna." *Quaternary Science Reviews* 29, nos. 19–20 (2010): 2591–2603.

Firestone, Richard B., Allen West, James P. Kennett, et al. "Evidence for an Extraterrestrial Impact 12,900 Years Ago That Contributed to the Megafaunal Extinctions and the Younger Dryas Cooling." *Proceedings of the National Academy of Sciences of the United States of America* 104, no. 41 (2007): 16016–16021.

Grün, Rainer, Stephen Eggins, Maxime Aubert, et al. "ESR and U-Series Analyses of Faunal Material from Cuddie Springs, NSW, Australia: Implications for the Timing of the Extinction of the Australian Megafauna." *Quaternary Science Reviews* 29, nos. 5–6 (2010): 596–610.

Koch, Paul L., and Anthony D. Barnosky. "Late Quaternary Extinctions: State of the Debate." *Annual Review of Ecology, Evolution, and Systematics* 37 (2006): 215–250.

Lorenzen, Eline D., David Nogués-Bravo, Ludovic Orlando, et al. "Species-Specific Responses of Late Quaternary Megafauna to Climate and Humans." *Nature* 479, no. 7373 (2011): 359–364.

Lyons, S. Kathleen, Felisa A. Smith, and James H. Brown. "Of Mice, Mastodons, and Men: Human-Mediated Extinctions on Four Continents." *Evolutionary Ecology Research* 6, no. 3 (2004): 339–358.

Miller, Gifford H., John W. Magee, Beverly J. Johnson, et al. "Pleistocene Extinction of *Genyornis newtoni*: Human Impact on Australian Megafauna." *Science* 283, no. 5399 (1999): 205–208.

Pacher, Martina, and Anthony J. Stuart. "Extinction Chronology and Palaeobiology of the Cave Bear (*Ursus spelaeus*)." *Boreas* 38, no. 2 (2009): 189–206.

Roberts, Richard G., and Barry W. Brook. "And Then There Were None?" *Science* 327, no. 5964 (2010): 420–422.

Roberts, Richard G., Timothy F. Flannery, Linda K. Ayliffe, et al. "New Ages for the Last Australian Megafauna: Continent-Wide Extinction about 46,000 Years Ago." *Science* 292, no. 5523 (2001): 1888–1892.

Steadman, David W., Paul S. Martin, Ross D.E. MacPhee, et al. "Asynchronous Extinction of Late Quaternary Sloths on Continents and Islands." *Proceedings of the National Academy of Sciences of the United States of America* 102, no. 33 (2005): 11763–11768.

Stuart, Anthony J., Pavel A. Kosintsev, Tom F.G. Higham, and Adrian M. Lister. "Pleistocene to Holocene Extinction Dynamics in Giant Deer and Woolly Mammoth." *Nature* 431, no. 7009 (2004): 684–689.

Stuart, Anthony J., and Adrian M. Lister. "Patterns of Late Quaternary Megafaunal Extinctions in Europe and Northern Asia." *Courier Forschungsinstitut Senckenberg* 259 (2007): 287–297.

Stuart, Anthony J., and Adrian M. Lister. "Extinction Chronology of the Cave Lion *Panthera spelaea*." *Quaternary Science Reviews* 30, nos. 17–18 (2011): 2329–2340.

Stuart, Anthony J., and Adrian M. Lister. Extinction Chronology of the Woolly Rhinoceros *Coelodonta antiquitatis* in the Context of Late Quaternary Megafaunal Extinctions in Northern Eurasia. *Quaternary Science Reviews* 51 (2012): 1–17.

Vartanyan, S.L., V.E. Garutt, and A.V. Sher. "Holocene Dwarf Mammoths from Wrangel Island in the Siberian Arctic." *Nature* 362, no. 6418 (1993): 337–340.

Veltre, Douglas W., David R. Yesner, Kristine J. Crossen, et al. "Patterns of Faunal Extinction and Paleoclimatic Change from Mid-Holocene Mammoth and Polar Bear Remains, Pribilof Islands, Alaska." *Quaternary Research* 70, no. 1 (2008): 40–50.

Webb, Steve. "Megafauna Demography and Late Quaternary Climatic Change in Australia: A Predisposition to Extinction." *Boreas* 37, no. 3 (2008): 329–345.

Wroe, Stephen, and Judith Field. "A Review of the Evidence for a Human Role in the Extinction of Australian Megafauna and an Alternative Interpretation." *Quaternary Science Reviews* 25, nos. 21–22 (2006): 2692–2703.

Other

Natural History Museum. "Dwarfing of Fossil Mammals on Mediterranean Islands." Accessed April 26, 2012. http://www.nhm.ac.uk/research-curation/departments/palaeontology/research/vertebrates/quaternary-mammals/dwarfing/.

Anthony John Stuart

North America (10,000–11,500 years ago)

Toward the end of the Pleistocene, or Ice Age, North America lost some 36 genera of mammals, either in the sense that they ceased to exist anywhere in the world after this time (30 genera) or that they no longer existed in North America even though they survived elsewhere. The causes of these extinctions have been the focus of an enduring debate in archaeology and paleontology.

The extinct mammals

Understanding this debate requires an understanding of the astonishing diversity of the mammals that were lost. It also requires an understanding of Late Pleistocene archaeology and climates in North America, but the mammals themselves must come first.

The cingulates

Today, North America supports only one species of cingulate, the nine-banded armadillo (*Dasypus novemcinctus*), and even it may not have arrived on the continent until the mid-nineteenth century. During the Pleistocene, however, southeastern North America not only had a now-extinct species of large armadillo—*D. bellus*, the beautiful armadillo—but also had two genera of cingulates of enormous size.

Pampatheres were armadillo-like in that they were enclosed in a flexible armor of bony scutes, but they differed sufficiently from armadillos to be placed in their own family. There are two genera of later Pleistocene pampatheres that may have occurred in North America, but only one is securely known from there. The northern pampathere (*Holmesina septentrionalis*) was some 6 feet (1.8 m) long and 3 feet (1 m) high and has been found in sites ranging from Kansas and North Carolina in the north to Texas and Florida in the south. The southern pampathere (*Pampatherium*) has been reported from only two sites, in Texas and Sonora, Mexico.

Simpson's glyptodont (*Glyptotherium floridanum*) is known primarily from near-coastal localities in Texas, Florida, and South Carolina. About 10 feet (3 m) long and 5 feet (1.5 m) tall, this animal had a turtle-like carapace, an armored tail and skull, massive limbs, and a pelvic girdle that was fused to its shell. From the settings in which its remains have been found, Simpson's glyptodont appears to have lived along lakes, streams, and marshes and may have been semiaquatic.

The sloths

Within the four genera of extinct North American ground sloths, the largest species was the Laurillard's ground sloth (*Eremotherium laurillardi*), which combined the height of a giraffe with the bulk of an elephant and is known from coastal plain environments from New Jersey south through Florida and west into Texas. The Shasta ground sloth (*Nothrotheriops shastensis*), which weighed about 330 pounds (150 kg) was the smallest of the North American sloths and is known only from western North America. Jefferson's ground sloth (*Megalonyx jeffersonii*) was the most widespread of them all, distributed from Florida to Alaska, with those in Alaska appearing to date to the last interglacial. Harlan's ground sloth (*Paramylodon harlani*) was also widespread in North America, found from coast to coast. Distinguished in part by the small bones embedded in its skin, this was the most abundant sloth at the famous Rancho La Brea Tar Pits in southern California.

The carnivores

The dhole (*Cuon alpinus*) is a pack-hunting, highly carnivorous member of the dog family that is widespread (but dwindling) in Asia. During the Pleistocene, it was found from southwestern Europe to Alaska and the Yukon, with a single additional site known from northern Mexico.

Dire wolves (*Canis dirus*), by contrast, were broadly distributed in North America. These large canids, estimated to have weighed about 130 pounds (60 kg), seem to have been pack hunters capable of taking prey that weighed well in excess of 1,000 pounds (455 kg). Dire wolves do not appear in Table 1 because the genus to which they belong includes the wolves and coyotes and thus still exists in North America.

Table 1 also excludes the giant American lion, (*Panthera leo atrox* [or *Panthera P. atrox*]), because the jaguar (*P. onca*) also still occurs in North America. Huge lions were not uncommon in North American open environments during the Pleistocene. Weighing around 950 pounds (430 kg), over twice the mass of the African lion, the giant American lion was the largest cat in North America.

A reconstruction of a *Glyptotherium* at the Arizona Museum of Natural History. Courtesy of Thomas H. Wilson/Arizona Museum of Natural History.

That species was not, however, the only huge member of the cat family to be found in North America. The famous sabertooth (*Smilodon fatalis*) weighed an estimated 850 pounds (385 kg) and ranged from coast to coast. The less well-known scimitar cat (*Homotherium serum*), which was found from Alaska to Texas and Florida, weighed in at 400 pounds (180 kg), while the American cheetah (*Miracinonyx trumani*), closely related to the cougar (*Puma concolor*), weighed an estimated 150 pounds (70 kg).

The largest Late Pleistocene carnivore, however, was the giant short-faced bear (*Arctodus simus*), the biggest of which may have reached 2,000 pounds (910 kg). Found from Pennsylvania to California, south into Mexico and north to Alaska and the Yukon, these long-limbed bears were highly carnivorous, with the meat they consumed perhaps obtained largely by scavenging.

The second extinct genus of North American bear, *Tremarctos*, exists today in the form of the spectacled bear (*T. ornatus*), which occupies the mountains of northwestern South America. The North American Pleistocene form was more powerfully built and larger, with an estimated weight of 300 pounds (135 kg). Known from Texas to South Carolina and Florida, it appears to have been omnivorous.

Harlan ground sloth (*Paramylodon harlani*) from the La Brea Tar Pits. Courtesy of D.K. Grayson.

Not all of the now-extinct North American Pleistocene carnivores were huge. The short-faced skunks (*Brachyprotoma*), whose later Pleistocene distribution ranged from eastern North America to the Yukon, were roughly equivalent in size to spotted skunks (*Spilogale*), the genus to which they appear to be most closely related.

Order and Family	Genus	Common Name
Cingulata		
Pampatheriidae	*Pampatherium*	Southern pampathere
	Holmesina	Northern pampathere
Glyptodontidae	*Glyptotherium*	Simpson's glyptodont
Pilosa		
Megalonychidae	***Megalonyx***	Jefferson's ground sloth
Megatheriidae	*Eremotherium*	Rusconi's ground sloth
Nothrotheriidae	***Nothrotheriops***	Shasta ground sloth
Mylodontidae	*Paramylodon*	Harlan's ground sloth
Carnivora		
Mustelidae	*Brachyprotoma*	Short-faced skunk
Canidae	*Cuon**	Dhole
Ursidae	*Tremarctos**	Florida cave bear
	Arctodus	Giant short-faced bear
Felidae	***Smilodon***	Sabertooth
	Homotherium	Scimitar cat
	Miracinonyx	American cheetah
Rodentia		
Castoridae	***Castoroides***	Giant beaver
Caviidae	*Hydrochoerus**	Holmes's capybara
	Neochoerus	Pinckney's capybara
Lagomorpha		
Leporidae	*Aztlanolagus*	Aztlan rabbit
Perissodactyla		
Equidae	***Equus****	Horses
Tapiridae	***Tapirus****	Tapirs
Artiodactyla		
Tayassuidae	***Mylohyus***	Long-nosed peccary
	Platygonus	Flat-headed peccary
Camelidae	***Camelops***	Yesterday's camel
	Hemiauchenia	Large-headed llama
	Palaeolama	Stout-legged llama
Cervidae	*Navahoceros*	Mountain deer
	Cervalces	Stag-moose
Antilocapridae	*Capromeryx*	Diminutive pronghorn
	Tetrameryx	Shuler's pronghorn
	Stockoceros	Pronghorns
Bovidae	*Saiga**	Saiga
	Eucheratherium	Shrub ox
	Bootherium	Harlan's muskox
Proboscidea		
Gomphotheriidae	*Cuvieronius*	Cuvier's gomphothere
Mammutidae	***Mammut***	American mastodon
Elephantidae	***Mammuthus***	Mammoths

The extinct Late Pleistocene mammals of North America. Genera marked with an asterisk live on elsewhere; those in bold have radiocarbon dates that fall between 12,000 and 10,000 years ago. Reproduced by permission of Gale, a part of Cengage Learning.

Sabertooth cat (*Smilodon fatalis*) skeleton at Panhandle-Plains Historical Museum, Amarillo, Texas. © Lana Sundman/Alamy.

The rodents

The giant beaver (*Castoroides ohioensis*) ranged from New York to Florida in the east and to Alaska in the far northwest but appears to have been most common in the Great Lakes region. Giant beavers were not dam builders, even though their preferred habitat appears to have been lakes, ponds, marshes, and swamps. A 2002 analysis suggests that they weighed some 170 pounds (75 kg), but, because these animals were the size of black bears (*Ursus americanus*), this estimate may be incorrect, and others suggest the weight to have been closer to 350 pounds (160 kg).

The two extinct North American capybaras—southeastern in distribution—are both related to the world's largest living rodent, *Hydrochoerus hydrochaeris* of central and northern South America, which weighs about 130 pounds (60 kg). The extinct Holmes's capybara (*H. holmesi*) was slightly larger than this, but Pinckney's capybara (*Neochoerus pinckneyi*) was perhaps half again larger.

The lagomorphs

The only genus of lagomorph known to have been lost during the later Pleistocene was *Aztlanolagus*, which included

The capybara (*Hydrochoerus hydrochaeris*). Photo by R.D. Lord, courtesy of the Mammal Images Library.

a single species, the small Aztlán rabbit (*A. agilis*). Fossils of this species have been found from eastern Nevada through New Mexico and Texas south into central Mexico.

The perissodactyls

Horses were widespread in North America but especially common in the west, including the far northwest. Tapirs, of at least two species, were also reasonably common, known from Pennsylvania to California, with the largest number of sites found in eastern North America.

The artiodactyls

Today, the collared peccary (*Tayassu tajacu*) ranges north into the southwestern United States and Texas, but it appears to be a very recent arrival, and peccaries were far more widespread in North America during the Late Pleistocene. The long-nosed peccary (*Mylohyus nasutus*) apparently flourished in wooded environments and was primarily eastern in distribution. The flat-headed peccary (*Platygonus compressus*), by contrast, was distributed from coast to coast and seemed to prefer more open settings. Unlike the long-nosed peccary, it also seems to have been gregarious, judging from the discovery of "fossil herds."

Toward the end of the Pleistocene, North America supported three members of the camel family. The western camel (*Camelops hesternus*), also known as yesterday's camel, looked something like a longer-legged, narrower-headed version of the dromedary (*Camelus dromedarius*) and was very common in the western half of North America. The large-headed llama (*Hemiauchenia macrocephala*) was widespread in southern North America, with sites known as far north as Iowa and Idaho. The stout-legged llama (*Palaeolama mirifica*) has been reported from South Carolina and Florida west to southern California, but most records are from the southeast. As its common name suggests, this species had relatively short and robust limbs, perhaps an evolutionary response to the need to escape predators in closed to semi-closed habitats.

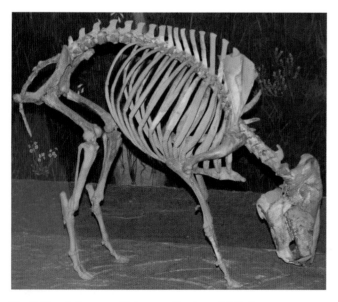

The flat-headed peccary *Platygonus*. Courtesy of D.K. Grayson.

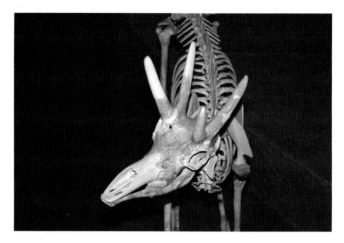

The extinct pronghorn *Stockoceros*. Courtesy of D.K. Grayson.

Mountain deer (*Navahoceros fricki*), whose closest relatives appear to have been reindeer and caribou (*Rangifer tarandus*), had short, robust limbs and fairly simple antlers and were found from southeastern California to the Plains. The elk-moose or stag-moose (*Cervalces scotti*) is far better known, with sites found from southern Canada and the eastern and central United States into the Yukon and Alaska. The contexts in which it has been found suggest that this animal may have been similar to the moose (*Alces alces*) not just in size but also in habitat preference.

There is a single species of Antilocapridae in North America today, the pronghorn (*Antilocapra americana*), a speedy animal with two laterally compressed horn cores. During the Late Pleistocene, this animal existed alongside three other genera of the same family, all of which were characterized by having four, rather than two, horns. The smallest of these was the diminutive pronghorn (*Capromeryx*), known from California to Texas. This species was 1.5 feet (0.5 m) tall at the shoulder, with estimates for the weight of this long-limbed antilocaprid falling at about 25 pounds (11 kg). The other two extinct antilocaprids were closer to the pronghorn in size: *Stockoceros*, known from Late Pleistocene contexts in the southwest and Texas, and *Tetrameryx*, reported from Texas, New Mexico, and southern Nevada.

Three genera of mammals related to cattle and sheep became extinct in North America toward the end of the Pleistocene. Of these, the saiga (*Saiga tatarica*) lives on in the arid steppes of Eurasia. During the later Pleistocene, however, it was to be found as far southwest as northern Spain and as far northeast as Alaska, the Yukon, and the Northwest Territories. The helmeted muskox (*Bootherium bombifrons*), also known as Harlan's muskox, is known from much of unglaciated North America; only the far southeast and far southwest appear to have been without it. Compared to the extant muskox (*Ovibos moschatus*), the helmeted muskox had longer legs and stood taller but was shorter from head to tail and had shorter pelage (known from preserved material from the far north). The horn cores of male, but not female, helmeted muskoxen fused in the midline, unlike those of *Ovibos*, providing the animal's common name. The shrub-ox (*Euceratherium collinum*) is characterized by horn cores that arise from near the rear edge of the frontal bones, sweep up and back, then out and forward, and curl upward near their tips. The skulls and neck vertebrae of shrub-oxen imply that, like bighorn sheep (*Ovis canadensis*), they were head-butters; their remains have been found from California to Iowa and south into Mexico.

The mountain goat (*Oreamnos americanus*) is doing well in northwestern North America, but Harrington's mountain goat (*O. harringtoni*) failed to survive the end of the Pleistocene. Sites that have provided the remains of this animal are known from the central Great Basin south through the Colorado Plateau into northern Mexico and as far east as southeastern New Mexico. Harrington's mountain goat was about 30 percent smaller than its modern counterpart, had a narrower face with thinner and smaller horns, and robust feet suitable for negotiating rough terrain.

The proboscideans

Gomphotheres are related to elephants and, among the extinct forms, closer to mastodons than to mammoths. They differ from mastodons in the form of their tusks and in that the cusps of their molar teeth are much more complex. Primarily Central and South American in distribution during the Late Pleistocene, one genus, *Cuvieronius*, was also found on the southern edge of North America during the very late portion of the Pleistocene.

The American mastodon (*Mammut americanum*) was widespread in unglaciated North America, found from coast to coast and from Alaska into Mexico, but was particularly abundant in the woodlands and forests of eastern North America; the species is also known from elevations as high as 9,800 feet (3,000 m). With shoulder heights of 6.5 feet (2 m) to 10 feet (3 m), American mastodons weighed an estimated 6,000 pounds (2,700 kg).

Most experts recognize two species of large Late Pleistocene North American mammoths, the Columbian mammoth (*Mammuthus columbi*) and the woolly mammoth (*M. primigenius*). There is also a diminutive form, the pygmy mammoth (*M. exilis*), known from the Channel Islands off the coast of California. Derived from *M. columbi*, the pygmy mammoth had shoulder heights that varied from about 4 feet (1.2 m) to 8 feet (2.4 m), compared to shoulder heights that ranged from about 7.5 feet (2.3 m; woolly mammoth females) to 13 feet (4 m; Columbian mammoth males) for the other species.

Some related changes

The mammals were not alone in suffering such extinctions. Some twenty genera of birds were also lost during the North American Late Pleistocene. While nine of these were predators or scavengers whose extinction may have been driven by the loss of the mammals, the others ranged from

Columbian mammoth (*Mammuthus columbi*) skeleton with tusks from La Brea Tar Pits, Page Museum, Los Angeles. © Martin Shields/Alamy.

storks and flamingoes to shelducks and jays. There was even a species of tree—the Critchfield's spruce (*Picea critchfieldii*)—that disappeared as the North American Pleistocene ended.

At the same time as these extinctions were occurring, other animals were undergoing often-massive distributional changes. Caribou, for instance, are known from as far south as northern Mississippi during the Late Pleistocene and muskoxen from as far south as Tennessee. Very small mammals, which underwent no generic extinctions, also moved dramatically across space as the Pleistocene ended. For instance, Late Pleistocene Tennessee was home to both the taiga vole (*Microtus xanthognathus*) and the heather vole (*Phenacomys intermedius*). Today, the taiga vole is found no further south than central Alberta and the heather vole no further south than the Canadian border in eastern North America.

Similarities and differences

With one exception, not much seems to tie these animals together, other than the fact that they lived in North America and became extinct. Their geographic distributions, for example, often differed widely. Some, such as the mammoths and tapirs, were found from coast to coast. Others, such as the capybaras and glyptodonts, were confined to the south. The saiga is known only from Alaska and the Yukon in North America; the shrub ox, only from the west.

The evolutionary origins of these animals vary as well. Mammoths, the extinct deer, and saiga originated in the Old World, making their way into North America via the Bering Land Bridge, the terrestrial connection between Siberia's Chukchi Peninsula and Alaska that was submerged around 11,000 years ago. The sloths and glyptodonts originated in South America, from which they made their way northward. And some of these animals originated in North America itself—the horses and camels are prime examples. The Old World has surviving horses and camels only because they, too, used the Bering Land Bridge to explore new worlds.

In fact, the only obvious thing that the extinct North American forms tend to share is their size: Nearly all were large. Of the 36 genera that were lost, only four were characterized by individuals that weighed less than 100 pounds (45 kg): the Aztlán rabbit, the diminutive pronghorn, the dhole, and the saiga, whose weight ranges above 100 pounds but averages less. All the others were larger—and often vastly larger.

The timing of the extinctions

There are two obvious questions to ask about these extinctions: When did they occur, and what caused them? Well-corroborated answers have not been found for either.

For years, it has been assumed that all of the extinctions happened around 11,000 years ago. (As is routine in the Pleistocene extinctions literature, all dates in this entry are in radiocarbon years; 11,000 years ago equates to about 13,000 calendar years ago.) This assumption has been made for several inappropriate reasons. First, by the late 1960s, it had become clear that six of these animals (the camel, horse, tapir,

mammoth, mastodon, and Shasta ground sloth) succumbed between 12,000 and 10,000 years ago. Second, it had become clear that this was a time of substantial climatic change. Third, it had become clear that the earliest well-corroborated evidence for people in North America, the Clovis culture, dated to around 11,000 years ago. Because almost all scientists agreed that either climatic change or newly arrived human hunters (Clovis), or some combination of the two, had caused the extinctions, these scientists were content to assume that all the extinctions had happened around 11,000 years ago. This was the case even though only a small fraction of the losses could be shown to have occurred at about this time, and it remained fully possible that population declines had begun long before this date.

Today, 16 of the 36 genera of extinct North American mammals can be shown to have lasted until between 12,000 and 10,000 years ago (see Table 1). It has also been shown that the more rare the animal, the less likely it is to have been dated to this 2,000-year interval. In fact, statistical analysis has shown that all 36 genera may have become extinct exactly when scientists have long assumed that they became extinct—about 11,000 years ago (Faith and Surovell 2009).

Nevertheless, that the extinctions may have all occurred at this time does not mean that they did. The youngest radiocarbon date for Laurillard's ground sloth falls at 39,000 years ago; the youngest dates for the Aztlán rabbit fall prior to the last glacial maximum, about 18,000 years ago. Scientists have a long way to go before they can know with any certainty when the last populations of these animals were lost.

The problem, however, is far more complex than this. Scientists are less interested in knowing when the last populations of an ancient species succumbed than they are in knowing what caused the losses in the first place. That requires knowing when populations in different parts of a species' range began to decline, not when that decline ended in extinction. The radiocarbon chronology for North America that scientists have compiled is not good enough to do this for any of the species involved.

Some things are known. It is known that at least some individuals of at least 16 genera lasted to between 12,000 and 10,000 years ago. There are some suggestions, from ancient DNA extracted from sediments in the interior of Alaska, that the mammoth and horse may have lasted in this area until as recently as about 7,000 years ago (Haile et al. 2009). This, though, stands as an intriguing exception to the general rule. The bones of these extinct mammals are not securely known from any North American archaeological or paleontological site that is less than 10,000 years old. The causes of the extinctions clearly lie prior to this time, though how much prior cannot be said for any of the animals that were lost.

Explaining the extinctions

Historically, there have been two prime, and hotly debated, explanations for the North American extinctions: that they were caused by the arrival of human hunters or that they were caused by massive climatic change. In the absence of a strong extinctions chronology, neither has much going for it. A third explanation maintains that the extinctions were caused by an extraterrestrial impact in northern North America about 11,000 years ago (Firestone et al. 2007). This hypothesis was quickly shown to be incorrect (Pinter et al. 2011) and does not need to be discussed further.

The earliest well-established archaeological phenomenon in North America is known as Clovis and dates to between about 11,200 and 10,800 years ago. Marked by a distinctive projectile point style that can be recognized wherever it is found, Clovis is best known from a series of sites in the Great Plains and the southwest. Similar projectile points are known from coast to coast. Clovis points have been found so tightly associated with the remains of mammoths and mastodons that there can be no doubt that Clovis peoples were hunting these now-extinct giants.

Those who pin the North American late Ice Age mammal extinctions on people choose Clovis hunters as the culprit. The ancestors of these people, the argument goes, crossed the Bering Land Bridge sometime before 11,200 years ago, then worked their way south of glacial ice, perhaps through the corridor that had opened up in central Canada between the western and eastern ice masses by about 11,500 years ago. Once south of glacial ice (roughly the US–Canada border), they encountered a wide array of large, edible mammals that had never before seen a human hunter. Soon, all had been driven to extinction by Clovis predators.

This explanation was first promoted in a powerful way by the American paleoecologist Paul S. Martin (1967, 1984, 2005) and is known by the name he gave it: Pleistocene overkill. As intriguing as this explanation is, it is subject to various debilitating problems. Perhaps most obviously, of the 29 genera of herbivores thought to have been driven to extinction in this fashion (carnivore extinctions are assumed to have followed the loss of their prey), only mammoths and mastodons have ever been found in a well-corroborated Clovis kill context (Grayson and Meltzer 2002).

In addition, two sites leave no doubt that Clovis people were not the first to have entered the Americas (Meltzer 2009). The first of these is Monte Verde, located near the coast in southern Chile and dated to about 12,500 years ago. The second is the Paisley Caves, located in south-central Oregon within the Great Basin. One of these caves provided well-preserved human coprolites (desiccated human fecal material) that contain ancient Native American DNA and that date to about 12,300 years ago. Because of these sites, few now think that Clovis people were the first to enter the Americas.

Perhaps, though, the pre-Clovis peoples who created these sites are not relevant to understanding the causes of North American late Ice Age extinctions. First, neither of these sites suggests adaptations that involved the hunting of now-extinct large mammals. Second, many archaeologists think that there were multiple early human entries into the Americas, one coming along the coast, the other through the interior. This suggestion has support from both archaeology and the

distribution of genetic markers across groups of modern native peoples (Perego et al. 2009). For a variety of reasons, both Monte Verde and Paisley may reflect coastal migrants, with adaptations initially pitched toward coastal resources, whereas Clovis may reflect interior migrants, with adaptations pitched toward such terrestrial resources as large mammalian herbivores.

Either way, it is certainly true that the earliest well-corroborated evidence scientists have for people hunting large, now-extinct mammals in North America is provided by Clovis sites that date to about 11,000 years ago. There is absolutely no well-corroborated evidence, however, that these people hunted anything beyond mammoths and mastodons. Most archaeologists doubt that such large mammals provided a significant portion of Clovis diet, unlike the overkill hypothesis, which routinely assumes that this was pretty much all they ate (Meltzer 2009).

Climatic change explanations also have their weaknesses. An obvious possible explanatory mechanism in this realm is provided by the climatic event known as the Younger Dryas. This starkly cold episode takes its name from a small, cold-loving herbaceous plant, the mountain avens (genus *Dryas*), which, in northwestern Europe, marks the replacement of forest by cold-adapted plants shortly after 11,000 years ago. This vegetational change was just one of many markers, in the Northern Hemisphere and elsewhere, of a cold snap of glacial intensity that lasted from about 10,600 to 10,100 years ago. For reasons that are not yet fully understood, the Younger Dryas ended much as it began: quickly.

So perhaps this switch-like, dramatically harsh climatic episode caused the demise of so many North American mammals. There are, though, many reasons to doubt this. The end of the most recent glaciation in North America was just that—the end of the most recent one. All of the now-extinct mammals had survived earlier, potentially similar, glacial terminations. While some scientists think that only the last termination saw a Younger Dryas–like interval, others think that such an event may have marked earlier ones as well. In addition, nothing similar to the North American extinctions occurred in Eurasia (see below), the very place where the Younger Dryas was first detected.

Finally, though, there is the most important issue of all. Scientists' understanding of the chronology of the North American extinctions is so weak that it is impossible to correlate it convincingly with any well-marked and well-dated event, whether that event is climatic change or the arrival of people.

Lessons from extinctions elsewhere

North American Late Pleistocene extinctions cannot be properly examined in a global biotic vacuum. South America saw Late Pleistocene extinctions that were every bit as pronounced as the North American ones. Unlike the situation for North America, where the classification of the extinct mammals is no longer heavily debated at the genus level, South American paleontologists often disagree about how to classify all the extinct Late Pleistocene mammals known from this vast area. Some, for instance, think that four genera of gomphotheres were lost (Lucas and Alvarado 2010); others insist there were only two (Prado et al. 2005). No matter what the exact count of extinct genera, almost all of the species were large. Of the roughly 52 genus-level mammalian extinctions in South America, the members of 47 would have weighed more than 100 pounds (45 kg). In both North and South America, about 80 percent of the mammalian genera whose members weighed more than 100 pounds were lost (45 kg). This is the case even though the kinds of animals involved were often distinctly different. Something causal must tie these losses together, but what that is has yet to be determined.

The South American extinctions share another attribute with the North American ones: Scientists do not have a particularly good idea as to when they occurred. Only 14 of the genera that were lost in South America have secure radiocarbon dates showing that they lasted beyond 12,000 years ago. In contrast with the North American case, however, it does seem that some of the South American forms lived beyond 10,000 years ago. Most notably, there are secure radiocarbon dates for a huge ground sloth (*Megatherium*) that fall at about 8,000 years ago and for a 1,500-pound (680-kg) glyptodont (*Doedicurus*) at about 7,500 years ago (Hubbe et al. 2007).

Ecologists have learned an important—and simple—lesson since about 1960. Every species has its own needs and responds in its own way to environmental change. This concept applies not just to modern organisms but also to extinct Pleistocene ones, no matter where the extinctions occurred. In fact, it is reasonable to argue that the continuing struggle to understand the causes of North American extinctions stems in part from so many scientists having viewed the problem at the community,

not the individual species, level (Grayson 2007). It is precisely this mistake that has been made by assuming that because some of the last populations of some of the extinct mammals occurred around 11,000 years ago, all of the populations of all of the mammals must have been lost at that time. That species respond in an individualistic way to the challenges nature throws at them means that scientists are not likely to understand the causes of North American Pleistocene extinctions until they have, at the least, far more precise information on the histories of all the mammals involved, across space and through time.

Eurasia provides a powerful example. Here, thanks to chronologies that are far stronger than any that are available for the Americas, it has become well-recognized that a variety of species, including woolly rhinoceros (*Coelodonta antiquitatis*), woolly mammoth, giant deer (*Megaloceros giganteus*), and muskox, had distinctly different histories, with different forms undergoing changes in abundance, and extinction, at different times in different places (Stuart et al. 2004; Stuart and Lister 2007; Lorenzen et al. 2011). Where compelling correlations with detailed climatic records or archaeological events are possible, those correlations are strongest for climate. Although a potential human role cannot always be ruled out, in some cases humans could not have been involved. The giant deer, for instance, became extinct in Ireland toward the very end of the Pleistocene even though there were no people there at the time.

What may or may not have happened in Eurasia, however, cannot reveal what may or may not have happened in North America. To determine this, far better population histories are needed for all of the mammals that were lost, from a far broader sample of the regions they occupied.

Resources

Books

Grayson, Donald K. "Late Pleistocene Faunal Extinctions." In *Handbook of North American Indians*, vol. 3, *Environment, Origins, and Population*, edited by Douglas H. Ubelaker, Bruce D. Smith, Dennis J. Stanford, and Emőke J.E. Szathmáry. Washington, DC: Smithsonian Institution Press, 2006.

Martin, Paul S. "Prehistoric Overkill." In *Pleistocene Extinctions: The Search for a Cause*, edited by Paul S. Martin and Herbert E. Wright Jr. New Haven, CT: Yale University Press, 1967.

Martin, Paul S. "Prehistoric Overkill: The Global Model." In *Quaternary Extinctions: A Prehistoric Revolution*, edited by Paul S. Martin and Richard G. Klein. Tucson: University of Arizona Press, 1984.

Martin, Paul S. *Twilight of the Mammoths: Ice Age Extinctions and the Rewilding of North America*. Berkeley: University of California Press, 2005.

Meltzer, David J. *First Peoples in a New World*. Berkeley: University of California Press, 2009.

Periodicals

Faith, J. Tyler, and Todd A. Surovell. "Synchronous Extinction of North America's Pleistocene Mammals." *Proceedings of the National Academy of Sciences of the United States of America* 106, no. 49 (2009): 20641–20645.

Firestone, Richard B., Allen West, James P. Kennett, et al. "Evidence for an Extraterrestrial Impact 12,900 Years Ago That Contributed to the Megafaunal Extinctions and the Younger Dryas Cooling." *Proceedings of the National Academy of Sciences of the United States of America* 104, no. 41 (2007): 16016–16021.

Grayson, Donald K. "Deciphering North American Pleistocene Extinctions." *Journal of Anthropological Research* 63, no. 2 (2007): 185–213.

Grayson, Donald K., and David J. Meltzer. "Clovis Hunting and Large Mammal Extinction: A Critical Review of the Evidence." *Journal of World Prehistory* 16, no. 4 (2002): 313–359.

Haile, James, Duane G. Froese, Ross D.E. MacPhee, et al. "Ancient DNA Reveals Late Survival of Mammoth and Horse in Interior Alaska." *Proceedings of the National Academy of Sciences of the United States of America* 106, no. 52 (2009): 22352–22357.

Hubbe, Alex, Mark Hubbe, and Walter Neves. "Early Holocene Survival of Megafauna in South America." *Journal of Biogeography* 34, no. 9 (2007): 1642–1646.

Lorenzen, Eline D., David Nogués-Bravo, Ludovic Orlando, et al. "Species-Specific Responses of Late Quaternary Megafauna to Climate and Humans." *Nature* 479, no. 7373 (2011): 359–364.

Lucas, Spencer G., and Guillermo E. Alvarado. "Fossil Proboscidea from the Upper Cenozoic of Central America: Taxonomy, Evolutionary, and Paleobiogeographic Significance." *Revista Geológica de América Central*, no. 42 (2010): 9–42.

Perego, Ugo A., Alessandro Achilli, Norman Angerhofer, et al. "Distinctive Paleo-Indian Migration Routes from Beringia Marked by Two Rare mtDNA Haplogroups." *Current Biology* 19, no. 1 (2009): 1–8.

Pinter, Nicholas, Andrew C. Scott, Tyrone L. Daulton, et al. "The Younger Dryas Impact Hypothesis: A Requiem." *Earth-Science Reviews* 106, nos. 3–4 (2011): 247–264.

Prado, José Luis, Maria Teresa Alberdi, Beatrice Azanza, et al. "The Pleistocene Gomphotheres (Proboscidea) from South America." *Quaternary International* 126–128 (2005): 21–30.

Stuart, Anthony J., and Adrian M. Lister. "Patterns of Late Quaternary Megafaunal Extinctions in Europe and Northern Asia." *Courier Forschungsinstitut Senckenberg* 259 (2007): 287–297.

Stuart, Anthony J., Pavel A. Kosintsev, Tom F.G. Higham, and Adrian M. Lister. 2004. "Pleistocene to Holocene Extinction Dynamics in Giant Deer and Woolly Mammoth." *Nature* 431, no. 7009 (2004): 684–689.

Donald K. Grayson

Extinction of the Eurasian megafauna

During his circumnavigation as naturalist onboard HMS *Beagle* from 1831 to 1836, Charles Darwin undertook numerous excursions to many parts of the South American mainland. Darwin was baffled by the enormous diversity of plant and animal life, yet the entire continent seemed devoid of large mammals. Despite the presence of virtually every sort of habitat in every type of climate from tropical to subarctic, the largest animal on the entire 6,880,000-square-mile (17,830,000-km^2) continent was the lowland tapir (*Tapirus terrestris*), which reaches a weight of only around 440 pounds (200 kg). Why was this? Where were all the large mammals? North America, too, seemed severely impoverished compared to the big game of southern Asia and most of sub-Saharan Africa. It is now known that large animals had in fact been in North and South America only recently. During the ice ages, North and South America, as well as Europe and northern Asia, had been home to an amazing bestiary of large mammals, collectively known as the megafauna, which included elephants, rhinoceroses, horses, camels, deer, ground sloths, lions, saber-toothed cats, bears, wolves, and hyenas. But somehow they have all disappeared with only remnants of what was once one of the richest mammalian faunas ever. The disappearance of the ice-age megafauna has been one of the most debated and discussed scientific issues of the twentieth and early twenty-first centuries.

Climate shifts at the end of the Pleistocene

Toward the end of the last ice age, northern Asia, Europe, and North and South America lost almost every species of large mammal weighing above 1 metric ton (1.1 tons), as well as most of the fauna of medium-sized mammals, and significantly, they were not replaced by their relatives from tropical Asia or Africa. This change represents one of the most dramatic and rapid declines in large animal diversity ever recorded, and it happened in the course of only a few thousand years. The question of what drove the megafauna to extinction has been the subject of thousands of research papers and books, as well as decades of debate.

The term *Eurasia* normally refers to all of Europe and Asia combined, but in the context of ice-age extinctions, it refers only to Europe and the central and northern parts of Asia. Southern Asia was and is characterized by subtropical and tropical climates, and this region appears to have undergone far fewer changes in climate and habitats, and as a result it lost far fewer species than the rest of Eurasia. Today, elephants, rhinoceroses, a variety of large buffalo (e.g., gaur [*Bos gaurus*], kouprey [*B. sauveli*], anoa [*B. quarlesi* and *B. depressicornis*], yak [*B. mutus*], big cats (e.g., lion [*Panthera leo*], tiger [*P. tigis*], leopard [*P. pardus*], snow leopard [*P. uncia*], clouded leopard [*Neofelis nebulosa*]), bears (sloth bear [*Melursus ursinus*], Asiatic black bear [*U. thibetanus*], Sun bear [*Helarctos malayanus*], brown bear [*U. arctos*], giant panda [*Ailuropoda melanoleuca*]), and many other large mammals still live in southern Asia.

These Eurasian extinctions happened toward the end of the geological period known as the Pleistocene, which began 2.558 million years ago and ended around 11,500 years ago. Unlike the previous geological epochs of the Cenozoic, the Pleistocene was an age of rapidly changing climates, during which ice ages, called glacials, which usually lasted 100,000 to 150,000 years, were interrupted by warm periods, known as interglacials, which lasted around 10,000 to 15,000 years. Ice-core drillings in the Arctic and Antarctic during the last decades of the twentieth century and early decade of the twenty-first century have demonstrated, however, that the climate was much more varied than previously realized (see Figure 1). Warm periods were not uniformly warm, nor were cold periods uniformly cold. Temperatures fluctuated constantly, and even in cold glacial periods there were dramatic shifts between true ice-age periods, called *stadials*, when huge glaciers began building up in northern and southern parts of the globe, and warmer periods, called *interstadials*, when temperatures rose and humidity levels increased. During the glacials, huge open plains dominated as a variety of grasses spread across much of northern and central Europe and northern Asia, whereas deciduous forests retreated to the south, and the vast open plains of eastern Asia turned to tundra. During warmer periods, the open, grassy plains vanished in many areas, and forests spread. This happened many times during the Pleistocene, and although a number of large mammals disappeared during this time, wholesale extinctions happened only in the last few thousand years of the Pleistocene.

For the past 11,500 years Earth has been in a warm interglacial period known as the Holocene, although there have also been numerous cold spells during this time. All the

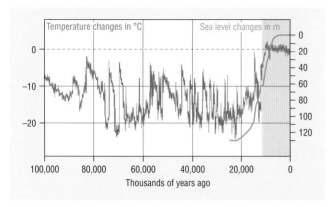

Figure 1. The Eemian–Weichselian transition was marked by a drop in sea levels as temperatures dropped and glaciers began building up. Glaciation reached its peak, the glacial maximum, around 22,000 to 25,000 years ago, where sea levels were nearly 400 feet (130 m) below present levels. From around 15,000 years ago a series of rapid climate shifts marked the end of the Tarantian. The Tarantian is colored blue, and the Holocene is colored green. Reproduced by permission of Gale, a part of Cengage Learning.

megafauna had died out either by the Pleistocene–Holocene boundary or in the Early Holocene. The previous interglacial was the Eemian, which lasted from around 130,000 to 114,000 years ago. The beginning of the Eemian interglacial also marks the beginning of the last geological stage of the Pleistocene, known as the Tarantian, and it is during this geological stage that the megafaunal extinctions took place. After the Eemian, the climate shifted back to an ice age known as the *Weichselian*, which lasted until the Tarantian–Holocene boundary. The shift from Eemian to Weichselian was marked by a sudden drop in sea level, as temperatures plummeted and huge inland glaciers began to form, and forests were replaced by steppe or tundra in many places. During the Middle Weichselian (75,000–25,000 years ago), enormous glaciations built up in the north, but there were also warmer, interstadial periods. Between 25,000 and 13,000 years ago the glaciers were at their largest, after which they slowly retreated until melting began in earnest during the first part of the current interglacial, the Holocene.

The latest Pleistocene saw great climatic changes affecting much of the world, and between 15,000 and 10,000 years ago the transition from a glacial to interglacial interval occurred, with a rise in global average temperature of no less than 10.8°F (6°C). Starting around 15,000 years ago, a number of dramatic climate shifts occurred in rapid succession, and it was during this final part of the Weichselian that the Eurasian megafauna began its rapid decline. This interval began with a cold period known as the Oldest Dryas, which lasted from around 19,000 to 14,700 years ago. At this time much of Eurasia was dominated by open, grassy plains and conifer forests around rivers, whereas deciduous and evergreen hardwood forests were confined to the continents' southern parts. This period was followed by a warmer interstadial period, the Bølling, which lasted from around 14,700 to 14,000 years ago. Then followed another brief stadial period, the Older Dryas (14,000–13,700 years ago), then another interstadial, the Allerød (13,700–13,000 years ago), and finally

temperatures dropped dramatically at the onset of the Younger Dryas stadial (13,000–11,600 years ago), which marks the end of the Tarantian and the end of the Pleistocene.

At the transition from the Younger Dryas to the Holocene, temperatures climbed dramatically. In some areas of the world (e.g., Greenland) temperatures may have risen as much as 12.6°F (7°C) within just a few decades. To any biologist, such dramatic and rapid climate shifts are a red flag, likely spelling doom to animals that are specialized to a given kind of habitat because in such circumstances climates and habitats change over and over again during a short period of only a few thousand years. Most of the ice-age megafauna were specialized animals; some were specialized to cold, open habitats, others to more temperate, wooded habitats. But specialists they were, and this fact holds important clues to their disappearance.

What caused the megafaunal extinction?

There are two competing hypotheses for the disappearance of the Eurasian megafauna at the end of the Pleistocene: climate change and human hunting. Some have suggested more exotic causes, such as disease and even the impact of a meteor or comet, but these hypotheses lack evidence and are not taken seriously by most scientists. There is no doubt that climate changes greatly altered the natural ecosystems of the latest Pleistocene. Climate change and thus ecosystem changes have been the most durable and popular explanations for the megafaunal extinctions, and these are well understood mechanisms of extinction throughout the geological record. Some 40,000 years ago, however, modern humans entered Eurasia. Although Neanderthals had lived in Europe for several hundred thousands of years, modern humans could have been more mobile, more organized, and more efficient hunters. There is some evidence that megafaunal extinctions closely followed the spread of modern humans in other parts of the world, and it is certain that humans did hunt the megafauna, as evidenced by numerous finds of bones with cut marks and other signs of having been butchered and processed by human hunters. Humans also drew on cave walls, and many cave paintings depict hunting scenes of, for instance, mammoths and horses. Nevertheless, the number of human settlements in Eurasia appears to have been far too small to have caused the disappearance of the megafauna. Also, it is by no means certain that the extirpation of the megafauna in North and South America happened for exactly the same reasons as in Eurasia.

A particularly interesting study was published in 2008 by Diana Pushkina and Pasquale Raia. They examined the diversity and abundance of large mammal species across Eurasia at sites known to have been inhabited by humans and sites where humans did not live. They also looked at human settlements to try to establish which animals they actually hunted. Surprisingly, there was little correlation between the presence and abundance of many species and human occupation. Several of the extinct species, such as the woolly mammoth (*Mammuthus primigenius*), woolly rhinoceros (*Coelodonta antiquitatis*), and Irish elk (*Megaloceros giganteus*), appear to have been hunted only infrequently. In contrast, humans often hunted reindeer (*Rangifer tarandus*), roe deer (Capreolus capreolus), red deer (*Cervus elaphus*), saiga antelope (*Saiga tatarica*), horses, and wild boar; animals that are still present. The carnivore fauna also did not appear to have been affected to any large extent by human presence. The only exception to the above was the steppe bison, which was hunted frequently and went extinct. This pattern suggests that the extinctions were driven by climate changes causing disruptions of the natural ecosystems and fragmentation of habitats, with large mammals not being able to sustain large enough populations to survive for longer periods of time.

That the really large mammals all died out holds important clues. Large animals require a lot of food, and they are, by default, specialists as they usually evolved to live in specific environments and consume specific foods. A given area of land may support a few dozen roe deer, but will support only a single mammoth. Accordingly, large animals need large areas in order to have large, genetically viable populations. Also, large mammals reproduce much more slowly than small mammals. If populations undergo rapid decline their slow reproduction may not be able to replenish what is lost before the entire population dies out (see Figure 2). The average number of young produced by an individual is a function of how many young are produced in a single litter and how many litters are produced per year. In big cats a female usually has only a single litter every two to three years, but often has three to four cubs per litter. Rhinoceroses, hippopotamuses, and elephants usually have only a single calf, and long gestation times and extended parental care means that it takes several years between births. Small mammals such as rodents usually have lots of young per litter and several litters per year. As such, the average yearly reproductive output dramatically decreases with body size.

Given these factors, the current state of knowledge seems to indicate that the disappearance of most of the Eurasian megafauna—in contrast with some medium-sized mammals

such as wolves, brown bears, and many deer, which survive to the present day—is far more complicated than a simple division into either climate change or human hunting. It is now known that the disappearance of many species was a drawn-out process that spanned centuries, even millennia, and that species underwent several declines and subsequent expansions of their range before they finally went extinct. It seems almost certain that climate change was the primary agent, but the biology of the species in question holds vital clues to understanding why some survived whereas others did not.

Proboscideans

Eurasia was home to several species of elephants during the Pleistocene, the most famous of which is, of course, the woolly mammoth. The straight-tusked elephant (*Palaeoloxodon antiquus*) and the woolly mammoth were both true elephants, members of the family Elephantidae, as indicated by their having tusks only in the cranium and not the mandible as well, and by their high-crowned molars with transverse grinding lamellae.

The straight-tusked elephant was a close relative of the modern Asian elephant (*Elephas maximus*), and some scientists believe it should be included in the genus *Elephas*. This was an enormous creature; a large bull stood 13 feet (4 m) in shoulder height and weighed nearly 11 tons. The straight-tusked elephant was adapted for warmer conditions and preferred open-woodland habitats where it fed on a mixture of browse, shrubs, and grasses. It was widely distributed across much of western and northern Europe in the Eemian interglacial, but when temperatures gradually began dropping during the beginning of the Weichselian glacial, it retreated to southern Europe. In northern Europe it had lived in temperate forests, but in much of southern Europe it lived in Mediterranean evergreen forest. *Palaeoloxodon* continued to live in southern Europe until around 35,000 year ago, when the spread of grasslands contracted the forest habitats, causing the straight-tusked elephant to disappear from the European mainland.

During its long reign in southern Europe, however, the straight-tusked elephant repeatedly spread out onto many of the Mediterranean islands, such as Cyprus, Crete, Malta, Sardinia, Sicily, and Tilos, which were connected to the mainland by land bridges owing to a global drop in sea level during the glacials. On these islands, after they were cut off from the mainland by rising temperatures and thus rising sea levels during the interstadials, the gigantic straight-tusked elephant quickly reduced in body size owing to the smaller habitats that provided less food resources. These dwarf elephants stood less than 5 feet (1.5 m) in shoulder height and weighed only around 440 pounds (200 kg), or about 2 percent of the weight of the large bulls from the mainland. These island populations went extinct much later than those of the mainland. The straight-tusked elephant persisted on Crete until around 18,000 years ago, whereas they lived right up to the Pleistocene–Holocene transition on Cyprus, around 11,000 years ago. On Tilos, they survived much longer and went extinct only well into the Holocene around 4,000 years ago. Their continued presence on isolated islands is a conundrum, because if climate changes caused their extinction

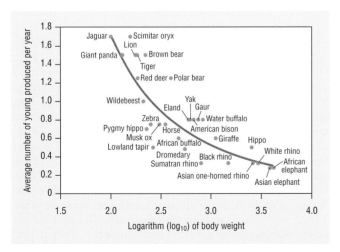

Figure 2. A strong relationship exists between body weight and the average number of young produced per year. Shown here are larger mammals (above 220 pounds [100 kg]), where the correlation is 86 percent. Body weight spans a factor of 40 from 220 pounds (jaguar) to over 4.4 tons (African elephant) and has been transformed into natural logarithms (Log₁₀) to produce a clearer figure. Reproduced by permission of Gale, a part of Cengage Learning.

Numerous woolly mammoth (*Mammuthus primigenius*) carcasses have been found frozen in the permafrost of Siberia. One of the most famous was the six-month old male nicknamed Dima, found near the Kolyma River in northeastern Siberia in 1977. In life he would have been around 3 feet (1 m) tall at the shoulders and weighed around 220 pounds (100 kg). Radiocarbon dating showed that he died about 40,000 years ago. Today he is on display at the Zoological Museum in the Russian city of St. Petersburg. © Vova Pomortzeff/Alamy.

on the mainland, the smaller, isolated habitats on islands should have made them equally or even more vulnerable. Perhaps the vegetation did not change markedly on the islands. At any rate, the extinction pattern of the straight-tusked elephant is complicated and evidently had nothing to do with human hunting.

The woolly mammoth is arguably the most famous ice-age mammal, and researchers know more about this species than any other extinct animal. Bones from thousands of individuals have been recovered from across much of Europe (except the southernmost part) and across most of northern Eurasia. The woolly mammoth also spread across the Bering Land Bridge into North America. It is not known only from skeletons. Several dozen individuals have been found frozen in the permafrost with much of the soft tissues of these bodies more-or-less intact.

The woolly mammoth evolved around 300,000 years ago in Europe and was the most cold-adapted of all the mammoths. It was smaller than many of the previous mammoth

species, which lived in warmer conditions, and was about the size of a modern Asian elephant. Bulls stood around 10 to 11 feet (3 to 3.4 m) at the shoulder and weighed around 5.5 to 6.5 tons, whereas cows were smaller, 8.5 to 9 feet (2.7 to 2.9 m) and around 3.5 to 4 tons. The woolly mammoth had a characteristically domed head and a hump above the shoulders. It was covered in multiple layers of tough, coarse hair. Underneath was a more woolly undercoat of shorter hairs covered by an outer coat of up to 3.3-foot-long (1-m), shaggy hair. Some mammoths had dark or almost black hair, whereas others' hair was more ginger, but the bright red-orange color often seen in drawings and found on some of the frozen mummies is an artifact of preservation. The skin of the woolly mammoth was similar to that of modern elephants, but contained more sebaceous glands which secreted fatty substances into the fur to keep it in good condition. The animal also had up to 3 inches (8 cm) of fat under the skin and even had a flap of skin protecting its rectum from the cold. Its ears were much smaller than those of modern elephants, so as not to lose heat. The trunk was very similar to that of an Asian elephant, but the tip was different. Modern elephants have either an upper and a lower (African elephant) or just an upper (Asian elephant) fingerlike projection at the tip of the trunk, which is used to pick up and manipulate objects. Woolly mammoths had a wide upper finger and a large flap for a lower finger, an adaptation for plucking grass, their main stable diet. The mammoth group is distinguished from other elephants by their spiral tusks, curving along their length and also around their axis. Woolly mammoths had huge tusks of up to 13 to 16.5 feet (4 to 5 m), which were often worn underneath, suggesting that the animals used them to clear the ground of snow in winter to expose the vegetation.

Mammoths were strongly adapted for open, grassy plains in steppe-tundra habitats, where they roamed in large herds. Humans occasionally hunted mammoths, but had little, if any, part to play in their extinction in Eurasia. During the cold stages of the last glacial, much of Eurasia was covered in grassy plains, and conifer forests grew along riverine valleys. This was mammoth country; the woolly mammoth appears to have been extremely abundant in this region, with its remains being recovered across an enormous geographic range (see Figure 3). However, from an estimated 2,970,000 square miles (7,700,000 km^2) around 40,000 years ago, suitable mammoth habitat underwent a rapid and dramatic decline around 14,000 years ago during the relatively warm Allerød interstadial. During this time forests and open-

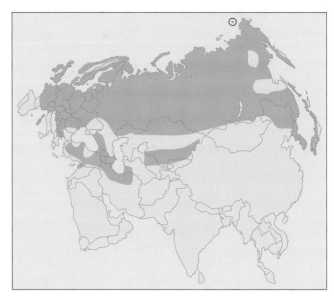

Figure 3. The geographical distribution of the woolly mammoth (*Mammuthus primigenius*) around 30,000 years ago. The mammoth was also widely distributed in North America. Wrangel Island (which has an area of 2,900 square miles [7,600 km^2]) off the coast of northeastern Siberia, where mammoths survived for millennia after their extinction on the Eurasian continent, is marked with a red circle. Reproduced by permission of Gale, a part of Cengage Learning.

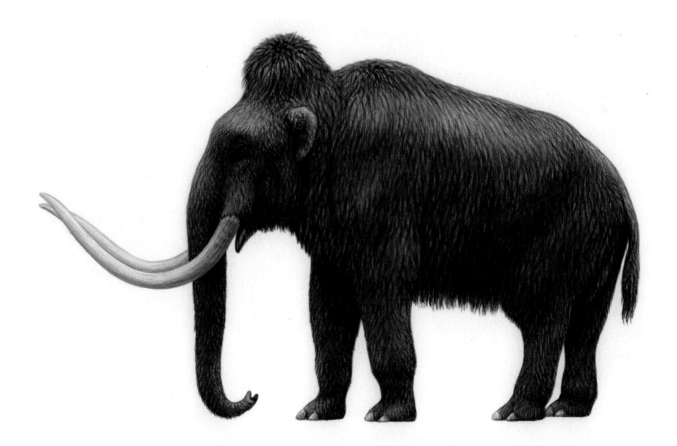

The woolly mammoth (*Mammuthus primigenius*) was about the size of a modern Asian elephant (*Elephas maximus*), but had a more sloping back and a domed head. Illustration by Francois Desbordes.

woodland habitats replaced grasslands across most of the woolly mammoth's former range. Mammoths were found only in the extreme north on the Taymyr Peninsula in northern Russia and along the coastal areas of the Arctic Sea. The Allerød interstadial was followed, however, by the brief but very cold and dry Younger Dryas, when much of the forests retracted and grasslands spread again. The woolly mammoth also expanded its range into northwestern Europe. This last cold period was replaced by the current interglacial, the Holocene, around 11,500 years ago, at which point the woolly mammoth disappeared.

Like the straight-tusked elephant, however, dwarf mammoths with shoulder heights of only around 6.5 feet (2 m) continued to live on some of the islands off the Russian coast in the Arctic Sea, such as Wrangel Island, where they went extinct as recently as around 4,000 years ago, possibly in connection with human arrival on the island. It seems likely that the disappearance of the mammoths from the Eurasian mainland was primarily a result of climate changes that eliminated much of its habitat. But it is doubtful this is the full story. Although the tundra areas in the north were more boggy and had less plant productivity, thus making them unsuitable for woolly mammoths, it is doubtful that all of the suitable habitats disappeared, and theoretically mammoths could still have persisted in isolated refugia, perhaps almost up to the present day. Maybe the slow reproduction of mammoths played an important part. The dramatic shifts in climate during the last part of the Tarantian caused mammoth populations to crash, and it is possible they reproduced too slowly to be able to recover. Also, with dramatic declines in populations, woolly

Figure 4. The geographical distribution of the woolly rhinoceros (*Coelodonta antiquitatis*) around 30,000 years ago. Reproduced by permission of Gale, a part of Cengage Learning.

mammoths would have been vulnerable to hunting by humans. The details of what caused the last mammoths to disappear entirely continue to be debated, and perhaps there simply was a certain element of chance involved. Other ice-age mammals, such as reindeer, also underwent massive population declines, yet they survived.

The woolly rhinoceros (*Coelodonta antiquitatis*) was a huge, powerfully built animal, resembling the extant white rhinoceros (*Ceratotherium simum*) in size and stature, but it had a thick woolly coat to protect it from the cold. Illustration by Francois Desbordes.

Eurasian rhinoceroses

Today, the rhinoceros is associated with African savannas and Asian rain forests. However, Eurasia had several species of rhinoceros, including *Stephanorhinus*, which died out in the Late Pleistocene around half a million years ago, and the gigantic *Elasmotherium*, which at 4 to 5 tons was even larger than the modern white rhinoceros (*Ceratotherium simum*) and died out as recently as 50,000 years ago. Next to the woolly mammoth, the most impressive of the Eurasian Weichselian megafauna was the woolly rhinoceros. Although less common than mammoths, woolly rhinoceroses have also been found not only as skeletons but as frozen mummies, and they are also common as cave paintings. The woolly rhinoceros was similar in size to the modern white rhinoceros, standing almost 6.5 feet (2 m) at the shoulder and weighing around 2.2 to 3.5 tons. It had two long, curved horns made of keratin—a short one above the eyes and a longer one at the tip of the snout of up to 3.3 feet (1 m) in length. This rhinoceros was covered in thick, coarse fur and had smaller ears than modern rhinoceroses as adaptations to cold environments. It probably used its horns for fighting, like the modern rhinoceroses, but also to clear the snow from low-growing vegetation in winter.

Like the woolly mammoth, the woolly rhinoceros lived on the cold, open treeless steppe and tundra areas dominated by a variety of grasses and low shrubs, which formed its main diet. The fluctuating stadial and interstadial cycles at the end of the Tarantian caused these habitats to expand and contract, and as for the woolly mammoth, the population size and geographic range of the woolly rhinoceros expanded and contracted with them (see Figure 4). At the end of the Tarantian, the open treeless steppe and tundra habitats vanished from most of mainland Eurasia, and so did the woolly rhinoceros. Being a large mammal, the woolly rhinoceros must have been a slow breeder, which probably played an important part in its extinction. It has been suggested that isolated populations of the woolly rhinoceros may have survived into the early Holocene in western Siberia, but this theory is still controversial.

Deer and relatives

Many deer species found in Eurasia during the last ice age—uch as roe deer, red deer, sika (*Cervus nippon*), fallow deer (*Dama dama*), moose (*Alces alces*), reindeer, and saiga antelope—did not go extinct. Most of these species are associated either with temperate forests or open-woodland habitats, and this fact holds vital clues as to why they survived the extinctions. Forest habitats were present all throughout the Tarantian in at least some part of the species' ranges, and with the exception of the moose, deer are medium-sized mammals, which have considerably higher breeding rates than, for instance, mammoths or rhinoceroses. The saiga antelope is adapted for cold, open, grassy plains and steppes, and in the Late Pleistocene it was widely distributed across much of Eurasia, but today it is confined to the open-steppe areas in central Asia. Reindeer are also adapted to cold, open steppe or tundra, and they underwent a rapid and dramatic population decline of around 85 percent at the end of the Tarantian, but they did not go

extinct, probably because of high reproductive rates. By this time humans had also domesticated the reindeer, which gave the species added protection from extinction. In contrast, the musk ox (*Ovibos moschatus*) is adapted to cold, open plains, or tundra, and has a lower reproductive rate. It went extinct in Eurasia, but had already migrated across the Bering Strait into North America, and, thus, it is still present in the Canadian Arctic and Greenland.

The most famous Eurasian cervid (scientifically deer are called *cervids*) was the great Irish elk or giant deer, famous for the enormous antlers which, in adult stags, could span more than 11.5 feet (3.5 m) in width. This was one of the largest cervid species, standing over 6.5 feet (2 m) tall at the shoulders and weighing around 1,100 pounds (500 kg). The Irish elk had been widespread across ice-age Eurasia for around 400,000 years before going extinct. It was an inhabitant of the open mixed-deciduous woodlands and was not a forest dweller as were most Eurasian deer. It was also not found on the treeless grassy steppes. Until around 20,000 years ago, these elk were common and widespread across much of Europe and central Asia, but then they vanished from large parts of Europe during the cold Oldest Dryas period. For millennia it appears to have been absent from most of its former range, probably because the cold stadial climate caused the open-woodland habitats to be replaced by tundra or steppe in most areas. During the Bølling and Allerød interstadials it again expanded its range as open woodland became more common. By then it was restricted to northwestern Europe, but during the Younger Dryas the populations apparently collapsed, and the Irish elk disappeared. This extinction coincided with the cold phase immediately prior to the Holocene, a period during which open woodland was replaced with treeless steppes.

But the Irish elk did not die out at the Pleistocene–Holocene boundary, as was once believed. In the Ural region and western Siberia it continued to live on until around 8,000 years ago or perhaps a little later. It would appear that this region underwent less dramatic climate changes during the Younger Dryas and that open-woodland and shrub habitats persisted. The Holocene saw a dramatic increase in temperature, and forests spread across Europe, preventing a recolonization by the elk. The composition of the vegetation appears also to have been important. Deer shed their antlers every year and in the Irish elk, these antlers could weigh about 60 to 90 pounds (30 to 40 kg). This antler size implies that stags must have fed on vegetation with a high content of calcium and phosphate in order to obtain sufficient minerals to grow such enormous structures. Thus, the last surviving populations of western Asia might have gone extinct not because the open-woodland habitats vanished completely, but because small changes in the vegetation meant that the food available for the elks was simply too poor in minerals.

Around 900,000 years ago, the steppe bison (*Bison priscus*) evolved in southern Asia and then spread out across Eurasia. It was a huge, sturdy animal, standing around 6 feet (1.8 m) in shoulder height and weighing around a ton. Around 300,000 years ago it crossed the Bering Land Bridge into North America. The steppe bison is the likely ancestor of the

Eurasian bison or wisent (*Bison bonasus*) and the American bison (*Bison bison*), both of which are still living. Like its American descendant, the steppe bison was adapted to a grass-eating diet and roamed the open treeless plains in huge numbers. In contrast, the Eurasian wisent is a forest dweller, browsing on grasses. The steppe bison also lived in tundra areas, so it was probably adapted for foraging on a wider variety of plants and in more sparse environments than the American bison.

Analyses of the steppe bison's anatomy have indicated that it probably did not migrate long distances, unlike the American bison. The range of the steppe bison, because of the species' adaptation to open plains, expanded and contracted with the stadial and interstadial periods. During the last part of the Weichselian, it appears to have been hugely abundant in North America, and competition with the steppe bison is thought to be part of the reason why many ice-age mammals went extinct in North America. In Eurasia the steppe bison was not as dominant, but it was still a common species, and population sizes expanded, starting around 30,000 to 20,000 years ago with the cold climate and the spread of the open plains. Unlike other Eurasian megafaunal species, however, the steppe bison is the only species that went extinct and was also frequently hunted by humans. The disappearance across most of Eurasia of the open grass and tundra steppes during the Pleistocene–Holocene transition undoubtedly led to a population crash, but a growing human population and intensified hunting may have been a factor in the steppe bison's ultimate extinction.

The large carnivores

Some of the large carnivores of the Weichselian, such as the wolf (*Canis lupus*), lynx (*Lynx lynx*), wolverine (*Gulo gulo*), and brown bear (*Ursus arctos*), did not go extinct, and neither did the foxes, weasels, and other smaller carnivores. Notably, these smaller species either can live in a wide variety of habitats, and thus are generalists, or are associated with forest-type environments. Late Pleistocene Eurasia was home to a number of spectacular carnivores normally associated with East Africa, such as hyenas and lions, as well as several species of large bears. The most remarkable was the huge cave bear (*Ursus spelaeus*), which vanished around 28,000 years ago during the last glacial maximum. Several ice-age carnivores, such as the cave lion and the cave hyena, have had the adjective *cave* added to their vernacular name, but this addition does not mean that these animals lived in caves to any larger extent than their modern counterparts do. The term has been added because the skeletons of these species are often found in caves, as the conditions in caves are often more favorable for preservation of skeletal remains.

The cave bear was probably an exception. Skeletal remains of cave bears have been found almost all over Europe and into western Asia, for instance, the Caucasus (see Figure 5). It also lived in the southern parts of Great Britain, which was connected to mainland Europe during the Weichselian owing to a huge drop in sea levels of up to 370 feet (120 m), as gigantic amounts of water were bound in the inland glaciers. Scandinavia and northern Great Britain, however, were

Figure 5. The geographical distribution of the cave bear (*Ursus spelaeus*) around 35,000 years ago. Reproduced by permission of Gale, a part of Cengage Learning.

covered by glaciers, and no cave bears are known from these areas. The cave bear appears to have been particularly numerous across central Europe, and hundreds of finds are known from numerous caves. In many cases it appears that these bears had dwelled in the caves for many generations, because the caves contain multiple sediment layers with cave bear bones, collectively representing hundreds or even thousands of years.

The cave bear typically inhabited forested areas and appears not to have lived in open grasslands, much like its close modern relative, the brown bear. It was one of the largest bears ever, and males would have rivaled the largest extant brown bears and weighed 1,000 to 1,100 pounds (450 to 500 kg), whereas females were smaller at 440 to 550 pounds (200 to 250 kg). One notable characteristic of the cave bear was its huge skull, featuring very large, flat molar teeth that were well adapted for grinding and chewing; these teeth often became heavily worn in older adults, a pattern that suggests cave bears had a largely herbivorous diet consisting primarily of tough, fibrous plants. Geochemical analyses of their bones has revealed that cave bears did indeed mainly ingest plants rather than meat although, as with all large ursine bears, they probably also scavenged and ate small mammals and insects when they had the chance.

The reason for the extinction of the cave bear continues to be disputed, but its higher dietary specialization may have made it more vulnerable to environmental changes caused by fluctuating temperatures than the more omnivorous brown bear. Cave bears may also have been more dependent on caves for hibernation during the winter. Brown bears are known to be less discriminate and will use caves; holes under logs or stumps, which they will further excavate; or simply thick brush as winter dens. Humans were once invoked as the primary agent for the disappearance of the cave bear, but human populations in Europe some 30,000 years ago were too small to have caused the species' complete extinction by hunting. Humans also used caves for shelter, however, and perhaps the influx of modern humans into Eurasia caused caves once occupied by cave bears for millennia to be taken over by humans. This change may have led to a crash in cave bear populations, with environmental conditions finally causing them to disappear.

The cave hyena is not strictly an extinct species, because it still lives in East Africa under the name spotted hyena (*Crocuta crocuta*). Therefore, scientists have much better chances of researching the ecology and possible disappearance of the cave hyena than many other ice-age mammals, because populations of the same species still exist. The Eurasian cave hyena was a distinct subspecies (*Crocuta crocuta spelaea*) and has been found across Eurasia south of the glaciers, from northern China to Spain, and it appears to have been particularly common in parts of northeastern Asia. The cave hyena was somewhat larger than its modern relative, weighing perhaps as much as 200 pounds (90 kg), whereas modern spotted hyenas rarely exceed 150 pounds (70 kg). Spotted hyenas are powerful hunters, living in large packs led by one or several dominant females. They have massive jaws and enormously strong teeth, so they can shear flesh with their molars and their premolars can break even large bones. Spotted hyenas are anything but scavengers as was once believed. They do obtain a substantial portion of nutrients from scavenging, but so do lions. In fact, lions usually steal more kills from spotted hyenas than vice versa. Although it is still being debated, cave hyenas probably also hunted in packs, but it is less certain whether they lived in large clans like their extant relatives. What seems certain, however, is that they were dependent on the open, grassy plains that covered much of Eurasia during stadials.

Spotted hyenas often use holes in the ground as dens and shelter for their pups, and cave hyenas often used caves as dens, bringing food back to feed their pups. Analyses of bones from such caves indicate that Przewalski's horses (*Equus ferus przewalskii*) were their preferred prey; modern spotted hyenas regularly feed on zebras. The cave hyena also hunted reindeer and, perhaps more surprisingly, the huge woolly rhinoceros. They appear to have hunted neither forest-dwelling deer such as roe deer or red deer nor the large Irish elk and steppe bison. Cave hyenas probably also stole kills from wolves, cave lions, and Neanderthal man. Toward the end of the Weichselian the vast grassy plains began to dwindle and were replaced by tundra in the north and forest or open mixed woodland toward the south. Cave hyenas were poorly adapted for these kinds of habitats, and much of their prey base also disappeared. Intensified competition with wolves, which thrive in open-woodland habitats, may also have been a factor in their decline. Cave hyena populations crashed, and the species began disappearing across much of its former vast geographic range, finally going extinct in Eurasia around 12,000 to 11,000 years ago.

The cave lion (*Panthera spelaea*) was a large, powerful, lionlike cat slightly larger and more robust in build than

Figure 6. A skull of a cave lion (*Panthera spelaea*) and redrawn images of cave lions from cave paintings from the famous Chauvet-Pont-d'Arc Cave, located in southern France. The skull is darkened from being buried for millennia in cave sediments. It was probably from an old male, because it is very large and the teeth are rather worn. Discovered in 1994, Chauvet Cave is the largest treasure trove ever discovered of cave paintings of a large variety of Eurasian ice-age mammals. The paintings were made around 30,000 to 32,000 years ago. Reproduced by permission of Gale, a part of Cengage Learning.

modern lions. It has traditionally been considered to have been a subspecies of the modern lion (*Panthera leo*), but more likely it was a separate species adapted for life in open or open-woodland habitats in temperate climate zones. Large lionlike cats have been present in Europe for at least 750,000 years, but it is uncertain if the earlier forms were cave lions or belonged to a different species. The cave lion was common in much of Eurasia, although its geographic range appears to have been smaller than that of the woolly mammoth, for instance, and most finds have been made in central Europe. The cave lion was an iconic species for ice-age humans, and cave paintings of lions have been found at several sites (see Figure 6). The cave lion probably hunted in groups, as indicated in several cave paintings, but the structure of these groups is unknown. It is not known if a group consisted mainly of females and their young led by one or a few males, as is common in modern lions.

Cave lions probably did not live in caves, but may have given birth in them because caves were secure places compared to open areas. These lions, however, often ventured into caves such as those occupied by cave bears. This has led to inferences that cave lions sometimes preyed on cave bears. Their main prey would have been horses, bison, deer, and young mammoths, but reindeer appear to have been their most important prey item, especially shortly before the lions disappeared. Although cave lions probably also took smaller prey, it is clear that they needed a stable prey base of large herbivores like their modern relatives. As with modern cats this would have been particularly important for females with cubs, since a female with cubs needs a lot of food to feed herself and to produce enough milk to feed a litter of three to four cubs. As such, the disappearance of the cave lion seems less mysterious than that of many of the large herbivores. With the dramatic decline and ultimate disappearance of most of the larger herbivores toward the end of the last glacial, cave lions had no hope of surviving in shifting climates and habitats where smaller, mainly forest-adapted deer were the only common herbivores left. Around 12,000 to 11,000 years ago, the cave lion disappeared from Eurasia.

Resources

Books

Clottes, Jean. *Chauvet Cave: The Art of Earliest Times.* Translated by Paul G. Bahn. Salt Lake City: University of Utah Press, 2003.

Curtis, Gregory. The Cave Painters: Probing the Mysteries of the World's First Artists. New York: Knopf, 2006.

Glen, William, ed. The Mass Extinction Debates: How Science Works in a Crisis. Stanford, CA: Stanford University Press, 1994.

Guthrie, R. Dale. *Frozen Fauna of the Mammoth Steppe: The Story of Blue Babe.* Chicago: University of Chicago Press, 1990.

Kurtén, Björn. *Pleistocene Mammals of Europe.* London: Weidenfeld and Nicolson, 1968.

Kurtén, Björn. *The Age of Mammals.* London: Weidenfeld and Nicolson, 1971.

Kurtén, Björn. *The Cave Bear Story: Life and Death of a Vanished Animal.* New York: Columbia University Press, 1976.

Lister, Adrian, and Paul Bahn. *Mammoths: Giants of the Ice Age.* Rev. ed. Berkeley: University of California Press, 2007.

Martin, Paul S., and Richard G. Klein. *Quaternary Extinctions: A Prehistoric Revolution.* Tucson: University of Arizona Press, 1984.

Owen-Smith, R. Norman. *Megaherbivores: The Influence of Very Large Body Size on Ecology.* Cambridge, U.K.: Cambridge University Press, 1988.

Shoshani, Jeheskel, and Pascal Tassy, eds. *The Proboscidea: Evolution and Palaeoecology of Elephants and Their Relatives.* Oxford: Oxford University Press, 1996.

Stewart, Kathlyn M., and Kevin L. Seymour, eds. *Palaeoecology and Palaeoenvironments of Late Cenozoic Mammals.* Toronto, Ont., CA: University of Toronto Press, 1996.

Turner, Alan, and Mauricio Antón. *The Big Cats and Their Fossil Relatives: An Illustrated Guide to Their Evolution and Natural History.* New York: Columbia University Press, 1997.

Ward, Peter D. *The Call of Distant Mammoths: Why the Ice Age Mammals Disappeared.* New York: Springer-Verlag, 1998.

Whitney-Smith, Elin. *The Second-Order Predation Hypothesis of Pleistocene Extinctions: A System Dynamics Model.* Saarbrüken, Germany: VDM Verlag Dr. Müller, 2009.

Periodicals

Barnosky, Anthony D., Paul L. Koch, Robert S. Feranec, Scott L. Wing, Alan B. Shabel. "Assessing the Causes of Late Pleistocene Extinctions on the Continents." Science 306, no. 5693 (2004): 70–75.

Burney, David A. "Recent Animal Extinctions: Recipes for Disaster." American Scientist 81, no. 6 (1993): 530–541.

Nogués-Bravo, David, Jesús Rodríguez, Joaquín Hortal, Persaram Batra, and Miguel B. Araúco. "Climate Change, Humans, and the Extinction of the Woolly Mammoth." *PloS Biology* 6, no. 4 (2008): e79.

Pacher, Martina, and Anthony J. Stuart. "Extinction Chronology and Palaeobiology of the Cave Bear (*Ursus spelaeus*)." *Boreas* 38, no. 2 (2009): 189–206.

Pastor, John, and Ron A. Moen. "Ecology of Ice-Age Extinctions." *Nature* 431, no. 7009 (2004): 639–640.

Pushkina, Diana, and Pasquale Raia. "Human Influence on Distribution and Extinctions of the Late Pleistocene Eurasian Megafauna." *Journal of Human Evolution* 54, no. 6 (2008): 769–782.

Per Christiansen

South American native mammals

South America is home to a fascinating and diverse mammal assemblage, ranging from mouse opossums, sloths, and armadillos to marmosets, vampire bats, and tapirs to distinctive canids, cats, weasels, and rodents. Brazil alone is home to over 650 mammal species—the highest count of any country in the world, representing about 12 percent of the global total (Brito et al. 2009). Many of these animals are endemic, and many are restricted to such regions as the Amazonian rain forest, the Atlantic Forest, the Andean highlands and mountains, and the grasslands of Argentina and the surrounding countries. Unfortunately, many of these species are being affected by rapid habitat loss, degradation, and fragmentation, as well as associated problems resulting from human encroachment. The conservation statuses of South America's mammals, and the threats that face them, became better understood in the late twentieth and early twenty-first centuries, though as of 2012 a basic deficiency of knowledge remains a problem for many. Major efforts are required to maintain healthy, genetically diverse populations of some of the larger species, and smaller ones will probably best benefit from the conservation of habitat blocks. While many conservation efforts are underway across South America, habitat loss, degradation, and fragmentation make it likely that some endangered species will be lost in the coming decades.

The background to South America's modern mammal fauna

South America's isolation for much of the 65 million years of the Cenozoic resulted in the evolution of a highly distinctive, insular fauna. Several now-extinct, unique, and often spectacular mammal groups inhabited the continent during the Cenozoic, including several ground sloth lineages; the diverse, herbivorous notoungulates; the often cursorial litopterns; and the bizarre, tapir- or proboscidean-like astrapotheres and pyrotheres. Ground sloths are xenarthrans; that is, part of the group that also includes armadillos and anteaters, but the affinities of notoungulates, litopterns, astrapotheres and pyrotheres within the placental mammal radiation remain uncertain. These many morphologically and ecologically diverse herbivore species would have formed complex communities, with species feeding—on all kinds of vegetation—from ground level to several meters up. Several lineages of omnivorous and carnivorous marsupials known as borhyaenoids lived alongside these placental mammals. Borhyaenoids superficially resembling martens (hathlyacynids), civets (prothylacinids) , hyenas (borhyaenids), bears (proborhyaenids) and even saber-toothed cats (thylacosmilids) lived across South America between approximately 60 and 4 million years ago. Hystricognath rodents and platyrrhine monkeys arrived on the continent approximately 30 million years ago following rafting across the Atlantic from Africa. Especially large rodents—the largest of which (*Phoberomys pattersoni*) may have weighed as much as 1,540 pounds (700 kg)—evolved in South America during the Miocene, Pliocene, and Pleistocene.

Approximately 4 million years ago, the formation of the Isthmus of Panama allowed many North American mammal groups to move south and some South American mammal groups to move north (though note that some groups had already been making the crossing, presumably by swimming). South America's modern carnivore and hoofed mammal lineages have their origins in this event, and rapid diversification resulted in the evolution of such distinctive forms as the maned wolf (*Chrysocyon brachyurus*), the bush dog (*Speothos venaticus*), the Andean cat (*Leopardus jacobita*), and the South American camelids.

As recently as the Late Pleistocene (approximately 12,000 years ago), South America was inhabited by more than 50 large terrestrial mammal species, including ground sloths, glyptodonts, pampatheres, mastodonts, litopterns, toxodonts, camelids, horses, bears, cats, and canids. While several of these groups are familiar, others are wholly extinct and include species unlike any living mammals. Glyptodonts and pampatheres were heavily armored, herbivorous relatives of armadillos, sometimes with mace-like tails and sometimes reaching body weights of over 4,400 pounds (2,000 kg). Litopterns were superficially horse-like and camel-like browsers and grazers, whereas toxodonts were stout-bodied herbivorous notoungulates, the largest of which were hippo-like in size. Humans had colonized at least part of South America by approximately 11,000 years ago. Archaeological evidence shows that they hunted many large mammals and that some of the species concerned were still present during the Early Holocene, approximately 7,000 years ago. Several

South America. Reproduced by permission of Gale, a part of Cengage Learning.

factors probably made these animals vulnerable to extinction at the hands of human hunters, including their specialization for life in open habitats and slow reproductive turnover (Cione et al. 2009). The extinction of most native large carnivores probably resulted from the decline in their prey species.

This selective extinction left Holocene South America almost entirely devoid of megaherbivores, a unique situation given the continent's long history of such animals. It has been argued that some modern South American ecosystems are filled with floral anachronisms: plants that coevolved with large herbivores such as mastodonts and sloths, but now produce redundant fruits too large to be successfully exploited by surviving species (Barlow 2000). In view of these extinctions, it should be understood that even the South America of pre-Columbian times had been strongly affected by human hunting and habitat change and that the modern South American mammal fauna is already depauperate.

Furthermore, archaeological evidence shows that large regions of Amazonia were modified by burning, agricultural activities, and the construction of settlements within the past 3,500 years (Willis et al. 2004). Dry climatic conditions that persisted during part of the Holocene also mean that forest cover in the Andes and elsewhere was more extensive in recent centuries than it was just 4,000 years ago (Mayle and Power 2008). The idea that many South American habitats represent pristine, ancient ecosystems can therefore no longer be regarded as accurate. While these data indicate that human impacts were more severe in some areas and ecosystems than conventionally thought, they also indicate that such habitats can recover if conditions allow.

Several major, pervasive conditions pose challenges for the health and future persistence of the remaining mammal species across the continent. The largest problem is habitat loss, fragmentation, and degradation, which is occurring at a

pace and scale exceeding that of the past. Habitats across South America are affected by logging and the removal of wood for charcoal, by the encroachment of areas used for soybean and cattle farming, by destruction associated with gold mining, and by the construction of roads and dams. The destruction and fragmentation of the Amazonian rain forest is well known. However, Brazil's dry caatinga forest; the tropical cerrado savanna habitat, also in Brazil; the hot, semiarid Chaco woodland of Bolivia, Paraguay, Brazil, and Argentina; and other habitat types across the continent are important diversity centers for endemic mammals and are similarly affected.

Additional problems that affect native carnivores especially include depredation, competition, and disease transmission from domestic dogs. Some large species, including the maned wolf, the giant anteater (*Myrmecophaga tridactyla*), and the mountain or Andean tapir (*Tapirus pinchaque*), suffer high mortality on roads. Illegal killing for meat and body parts threatens some species. Mammal groups of special conservation concern in South America include cats, canids, primates, and certain tapir, peccary, and deer species.

Xenarthrans

Xenarthrans—the sloths, armadillos, and anteaters—are ancestrally South American, and the majority of living species occur there. Many are famously elusive and rarely seen, but at least some are known to be threatened by hunting and habitat loss. These problems affect the largest and most spectacular living members of the group: the giant armadillo (*Priodontes giganteus*) and the giant anteater, both of which can weigh in excess of 66 pounds (30 kg). Giant anteaters remain widespread, but their vulnerability to being traffic casualties puts them at special risk in countries with expanding road networks. They are also poor at escaping from the fires that sometimes affect the grassland habitats they inhabit. They are regarded as locally extinct in Uruguay and parts of Brazil.

Armadillos are often elusive because of their nocturnal habits, burrow-dwelling lifestyle, and small size. Low population densities may be the reason for the apparent rarity of some species, and it is uncertain whether these rare species are in decline. Nevertheless, several species are of particular conservation concern. The giant armadillo, the southern long-nosed armadillo (*Dasypus hybridus*), the Brazilian three-banded armadillo (*Tolypeutes tricinctus*), and the southern three-banded armadillo (*T. matacus*) are threatened by hunting and have been extirpated across parts of their ranges. Indeed, the Brazilian three-banded armadillo has been so hard to locate that it was thought extinct prior to its rediscovery in the 1990s.

Among sloths, the maned sloth (*Bradypus torquatus*) is perhaps of most conservation concern, with a declining population restricted to the Atlantic coastal rain forest of eastern Brazil (Hirsch and Chiarello 2012). Rather than being a continuous population, this population may actually represent three separate phylogenetic units. Logging, charcoal burning, and encroaching agriculture are destroying and fragmenting the Atlantic coastal rain forest. Corridors

The giant anteater (*Myrmecophaga tridactyla*) is the only terrestrial member of an otherwise arboreal radiation of xenarthan mammals (though it still retains a limited climbing ability). Sadly, this spectacular species is threatened by habitat fragmentation, its vulnerability to road traffic and other problems. © Frans Lanting Studio/Alamy.

connecting the remaining areas of rain forest habitat need to be established before this important region becomes too fragmented to support its many endemic species. In contrast to the maned sloth, other sloth species are generally widespread and often extremely abundant.

Hoofed mammals

Three of the world's four tapir species occur in South America. Large size, prized meat, slow reproductive turnover, and a reliance on habitats prone to destruction or degradation put them at special risk. They are also easy to locate because of their prominent trails and are easily found by dogs and by humans who can imitate their whistles. All are threatened by hunting and habitat loss. The Brazilian tapir (*Tapirus terrestris*) is still found over a comparatively large area (extending from northern Colombia to northern Argentina and Paraguay) and is still regarded as common within parts of this range. The mostly Central American Baird's tapir (*T. bairdii*) has a smaller, more fragmented range. The mountain tapir is declining as its habitat is fragmented by encroaching agriculture, including poppy growing for the drug trade, and an expanding human population. Armed conflict in parts of its

range, illegal poaching for its skin and body parts, and collisions with road traffic represent additional problems.

Like the Brazilian tapir, two of South America's peccaries—the collared peccary (*Pecari tajacu*) and the white-lipped peccary (*Tayassu pecari*)—are enormously widespread and locally common across the northern part of the continent. However, the Chacoan peccary or tagua (*Catagonus wagneri*), endemic to parts of Paraguay, Bolivia, and Argentina, is endangered as a result of a history of hunting and a reduction of favored habitat. The Argentine Chaco inhabited by this species is being rapidly replaced by cattle pasture.

The status of the white-lipped peccary is also a cause for concern. This species has undergone a major range-wide decline since 1900, the local extinction of some populations has been reported, and it is estimated that the species is now absent from over 20 percent of the area inhabited in the early twentieth century (Altrichter et al. 2011). Given the ecological importance of the white-lipped peccary as an ecosystem engineer (a species that plays a key role in shaping an environment by manipulating soils, spreading plant seeds, and so on), its decline could result in serious habitat degradation across the continent. A wide-ranging species, it

Modern sloths, like this maned three-toed sloth (*Bradypus torquatus*), are the only surviving members of a substantially more diverse group of xenarthrans. Relatively large-bodied (though not gigantic) terrestrial and semi-arboreal sloths inhabited South America and the Caribbean until about 4,000 years ago. Human persecution was almost certainly integral to their disappearance. © Mark Moffett/Minden Pictures/Corbis.

is a key component of some rain forest, dry forest, and scrub habitats. Populations in arid habitats seem to have been affected most severely.

Habitat loss and fragmentation resulting from changing land use are among the documented causes of peccary decline. In the Brazilian Pantanal, the conversion of the surrounding highland to cash crop pastureland and the partial replacement of cerrado forest to cattle pasture have contributed to this decline. It is also possible that peccary decline has been caused by diseases transmitted by domestic livestock. It has been suggested that the creation and maintenance of habitat corridors that link mosaics of large, continuous habitat types will help maintain white-lipped peccary populations in the future.

Of the wild South American camelids, all remain abundant, despite enormous declines. According to some estimates, over 50 million guanacos (*Lama guanicoe*) inhabited Patagonia alone a few centuries ago, and vicuña (*Vicugna vicugna*) also numbered in the millions at that time. The valuable fleece of the vicuña resulted in nonsustainable hunting and a decline to approximately 6,000 animals by the 1970s. A conservation initiative in which local people were paid to protect vicuña and also harvest their fleeces, coupled with the controlled trade of these products through the Convention on International Trade in Endangered Species of Wild Fauna and Flora (CITES), prompted the recovery of populations to approximately 350,000. Nevertheless, the health of this species is dependent on continued conservation, because poaching and habitat loss continue to be threats.

Finally, among terrestrial hoofed mammals, South America is home to approximately 17 deer species. These include the

marsh deer (*Blastocerus dichotomus*), the highland-dwelling huemuls (genus *Hippocamelus*), the forest-dwelling brockets (genus *Mazama*), and the tiny pudus (genus *Pudu*). Endemic subspecies of white-tailed deer (Odocoileus virginianus) also occur in northern South America.

Several of these forms are in danger of extinction. The South Andean or Chilean huemul (*Hippocamelus bisulcus*) persists as approximately 600 animals in Argentina and 1,500 in Chile. Some estimates suggest that a decline of about 99 percent relative to the numbers present in pre-Columbian times has occurred, and the decline appears to be continuing. The Argentine population is fragmented into about 60 small, isolated groups. This reduction has partly been caused by hunting, domestic dog depredation, and habitat loss and fragmentation stemming from agricultural encroachment (Povilitis 1998). Ecological competition from alien red deer (*Cervus elaphus*) and vulnerability to diseases carried by sheep may also be contributing to the huemul's decline.

South America is inhabited by two or three aquatic relatives of terrestrial hoofed mammals: the Amazonian or South American manatee (*Trichechus inunguis*) and the Amazon River dolphin or boto (*Inia geoffrensis*). Bolivian botos are regarded by some experts as distinct enough from those of the Amazon and Orinoco river basins to be recognized as a distinct species, *Inia boliviensis*. The conservation status of all boto populations is uncertain, and more research is needed. The status of the manatee is also not well understood, mostly because of its secretive lifestyle and poor population data, but it is suspected to be in decline. Mostly this decline is due to continuing illegal hunting involving specially designed harpoons and nets. Habitat deterioration, pollution, and population fragmentation caused by damming are increasingly important threats to manatees.

Carnivores

South America includes an assemblage of relatively poorly known, often specialized carnivore species, most notably the Andean or spectacled bear (*Tremarctos ornatus*) and a list of endemic canids and felids. The conservation status of certain species is not well understood, population changes are not being adequately monitored, and some species seem to occur at naturally low densities.

A number of small carnivore groups, namely mustelids, procyonids, and mephitids (skunks), occur on the continent. Some members of these groups are poorly known in terms of population sizes and trends, and their apparent restriction to small areas puts them at extinction risk. The Colombian weasel (*Mustela felipei*), for example, is known only from ten cloud forest localities in the Cordillera Central of Colombia and Andean Ecuador and is likely threatened by the deforestation occurring in these regions. The marine otter (*Lontra felina*), an endemic of the coasts of Peru, Chile, and southern Argentina, is in decline due to human overfishing, illegal killing by fishermen, and deaths caused by domestic dogs. The fragmented populations total less than 1,000 animals (Hunter and Barrett 2011). The giant otter (*Pteronura brasiliensis*)—the world's largest mustelid after the sea otter (*Enhydra lutris*)—previously suffered from extensive hunting for its pelt but is currently seriously endangered by habitat destruction, fish depletion, and pollution caused by increasing human activity around the rivers it prefers.

One of the region's most distinctive canids, the maned wolf, is threatened by many factors (Soler et al. 2005) and is already extinct in Uruguay. Its favored habitat has been reduced by the conversion of land for agriculture. Direct persecution by people, deaths following conflicts with domestic dogs, and collisions with motor vehicles represent serious threats, and the use of its body parts in medicine and ritual also results in the killing of individuals. Conservation interest in this bizarre, omnivorous canid has resulted in legal protection in Brazil and Argentina, but specific action tailored to the ecology of the maned wolf has yet to be implemented. Significant population declines are expected to occur because of continuing habitat loss and degradation.

The status of another highly distinctive South American canid, the short-limbed bush dog, is difficult to determine because of its natural rarity and elusive, nomadic nature. There are indications, however, that habitat loss and fragmentation, reduction in prey, and domestic dog depredation have resulted in population declines, and the species is also known to be susceptible to diseases transmitted by domestic dogs. Bush dogs are social predators of savannas and diverse wooded habitats and require healthy populations of small and mid-sized vertebrate prey. The persistence of both maned wolves and bush dogs depends on the protection of large, connected areas of grassland, woodland, and shrub-dominated lowland where domestic dogs and vehicle traffic are minimal or absent. The presence of bush dogs in disturbed areas (Oliveira 2009) suggests, however, that they are more adaptable in the face of habitat change than usually thought.

The poorly known short-eared dog (*Atelocyon microtis*) is a forest-dependent canid, endemic to western Amazonia. Sleek fur, a streamlined form, and reports of it swimming suggest a semiaquatic lifestyle and reliance on watercourses (Hunter and Barrett 2011). Like other wild canids worldwide, it is known to be at risk from diseases transferred from domestic dogs and also from habitat loss.

Several medium-sized, fox-like canids inhabit South America, including the crab-eating fox (*Cerdocyon thous*) and the zorros (genus *Lycalopex*). Some, including the hoary fox (*L. vetulus*), the pampas fox (*L. gymnocercus*), and the crab-eating fox, seem tolerant of habitat degradation and remain

One of the world's most distinctive canids, the bush dog (*Speothos venaticus*) is a close relative of the maned wolf (*Chrysocyon brachyurus*). Both may have evolved in North or Central America before invading South America. Bush dogs live in packs, thought to consist of a breeding pair and their grown offspring. © Amazon-Images MBSI/Alamy.

widespread and common. The pampas fox, however, is heavily persecuted because of its perceived predation on livestock. Darwin's fox (*L. fulvipes*) is critically endangered, with two remnant, Chilean populations (one on Chiloé Island) containing fewer than 250 individuals. The assumption that Darwin's fox is a subspecies of the Argentine gray fox or chilla (*L. griseus*) has contributed to a lack of conservation or captive breeding efforts. The two surviving Darwin's fox populations are relicts, stranded on habitat islands surrounded by agricultural land and, on the mainland, severely at risk from domestic dogs. This species, however, has increasingly been reported from secondary forests, suggesting that it may be adaptable in the face of habitat change. Its long-term survival perhaps depends on reduced interaction with domestic dogs.

South America's only native bear, the Andean or spectacled bear, is most usually associated with humid Andean forests, but it also occurs in other wooded habitats at elevation as well as in coastal scrub forest in northwestern Peru. The risks facing this species are similar to those affecting other South American carnivores: habitat loss and fragmentation, persecution stemming from its supposed livestock-killing behavior, and illegal hunting. Its fur, meat, and other body parts are used by people. It is estimated that the bear has now disappeared from about 60 percent of its range, and its restriction to small scattered populations places many populations at risk of extinction (Kattan et al. 2004). As is the case with many South American mammals, a lack of good information on population numbers and trends hinders an understanding of conservation priority. Some protected areas do not contain suitable bear habitat or are too small. Habitat corridors that link populations are crucial to the conservation of this unique species. The spectacled bear is the only living member of a bear group that was distributed throughout the Americas in the Pliocene and Pleistocene.

Cats

South America is home to about 10 cat species (the exact number is debatable because the taxonomic validity of some forms remains controversial). While several are endemic, the jaguar (*Panthera onca*), puma (*Puma concolor*), jaguarundi (*Puma yagouaroundi*), ocelot (*Leopardus pardalis*), and margay (*L. wiedii*) also occur in Central and North America, and the oncilla (*L. tigrinus*) has a relict population in Costa Rica and Panama. Of all South American mammals, cats are perhaps the best known for being endangered by hunting, habitat loss, and persecution resulting from their depredations on domestic animals. Extinction risk varies considerably among these species. Some, such as the jaguar, the puma, the jaguarundi, and Geoffroy's cat (*Leopardus geoffroyi*) are sufficiently widely distributed to (theoretically) ensure survival for some time. Others, however, are increasingly restricted to small, threatened regions or habitat types and are at risk of extinction in the early decades of the twenty-first century.

Jaguars inhabit savannas, deserts, scrubland, swamps, and forests across the continent. It is estimated, however, that the species is now absent from about 45 percent of its historic range, and the jaguar is now extinct in Uruguay. Similarly, the puma, while occurring in diverse habitats across the continent,

Largest of South America's cats, the jaguar (*Panthera onca*) is now absent from about 45 percent of its original range. Like other large cats elsewhere in the world, its survival and persistence depends on the creation of habitat corridors that allow gene flow between otherwise isolated populations. © Amazon-Images/Alamy.

is reckoned by some estimates to have been extirpated from about 40 percent of its original range. Like other large carnivores, these cats require large habitat tracts and are at great risk of local extinction when restricted to habitat fragments. Furthermore, genetic studies show that modern jaguars do not exhibit any strong geographic structure in variation (Eizirik et al. 2001). Combined, these details emphasize the need for conservation strategies that promote jaguar movement across large areas (Sanderson et al. 2002).

Among the smaller cats, the margay and ocelot are highly reliant on forest habitat and hence are threatened by its destruction. Both are heavily spotted cats of forested habitats; the ocelot also occurs in scrublands whereas the margay is an especially adept tree-climber reliant on dense lowland forests. Geoffroy's cat is an adaptable cat of forests, grasslands, and marshes, notable for its numerous small spots, whereas the colocolo or pampas cat (*Leopardus colocolo*) is a confusingly variable animal that occurs across desert, grassland, and woodland habitats but does not inhabit rain forest. Some experts suggest that the colocolo represents a complex of several related species. Both Geoffroy's cat and the colocolo, as traditionally conceived, appear to remain widespread and relatively abundant. The status of the oncilla (*L. tigrinus*), a small, slender, spotted cat that occurs in savanna, scrub, and woodland, is poorly known. The guiña or kodkod (*L. guigna*), another small, spotted cat, is a major cause for concern. Highly restricted in range, dependent on dense temperate forest, and showing a habit of preying on domestic chickens when the opportunity allows, this species is declining as a result of habitat loss and fragmentation and illegal persecution. The Andean cat, a thick-tailed, blotch-marked, montane specialist that mostly preys on rodents, appears to be naturally rare. Unfortunately, it is beleaguered by problems, and conservation initiatives need to be implemented across its range. Habitat destruction caused by mining and farming as well as depletion of its chinchilla prey represent problems for this species. Hunting for use in the harvest ceremonies of local people represents an additional major threat.

Marsupials, insectivores, bats, and rodents

South America's small mammal fauna includes a rich assemblage of opossum and opossum-like marsupials, bats, and rodents. Lagomorphs and shrews also occur but at low diversity. While certain species within these groups have been the subjects of specific conservation concern, the same problem affects many of them: They are simply not well known enough—taxonomically (Brito 2004), ecologically, or in terms of population trends—to be assessed in terms of conservation priority. Among rodents, for example, a long list of species belonging to all the native groups are known from one, two, or a few specimens found in just one restricted area. Inadequate sampling and low population densities may obscure their true status. Nevertheless, given that many are unique to habitats or regions subject to fragmentation, degradation, or loss, it is likely that some or many are declining and endangered. Similarly, the conservation status of many South American bat species is poorly known, and the inclusion of several as targets of conservation concern has been essentially arbitrary

Among South American marsupials, such opossums as the Handley's slender mouse opossum (*Marmosops handleyi*) of Colombia are regarded as endangered because of their small range, their apparent rarity, and the loss of their habitat as a result of agricultural encroachment. Small ranges and natural rarity may mean that some opossums are prone to extinction. Indeed, the red-bellied gracile mouse opossum (*Cryptonanus ignitus*), named in 2002, is already believed extinct because the region in which it was collected (Yuto, in Jujuy, Argentina) was converted to agricultural use during the 1960s—that is, some time prior to the official recognition of this form as a new species (Díaz et al. 2002).

Also inhabiting the South American tropics are the poorly known shrew opossums or caenolestoids. Again, the data are lacking, but some shrew opossums are suspected of being in decline and in danger from habitat loss. Finally, the monito del monte (*Dromiciops gliroides*) is restricted to cool forests in Chile and adjacent areas of Argentina, where it is considered threatened because of the loss of habitat resulting from the spread of conifer plantations. The presence of black rats (*Rattus rattus*) may also represent a threat to this small arboreal marsupial. This species is the only living member of Microbiotheria, a group otherwise known only as fossils and thought to be more closely related to Australasian marsupials than to any other South American forms. The conservation of the monito del monte should be considered a priority in view of its uniqueness.

Among South American shrews, the Darién small-eared shrew (*Cryptotis mera*) and Enders's small-eared shrew (*C. endersi*) are both restricted to small areas of fragmented evergreen forest in Colombia and Panama; both are believed to be declining as a result of deforestation. Little is known of the ecological preferences and biology of these species. The conservation status of several other South American shrews is unknown.

Given the sheer number of South American rodent species and the deficiency in knowledge of their taxonomic and conservation status, it is probably most appropriate to concentrate on higher taxa and on habitats and regions known to be diversity hot spots (Amori and Gippoliti 2003). Nevertheless, the South American rodent species that are the focus of special attention include the fossorial giant rat (*Kunsia fronto*), the two chinchillas (*Chinchilla chinchilla* and *C. lanigera*), the Bolivian chinchilla rat (*Abrocoma boliviensis*), the Pacific degu (*Octodon pacificus*), and the Santa Catarina's guinea pig (*Cavia intermedia*). Other poorly known species that might require special conservation concern include the owl's spiny rat (*Carterodon sulcidens*), and the Bahian giant tree rat (*Phyllomys unicolor*) (Costa et al. 2005). The candango mouse (*Juscelinomys candango*), named as a new species in 1965, has not been recorded since 1960, and the area of Brazil in which it was discovered is now entirely urbanized; repeated attempts to discover it have been unsuccessful, and it is now thought to be extinct. Among South American porcupines, several are known from a small number of specimens or from a small area, and these species may or may not be in critical decline. The bristle-spined or thin-spined porcupine (*Chaetomys subspinosus*) from the Brazilian Atlantic Forest is believed to be threatened by habitat loss, and the brown hairy dwarf porcupine (*Coendou vestitus*) was originally collected from an area (Cundinamarca, Colombia) that has since been extensively deforested.

Like other South American mammals, many of these rodents are threatened by the loss, fragmentation, and degradation of their habitat. Viscachas (represented by the two genera *Lagostomus* and *Lagidium*) are superficially lagomorph-like, thick-furred relatives of chinchillas. Populations discovered in the Ecuadorian Andes in the early twenty-first century compete with domestic cattle and are threatened by humans using fire to maintain and control crop fields and pasture (Werner et al. 2006). Similarly, declines in the two chinchilla species have been caused not only by habitat loss resulting from mining and livestock grazing but also by nonsustainable hunting for the fur trade. At the height of the chinchilla trade (exports to North America and Europe peaked during the late 1800s), about two million pelts were exported over a 5-year period. The short-tailed chinchilla

The long-tailed chinchilla (*Chinchilla lanigera*) of Chile and Peru, one of two living chinchilla species. Chinchillas are famous for their dense fur where as many as 60 individual hairs grow from each follicle. Hunting and habitat loss have severely affected chinchillas as well as other South American highland rodents. © Luciano Candisani/Minden Pictures/Corbis.

(*Chinchilla chinchilla*) became extinct in Peru and possibly in Bolivia as a result of these pressures.

Primates

Approximately 100 to 140 endemic primate species, all belonging to the platyrrhine or New World monkey clade, inhabit South America. They include the marmosets and tamarins, night monkeys, titis, capuchins, squirrel monkeys, sakis, uakaris, howler monkeys, muriquis, and spider monkeys. This enormous radiation, varying in body size from 4.2 ounces (120 g) to over 26 pounds (12 kg), encompasses a diversity of behaviors and bodily structures unique to the continent.

Views differ considerably as to which South American primate populations are worthy of recognition at species level, and there has been an enormous increase in the proposed number of species. Regardless of taxonomy, many populations are highly threatened or endangered, some critically so. Species especially worthy of conservation attention include the black-faced lion tamarin (*Leontopithecus caissara*), the pied tamarin (*Saguinus bicolor*), the black bearded saki (*Pithecia satanas*), the buff-headed or yellow-breasted capuchin (*Cebus xanthosternos*), the blond capuchin (*Cebus flavius*), the brown or variegated spider monkey (*Ateles hybridus*), the two muriqui species (*Brachyteles arachnoides* and *B. hypoxanthus*), and the yellow-tailed or Hendee's woolly monkey (*Oreonax flavicauda*). These include Amazonian forest species as well as Atlantic Forest endemics. Two titis, Barbara Brown's titi (*Callicebus barbar-abrownae*) of the Brazilian caatinga and Coimbra Filho's titi (*Callicebus coimbrai*) of the Atlantic coastal forest, are critically endangered, with populations in the hundreds.

Many South American primates are affected by the same problems of habitat loss, degradation, and fragmentation that are causing declines in mammals across the region

(Mittermeier et al. 1989). This is especially problematic for some of the smaller monkeys, because their ranges can be tiny. As one example, the pied tamarin is restricted to forest patches located around the periphery of the Brazilian state of Manaus, where urban growth and habitat destruction have removed the tamarin's original habitat. Ecological competition from another species, the red-handed tamarin (*Saguinus midas*), poses a new problem for this species. Somewhat similarly, competition and even hybridization from two hardy, artificially introduced marmoset species—the common marmoset (*Callithrix jacchus*) and the black-tufted-ear marmoset (*C. penicillata*)—now affect the survival of local marmoset species in the Atlantic Forest.

Some specific problems of human persecution have also affected South American primate populations. Prior to the passage of the U.S. Endangered Species Act of 1973, enormous numbers of marmosets and tamarins were captured and exported to the United States as pets and laboratory animals. This was believed to be beneficial to the countries where the animals are native, because it was thought that these monkeys acted as reservoirs of malaria and yellow fever. The numbers of animals involved are staggering: 30,000 to 40,000 cotton-top tamarins (*Saguinus oedipus*), for example, are believed to have been exported to the United States during the 1960s and 1970s.

Large monkeys in particular suffer from human hunting. Low-level, traditional-style hunting practiced by some local people can probably be sustained, but commercial hunting involving firearms is not sustainable in view of the slow reproductive turnover of these primates. The hunting of some monkeys occurs in an opportunistic fashion in areas where people illegally harvest trees. This is especially a problem for the southern muriqui (*Brachyteles arachnoides*) in the Atlantic Forest where the removal of the economically important, endemic palm *Euterpe edulis* means that palm harvesters—known as *palmiteiros*—exert continuous, persistent hunting pressure on these monkeys.

The northern muriqui (*Brachyteles hypoxanthus*)) is South America's largest native primate and has a distinctive egalitarian social system where males do not compete over females. Sadly, it is critically endangered, with less than 1,000 individuals existing in several isolated sub-populations. © Andre Seale/Alamy.

Some South American primates have received a large amount of conservation attention, partly because of their observability and a comparatively large number of interested researchers. The critically endangered northern muriqui (*Brachyteles hypoxanthus*), reduced to less than 1,000 individuals split into more than a dozen Atlantic Forest populations, has been the subject of a sustained effort to increase public awareness of its plight and to involve local landowners and other people in its conservation.

Overall concerns and conservation needs

Habitat fragmentation, destruction, and degradation is evidently increasing in pace and extent across South America, as it is worldwide. This is the number one cause of species decline and potential loss among South American mammals. Habitat fragmentation caused by road building, damming, the creation of cash-crop plantations, and the conversion of land for agriculture is potentially disastrous for the survival of many larger mammals across the region, because the creation of isolated subpopulations weakens the genetic health of species as a whole and destroys their ability to resist change and persist at viable levels. Furthermore, fragmented forest enclaves are subject to rapid shifts in plant community structure and increased tree mortality resulting from desiccation stress, wind turbulence, liana infestation, increased vulnerability to fire, and other factors (Laurance et al. 1998). They cannot, therefore, persist as healthy habitat islands for mammals and other large vertebrates.

Of additional concern is that the removal or loss of species in these fragmented habitats (including plants, invertebrates, and vertebrates) through illegal harvesting, hunting, and factors related to climate change is further degrading the health of the habitats as a whole. There is also evidence that alterations to the Amazonian forest composition are occurring as a consequence of climate change, with assemblages of slow-growing sub-canopy tree species declining in importance and fast-growing canopy and emergent forms becoming more important (Laurance et al. 2006). It is unknown what effect these community changes will have on endemic mammals and other animals.

In view of ongoing habitat fragmentation, the protection of areas linking large and sustainable habitat patches must be considered a priority. The giant anteater, maned wolf, bush dog, jaguar, and white-lipped peccary use woodlands, grasslands, and shrub-dominated regions rather than one habitat alone, and habitat mosaics are more likely to preserve biodiversity and healthy populations of large-bodied species than forest or grassland patches alone. Large collapses seen in widespread species that were not previously considered threatened, such as the white-lipped peccary, have shown the importance of habitat patches connected via safe corridors. Efforts to create such corridors are underway in many areas, including the Ipanema/Caratinga/Sossego Biodiversity Corridor in the Brazilian Atlantic Forest, the Bogotá Conservation Corridor in Colombia, and the North Andean Conservation Corridor in Colombia and Venezuela. At the time of writing, the ambitious Jaguar Corridor Initiative seeks to connect wild spaces from Mexico all the way to Argentina.

A number of other conservation problems require attention across South America. European hares (*Lepus europaeus*), first introduced to Argentina in 1888, have colonized an enormous part of the continent, including virtually the whole of Argentina and Uruguay, much of Chile, and parts of Bolivia, Brazil, Peru and Paraguay. The numbers are so great that more than 2.5 million Argentinean hares were shot annually for their meat during the 1980s (Bonino et al. 2010). In Tierra del Fuego, huge numbers of North American beavers (*Castor canadensis*)—introduced during the 1940s for the fur trade—are causing ecological devastation and urgently require control. As of 2011, the estimated population exceeded 200,000. The interaction of domestic dogs with wild carnivores is a serious problem for native canids, cats, and other species. It is hoped that efforts will be made to exclude domestic dogs from national parks inhabited by endangered native carnivores and to minimize the impact of domestic dogs and the diseases they transmit to wild carnivores. Efforts to minimize the danger that road traffic presents to large mammal species should also be considered a priority in some areas.

Resources

Books

Barlow, Connie. *The Ghosts of Evolution: Nonsensical Fruit, Missing Partners, and Other Ecological Anachronisms*. New York: Basic Books, 2000.

Cione, Alberto L., Eduardo P. Tonni, and Leopoldo Soibelzon. "Did Humans Cause the Late Pleistocene–Early Holocene Mammalian Extinctions in South America in a Context of Shrinking Open Areas?" In *American Megafaunal Extinctions at the End of the Pleistocene*, edited by Gary Haynes, 125–144. Dordrecht, Netherlands: Springer, 2009.

Eisenberg, John F. *Mammals of the Neotropics*. Vol. 1, *The Northern Neotropics: Panama, Colombia, Venezuela, Guyana, Suriname, French Guiana*. Chicago: University of Chicago Press, 1989.

Eisenberg, John F., and Kent H. Redford. *Mammals of the Neotropics*. Vol. 3, *Ecuador, Bolivia, Brazil*. Chicago: University of Chicago Press, 1999.

Emmons, Louise H. *Neotropical Rainforest Mammals: A Field Guide*. 2nd ed. Chicago: University of Chicago Press, 1997.

Gardner, Alfred L., ed. *Mammals of South America*. Vol. 1, *Marsupials, Xenarthrans, Shrews, and Bats*. Chicago: University of Chicago Press, 2007.

Hunter, Luke, and Priscilla Barrett. *Carnivores of the World*. Princeton Field Guides. Princeton, NJ: Princeton University Press, 2011.

Macdonald, David, ed. *The New Encyclopedia of Mammals*. Oxford: Oxford University Press, 2001.

Nowak, Ronald M. *Walker's Mammals of the World*. 6th ed. 2 vols. Baltimore: Johns Hopkins University Press, 1999.

Redford, Kent H., and John F. Eisenberg. *Mammals of the Neotropics*. Vol. 2, *The Southern Cone: Chile, Argentina, Uruguay, Paraguay*. Chicago: University of Chicago Press, 1992.

Periodicals

Altrichter, Mariana, Andrew Taber, Harald Beck, et al. "Range-Wide Declines of a Key Neotropical Ecosystem Architect, the Near Threatened White-Lipped Peccary *Tayassu pecari*." *Oryx* 46, no. 1 (2011): 87–98.

Amori, Giovanni, and Spartaco Gippoliti. "A Higher-Taxon Approach to Rodent Conservation Priorities for the 21st Century." *Animal Biodiversity and Conservation* 26, no. 2 (2003): 1–18. Accessed June 13, 2012. http://w3.bcn.es/fitxers/icub/museuciencies/abc262pp118.451.pdf.

Bonino, Never, Daniel Cossíos, and Joao Menegheti. "Dispersal of the European Hare, *Lepus europaeus* in South America." *Folia Zoologica* 59, no. 1 (2010): 9–15.

Brito, Daniel. "Lack of Adequate Taxonomic Knowledge May Hinder Endemic Mammal Conservation in the Brazilian Atlantic Forest." *Biodiversity and Conservation* 13, no. 11 (2004): 2135–2144.

Brito, Daniel, Leonardo C. Oliveira, Monik Oprea, and Marco A. R. Mello. "An Overview of Brazilian Mammalogy: Trends, Biases, and Future Directions." *Zoologia* 26, no. 1 (2009): 67–73.

Costa, Leonora Pires, Yuri Luiz Reis Leite, Sérgio Lucena Mendes, and Albert David Ditchfield. "Mammal Conservation in Brazil." *Conservation Biology* 19, no. 3 (2005): 672–679.

Díaz, M. Mónica, David A. Flores, and Rubén M. Barquez. "A New Species of Gracile Mouse Opossum, Genus *Gracilinanus* (Didelphimorphia: Didelphidae), from Argentina." *Journal of Mammalogy* 83, no. 3 (2002): 824–833.

Eizirik, Eduardo, Jae-Heup Kim, Marilyn Menotti-Raymond, et al. "Phylogeography, Population History, and Conservation Genetics of Jaguars (*Panthera onca*, Mammalia, Felidae)." *Molecular Ecology* 10, no. 1 (2001): 65–79.

Hirsch, André, and Adriano Garcia Chiarello. "The Endangered Maned Sloth *Bradypus torquatus* of the Brazilian Atlantic Forest: A Review and Update of Geographical Distribution and Habitat Preferences." *Mammal Review* 42, no. 1 (2012): 35–54.

Kattan, Gustavo, Olga Lucía Hernández, Isaac Goldstein, et al. "Range Fragmentation in the Spectacled Bear *Tremarctos ornatus* in the Northern Andes." *Oryx* 38, no. 2 (2004): 155–163.

Laurance, William F., Leandro V. Ferreira, Judy M. Rankin-de Merona, and Susan G. Laurance. "Rain Forest Fragmentation and the Dynamics of Amazonian Tree Communities." *Ecology* 79, no. 6 (1998): 2032–2040.

Laurance, William F., Henrique E.M. Nascimento, Susan G. Laurance, et al. "Rapid Decay of Tree-Community Composition in Amazonian Forest Fragments." *Proceedings of the National Academy of Sciences of the United States of America* 103, no. 50 (2006): 19010–19014.

Mayle, Francis E., and Mitchell J. Power. "Impact of a Drier Early–Mid-Holocene Climate upon Amazonian Forests." *Philosophical Transactions of the Royal Society* B 363, no. 1498 (2008): 1829–1838.

Mittermeier, Russell A., Warren G. Kinzey, and Roderic B. Mast. "Neotropical Primate Conservation." *Journal of Human Evolution* 18, no. 7 (1989): 597–610.

Oliveira, Tadeu G. de. "Distribution, Habitat Utilization, and Conservation of the Vulnerable Bush Dog *Speothos venaticus* in Northern Brazil." *Oryx* 43, no. 2 (2009): 247–253.

Povilitis, Anthony. "Characteristics and Conservation of a Fragmented Population of Huemul *Hippocamelus bisulcus* in Central Chile." *Biological Conservation* 86, no. 1 (1998): 97–104.

Sanderson, Eric W., Kent H. Redford, Cheryl-Lesley B. Chetkiewicz, et al. "Planning to Save a Species: The Jaguar as a Model." *Conservation Biology* 16, no. 1 (2002): 58–72.

Soler, Lucía, Jean Marie Carenton, Diego Birochio, et al. "Problems and Recommendations for the Conservation of the Maned Wolf in Argentina." *Endangered Species Update* 22, no. 1 (2005): 3–9.

Werner, Florian A., Karim J. Ledesma, and Rodrigo Hidalgo B. "Mountain Vizcacha (*Lagidium* cf. *peruanum*) in Ecuador: First Record of Chinchillidae from the Northern Andes." *Mastozoología Neotropical* 13, no. 2 (2006): 271–274.

Willis, Kathy J., L. Gillson, and T.M. Brncic. "How 'Virgin' Is Virgin Rainforest?" *Science* 304, no. 5669 (2004): 402–403.

Darren Naish

· · · · ·

Australia (50,000 years ago)

The continent of Australia supports between 7 percent and 10 percent of the world's species. This high species richness is a result of the vast land area of Australia and its nearby islands (approximately 2.97 million square miles [7.69 million km^2]); the broad range of terrestrial environments encompassed by this area, from tropical rain forest to deserts and cool temperate forests; and the diversity of temperate to tropical marine environments around the long Australian coast. For several major groups of plants and animals, Australia is a global center of diversity. For example, the number of reptile species in Australian deserts is higher than in desert environments elsewhere in the world, and Australia's coasts have 57 percent of the world's mangrove species (State of the Environment Committee 2011).

Not only is Australia's biota rich in species, those species are also highly distinctive. Prior to about 60 million years ago, Australia was connected to the other southern continents as part of the supercontinent Gondwanaland. At about that time the Australian continental plate began to break away and drift northwards into isolation, and by 45 million years ago it was completely separated from the other continents (White 2006). Only within the last few million years has Australia approached Asia closely enough to allow some interchange of northern Australian and Asian species to occur. Because of this long period of separate evolution, Australia has many groups of plants and animals that are found nowhere else in the world. For example, Australia has the world's highest number of endemic plant families (16), a frog fauna in which 94 percent of the species are endemic, and, along with New Guinea, all of the world's monotremes (the platypus and echidnas, which together form a sister group to all other mammals). Australia and New Guinea combined have about two-thirds of the world's marsupial species, all of which are endemic to the region. Because of this exceptional combination of species richness and uniqueness, Australia is regarded as one of the world's "megadiverse" countries (State of the Environment Committee 2011).

Unfortunately, Australia has also suffered a higher rate of extinction than other continents. This entry reviews the chronicle of extinctions in the Australian biota from 50,000 years ago until the present. The date of 50,000 years ago is significant because it was about then, or soon after, that people first settled in Australia (Hiscock 2008). Therefore, this review is concerned primarily with extinctions related to the impacts of people, both in prehistory and the modern era. The extinctions can be considered as three distinct waves that unfolded in the Late Pleistocene, in the mid-Holocene, and since the beginning of the nineteenth century.

For part of this period, there was a broad land connection between Australia and New Guinea to the north. This land bridge was flooded as sea levels rose about 10,000 years ago. This review focuses on the continent of Australia and its continental islands (Tasmania and Kangaroo Island, plus many smaller islands) as defined by the present-day coastline, but makes passing reference to extinctions in New Guinea during the period when it was joined to Australia and also notes extinctions on islands (such as Norfolk and Lord Howe Islands) that are beyond the Australian continental shelf but are now part of Australia's national territory.

Late Pleistocene extinctions: Disappearance of the megafauna

The Pleistocene is the epoch of geological history that includes most of the last 2 million years, up to 10,000 years ago (the beginning of the Holocene). Pleistocene climates were generally cooler and drier than that of today and were dominated by the temperature cycles that caused the succession of "ice ages." Australia remained almost ice-free during the Pleistocene, but the open habitats created by a generally cool and dry climate favored the evolution of large size among vertebrate animals. These giants are collectively termed the Pleistocene megafauna (Johnson 2006; Long et al. 2002).

Most of the Pleistocene megafauna were very large marsupials, of which there were 50 or more species. These included a diverse suite of herbivores, dominated by kangaroos adapted for feeding on tough plant material, especially the leaves of shrubs and small trees. The biggest of these kangaroos was *Procoptodon goliah*, a species that weighed up to 550 pounds (250 kg) and stood close to 10 feet (3 m) tall. The many extinct species of kangaroos related to *P. goliah* show a variety of distinct adaptations in their teeth and body form that indicate complex ecological communities in which many related species shared the same habitats by occupying different foraging niches. The largest living kangaroos have average weights under 88 pounds (40 kg) and feed mainly on grasses.

The largest marsupial herbivores of the Pleistocene were the diprotodontids, a family that included species ranging in size from small cattle up to *Diprotodon optatum*, which weighed around 6,000 pounds (2,700 kg); and the palorchestids, animals with short trunks like tapirs' that evidently fed on coarse plant material such as twigs and bark. The marsupial herbivores of the Pleistocene also included several large and one small wombat (six extinct species, as well as the three surviving wombat species). The giant wombat (*Phascolonus gigas*) weighed perhaps 440 pounds (200 kg) and had a body form suggesting that, like the living wombats, it dug extensive burrow systems; if so, it may have been the largest burrowing mammal ever.

Some of the Pleistocene kangaroos—several species of *Propleopus*, related to the living musky rat-kangaroo (*Hypsiprymnodon moschatus*), a tiny rain-forest-dwelling omnivore and frugivore—were evidently carnivorous, with teeth and skeletal remains suggesting they were broadly similar in their adaptations and habits to dogs, bears, or wolverines. The most formidable Pleistocene predator was the marsupial lion (*Thylacoleo carnifex*). The Family Thylacoleonidae, of

which the marsupial lion was the largest, most specialized, and last member, was related to the herbivorous diprotodonts and wombats. Its herbivorous ancestry notwithstanding, the marsupial lion was an extraordinarily well-armed predator. In place of the elongated canines used by most carnivores to pierce and grasp their prey, the marsupial lion had incisors modified as pointed canine-like teeth. All its cheek teeth other than the third premolars were reduced, and the third premolars were enlarged as enormous slicing blades. These were probably used to rapidly scissor through hide and flesh, causing massive trauma to prey and a rapid kill. A functional analysis indicated that the marsupial lion was able to exert exceptional bite force, much greater (relative to body mass) than the African lion, and showed that its robust skull was able to resist high external forces that would be generated by struggling prey (Wroe 2008). These animals also had powerful forelegs with large paws, each with an enlarged thumb tipped with a long claw that would have enabled it to grasp and tear large prey. All these features suggest that the marsupial lion regularly killed prey well above its own body weight and was probably a significant predator of all but the largest herbivores in Australia's Pleistocene megafauna.

The extinct megafauna also included some large monotremes, reptiles, and birds. One of the monotremes, the

Diprotodon optatum, the largest marsupial ever to have lived. This species weighed around 6,000 pounds (2,700 kg) and was widespread in Australia until it went extinct between 50,000 and 40,000 years ago. Illustration by Francois Desbordes.

long-beaked echidna (*Zaglossus hacketti*), reached a weight of perhaps 66 pounds (30 kg), almost ten times the weight of the living short-beaked echidna (*Tachyglossus aculeatus*). The genus *Zaglossus* now survives only in remote high-altitude forest in New Guinea, and the living species have a maximum weight of about 33 pounds (15 kg). The reptilian fauna included several large carnivores: the giant goanna (*Megalania priscus*), a large terrestrial crocodile (*Quinkana fortirostrum*), and a giant serpent (*Wonambi naracoortensis*). Australian paleontologists have debated the extent to which these reptiles were ecologically significant as predators, in comparison with mammals such as the marsupial lion; in a 2002 article, Stephen Wroe made a strong case for mammals as the dominant predators of Pleistocene Australia. In addition to these carnivorous reptiles, two species of herbivorous land turtles of the genus *Meiolania* lived in Australia. The best-known species, *M. platyceps*, was more than 6.6 feet (2 m) long and bore horns on its head and a bony club on its tail. The largest bird in Pleistocene Australia was *Genyornis newtoni*, a flightless bird perhaps three times the weight of the emu. This was the last representative of the dromornithids, a family known as mihirungs (after a giant bird from Aboriginal mythology) and related to geese.

Extinctions in the Late Pleistocene were strongly size selective, removing all species above an average body mass of about 88 pounds (40 kg) in both Australia and New Guinea (see Figure 1). In this respect, the Australian megafaunal extinctions resemble the wholesale extinctions of very large vertebrates that happened in other parts of the world during the last 50,000 years (Koch and Barnosky 2006). In Australia their effect was to remove about one-fourth of all marsupials and half the monotremes, with

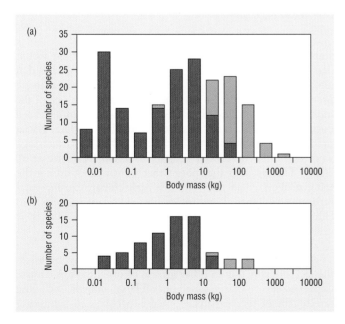

Figure 1. Body-mass distributions of land mammals (excluding bats) that were present in (a) Australia and (b) New Guinea during the Pleistocene, distinguishing species that survived into the Holocene (orange shading) from those that did not (beige). Reproduced by permission of Gale, a part of Cengage Learning.

significant, but proportionally much smaller, impacts on reptiles and birds.

Hypothesized causes of the Australian Pleistocene extinctions

The question of what caused these extinctions has generated one of the longest and, at times, most intense debates in Australian science (Johnson 2006). The three contending hypotheses are: (1) the arrival of people caused the extinction of large vertebrates as a result of overhunting; (2) fires set by early human populations caused such dramatic environmental change that they destroyed the habitat of large animals and indirectly caused their extinction; or (3) a deteriorating climate caused a gradual decline in megafaunal abundance and diversity, culminating in the final disappearance of these animals during the last glacial cycle.

Between 2000 and 2012, most evidence supported the so-called overkill hypothesis—that hunting by newly arrived human populations was responsible. This hypothesis argues that megafaunal populations were particularly susceptible to overhunting, mainly because large-bodied animals typically have low reproductive rates so that even quite low rates of hunting can hold mortality rates above population replacement rates. Also, large animals typically live on the ground and in open habitats, and this means the megafauna would have been highly exposed to human hunters. The arrival of people would, therefore, cause a differential loss of the largest animals. Other relevant factors are that people might have targeted larger animals because they provided a greater reward, and those animals could well have been naive to human hunters, but these postulates are not essential to the overkill model.

The timing of these extinctions is crucial to testing hypotheses of extinction. Thus, a key question concerns whether the megafaunal extinctions were correlated with human arrival. The date of the arrival of people in Australia has been almost as controversial as the cause of the megafaunal extinctions, but a reasonable consensus view of the archaeological evidence is that people were definitely in Australia by 45,000 years ago and had occupied the whole continent by 40,000 years ago, possibly several thousand years earlier (Hiscock 2008). Several comprehensive dating studies have shown that megafaunal species disappeared around 45,000 years ago (see Figure 2; Grün et al. 2008; Miller et al. 2005; Roberts et al. 2001; Rule et al. 2012). This evidence supports a link between human arrival and mass extinction, but it remains controversial because for many species few or no fossils have been dated precisely, so it is not possible to show they were extant when humans reached Australia. This leaves open the possibility that the greater part of the disappearance of the megafauna resulted from a prolonged series of extinctions rather than a synchronized event precipitated by human arrival (Field et al. 2008). Evidence for extinction close to 45,000 years ago is strongest for species that are common as fossils and for which researchers have amassed large collections of dates (Johnson 2006).

Another point of contention with regard to the overkill hypothesis concerns the archaeological evidence for the hunting of extinct megafauna. There is little such evidence, and no accumulations of bones at ancient occupation sites that could indicate butchery of large numbers of extinct animals. This could be explained, however, by two factors. First, although there is sufficient archaeological evidence to make it clear that people were present in Australia when the megafauna went extinct, there still are few archaeological sites from the first few thousand years of human settlement of Australia, and thus few opportunities to find evidence of hunting. Second, because of the great age of those sites, most contain very little material to reveal what the people who made them were doing. It is possible that those people were hunting large animals, but that the evidence of that activity has not survived (Johnson 2006).

The fire hypothesis envisages that fires set by newly arrived people simplified the structure and composition of vegetation, and in particular removed a complex layer of shrubs and small trees that was important to large browsing animals (Miller et al. 2005). This hypothesis has been refuted by three pieces of research. First, a meta-analysis of data on changes in fire regimes in prehistoric Australia (indicated by charcoal in sediments) concluded that human arrival did not cause a consistent increase in burning across the continent (Mooney et al. 2010). Second, a reconstruction of ecological changes related to megafaunal extinction at a well-studied site (Lynch's Crater in tropical northeastern Australia), where burning evidently did increase soon after human arrival, showed that this increase followed megafaunal extinction rather than preceding it as required by the fire hypothesis (Rule et al. 2012). Third, detailed ecological reconstructions of extinct species show that some of those that went extinct around 45,000 years ago actually occupied nonflammable vegetation types and so ought not to have been exposed to the effects of increased burning elsewhere, assuming such increases happened (Prideaux et al. 2009).

The climate hypothesis suggests that the megafaunal extinctions were a result of a long-term deterioration of the climate of Australia, superimposed on the glacial climate cycles (each of which lasted about 130,000 years) and extending over the last few hundred thousand years. This, it is argued, caused a gradual reduction in abundance and distribution of megafaunal species, while also generating some abrupt declines during episodes of severe climate stress at the extreme points of the glacial cycles. The climate trend was toward generally cooler, and therefore drier conditions, which reduced the productivity of vegetation and also constrained the distribution of large animals by reducing the availability of water. The latest of the episodes of severe stress was the last glacial maximum, which occurred between 35,000 and 12,000 years ago. This picture of gradual decline of the megafaunal assemblage, with the extinction of the last populations of the last species occurring during the last glacial maximum, is not supported by the dating studies referred to above and illustrated in Figure 2, which suggest abrupt extinction around 45,000 years ago. At that time the climate of Australia was relatively benign (Johnson 2006). Two records of relative abundance of megafauna through the last 130,000 years—large collections of eggshells of the extinct giant bird *Genyornis newtoni* (Miller et al. 2005) and spores of dung fungi indicating activity of large herbivores (Rule et al. 2012)—suggest no effects of climate change on megafaunal

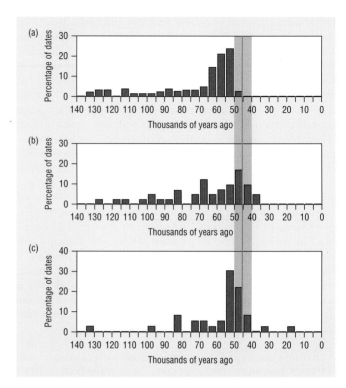

Figure 2. Distribution of dates on remains of extinct megafaunal taxa from the last glacial cycle, from three studies: (a) dates on egg shells of the bird *Genyornis newtoni* (Miller et al. 2005); (b) direct dates on teeth of marsupial megafauna (Grün et al. 2008); and indirect dates, from sediments associated with remains, or marsupial megafauna (Roberts et al. 2001). The vertical line is the estimate of the extinction date for all species combined, with the 95% confidence range shaded, estimated by Roberts et al. (2001) from their data using a statistical procedure that takes account of measurement error of individual dates and sampling error across the whole collection of dates. Reproduced by permission of Gale, a part of Cengage Learning.

abundance during that interval, while also confirming abrupt decline around 45,000 years ago. The megafaunal species that disappeared during the last glacial cycle had survived many previous glaciations. The last glacial maximum was one of the more severe episodes of aridity that Australia experienced during the Pleistocene, but not uniquely so. The glacial maxima that occurred around 440,000 and 635,000 years ago were of similar or greater severity (Masson-Delmotte et al. 2010). The fact that the species that went extinct during the last glacial cycle had survived those early climate extremes and remained widespread and (evidently) abundant suggests that climate stress was not a significant part of the cause of their extinction.

Another piece of evidence against the climate hypothesis is the lack of extinction in other groups of organisms. A climate impact so severe that it wiped out all large animals, including one-fourth of the marsupials, ought to have had significant impacts on other groups as well. Allowing for the poor fossil record in many groups, there is no evidence for this. In particular, the contrast in plant extinction between the Early and Late Pleistocene is instructive. At the beginning of the Pleistocene there was a pulse of plant extinctions, especially at regional scales, which can be attributed to the onset of the climate cycles that characterize the Pleistocene and that brought episodes of drier and cooler conditions exceeding those of the preceding Pliocene (Sniderman 2011). But no comparable pulse of plant extinctions has been detected near the end of the Pleistocene, when the megafauna went extinct (Kershaw et al. 2000).

Holocene extinctions: Marsupial carnivores on mainland Australia

Scientists have not uncovered evidence of any species that went completely extinct during the Holocene (i.e., from the beginning of the current interglacial approximately 10,000 years ago) to European settlement of Australia in the late eighteenth century. Three species, however, went extinct on mainland Australia and persisted only on the large island of Tasmania: the thylacine or Tasmanian tiger (*Thylacinus cynocephalus*), the Tasmanian devil (*Sarcophilus harrisii*), and the Tasmanian native hen (*Gallinula mortierii*). These extinctions are significant because of the ecological implications of the loss of two large carnivores from mainland Australia and because the extinction of the thylacine from the mainland set the stage for the complete extinction of that species in recent times.

In the Late Pleistocene both the thylacine and the Tasmanian devil were widespread over mainland Australia and were significant members of the carnivore assemblage. The thylacine was a medium-large marsupial (weighing about 66 pounds [30 kg]) with classically wolflike features of the skull and teeth: a long face, well-developed canines, and molars highly modified to function as shearing blades. Reconstruction of its diet by interpretation of its morphology and the piecing together of fragmentary observational evidence of Tasmanian animals suggests that the thylacine was a specialized ambush predator that killed live prey, mostly

A Tasmanian tiger, or thylacine (*Thylacinus cynocephalus*) in captivity, circa 1930. The species is believed to have become extinct in the early twentieth century. © Popperfoto/Getty Image.

smaller than itself—typically medium-sized kangaroos and wallabies. The Tasmanian devil is also a predator, but by virtue of its robust and powerful jaws and broad molars, it is also a formidable scavenger capable of devouring all parts of the carcasses of mammals much larger than itself.

The thylacine went extinct on mainland Australia just over 3,000 years ago; the Tasmanian devil disappeared at around the same time or perhaps more recently (the youngest dated remains of Tasmanian devils on mainland Australia are 400 years old, but that date is considered unreliable). The earlier date of 3,000 years ago coincides with the arrival in Australia of the dingo (*Canis lupus dingo*). The dingo was brought to Australia by Asian seafaring people perhaps 4,000 years ago. It was a semi-domesticated animal little modified from ancestral wolf populations, and it quickly established wild populations across the whole of mainland Australia but did not reach Tasmania. Both a predator and scavenger, the dingo would have posed a competitive challenge to both the thylacine and the Tasmanian devil. For these reasons, the idea that the dingo caused the extinction of the thylacine and the Tasmanian devil on mainland Australia is widely accepted.

There is, however, an alternative hypothesis (Johnson 2006). Evidence exists for a dramatic increase in the human population about 4,000 years ago, an increase associated with improvements in hunting technology, which happened on mainland Australia, but not Tasmania. The effect of this population change would have been to impose higher hunting pressure on both the thylacine and the Tasmanian devil, as well as on their prey. The combined effects might have caused or at least contributed to their extinction. Dingoes would have been buffered from these pressures because, in addition to existing in the wild, they were also maintained as companion animals by aboriginal people. Further research will be needed to determine which of these two hypotheses is correct, but there is intriguing support from ancient rock art for the idea that humans hunted thylacines to extinction. Thylacines are commonly depicted in rock art in northern Australia, and some painting show that people hunted them. Dingoes were also painted, but they appear in art styles that are thought to be more recent than those that include thylacines, suggesting that at least in the far north thylacines disappeared before dingoes arrived

Recent extinctions

Animal species that have gone extinct in Australia since the early nineteenth century, as recognized under the Commonwealth of Australia's Environment Protection and Biodiversity Conservation Act of 1999, are listed in Table 1. Just as the extinctions of the Late Pleistocene were highly selective, removing all the largest vertebrates, so these recent extinctions have also been highly unbalanced. By far the highest rate of loss has been among mammals, and there have been four frog extinctions. The mammal and frog extinctions have been concentrated on mainland Australia, whereas only one bird and no reptile species have gone extinct from the mainland. Another eight bird species have been lost from islands. Only one extinction of an invertebrate animal is recognized, although it is quite likely that other invertebrates have gone

Scientific name	Common name
Mammals:	
Caloprymnus campestris	Desert rat-kangaroo
Chaeropus ecaudatus	Pig-footed bandicoot
Conilurus albipes	White-footed rabbit-rat
Lagorchestes asomatus	Central hare-wallaby
Lagorchestes leporides	Eastern hare-wallaby
Leporilus apicalis	Lesser stick-nest rat
Macropus greyi	Toolache wallaby
Macrotis leucura	Lesser bilby
Notomys amplus	Short-tailed hopping-mouse
Notomys longicaudatus	Long-tailed hopping-mouse
Notomys macrotis	Big-eared hopping-mouse
Notomys mordax	Darling Downs hopping-mouse
Nyctophilus howensis	Lord Howe long-eared bat*
Onychogalea lunata	Crescent nail-tail wallaby
Perameles eremiana	Desert bandicoot
Pipistrellus murrayi	Christmas Island pipistrelle*
Potorous platyops	Broad-faced potoroo
Pseudomys gouldii	Gould's mouse
Rattus macleari	Maclear's rat, Christmas Island rat*
Rattus nativitatis	Christmas Island rat, bulldog rat*
Thylacinus cynocephalus	Thylacine
Birds:	
Aplornis fusca	Tasman starling*
Dromaius ater	King Island emu*
Dromaius baudinianus	Kangaroo Island emu*
Gerygone insularis	Lord Howe warbler*
Nestor productus	Norfolk Island kaka*
Porphyrio albus	White gallinule*
Psephotus pulcherrimus	Paradise parrot
Zosterops alboqularis	White-chested white-eye*
Zosterops stenuus	Robust white-eye*
Frogs:	
Rheobatrachus silus	Southern gastric-brooding frog
Rheobatrachus vitellinus	Northern gastric-brooding frog
Taudactylus acutirostris	Sharp-snouted day frog
Taudactylus diurnus	Southern day frog
Invertebrates:	
Hypolimnus pedderensis	Lake Pedder earthworm

*Species that occurred only on islands.

Table 1. Recently extinct species of Australian animals (Source: Environment Protection and Biodiversity Conservation Act 1999 List of Threatened Fauna,;the Christmas Island Pipistrelle went extinct in 2009, and has not yet been added to the EPBC list). Reproduced by permission of Gale, a part of Cengage Learning.

extinct without notice. That one case was an earthworm (*Hypolimnus pedderensis*) known only from the shores of Lake Pedder in southwestern Tasmania, an animal not seen since its habitat was inundated by the creation of a hydroelectric project at the lake in 1972.

Mammals

The most celebrated of the recent extinctions was of the thylacine in Tasmania. The last confirmed record of the thylacine in the wild was made in 1930, and the last captive animal died in 1936 (ironically, the same year that the species was given legal protection), but the species' fate was probably sealed around 1905 when bounty records indicate that numbers dropped to critically low levels. There could have been three main causes for that decline: hunting by people, which was motivated by the perception that thylacines threatened sheep (although historical research suggests this was not true) and encouraged by a bounty scheme; reduction of prey populations stemming from the expansion of the sheep industry; and a disease outbreak. Of these, there is least support for disease, because there is no direct evidence for a disease outbreak at the time. It seems most likely that a combination of the direct effects of persecution of thylacines and the indirect effects of suppression of their prey populations accounts for the extinction of the thylacine.

Apart from the thylacine, the other mammal extinctions followed a remarkably consistent pattern (Johnson 2006). They predominantly affected species of medium body size, between about 1.2 ounces (35 g) and 12 pounds (5.5 kg), and mostly from relative dry and open habitats of the arid, semiarid, and temperate zones of southern Australia (see Table 1). The species most likely to be affected were ground-dwelling forms that did not shelter in rock piles or burrows. There was a tendency for small species to have gone extinct earlier, beginning in the 1850s, than larger species, many of which disappeared during the 1950s and 1960s, and the extinctions tended to progress from the southeastern to the western portions of the continent, before finally reaching the central and western deserts.

All these factors implicate two invasive predators, the feral cat (*Felis catus*) and the red fox (*Vulpes vulpes*), as the main causes of extinction. This conclusion is supported by several other lines of evidence. In many cases, species that declined on mainland Australia survived on nearshore islands without cats or foxes (and are therefore not listed among the casualties in Table 1); the control of cats and foxes on mainland Australia has allowed the recovery of threatened species similar to those that went extinct; and in the absence of effective control, attempts to reintroduce such species have failed because of high rates of predation by foxes and cats.

Why have foxes and cats had such a devastating impact on native Australian mammals? There appear to be several reasons. First, Europeans introduced not just foxes and cats to Australia, but also additional prey species, such as the house mouse (*Mus musculus*) and (most importantly) the European rabbit (*Oryctolagus cuniculus*). These are highly fecund species capable of living at very high densities. They can thus maintain fox and cat populations at abundances so high that they overwhelm native species that are typically not so abundant. Medium-sized predators such as foxes and cats are typically strongly suppressed by larger predators. But on mainland Australia the larger native predators, the thylacine and the Tasmanian devil, were removed in the Holocene and replaced by the dingo. There is good evidence that dingoes do suppress foxes and cats, but dingoes were heavily controlled and eliminated in some areas to protect livestock, allowing fox and cat populations to reach very high density. Third, the native mammals of Australia may have been intrinsically vulnerable to predation by foxes and cats. Marsupials and the native rodents of Australia tend to have low reproductive rates, lower than that of foxes and cats. Thus, the capacity of their populations to replace animals killed by predators is comparatively low. It is possible that to some extent Australia's marsupials and rodents were behaviorally naive to placental predators, although this seems less likely, as they had previously been exposed to predation by the dingo as well as by other marsupial predators. A final factor that could well have amplified the impacts of foxes and cats is modification of habitat by fire and livestock grazing. While habitat destruction per se was not a significant factor in most of these mammal extinctions, the simplification of habitat by fire and grazing removed cover and small-scale refuges from predators, thereby increasing the exposure of small and medium-sized mammals to predation.

Frogs

The four extinct frogs have all disappeared since 1970. The loss of the two species of gastric-brooding frogs is especially to be lamented, as they possessed one of the most remarkable breeding adaptations known for any amphibian. During reproduction, the female swallowed her fertilized eggs or early larvae, which then secreted hormones to suppress production of digestive secretions in the stomach so that it functioned as a brood pouch. Fully developed metamorphs then emerged from the mother's mouth after six or seven weeks, whereupon the stomach resumed its function as a digestive organ. The cause of extinction of these four species is not completely certain, but it is highly likely to have been the chytrid fungus *Batrachochytrium dendrobatidis*.

The life history and impacts on frogs of *B. dendrobatidis* are understood from study of its interaction with species that have declined as a result of infection but not gone extinct (Alford 2010). The fungus lives in the epidermis of amphibians. It spreads from host to host by dispersal of zoospores in water; in addition, the intensity of infection on individual hosts is increased by zoospores that reinvade the epidermis. There is no intermediate host, and the fungus has a broad host range. Thus, the species that are able to tolerate the infection act as environmental reservoirs of the disease and spread it to highly susceptible species. It causes death in susceptible species by upsetting the balance of electrolytes and causing heart failure (Voyles et al. 2009). A widely distributed organism, *B. dendrobatidis* has caused severe population crashes and extinctions in other parts of the world, especially Central America. The extent to which its global distribution is natural or the result of recent expansion is unclear. It seems likely, however, that it introduced to Australia within the past 30 years.

The impact of *B. dendrobatidis* on Australian frogs is strongly dependent on environmental conditions. In the tropical forests of northeastern Queensland, seven species have suffered severe declines above an elevation of 1,300 feet (400 m), but low-elevation populations of the same species have been unaffected; in addition, prevalence has been shown to increase in winter. This is attributable to the direct effects of temperature on the growth and reproduction of the fungus, which means that it grows best between 50 and 77°F

(10–25°C). Consequently, frog species that live in streams in cool environments are most likely to decline because of *B. dendrobatidis*.

Birds

The only bird extinction from mainland Australia is that of the paradise parrot (*Psephotus pulcherrimus*). This beautiful parrot was once quite widely distributed in open woodlands in central and southern Queensland. It fed on grass seeds and reproduced by excavating tunnels into termite mounds where eggs were laid and brooded (presumably, the birds selected termite mounds because of the optimal temperature they provided). The species was last recorded in the 1920s. Collecting for aviaries may have contributed to its decline, but loss of food resources resulting from livestock grazing and fire, and perhaps habitat destruction, was also important. The disappearance of the paradise parrot is one instance of a general decline of seed-eating, ground-dwelling birds from the subtropical and tropical woodlands and grasslands of northern Australia, although none of the other birds has yet gone extinct. The Kangaroo Island and King Island emus were distinctive, small emus that lived on large continental islands close to the Australian coastline. Both emus went extinct soon after those islands were occupied by Europeans. Hunting for food caused the extinction of the King Island emu (*Dromaius ater*) and contributed to the extinction of the Kangaroo Island emu (*Dromaius baudinianus*), although it is thought that destruction of habitat by fire also contributed in the latter case. A similar emu occurred on Tasmania and was also hunted to extinction very soon after British settlement, but that taxon is classified as a subspecies of the living emu of mainland Australia (*Dromaius novaehollandiae*). The other extinct bird species occurred on Norfolk Island or Lord Howe Island. These islands lie in the Pacific Ocean beyond the Australian continental shelf, but are administered as part of Australia's national territory. Neither island had been settled permanently by Polynesians, and, like other pristine Pacific Islands, each had a complement of endemic species of land birds. The extinction of these birds stemmed from a combination of hunting by people (in the case of the larger species) and impacts of invasive species, especially rats and feral pigs.

Overall characteristics of the extinctions

Australia has lost many species since the first arrival of Aboriginal people around 50,000 years ago. There have been two general features of this history of extinctions. First, they have been highly selective, removing distinct components of the original fauna with remarkable precision. The Pleistocene extinctions removed all large land-dwelling vertebrates, in the process eliminating many large herbivorous mammals; the ecological effects of this removal of herbivores on vegetation dynamics are not well understood, but they are likely to have been profound (Rule et al. 2012). In the Holocene that process was extended by the elimination of midsized predators from mainland Australia. The major losses from recent times were of medium-sized ground-dwelling mammals from drier habitats. Across large tracts of inland Australia, essentially all such mammals vanished. This probably also had significant ecological effects, because this assemblage included many species that foraged by turning over soil, promoting soil health and dispersing seeds and the spores of mycorrhizal fungi as they did so (Johnson 2006).

The second characteristic feature is the extent to which extinctions have followed from the direct effects of invasive species, as predators or (in the case of frogs) disease organisms. This includes people as predators during the Pleistocene and to a lesser extent in recent times, as well as predators introduced by people, especially the red fox and the feral cat. The high selectivity of extinctions in Australia is largely a function of the native species' differential susceptibility and exposure to those predators (and disease). The recent impact of invasive species on the native biota has been far greater in Australia than on any other continent, but the impacts of other processes, especially habitat destruction, is likely to increase in the future, given that many species now listed as endangered or vulnerable are threatened by habitat loss.

Resources

Books

Alford, Ross A. "Declines and the Global Status of Amphibians." In *Ecotoxicology of Amphibians and Reptiles*, 2nd ed., edited by Donald W. Sparling, Greg Linder, Christine A. Bishop, and Sherry K. Krest, 13–45. Boca Raton, FL: CRC Press; Pensacola, FL: Society of Environmental Toxicology and Chemistry, 2010.

Hiscock, Peter. *Archaeology of Ancient Australia*. London: Routledge, 2008.

Johnson, Chris. *Australia's Mammal Extinctions: A 50,000 Year History*. Cambridge, U.K.: Cambridge University Press, 2006.

Kershaw, A.P., P.G. Quilty, B. David, et al. "Palaeobiogeography of the Quaternary of Australia." In *Palaeobiogeography of Australasian Faunas and Floras*, edited by A.J. Wright, G.C. Young, J.A. Talent, and J.R. Laurie, 471–515. Canberra: Association of Australasian Palaeontologists, 2000.

Long, John A., Michael Archer, Timothy Flannery, and Suzanne Hand. *Prehistoric Mammals of Australia and New Guinea: One Hundred Million Years of Evolution*. Sydney: University of New South Wales Press, 2002.

State of the Environment Committee. *Australia State of the Environment 2011*. Canberra: Commonwealth of Australia, Department of Sustainability, Environment, Water, Population, and Communities, 2011.

White, M.E. "Environments of the Geological Past." In *Evolution and Biogeography of Australasian Vertebrates*, edited by J.R. Merrick, M. Archer, G.M. Hickey, and M.S.Y. Lee, 17–50. Oatlands, Australia: Auscipub, 2006.

Periodicals

Field, Judith, Melanie Fillios, and Stephen Wroe. "Chronological Overlap between Humans and Megafauna in Sahul (Pleistocene Australia–New Guinea): A Review of the Evidence." *Earth-Science Reviews* 89, nos. 3–4 (2008): 97–115.

Grün, R., R. Wells, S. Eggins, et al. "Electron Spin Resonance Dating of South Australian Megafauna Sites." *Australian Journal of Earth Sciences* 55, nos. 6–7 (2008): 917–935.

Koch, Paul L. and Anthony D. Barnosky. "Late Quaternary Extinctions: State of the Debate." *Annual Review of Ecology, Evolution, and Systematics* 37 (2006): 215–250.

Masson-Delmotte, V., B. Stenni, K. Pol, et al. "EPICA Dome C Record of Glacial and Interglacial Intensities." *Quaternary Science Reviews* 29, nos. 1–2 (2010): 113–128.

Miller, Gifford H., Marilyn L. Fogel, John W. Magee, et al. "Ecosystem Collapse in Pleistocene Australia and a Human Role in Megafaunal Extinction." *Science* 309, no. 5732 (2005): 287–290.

Mooney, S.D., S.P. Harrison, P.J. Bartlein, et al. "Late Quaternary Fire Regimes of Australasia." *Quaternary Science Reviews* 30, nos. 1–2 (2010): 28–46.

Prideaux, Gavin J., Linda K. Ayliffe, Larisa R.G. DeSantis, et al. "Extinction Implications of a Chenopod Browse Diet for a Giant Pleistocene Kangaroo." *Proceedings of the National Academy of Sciences of the United States of America* 106, no. 28 (2009): 11646–11650.

Roberts, Richard G., Timothy F. Flannery, Linda K. Ayliffe, et al. "New Ages for the Last Australian Megafauna: Continent-Wide Extinction about 46,000 Years Ago." *Science* 292, no. 5523 (2001): 1888–1892.

Rule, Susan, Barry W. Brook, Simon G. Haberle, et al. "The Aftermath of Megafaunal Extinction: Ecosystem Transformation in Pleistocene Australia." *Science* 335, no. 6075 (2012): 1483–1486.

Sniderman, J.M. Kale. "Early Pleistocene Vegetation Change in Upland South-Eastern Australia." *Journal of Biogeography* 38, no. 8 (2011): 1456–1470.

Voyles, Jamie, Sam Young, Lee Berger, et al. "Pathogenesis of Chytridiomycosis, a Cause of Catastrophic Amphibian Declines." *Science* 326, no. 5952 (2009): 582–585.

Wroe, Stephen. "A Review of Terrestrial Mammalian and Reptilian Carnivore Ecology in Australian Fossil Faunas, and Factors Influencing Their Diversity: The Myth of Reptilian Domination and Its Broader Ramifications." *Australian Journal of Zoology* 50, no. 1 (2002): 1–24.

Wroe, Stephen. "Cranial Mechanics Compared in Extinct Marsupial and Extant African Lions Using a Finite-Element Approach." *Journal of Zoology* 274, no. 4 (2008): 332–339.

Other

Commonwealth of Australia. Department of Sustainability, Environment, Water, Population, and Communities. "Environment Protection and Biodiversity Conservation (EPBC) Act List of Threatened Fauna." Last modified November 26, 2009. http://www.environment.gov.au/cgi-bin/sprat/public/publicthreatenedlist.pl.

Chris Johnson

Madagascar (2,000 years ago)

The story of extinction in Madagascar is an important part of the global story of extinction in the wake of human migration. Giant birds, reptiles, and mammals, including many species of large lemurs, have disappeared from Madagascar over the last 2,000 years following human arrival. As the last place on Earth where a large-animal fauna (megafauna) with continental-level diversity disappeared after human colonization, this California-sized island in the western Indian Ocean provides important clues to how humans can transform ecosystems quickly, whether that is their intention or not.

Debate continues as to the possible role factors other than humans, such as climate and disease, may have played in Madagascar's catastrophic extinction. More recent topics for research and debate have focused on how and when humans might have caused these extinctions, which human-related factors were most important, and how they affected each other. After human arrival just over two millennia ago, the fossil record shows a decline of the large animals, probably through human hunting over the ensuing centuries. This decline is followed by an increase in the density of flammable vegetation. Subsequently, fires and species introduced by humans took advantage of the new disturbance, finishing the job over the ensuing centuries. By the time the first Europeans recorded observations in Madagascar in the early 1600s, it is likely that most of the larger animals were already extinct. Extinction and environmental change continue to plague this beleaguered island nation, regarded as one of the most important biodiversity hotspots on the planet.

The world's strangest giants

In modern times the world's largest oceanic island, Madagascar has apparently not been in contact with a continent in the last 100 million years or more of its geological history. Since its initial separation from Africa with the breakup of Gondwana, Madagascar has become increasingly isolated, with India fully separating from the other side. Both the increasing distance to Africa and the decreasing likelihood of transport by winds and sea currents from that direction (Samonds et al. 2012) have profoundly isolated "the eighth continent," as Madagascar is sometimes called. Madagascar's animals were a remarkable bestiary that included giant tortoises, elephant birds, pygmy hippopotami, and a great diversity of large lemurs unlike any other living creatures. Apparently ruminants and other familiar hoofed animals of Africa never reached Madagascar. Instead many of Madagascar's animals came by swimming, flying, or rafting on floating debris. Among mammalian carnivores, only the mongoose is represented. Members of this family are usually small, but one extinct species, *Cryptoprocta spelaeo*, reached the size of some of the larger cats of continental areas. The elephant birds included the largest bird that ever lived, the half-ton *Aepyornis*, and a smaller genus, *Mullerornis*, about the size of its African cousin, the ostrich (*Struthio camelus*). Scientists recognize two to four species of hippopotami, ranging from one the size of the African hippopotamus down to cow-sized pygmies. One of the strangest animals of all was *Plesiorycteropus*, a unique creature, vaguely similar to the aardvark of Africa, but regarded by the paleontologist Ross D.E. MacPhee as quite distinct and perhaps related to long-extinct condylarths. In a 1994 work, MacPhee placed this extinct oddity in an order of mammals, Bibymalagasia, all by itself.

For primatologists, Madagascar's megafaunal extinction is the world's most special case, as it is the only megafaunal extinction on the planet that was dominated, in terms of diversity at least, by large primates. As of 2012, paleontologists recognized 18 species of extinct lemuroids that were larger than any living lemur. These fall into five families and include animals too strange to imagine fully. *Archaeoindris* was gorilla-sized and probably fed like a ground sloth, breaking down limbs and small trees. *Megaladapis* was a careful climber like the orangutan. Its elongated face and very long fingers and toes would have perhaps suited the slow lifestyle of the Australian koala, but this was a creature ranging to the size of a female gorilla. *Palaeopropithecus* also has no living analog. This chimpanzee-sized lemuroid probably hung upside down like a tree sloth. One of the strangest of all would have been the giant aye-aye (*Daubentonia robusta*), a much larger version of the living creature with its long middle finger and huge incisors. One living analog is the Australian marsupial *Dactylopsila*, the striped possum (a peturid) with an elongate fourth manual digit and gnawing incisors. It is believed that,

Megaladapis was a giant lemur that probably moved slowly about in the trees in a manner similar to the koala of Australia. Illustration by Francois Desbordes.

through their remarkable adaptations, living aye-ayes (*Daubentonia madagascariensis*) occupy the same niche filled by large woodpeckers in other lands. Each of these remarkable giant lemurs was likely to have exhibited behaviors and filled ecological niches about which scientists know almost nothing.

Archaeolemur and the related *Hadropithecus* were two genera of large lemurs that might be thought of as somewhat like large monkeys or baboons. Remains of these so-called monkey lemurs have been studied by collaborating primatologists with fascinating results. Reconstructions of the hands and feet of *Archaeolemur*, for instance, by William L. Jungers and colleagues (2005) have shown them to have been capable of efficient locomotion both on the ground and in the trees, somewhat paralleling humanity's own distant ancestors. Studies of their coprolites—fossil fecal pellets—shows that they were omnivorous, feeding on fruits and seeds, insects, frogs, and very small mammals (Burney et al. 1997). *Archaeolemur* may have frequented caves, like humanity's own ancestors.

Hadropithecus, the other genus of monkey lemurs, has always been somewhat mysterious to primatologists, both because of the lack of some skeletal elements in collections and also because it possessed sheep-like molars, high-ridged

teeth suggesting a specialized diet of coarse plants. Missing parts of the skeleton, and even missing parts of the skull that serves as the museum-type specimen, were found in 2004 in Andrahomana Cave on the southeastern coast of the island (Godfrey et al. 2006). A team of experts in computer simulation were able to put the newly found pieces of the skull back together in cyberspace (Ryan et al. 2008). Studies of

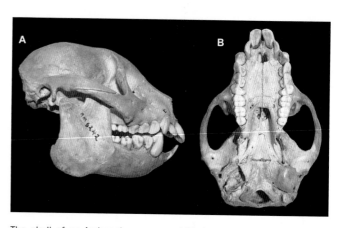

The skull of an Archaeolemur, one of Madagascar's now-extinct giant lemurs. Courtesy of William Jungers.

the tooth wear and the isotopic composition of *Hadropithecus* fossils have suggested that, indeed, this strange creature was capable of eating very hard or coarse plant materials. Among human ancestors, robust australopithecines may have occupied a similar niche in the African savannas, across the Mozambique Channel.

It is not easy to imagine the landscapes that all these strange giants shaped with their grazing and browsing. Hungry tortoises must have made short work of vegetation within reach, half-ton elephant birds could have reached over 10 feet (3.3 m) into the trees for their browse, and hippos were key grazers as they are within African savannas and riverine forests.

Paleoecologists, who study past environments through excavation and sediment coring, are trying to reconstruct the setting and events of this lost time by indirect evidence and reference to related species and systems elsewhere. By carefully collecting each layer of sediment, scientists can date the materials and identify even microscopic fossils for clues to past environments and changes that occurred before and after human presence. Studies of fossil pollen, charcoal particles, and other microfossil evidence (summarized in Burney et al. 2004) show that the environments of Madagascar's interior before human arrival were diverse systems with a mixture of trees, shrubs, and grasses most similar to mosaic-like environments in parts of northern Madagascar today, where riverine forest, wooded grasslands, and shrub vegetation share the landscape. Much of Madagascar's interior in the early twenty-first century is an eroding grassland with few native species, but central fossil sites such as Ampasambazimba contain seeds and pollen of a wider range of plants when the big animals were there. So what happened?

What happened and why it matters

Humans had found most of the larger islands on Earth well before they found Madagascar. The human species spread from its native homeland—Africa and southern Asia—by reaching Australia and New Guinea over water roughly 50 millennia ago. By 35,000 years ago, people had spread northward in Eurasia and onto Japan, probably over sea ice from Siberia. At the end of the last ice age roughly 15,000 years ago, or perhaps earlier according to some scientists, humans made it to the Americas. Amerindians reached the large remote islands of the West Indies 5,000 years ago or more. Sailors were wandering the Mediterranean perhaps 4,000 years ago, discovering strange new islands with bizarre creatures, as recounted in the *Odyssey* of Homer and other ancient texts. Toward the end of this first period of expansion to islands, Southeast Asian people of the Austronesian language group spread out via advanced sailing craft from the Indonesian region eastward and westward. To the east, humans spread all the way across the Pacific Ocean to reach, less than 1,000 years ago, remote eastern Polynesia, Hawaii, Rapa Nui (Easter Island), and finally New Zealand well to the south. They spread westward around the north end of the Indian Ocean to find, perhaps around the fourth century BC—or maybe later according to some archaeologists—the great island of Madagascar.

Mother and baby Coquerel's sifaka (*Propithecus coquereli*) near Anjohibe Cave in northwestern Madagascar. Courtesy of David Burney.

Much of what happened over the next two millennia has been reconstructed almost entirely from indirect sources: the layers of sediments in lake bottoms, caves, and archaeological sites, and the bones, shells, and dung of the lost creatures themselves, preserved as subfossils, or fossils that are only centuries to a few millennia old and are largely unaltered by subsequent aging processes. With great care in dating these materials by radiocarbon and occasionally other methods, it has been possible to reconstruct a sequence of events in Madagascar that has been variously explained by hypotheses of human overhunting, fire, disease, and climate change. At least 14 major hypotheses have been put forward globally to explain the catastrophic extinction of large animals on the continents and islands over the last 50 millennia.

A place "beyond the Mountains of the Moon" (East Africa) that supported giant birds was mentioned by the Greek historian Herodotus about 400 BC, recounting stories he had heard from Egyptian priests. Radiocarbon dates on bones of extinct megafauna apparently butchered by humans date to about the same time. Also about this time pollen records begin to show the presence of a plant (*Cannabis*) believed to have been introduced by people from Southeast Asia. By no coincidence, the native language of the present-day Malagasy people of Madagascar is most similar to languages spoken in the highlands of modern Indonesia. Thus, as Austronesian speakers spread from Indonesia eastward into Polynesia, they apparently also crossed to the other side of the Indian Ocean and colonized the huge island of Madagascar. There they found a megafauna that undoubtedly had no initial fear of them, because humans were not previously known there as predators. Although actual human occupation sites are known only from the archaeological record of the first millennium, butchered bones of extinct giant lemurs and hippos suggest the megafauna was being affected by early hunters even earlier (Burney et al. 2004).

Sediment cores for pollen analysis contain other clues to prehistoric human activity, and paleoecologists and their collaborators have conducted this type of work since the early 1980s in areas of Madagascar ranging from the driest parts of the southwest to some of the wettest rain-forest areas of the island's north and east sides. A transect of sediment cores from some of the driest areas of Madagascar to some of the wettest, spanning the entire length of the island, was analyzed for evidence of fires. Although some areas in the highlands were burning even before human arrival through natural agencies such as lightning and volcanic activity, nearly all sites show a big increase in charcoal that apparently corresponds with the local advent of human settlers. The pattern is one of increasing fires in the driest areas of southwestern Madagascar, where the early butchery evidence has been dated, with the fires spreading into the interior and up the west coast over the next thousand years or so. One of the most significant clues as to what happened, and in what order, came from studies of the occurrence of fungal spores (*Sporormiella*) in these sediment cores (Burney et al. 2003). These types of fungal spores have been shown to occur in North America, New Zealand, and elsewhere at high rates in areas with many large animals, because the fruiting bodies grow on dung. In sites around the world where these spore studies have been done, it has been shown that a precipitous drop in their abundance in sediments correlates with the decline of large mammals. Interestingly, the spore values often go up again once people have settled in a new land and stocked it with cattle, sheep, and goats.

In Madagascar, combining and comparing dated information from pollen, as an index of vegetation; charcoal particles to indicate fires; and dung fungus spores to indicate large animal presence (or absence) have consistently yielded the following scenario: People arrive just over two millennia ago. Within a few centuries, large animals decline in areas with humans, and soon after, within just a few decades perhaps, fires sweep through the landscape. Nonnative species appear

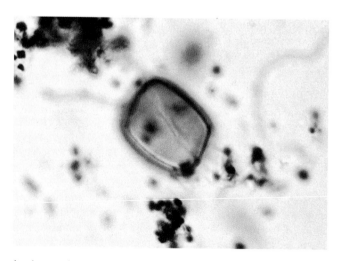

A microscopic spore of *Sporormiella*, a fungus that commonly grows on the dung of large animals and has been used to detect the presence of megafauna in the fossil record of Madagascar and many other places. Courtesy of David Burney.

in the record, many plant species decline, and much of the interior eventually becomes a steppe grassland with few native woody plants. Rain forests become increasingly fragmented, and archaeological sites show fairly large human populations in favorable areas by approximately 1,000 years ago. Fungal spores once again appear in many sites, as people introduce livestock to landscapes now missing the huge biomass of giant tortoises, pygmy hippopotami, elephant birds, and large lemurs that the fossil record shows once occupied this land.

It is not hard to imagine how a kind of negative synergy might occur in places such as Madagascar during this unique transition. A world that evolved for more than 100 million years in isolation from key evolutionary developments on the continents (including the appearance of ruminant mammals; keystone predators such as big cats, bears, and wolves; and that ultimate ecological game-changer, *Homo sapiens*) changed in a few centuries or less to a world dominated by humans and their camp followers. Various major factors have been put forward to explain the so-called mysterious extinction of Madagascar's megafauna, and that of many other places, and these factors interacted in a variety of ways to make the situation worse. A key set of interactions has been identified to show how human effects may be amplified.

The simple model begins with the rapid decline of native megafauna (or in any case the biggest animals on a remote island) at the onset of human predation and that of predatory creatures that come along with humans, such as dogs and cats. Some smaller introduced creatures, such as rats and even ants, may have quite dramatic effects on plants and animals of all sizes as well. In any case, the ecosystems were probably maintained in at least one key way primarily by the largest herbivores on the scene, which shape the landscape with their grazing and browsing. At high densities such as probably occurred in the complete absence of humans, these herbivores would have greatly reduced fire danger most of the time by keeping many habitats relatively open and free of flammable litter near the ground. With the demise of these keystone megaherbivores, grasslands, shrublands, woodlands, and even relatively dense forest would become extremely vulnerable to sudden transformations by destructive fires, one of the greatest forces for landscape change available to both nature and premodern humans. Within a matter of decades, vegetation in many precipitation regimes (perhaps all but the wettest and driest) can change with the impact of fires from woody to almost entirely grass or from a complex mosaic of vegetation types to an often monotonous system of aggressive fire-tolerant plants dominated by nonnative invasive plants that came with humans and their livestock.

Critics of this reconstruction often focus on the indirect nature of the evidence, and rightly so as such indirect indicators of past conditions as fossil pollen, charcoal particles, dung fungus spores, cut bones of extinct animals, and even archaeological sites must be interpreted in terms of complex natural processes and intricate timings not always easy to tease out from buried sediments. A question worth asking, however, is whether this synergistic pattern of human arrival, megafaunal decline, increased fire incidence, and biological invasion has occurred in any similar place, but with eye-witnesses.

Rodrigues Island as Rosetta stone

Jared Diamond (1984) has suggested that studying historical extinctions might be one way to better understand extinction events thought to have been precipitated by human arrival to continents and islands in prehistoric times. Scientists could learn much, he pointed out, by comparing prehistoric records for human arrival to the historical record for the small number of very remote islands around the world that were apparently never colonized prehistorically, such as the Galápagos and Mascarene Islands. What people describe in these early records could in that sense serve as a Rosetta stone for interpreting the less direct fossil evidence. Early naturalists in these so-called last places captured

moments in the earliest stages of transformation as historical events they claim to have witnessed, and this scant literature provides some remarkable insights that almost certainly have relevance to interpreting and perhaps even calibrating events known from places such as Madagascar only as fossils and artifacts.

Rodrigues Island lies about 940 miles (1,520 km) east of Madagascar in a particularly lonely stretch of the southwestern Indian Ocean. There are no larger islands to the east and north for thousands of miles, and Mauritius, one of the other two inhabited islands of the Mascarenes, is over 400 miles (650 km) away to the west. Rodrigues is an ancient piece of volcanic rock, possibly up to 10 million years old, 41.7 square miles (108 km^2) in extent. In the early 2000s, it has 38,000 Creole inhabitants, descendants of all the people who have come there and stayed since it was first successfully colonized in the eighteenth century. A member of the very first known human colonization attempt, François Leguat, was there from

The tiny remote island of Rodrigues, in the middle of the Indian Ocean, may provide a kind of "Rosetta Stone" for understanding late prehistoric events in Madagascar. The first settlers arrived in the historical period and recorded the transformation of the island. Loss of the large herbivores due to overhunting led to the accumulation of flammable plant biomass that fueled destructive fires. This large-scale disturbance provided ample opportunity for invasive alien species and resulted in ecological crisis and catastrophic extinction. Courtesy of Julian Jume and Francois Leguat Reserve Rodrigues.

1691 to 1693 and wrote his detailed memoirs years later, after the colony was abandoned. Other Frenchmen visited for several months at roughly 30-year intervals until the first governor began keeping his own records, including natural history observations, in 1795 (for a summary of the island's history, see Cheke and Hume 2008).

In the space of one century, these four literate naturalists describe a sequence of events that perfectly match the fossil record of Madagascar. The fauna is much simpler, of course, owing to Rodrigues Island's isolation from any source of colonizing animals. Two species of giant tortoise (*Cylindraspis* spp.), one evolved to reach up to about 4 feet (1.2 m) into the vegetation, were the primary large herbivores. A giant flightless pigeon called the Rodrigues solitaire (*Pezophaps solitaria*), bigger even than its famous cousin, the dodo of Mauritius, stood perhaps 3.3 feet (1 m) tall. Geckos and other lizards had diversified into many large and small forms, and seabirds nested in large numbers on the island and its offshore islets on a reef platform several times the size of the island itself. Leguat said that the tortoises were in places so numerous he could run 100 paces on their backs without touching the ground and that birds showed no fear at all of humans and could be knocked down with a stick to provide abundant food. The vegetation was open and parklike, with a diverse woodland of native trees and shrubs and a clear understory. No wildfires were recorded by him or the next visitor to report in, Julien Tafforet in 1725, as tortoises ate anything within reach that might otherwise burn. In the mid-eighteenth century, a few French families and their slaves settled on the island and began catching the tortoises and selling them to ships as a ready meat supply. They kept records, showing that in less than four decades they caught over 280,000 tortoises, eventually nearly wiping them out by the time of the third reporting visitor, Alexandre-Guy Pingré. In 1761 he noted that tortoises were scarce and that he did not see a solitaire. Huge fires swept through the thick underbrush periodically, and natural vegetation was in decline.

Philibert Marragon, the first governor of the tiny colony, described in 1795, just a century after the first colonist arrived, a completely transformed land with no large native animals, scarcely any native forest, eroding soils, and still only a tiny human population. A small number of humans, in a matter of decades, had changed Rodrigues forever, leaving a legacy of extinction, environmental change, and biological invasion. Every bit of the process was described in detail by eyewitnesses, providing an experimental microcosm of change in the wake of human arrival, truly a Rosetta stone for interpreting prehistoric extinction on Madagascar and other landmasses. The primary obstacle to fully accepting this model for islands larger than Rodrigues and potentially even continents such as the Americas and Australia is one of scale—can the dynamic on larger areas with more diverse (and larger) faunas be this simple? Some scientists have argued to the contrary, but at this point no clear cutoff in terms of area and biological diversity has been identified. Paleoecological studies in North America, Australia, and a host of islands (reviewed in Burney and Flannery 2005; see also Rule et al. 2012) have shown this same sequence, starting with a decline of fungus spores associated with large animals, followed by a large increase in charcoal, and ending with extinction, vegetation change, and biological invasion.

Extinction continues

The extinction catastrophe in Madagascar removed all large native animals weighing over about 22 pounds (10 kg) except the crocodile and large boid snakes. The largest living lemur species, the indri (*Indri indri*), is smaller than any of the 18 extinct subfossil lemurs. Thus, the original pattern of extinction in Madagascar following human arrival was similar to that of the Americas, Australia, and other lands, as Paul S. Martin first pointed out in the 1960s, in which presumably human-caused extinctions shortly after arrival primarily affected the largest herbivores and carnivores (Martin 2005). The history of these continents and such islands as Japan, the West Indies, and Hawaii shows graphically that second and even third waves of extinction often occur when successive groups arrive and bring new technology and new ways of life. Clearly, Madagascar is experiencing another phase in the extinction catastrophe that began two millennia ago. Like cases throughout the world, these more recent extinctions are far less partial to large animals, if any remain. Human overpopulation and resultant deforestation, biological invasions, and climate change are potentially more like the five or more great mass extinctions of remote geological time, before humans, which often affected large and small animals alike, as well as plants and entire ecosystems.

Madagascar continues to pose one of the conservation world's greatest challenges and one with the most at stake in terms of the global stock of biodiversity. Malagasy conservationists and their partner organizations from around the world must cope with the difficult economic situation in one of the world's poorest countries. Deforestation continues to pose a threat, and many species of lemurs, birds, reptiles, plants, and other organisms remain on the verge of extinction.

Backed by only meager resources, conservation projects have struggled to protect some of the last vestiges of rare plant communities, from rain forests to semideserts. One of the greatest unmitigated threats is fire, as even moderately wet areas of Madagascar may have lengthy dry seasons, and Malagasy pastoralists set fires to produce a green flush for hungry livestock. Ocean-atmospheric dynamics on decadal scales (e.g., the El Niño Southern Oscillation), coupled with global climate changes, mean that drought can affect almost any part of Madagascar. Rare species such as the plowshare tortoise (*Geochelone yniphora*) are under continued threat from poachers who sell them through an illicit pet trade. Rare forest trees are being exploited for timber, and people are desperately poor.

Innovative conservation strategies are needed that integrate community development with sustainable management of natural resources. Some projects, such as the establishment of Ranomafana National Park, have included ecotourism as a component of a broader community-based strategy that seeks to address the needs of people, wildlife, and watersheds. The hope in this and many other projects throughout the island is to preserve biodiversity while helping people.

Grandidier's baobab (*Adansonia grandidieri*) is the largest of the Madagascar baobabs. © imagebroker/Alamy.

A lesson for the world

What happened in Madagascar over the last two thousand years is in many ways unique, but it is also part of a larger picture. In a general way, the patterns observed by paleoecologists working in Madagascar serve to remind scientists and conservationists everywhere that individual threats to ecosystems actually work together synergistically to have a much bigger combined impact. This collision of powerful forces in nature, precipitated by human arrival to a new land, can overwhelm natural systems on many scales, and indeed often has. The case of Madagascar, especially when calibrated against historical events in the nearby Mascarene Islands, shows that initial human impact proceeds through three stages, beginning with the easy overexploitation of a fauna completely naive to the danger posed by human hunters. The collapse of the large herbivore component of the ecosystem leads to vegetation changes that start with the accumulation of underbrush, followed by devastating fires. In the last stage, the stage that modern Madagascar and indeed the entire world is in today, simpler vegetation communities struggle against invasive species, and many organisms in all size ranges are in decline or already extinct.

If the critical transition in landscapes, from a prehuman to posthuman mode, is driven by a negative synergy between the loss of large herbivores and the accumulation of flammable vegetation, one might ask whether this phase shift could be reversed to some extent by reintroducing large herbivores. Clearly in Madagascar, the presence of domestic livestock does not automatically prevent fires, as a range imbalance from overgrazing leads to unpalatable brush

The Madagascar ploughshare tortoise (*Geochelone yniphora*) is the world's rarest tortoise, restricted to a few hundred wild individuals in a small area of the western part of the island. Courtesy of David Burney.

In Madagascar's central plateau, extensive rice terraces produce the island's staple crop. Courtesy of David Burney.

encroachment in some habitats. In addition, pastoralists like to encourage fire to generate fresh growth for cattle. Christian A. Kull points out in his 2004 book on the subject, *Isle of Fire*, that huge areas of Madagascar and other lands are burned almost every year because the standing biomass is too dry or unpalatable for livestock.

In a 2010 study, Christine J. Griffiths and colleagues showed that tortoises have multiple beneficial effects on landscapes in

Mauritius and Rodrigues when used in restoration projects. Giant tortoises have been technically extinct in the Mascarenes since the end of the eighteenth century, but as a result of advice in a letter written by Charles Darwin in 1874 to the governor of Mauritius, some giant tortoises then surviving only on a single uninhabited island of the Indian Ocean—Aldabra Atoll—were moved to other islands of the region where they had formerly thrived, such as Mauritius and the Seychelles. These rewilded tortoises—believed to be identical to Madagascar's largest extinct species—today number in the thousands on other Indian Ocean islands. On both Rodrigues Island and uninhabited Round Island off Mauritius, hundreds of rewilded giant tortoises are employed in restoration projects in which they eat invasive nonnative plants in preference to the native plants, which coevolved with tortoises and are presumably defended from overgrazing by their leaf characteristics.

Ecosystem services provided by the tortoises in these projects include weed control, fire prevention, recycling of nutrients in their dung, and enhanced germination of native tree seeds as a result of passing through the gut of a giant reptilian herbivore. Projects of this type are under consideration for Madagascar. Although such projects are hardly likely to solve all or even most of Madagascar's conservation problems, they do give hope and create interest in conservation. The results demonstrate convincingly the important link, first inferred in the fossil record of Madagascar, between large herbivores, fire, landscape transformation, and extinction.

Resources

Books

Burney, David A. *Back to the Future in the Caves of Kaua'i: A Scientist's Adventures in the Dark.* New Haven, CT: Yale University Press, 2010.

Cheke, Anthony, and Julian Hume. *Lost Land of the Dodo: The Ecological History of Mauritius, Réunion, and Rodrigues.* New Haven, CT: Yale University Press, 2008.

Kull, Christian A. *Isle of Fire: The Political Ecology of Landscape Burning in Madagascar.* Chicago: University of Chicago Press, 2004.

Martin, Paul S. *Twilight of the Mammoths: Ice Age Extinctions and the Rewilding of America.* Berkeley: University of California Press, 2005.

Periodicals

Burney, David A., Lida Pigott Burney, Laurie R. Godfrey, et al. "A Chronology for Late Prehistoric Madagascar." *Journal of Human Evolution* 47, nos. 1–2 (2004): 25–63.

Burney, David A., and Timothy F. Flannery. "Fifty Millennia of Catastrophic Extinctions after Human Contact." *Trends in Ecology and Evolution* 20, no. 7 (2005): 395–401.

Burney, David A., Helen F. James, Frederick V. Grady, et al. "Environmental Change, Extinction, and Human Activity: Evidence from Caves in NW Madagascar." *Journal of Biogeography* 24, no. 6 (1997): 755–767.

Burney, David A., Guy S. Robinson, and Lida Pigott Burney. "*Sporormiella* and the Late Holocene Extinctions in

Madagascar." *Proceedings of the National Academy of Sciences of the United States of America* 100, no. 19 (2003): 10800–10805.

Diamond, J.S., 1984. Historic extinctions: A Rosetta Stone for understanding prehistoric extinctions. In, P.S. Martin and R.G. Klein, (Eds.) *Quaternary Extinctions: A Prehistoric Revolution.* University of Arizona Press, Tucson.

Godfrey, Laurie R., William L. Jungers, David A. Burney, et al. "New Discoveries of Skeletal Elements of *Hadropithecus stenognathus* from Andrahomana Cave, Southeastern Madagascar." *Journal of Human Evolution* 51, no. 4 (2006): 395–410.

Griffiths, Christine J., Carl G. Jones, Dennis M. Hansen, et al. "The Use of Extant Non-indigenous Tortoises as a Restoration Tool to Replace Extinct Ecosystem Engineers." *Restoration Ecology* 18, no. 1 (2010): 1–7.

Jungers, William L., Pierre Lemelin, Laurie R. Godfrey, et al. "The Hands and Feet of Archaeolemur: Metrical Affinities and Their Functional Significance." *Journal of Human Evolution* 49, no. 1 (2005): 36–55.

MacPhee, Ross D.E. "Morphology, Adaptations, and Relationships of *Plesiorycteropus* and a Diagnosis of a New Order of Eutherian Mammals." *Bulletin of the American Museum of Natural History*, no. 220 (1994): 1–214.

Rule, Susan, Barry W. Brook, Simon G. Haberle, et al. "The Aftermath of Megafaunal Extinction: Ecosystem Transformation in Pleistocene Australia." *Science* 335, no. 6075 (2012): 1483–1486.

Ryan, Timothy M., David A. Burney, Laurie R. Godfrey, et al. "A Reconstruction of the Vienna Skull of *Hadropithecus stenognathus*." *Proceedings of the National Academy of Sciences of the United States of America* 105, no. 31 (2008): 10699–10702.

Samonds, Karen E., Laurie R. Godfrey, Jason R. Ali, et al. "Spatial and Temporal Arrival Patterns of Madagascar's Vertebrate Fauna Explained by Distance, Ocean Currents, and Ancestor Type." *Proceedings of the National Academy of Sciences of the United States of America* 109, no. 14 (2012): 5352–5357.

Other

Madagascar National Parks. "Ranomafana." Accessed June 5, 2012. http://www.parcs-madagascar.com/fiche-aire-protegee_en.php?Ap=26.

David A. Burney

New Zealand (500 years ago)

The indigenous fauna of New Zealand is characterized by a great diversity of unique and unusual species. Over millions of years the combined effects of prolonged geographic isolation, a lack of nonvolant mammals, and a geologically dynamic landscape effectively led to an independent experiment in evolution—one that the author Jared Diamond referred to as being "as close as we will get to the opportunity to study life on another planet" (Diamond 1990, 3). The so-called alien oddities of New Zealand include mammal-like birds (e.g., kiwi [*Apteryx* spp.]), a large flightless nocturnal parrot (the kakapo [*Strigops habroptilus*]), giant flightless insects (e.g., giant weta [*Deinacrida* spp.]), viviparous lizards, terrestrial bats (*Mystacina* spp.), and the world's largest eagle (Haast's eagle [*Aquila moorei*]). New Zealand's fauna is also notable for several ancient lineages that persisted in this isolated part of the world long after they had become extinct elsewhere. Examples include the New Zealand wrens (Acanthisittidae) and sphenodontine reptiles (tuatara [*Sphenodon* spp.]; Gibbs 2006).

Although many of New Zealand's unique species can still be seen today, a large number have now been relegated to the pages of history. A major extinction event, beginning in the thirteenth century, led to the loss of about one-fourth of all the native terrestrial bird species (including nine species of the large ratite moa) as well as frogs, reptiles, a freshwater fish, a bat, and an unknown number of invertebrates. New Zealand, therefore, shares the fate of many other landmasses around the world that have suffered from major extinction events during the past 50,000 years. However, while elsewhere there is vigorous debate as to the relative contributions of human settlement and contemporaneous climate change in causing these extinction events, in New Zealand the situation is clear. New Zealand was one of the last major landmasses on Earth to be settled by humans, and it was settled during a period of relative climatic stability. On New Zealand, extinctions have undeniably been driven by human impacts.

The first people

The settlement of eastern Polynesia from approximately 1,000 to 700 years ago was the last great movement of modern humans to unsettled lands. The first stage of this migration occurred around 1,000 years ago with people from the Tonga/Samoa area moving into the Society Islands region. After a pause of approximately 200 years, a rapid expansion occurred from this region out to the numerous island groups between Hawaii, Rapa Nui (Easter Island), and New Zealand (Wilmshurst et al. 2011). The reasons why this mass migration took place are not clear, but the repercussions were felt by numerous animal populations throughout the island archipelagos of eastern Polynesia (Steadman 2006). Three factors associated with the initial human settlements appear to have been the main contributors to the extinction of native fauna in New Zealand: hunting, predation, and habitat loss.

Hunting

Chicken bones are common in nearly all early settlement sites across the Pacific (Steadman 2006), reflecting the importance of these birds as a food resource for the colonizers of eastern Polynesia. A few chicken bones have been excavated from early New Zealand archaeological sites, although their exact context and age remains unverified (Scofield et al. 2003). It would appear, however, that if chickens were brought to New Zealand they died out quickly, as the plentiful supply of large flightless native birds would have provided an alternative and more than sufficient food supply for the first generations of settlers.

Archaeological deposits at Wairau Bar, on New Zealand's South Island, provide a unique glimpse into the lives and hunting practices of some of the earliest settlers. A combination of archaic artifacts and radiocarbon dating evidence suggests the people found buried at this site represent either a founding population of New Zealand or their immediate descendants (Knapp et al. 2012). The DNA sequenced from the bones of 11 individual moa excavated from a single umu (earth oven) at Wairau Bar showed a bias toward males (with a male-to-female ratio of 2.7 to 1), despite similar data from natural bone deposits showing that moa populations were strongly biased toward females (with a mean male-to-female ratio of 1 to 2.4; Oskam et al. 2012). It is likely that in moa, as with extant ratites, it was the males who incubated the eggs, and so they may have represented much easier targets for hunting parties. The concentrated raiding of moa nests is also evident in the Wairau Bar deposit. Radiocarbon dating by Thomas Higham and colleagues (1999) indicates that moa eggshell fragments from Wairau Bar were all collected within a very short time period (late thirteenth century, c. 1285–1300). Genetic analysis

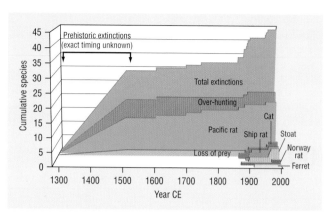

Terrestrial vertebrate species extinctions in New Zealand since human settlement (thirtheenth century CE). The blue plot represents the cumulative total number of extinct species, and the other plots represent the cumulative total number of extinctions in which each different factor/predator has been directly implicated. Reproduced by permission of Gale, a part of Cengage Learning.

identified 152 moa eggshell fragments from the site as having come from a minimum of 53 individual eggs, which Charlotte L. Oskam and colleagues suggested was likely to represent "a considerable proportion of the total reproductive output of moa in the area" (2012, 41).

Targeted hunting of moa at all life stages, from eggs to adults, was undoubtedly the major cause of their rapid extinction. The impact of such intense hunting would have been particularly exacerbated by their K-selected life strategy (large body size, low reproductive rate, and prolonged juvenile stage)—a trait that moa shared with many other large New Zealand bird species. Research on the ontogeny of moa (using ages determined by an analysis of lags or growth rings in the bones) has shown that, considering the time required for these birds to mature and establish territories, they probably did not begin breeding until eight to ten years of age. Once they started breeding, female moa would have laid one to two eggs each year. Modeling based on such a life strategy has shown that, with an estimated starting population size of 158,000, it would have required only very low harvest rates for the birds to have become extinct in less than 200 years (Holdaway and Jacob 2000).

Analysis of moa bones from a large prehistoric settlement site at the mouth of the Shag River, on the east coast of the South Island, has shown that, as moa populations declined, major changes occurred in human foraging patterns (Nagaoka 2005). Some parts of the moa carcasses were used to a much higher degree, as indicated by an increase in the relative frequency of high-utility elements (i.e., bones that held the most meat, such as femora and tibiotarsi) and increased fragmentation of bones (to access marrow and grease from the bones). As local prey populations began to decline, however, it became necessary for hunting parties to travel farther afield to find moa, and the field processing of carcasses became commonplace. The lower-utility parts of moa carcasses, such as those related to feet and necks, were removed at the kill site and not transported back to the settlement. This is shown by a decline in the relative frequency of elements related to these body parts (i.e.,

phalanges, tarsometatarsi, tracheal rings, and caudal vertebrae) in younger midden (refuse heaps containing the remains of consumed animals) layers. There is evidence that, as moa became harder to find, the human population increased its dependence on local coastal marine resources such as the New Zealand fur seal (*Arctocephalus forsteri*; Nagaoka 2006) and marine birds. Just a single marine species (the spotted shag, [*Stictocarbo punctatus*]) was represented in the 10 most common bird species in the late-thirteenth-century Wairau Bar site, yet five are represented in the later (mid-fourteenth century) midden layers at Shag River: the spotted shag, the blue penguin (*Eudyptula minor*), the giant petrel (*Thalassarche* sp.), the Waitaha penguin (*Megadyptes waitaha*), and the Stewart Island shag (*Leucocarbo carunculatus*).

Despite overwhelming evidence that the majority of moa disappeared within the first 200 years of human settlement, it seems likely that a few individuals or isolated populations may have persisted longer in some remote mountainous regions of New Zealand. Spores of *Sporormiella* (a fungus that lives on herbivore dung) preserved in a soil core from the Murchison Mountains in Fiordland suggest that a local decline of large herbivores (possibly including moa) did not occur until the late seventeenth to early eighteenth century (Wood et al. 2011). Yet there are certainly no confirmed sightings or radiocarbon-dated specimens, suggesting that moa survived into the post-European-settlement era (after 1769).

Population decline through hunting pressure was obviously not confined just to moa. Bones of other large flightless birds such as the New Zealand geese (*Cnemiornis* spp.) and adzebills (*Aptornis* spp.) occur in middens, and these species appear to have become extinct just as rapidly as moa did. In a detailed analysis of bird remains from Wairau Bar, published in 2003, R. Paul Scofield and colleagues showed that just 36.5 percent

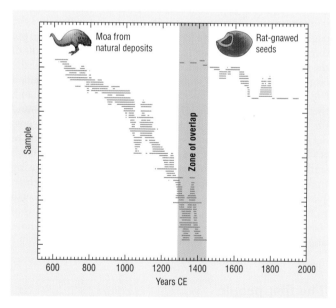

Ages of radiocarbon dated moa remains (bones, coprolites and eggshell <1,500 BP) and rat-gnawed seeds (a proxy for human settlement) from New Zealand. The ages are ordered from oldest to youngest, and shown as 1σ confidence ranges of calibrated dates. Reproduced by permission of Gale, a part of Cengage Learning.

Naturally desiccated head of an upland moa (*Megalapteryx didinus*) from a dry cave in southern New Zealand. © The Natural History Museum, London.

of 3,019 excavated bones were attributable to moa. The remaining 63.5 percent comprised bones from fifty additional bird species. The most common species represented was the New Zealand pigeon (*Hemiphaga novaezealandiae*). It is also interesting to note that only four of the ten most common birds represented in the site eventually became extinct: the coastal moa (*Euryapteryx curtus*), the black swan (*Cygnus atratus*, which became extinct but has since recolonized from Australia), the eastern moa (*Emeus crassus*), and the New Zealand coot (*Fulica prisca*). At the nearby Lake Grassmere, however, adjacent natural dune and midden bone deposits provide the opportunity to determine the natural bird community and therefore assess which species were most intensively hunted. In an analysis of these deposits published in 2002, Richard P. Duncan and colleagues found a significant relationship between those bird species that were intensively hunted and their probability of extinction.

As might be expected, the likelihood that a bird species was selectively hunted (and thereby had a higher chance of extinction) appears to be linked to body size and the ability to escape hunters. In a traits-based approach to understanding patterns of prehistoric extinction in New Zealand, the results of which were published in 2003, Derek A. Roff and Robin J. Roff determined that the probability of bird extinction was correlated positively with body mass and negatively with volancy (flying ability). Both of these can be explained by hunting in that large flightless birds would be easier to catch and provide more food value.

The relatively recent settlement date of New Zealand means that a large number of bone-rich midden sites still exist. As a result, the diets of the early settlers and the species that were driven to extinction through hunting are relatively well understood. Yet surprise discoveries are still being made. In 2009 Sanne Boessenkool and colleagues extracted DNA from supposed yellow-eyed penguin (*Megadyptes antipodes*) bones from early South Island middens and revealed them to actually be a distinct species, which was

named the Waitaha penguin (*Megadyptes waitaha*) after the first Polynesian tribe that occupied the South Island. The evidence seems to suggest that the Waitaha penguin was hunted to extinction and that yellow-eyed penguins from the subantarctic islands south of New Zealand moved north to recolonize the vacant niche. This discovery raises questions as to whether any other cryptic extinct species may remain to be identified in New Zealand archaeological deposits. One such extinction that has been recognized in the past is that of the black swan. Bones of the black swan are common in prehistoric lake, coastal, and midden deposits, and the species had been exterminated from New Zealand by the time of European settlement. However, black swans arrived naturally from Australia and reestablished a breeding population in New Zealand during the 1860s, and as a species now protected by law the black swan is once again a common bird throughout the country.

Predation

The first Polynesian settlers to arrive in New Zealand brought with them two commensal (living in close association

Palaeontologist Trevor Worthy sits beside the bones of a female South Island giant moa (*Dinornis robustus*) that fell to it's death in a New Zealand cave hundreds of years ago. Courtesy of Jamie Wood.

with humans) mammals. One was a medium-sized dog (*Canis familiaris*) known as the kuri, or the Polynesian dog. The impact of the kuri on the native fauna is not well understood. The kuri may have played a role in the hunting practices of the Polynesian settlers, but despite kuri bones being relatively common in archaeological middens (they also served as an important food source for the early settlers), their absence from natural bone deposits, such as caves, suggests that the kuri probably did not become feral.

In contrast, the other commensal mammal, the kiore or Pacific rat (*Rattus exulans*), rapidly spread throughout the country. Evidence for this species attaining a widespread distribution across New Zealand very soon after human arrival (approximately AD 1280) centers on radiocarbon dating of rat-gnawed seeds excavated from swamps and flood deposits (Wilmshurst et al. 2008). Predation by the kiore is implicated in the extinction of several frog, reptile, and small flightless and ground-nesting bird species that disappeared soon after initial human settlement (see Table 1). The kiore is also likely to have contributed to the extinction of large terrestrial invertebrates such as the extinct beetle *Waitomophylax worthyi* (Leschen and Rhode 2002).

The number of New Zealand bird extinctions directly attributable to the kiore is second only to those attributed to human hunting. When the extinction of frogs, reptiles, and an unknown number of invertebrate species is also considered, however, the kiore can be regarded as perhaps the single greatest cause of recent extinctions in New Zealand. It should be noted, of course, that the kiore would never have gotten to New Zealand if it were not for humans transporting them there in the first place.

Some New Zealand birds bore evolutionary traits that made them particularly susceptible to the novel threat of mammalian predation. Prior to human arrival, New Zealand was a land with few large terrestrial predators. Except for the flightless adzebills, which bone isotope analyses suggest may have preyed on ground-based birds and reptiles, most large predators were aerial. These included a giant eagle, a giant harrier, a falcon, and two owls. With no mammalian terrestrial predators, the requirement for flight was reduced, and many New Zealand birds lost the ability all together. In addition, those that could fly often fed or nested on the ground. These factors, combined with an overall naivety to the danger posed by predatory mammals, made New Zealand birds particularly susceptible to both hunting by humans and predation by the kiore. In their 2003 study, Roff and Roff found that the probability of extinction for small (less than 8.25 pounds [3.75 kg]) New Zealand flightless bird species was negatively correlated with body size. This is likely to be directly related to the impact of kiore predation, with the kiore having a greater impact on smaller birds and their smaller eggs.

Another aspect of New Zealand birds that made them particularly susceptible to mammalian predators was their smell. Predatory birds rely mostly on visual acuity to detect their prey, and therefore the selection pressures within natural New Zealand ecosystems meant that becoming nocturnal (to avoid the diurnal eagle and hawk) and/or developing camouflaged plumage were beneficial adaptations. The kiwi and kakapo are both examples of New Zealand birds that had adopted these mechanisms of predator avoidance. Strong smells were not selected against, and many such native birds evolved musky scents as a means of territory marking and communication, making them easily detected by predatory mammals with their keen sense of smell.

Habitat loss

At the time of human settlement New Zealand was a land of forests, with approximately 85 percent to 90 percent of the country covered by tall woody vegetation. On their arrival, Polynesian settlers quickly set about clearing the forests using fire in order to establish land for habitation and gardening and to promote growth of the early successional bracken fern (*Pteridium esculentum*), one of their main sources of starch. The signature of the widespread use of fire to modify vegetation at this time is clear to see in the numerous sediment cores from across the country that show the same typical pattern. In these cores pollen assemblages dominated by native forest trees are replaced by assemblages of grass pollen and bracken spores, coincident with a dramatic rise in the abundance of charcoal fragments. The timing of this change varied slightly between regions, but most South Island sites experienced significant burning within the first 200 years of settlement (McWethy et al. 2010). During this time approximately 40 percent of New Zealand's forests were cleared by burning.

The loss of forest habitat almost certainly reduced the populations of some animal species. This would have magnified the other pressures of hunting and predation, but with so much forest remaining it is doubtful that habitat loss was the sole cause for any extinctions. In some instances fauna, such as the New Zealand quail (*Coturnix novaezelandiae*), may even have benefited from the increased distribution of open habitat.

Other causes of the prehistoric extinctions

Despite the overwhelming evidence implicating humans and their commensals, it has been postulated that other factors may also have partly contributed to the postsettlement extinctions of some New Zealand fauna. For example, it has been suggested that the populations of some species may already have been in natural decline at the time of human settlement as a result of climatic, environmental (e.g., volcanic eruption), or biological (e.g., disease) factors. Although such hypotheses can be difficult to test, the application of Bayesian skyline plot analysis allows past population trends to be reconstructed using the relationship between effective population size and genetic diversity. By recovering ancient DNA sequences from radiocarbon-dated bones, Nicolas J. Rawlence and colleagues, in a study published in 2012, found no evidence for a presettlement decline in *Pachyornis* moas; these researchers instead found that the populations of these birds had remained relatively stable during the 50,000 years preceding human settlement.

Species		Fossil distribution	Extinction period CE	Causes (likely predators listed^)
Fish				
New Zealand grayling	*Prototroctes oxyrhynchus*	—	2 (1920–30s)	Introduced trout species
Frogs				
Aurora frog	*Leiopelma auroraensis*	SI	1	Pacific rat
Markham's frog	*Leiopelma markhami*	NI, SI	1	Pacific rat
Waitomo frog	*Leiopelma waitomoensis*	NI	1	Pacific rat
Reptiles				
Kawekaweau	*Hoplodactylus delcourti*	?*	2 (c.1800s)	European rats?
Northland skink	*Cyclodina northlandi*	NI	1	Pacific rat
Birds				
Little bush moa	*Anomalopteryx didiformis*	NI, SI, ST	1	Over-hunting
Upland moa	*Megalapteryx didinus*	SI	1	Over-hunting
Mantell's moa	*Pachyornis. geranoides*	NI	1	Over-hunting
Heavy-footed moa	*P. elephantopus*	SI, ST	1	Over-hunting
Crested moa	*P. australis*	SI	1	Over-hunting
Eastern moa	*Emeus crassus*	SI	1	Over-hunting
Coastal moa	*Euryapteryx curtus*	NI, NO, SI	1	Over-hunting
North Island giant moa	*D. novaezealandiae*	NI, NO	1	Over-hunting
South Island giant moa	*Dinornis robustus*	SI, SO, ST	1	Over-hunting
New Zealand quail	*Coturnix novaezelandiae*	NI, NO, SI	2 (c.1870s)	Norway rat, cat
Black swan	*Cygnus atratus*	NI, SI	1 (but re-established during 2)	Over-hunting
North Island goose	*C. gracilis*	NI	1	Over-hunting
South Island goose	*Cnemiornis calcitrans*	SI	1	Over-hunting
Scarlett's duck	*Malacorhynchus scarletti*	NI, SI	1	Over-hunting, Pacific rat
New Zealand blue-billed duck	*Oxyura vantetsi*	NI, SI	1	Over-hunting, Pacific rat
New Zealand musk duck	*Biziura delautouri*	NI, SI	1	Over-hunting, Pacific rat
New Zealand merganser	*Mergus australis*	NI, SI, ST	1#	Over-hunting, Pacific rat
Finsch's duck	*Chenonetta finschi*	NI, SI	1	Over-hunting, Pacific rat
Waitaha penguin	*Megadyptes waitaha*	SI, ST	1	Over-hunting

Table 1. Terrestrial and freshwater vertebrate species from New Zealand that have become extinct since human settlement in the thirteenth century. Distributions: NI, North Island; NO, northern offshore islands; SI, South Island; SO, southern offshore islands; ST, Stewart Island. Extinctions on the Chatham and subantarctic island groups are not included. Extinction periods are: 1, Post-settlement pre-European era (thirteenth century until 1769 CE); 2, Post-European era (1769 CE until present). Approximate dates of extinction or last credible sightings are given where known. Data are largely from Worthy & Holdaway (2002), Tennyson & Martinson (2006) and OSNZ checklist committee (2010). Reproduced by permission of Gale, a part of Cengage Learning.

A second wave of settlement

The high rate of extinction that immediately followed the first settlement of New Zealand probably slowed between the sixteenth and eighteenth centuries, although several extinctions are still likely to have occurred during this time. Bayesian modeling of radiocarbon dates for Finsch's duck (*Chenonetta finschi*) suggest that this once-common and widespread bird species may have persisted until the seventeenth or eighteenth century. Possible sightings during the nineteenth century provide the tantalizing possibility that birds such as the North Island takahe (*Porphyrio mantelli*), the New Zealand owlet-nightjar (*Aegotheles novaezealandiae*), and even Haast's eagle may also have survived into relatively recent times. But a new wave of settlement would soon expose New Zealand's fauna to a

Species		Fossil distribution	Extinction period CE	Causes (likely predators listed^)
Birds				
Scarlett's shearwater	*Puffinus spelaeus*	SI	? (sixteenth–nineteenth centuries)	Pacific rat
New Zealand little bittern	*Ixobrychus novaezelandiae*	NI, SI	2	Rats, mustelids, cat
Eyles' harrier	*Circus teauteensis*	NI, SI	1	Loss of prey
Haast's eagle	*Aquila moorei*	SI	1	Loss of prey
North Island adzebill	*A. otidiformis*	NI	1	Over-hunting
South Island adzebill	*Aptornis defossor*	SI	1	Over-hunting
Snipe rail	*Capellirallus karamu*	NI	1	Pacific rat
Hodgen's waterhen	*Gallinula hodgenorum*	NI, SI	1 (1700s)	Pacific rat
North Island takahe	*Porphyrio mantelli*	NI	2 (c.1890s)	Over-hunting, mustelids or cat?
New Zealand coot	*Fulica prisca*	NI, SI	1	Over-hunting, Pacific rat?
North Island snipe	*Coenocorypha barrierensis*	NI, NO	2 (c.1870s)	Pacific rat, cat
South Island snipe	*Coenocorypha iredalei*	SI, SO, ST	2 (1960s)	Rats, cat
Laughing owl	*Sceloglaux albifacies*	NI, SI, ST	2 (1914–1920)	Mustelids, rats, cat
New Zealand owlet-nightjar	*Aegotheles novaezealandiae*	NI, SI		Pacific rat
Bush wren	*Xenicus longipes*	NI, SI, SO, ST	2 (c.1972)	Rats, cat
Lyall's wren	*Traversia lyalli*	NI, SI, SO	2 (1890s)	Pacific rat, cat
North Island stout-legged wren	*Pachyplichas jagmi*	NI	1	Pacific rat
South Island stout-legged wren	*Pachyplichas yaldwyni*	SI	1	Pacific rat
Long-billed wren	*Dendroscansor decurvirostris*	SI	1	Pacific rat
South Island kokako	*Callaeas cinerea*	SI, ST	2 (c.1960s)	Rats, mustelids, possum, cat
Huia	*Heteralocha acutirostris*	NI	2 (c.1907–1920s)	Over-hunting, rats, mustelids, cat
North Island piopio	*Turnagra tanagra*	NI	2 (c.1900)	Rats
South Island piopio	*Turnagra capensis*	SI, SO, ST	2 (c. early 1900s)	Rats, mustelids, cat
New Zealand raven	*Corvus antipodum*	NI, SI, ST	1	Over-hunting?, Pacific rat
Mammals				
Greater short-tailed bat	*Mystacina robusta*	NI, SI, SO, ST	2 (c.1967)	Rats

*The single mounted specimen of this large gecko in the collections of the Muséum d'histoire naturelle de Marseille, France, has been attributed to *Hoplodactylus*; a genus endemic to New Zealand. However there are no known fossil bones of the species from New Zealand and so a determination of its exact relationships and place of origin will probably require molecular analysis of the specimen.

#Persisted on subantarctic islands until early twentieth century.

^Based on Tennyson & Martinson 2006

Table 1 continued. Reproduced by permission of Gale, a part of Cengage Learning.

novel suite of predators from different regions of the world and once again lead to an increase in the extinction rate.

The first known sighting of New Zealand by a European explorer was made by the Dutch navigator and explorer Abel Tasman in 1642. Following a hostile encounter with local Maori while anchored in Golden Bay on the northern coast of the South Island, Tasman departed before actually stepping foot on New Zealand soil.

The first official European landing in New Zealand was made by James Cook, a British lieutenant who sailed to New Zealand onboard the *Endeavour*. In 1769 and 1770 his crew made multiple landings on different parts of the New

Zealand coastline, and it is likely that ship rats (*Rattus rattus*) may have first gotten ashore at this time. Over the next few years Cook revisited New Zealand, and at the same time French ships were also visiting. The first recorded deliberate introduction of European animals into New Zealand took place in 1773 at Dusky Sound, Fiordland, when Cook released geese, sheep, chickens, and goats, and a French captain released pigs. Cook also had cats onboard his ship to control rats, one of which "regularly took a walk in the woods every morning and made great havoc among the little birds" (Forster 1777, 128). By the early nineteenth century whalers and sealers plied the coasts of New Zealand, and it is thought that feral populations of cats may have become established at this time (Fitzgerald 1990). Although cats are implicated in the extinction of several bird species, the most notable was that of the Lyall's (or Stephens Island) wren (*Traversia lyalli*). One of just a handful of known flightless passerines, this small terrestrial insectivore was eliminated from mainland New Zealand when humans first arrived with the cause being predation by the kiore. The 380-acre (154-ha) Stephens Island, off the northern tip of the South Island, became the species' last refuge. Unfortunately, by the 1890s a significant population of cats had accrued on the island. A construction worker associated with the island's lighthouse was the first to report seeing Lyall's wren, but by the time news reached the scientific community, and naturalist Henry Travers visited the island in search of the bird in 1895, it had become extinct. The only specimen Travers was able to obtain was one that had been caught by a cat (Tennyson and Martinson 2006).

Over the next century, a number of exotic species were released in New Zealand intentionally. In the late nineteenth century this practice was largely performed by so-called acclimatization societies, whose aims were to introduce animals that were seen as being beneficial, either aesthetically, for sport (i.e., hunting), or to make the New Zealand fauna more similar to that which the largely British immigrants were familiar with (see Thomson 1922 for a discussion of these introductions). Many were unsuccessful (e.g., the long-nosed potoroo [*Potorous tridactylus*] and the western gray squirrel [*Sciurus griseus*]), but others multiplied and spread throughout the country.

Against the protests of ornithologists, mustelids (ferrets, stoats, and weasels) were introduced to New Zealand from the 1860s to the 1880s in an attempt to control the booming rabbit (*Oryctolagus cuniculus*) populations that were destroying agricultural grasslands. Stoats (*Mustela erminea*), as indiscriminant killers, quickly became one of the worst exotic predators New Zealand's native fauna had to confront. Stoats are likely to have brought about the extinction of several bird species that had managed to persist despite the onslaught from rats and cats, such as the New Zealand little bittern (*Ixobrychus novaezelandiae*), the laughing owl (*Sceloglaux albifacies*), the South Island kokako (*Callaeas cinerea*), and the South Island piopio (*Turnagra capensis*).

Hunting and specimen collecting are also likely to have contributed to the extinction of some New Zealand bird species. In no case is this better exemplified than that of the huia (*Heteralocha acutirostris*). Huia tail feathers were long

Moa bones in a Maori midden, northern South Island of New Zealand. Courtesy of Quinn Berentson.

Male and female huia (*Heteralocha acutirostris*), which became extinct following European settlement of New Zealand (painted by Johannes G. Keulemans). © The Natural History Museum/The Image Works.

prized by Maori for whom the wearing of the feathers in the hair signified high rank. But the tail feathers quickly became a popular item in European fashion as well after the Duke of York was presented with a huia feather during his visit to New Zealand in 1901. As huia populations declined as a result of predation and forest clearance, the demand for huia specimens increased. This was driven not only by hunting for feathers but also by scientific institutions and museums around the world seeking to secure specimens of the huia before the bird became extinct. The final huia was officially sighted in 1907, although credible reports indicate the species may have persisted until the mid-1920s (Tennyson and Martinson 2006).

The New Zealand grayling or upokororo (*Prototroctes oxyrhynchus*) is the only freshwater fish known to have become extinct in New Zealand, although several other species are now very rare. The upokororo was abundant in New Zealand waterways until its population began to decline in the 1870s. The exact cause of its decline is not known, although, as with several other native New Zealand fish (particularly in the Galaxiidae family), it is likely to have been impacted severely by the introduction of predatory trout species. Watershed modification through the removal of forest and exploitation by humans has also been suggested as a factor contributing to the upokororo's demise. The species was restricted to a

few isolated localities by the 1920s and probably disappeared sometime during the following decade.

The most recent vertebrate extinctions in New Zealand followed the arrival of shipborne rats on Big South Cape Island. The island, located near the southern tip of Stewart Island, is a place of traditional harvest of titi (the muttonbird or sooty shearwater [*Puffinus griseus*]). The first rats likely arrived on the island accidentally around the 1950s or 1960s, with cargo brought for the annual titi harvest. By 1964 their numbers had irrupted, and rats swept across the island devouring everything in their path like a plague. Unfortunately, the island was the last stronghold for several species of bird, including the bush wren (*Xenicus longipes*), the South Island snipe (*Coenocorypha iredalei*), and the South Island saddleback (*Philesturnus carunculatus*), as well as the greater short-tailed bat (*Mystacina robusta*). Wildlife managers, realizing the peril these species faced, acted quickly by attempting to translocate them to predator-free islands. This management successfully saved the South Island saddleback from extinction. However, the bush wren, the South Island snipe, and the greater short-tailed bat were not as fortunate.

There have also been some stories of good fortune. Only four specimens of the South Island takahe (*Porphyrio hochstetteri*), a large flightless rail, had been collected between 1850 and 1898. For the next 50 years the species was thought to be extinct. But in 1948 a small population was discovered living in an alpine valley in the Murchison Mountains, Fiordland. Since then active management has increased the population of takahe to approximately 260 birds. Subsequently, the New Zealand storm petrel (*Oceanites maorianus*) was also rediscovered. Known only from a handful of museum specimens collected in the mid-nineteenth century, the species was assumed to be extinct. Then, in 2003, birds matching the description of the New Zealand storm petrel were sighted in the Hauraki Gulf in northern New Zealand. Their identity was confirmed in 2011 through a DNA comparison of captured individuals with the museum specimens. Active management to protect the birds from predation awaited as of 2012 the discovery of their breeding sites.

Lessons for the future

Many species in New Zealand's native fauna continue to be driven toward extinction by introduced predators. Within the short period between 2002 and 2005 the conservation status of forty New Zealand bird species declined, with a similar trend seen in other native fauna (Scofield et al. 2011). As populations decline, the potential increases for stochastic events—such as periods of extreme climate or rat and stoat plagues induced by beech (*Nothofagus* spp.) seed masting (synchronised seeding events which result in a vast amount of seed being produced at the same time)—to bring about the end of a species. Unintended consequences of pest control, such as prey switching, are also a problem. Through experiences gained in unfortunate circumstances, New Zealand has become a world leader in conservation management, pest management, and ultimately the prevention of further extinctions.

Irish immigrant Richard Henry is credited with pioneering practical conservation in New Zealand. From 1894 to 1908 he was the appointed government caretaker of Resolution Island, a designated predator-free island reserve in southern Fiordland. During his first few years on the island, Henry developed equipment and methods for the translocation of rare bird species from the nearby mainland. Henry caught and transferred hundreds of kakapo and kiwi before his enterprise was tragically halted by the invasion of stoats on Resolution Island in 1900. But Henry's methods of moving fauna to predator-free sanctuaries would become a standard conservation tool used in New Zealand throughout the twentieth century and into the twenty-first.

Since the 1940s there have been several instances in which great effort and ingenuity has rescued New Zealand bird species from the brink of extinction. After the takahe was rediscovered in 1948, a small number of eggs were collected and brought down from the mountainous valley, with a bantam hen used to sit on the eggs while they were transported. This would be the start of a program that continued into the early twenty-first century, through which chicks are raised in captivity before being re-released back into managed refuges once they are big enough to defend themselves against predators. Another innovation of this program was the use of takahe puppets to feed the chicks, so that they did not become imprinted on humans. Another success story is that of the Chatham Island robin (*Petroica traversi*). In 1980 the species was precariously on the edge of extinction, with just five known individuals (and a single female) remaining. The eggs from the first clutch of this bird each year were carefully removed and placed in nests of Chatham Island tomtits (*Petroica macrocephala chathamensis*), which acted as surrogate parents and raised the chicks (a method known as cross-fostering).

New Zealand has also become renowned internationally for its expertise in island pest eradications. New Zealand has approximately 220 offshore islands larger than 12.4 acres (5 ha) in size, of which 42 percent are classified as nature reserves. By 2005 a total of 218 successful eradications of seventeen exotic mammal species had been performed on these islands (Clout and Russell 2006). In June 2001 a New Zealand team managed to eradicate the Norway rat (*Rattus norvegicus*) from the subantarctic Campbell Island, which at 43.6 square miles (11,300 ha) made it, at that time, the largest

island on which pest eradication had been accomplished successfully. Since then New Zealand expertise has been called in to assist with pest eradications on many other islands, including two large-scale projects on Macquarie Island (49.7 square miles [12,870 ha]) and South Georgia (1,505 square miles [390,000 ha]). In 2009 a feasibility analysis began for the eradication of rats from Stewart Island (674 square miles [174,600 ha]) at the southern tip of New Zealand, and there are also presently discussions beginning regarding the potential for the eradication of exotic predators from the entire country (Priestley 2012).

During the past 500 years, New Zealand has suffered from an unprecedented rate of faunal extinctions. These extinctions occurred in two main phases: the first beginning with the initial Polynesian settlement (mid- to late thirteenth century) and the second following European settlement (late eighteenth century).

New Zealand was one of the last major landmasses on Earth to be settled by humans, at a time of relatively stable climate conditions. This, combined with the abundance of midden deposits containing bones of large extinct birds, means that New Zealand is globally regarded as providing the best evidence in support of the overkill or blitzkrieg extinction hypotheses. But it is clear that predatory mammals, both those brought by Polynesians (particularly the kiore) and Europeans (particularly rats, cats, and mustelids), have also been major causes of extinctions in New Zealand's mammal-naive fauna. The extent of their impact, especially with respect to invertebrate extinctions, probably remains to be fully realized.

The wider ecological implications of New Zealand's extinctions are also only beginning to be understood. With such a long evolutionary history in isolation from other landmasses, close relationships have formed between many New Zealand plant and animal species. Questions remain as to how disruptions to these ecological links may have affected such processes as nutrient cycling, seed dispersal, and pollination. Despite its legacy of vanished fauna, experience gained in pest eradication and managing species declines in New Zealand over the past century is now proving beneficial in helping to prevent further extinctions not just in New Zealand but across the world.

Resources

Books

Clout, Mick N., and James C. Russell. "The Eradication of Mammals from New Zealand Islands." In *Assessment and Control of Biological Invasion Risks*, edited by Fumito Koike, Mick N. Clout, Mieko Kawamichi, et al., 127–141. Kyoto, Japan: Shoukadoh Book Sellers; Gland, Switzerland: IUCN—the World Conservation Union, 2006.

Diamond, Jared M. "New Zealand as an Archipelago: An International Perspective." In *Ecological Restoration Of New Zealand Islands*, edited by David R. Towns, Ian A.E. Atkinson,

Charles H. Daugherty, 3–8. Wellington, New Zealand: Department Of Conservation, 1990.

Fitzgerald, B.M. "House Cat." In *The Handbook of New Zealand Mammals*, edited by Carolyn M. King, 330–348. Auckland, New Zealand: Oxford University Press, 1990.

Forster, Georg. *A Voyage Round the World, in His Britannic Majesty's Sloop, Resolution*. London: B. White, 1777.

Gibbs, George. *Ghosts of Gondwana: The History of Life in New Zealand*. Nelson, New Zealand: Craig Potton, 2006.

OSNZ (Ornithological Society of New Zealand). Checklist Committee. *Checklist of the Birds of New Zealand, Norfolk and Macquarie Islands, and the Ross Dependency, Antarctica.* 4th ed. Wellington, New Zealand: Te Papa Press, 2010.

Steadman, David W. *Extinction and Biogeography of Tropical Pacific Birds.* Chicago: University of Chicago Press, 2006.

Tennyson, Alan J.D., and Paul Martinson. *Extinct Birds of New Zealand.* Wellington, New Zealand: Te Papa Press, 2006.

Thomson, George M. *The Naturalisation of Animals and Plants in New Zealand.* Cambridge, U.K.: Cambridge University Press, 1922.

Worthy, Trevor H., and Richard N. Holdaway. *The Lost World of the Moa: Prehistoric Life of New Zealand.* Christchurch, New Zealand: Canterbury University Press, 2002.

Periodicals

Boessenkool, Sanne, Jeremy J. Austin, Trevor H. Worthy, et al. "Relict or Colonizer? Extinction and Range Expansion of Penguins in Southern New Zealand." *Proceedings of the Royal Society* B 276, no. 1658 (2009): 815–821.

Duncan, Richard P., Tim M. Blackburn, and Trevor H. Worthy. "Prehistoric Bird Extinctions and Human Hunting." *Proceedings of the Royal Society* B 269, no. 1490 (2002): 517–521.

Higham, Thomas, Atholl Anderson, and Chris Jacomb. "Dating the First New Zealanders: The Chronology of Wairau Bar." *Antiquity* 73, no. 280 (1999): 420–427.

Holdaway, Richard N., and Chris Jacomb. "Rapid Extinction of the Moas (Aves: Dinornithiformes): Model, Test, and Implications." *Science* 287, no. 5461 (2000): 2250–2254.

Knapp, Michael, K. Ann Horsburgh, Stefan Prost, et al. "Complete Mitochondrial DNA Genome Sequences from the First New Zealanders." *Proceedings of the National Academy of Sciences of the United States of America* 109, no. 45 (2012): 18350–18354.

Leschen, Richard A.B., and Brigit E. Rhode. "A New Genus and Species of Large Extinct Ulodidae (Coleoptera) from New Zealand." *New Zealand Entomologist* 25, no. 1 (2002): 57–64.

McWethy, David B., Cathy Whitlock, Janet M. Wilmshurst, et al. "Rapid Landscape Transformation in South Island, New Zealand, following Initial Polynesian Settlement." *Proceedings of the National Academy of Sciences of the United States of America* 107, no. 50 (2010): 21343–21348.

Nagaoka, Lisa. "Declining Foraging Efficiency and Moa Carcass Exploitation in Southern New Zealand." *Journal of Archaeological Science* 32, no. 9 (2005): 1328–1338.

Nagaoka, Lisa. "Prehistoric Seal Carcass Exploitation at the Shag Mouth Site, New Zealand." *Journal of Archaeological Science* 33, no. 10 (2006): 1474–1481.

Oskam, Charlotte L., Morton E. Allentoft, Richard Walter, et al. "Ancient DNA Analyses of Early Archaeological Sites in New Zealand Reveal Extreme Exploitation of Moa (Aves: Dinornithiformes) at All Life Stages." *Quaternary Science Reviews* 52 (2012): 41–48.

Priestley, Rebecca. "Sir Paul Callaghan's Crazy Idea." *New Zealand Listener* April 14, 2012. Accessed November 20, 2012. http://www.listener.co.nz/current-affairs/science/sir-paul-callaghans-crazy-idea.

Rawlence, Nicolas J., Jessica L. Metcalf, Jamie R. Wood, et al. "The Effect of Climate and Environmental Change on the Megafaunal Moa of New Zealand in the Absence of Humans." *Quaternary Science Reviews* 50 (2012): 141–153.

Roff, Derek A., and Robin J. Roff. "Of Rats and Maoris: A Novel Method for the Analysis of Patterns of Extinction in the New Zealand Avifauna before European Contact." *Evolutionary Ecology Research* 5, no. 5 (2003): 759–779.

Scofield, R. Paul, Ross Cullen, and Maggie Wang. "Are Predator-Proof Fences the Answer to New Zealand's Terrestrial Faunal Biodiversity Crisis?" *New Zealand Journal of Ecology* 35, no. 3 (2011): 312–317.

Scofield, R. Paul, Trevor H. Worthy, and Heidi Schlumpf. "What Birds Were New Zealand's First People Eating?— Wairau Bar's Avian Remains Re-examined." *Records of the Canterbury Museum* 17 (2003): 17–35.

Wilmshurst, Janet M., Atholl J. Anderson, Thomas F.G. Higham, and Trevor H. Worthy. "Dating the Late Prehistoric Dispersal of Polynesians to New Zealand Using the Commensal Pacific Rat." *Proceedings of the National Academy of Sciences of the United States of America* 105, no. 22 (2008): 7676–7680.

Wilmshurst, Janet M., Terry L. Hunt, Carl P. Lipo, and Atholl J. Anderson. "High-Precision Radiocarbon Dating Shows Recent and Rapid Initial Human Colonization of East Polynesia." *Proceedings of the National Academy of Sciences of the United States of America* 108, no. 5 (2011): 1815–1820.

Wood, Jamie R., Janet M. Wilmshurst, Trevor H. Worthy, and Alan Cooper. "*Sporormiella* as a Proxy for Non-mammalian Herbivores in Island Ecosystems." *Quaternary Science Reviews* 30, nos. 7–8 (2011): 915–920.

Jamie R. Wood

· · · · ·

Mediterranean islands (9,000 years ago)

Regarded by some as an ocean in miniature surrounded by the African and Eurasian plates, as well as a few microplates, the Mediterranean Sea has experienced a long and complex geological history. This history has been characterized by orogenic processes and widespread extensional tectonics, causing repeated separations/connections of microplates or isolations/connections of insular and mainland territories (Carminati and Doglioni 2005). Like other seas landlocked between contiguous plates, the Mediterranean was, and is, rich in islands. The fossil records of these islands testify to the existence of an exceptional level of biodiversity and a high rate of regional endemism both in the past and at present (Grill et al. 2007). Since the beginning of the evolution of the Mediterranean basin, more than 50 million years ago, endemic faunas from islands, which had either a difficult or no connection with a mainland, have represented an important part of the basin's faunal successions. These successions comprise a long and complex history of faunal turnovers and species originations, extinctions, dispersals, and competitive exclusions. This history documents the evolution of insular paleocommunities through time, up to the modern composition and diversity.

The Mediterranean is thus an especially favorable natural laboratory for biogeographic and paleobiogeographic studies, as well as an inexhaustible source of data suitable for testing evolutionary models, including extinction patterns of the vertebrate fauna, especially mammals. Most of the Pleistocene endemic insular species (including dwarf elephants, mammoths, hippopotamuses, and deer; giant murids, voles, and owls; and strange, extremely specialized wild dogs) disappeared either by the end of the Pleistocene or during the early Holocene. Nearly all the Mediterranean insular mammals were not descendants of previous island settlers, whereas most of the herpetofauna was. A better understanding of the causal factors behind the extinction of these endemic small and large mammals—and a more precise unraveling of the extent to which each factor affected the timing and mechanisms of faunal demise on each island—is of great importance. This is especially true in light of the debated issues on whether the ongoing Holocene extinctions of insular species are principally caused, either directly or indirectly, by human activity or by climate action and environmental changes.

The Pleistocene insular faunas from the Mediterranean islands

A complex and still not completely understood combination of paleogeographic evolution, geodynamic and tectonic action, and glacio-eustatic sea-level changes has resulted in each Mediterranean island or archipelago having its own unique history. The differences extend to antiquity, area, physical geography, duration and extent of isolation, number of dispersal events and faunal turnovers, and, ultimately, the timing of human colonization.

The time of isolation of some Mediterranean islands extends over hundreds of thousands or even millions of years (e.g., the Balearic Islands). Some islands, although sitting on continental selves, were isolated from the mainland permanently (e.g. the Balearic Islands, Crete, and Cyprus), whereas others were not separated from the mainland by a great barrier or were connected by a filtering corridor (e.g., Sicily during the Late Pleistocene). Still others changed their status from complete isolation to a temporary connection to the continent to being part of a large archipelago, and then once again to being isolated (e.g., Sardinia). Isolation through time affected successful dispersals and colonization by island settlers strongly. This also affected the composition of insular faunas (from impoverished but balanced, to disharmonic, strongly impoverished, and highly unbalanced with respect to faunas inhabiting similar, but not isolated, continental ecosystems), the competitive coevolutionary dynamics, the community equilibrium, and, finally as a result of the preceding, the extinction patterns of insular terrestrial species. Depending on the paleogeographic control of dispersals and the characteristics of island settlers, endemic faunas differed from one another in taxonomic composition and structure, either from island to island or on the same island through time. Nevertheless, the extinctions of insular mammalian species exhibit some common patterns, as shown in the following overview of the faunal history of the Mediterranean islands that yielded the most intriguing endemic species.

The Balearic Islands

The Balearic Islands, which include the eastern Gymnesics (including Majorca and Minorca) and the western Pityusics archipelagos, were completely isolated from the mainland

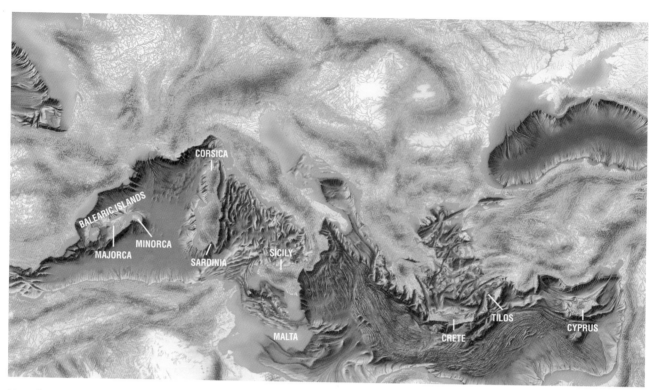

Map of the Mediterranean Islands. Reproduced by permission of Gale, a part of Cengage Learning.

probably since the end of the Messinian Salinity Crisis, after the Pliocene flooding about 5.3 million years ago. From that moment the Balearics acted as oceanic-like islands, and terrestrial vertebrates changed only through in situ evolutionary processes.

The most representative mammalian lineages characterizing the Gymnesic Islands are those of the bovid *Myotragus* (including up to six chronospecies: *M. palomboi* [recte *M. palomboae* according to the International Commission on Zoological Nomenclature (ICZN)], *M. pepgonellae*, *M. antiquus*, *M. kopperi*, *M. batei*, and *M. balearicus*), the glirid rodent *Hypnomys*, and the soricid *Nesiotites*. The herpetofauna includes small-sized lizards (the Early Pleistocene Lacertidae *Podarcis* sp. and the still living *Podarcis lilfordi*, present since the Middle Pleistocene) and two amphibians (the Early Pleistocene large-sized frog *Discoglossus* sp. and the still living Majorcan midwife toad [*Alytes muletensis*]; see bibliography in Bover et al. 2010).

The first representatives of mammalian lineages were already present on Majorca in the Early Pliocene; the last ones went extinct on Majorca and Minorca between 5,000 and 4,000 years ago (Bover et al. 2008, 2010). During the long period of isolation in an environment free of predators and competitors, *Myotragus* acquired peculiar features (including stereoscopic vision. monophyodontic dentition with only two lower perennial-growth incisors and a reduced number of premolars, and extremely short and robust, but not very flexible, limb bones) typically displayed by the terminal species of the lineage, *Myotragus balearicus*, which has been present on the Gymnesics since the Middle Pleistocene.

Although climate changes in some phases likely affected the evolution of *Myotragus*, it has been suggested that climate change did not cause its extinction. Pere Bover and Josep

Artistic reconstruction of the landscape of Tilos during the early Holocene showing the dwarfed *Palaeoloxodon tiliensis* and the turtle *Testudo marginata*. Illustration by Stefano Maugeri.

Artistic reconstruction of the landscape of Mallorca around 5,000 years ago, showing the extinct bovid *Myotragus balearicus* and shrew *Nesiotites hidalgo*, and the still living lizard *Podarcis lilfordi*. Illustration by Stefano Maugeri.

Antoni Alcover, reviewing all the available evidence in a 2003 study, concluded that this hypothesis is weakly supported by paleoclimatic and paleontological data, as Late Pleistocene–early Holocene climate changes did not affect the composition of the fauna and that their effect on the vegetation was dampened by the influence of the surrounding sea.

These two paleontologists considered a human-caused extinction of *Myotragus balearicus* to be the most plausible hypothesis. They based their conclusions on the following evidence and lines of reasoning:

- The Balearic Islands, because of their degree of isolation, were colonized by humans very recently in comparison with less isolated Mediterranean islands. Until 2001, human arrival was assumed to have occurred 8,000 years ago (or even earlier), but a reappraisal of the evidence has pinpointed a human arrival in Majorca at around 5,000 years ago, although permanent settlement probably occurred later, perhaps about 4,300 years ago (Bover and Alcover 2003).

- *M. balearicus* was present at least on the larger Gymnesic Islands at the time of the first human arrival, but it disappeared not long after.

- The dramatically short time of coexistence of humans and endemic bovids suggests that the sitzkrieg extinction model (slow, drawn-out extinction events, caused by secondary human effects, such as deforestation, anthropogenic burning, and general habitat modification) should be eliminated as a possibility for *M. balearicus*.

- The lack of compelling archaeological evidence documenting contact between humans and *M. balearicus* cannot be explained by mutual avoidance because of *M. balearicus*'s limited agility in escaping and the supposed tameness of this species.

Consequently, Bover and Alcover (2003) concluded that the rapid extinction of *M. balearicus* following human arrival is consistent with either the blitzkrieg (which might have left no traces in the archaeological record) or hyperdisease models. Whereas data on *M. balearicus* are abundant, if not fully conclusive as to the factors driving its extinction, information on the demise of the other two Majorcan and Minorcan species, *Hypnomys morpheus* and *Nesiotites hidalgo*, is scarce and more difficult to interpret. In a 2008 article, Bover and Alcover dismissed climate change, predation by invasive species, competition with newcomers, and habitat deterioration as possible causes for their extinction, suggesting that the introduction of diseases is the most reasonable explanation. Just after human arrival, the Balearic wild mammalian fauna shifted from three species (*M. balearicus*, *H. morpheus*, and *N. hidalgo*) to only two (the garden dormouse *Eliomys quercinus* and the wood mouse *Apodemus sylvaticus*), with the latter two introduced by humans together with four or five domesticated mammals (sheep, goats, cows, pigs, and probably also dogs).

Corsica and Sardinia

The Corsica–Sardinia massif has had a complex paleogeographic history mirrored, in particular, by different degrees of endemism and biodiversity characterizing the Sardinian faunas, which bear witness to a series of dramatic extinctions throughout most of the Cenozoic. During the last 5 million years, the Corsica–Sardinia massif maintained a permanent isolation from the continent—although the two islands were temporarily connected to each other during the most important phases of sea-level drop—and new settlers mainly reached the islands by either sweepstake or passive dispersal. Colonization events, especially by jump dispersal of large mammals, were sometimes facilitated during glacial phases that produced a lowering of sea level. A 2009 study by Maria Rita Palombo indicated that the mammalian fauna of Sardinia can be split into two major complexes: the *Nesogoral* faunal complex (Late Pliocene–Early Pleistocene) and the *Microtus* (*Tyrrhenicola*) faunal complex (late Early Pleistocene–early Holocene). A significant faunal turnover marks the transition between these complexes, although individual appearance and disappearance bioevents characterized each complex. For instance, during the time of *Nesogoral* complex (including the endemic midsized bovids *Nesogoral* and *Asoletragus*, the dwarf monkey *Macaca majori*, the small suid *Sus sondaari*, three carnivores—a hunting hyena *Chasmaporthetes melei* and two mustelids—and archaic small mammals), the renewal mainly relies on evolution within small mammal endemic lineages.

Conversely, at the transition to the *Microtus* (*Tyrrhenicola*) complex, several species disappeared. These extinctions were probably driven by internal dynamics and competition with new settlers, but climatic and environmental changes possibly contributed to the faunal turnover, negatively affecting the most specialized endemic species and favoring species with a broader niche. This caused the faunal diversity to drop, despite the arrival, mainly by the end of the Early Pleistocene, of new settlers, such as the ancestors of the wild dog *Cynotherium sardous*, the deer *Praemegaceros cazioti*, and the large vole *Microtus (Tyrrhenicola) henseli*. The presence of these species—together with the appearance of advanced murids and insectivores in preexisting phyletic lineages, endemic otters, and the dwarf Sardinian mammoth *Mammuthus lamarmorai*—typified the Late Pleistocene Sardinian and Corsican faunas.

Most of the endemic species, except for the mammoth and otters, last occurred in the Holocene. Among the most common large mammals, the wild dog *Cynotherium sardous* seems to have disappeared first (with a latest presence in Sardinia traced to Corbeddu Cave [eastern Sardinia], about 11,350 years ago), followed by *Praemegaceros cazioti* (with a latest presence in Corsica at Teppa di U Lupinu Cave [Haute-Corse, France], about 8,500 years ago, and in Sardinia at Juntu Cave [North Eastern Sardinia], about 7,000 years ago). Across the Pleistocene–Holocene boundary two endemic birds of prey disappeared from the massif: the large eagle *Aquila nipaloides*, last recorded during the last glacial, and the small endemic eagle-owl *Bubo insularis*, recorded until about 8,500 years ago in Corsica. In Sardinia, the disappearance of small mammals is curiously related to the relative frequency and abundance of each species in the Late Pleistocene faunal assemblages. The mole *Talpa tyrrhenica* and the shrew *Asoriculus similis* first disappeared in the early Holocene; the larger high-crowned murid *Rhagamys orthodon* has been reported in a few Neolithic localities; the vole *Microtus (Tyrrhenicola) henseli* is recorded until the late Bronze Age; whereas the pika *Prolagus sardus* survived until historical times, and it seems to have played an important role as an edible resource for the Mesolithic people of Sardinia (Wilkens and Delussu 2002).

The actual time of arrival of human settlers in Sardinia has been debated (e.g., Sondaar 2000; Mussi 2001), but there is little doubt about the existence of a human presence on the Corsica–Sardinia massif during the Mesolithic. In Corsica, the data attest to a human presence at least by 11,000 years ago (Vigne 2000), although some evidence suggests a human presence by about 17,000 BC (Salotti et al. 2008). In Sardinia, the oldest human remains (with an inferred age of about 20,000 years ago) are a phalanx found in Corbeddu Cave and a mandible from the same cave dated at 8,750 +/− 104 years ago (Klein Hofmeijer et al. 1987). In 2009 Rita T. Melis and colleagues reported finding two sepultures dated at about 8,500 years ago at S'Omu e s'Orku (in western Sardinia).

Although a noticeable turnover followed the arrival of Mesolithic humans and their accompanying fauna, the long survival of deer and small mammals suggests that prehistoric humans had only a minimal direct role in the demise of endemic mammals from the Corsica–Sardinia massif. By contrast, the introduction of alien species altered the equilibrium of insular

Artistic reconstruction of the landscape around Corbeddu cave about 12-11,000 years ago showing the wild dog *Cynotherium sardous*, the deer *Praemegaceros cazioti*, the murid *Rhagamys orthodon*, the large vole *Microtus (Thyrrenicola) henseli*, the shrew *Asoriculus henseli*, the pika *Prolagus sardus* and the large eagle *Aquila nipaloides*. Illustration by Stefano Maugeri.

communities, especially during a period characterized by climate oscillations.

Sicily and Malta

The large island of Sicily, now separated from the southern Italian mainland by the narrow Strait of Messina (1.9 to 3.2 miles [3.1 to 5.1 km] in width), came into existence in the earliest Pleistocene. Although it is a relatively young insular system, Sicily has had a complex faunal history (Masini et al. 2008), which has mainly been driven by the paleogeographic setting of the area, which controlled species dispersal. During the Early and possibly the earliest Middle Pleistocene, Sicily consisted of separated islands that had been colonized by a few species through sweepstake and passive dispersals as shown by the extremely impoverished and strongly disharmonic fauna of the so-called *Elephas falconeri* faunal complex. This complex included, together with a quite rich herpetofauna, birds and bats and mostly small mammals, whereas the only large mammals were the dwarf elephant *Palaeoloxodon falconeri* (the smallest in the Mediterranean islands, with a height at the shoulder of no more than 39 inches [100 cm]) and the endemic otter *Lutra trinacriae*. By the late Middle Pleistocene the filtering action of the barrier had decreased, allowing a number of large mammals (lion, spotted hyena, wolf, brown bear, elephant, boar, hippopotamus, fallow and red deer, bison, and auroch) to reach Sicily, but not any small mammals. As a result, *P. falconeri* disappeared, and a new, deeply renewed and more balanced fauna (the "*Elephas*" *mnaidriensis* faunal complex) inhabited the island, including some species identical to or only somewhat modified from the congeneric/conspecific taxa of the Italian Peninsula. With the climate growing colder, forecasting the onset of the last glacial, a few species arrived, such as the Savi's Pine Vole *Microtus* [*Terricola*] ex gr. *M. savii*, the wood mouse *Apodemus sylvaticus*, the ancestor of the extant shrew *Crocidura sicula*, the red fox, and possibly the slender horse *Equus hydruntinus*, whose cursorial body proportions suggest an adaptation to semiarid conditions.

Climate change and competition with the newcomers likely caused the extinction of some endemic species, such as the otter, the fallow deer, the hippopotamus, and all the small mammals. The middle-sized elephant *Palaeoloxodon mnaidriensis*, the endemic red deer, and the endemic bovids survived at least until about 32,000 years ago, as documented by remains found in San Teodoro Cave, the fossil record of which is the epitome of the new San Teodoro Cave–Pianetti faunal complex (see Bonfiglio et al. 2008). These Middle and Late Pleistocene extinction events cannot be attributed to direct or indirect actions by humans. Indeed, researchers have demonstrated that there is no compelling evidence of a human presence in Sicily before 17,500 years ago (Mannino and Thomas 2007; Martini et al. 2007), a date considered to be the last possible date for the colonization of Sicily by Upper Paleolithic humans. The latter possibly entered the island during the Last Glacial Maximum by crossing a temporarily emerged land bridge, as the terrestrial mammals did, some of which could have been introduced accidentally or intentionally on the island by humans. These species characterize the so-called Castello faunal complex, which

includes the faunal assemblages recorded in Sicily from about 20,000 to 11,000 years ago (Masini et al. 2008). The fauna of this complex is largely new, because endemic herbivores and large predators are no longer recorded, while a number of continental mammals appeared. Nonetheless, the Castello complex represents only a subset of the fauna inhabiting the Italian Peninsula at the time, as several species widely spread over the peninsula (e.g., the mountain ungulates chamois and ibex, as well as a number of rodents) did not reach Sicily, probably because of a severe filtering effect of the land bridge.

The history of the fauna of Malta, although still not known in detail (see Hunt and Schembri 1999), roughly matches that of Sicily. Malta's endemic fauna likely disappeared by the Last Glacial Maximum, when a number of species arrived by crossing the narrow land bridge that connected the Maltese island system to Sicily at that time. These arrivals are documented in the red earth layers of Ghar Dalam Cave, which date back to about 18,000 years ago (see Zammit-Maempel 1989). In these layers, carnivores (e.g., *Canis lupus*, *Vulpes vulpes*, *Ursus* cf., *U. arctos*) are recorded for the first time together with other large mammals (e.g., *Sus scrofa*, *Cervus elaphus*, *Bos primigenius*) and small mammals (e.g., *Microtus* [*Terricola*] ex gr. *M. savii*, *Crocidura* sp.), which were present at the same time in Sicily. Other species, such as the Western European hedgehog *Erinaceus europaeus*, the wood mouse *Apodemus sylvaticus*, and brown hare *Lepus europaeus*, apparently did not enter Malta, whereas the presence on this island of *Equus hydruntinus* remains uncertain.

Overall, it seems that the progressive extirpation of endemic mammals in Sicily and Malta was mainly a result of competition with alien species, though the arrival of Epigravettian Upper Paleolithic humans on Sicily had an impact on the fauna. Conversely, human action cannot be claimed as the cause of extinction of endemic mammals in Malta because the island was first occupied about 7,400 years ago by humans coming from Sicily (see Marriner et al. 2012).

Crete

The paleogeographic history of Crete is fairly complex and includes phases of connection to the mainland, fragmentation into small islands, and submersion. The island acquired its present status only in the Early Pleistocene (see Van der Geer et al. 2010). Since that time the island has been inhabited by strongly impoverished, disharmonic faunas and colonized by overseas sweepstake dispersal of small mammals, elephants, hippopotamuses, deer, and otters. The extremely rich Cretan fossil record found in a number of localities of different ages makes this island a valuable case history for colonization, adaptation, originations, and extinction of island endemic ungulates and rodents. Based on murids, two main faunal complexes have been defined: the *Kritimys* zone (Early–early Middle Pleistocene; split into the *K.* aff. *K. kiridus*, *K. kiridus*, and *K. catreus* subzones) and the *Mus* zone (late Middle–Late Pleistocene; split into the *Mus bateae* and *M. minotaurus* subzones), with a minimal overlap between the two. The *Kritimys* complex includes, in addition to the Cretan rat, a tortoise (*Clemmys* cf. *C. caspica*) and a frog (*Rana* cf. *R. ridibunda*), the pygmy mammoth *Mammuthus creticus*, and the

Artistic reconstruction of the landscape around San Teodoro cave (Sicily) about 30,000 year ago, showing the still surviving endemic large mammals *Palaeoloxodon mnaidriensis, Bos primigenius siciliae, Bison priscus siciliae,* and *Cervus elaphus siciliae* (the skull on the right), along with *Ursus arctos, Canis lupus, Crocuta crocuta* as well as the slender horse *Equus hydruntinus,* claimed to be present since that time, but which entered the island for sure by crossing a land bridge temporarily connecting Sicily to Calabria during the Last Glacial Maximum. Small mammals, first recorded at that time, are (from the left to the right) *Erinaceus europaeus, Apodemus* cf. *A. silvaticus Microtus* (*Terricola*) gr. *M. savii,* and *Crocidura* cf. *C. sicula.* Illustration by Stefano Maugeri.

dwarf hippopotamus *Hippopotamus creutzburgi.* The faunas of the *Mus* complex were characterized by the occurrence of no fewer than eight different, highly specialized deer (*Candiacervus ropalophorus, C.* sp. IIa, b, and c, *C. cretensis, C. rethymnensis, C. dorothensis, C. major;* De Vos 1996), whose height at the shoulder ranged from about 16 to 65 inches (40 to 165 cm). Cretan deer likely originated by a radiative evolution from a common ancestor, which probably colonized Crete when some elements of the *Kritimys* complex (e.g., *H. creutzburgi*) were still present. Other typical taxa accompanying the mice were an elephant quite reduced in size (*Palaeoloxodon creutzburgi*), an otter (*Lutrogale cretensis*), a white-toothed shrew (*Crocidura zimmermanni*), and a tortoise (*Testudo marginata cretensis*).

Archaeological evidence indicates that this fauna, with the exception of the still surviving *C. zimmermanni,* probably disappeared by the arrival of Neolithic humans on the island. For instance, in the Aceramic Neolithic level of Knossos, the mammalian fauna includes only alien species introduced by humans, such as the agrimi *Capra aegagrus,* the goat *Capra hircus,* the dog *Canis familiaris,* and ancestors of the extant endemic mustelids *Martes foina bunites* and *Meles meles arcalus.* Some features of the Knossos mustelids are somewhat similar to those of their Near Eastern relatives, which suggests an

eastern origin of the human settlers and their accompanying fauna (Evans 1968).

Two lines of evidence have provided clues to the timing and driving forces behind the extinction of the Cretan endemic deer. First, although evidence suggests that the largest deer were not present during the last glacial, some amino acid racemization datings (Reese et al. 1996) indicate that the smallest deer survived for a longer time—47,000 years +/– 20 percent at Gerani 2 and 21,000 years +/– 20 percent at Simonelli I. Second, human exploitation of coastal and estuarine wetland resources on Crete took place in both the Pleistocene and the early Holocene, and two separate human groups inhabited the island—one in the latest Middle to Late Pleistocene (since about 130,000 years ago) and the other in the Late Pleistocene to Early Holocene (about 11,000 to 9,000 years ago). This hypothesis is based on a number of artifacts found in 28 sites of the Cretan Plakias region. In some sites the characteristic of Cretan artifacts are consistent with those of Lower Paleolithic artifacts found in the Greek mainland, whereas they are consistent with those of the Mesolithic at other sites (Strasser et al. 2010). Accordingly, the hypothesis that humans had any impact on the selective, diachronous extinction of some Cretan deer species cannot be rejected.

Tilos

The small island of Tilos is located between Rhodes and Kos about 12.5 miles (20 km) from the nearest point of the Anatolian peninsula. In spite of the short distance from the mainland, Tilos was inhabited during the late Quaternary by two monospecific endemic mammalian faunas, both of which have been documented by fossil remains discovered in Charkadio Cave in the early 1970s. The first consists of deer, which inhabited the island about 140,000 years ago, whereas the second includes the endemic dwarf elephant *Palaeoloxodon tiliensis* (redescribed in 2007 by Theodorou et al.), one bat (*Myotis blythii*), a tortoise (*Testudo marginata*), and various unstudied birds. The dwarf elephants, which stood no more than 5 feet (1.5 m) at the shoulder, lived on the island from about 45,000 to 3,500 years ago. Their ancestor possibly arrived during the beginning of the last glacial, about 90,000 years after the extinction of the Telian deer.

It is generally accepted that the extinction of the deer was related to the volcanism that around 10,000 years ago turned Tilos into an uninhabitable island, but the forces that drove the extinction of *Palaeoloxodon tiliensis* are still unknown. The Telian elephant species was the last European straight-tusked elephant. It coexisted on the island with post-Paleolithic humans (Bachmayer et al. 1984) and was perhaps known by Egyptian people (Masseti 2001). Dwarf elephants survived on Tilos at least until the beginning of the Aegean Bronze Age, in spite of hunting by humans and competition with introduced animals (e.g., the Eastern European hedgehog *Erinaceus*

Artistic reconstruction of the landscape around Simonelli cave (Crete) about 20000 years ago, showing the extinct endemic deer *Candiacervus*, giant owl *Athena cretensis* and murid *Mus minotaurus* (on the right), and the still living insectivore *Crocidura zimmermanni* (on the left). Illustration by Stefano Maugeri.

concolor, the lesser white-toothed shrew *Crocidura suaveolens*, the rabbit *Oryctolagus cuniculus*, the broad-toothed field mouse *Apodemus mystacinus*, the roof rat *Rattus rattus*, and the house mouse *Mus domesticus*). Therefore, other factors have to be taken into account to explain their extinction (Theodorou et al. 2007). These factors include the eruptions, earthquakes, and destruction of grasses related to the activity of the Nisiros and Santorini volcanoes; the pollution of water by volcanic tuffs; and, possibly, the reduction of the island's surface area stemming from the progressive rise of sea level following the Last Glacial Maximum.

Cyprus

Cyprus is one of the most geologically and biogeographically isolated Mediterranean islands, separated from the southern seaboard of Anatolia and the Syro-Palestinian littoral by two deep submarine features, the Adana Trough and the Latakia Basin. Even at times of the maximum glacio-eustatic drop of the sea level, Cyprus remained separated from the mainland by at least 18.5 to 25 miles (30 to 40 km) of deep water. During the Late Pleistocene the island was inhabited consistently by an oligotypic, strongly disharmonic mammalian fauna, including a dwarf hippopotamus (*Phanourios minor*, a browser, with a unguligrade locomotion, about 30 inches [76 cm] tall, weighing about 440 pounds [200 kg]) and an elephant (*Palaeoloxodon cypriotes*), which were later joined at the Pleistocene to Holocene transition by a small carnivore (*Genetta plesictoides*) and an endemic mouse (*Mus cypriacus*) that colonized the island by overseas sweepstake dispersals. This fauna also included some bats, such as the fruit bat *Rousettus aegyptiacus*, and birds. Reliable dates of *P. minor* bone collagen testify to the presence of this hippopotamus on Cyprus at least between about 22,000 and 3,700 years ago (Reese 1995), although its extreme specialization would suggest a Middle, or perhaps even Early Pleistocene, colonization (Van der Geer et al. 2010).

Humans and their accompanying fauna arrived on the island toward the end of the Pleistocene or at the beginning of the Holocene, but whether an overkill or blitzkrieg model can explain the megafaunal extinction is still a matter of heated debate. The arguments entirely focus on the findings at Aetokremnos, a collapsed rock shelter on the Akrotiri Peninsula that shows evidence of one of the oldest human occupations of the western Mediterranean islands. The oldest cultural level, known as the Aetokremnos phase, is Late Epipaleolithic (11,000–10,000 years ago) and contains blackish hippopotamus bones, which are absent in the younger Neolithic layers, which in turn contain burnt shells and bird and hare bones. Alan H. Simmons (1999) regarded this evidence as proof of the demise of the endemic megafauna not long after the first human presence on the island. Nevertheless, many researchers remain skeptical about the overkill hypothesis at this site, stressing the absence of clear evidence of marrow extraction, butchering, and burning of the hippopotamus bones. But the hypothesis that the Epipaleolithic people of Aetokremnos might have played any role in the extinction of the Pleistocene megafauna, likely in combination with other factors such as climatic and environmental change, competition with introduced species, and possibly disease, cannot be ruled out (see Knapp 2010 for a discussion).

Artistic reconstruction of the landscape of Cyprus at the end of the Pleistocene showing the pigmy hippopotamus (*Phanourios minor*), the dwarf elephant (*Palaeoloxodon cypriotes*), the genet (*Genetta plesictoides*) and the great bustard (*Otis tarda*). Illustration by Stefano Maugeri.

Causes of the demise of the endemic mammals of the Mediterranean islands

Intrinsic and extrinsic factors associated with extinction and extinction risk have been scrutinized and debated by a number of paleontologists and neontologists. A special emphasis has been given to causal mechanisms leading to Late Pleistocene/early Holocene megafaunal extinctions across continents (e.g., Martin and Klein 1989; MacPhee 1999; Haynes 2009; Johnson 2009; Nogués-Bravo et al. 2010; Barnosky et al. 2011; Field and Wroe 2012; Prescott et al. 2012). It is not a simple task to delineate a general model to explain the dramatic global demise of mammals during the late Quaternary. All in all, extinctions result from a complex interaction of several biotic and abiotic factors, whose synergetic dynamics could vary depending on body size, basal metabolism and correlated characteristics, preferred habitat, population density, dietary specialization, reproductive rate, and the degree of tolerance to environmental changes of each species. Extinctions also depend on the response of individuals and species to perturbations of coevolutionary equilibrium characterizing each paleocommunity. In Europe, compelling evidence suggests that climate changes were a major factor in promoting some extinctions of less ecologically flexible species, while human impacts likely either increased the risk of extinction or accelerated the process of extinction for species already stressed by environmental change, range contraction, and reduction of population density and size.

What about Mediterranean insular species? The classic equilibrium theory of Robert H. MacArthur and Edward O.

Wilson (1967) assumes that, on islands, the number of species will be augmented continuously by the arrival of new species from mainland sources and continuously depleted by extinctions. The immigration rate is obviously a function of isolation, which can change in response to geodynamic events, glacio-eustatic fluctuations of sea level, or a combination of both. MacArthur and Wilson regarded extinctions on islands as a function of population size, although they were not explicit about the mechanism leading to the species turnover. They nevertheless recognized the importance of competitive replacement, whose dynamics mostly depend on available niches and the resources and surface area of the island (along with the biological characteristics of resident and immigrant species).

Keeping in mind this theoretical base, as well as the fact that in situ speciation events could increase the number of resident species, the regional effects of global climate changes, and the paleontological and archaeological evidence from the Mediterranean islands, the remainder of this entry delves into the extinction events in an attempt to disentangle the causal mechanisms of the diachronous demise of the Pleistocene species across islands and on the same island throughout time. Three main factors, (1) remarkable environmental changes, (2) biological interaction between previously resident and new species, and (3) human settlement on the islands, have been claimed as key causal mechanisms to explain the extirpation of the autochthonous insular vertebrates, although the discussion has mainly been focused on mammals.

Environmental and climate changes

In regard to dramatic environmental changes, including stochastic events and natural disasters, such as hurricanes, volcanic eruptions, and tsunami (see Whittaker and Fernández-Palacios 2007), only the latter would have directly and sharply led to the extinction of some Mediterranean insular species, but there is not any compelling evidence of either significant volcanic activity or tsunami deposits in the latest Pleistocene to early Holocene of the Mediterranean islands. The volcanic eruption claimed to have caused the extinction of Tilos deer is much older (dating to about 140,000 years ago; Theodorou et al. 2007), and the hypothetical tsunami deposits claimed to be present at Su Nuraxi nuragic archaeological site (Barumini, Sardinia) are much more recent, dating to about 1300 to 1100 years ago.

The role of global climate changes in promoting faunal evolution and turnovers is a hot topic in paleobiology and paleoecology. Evidence from the Mediterranean suggests that climate itself had only a trivial impact on extinctions in isolated systems. For instance, the long stability of the Majorcan fauna indicates that no faunal turnovers occurred in the Gymnesic Islands during most of the Pleistocene in the face of the changes in physical parameters (e.g., average temperature and precipitation) related to glacial–interglacial cycles. On the contrary, the endemic large and small mammals of the Gymnesics became extinct during nearly stable climate periods.

In Sardinia and possibly Crete, no apparent correlation occurs, at least during the Middle and Late Pleistocene, between climatic oscillations and changes in faunal diversity.

Climate changes likely had an indirect effect on the persistence instead of the extinction of insular species—especially of large mammals—by providing easier conditions for dispersal during periods of sea-level lowering. This was, for instance, the case with the younger faunal history of Sicily, an island close to and separated from the continent by a rather narrow sea barrier that turned into, at least during the Last Glacial Maximum, a temporarily emerged land bridge as a result of the combined effect of tectonic action and a glacio-eustatic sea-level drop.

Competition with new/alien species

The reconstruction of the faunal histories of Mediterranean islands shows that, without exchanges with the mainland, the insular paleocommunities inhabiting oceanic-like islands were not significantly affected by changes in physical parameters, such as temperature and precipitation. Conversely, climate changes and changing paleogeography likely triggered discrete dispersals from the mainland toward islands not far from the continent, promoting new competitive dynamics inside the insular communities indirectly. Although compositional changes also depend on species interaction (intra- and interspecific competition) and coevolution, which allows origination and extinction events to also occur during periods of complete isolation and of no arrival of new settlers (e.g., the asynchronous disappearances of Cretan endemic deer), the disruption of paleocommunity equilibrium by the introduction of alien species was a key causal factor in promoting extinctions. A number of insular taxa, indeed, went extinct after the successful establishment of new, alien species. It is worth noting that because carnivores—especially top predators—are usually absent or rarely recorded in insular faunas, and on islands a broad spectrum of niches was available because of the low biodiversity of insular ecosystems, most of the endemic mammals underwent structural and morphofunctional modifications related to a change in their adaptive strategy. For endemic mammals, the absence of powerful predators entailed the loss of their ability to escape predators of potential prey, while also enabling them to more efficiently exploit resources through a more parsimonious use of energy in a predator-free environment. This makes insular mammals particularly vulnerable to predation and enhances their extinction risk.

Nonetheless, some insular mammals became extinct as a result of the arrival of competitors from the mainland. Researchers are led to ask why, then, were and are insular species so vulnerable to alien species introductions. Insular faunas have a particular composition and structure, being impoverished and unbalanced, and feature particular interactions among species. As stated above, frequently functional types of endemic insular mammals significantly differ not only from those of their ancestors but also from those of species that on the mainland occupy, in a more efficient way, a niche similar to what the insular settlers have been adapted to. Consequently, the specialized resident species could not successfully compete with more generalist, efficient newcomers and ultimately were extirpated. This might be the case, for instance, with the deer *Praemegaceros cazioti* on Sardinia. During Nuragic times it was not actively hunted by

humans, but had to compete with alien species that humans had introduced, such as red deer, goats, and sheep.

The introduction of alien species could also mean the introduction of lethal diseases (or the hyperdisease hypothesis; see MacPhee and Marx 1997), which has been argued as a cause of insular extinctions of animals having no immunity to pathogens that arrived on the island with humans and their accompanying fauna. Although this hypothesis is difficult to test in fossil species, and it not clear how pathogens could actually have sent different species to extinction under natural conditions, the disease hypothesis emerges as the most reasonable way to explain extinctions when other causes—climate, competition with invasive species, predation, and human direct impact—fail to be proven, such as the case of small mammals from the Gymnesic Islands.

Human settlement

Prehistoric human hunting undoubtedly dramatically modified insular animal communities, but how many extinctions were actually directly caused by human hunting is still an unanswered question. Because of the relatively little archaeological evidence for dramatic human hunting of endemic mammals, a blitzkrieg extinction model (sudden extinctions caused by rapid human population growth and the extensive human hunting of the large game of naive faunas) is hardly valid for the Mediterranean, except perhaps for the Balearic *Myotragus* (see above). But a combination of the sitzkrieg (human indirect actions) and overkill (protracted human hunting and correlated effects) hypotheses could be more realistic.

Looking at the Mediterranean, the only evidence of active hunting of large endemic mammals is provided by fossil records on Cyprus. On Crete, humans were hardly responsible for the extinction of the smallest endemic deer. In fact, assuming a long persistence of Paleolithic humans who arrived on Crete about 130,000 years ago, they would have had some impact on large deer, but they surely coexisted for thousands of years with the small ones. The small Cretan deer apparently did not last until the arrival of Late Epipaleolithic/Neolithic humans. Therefore, it appears that neither Paleolitic nor Neolithic humans were responsible for their extinction. In Sicily, the endemic fauna, except for the still living *Crocidura sicula*, disappeared before the first settlement of Upper Paleolithic humans, given the lack of a human presence on the island before about 17,000 years ago. Accordingly, the turnover between fauna that still yielded few endemic mammals (the San Teodoro Cave—Pianetti faunal complex) and fully continental fauna (the Castello faunal complex) was triggered by the arrival of continental alien species at the time of the dramatic climate variations of Last Glacial Maximum.

On the Corso-Sardinian massif, where humans were present possibly by about 20,000 years ago and definitely by about 9,000 years ago, their hunting did not cause the direct extinction of either deer or pikas (which likely were an important component in the diet of Mesolithic humans) or the indirect extinction of the wild dog *Cynotherium*, by usurping its preferred, tremendously abundant prey. Nonetheless, human impacts acted not only by introducing exotic species (which could have significantly disrupted the equilibrium, competing with or even preying on resident endemic species) and virulent diseases but also through the fragmentation of viable habitat during critical climate periods.

Overall, based on the main data, it appears that although different combinations of factors likely triggered the disappearance of Late Quaternary megafauna from the Mediterranean islands, there is little question that prehistoric humans were a key component in the demise of most of the Pleistocene insular mammals, but their impact, mainly an indirect one, differed from island to island. Moreover, it appears that the combined action of direct and indirect human and climate action affected large and small mammals differently. As stressed by Lawrence R. Heaney in a 1984 article, large mammals are confirmed to have a greater likelihood of extinction than small mammals, given that the latter are in general less liable to extinction in the face of changing ecosystem dynamics, as is the case with herpetofauna as well. Indeed, the only endemic Pleistocene species, other than lizards, still inhabiting Mediterranean islands are the Sicilian and Cretan insectivores *Crocidura sicula* and *C. zimmermanni*, respectively, and the Cypriot mouse *Mus cypriacus*.

Much work has still to be done to improve the scientific community's knowledge of the extinction dynamics in isolated ecosystems. But there is no doubt that the demise of endemic Pleistocene species from the Mediterranean islands led to the disappearance of exclusive, astonishing biomes nowhere present today.

Resources

Books

Barnosky, Anthony D., Marc A. Carrasco, and Russell W. Graham. "Collateral Mammal Diversity Loss Associated with Late Quaternary Megafaunal Extinctions and Implications for the Future." In *Comparing the Geological and Fossil Records: Implications for Biodiversity Studies*, edited by A.J. McGowan and A.B. Smith, 179–190. Geological Society Special Publication 358. London: Geological Society, 2011.

Carminati, Eugenio, and Carlo Doglioni. "Mediterranean Geodynamics." In *Encyclopedia of Geology*, edited by Richard C. Selley, L. Robin M. Cocks, and Ian R. Plimer, Vol. 2, 135–146. Amsterdam: Elsevier, 2005.

De Vos, John. "Taxonomy, Ancestry, and Speciation of the Endemic Pleistocene Deer of Crete Compared with the Taxonomy, Ancestry, and Speciation of Darwin's Finches." In *Pleistocene and Holocene Fauna of Crete and Its First Settlers*, edited by David S. Reese, 111–124. Monographs in World Archaeology 28. Madison, WI: Prehistory Press, 1996.

Diamond, Jared M. "Historic Extinctions: A Rosetta Stone for Understanding Prehistoric Extinctions." In *Quaternary Extinctions: A Prehistoric Revolution*, edited by Paul S. Martin and Richard G. Klein, 824–862. Tucson: University of Arizona Press, 1984.

Grill, Andrea, Paolo Casula, Roberta Lecis, and Steph Menken. "Endemism in Sardinia." In *Phylogeography of Southern European Refugia: Evolutionary Perspectives on the Origins and Conservation of European Biodiversity*, edited by Stephen Weiss and Nuno Ferrand, 273–296. Dordrecht, Netherlands: Springer, 2007.

Haynes, Gary, ed. *American Megafaunal Extinctions at the End of the Pleistocene*. Dordrecht, Netherlands: Springer, 2009.

Hofmeijer, Gerard Klein. *Late Pleistocene Deer Fossils from Corbeddu Cave: Implications for Human Colonization of the Island of Sardinia*. BAR International Series 663. Oxford: British Archaeological Reports/John and Erica Hedges, 1997.

Hunt, Christopher O., and Patrick J. Schembri. "Quaternary Environments and Biogeography of the Maltese Islands." In *Facets of Maltese Prehistory*, edited by Anton Mifsud and Charles Savona Ventura, 41–75. Malta: Prehistory Society of Malta, 1999.

Le Pichon, Xavier. "Les fonds de la Méditerranée". Scale 1:4,250,000. Paris: Hachette, 1987.

MacArthur, Robert H., and Edward O. Wilson. *The Theory of Island Biogeography*. Princeton, NJ: Princeton University Press, 1967.

MacPhee, Ross D.E., ed. *Extinctions in Near Time: Causes, Contexts, and Consequences*. New York: Kluwer Academic/Plenum, 1999.

MacPhee, Ross D.E., and Preston A. Marx. "The 40,000-Year Plague: Humans, Hyperdisease, and First-Contact Extinctions." In *Natural Change and Human Impact in Madagascar*, edited by Steven M. Goodman and Bruce D. Patterson, 168–217. Washington, DC: Smithsonian Institution Press, 1997.

Martin, Paul S., and Richard G. Klein, eds. *Quaternary Extinctions: A Prehistoric Revolution*. 2nd ed. Tucson: University of Arizona Press, 1989.

Martini, Fabio, Domenico Lo Vetro, André C. Colonese, et al. "L'Epigravettiano finale in Sicilia." In *L'Italia tra 15,000 e 10,000 anni fa: Cosmopolitismo e regionalità nel Tardoglaciale*, edited by Fabio Martini, 209–254. Florence, Italy: Museo Fiorentino di Preistoria "Paolo Graziosi," 2007.

Masseti, Marco. "Did Endemic Dwarf Elephants Survive on Mediterranean Islands up to Protohistorical Times?" In *The World of Elephants: Proceedings of the 1st International Congress*, edited by Giuseppe Cavarretta, Patrizia Gioia, Margherita Mussi, and Maria R. Palombo, 402–406. Rome: Consiglio Nazionale delle Ricerche, 2001.

Melis, Rita T., Margherita Mussi, Rosalba Floris, et al. "Popolamento e ambiente nella Sardegna centro occidentale durante l'Olocene antico: Primi risultati." In *Atti della XLIV Riunione Scientifica: La preistoria e la protostoria della Sardegna—Cagliari November 22–28, 2009*. Florence, Italy: Istituto di preistoria e protostoria, in press.

Mussi, Margherita. *Earliest Italy: An Overview of the Italian Paleolithic and Mesolithic*. New York: Kluwer Academic/Plenum, 2001.

Simmons, Alan H. *Faunal Extinction in an Island Society: Pygmy Hippopotamus Hunters of Cyprus*. New York: Kluwer Academic/Plenum, 1999.

Van der Geer, Alexandra, George Lyras, John de Vos, and Michael Dermitzakis. *Evolution of Island Mammals: Adaptation and Extinction of Placental Mammals on Islands.* Chichester, U.K.: Wiley-Blackwell, 2010.

Vigne, Jean-Denis. "Les débuts néolithiques de l'élevage des ongulés au Proche-Orient et en Méditerranée: Acquis récents et questions." In *Premiers paysans du monde: Naissances des agricultures,* edited by Jean Guilaine, 143–168. Paris: Éditions Errance, 2000.

Whittaker, Robert J., and José Maria Fernández-Palacios. *Island Biogeography: Ecology, Evolution, and Conservation.* 2nd ed. Oxford: Oxford University Press, 2007.

Wilkens, Barbara, and Fabrizio Delussu. "Les mammifères sauvages de la Sardaigne: Extinctions et nouvelles arrivées au cours de l'Holocène." In *Mouvements ou déplacements de populations animales en Méditerranée au cours de l'Holocéne,* edited by Armelle Gardeisen, 23–31. BAR International Series 1017. Oxford: John and Erica Hedges/Archaeopress, 2002.

Zammit-Maempel, George. *Ghar Dalam Cave and Deposits.* Malta: Selbstverl, 1989.

Periodicals

Bachmayer, Friedrich, Nikos K. Symeonidis, and Helmuth Zapfe. "Die Ausgrabungen in der Zwergelefantenhöhle der Insel Tilos (Dodekanes, Griechenland) im Jahr 1983." *Sitzungsberichten der Österreichischen Akademie der Wissenschaften, Mathematisch-naturwissenschaftliche Klasse* Abteilung 1, 193, nos. 6–10 (1984): 321–328.

Bonfiglio, Laura, Daniela Esu, Gabriella Mangano, et al. "Late Pleistocene Vertebrate-Bearing Deposits at San Teodoro Cave (North-Eastern Sicily): Preliminary Data on Faunal Diversification and Chronology." Quaternary International 190, no. 1 (2008): 26–37.

Bover, Pere, and Josep Antoni Alcover. "Understanding Late Quaternary Extinctions: The Case of Myotragus balearicus (Bate, 1909)." Journal of Biogeography 30, no. 5 (2003): 771–781.

Bover, Pere, and Josep Antoni Alcover. "Extinction of the Autochthonous Small Mammals of Mallorca (Gymnesic Islands, Western Mediterranean) and Its Ecological Consequences." Journal of Biogeography 35, no. 6 (2008): 1112–1122.

Bover, Pere, Josep Quintana, and Josep Antoni Alcover. "Three Islands, Three Worlds: Paleogeography and Evolution of the Vertebrate Fauna from the Balearic Islands." *Quaternary International* 182, no. 1 (2008): 135–144.

Bover, Pere, Josep Quintana, and Josep Antoni Alcover. "A New Species of *Myotragus* Bate, 1909 (Artiodactyla, Caprinae) from the Early Pliocene of Mallorca (Balearic Islands, Western Mediterranean)." *Geological Magazine* 147, no. 6 (2010): 871–885.

Evans, John D. "Knossos Neolithic. Part II: Summary and Conclusion." *Annual of the British School at Athens* 63 (1968): 267–276.

Field, Judith, and Stephen Wroe. "Aridity, Faunal Adaptations, and Australian Late Pleistocene Extinctions." *World Archaeology* 44, no. 1 (2012): 56–74.

Gippoliti, Spartaco, and Giovanni Amori. "Ancient Introductions of Mammals in the Mediterranean Basin and Their Implications for Conservation." *Mammal Review* 36, no. 1 (2006): 37–48.

Heaney, Lawrence R. "Mammalian Species Richness on Islands on the Sunda Shelf, Southeast Asia." *Oecologia* 61, no. 1 (1984): 11–17.

Johnson, Christopher. "Megafaunal Decline and Fall." *Science* 326, no. 5956 (2009): 1072–1073.

Klein Hofmeijer, Gerard, Paul Y. Sondaar, Carl Alderliesten, Klaas van der Borg, and Arie de Jong. "Indications of Pleistocene Man on Sardinia" *Nuclear Instruments and Methods in Physics Research* 29 (1987): 166–168.

Knapp, A. Bernard. "Cyprus's Earliest Prehistory: Seafarers, Foragers, and Settlers." *Journal of World Prehistory* 23, no. 2 (2010): 79–120.

Mannino, Marcello A., and Kenneth D. Thomas. "New Radiocarbon Dates for Hunter-Gatherers and Early Farmers in Sicily." *Accordia Research Papers* 10 (2007): 13–34.

Marriner, Nick, Timothy Gambin, Morteza Djamali, et al. "Georchaeology of the Burmarrad Ria and early Holocene Human Impacts in Western Malta." *Palaeogeography, Palaeoclimatology, Palaeoecology* 339–341 (2012): 52–65.

Masini, Federico, Daria Petruso, Laura Bonfiglio, and Gabriella Mangano. "Origination and Extinction Patterns of Mammals in Three Central Western Mediterranean Islands from the Late Miocene to Quaternary." *Quaternary International* 182, no. 1 (2008): 63–79.

Masseti, Marco, and Maurizio Sarà. "Non-volant Terrestrial Mammals on Mediterranean Islands: Tilos (Dodecanese, Greece), a Case Study." *Bonner zoologische Beiträge* 51, no. 4 (2003): 261–268.

Nogués-Bravo, David, Ralf Ohlemüller, Persaram Batra, and Miguel B. Araújo. "Climate Predictors of Late Quaternary Extinctions." *Evolution* 64, no. 8 (2010): 2442–2449.

Palombo, Maria Rita. "Biochronology, Paleobiogeography, and Faunal Turnover in Western Mediterranean Cenozoic Mammals." *Integrative Zoology* 4, no. 4 (2009): 367–386.

Prescott, Graham W., David R. Williams, Andrew Balmford, et al. "Quantitative Global Analysis of the Role of Climate and People in Explaining Late Quaternary Megafaunal Extinctions." Proceedings of the National Academy of Sciences of the United States of America 109, no. 12 (2012): 4527–4531.

Salotti, Michelle, Antoine Louchart, Salvador Bailon, et al. "A Teppa di U Lupinu Cave (Corsica, France): Human Presence since 8500 Years BC, and the Enigmatic Origin of the Earlier, Late Pleistocene Accumulation." *Acta zoological cracoviensia* 51A, nos. 1–2 (2008): 15–34.

Sondaar, Paul Y. "The Island Sweepstakes: Why Did Pygmy Elephants, Dwarf Deer, and Large Mice Once Populate the Mediterranean?" *Natural History* 95, no. 9 (1986): 50–57.

Sondaar, Paul Y. "Early Human Exploration and Exploitation on Islands." *Tropics* 10, no. 1 (2000): 203–230.

Strasser, Thomas F., Eleni Panagopoulou, Curtis N. Runnels, et al. "Stone Age Seafaring in the Mediterranean: Evidence from the Plakias Region for Lower Palaeolithic and Mesolithic Habitation of Crete." *Hesperia* 79, no. 2 (2010): 145–190.

Theodorou, George, Nikolaos Symeonidis, and Elizabeth Stathopoulou. "*Elephas tiliensis* n. sp. from Tilos Island

(Dodecanese, Greece)." *Hellenic Journal of Geosciences* 42 (2007): 19–32. Accessed August 10, 2012. http://www. hellenjgeosci.geol.uoa.gr/42/19-32.pdf.

Other

Reese, David S. *The Pleistocene Vertebrate Sites and Fauna of Cyprus.* Bulletin of the Geological Survey Department of Cyprus 9. Nicosia, Cyprus: Ministry of Agriculture, Natural Resources, and Environment, Geological Survey Department, 1995.

Reese, David S., Giorgio Belluomini, and M. Ikeya. "Absolute Dates for the Pleistocene Fauna of Crete." In *Pleistocene and Holocene Fauna of Crete and Its First Settlers*, edited by David S. Reese, 47–51. Monographs in World Archaeology 28. Madison, WI: Prehistory Press, 1996.

Maria Rita Palombo
Roberto Rozzi

• • • • •

Modern biodiversity crisis

Biological diversity includes the variation in populations, species, and ecosystems around the world. Biological diversity is documented by the description of species and the categorization of ecological regions and in terms of species interactions, the evolution of species and ecosystems over time, and the mutual interactions between landscapes and life. A dynamic network of interactions between organisms and their environments occurs over daily to seasonal to millennial timescales. These ecological interactions are the basis for ecosystem services that support fundamental aspects of the quality of life for human societies. Just as scientists are learning how biological diversity contributes to these ecosystem services, a veritable mountain of evidence now demonstrates the current and impending losses of species and ecosystems worldwide.

The modern biodiversity crisis refers to the decline and extinction of populations and species and the degradation of ecosystems as a result of human activities. Extinction is the termination of an evolutionary lineage by the death of individuals in all populations of that species. A species becomes doomed to extinction long before the last individual dies, when the number of individuals is too small to maintain a viable population size and the death rate exceeds the birth rate. Degradation of ecosystems includes changes in interactions between organisms and their environments such that ecological processes become less efficient. Extinction is the natural fate of all species eventually, but the scope and rate of extinction during recent human history have increased to the point that many scientists are sounding the alarm that Earth is on the verge of a sixth mass extinction, following the so-called *Big Five* that have already been documented. Ongoing efforts to document the conservation status of species and ecosystems, to understand the causes and consequences of their decline, and to safeguard and improve the prospects for their persistence involve scientists, policymakers, and concerned citizens around the world. The general threats to biodiversity are no mystery—they involve habitat transformation, overharvesting, pollution, invasive species, and climate change. The underlying drivers of these threats involve the ordinary activities of individuals and societies as they affect both naturally rare and common species. For a historical perspective, the fossil record provides estimates of the frequency and magnitude of extinction rates across Earth history. Case studies illustrate the various proximate causes of species decline.

Most of the examples and case studies presented below involve estimates of changes in species diversity or ecosystem composition over time. For examples from the distant past, estimates of species diversity at the local scale are based on the remains of species in fossil assemblages or archaeological concentrations. At the regional to global scales, diversity is estimated both from fossil remains occurring in specific time intervals as well as from lineages represented in fossil assemblages before and after but absent from the specific time interval. Ecosystem composition is assessed from the number and kinds of species present as well as from the sedimentary environment (e.g., river, marsh, cave, coral reef). In modern ecosystems, the estimates of species diversity are based on surveys by taxonomic specialists in particular locations and from prior information about species' geographic ranges. Ecosystem composition is assessed in terms of the species present and the biophysical attributes of the local environment, such as the depth of soil horizons and the rates of nutrient cycling.

The historical context

Change characterizes natural systems over short and long timescales. When does change constitute a crisis? For the modern biodiversity crisis, recognition that native species and ecosystems are being disrupted across a wide range of terrestrial and aquatic ecosystems and across many different taxonomic groups has involved comparisons with historical baselines. Baselines from recent human history, even from several millennia ago, represent important benchmarks in environmental history (for example, the local animals during the time of the early Egyptian pharaohs, the primeval forest of New England before European colonization, North Atlantic coastal fisheries before trawling began), but even these ecosystem conditions have been substantially altered by human activities. The ultimate historical baseline is the fossil record of the history of life.

The geohistorical record of fossils and chemical traces of photosynthesis indicate that life on Earth is almost 4 billion years old. The fossil record of multicellular organisms with external or internal skeletons spans about the last 540 million years, and it is

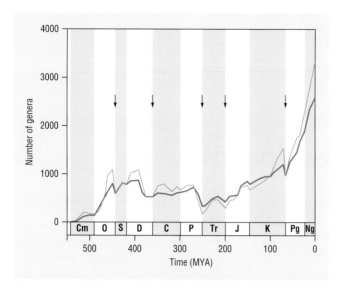

Figure 1. Marine diversity over 540 million years of Earth history. Diversity was compiled at the genus level (just above the species level in the taxonomic hierarchy) because fossils were most reliably identified at this level for a global dataset. Heavy brown line indicates current data, thin blue line indicates an older dataset. Arrows indicate the five mass extinctions documented in the fossil record (Alroy et al. 2008). Abbreviations: Cm, Cambrian; O, Ordovician; S, Silurian; D, Devonian; C, Carboniferous; P, Permian; Tr, Triassic; J, Jurassic; K, Cretaceous; Pg, Paleogene; Ng, Neogene. Reproduced by permission of Gale, a part of Cengage Learning.

over this time span that paleontologists have estimated the average duration of animal or plant species, the global diversity for standardized time intervals, and the distribution of ecosystems in relation to changing configurations of continents, ocean basins, and paleoclimates. This record provides a critical perspective on normal or background rates of extinction and on what constitutes a crisis in the history of life.

The fossil record of marine diversity over the last 540 million years (see Figure 1) is based on documented occurrences of marine invertebrates, which include many groups with fossilizable skeletons (Alroy et al. 2008). Marine diversity rose from the initial appearance of multicellular invertebrates just prior to the Cambrian period to about 900 genera in the Ordovician. Fluctuations around this value persisted until about 250 million years ago. At the end of the Permian, diversity dropped substantially, and it then increased from the Triassic to the present day. After correcting for variation in sample size among time intervals, the increase in diversity from the Triassic to the present day is less steep than it appears in Figure 1.

Five intervals are marked by substantial, geologically rapid declines in diversity (arrows in Figure 1). During these intervals, 35 percent to 76 percent of marine genera disappeared, with an estimated loss of 75 percent to 95 percent of marine species (Jablonski 1991). These mass extinctions were geographically, taxonomically, and ecologically broad. Between these intervals, extinction was approximately matched or exceeded by the appearance of new lineages, so that global marine diversity hovered around a narrow range of values or slowly increased

over time. In addition to the five mass extinctions, minor fluctuations in global marine diversity, as well as in similar data sets for terrestrial vertebrates and plants, occurred many times in the past. Minor extinction episodes and the ecosystem transformations that accompanied them happened in restricted geographic regions or involved a small proportion of species.

In order to compare extinction rates from different episodes in Earth history or with data from modern ecosystems, it is necessary to account for the time interval over which rates are measured and the number of species (or genera) present during that interval. Consequently, rates are often expressed as the number of extinctions per species present per unit of time (e.g., number of extinctions per million species per million years). Documented extinction rates compiled over many time intervals in the fossil record indicate that mass extinction rates are not a separate class unto themselves but rather are the highest rates from a distribution with a much lower average extinction rate. These lower average rates are typical over most of the fossil record. For example, estimates of the long-term extinction rate for mammals range from 0.1 to 1 species per thousand species per millennium (Millennium Ecosystem Assessment 2005).

Such estimates of the background extinction rate are uncertain for reasons pertaining to the nature of the fossil record. One reason relates to fossil preservation. The fossil record comes from sedimentary rocks that accumulate in depositional environments mainly at low elevations, such as lakes, coastal rivers, and marine settings. Continuing with mammals as an example, the diversity of terrestrial mammals today is greatest in regions of complex topography with strong environmental gradients. Estimates from the fossil record are thus based on species from a selective subset of habitats that mammals actually occupy. It is not known whether the longevities and extinction rates of mammals from depositional ecosystems differ systematically from those of species in nondepositional environments. Another consideration is that the recognition of ancient mammals from fossilized bones and teeth may underestimate the number of species that were actually present if species-level differences were based on traits that do not readily fossilize, such as behavior or coloration. In addition, the empirical estimates of extinction rates for mammals in the fossil record do not necessarily apply to other major groups of animals. Thus, it is important to bear in mind these caveats when comparing background extinction rates among groups from the fossil record and with estimates of extinction rates in recent times.

Mass extinctions are distinctive in three respects from extinction episodes of lower intensity. First is the large magnitude of effect: The cumulative loss of species during a mass extinction amounts to a major fraction of the global biodiversity for that time period, with estimates exceeding 75 percent of marine animal species. Second, the losses are global in scope, affecting terrestrial and marine ecosystems all over the world. Third, the recovery period, when the rate of origination of new species exceeds the extinction rate, takes millions of years.

The fossil record also demonstrates that species are capable of persisting through remarkable perturbations to landscapes and ecosystems. For example, during the glacial–interglacial cycles of the last 2.5 million years, massive continental ice sheets expanded to cover almost half of North America when climatic conditions were generally cool and dry, compared to the short interglacial periods (such as the present day) when climatic conditions were warm and moist. The geographic ranges of continental species shifted southward, or to lower elevations for montane species, sometimes by hundreds of miles. Geographic ranges were compressed into much smaller areas than most species occupy today, and many species occurred with animals or plants that they do not live with today. During the rapid transitions from glacial to interglacial conditions, continental glaciers melted away and montane glaciers retreated to high peaks. The ranges of many species shifted northward, or to higher elevation in mountain ranges, during a few thousand years and then stabilized geographical-ly. These migrations occurred more than twenty times in the last 2.5 million years, yet extinction rates were low for both terrestrial and marine species, with the exception of one time period. Between 10,000 and 50,000 years ago, most of the large terrestrial vertebrates from four continents disappeared. The coincidence of these extinctions with the presence or arrival of tool-using humans supports the hypothesis that human hunting played a major role in these Late Quaternary extinctions (Koch and Barnosky 2006).

The fossil record provides a critical perspective on the intensity and pace of change in biological diversity. While the extinction of lineages is a fundamental process in the history of life, for most of Earth history, extinction rates have been low to moderate—slow enough as to escape notice during single human generations. Uncommonly, many species disappeared during a geologically brief time span (thousands to hundreds of thousands of years), either regionally or globally, in association with changing environmental conditions, such as rising or falling sea levels or global cooling or warming. Rarer still were mass extinctions, in which a large fraction of the world's species and many ecosystems were decimated during a geologically brief period. Recovery took millions of years, and the survivors that proliferated were a different cast of taxonomic groups compared to those that were dominant before.

When did the modern biodiversity crisis begin?

Human impacts on species and ecosystems began thou-sands of years ago. There is a tendency to anchor the modern biodiversity crisis within the last 500 years, when colonial expansion and industrialization became dominant global influences. However, sediments and fossils, including human artifacts, provide evidence of widespread human impacts much earlier. Highlights of these early impacts include

1.   the Late Quaternary extinctions of large terrestrial vertebrates,

2.   the extinction of hundreds of birds on Pacific islands,

3.   soil erosion in early agricultural societies, and

4.   the early stages of the exploitation of marine and freshwater ecosystems.

Before 10,000 years ago, terrestrial ecosystems were inhabited by many kinds of large mammals, birds, and reptiles that are now extinct (Koch and Barnosky 2006). Mammoths, mastodons, ground sloths, tall camels, saber-toothed cats, giant marsupials and lizards, and huge flightless birds mingled with the animals that are found in today's world. This megafauna (species with adult body weights greater than 100 pounds [44 kg]) had been a characteristic feature of most terrestrial ecosystems for millions of years, with many ecosystems supporting ten or more species with adult body weights greater than 1,100 pounds (500 kg) in addition to the full complement of small and medium-sized animals. No ecosystem today has such high megafaunal diversity.

The Late Quaternary extinctions were dominated by the loss of large vertebrates, although a few smaller species also disappeared. The tally of species that became extinct differed on each continent but everywhere included large mammals, reptiles, and birds. The potential causes of this wave of extinctions include environmental change, human hunting, and various interactions of these factors. Environmental hypotheses emphasize that these extinctions occurred during the transition from glacial to interglacial conditions, when global and regional climate and vegetation were changing rapidly. Yet the species that succumbed to extinction had all survived many previous glacial–interglacial transitions, when no extinction pulses occurred. What was distinctive about the last glacial to interglacial transition were the small groups of modern *Homo sapiens* rapidly expanding in geographic range, across both continents and islands. The archaeological record makes clear that these aboriginal peoples were adept hunters, and some groups were skilled in the use of fire. The synchrony between human arrival times and the disappear-ance of megafaunal species, the evidence of hunted individuals (for example, mammoths and mastodons by Paleo-Indians in the Great Lakes region of North America), and simulations evaluating the impacts of low but steady human hunting rates on particular large species all suggest that human activities, even with a low level of technology, played a significant role in the Late Quaternary extinctions, sometimes in combination with rapid environmental change (Koch and Barnosky 2006).

Another striking example of aboriginal human impacts involves native birds on Pacific islands (Steadman 2006). The vast expanse of the Pacific known as Oceania hosts 25,000 islands. Most lie within 25 degrees north or south of the equator and occur as archipelagos extending east of the Philippines, New Guinea, and Australia. Many of these islands were colonized prehistorically by aboriginal Pacific islanders, as early as 35,000 years ago for islands near New Guinea, and between 3,000 and 1,000 years ago for more remote islands. Although the floras and faunas of these islands were naturally depauperate compared to those of the nearest mainland, the remoteness of many islands promoted the evolution of unique species. More than 1,100 species of birds, including songbirds (passerines), other land birds (such as rails, pigeons, parrots, and hawks), and seabirds inhabited these islands prior to human colonization. Some island birds, such as the famous dodo—a large, flightless pigeon—lost the ability to fly, and many species were naive in terms of predators or hunters.

	Landbirds		Seabirds	All species
	Non-Passerines	Passerines		
Living species	209	216	50	475
Documented extinct species	105	10	6	121
Estimated extinct species	561–1696	82	16	659–1794
Estimated total species	770–1905	298	66	1134–2269

Table 1. Documented and estimated number of living and extinct species of landbirds and seabirds from tropical Oceania. Estimated extinctions are based on the incompleteness of the archeological record. Data from Steadman (2006). Reproduced by permission of Gale, a part of Cengage Learning.

Two-thirds of these species have become extinct since human arrival. This calculation is derived from documented extinctions (fossil bones of species with no modern counterpart) and from estimates of the incompleteness of the archaeological record of bird remains (see Table 1). Extinctions resulted from hunting, habitat transformation, and introduction of exotic predators, such as pigs, cats, and dogs, as well as novel diseases. The same processes affect many island species today.

Early agricultural societies from many regions left signatures of ecosystem transformation. Deforestation and plowing, especially on steep slopes, led to the rapid accumulation of hill-slope sediments in valley bottoms. Although agricultural practices varied widely, early writings in ancient Greek, Roman, and Chinese societies noted the importance of maintaining soils on hill slopes. However, many an empire went into decline soon after episodes of massive soil erosion from its agricultural landscapes (Montgomery 2007). Exposed foundations of ancient houses, cisterns, and roads, as well as layers of mud in adjacent rivers, attest to sustained periods of soil erosion. Geological detective work has shown that erosion rates followed periods of empire expansion and intensification of agricultural practices (such as more frequent plowing, expanding farm size, and increasing yields) rather than regional climatic changes. The impacts on biodiversity increased as more intensive agriculture expanded over the landscape. Once the original vegetation was cleared and fields were plowed, erosion transferred soil into lowlands, choking rivers with mud. In Italy, to cite one example, cities such as Ravenna and Ostia were seaports during the days of the Roman Empire became now landlocked by freshly eroded sediments piled up in river deltas. These processes transformed both upland habitats and river ecosystems. In some cases, once-clear rivers that had supported abundant fish and mollusk populations became malaria-infested marshes. Such changes occurring locally and regionally several thousand years ago altered floras, faunas, soils, and microclimates, creating less productive ecosystems that have not reverted to their prior state. In North Africa, Iraq (Mesopotamia), and the Indus Valley in Pakistan, diverse woodland and riparian ecosystems were transformed by deforestation, overgrazing, and erosion into extensive semideserts with reduced plant and animal diversity. These changes at the regional scale thousands of years ago foreshadowed the transformations occurring at the global scale today.

The ecosystem-level consequences of hunting aquatic species for subsistence or commerce also extend well back in time. In the Aleutian Islands of the North Pacific, bones and shells in stratified middens document the changes in coastal and marine prey species harvested by aboriginal Aleuts over about 2,000 years (Simenstad et al. 1978). Over time, the frequency of sea otter, harbor seal, and marine fish bones have been found to be inversely correlated with the frequency of shells from sea urchins, limpets, and other large invertebrates. The alternation of vertebrate-dominated and invertebrate-dominated midden assemblages reflected a shift in the structure of nearshore marine ecosystems, depending on the presence or absence of a keystone predator—the sea otter. Sea otters prey on sea urchins and other large invertebrates, thereby reducing the grazing intensity of these invertebrates on macroalgae (kelp and other seaweeds). Healthy kelp forests support a high diversity of crustaceans, nearshore fishes, and marine mammals. When sea otters are absent, their prey populations prosper and graze kelp and other seaweeds, leading to ecosystems dominated by sea urchins, limpets, bivalves, asteroids, and octopus. The midden record suggests that Aleut communities had at times overexploited sea otters, causing the nearshore ecosystem to become dominated by large invertebrates. These northern aquatic ecosystems have alternated between two states, depending on the presence or absence of sea otters.

In other cases, the changes in aquatic ecosystems have involved directional transformation (Roberts 2007). Both written and archaeological records from European towns and cities show that fish were an important human food source during the first millennium Common Era. Prior to the eleventh century, fish bones from British excavations were primarily from freshwater or migratory species that spend part of their lives in rivers and streams. An abrupt change occurred during the eleventh century in Britain and slightly later in northern Europe, when marine species replaced river species in the bone piles left near households and taverns. Analyses of written records and bone assemblages have revealed that environmental changes on land, including soil erosion following some forms of agriculture, the damming of rivers for mill power, and the intensive harvesting of fish species, caused fish stocks to plummet across much of northern Europe. Dams and barrier nets made it easier for fishers to deplete resident and migrating fish populations. In response to declines in freshwater fish stocks, early aquaculture practices arose, first in France and then throughout Europe. Many fishers went to sea, where abundant populations of marine species were then found close to shore. So large were early harvests of marine species, including cod, herring, and pollack, that they dominated trade relations among European nations and contributed to the colonial expansions of the major European powers. By the 1600s, declines in the adult body size of intensively fished populations were noted from ship logs and travelers' reports. The sense of unlimited abundance of fish populations, noted by generations of fishers and seafarers, turned out to be a series of shifting baselines of gradually declining natural populations. Initially, populations declined or were extirpated locally (e.g., within a single watershed or marine region), and over centuries, extirpations extended across much larger regions. Historical records show that each generation of fishers considered the fish stocks of their parents' generation to be the so-called normal baseline of natural populations. One thousand years of hindsight has revealed the deterioration of these baselines from natural or sustainable conditions.

These historical perspectives, anchored in the fossil record, the archaeological record, and written history, indicate that the species and ecosystems of today were transformed or impoverished long before biologists began to document the world's biodiversity and its conservation status.

The biodiversity crisis today

The number of described species on Earth today has risen rapidly since the 1980s as a result of increased expertise, resources, and tools for documenting modern biodiversity and its ecosystem services. About 2 million species have been described, with an estimated 3 million to 28 million species as yet undescribed. Ironically, even as humanity's knowledge of global biodiversity has expanded, the rate of species extinction and the decline of ecosystem services have increased. Human activities now dominate most regions of the world, resulting in a wide range of impacts on terrestrial and marine ecosystems. Conservation efforts and environmental awareness are also increasing around the world and have accomplished notable successes in restoring endangered species to healthy population levels in the wild. But losses are rising in all groups of animals and plants that have been systematically evaluated.

Human activities are now a global force of landscape evolution, ecosystem transformation, and climate modification, as well as biodiversity decline. One indicator of the magnitude of impact is the amount of global primary production that is dominated by human activities. Net primary production (NPP) is the amount of biomass from photosynthesis that is available to plant consumers; essentially, it is the base of the consumer food web. Human activities alter NPP either by harvesting it directly or by modifying positively or negatively the productive capacity of natural ecosystems. In a 2007 global assessment, Helmut Haberl and colleagues found that human activities have appropriated about one-fourth (23.8%) of global NPP from terrestrial ecosystems. Over half of this human domination comes from harvest, about 40 percent results from changes in productivity, and the rest is attributable to fires. All continents except Antarctica contain some regions with moderate to high appropriation of NPP by human activities (see Figure 2). Only immense deserts, high mountains and plateaus, and very cold regions show no measurable effect.

Other global impacts of human activities include the increasing concentration of greenhouse gases in the atmosphere and the associated climatic changes; the erosion and redeposition of sediment from the land to rivers, lakes, and oceans; the acidification of the oceans; and rising sea levels (Zalasiewicz et al. 2010). These influences are being recorded in sedimentary deposits and organic remains that will become the fossil record of the future.

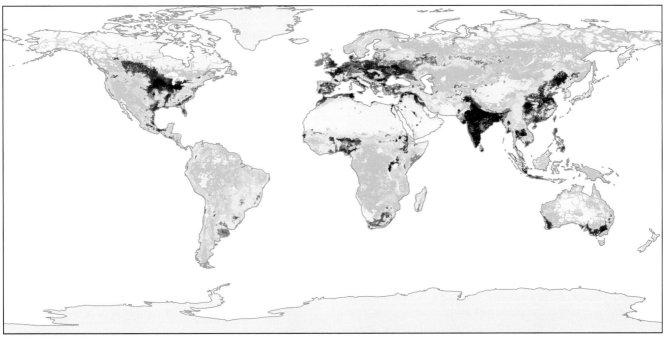

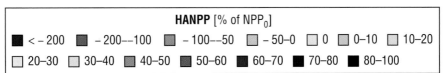

Figure 2. Human appropriation of NPP as a proportion of total NPP in native ecosystems. Blue shading signifies that human activities have increased NPP over its natural rate; green, yellow, and red shading signifies that human activities have reduced NPP in comparison with the levels in natural ecosystems (Haberl et al. 2007). Reproduced by permission of Gale, a part of Cengage Learning.

The Red List

Human activities impact many species throughout the world through both targeted exploitation and incidental effects, such as clearing of forest for farming. The ongoing tally of species made vulnerable by human activities is kept in the Red List of Threatened Species, a database of global biodiversity in which species are classified according to their conservation status, ranging from extinct to not vulnerable (Vié et al. 2009). The International Union for Conservation of Nature (IUCN) coordinates the monitoring and assessment processes; hundreds of taxonomic specialist groups work with the IUCN to keep the Red List up to date. As of 2012, 31 percent of the species assessed for their conservation status are threatened with extinction (IUCN Red List 2012). For several reasons, the assessments for the Red List have focused on particular taxonomic groups—initially, groups with species of economic or cultural importance (for example, mammals and birds); more recently, groups documented well enough by taxonomic specialists and ecologists to be assessed in their entirety (mammals, birds, amphibians, conifers, and others); groups that convey significance about entire ecosystems (such as corals), and groups that are experiencing intensive threats in recent years (such as cartilaginous fishes, the sharks and rays). The estimated level of threat for the assessed groups should not be interpreted as the level of threat for all of global biodiversity today (this level could be higher or lower). Rather, the estimates provide a consistent measure over time of changes in the status of well-documented groups from a wide range of global habitats and form the basis for the assessment of changes in exploitation as well conservation successes.

Each species reviewed for the Red List is placed in one of eight categories (Mace et al. 2008). *Extinct* means that the species no longer exists in the wild or in captivity and is not likely to be persisting undetected. *Extinct in the wild* applies to species that survive only in captive breeding programs. Such species could potentially be restored in the wild but would remain vulnerable there for many generations. These extinctions have been assessed over the last 500 years and do not include earlier extinctions. *Critically endangered*, *endangered*, and *vulnerable* represent different degrees of extinction risk by criteria such as rate of population decline, amount of original geographic range occupied, and population size. The critically endangered category includes species assessed as *probably extinct*. *Near threatened* applies to species that are close to meeting criteria for one of the three threatened categories. *Least concern* refers to species that do not qualify for any of the threatened categories. Finally, *data deficient* species are those that are not well documented in the wild. Such species could be in various high-risk categories but have not been studied enough for a reliable assessment. Over time, the assessment of a species can change its assignment to greater or lesser degrees of risk, depending on changes in its population size and geographic range, as well as in the factors that affect its persistence.

As of 2012, 742 animal species and 122 plants species had been declared extinct or extinct in the wild (with a large majority in the extinct category) from a total of 63,819 species assessed (IUCN Red List 2012). Expressed in terms of the number of extinctions per 1,000 species present, the value is 13.5 extinctions per 1,000 species. For mammals, the value is 14.4 extinctions

per 1,000 species. In comparison to the long-term average extinction rate of 0.1 to 1 species per 1,000 species per 1,000 years for mammals estimated from the fossil record, the current extinction rates are one to two orders of magnitude greater. In evaluating this comparison, it is necessary to bear in mind the biases inherent in estimating risk and extinction over the last few centuries as well as from the fossil record. As mentioned earlier, the fossil record at best preserves organic remains in depositional ecosystems and leaves large portions of the biosphere undocumented. Only rarely are soft-bodied organisms, which comprise most of modern animal and plant diversity, preserved as fossils. Evaluations of modern biodiversity have their own biases. The most diverse taxonomic groups (e.g., insects, flowering plants) are incompletely described and even more incompletely assessed in terms of their conservation status (with less than 3 percent so assessed). Even the assessed groups have a substantial proportion of species categorized as data deficient (see Figure 3). Different comparative methodologies can minimize the effects of these disparities, but it is important not to overextend these estimates of risk and extinction (Barnosky et al. 2011).

The proportion of current species that are threatened varies considerably among the major groups of vertebrates, several groups of invertebrate animals, and three kinds of plants (see Figure 3; Hoffmann et al. 2010). Some groups have

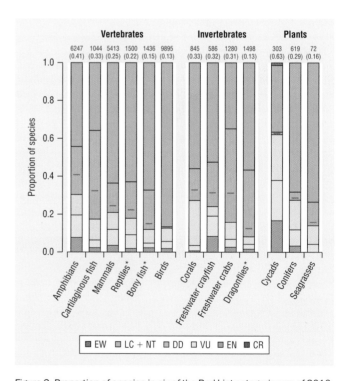

Figure 3. Proportion of species in six of the Red List categories as of 2010. Horizontal red lines show the estimated proportion of threatened species. Estimates are based on species placed in the critically endangered (CR), endangered (EN) and vulnerable (VU) categories as well as the same proportion of species from the data-deficient (DD) category. Numbers above each bar represent the number of living species assessed in each group; asterisks denote groups for which a randomized sample was assessed (Hoffman et al. 2010). Reproduced by permission of Gale, a part of Cengage Learning.

been assessed in their entirety, whereas for others a representative subset has been evaluated. The large number of data-deficient species in each group is likely to include species ranging from critically endangered to least concern status. Among vertebrates, the proportion of species that are threatened ranges from 40 percent for amphibians to 13 percent for birds. Among invertebrate groups, the proportion ranges from 33 percent for corals to 13 percent for dragonflies. For the plants assessed, the proportion ranges from 63 percent for cycads to 16 percent for sea grasses. These estimates do not include extinct species. Geographically, the highest proportion of vertebrate species that are threatened occurs in the tropics, even after accounting for the high tropical species diversity of all vertebrate groups. The greatest number of threatened mammals occurs in the Indo-Malayan region, amphibians show the greatest declines in the Neotropics (biogeographic region that extends southward from the Tropic of Cancer and includes southern Mexico, Central and South America, and the West Indies), and birds are more threatened in Oceania even now than in other regions of the world. A higher proportion of bird and mammal species used for food or medicine is threatened with extinction than species not impacted in these ways (Vié et al. 2009).

Over time, more species have increased than decreased in vulnerability to extinction, even though conservation efforts have improved the status of some species considerably. The Red List index summarizes changes in species assessments for all species in a particular group as a consequence of improving or deteriorating conservation status for the group as a whole. This index takes into account changes in the number of species described or assessed but depicts changes only among Red List categories. Between 1980 and 2010, the index declined for four animal groups, signifying that a greater proportion of species had become threatened (Secretariat of the Convention on Biological Diversity 2010). Mammals, birds, amphibians, and reef-building corals have been assessed in their entirety. Corals showed the steepest drop as a result of increased bleaching and disease since 1996, associated with warming ocean temperatures and pollution. In coral reefs around the world, live coral cover has decreased by 50 percent to 93 percent. Macroalgae (seaweeds) have increased as corals have declined, further accelerating the decline of the remaining live coral by overgrowing them (Jackson 2008). Amphibians face a host of threats. Since 1980, eleven species have become extinct and another 120 are probably extinct (Vié et al. 2009). The interaction of warming temperatures with disease agents, such as the chytrid fungus, has been particularly devastating for tropical amphibians (Wake 2012). Birds and mammals have shown slower but steady rates of decline.

In addition to species, another important aspect of biodiversity consists of genetically distinct populations that may differ in terms of morphology, behavior, and life-history traits through adaptation to geographically varying environmental conditions. For example, in a 2003 study, Ray Hilborn and colleagues found that sockeye salmon (*Oncorhynchus nerka*) populations that spawn in seven different watersheds draining into Bristol Bay, Alaska, showed differences in adult body size and shape, spawning times, egg size, and the time that young fish spent in freshwater before migrating to the sea. Each population responded to the substrates, predators, and microclimates of its own watershed. The differences in the timing of decade-scale environmental changes among the watersheds resulted in the distinct sockeye populations increasing and decreasing at different times among the watersheds. The aggregate behavior of all these population histories, with some populations increasing as others were decreasing, has supported a robust salmon fishery in this area for over a century.

The assessment of marine diversity lags behind that of terrestrial and freshwater diversity because of the challenges of monitoring species over vast expanses of ocean. Nonetheless, concerted efforts to document the status of critical habitats, species, and regions have revealed alarming degradation of marine ecosystems (Jackson 2008). Except for remote, uninhabited oceanic islands and regions of the open ocean rarely visited by fishing vessels, no areas of the world's oceans are of least concern." Coral reefs around the world are critically endangered, as are estuaries and coastal seas. For the latter, which provide significant food resources for many coastal nations, the global degradation of the eighty species or species groups assessed ranged from 39 percent to 91 percent. Continental-shelf ecosystems are endangered, and open-ocean ecosystems are threatened. In tropical, temperate, and polar regions, many commercially important fish and invertebrate stocks are severely depleted. By 2003, 29 percent of fished species were considered to have collapsed (defined as catches dropping below 10 percent of the recorded maximum), with a cumulative 65 percent of fished species having collapsed between 1950 and 2003 (Worm et al. 2006). Some of these collapsed species have since recovered following the implementation of protective measures. In addition to the decline of stocks (regional populations) or species to the point of commercial extinction (such as the North Atlantic cod) and biological extinction (such as the Caribbean monk seal), major ecosystem changes are occurring throughout the oceans.

A global assessment of human impacts on marine ecosystems based on scores derived from seventeen anthropogenic activities and their ecological impacts across twenty marine ecosystems (see Figure 4) revealed some level of impact throughout the oceans (Halpern et al. 2008). Ecosystems affected by multiple impacts (such as fishing pressure, high inputs of nitrogen or phosphorus, warming temperatures) had higher impact scores. Over one-third of the oceans showed medium to very high impact scores, with the areas of highest impact in coastal regions and continental shelves. The areas of low impact occurred predominantly at high latitudes where sea ice limits fishing access. Ecosystems with the highest cumulative impact scores included coral reefs, mangroves, rocky reefs, and continental shelves.

Is a sixth mass extinction already underway? These summaries of global biodiversity on land and at sea illustrate the taxonomic, geographic, and ecological dimensions of the modern biodiversity crisis. While some groups have become highly threatened quite recently (e.g., Neotropical amphibians), others have been

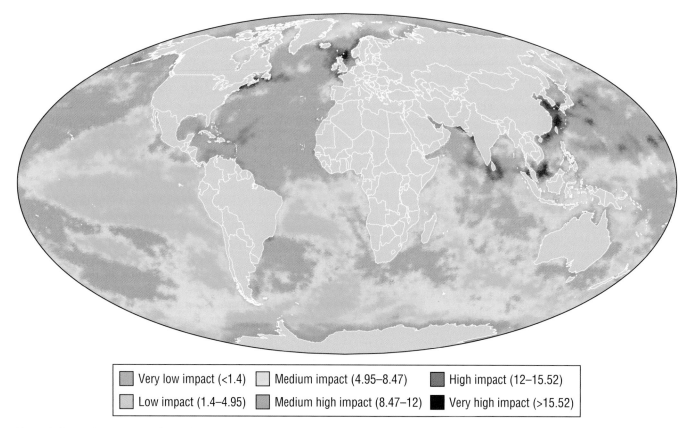

| ■ Very low impact (<1.4) | ☐ Medium impact (4.95–8.47) | ■ High impact (12–15.52) |
| ☐ Low impact (1.4–4.95) | ■ Medium high impact (8.47–12) | ■ Very high impact (>15.52) |

Figure 4. Human impact scores for marine ecosystems, with impact scores based on documented human activities in different regions of the world's oceans. Many regions exhibit medium-high to very high impacts, with the highest impacts occurring in coastal regions and continental shelves (modified from Halpern et al. 2008). Reproduced by permission of Gale, a part of Cengage Learning.

declining steadily for centuries (for example, mammals and birds hunted for food or medicine). The scale and scope of extinct and threatened species and ecosystems match those of the five mass extinctions of the fossil record. In terms of the number of documented extinctions in the last 500 years, the extinction rate exceeds the long-term average but has not yet reached the levels estimated for mass extinction (Barnosky et al. 2011). However, if the species currently listed as critically endangered, endangered, and vulnerable on the Red List all become extinct in the next few centuries and still more species become threatened at the same rate, then these losses would usher in a sixth mass extinction. Even the current state of loss, decline, and degradation is undermining ecosystem processes and human wellbeing.

Ecosystem processes and services

The modern biodiversity crisis was initially recognized and documented as a loss of populations and species. Another important dimension comprises changes to ecological processes as a consequence of losses of native biodiversity or gains from invasive species. Every individual, population, and species engages in many ecological interactions, such as acquiring food, processing nutrients, and competing for various resources essential for survival and reproduction. Many organisms also modify the landscapes that they inhabit.

Thus, species are significant not only as evolutionary lineages but also as ecological actors and members of ecosystems. Ecosystems are assemblages of species interacting with their physical context (including the substrate, local climate, and local hydrology) at the landscape scale. Ecosystem processes, such as photosynthesis and decomposition, regulate the flow of energy and matter within and beyond the ecosystem. In healthy natural ecosystems, the diversity of species, from microbes to megafauna, corresponds to a diversity of ecological roles. Some species have quite similar roles, and this redundancy tends to ensure the continuity of ecosystem processes over time even as populations of different species fluctuate seasonally or from year to year. Many ecosystem processes deliver direct benefits to human societies; these benefits are called ecosystem services (Millennium Ecosystem Assessment 2005).

Research in experimental ecosystems, natural ecosystems, and theory has confirmed the importance of biological diversity per se (many kinds of species and ecological roles) to ecosystem processes (Cardinale et al. 2012). Biomass production and nutrient cycling show more stability over time in more diverse ecosystems. Particular species that are highly productive as well as a diverse species pool with complementary ecological roles contribute to highly productive ecosystems, such as tropical forests, prairies, and kelp forests. Loss of species diversity reduces the efficiency of photosynthesis, energy flow through food webs, and nutrient cycling across a wide variety of ecosystems. Furthermore, increasing losses of species accelerate the decline in ecosystem processes.

A dramatic illustration of changes in ecosystem processes following a decline in diversity involves the loss of apex predators (those at the top of the consumer food web) in many terrestrial and aquatic ecosystems. The loss of these predators leads to increases in populations of their prey, which in turn consume more of their prey, with effects cascading throughout food webs. In some natural ecosystems, entire predator trophic levels have been eliminated by overharvesting, resulting in substantial changes in biomass and diversity at lower trophic levels (Estes et al. 2011). These regime shifts following the removal of apex predators have altered the vegetation structure of ecosystems by either suppressing or enhancing herbivores, depending on the length of the natural food chains. Changes in disease transmission, carbon storage, fire frequency, and water quality have followed from the loss or restoration of apex predators (see Figure 5).

The biological resources on which human life and wellbeing depend come mainly from ecosystem services. Human activities have drastically simplified many natural ecosystems to suit the consumption by human societies of food, fiber, building materials, and biofuels. This simplification has enhanced some ecosystem processes while diminishing others. For example, high-yielding monocultures of agricultural crops have high net primary production but are vulnerable to pests and diseases. In contrast, a diversity of crop plants reduces the prevalence of diseases and invasion by exotic species. Many ecosystem services are more stable or operate at higher rates with higher genetic or species diversity in the ecosystem, but some do not (Cardinale et al. 2012). For services that depend largely on wild species, such as wood production, soil nutrient mineralization, and fisheries yield, diversity tends to show a positive effect. In experimental studies, a loss of species diversity reduced biomass productivity and decomposition—two fundamental ecosystem processes (Hooper et al. 2012). The effect of reduced species diversity rivaled the effects of other major environmental perturbations, such as elevated carbon dioxide, increased ultraviolet radiation, drought, and agricultural pollutants. Despite uncertainties in scaling up experimental results to the ecosystem level, these results imply that a loss of biodiversity ranks with other drivers of global change in terms of altering ecosystem processes.

Finally, documented species losses and the proportion of species that are threatened (see Figure 3), as well as the known and anticipated ecosystem consequences of declining biodiversity, are a key part of the assessments contending that the global ecosystem is close to thresholds that will abruptly change to a different state than any that humans have ever experienced. Three interacting Earth systems—climate, biodiversity, and the global nitrogen cycle—are already perturbed by human activities to the point of undermining present-day global ecosystem processes (Rockstrom et al. 2009). The increase in anoxic zones (areas depleted of dissolved oxygen) in coastal marine areas, ocean acidification, melting

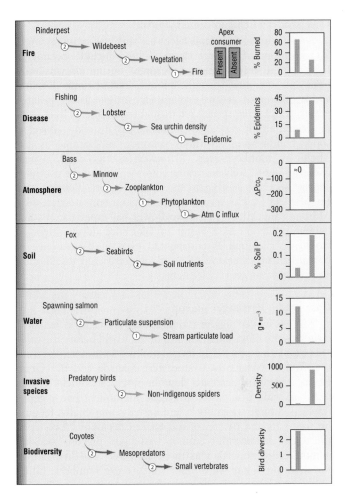

Figure 5. Effects of apex predators on fire frequency, disease transmission, carbon exchange with the atmosphere, soil nutrient storage, water quality, rate of species invasion, and maintenance of native biodiversity. On the right, blue bars represent data when predator is present, brown bars when predator is absent (Estes et al 2011). Reproduced by permission of Gale, a part of Cengage Learning.

glaciers, and increasing desertification are further indicators of widespread human impacts, even on regions where few people live (Barnosky et al. 2012). With no abatement in these impacts over the twenty-first century, a shift in the planetary state with unpredictable and unpleasant consequences for human societies becomes more feasible.

Victims and survivors

The modern biodiversity crisis affects species and ecosystems selectively. Species that are extinct or threatened preferentially include animals of large adult body size, animal or plant species with restricted geographic ranges, species that inhabit oceanic or montane islands, and species of commercial importance to human societies. These trends extend back in time to the Late Quaternary extinctions of large mammals, birds, and reptiles. Different human activities affect species with different ecological properties. The following five case studies illustrate the ecological traits that characterize many vulnerable species.

Case study: Michigan monkeyflower

Some animals and plants have very small geographic ranges for a variety of natural causes. These include occupying habitats that were formerly widespread during glacial conditions but are now spatially limited, having poor dispersal ability, being of recent evolutionary origin as an isolated population, having specialized habitat requirements that occur in geographically restricted circumstances, or being subject to some combination of these factors. An example is the Michigan monkeyflower, (*Mimulus michiganensis*), which occurs on moist soils near calcareous seeps along the northern shores of Lake Michigan and Lake Huron (Penskar and Higman 2001). The species is known from only a dozen sites. Although the Michigan monkeyflower has experienced some habitat reduction as a result of lakeshore development, it would be classified as endangered even in the absence of human impacts because of its highly restricted distribution and narrow habitat requirements.

Case study: Passenger pigeon

The passenger pigeon (*Ectopistes migratorius*) was by some estimates the most abundant bird in the world before the mid-1800s (Matthiessen 1959). Its geographic range extended from the Gulf of Mexico to southeastern Canada and the Great Plains. Historical accounts describe migratory flocks 1 mile (1.6 km) wide that passed overhead for hours. The pigeons traveled in huge social flocks and nested communally, feeding on tree nuts and berries of the eastern deciduous forests. In the early nineteenth century, passenger pigeons became commercially important as meat, fertilizer, and live targets for shooting. They were hunted in massive numbers, filling wagons and boxcars, even as witnesses spoke of unlimited abundance. Individual hunters boasted of killing more than 1 million birds. Killing pigeons on their nesting sites and deforestation over large expanses of their habitat were critical to their decline. Although legislation to reduce the hunting season or establish protected areas began as early as the 1850s in some states, the pigeon populations were so reduced by the 1890s that breeding in the wild and in captivity was unsuccessful. Sightings of small flocks of the birds continued until the early 1900s; the last individual died in the Cincinnati Zoo in 1914. The passenger pigeon had a large geographic range and spectacular abundance. Nonetheless, sustained overharvesting for commerce drove the bird to extinction in less than a century.

Case study: Northern sportive lemur

Lemurs are a unique evolutionary radiation of primates that occur only on the microcontinent of Madagascar. Primatologists recognize 50 to 100 species of extant lemurs ranging from tiny (the smallest primate, about 1 ounce [30 g]) to an adult body size of 20 pounds (9 kg). Different species occupy the various habitats of Madagascar, ranging from arid woodland to lush rain forest. Since the arrival of humans on Madagascar about 2,000 years ago, hunting and conversion of habitat for agriculture have caused the extinction of seventeen species of lemur, including giant lemurs that weighed about 440 pounds (200 kg). The remaining lemurs live on about 10 percent of the island. The northern sportive lemur (*Lepilemur septentrionalis*) is listed as critically endangered (Andrainarivo

et al. 2012) and is considered one of the world's twenty-five most endangered primates. It inhabits dry forest in northern Madagascar and is active at night, when it eats leaves, fruits, and flowers. Hunting and logging for charcoal production have reduced the populations of this species to fewer than 200 individuals distributed in remnant patches of forest (Mittermeier et al. 2009). A small geographic range and low rate of reproduction (females have one young per year) would make this species vulnerable to extinction even in the absence of human impacts. Habitat destruction and hunting have intensified the level of extinction risk.

Case study: Golden toad

The golden toad (*Bufo periglenes*) lived at high elevation in the cloud forests of Costa Rica. The last sighting of this toad occurred in 1989, even though more than 1,500 adults had been seen two years earlier (Crump 2000). Most of this species' range was in the Monteverde Cloud Forest Preserve, deemed one of the "seven wonders of Costa Rica" by popular vote. This small toad was brightly colored—unusual for toads—and was thought to live underground. During the breeding season, bright orange males congregated around puddles of water and mated there with red and black females. The golden toad occupied a restricted altitudinal range along steep montane temperature and moisture gradients. Only under a narrow range of climatic conditions would tadpoles survive to metamorphosis. Warming temperatures are implicated in increasing the susceptibility of golden toads to diseases, especially the chytrid fungus, *Batrachochytrium dendrobatidis*, which grows on the moist skin of infected amphibians. More than nearby deforestation, global warming has increased the elevation at which mist condenses on mountaintops during the dry season. In the Neotropics air temperatures have increased at rates three times higher than the global average, resulting in a significant increase in the number of dry days since 1980 (Pounds et al. 2006). Warmer years have been associated with upward shifts in elevation for birds and lizards, as well as the disappearance of many populations of frogs and toads from Central America to Peru. At the Monteverde preserve, the combination of warmer years and increased cloud cover at high elevation increased the areas favorable to growth of the chytrid fungus. The interaction of climate change and fungal disease is considered the cause of the extinction of the golden toad, which occurred in a well-protected habitat. This example illustrates that multiple factors can interact to cause irreversible population declines and that well-protected reserves provide no guarantee against extinction.

Case study: Atlantic cod

The Atlantic cod (*Gadus morhua*) is a bottom-feeding fish with a large geographic range along the coast and continental shelves from the middle latitudes of the North Atlantic to southern Greenland and northern Europe. Cod live at depths ranging from shallow river mouths to 2,000 feet (600 m), with most fish inhabiting the continental shelves between 500 and 650 feet (150 and 200 m; Cohen et al. 1990). Adults can grow to 6.5 feet (2 m) in length and can live for more than 20 years. Plankton, invertebrates, and smaller fish form the diet, and cod are significant predators of coastal and shelf ecosystems.

This gregarious species has both migratory and non-migratory populations, with migrations of up to 620 miles (1,000 km) in each direction. Females have legendary fecundity, with the average female producing 1 million eggs. A female weighing 33 pounds (15 kg) can produce 7.5 million eggs. Despite the large geographic range, high fecundity, and omnivorous feeding habits, several populations of Atlantic cod have crashed to commercial extinction, although other cod populations still sustain regular harvests.

The Atlantic cod fishery is many centuries old, with reports of Basque and Scandinavian vessels fishing the western North Atlantic in the first millennium AD. Cod were so numerous historically that they were caught with dip nets and reportedly slowed the progress of ships. Prior to the 1950s, cod landings in the northwestern Atlantic regularly exceeded 200,000 tons per year. Catches spiked in the mid-1950s for two decades, then fell sharply in the 1970s and again in the 1990s (see Figure 6). Changes in fishing technology were behind two of the trends (Pauly and Maclean 2003). The use of bottom trawls in the productive fishing grounds off the east coast of Canada began in the late 1800s; this innovation led to catches increasing to above 200,000 tons per year. Following World War II (1939–1945), the advent of factory ships and sonar greatly expanded the capacity of fishing fleets. Over the same period, population models of sustainable yields and elaborate management schemes were applied to North Atlantic fisheries. The spectacular catch increases, then crashes, and management failures led to what had been unthinkable for centuries—the complete closure to commercial fishing of formerly productive fishing grounds in parts of the northwestern Atlantic. In many of these regions, the cod populations have not yet recovered and remain at about 1 percent of their original abundance. The loss of these predators has caused cascading effects throughout the food web, with increases in populations of large invertebrates and small mid-water fish species; increases in gray seals, which feed

primarily on large invertebrates and small fish; decreases in large zooplankton, which are the prey of large invertebrates and small fish; and increases in phytoplankton, which are consumed by the zooplankton (Frank et al. 2005). Currently, the Red List status of the Atlantic cod is vulnerable (Sobel 1996). Other commercially harvested marine fish have followed a similar pattern, even as fishing effort has increased (Pauly and Maclean 2003). This example demonstrates that an abundant, widely distributed species of economic importance to human societies can be driven close to extinction over areas of its range where it was once spectacularly abundant, even under harvest management plans.

What the case studies illustrate

These examples illustrate some of the properties of species that elevate vulnerability to extinction. Species with very small geographic ranges—as part of their postglacial historical distribution (such as the Michigan monkeyflower and the golden toad) or from recent reductions in habitat (the northern sportive lemur)—are more vulnerable to extinction than species with large geographic distributions. Natural variation in extreme weather events or in reproduction can diminish population sizes below the capacity to rebound. Human impacts, including hunting, collecting, or habitat transformation, can accelerate the transition to extinction. Species with highly specialized requirements for growth (for example, the Michigan monkeyflower), reproduction (the golden toad), or other critical resources are also vulnerable, even over large geographic ranges. Specialized conditions are easily altered by habitat transformations, climate change, and invasive species (see below). Low reproductive rates (for example, the northern sportive lemur) also dispose species to vulnerability because their populations grow slowly, especially when reduced to low numbers. At very low numbers, individuals may have difficulty finding mates, may be vulnerable to inbreeding, and for social species (such as the passenger pigeon) may lack the group stimuli for normal behavior. Large body size in animals is correlated with large geographic ranges but low reproductive rates. Large animals (for example, baleen whales, elephants) are disproportionately represented among the extinct and threatened species on the Red List. Finally, apex predators are more vulnerable to extinction than species at lower trophic levels. Apex predators have been directly targeted for food (for example, sharks) and other extractive uses (for example, tigers) or have been targeted as competitors with human interests (gray wolf, African lion).

The cases also illustrate three important points about the management of threatened species. First, although protected areas have played an important role in maintaining threatened species and ecosystems, they cannot prevent extinction (consider the golden toad). Second, management plans for commercially harvested species may fail to maintain healthy populations (for example, Atlantic cod) because the population models are flawed, because data about harvests are grossly incomplete, or because the plans are overwhelmed by conflicting political and economic interests. Third, most threatened species are at risk from multiple interacting changes to their ecosystems, usually as a consequence of direct (for example, hunting) and indirect (for example, global climate change) impacts of human activities.

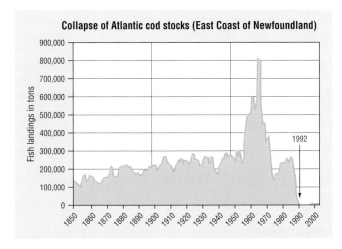

Figure 6. Landings of Atlantic cod from the productive banks east of Newfoundland. The stocks of cod declined to such low levels in the early 1990s that the Canadian government declared a moratorium on fishing these and other North Atlantic cod populations. From Millennium Ecosystem Synthesis Report (2005). Reproduced by permission of Gale, a part of Cengage Learning.

Which species are thriving during the modern biodiversity crisis? Three general categories of species are increasing in population size or geographic range. One category includes those species that are cultivated for human consumption. Through deliberate introduction and the channeling of resources to their survival, corn, wheat, and rice, as well as cattle, pigs, and sheep, now occupy vastly greater areas than their wild ancestors ever did. Second are species with broad habitat requirements, high reproductive rates, high dispersal ability, and a tolerance for disturbed environments. Examples include dandelions, house mice, cockroaches, coyotes, house sparrows, silver carp, and poison ivy. Some such species are expanding on their own, whereas others also belong in the third category—introduced species. Among the thousands of intentional and accidental introductions, most are unsuccessful in becoming established in the new location. Some that have become established displace native species and transform ecosystems through their rapid growth.

Causes

The modern biodiversity crisis is the consequence of interacting proximate and underlying causes. The proximate causes are widely recognized (Secretariat of the Convention on Biological Diversity 2010). The major pressures acting to reduce the population sizes, geographic ranges, or ecological interactions of wild species are habitat loss, fragmentation, and degradation; overharvesting; invasive species; pollution; and climate change.

Habitat loss, fragmentation, and degradation

Habitat loss, fragmentation, and degradation are the principal reasons that many species become threatened with extinction. These processes reduce the areas and connectivity of habitats that species are capable of occupying and thereby the available resource base. The consequence is that smaller, isolated populations become vulnerable to extinction. Fragmentation, the division of once continuous habitats into separated patches, has a significant influence on the persistence of populations because small areas of even pristine habitat cannot support large populations and will lose species over time (Perfecto et al. 2009). Habitat loss and fragmentation may create an extinction debt, whereby a species does not disappear immediately but is doomed to extinction because the remaining habitats are too small to support viable populations. In the Brazilian Amazon, where deforestation rates have been as high as 10,000 square miles (26,000 sq km) per year, most extinctions have yet to occur as populations adjust to habitat fragments. Simulated deforestation scenarios projected through the year 2050 suggest that nine vertebrate species will become extinct and another sixteen committed to extinction (Wearn et al. 2012).

On the continents, habitat conversion for agriculture is the major agent of habitat transformation for wild species. More than 30 percent of the global land area has been converted to agricultural uses (see Figure 2). The kind of agriculture makes a difference in terms of the impact on biodiversity. Large-scale monocultures, especially when maintained by synthetic fertilizers and biocides, displace more native biodiversity than

do small-scale diversified farms (Jackson and Jackson 2002). The impact of agriculture on biodiversity also depends on how similar the agricultural system is to the native ecosystem in terms of vegetation structure, as well as the distribution of farms and patches of native vegetation on the landscape. Although more than half of the world's people now live in cities, urban areas and infrastructure networks (such as highways and airports) represent a vastly smaller area of habitat transformation than that of agriculture. Agriculture has also affected freshwater ecosystems through nonrenewable water withdrawals as well as high nutrient loading.

In marine ecosystems, overharvesting is the major human impact (see below), and some harvesting methods cause massive damage to marine ecosystems. Chief among these is bottom trawling, which involves dragging of weighted nets along the seafloor in order to capture deepwater fishes or invertebrates. The increase in the size of trawl nets, the frequency of trawls over the same locations, and the geographic scope of trawling since World War II have decimated many benthic ecosystems of the continental shelves and even the deep sea. In addition, the dredging of sediments for shellfish in estuaries and shallow oceans has disrupted large areas of seafloor habitats and has thereby affected many non-target species (Jackson 2008).

Overharvesting

Overharvesting is the removal of individuals from wild populations at a rate greater than the rate of reproduction. If such removal occurs over successive generations, the populations can decline precipitously. The passenger pigeon and several populations of Atlantic cod exhibited effects of overharvesting during the decades prior to their population crashes. Even abundant, widespread species cannot survive continuous overharvesting for many generations. Exploitation is particularly unsustainable when it targets individuals during their reproductive period. Harvesting of wild species occurs for food, garments, medicinal uses, status objects, and medical research. Other species are hunted mainly to reduce or eliminate their presence (for example, the gray wolf over much of its geographic range). Many species can tolerate low to moderate harvests. But even aboriginal, subsistence harvests have caused the extinction of some species and are threatening the persistence of others today (such as the northern sportive lemur). Overharvesting is particularly acute in marine ecosystems but also affects a wide range of terrestrial species and ecosystems (Pereira et al. 2010). Indiscriminate harvesting methods, such as long gill nets and trawling in lakes or oceans, have captured vast numbers of non-target species, including not only fishes but also marine mammals, sea turtles, and seabirds that become entangled in fishing gear (Roberts 2007).

Invasive species

Invasive species, resulting from the intentional and accidental introductions of non-native species, establish vigorous populations beyond the reach of agriculture and human residences. In their introduced context, such species often lack the full set of ecological interactions that kept their populations at moderate levels in their original habitats. Invasive species can overwhelm populations in their new habitats as competitors (for example, house sparrows in the United States competing with native songbirds for nest sites), predators (such as the Nile perch, introduced into Lake Victoria that consumed to extinction more than 100 endemic cichlid fishes), or parasites (such as the chestnut blight, a fungus that killed mature chestnut trees after introduction into North America around 1900). Introduced plants can transform entire habitats by overgrowing or replacing native species. While many species have undergone natural geographic range expansions over long periods of time, including intercontinental range shifts, what is unusual in the last few centuries is the magnitude of introduced species resulting from trade and travel. In Hawaii, more than half of the flowering plants are introduced species. Invasive species contribute to the status of 49 percent of threatened species in the United States (Stein et al. 2000). In Europe, out of 11,000 documented alien species, about 10 percent have ecological impacts and about 12 percent have economic impacts (Secretariat of the Convention on Biological Diversity 2010).

Pollution

Pollution, including excessive nutrient loading from agricultural runoff, may originate locally and then travel through waterways, the atmosphere, and food webs to have distant impacts. Pollution includes synthetic chemicals and materials as well as ordinary compounds with strong bioactive effects that have increased in quantity and flow through human-managed as well as native ecosystems. Since the publication of Rachel Carson's *Silent Spring* in 1962, the effects of synthetic biocides on species and ecosystems have become well known, although not as well regulated. The discovery that low doses of widely used herbicides (such as atrazine and glyphosate) can cause developmental abnormalities (Hayes et al. 2002) or increased mortality (Relyea 2005) in frogs supports the claim that biocides are contributing to global declines in amphibians. Plastics in the form of microscopic particles and macroscopic debris are major contaminants of terrestrial and aquatic ecosystems. In the marine environment, plastics are widespread as floating mats of garbage and suspended particles (Moore 2008). Worldwide, 267 species of marine organisms have been affected by plastic debris, including rare and endangered sea turtles and seabirds. Ordinary compounds, including nitrogen and phosphorus, are the main ingredients in agricultural fertilizers. Human activities have now more than doubled the amount of reactive nitrogen that cycles through ecosystems compared to contributions from natural processes (Millennium Ecosystem Assessment 2005). The addition of the often limiting nutrients nitrogen and phosphorus to ecosystems tends to favor a few plant species that rapidly assimilate these nutrients and increase at the expense of many other species. Nitrogen deposition from the atmosphere may cause acid rain, which has killed large expanses of forest in the United States and Europe and has reduced insect and fish populations from acidified streams and lakes. In aquatic ecosystems, agricultural runoff and sewage have increased the concentration of nitrogen and phosphorus in rivers, lakes, and coastal seas, resulting in eutrophication and dead zones. These are areas in which the rapid growth of algae and photosynthetic bacteria, stimulated by an influx of nitrogen

and phosphorus, has resulted in an overload of decomposing organic matter. Microbial decomposers consequently increase and deplete the body of water of oxygen, making it uninhabitable for most animal life. The number of dead zones in coastal marine ecosystems has nearly doubled each decade since the 1960s (Diaz and Rosenberg 2008).

Climate change

Climate changes at both the global scale and the habitat scale are underway over much of the world. Earth's climate has ranged from ice covered to ice free over Earth history, and within the history of most living species it has fluctuated between glacial and interglacial conditions. Neither species nor ecosystems are strangers to climate change. But global climate change from anthropogenic causes is occurring more rapidly than is usual in combination with other acute stresses. Climate change is already causing shifts in species ranges and changing interactions among species. The extinction of the golden toad, which vanished after a particularly warm year, showed that high-elevation species are particularly at risk to changes in local climatic conditions, reductions in habitat area, and novel species interactions.

Rates of change in temperature and moisture vary widely around the world. The highest rates of warming are occurring at high latitudes, with sea ice decreasing in the Arctic and the calving of glaciers increasing in Antarctica. Montane and polar ecosystems are decreasing in area, and these decreases will likely accelerate over the course of the twenty-first century. Climate simulations suggest that novel climatic conditions, unlike any present today, may occur by the end of the twenty-first century, and some climatic conditions of today may disappear by then (Williams and Jackson 2007). Both novel and disappearing climates are most likely to occur across the tropics, where the greatest diversity of species lies.

Species' geographic ranges can shift poleward or toward higher elevation and have done so many times over the last 2 million years. But many species, especially long-lived plants, will have difficulty establishing new populations at the pace of projected climatic changes. Terrestrial species will face barriers to dispersal, including agricultural regions and urban areas, as well as novel competitors, predators, and diseases. The oceans have become warmer and slightly more acidic since the early twentieth century, as increased concentration of atmospheric carbon dioxide has led to more carbon dioxide dissolving in seawater, where it forms a weak acid. These changes are projected to increase over the next 100 years. Marine organisms with calcareous skeletons show varying responses to acidification. Reef-building corals and mollusks have shown decreased calcification under experimental exposure to acidic ocean water; in contrast, coccolithophores (calcareous phytoplankton) and photosynthetic bacteria have shown increased calcification rates in more acidic water (Doney et al. 2009). As regions of the ocean become undersaturated with respect to calcareous minerals (including calcite and aragonite), organisms with calcareous skeletons will relocate to shallower depths and lower latitudes—leading to potentially significant changes in the composition of marine ecosystems.

Laysan albatross chick that died after ingestion of floating plastic debris. © FLPA/Alamy.

Most threatened species and ecosystems are facing multiple pressures. Climate change and the chytrid fungus together caused the extinction of the golden toad. Deforestation and indigenous hunting have reduced the northern sportive lemur to critically low numbers. Pollution, ocean acidification, and overfishing are causing a decline in coral reefs. These interactive effects increase the risk of extinction and make the future composition of ecosystems and rates of ecosystem processes difficult to predict. The twenty-first century will be full of ecological surprises.

Underlying causes

These proximate causes are themselves the consequences, both intentional and unintentional, of common human activities practiced by societies around the world. Four of the underlying drivers are food and agriculture, trade, consumption, and policies and priorities. These drivers have overlapping spheres of influence both locally and globally. They are embedded in political institutions, cultural habits, powerful economic interests, and limited understanding of the consequences to human well-being of the loss of species and the degradation of ecosystems.

FOOD AND AGRICULTURE Among all human activities, food acquisition through agriculture and harvesting of wild species has the greatest impact on biodiversity (Badgley 1998). Agriculture is the principal cause of habitat transformation, and overharvesting is predominantly for food and feed. Many pollutants have their primary uses in agriculture (for example, biocides and fertilizers), many invasive species were initially

introduced in association with agriculture, and agriculture is a major contributor to greenhouse gas emissions. Changes in agricultural management to accommodate native biodiversity, changes in dietary habits to reduce the ecological footprint of everyday meals, and reductions in food waste to achieve greater delivery of food from current agricultural lands could alleviate acute pressures on threatened species and ecosystems around the world.

Northern sportive lemur (*Lepilemur septentrionalis*) of northern Madagascar. Karl Lehmann/Getty Images.

TRADE Biological resources have been the dominant component of economic exchanges locally and globally throughout human history. The extractive use of a species or habitat often has a low impact when the use is for local subsistence, but it often becomes unsustainable when the use is for commerce, especially for export. The extension of distance and supply links between harvester and consumer weakens the accountability for adverse impacts of consumption and trade. In today's highly globalized economy, supply chains often involve many links between the point of origin and the point of consumption. Such distant connections make it difficult for consumers to comprehend the consequences of their purchases, even when the consequences are quite ominous. Many threats to biodiversity are exported from developing countries to developed countries in the form of commodities (see Figure 7). The top 10 importers of commodities associated with threats to biodiversity are countries with relatively low biodiversity per unit area, whereas the top ten exporters of biodiversity threats are among the most biodiverse countries in the world (Lenzen et al. 2012). Developed countries have relatively low exports of biodiversity threats, partly because of strong environmental policies and partly because of their low native biodiversity in comparison to tropical regions. Consequently, their biodiversity impacts through trade are greater in other countries than at home. Many of the exported commodities involve agricultural, fishing, and forest products. More effective regulation, certification, and information—as in fair-trade products—would enhance the distribution of responsibilities along these trade routes.

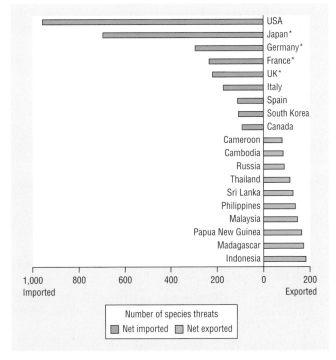

Figure 7. Top net importers and exporters of biodiversity threats. Asterisk signifies that a greater proportion of biodiversity threats occur through imports than domestically (Lenzen et al. 2012). Reproduced by permission of Gale, a part of Cengage Learning.

CONSUMPTION While the global human population doubled between 1970 and 2010, it is not projected to double again. In contrast, global economic activity multiplied seven times between 1950 and 2000 and is projected to grow by a factor of three to six times more by 2050 (Millennium Ecosystem Assessment 2005). The economic benefits of this growth have been unevenly shared, with the majority of the world's population living in poverty. The wealthy minority have high per-capita consumption rates, and their lifestyles dominate global trade. The poor majority have much lower per-capita consumption rates but often have direct negative impacts on species and ecosystems for subsistence or in livelihoods that export biodiversity threats. Most forms of consumption—food, water, energy, building materials, clothing, and luxury goods—have direct or indirect impacts on biodiversity. The ecological footprint index assesses the demand on biological capacity (biocapacity) of ecosystems required to support a particular pattern of consumption (Ewing et al. 2010). The cumulative ecological footprint of the human population has exceeded Earth's biocapacity since the 1970s. This means that current demands are met only by undermining the biocapacity of the future. Ecological footprints per capita vary widely among countries, as do the ecological footprints of major regions. North America, Europe, and Asia have greater ecological footprints than their region's biocapacity, a situation that is possible only because these regions are relying on the biological resources of other regions. In contrast, Latin America, Africa, and Oceania have greater biocapacity than their regional ecological footprints. The major challenge for a sustainable future is for all members of the global human community to achieve basic material well-being and access to social services without degrading Earth's biological capacity. This goal will require greater efficiencies of resource use as well as more equitable access to resources for those currently living in poverty.

POLICIES AND PRIORITIES Policies and regulations can be effective at reversing the decline of species or the degradation of ecosystems. For example, the ban on DDT in the United States, the moratorium on most whaling established by the International Whaling Commission, and the restrictions on international trade in endangered species from the Convention on International Trade in Endangered Species of Wild Fauna and Flora (CITES) have alleviated threats to particular species and ecosystems. In general, however, the rate of decline in species and ecosystems has exceeded the goals of most national and global agreements for reducing the magnitude of impacts (Secretariat of the Convention on Biological Diversity 2010). The world is facing unprecedented challenges in managing and protecting biodiversity sustainably and equitably.

Even though many biological resources are bought and sold, many aspects of their value are poorly represented in market prices (Millennium Ecosystem Assessment 2005). The monetary value of a product extracted and sold by an individual or company may fall far short of the value of the intact ecosystem. For example, the conversion of a mangrove wetland to aquaculture can provide direct economic benefits to a private company, but the estimated value of the original wetland for storm protection, coastal fisheries, and light harvesting of wood can greatly exceed the profits from shrimp farming. Furthermore, extractive uses of biological resources tend to benefit individuals or companies, whereas the costs of degraded ecosystems are more often borne by local communities or entire societies.

Species and ecosystems are part of the global commons shared by all people rather than being the property of individuals, businesses, or nations. The use, management, and protection of biodiversity are likewise the responsibility of the entire global community. It will take novel as well as traditional ways of thinking to manage species and ecosystems for their own integrity, for the continued delivery of ecosystem services, and for equity in access to the benefits of those services. Such planning is most likely to succeed when the parties involved all place high value on native species and ecosystems, when the monetary and non-monetary benefits of healthy ecosystems are shared equitably, when monitoring and communication are effective, and when the resource users are involved in developing the regulations about resource use (Ostrom and Nagendra 2006).

Is a sixth mass extinction inevitable?

Assessments of global biodiversity document that species extinctions are occurring at higher rates than have prevailed over most of Earth history. Many natural ecosystems are declining in area and in the services that they provide to human societies. Human activities have eliminated species and altered ecosystems locally for thousands of years. Now the cumulative effects have a global reach. Moreover, the loss of species and the significant extinction debts in many habitats, in combination with intensifying pressures, imply that extinctions are likely to accelerate over the course of the twenty-first century. Many individuals, organizations, and societies are deeply committed to the protection and appreciation of nature, and the current biodiversity crisis would be even worse in the absence of those efforts.

The causes of biodiversity loss are grounded in ordinary human activities, often driven by economic policies that fail to acknowledge the value and irreplaceable nature of species and ecosystem services. Yet there is nothing inevitable about the current patterns of resource use. Societies could, by changing vested interests and incentives, rescale their agriculture, trade, and consumption to match the capacity of Earth's biological resources and ensure equitable access. For terrestrial ecosystems, the most far-reaching changes would involve the conversion of large-scale, chemical-intensive monoculture agriculture to small-scale, agroecologically managed, biodiverse farms (Perfecto et al. 2009). Economic incentives for farming practices that support native biodiversity through both initiatives in civil society and changes in the structure of national agricultural subsidies and regulations could stimulate changes in trade and consumption. Such changes are already occurring on a small scale through wildlife-friendly certification programs for various foods and clothing. For marine ecosystems, stronger regulations and enforcement mechanisms for overharvested regions, species, and stocks, along with an increase in marine protected areas, could promote the recovery of species and ecosystems in formerly degraded regions (Worm et al. 2006). Efforts at all scales of governance, from the household to the city or district to global institutions, can have positive impacts. Education and experiences that increase the public understanding of biological diversity and ecosystem services can broaden support for more effective regulations and accelerate the adoption of lifestyle changes that have a smaller ecological footprint. All these changes would also promote more equitable access to Earth's biological resources. Such transformations would expand the opportunities to restore threatened species and degraded ecosystems. If the global community can meet this challenge, it will be an unprecedented measure of human achievement.

Resources

Books

Carson, Rachel. *Silent Spring*. Boston: Houghton Mifflin, 1962.

Cohen, Daniel M., Tadashi Inada, Tomio Iwamoto, and Nadia Scialabba. *Gadiform Fishes of the World (Order Gadiformes): An Annotated and Illustrated Catalogue of Cods, Hakes, Grenadiers, and Other Gadiform Fishes Known to Date*. FAO Species Catalogue, vol. 10. Rome: Food and Agriculture Organization of the United Nations, 1990.

Crump, Marty. *In Search of the Golden Frog*. Chicago: Chicago University Press, 2000.

Ewing, Brad, David Moore, Steven Goldfinger, et al. *The Ecological Footprint Atlas 2010*. Oakland, CA: Global Footprint Network, 2010. Accessed November 13, 2012. http://www. footprintnetwork.org/images/uploads/Ecological_Footprint_ Atlas_2010.pdf.

Jackson, Dana L., and Laura L. Jackson. *The Farm as Natural Habitat: Reconnecting Food Systems with Ecosystems*. Washington, DC: Island Press, 2002.

Matthiessen, Peter. *Wildlife in America*. New York: Viking Press, 1959.

Millennium Ecosystem Assessment Synthesis Report. Washington, DC: World Resources Institute, 2005.

Montgomery, David R. *Dirt: The Erosion of Civilizations*. Berkeley: University of California Press, 2007.

Pauly, Daniel, and Jay Maclean. *In a Perfect Ocean: The State of Fisheries and Ecosystems in the North Atlantic Ocean*. Washington, DC: Island Press, 2003.

Perfecto, Ivette, John Vandermeer, and Angus Wright. *Nature's Matrix: Linking Agriculture, Conservation, and Food Sovereignty*. London: Earthscan, 2009.

Roberts, Callum. *The Unnatural History of the Sea*. Washington, DC: Island Press/Shearwater Books, 2007.

Secretariat of the Convention on Biological Diversity. *Global Biodiversity Outlook 3*. Montreal: Secretariat of the Convention on Biological Diversity, 2010.

Steadman, David W. *Extinction and Biogeography of Tropical Pacific Birds.* Chicago: University of Chicago Press, 2006.

Stein, Bruce A., Lynn S. Kutner, and Jonathan S. Adams, eds. *Precious Heritage: The Status of Biodiversity in the United States.* Oxford: Oxford University Press, 2000.

Vié, Jean-Christophe, Craig Hilton-Taylor, and Simon N. Stuart, eds. *Wildlife in a Changing World: An Analysis of the 2008 IUCN Red List of Threatened Species.* Gland, Switzerland: IUCN, 2009.

Periodicals

Alroy, John, Martin Aberhan, David J. Bottjer, et al. "Phanerozoic Trends in the Global Diversity of Marine Invertebrates." *Science* 321, no. 5885 (2008): 97–100.

Badgley, Catherine. "Can Agriculture and Biodiversity Coexist?" *Wild Earth* 8, no. 3 (1998): 39–47.

Barnosky, Anthony D., Elizabeth A. Hadly, Jordi Bascompte, et al. "Approaching a State Shift in Earth's Biosphere." *Nature* 486, no. 7401 (2012): 52–58.

Barnosky, Anthony D., Nicholas Matzke, Susumu Tomiya, et al. "Has the Earth's Sixth Mass Extinction Already Arrived?" *Nature* 471, no. 7336 (2011): 51–57.

Cardinale, Bradley J., J. Emmett Duffy, Andrew Gonzalez, et al. "Biodiversity Loss and Its Impact on Humanity." *Nature* 486, no. 7401 (2012): 59–67.

Diaz, Robert J., and Rutger Rosenberg. "Spreading Dead Zones and Consequences for Marine Ecosystems." *Science* 321, no. 5891 (2008): 926–929.

Doney, Scott C., Victoria Fabry, Richard A. Feely, and Joan A. Kleypas. "Ocean Acidification: The Other CO_2 Problem." *Annual Review of Marine Science* 1 (2009): 169–192.

Estes, James A., John Terborgh, Justin S. Brashares, et al. "Trophic Downgrading of Planet Earth." *Science* 333, no. 6040 (2011): 301–306.

Frank, Kenneth T., Brian Petrie, Jae S. Choi, and William C. Leggett. "Trophic Cascades in a Formerly Cod-Dominated Ecosystem." *Science* 308, no. 5728 (2005): 1621–1623.

Haberl, Helmut, K. Heinz Erb, Fridolin Krausmann, et al. "Quantifying and Mapping the Human Appropriation of Net Primary Production in Earth's Terrestrial Ecosystems." *Proceedings of the National Academy of Sciences of the United States of America* 104, no. 31 (2007): 12942–12947.

Halpern, Benjamin S., Shaun Walbridge, Kimberly A. Selkoe, et al. "A Global Map of Human Impact on Marine Ecosystems." *Science* 319, no. 5865 (2008): 948–952.

Hayes, Tyrone B., Atif Collins, Melissa Lee, et al. "Hermaphroditic, Demasculinized Frogs after Exposure to the Herbicide Atrazine at Low Ecologically Relevant Doses." *Proceedings of the National Academy of Sciences of the United States of America* 99, no. 8 (2002): 5476–5480.

Hilborn, Ray, Thomas P. Quinn, Daniel E. Schindler, and Donald E. Rogers. "Biocomplexity and Fisheries Sustainability." *Proceedings of the National Academy of Sciences of the United States of America* 100, no. 11 (2003): 6564–6568.

Hoffmann, Michael, Craig Hilton-Taylor, Ariadne Angulo, et al. "The Impact of Conservation on the Status of the World's Vertebrates." *Science* 330, no. 6010 (2010): 1503–1509.

Hooper, David U., E. Carol Adair, Bradley J. Cardinale, et al. "A Global Synthesis Reveals Biodiversity Loss as a Major Driver of Ecosystem Change." *Nature* 486, no. 7401 (2012): 105–108.

Jablonski, David. "Extinctions: A Paleontological Perspective." *Science* 253, no. 5021 (1991): 754–757.

Jackson, Jeremy B.C. "Ecological Extinction and Evolution in the Brave New Ocean." *Proceedings of the National Academy of Sciences of the United States of America* 105, supp. 1 (2008): 11458–11465.

Koch, Paul L., and Anthony D. Barnosky. "Late Quaternary Extinctions: State of the Debate." *Annual Review of Ecology, Evolution, and Systematics* 37 (2006): 215–250.

Lenzen, M., D. Moran, K. Kanemoto, et al. "International Trade Drives Biodiversity Threats in Developing Nations." *Nature* 486, no. 7401 (2012): 109–112.

Mace, Georgina M., Nigel J. Collar, Kevin J. Gaston, et al. "Quantification of Extinction Risk: IUCN's System for Classifying Threatened Species." *Conservation Biology* 22, no. 6 (2008): 1424–1442.

Mittermeier, Russell A., Janette Wallis, Anthony B. Rylands, et al. "Primates in Peril: The World's 25 Most Endangered Primates, 2008–2010." *Primate Conservation* 24 (2009): 1–57.

Moore, Charles James. "Synthetic Polymers in the Marine Environment: A Rapidly Increasing, Long-Term Threat." *Environmental Research* 108, no. 2 (2008): 131–139.

Ostrom, Elinor, and Harini Nagendra. "Insights on Linking Forests, Trees, and People from the Air, on the Ground, and in the Laboratory." *Proceedings of the National Academy of Sciences of the United States of America* 103, no. 51 (2006): 19224–19231.

Pereira, Henrique M., Paul W. Leadley, Vânia Proença, et al. "Scenarios for Global Biodiversity in the 21st Century." *Science* 330, no. 6010 (2010): 1496–1501.

Pounds, J. Alan, Martín R. Bustamante, Luis A. Coloma, et al. "Widespread Amphibian Extinctions from Epidemic Disease Driven by Global Warming." *Nature* 439, no. 7073 (2006): 161–167.

Relyea, Rick A. "The Lethal Impact of Roundup on Aquatic and Terrestrial Amphibians." *Ecological Applications* 15, no. 4 (2005): 1118–1124.

Rockstrom, Johan, Will Steffen, Kevin Noone, et al. "A Safe Operating Space for Humanity." *Nature* 461, no. 7263 (2009): 472–475.

Simenstad, Charles A., James A. Estes, and Karl W. Kenyon. "Aleuts, Sea Otters, and Alternate Stable-State Communities." *Science* 200, no. 4340 (1978): 403–411.

Wake, David B. "Facing Extinction in Real Time." *Science* 335, no. 6072 (2012): 1052–1053.

Wearn, Oliver R., Daniel C. Reuman, and Robert M. Ewers. "Extinction Debt and Windows of Conservation Opportunity in the Brazilian Amazon." *Science* 337, no. 6091 (2012): 228–232.

Williams, John W., and Stephen T. Jackson. "Novel Climates, No-Analog Communities, and Ecological Surprises." *Frontiers in Ecology and the Environment* 5, no. 9 (2007): 475–482.

Worm, Boris, Edward B. Barbier, Nicola Beaumont, et al. "Impacts of Biodiversity Loss on Ocean Ecosystem Services." *Science* 314, no. 5800 (2006): 787–790.

Zalasiewicz, Jan, Mark Williams, Will Steffen, and Paul Crutzen. "The New World of the Anthropocene." *Environmental Science and Technology* 44, no. 7 (2010): 2228–2231.

Other

Andrainarivo, C., V.N. Andriaholinirina, A. Feistner, et al. "*Lepilemur septentrionalis.*" 2008. IUCN Red List of Threatened Species. Version 2012.1. http://www.iucnredlist.org/details/11622/0.

IUCN Red List (International Union for Conservation of Nature). "Summary statistics: Threatened species in past and present Red Lists. Accessed December 11, 2012. http://www.iucnredlist.org/about/summary-statistics. Note: see Table 1 at http://www.iucnredlist.org/documents/summarystatistics/2012_2_RL_Stats_Table_1.pdf.

Penskar, M.R., and P.J. Higman. "Special Plant Abstract for *Mimulus michiganensis* (Michigan Monkey-Flower)." Michigan Natural Features Inventory. 2010. Accessed December 7, 2012. http://mnfi.anr.msu.edu/abstracts/botany/Mimulus_michiganensis.pdf.

Sobel, J. "*Gadus morhua.*" 1996. IUCN Red List of Threatened Species. Version 2012.1. http://www.iucnredlist.org/details/8784/0.

Catherine Badgley

Historical extinctions (1600–1850)

With the exceptions of a few larger species, scientific knowledge concerning the scale of extinctions among protists, fungi, animals, plants, and other organisms, rests at the mercy of both a poorly mined and often fragmentary historical record and the interpretation of that record by a body of biodiversity scholars that declined exponentially in the twentieth century. The world's oceans serve as a model for how little is known about the depth and breadth of this deletion of species from communities, which suggests scientists' working assumption—that extinctions are rare among many groups of organisms—is not tenable.

The database

Knowledge of the world's biodiversity, and particularly those organisms within the oceans, between the 1600s and 1850s relies primarily on two sources: archeological and historical. Archeological and other prehistoric data are a potentially vast underexplored resource for understanding the presence, distribution, and relative abundance of many coastal shallow-water invertebrates, vertebrates, and sea-grasses from the seventeenth century to the early nineteenth century. These data consist of organic remains in shell middens (mounds), kitchen and other cooking sites, ornaments, and inscriptions and artwork. While zooarcheological (archaeozoological, or the study of faunal remains) records have effectively determined the historical distribution of species such as the Steller's sea cow (*Hydrodamalis gigas*), sea mink (*Neovison macrodon*), and great auk (*Pinguinus impennis*) (all now extinct), along with sea otters (*Enhydra lutris*) (having sustained extensive metapopulation extinctions), the archeological record—with local exceptions—remains largely untapped in terms of understanding the disappearance and altered abundances of marine invertebrates. Of historical data, two types are most relevant: written accounts and museum collections. The 1600s to early 1800s comprise a broad era bridging early efforts at taxonomy with the beginning years of standardized global nomenclature for the scientific names of species based upon the system popularized by the Swedish biologist Carl von Linnaeus in 1758.

Unfortunately, many of the animals and plants (especially smaller, soft-bodied organisms) observed in the natural world in this period often lack sufficient descriptive or illustrative detail to properly identify the taxa in question. Thus, scientists are unable to match them with extant species (exceptions include, on land, larger terrestrial animals and plants, and, for the ocean, historical drawings, dating back to ancient times, of such animals as marine birds, mammals, fish, shelled mollusks, and edible crabs). A typical supposition in analyzing the historical descriptions of species unable to be matched with living species is that the accounts are simply too coarse to permit recognition, but that such accounts almost certainly represent still living species. An alternative hypothesis is that a number of these early descriptions represent the only known records of some species that in fact became extinct at the hands of human endeavor.

A global database of pre-1800s holdings does not exist for the world's museum collections. Likewise, scientists have not systematically searched major natural history collections for such holdings of marine invertebrates or other organisms that have not been seen for over 200 years. A number of caveats would, inevitably, attend interpreting this potential database. As oceanographic exploration technology expanded, deeper waters were sampled. A large number of deep-sea species (those at depths of greater than 650 feet [200 m]) are represented by few individuals, many collected only once, and their subsequent absence in collections—for decades or centuries—cannot be taken as evidence of extinction. In contrast, museum collections hold substantial inventories of intertidal and shallow-water species, many potentially dating from the mid- to late-1700s, especially in Europe, the center of global collections for centuries. Few, if any, of these collections have been carefully analyzed for the presence of species (including epi- and endobiota) that have not been found since before the mid-1800s. The potential is profound for these collections to reveal the extinction of species that are presumed to be still alive, but that have, in fact, disappeared long ago, as well as to recover from preserved specimens of known extinctions also-extinct parasites and symbionts. Scientists know, for example, that the extinct Steller's sea cow had external and internal bionts (associated organisms) that in all likelihood went extinct as well.

A compelling modern-day example—one with particular relevance to this largely untapped historical record—of

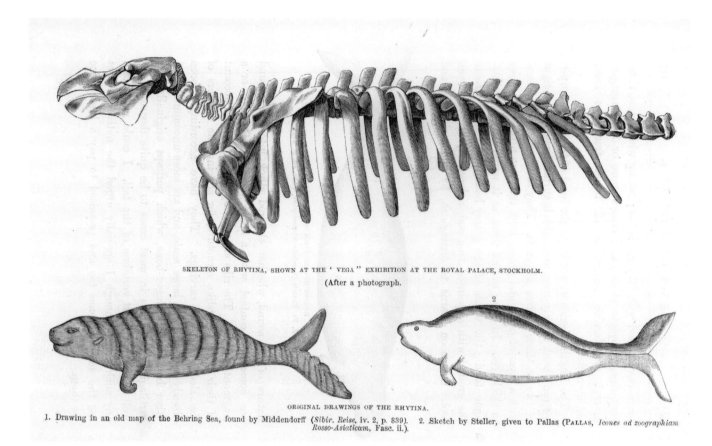

SKELETON OF RHYTINA, SHOWN AT THE ' VEGA '' EXHIBITION AT THE ROYAL PALACE, STOCKHOLM.
(After a photograph.

ORIGINAL DRAWINGS OF THE RHYTINA.

1. Drawing in an old map of the Behring Sea, found by Middendorff (*Sibir. Reise*, iv. 2, p. 839). 2. Sketch by Steller, given to Pallas (PALLAS, *Icones ad zoographiam Rosso-Asiaticam*, Fasc. ii.).

Steller's sea cow (*Hydrodamalis gigas*) was almost 6 meters long and is now extinct. Top: engraving of the skeleton of the sea cow. Bottom left: Engraving after an early map of the Behring Sea. Bottom right: Engraving of Steller's sketch. © UniversalImagesGroup/Getty Images.

the latter phenomenon is the long-overlooked extinction of a marine snail in the Northwest Atlantic Ocean. The marine limpet (*Lottia alveus*) was an abundant native gastropod living on the blades of the eelgrass (*Zostera marina*) from eastern Canada to New York, recorded commonly in the literature and in guides to seashells from the 1830s to the 1980s. A small (0.4 inch [10 mm] in length), but easily recognized mollusk, it was, for example, recorded in 1929 as "thousands of individuals readily accessible" (Proctor, 1933, p. 167) on the coast of Maine. Between 1930 and 1933, the eelgrass (the limpet's host plant) precipitously disappeared from much of the Atlantic seaboard due to a wasting disease. Remarkably, although the limpet continued to be cited in literature as common on eelgrass blades for decades after the disappearance of *Zostera*—in a region as densely populated with marine biologists as any in the world—no one noticed that the limpet had gone extinct. No living *L. alveus* have been collected since 1929, as revealed by thorough searches of museum collections and confirmed by extensive field surveys since the 1980s.

If a shelled mollusk that thrived at the foot of marine biological laboratories went extinct in the early twentieth century without notice (its extinction was announced in 1991), it is not difficult to imagine that a potential wealth of intertidal and shallow-water species went extinct before the

1800s, ushered in by over-extraction fisheries (which removed, for example, enormous expanses of oyster beds, and their rich associated communities). Contributing to further extinctions going virtually unnoticed include the profound forces of urbanization, industrialization and port and harbor development (which resulted in the extirpation of huge areas of salt marshes, supralittoral zones, and lagoons, all habitats highly vulnerable to the footprint of human expansion).

Susceptibility to extinction

Numerous aspects of the biology and ecology of a given species increase or decrease its vulnerability to extinction (see Table 1). These include the length and rate of population turnover; the type and frequency of reproduction; the capacity for population recovery or reestablishment, including dispersal distance and mobility, competitive ability, and colonizing ability; and the species' range and distribution, including habitat specificity and susceptibility to destruction, rarity, and trophic level occupied. Scientific knowledge (see below) of the biological and ecological attributes of older extinctions is such that it is difficult to retrospectively apply all of these criteria, but these characteristics may provide a basis for predicting which species are more vulnerable to extinction in the future.

Characteristics that make marine species vulnerable to extirpation and extinction	Vulnerability	
	High	Low
Population turnover (age, growth, biomass)		
Longevity	Long	Short
Growth rate	Slow	Fast
Natural mortality rate	Low	High
Production biomass	Low	High
Reproduction		
Reproductive effort	Low	High
Reproductive frequency	Semelparity (breeding once)	Iteroparity (breeding more than once)
Age or size at sexual maturity	Old or large	Young or small
Sexual dimorphism	Large difference in size between sexes	Does not occur
Sex change	Occurs (in particular protandry [a male stage followed by a female stage])	Does not occur
Spawning	In aggregations at predictable locations	Not in aggregations
Allee effects (reduction in density has significant impacts on reproduction)	Strong	Weak
Capacity for recovery (dispersal and competition)		
Dispersal	Short-distance	Long-distance
Competitive ability	Poor	Good
Colonizing ability	Poor	Good
Adult mobility	Low	High
Range, distribution, and habitat specificity and vulnerability		
Horizontal distribution	Nearshore	Offshore
Vertical depth range	Narrow	Broad
Geographic range	Small	Large
Patchiness of population within range	High	Low
Habitat specificity	High	Low
Habitat vulnerability to destruction by people	High	Low
Abundance	Rare	Abundant
Trophic level	High	Low

Table 1. Characteristics that make marine species vulnerable to extirpation and extinction. Reproduced by permission of Gale, a part of Cengage Learning.

Types of extinctions

While the focus has often been on worldwide extinctions, it is important to note that extinctions can and do occur on various spatial and regulatory scales, and the local consequences of such deletions can be profound.

Habitat extinction refers to the restriction of species to fewer habitats (realized niche) than previously occupied (fundamental niche) due, for example, to competition or predation from introduced species, or from the wholesale destruction of populations in one or more original habitats. An example is the restriction of the native California mudsnail (*Cerithidea californica*) to high intertidal salt marsh refugia due to the invasion of the non-native Atlantic mudsnail (*Ilyanassa obsoleta*), which displaced *Cerithidea* from open mudflats.

Local extinction refers to a species' spatial restriction within, for example, a large estuary (such as Chesapeake Bay

or Puget Sound) compared to its previous documented distribution.

Regional or provincial extinction refers to the extirpation of a species from a large part of its functional range (the total geographic area a species previously occupied, such as multiple provinces or an ocean basin); a great many examples exist, ranging from mollusks (such as the giant clam, *Tridacna crocea*) to fish.

Global extinction (as reviewed within this text) is the worldwide demise of a species, as assessed by consistent field surveys over many decades.

Functional extinction refers to the phenomenon whereby a species no longer plays a key regulatory role as a competitor, predator, or disturbing agent in a community or ecosystem, due to the scale of population reduction. Thus, many populations of whale species used to number in the hundreds of thousands to millions, but now exist in numbers that are a small fraction of their previous populations, and thus presumably are no longer able to regulate their prey or competitors at their historic scale.

Of no small value would be studies of select communities that attempt to assess the full scale of extinctions from habitat to functional levels.

The record of global marine extinctions

At least two species of birds, one mammal, one fish, and one mollusk went extinct in the years before 1850 (see Table 1). The year 1850 is used in this essay as a date of record for two reasons. First, the formal year of the last known living individual(s) represents a time long after the species had suffered vast population restriction and decline, which in the case of the species discussed here, would have been largely in the eighteenth and early nineteenth centuries. Second, were scientists to recognize species only known to have become extinct prior to 1800, then only one mammal would be discussed; a record, as noted above, that belies and obscures the real scale of extinctions. Thus, the other species included here, albeit persisting into the nineteenth century, serve as useful reminders and models that extinctions were proceeding in the 1600s and 1700s across a variety of taxa.

Great auk (*Pinguinus impennis*)

Archeological remains reveal that the flightless piscivorous great auk, standing slightly less than 3.3 feet (1 m) tall, historically occurred across much of the North Atlantic. It ranged in the western hemisphere south along the North American Atlantic coast and, in the eastern hemisphere, south to Spain, but by approximately AD 1000 was already restricted to the higher latitudes of Canada and northern Europe. Hunting for its meat, oil, and feathers finally reduced the population to Iceland, where—when the last verified individuals were found (in 1844)—they were promptly shot. (However, reports and rumors of great auks persisted for many decades thereafter.)

Great auk (*Pinguinus impennis*) specimen of extinct bird. © Field Museum Library/Getty Images.

Pallas's (or spectacled) cormorant (*Phalacrocorax perspicillatus*)

Little is known about this largely flightless cormorant that lived in the Commander (Komandorski) Islands of the Bering Sea. It was hunted to extinction for food and feathers.

Steller's sea cow (*Hydrodamalis gigas*)

Similar to the great auk of the North Atlantic, paleontological and archeological records indicate that this herbivorous North Pacific dugong occurred over a wide range before its eventual restriction to the Commander Islands. Huge (up to 30 feet [9 m] in length, with a weight of perhaps 11 tons [10 t]) and slow, it was immediately susceptible to hunting, with a remarkable rate of extirpation from its discovery in 1741 to the last recorded individuals in 1768.

Green wrasse (*Anampses viridis*)

This small labrid fish was first and last collected in 1839 on the island of Mauritius in the Indian Ocean. A member of a well-studied group of tropical fish, it is unlikely to have remained overlooked. The cause and timing of extinction is not known, but may be related to habitat loss.

Periwinkle snail (*Littoraria flammea*)

The only marine mollusk thought to have become extinct before 1850, this snail was collected a number of times in the early nineteenth century from the Chinese mainland, perhaps associated with once more-pristine mangrove habitats.

Other possible early extinctions

A gray whale species existed in the North Atlantic Ocean until about 1675, and appears to have been recorded (as the scrag or sandloegia) since at least the ninth century in Norse writings. While long held to have been the Atlantic population of the still-extant Pacific gray whale (*Eschrichtius robustus*), extant bones of the Atlantic gray whale await potential genetic analysis as of April 2012. Similarly, the Auckland Island Shore plover (*Thinornis rossi*) known from a single extant specimen collected in 1840, and thought by many scientists to be the still-living plover *Thinornis novaeseelandiae* from New Zealand, remains unexamined, as of April 2012, for potential DNA (deoxyribonucleic acid) mining. These two examples underscore that a further use of museum collections to determine whether what are presumed to have been allopatric but now extinct metapopulations of many extant species were in fact distinct species.

Vulnerability of known extinctions

Comparing Table 1 (vulnerable characteristics) to Table 2 yields a mixed picture. For example, the great auk and Steller's sea cow were likely long-lived (high vulnerability, per characteristics in Table 1), while the cormorant, wrasse, and limpet were almost certainly comparatively short-lived (ostensibly low vulnerability). The auk, sea cow, and possibly the cormorant were likely older or larger at sexual maturity (higher vulnerability), whereas the wrasse and limpet were likely young or small at sexual maturity (lower vulnerability). On the other hand, all likely occurred in some type of aggregation to spawn, all occurred nearshore, and all of these species were likely relatively habitat specific. While the characteristics of species that have become extinct may fall in either column, it is likely that certain phenomena—particularly those that closely interface with human pressures such as over-hunting and habitat decimation—are the more stringent Achilles' heels that lead to the final demise of species.

Species	Location	Last known living	Cause of extinction
Birds			
Pinguinus impennis (Great auk)	Atlantic: Iceland	1844	Hunting
Phalacrocorax perspicillatus (Pallas's cormorant)	Pacific: Commander Islands	ca. 1850	Hunting
Mammals			
Hydrodamalis gigas (Steller's sea cow)	Pacific: Commander Islands	1768	Hunting
Fish			
Anampses viridis (Green wrasse)	Indian Ocean: Mauritius	1839	Habitat loss?
Molluscs			
Littoraria flaminea (Periwinkle snail)	Pacific: China	<1840s	Habitat loss?

Table 2. Marine animals and plants known to have gone extinct before 1850. Reproduced by permission of Gale, a part of Cengage Learning.

The scale of potential extinction

No sponges, sea anemones, worms, crustaceans, marine insects, clams, mussels, bryozoans, or most other marine invertebrates are regarded as having become extinct. Only one obscure fish species is considered extinct, as noted above. Given the scale of modification of coastal environments over the past several thousand years, especially where dense human populations exist, it is inconceivable that these taxa have completely escaped global extinction—rather, it is more likely that scientists are simply unaware of what is now gone. Table 3 presents the probable minimum hypothetical scale of global extinctions. Summing these numbers, at least 190 species of marine protists and invertebrates may have gone extinct in the past several hundred years from marine environments that have been highly susceptible to destruction. This number is small compared to other estimates that concluded a minimum of 1,000 species must have gone extinct in coral reefs in recent centuries, given the percentage of reef loss due to severe habitat degradation. These numbers further support the proposition that museums hold a fundamentally unexplored window into the past.

Modern assessments of marine biodiversity

The population status of certain high-profile endangered species of marine mammals (such as the Hawaiian monk seal [*Monachus schauinslandi*], vaquita [*Phocoena sinus*], and North Atlantic right whale [*Eubalaena glacialis*], along with Beluga whales [*Delphinapterus leucas*]), certain corals (including species of the stony coral *Acropora*), and edible snails (such as the black and white abalones of the North American Pacific coast) is under regular and often close monitoring. The records of these species being compiled in the 2010s should provide a clear basis for the timing and geography of either their demise or recovery. This said, there are no globally standardized, regular quantitative monitoring programs for most species of invertebrates, fish, plants (seaweed and seagrasses), and other organisms for most coasts of the world. An understanding of the population sizes of heavily extracted finfish and shellfish populations often relies on incomplete data and numerous assumptions that may lead to estimates with order-of-magnitude ranges. Severely compromising scientific understanding of the distribution and abundance of a great many marine species is the decline in taxonomy and systematics, disciplines that are the *sine qua non* of measuring the pulse of biodiversity. Linked to this is the absence of thorough modern-day quantitative assessments of the status of marine life for most, or perhaps all, of the iconic estuaries of the world. Given the current state of biodiversity assessment in much of the world's oceans, it seems probable that many more species may go extinct without notice.

Taxon or guild	Habitat where extinctions may have occurred	Possible scale of extinctions (1*s* = <10 species; 10*s* = <100; 100*s* = 100 or more)
Meiofauna (psammofauna, interstitial fauna)	Estuarine muds and sands	100*s*
Symbionts, commensals, and parasites	Extinct birds, mammals, fish, and invertebrates	10*s*
Porifera (sponges)	Estuaries and shallow-water near-shore environments	1*s*
Cnidaria Hydrozoa (hydroids) and anthozoa (sea anemones)	Estuaries and shallow-water near-shore environments	1*s*
Platyhelminthes (flatworms)	Estuaries and shallow-water near-shore environments	10*s*
Annelida (worms) Oligochaeta	Supralittoral shores, lagoons, and estuaries	1*s*
Annelida (worms) Polychaeta	Estuaries and shallow-water near-shore environments	10*s*
Mystacocarida	Estuarine muds and sands	1*s*
Ostracoda (ostracodes)	Estuaries and shallow-water near-shore environments	1*s*
Copepoda (copepods)	Estuaries and shallow-water near-shore environments	10*s*
Isopoda (pill bugs) and **Amphipoda** (scuds)	Supralittoral shores, lagoons, and estuaries	10*s*
Decapoda (crabs and shrimps)	Estuaries and shallow-water near-shore environments	1*s*
Insecta Hemiptera (water boatmen), diptera (flies), dermaptera (earwigs), coleoptera (beetles)	Supralittoral shores, lagoons, and estuaries	10*s*
Chilopoda (marine centipedes)	Supralittoral shores	1*s*
Mollusca Bivalvia (clams, mussel, oysters) and gastropoda (snails)	Estuaries and shallow-water near-shore environments	10*s*
Bryozoa (moss animals)	Estuaries and shallow-water near-shore environments	1*s*
Chordata Ascidiacea (sea squirts)	Estuaries and shallow-water near-shore environments	1*s*
Chordata Pisces (fish)	Estuaries and shallow-water near-shore environments	1*s*
Algae (seaweeds)	Estuaries and shallow-water near-shore environments	10*s*

Table 3. Possible scale of extinctions of marine organisms. Reproduced by permission of Gale, a part of Cengage Learning.

Why should scientists care?

Why be concerned about which species are already gone? There are numerous compelling reasons to refine the opaque record of historic anthropogenic extinctions. Scientists and others are fundamentally concerned—for scientific, cultural, and even aesthetic reasons—about the global loss of distinct lineages. Critical to understanding the modern-day structure and function of communities and ecosystems is knowing whether key species, including potential ecological engineers and other regulatory species, have in fact been deleted from the systems whose evolutionary history are trying to be elucidated. Perhaps of greatest value is that a knowledge of which species have disappeared can provide science a measure of both what types of organisms and what types of communities are most vulnerable to extinction, permitting better focus of limited monitoring and assessment resources.

Resources

Books

Dulvy, Nicholas K. et al. "Holocene Extinctions in the Sea." In *Holocene Extinctions*, edited by Samuel T. Turvey, 129–150. Oxford: Oxford University Press, 2009.

Periodicals

Carlton, James T. "Apostrophe to the Ocean." *Conservation Biology* 12 (1998): 1165–1167.

Carlton, James T. "Neoextinctions of Marine Invertebrates." *American Zoologist* 33 (1993): 499–509.

Carlton, James T., Jonathan B. Geller, et al. "Historical Extinction in the Sea." *Annual Review of Ecology and Systematics* 30 (1999): 515–538.

Carlton, James T., Geerat J. Vermeij, et al. "The First Historical Extinction of a Marine Invertebrate in an Ocean Basin: The Demise of the Eelgrass Limpet *Lottia alveus*." *Biological Bulletin* 180 (1991): 72–80.

del Monte-Luna, Pablo et al. "Marine Extinctions Revisited." *Fish and Fisheries* 8 (2007): 107–122.

Dulvy, Nicholas K. et al. "Extinction Vulnerability in Marine Populations." *Fish and Fisheries* 4 (2003): 25–64.

Lotze, Heike K. et al. "Historical Baselines for Large Marine Animals." *Trends in Ecology and Evolution* 24 (2009): 254–262.

Proctor, William. 1933. "Biological survey of the Mount Desert region. Part V." Wistar Institute of Anatomy and Biology. Philadelphia, PA.

Reynolds, John D. et al. "Biology of Extinction Risk in Marine Fishes." *Proceedings of the Royal Society (B)* 272 (2005): 2337–2344.

Roberts, Callum M et al. "Extinction Risk in the Sea." *Trends in Ecology and Evolution* 14 (1999): 241–246.

James T. Carlton

Historical extinctions (1850–present)

Following the Pliocene-Pleistocene megafaunal extinctions is the Holocene epoch—the time interval between 12,000 years ago and the present. Though this has been a climatically stable interglacial period, it has nevertheless borne witness to a large number of extirpation events at a variety of geographic scales as well as the global extinction of over 800 species. Also, within this time interval the rates of extirpation/extinction appear to have increased progressively, though, to no small extent, this apparent trend is attributable, at least in part, to a combination of sampling factors along with the fact that historico-scientific records have become much more accurate over the last 200 years. Nevertheless, and as Samuel T. Turvey (2009) has noted, in many ways the climatic stability of the Holocene represents a more interesting, and certainly a more compelling, lens through which to view modern extinctions than either the climatically more variable Plio-Pleistocene or the—by comparison—terra incognito of the major ancient extinction intervals. Within the Holocene epoch many of the best data on the effect human societies have had on plant and animal populations come from the very last portion, which encompasses the Industrial Revolution—typically defined as the period from 1800 to the present. Among the 864 species currently listed as either extinct or extinct in the wild by the International Union for Conservation of Nature (IUCN; see Tables 1 and 2), 636 have specific estimates of extinction dates associated with their records. Of these, all but 33 (95%) have occurred since 1800. Moreover, this "industrialization interval" represents the time of the most intense human population increase; the repurposing of major expanses of land for agriculture; widespread deforestation; the proliferation of housing, physical infrastructure, and commercial development of all sorts; massively increased water use; routine fertilizer use; factory fishing; and, of course, global temperature change—all induced by the technology-related activities of a single species, *Homo sapiens*.

Careful analysis of these data is important for a variety of reasons. First, they provide unambiguous proof that extinction is not a phenomenon confined to the dim and distant past, but rather a prominent—even routine—feature of the modern world. These data also represents the cohort of extinct species for which scientists have the best, most complete, and most detailed ancillary data—data that, in many cases, can be used to identify the factors most likely to be responsible for many species' extinctions. In addition, researchers can query this

database and identify geographic regions, ecological biomes, and (to some extent) taxonomic groups that are susceptible to extinction as a result of human technological and/or social innovation. Finally, when coupled with assessments of the degree of risk that characterizes other, non-extinct populations and species, these data provide a means for estimating future extinction rates as well as the effect that extinction mitigation strategies may (or may not) have with regard to the ongoing efforts to conserve and protect the natural world.

As a final introductory point, it is important to emphasize that the database of species extinctions over the last 200 years represents, inevitably, only the tip of the biological iceberg when it comes to assessing the magnitude of the conservation problem implied by efforts to save modern species from extinction. In essence, these are the extinctions it was easy for researchers, natural historians, and the commercial exploiters of many of these species to notice either because the species' loss was so striking (e.g., the passenger pigeon [*Ectopistes migratorius*], the dodo [*Raphus cucullatus*], the Tasmanian tiger [*Thylacinus cynocephalus*]) or because a particular individual had a particular interest in a particular species or set of species (e.g., the gastropods *Amastra albolabris*, *Amastra cornea*, and *Amastra crassilabrum*, all of which the American malacologist Wesley Newcomb listed as extinct, as reported by J. T. Gulick in an 1858 monograph on Hawaiian land snails). Not included in the IUCN (and other) extinction list(s), but just as important from the standpoint of understanding how humans are affecting species' overall extinction susceptibility, are extinctions of localized populations (known as extirpations). These extirpations reduce overall geographic range, morphological diversity, and genetic diversity of a given species and increase its extinction susceptibility, but the species itself remains extant, often for long periods of time.

Indeed, global population size and geographic range reductions are used to place species in the three "endangered" IUCN categories: vulnerable (overall population size reduction of 30% to 50%), endangered (overall population size reduction of 50% to 70%), and critically endangered (overall population size reduction of 70% to 90%). The IUCN lists for these three categories are much larger than the lists of species totally extinct and extinct in the wild—10,103; 5,764; and 3,941 species, respectively. When combined, a total of over 20,000 species are known to have already become extinct (3%

Class	Extinct	EW	CR	ED	VU	NT	LR	DD	LC	Total
Mammalia	77	2	196	447	497	324	0	835	3,123	5,501
Aves	130	4	197	389	727	880	0	60	7,677	10,064
Reptilia	21	1	145	293	364	255	3	664	1,917	3,663
Amphibia	36	2	507	767	657	388	0	1,623	2,390	6,370
Cephalaspidomorphi	1	0	2	0	1	2	0	3	10	19
Myxini	0	0	1	2	6	2	0	30	35	76
Chondrichthyes	0	0	25	41	116	133	0	504	274	1,093
Actinopterygii	59	7	385	443	1,017	306	10	2,008	4,931	9,166
Sarcopterygii	0	0	1	0	1	0	0	0	3	5
Echinoidea	0	0	0	0	0	1	0	0	0	1
Arachnida	0	0	3	5	11	2	0	9	3	33
Chilopoda	0	0	0	0	1	0	0	0	0	1
Diplopoda	0	0	1	6	7	0	0	7	10	31
Crustacea	11	1	116	145	335	54	9	867	861	2,399
Insecta	59	1	103	190	483	203	3	925	1,930	3,897
Merostomata	0	0	0	0	0	1	0	3	0	4
Onychophora	0	0	3	2	4	1	0	1	0	11
Clitellata	1	0	1	0	4	2	0	0	0	8
Polychaeta	0	0	1	0	0	0	0	1	0	2
Bivalvia	33	0	63	47	39	56	5	169	268	680
Gastropoda	281	14	432	390	758	433	1	1,409	1,435	5,153
Cephalopoda	0	0	0	0	0	1	0	148	46	195
Enopla	0	0	0	0	2	1	0	3	0	6
Turbellaria	1	0	0	0	0	0	0	0	0	1
Anthozoa	0	0	6	23	202	175	0	147	289	842
Hydrozoa	0	0	1	2	2	1	0	2	8	16
Total	710	32	2,189	3,192	5,234	3,221	31	9,418	25,210	49,237

Abbreviations: EW, Extinct in the Wild; CR, Critically Endangered; ED, Endangered; VU, Vulnerable; NT, Near Threatened; DD, Data Deficient; LC, Least Concern.

Table 1. Current (2012) summary table for IUCN extinction risk categories for animal taxa. Reproduced by permission of Gale, a part of Cengage Learning.

of the total "at-risk" species) or to have suffered significant population local reductions (97% of the total "at-risk" species) since 1800, it is safe to assume that the true figure is at least an order of magnitude larger.

Sources of evidence

Since 1800, the standard of evidence for a species' existence has been the visual observation of a living individual, the visual observation of a recently dead individual such that there can be no question of the individual's having been alive some short time before, or secondhand evidence (e.g., scat or droppings, footprints, fur, DNA from tissue) that can be attributed to the species in question unambiguously. Any of these lines of evidence is sufficient to conclude that the species remains extant. However, there are often issues surrounding the reliability of secondhand reports of species sightings with regard to which species was observed, especially in cases in which the observer has not been trained in the species' taxonomy and identification. For example, local people may use the same colloquial name to refer to different species and/or different names to refer to the same species—especially when multiple observations originate from more than one cultural group. This phenomenon is by no means confined to lay naturalists. Even experienced, expert taxonomists can suffer from the same terminological complications through differences in training, differences in access to reference literature, and personal differences of opinion with regard to

Class	Extinct	EW	CR	ED	VU	NT	LR	DD	LC	Total
Anthocerotopsida	0	0	0	2	0	0	0	0	0	2
Bryopsida	2	0	12	13	7	1	0	3	3	41
Charophyaceae	0	0	0	0	0	0	0	3	8	11
Chlorophyceae	0	0	0	0	0	0	0	1	0	1
Coniferopsida	0	0	24	62	92	63	13	22	333	609
Cycadopsida	0	4	53	65	74	63	0	3	45	307
Florideophyceae	1	0	6	0	3	0	0	44	4	58
Ginkgoopsida	0	0	0	1	0	0	0	0	0	1
Gnetopsida	0	0	0	1	3	7	0	10	76	97
Isoetopsida	0	0	8	4	2	3	0	6	5	28
Jungermanniopsida	1	0	10	11	12	1	0	0	10	45
Liliopsida	3	4	190	345	459	158	16	276	724	2,175
Lycopodiopsida	0	0	1	2	8	1	0	0	1	13
Magnoliopsida	82	23	1,412	2,021	4,139	933	195	647	1,457	10,909
Marchantiopsida	0	0	1	3	2	0	0	4	1	11
Ophioglossopsida	0	0	0	0	0	0	0	0	3	3
Osmundopsida	0	0	0	0	0	0	0	0	1	1
Polypodiopsida	2	0	35	42	64	15	0	50	54	262
Sellaginellopsida	0	0	0	0	1	1	0	0	2	4
Sphagnopsida	0	0	0	0	2	0	0	0	0	2
Takakiopsida	0	0	0	0	1	0	0	0	0	1
Ulvophyceae	0	0	0	0	0	0	0	1	0	1
Total	91	31	1,752	2,572	4,869	1,246	224	1,070	2,727	14,582

Abbreviations: EW, Extinct in the Wild; CR, Critically Endangered; ED, Endangered; VU, Vulnerable; NT, Near Threatened; DD, Data Deficient; LC, Least Concern.

Table 2. Current (2012) summary table for IUCN extinction risk categories for plant taxa. Reproduced by permission of Gale, a part of Cengage Learning.

the limits of species variation. In some cases it is possible for contemporary investigators to mount searches of particular areas to look for signs of the species' presence based on alleged sightings. But if the sighting is recorded in a journal, diary, letter, technical article, or some other historical document, it may not be possible to accept or deny the correctness of an observation made at some point in the distant, or even in the recent, past. Moreover, even if confirmed, observations can testify only to a species' presence. In terms of extinction studies, what investigators really want to know is whether a species is absent from a locality, region, or worldwide.

Logically, there is no way of collecting data that prove a negative proposition—that a species is not present. This issue can be addressed, however, on the basis of probability estimates by asking the question how many times a negative result must be obtained before a negative proposition can be accepted as the one most likely to be correct, especially if a reasonably accurate model of a species occurrence pattern prior to extinction can be obtained or estimated. Typically the results of such estimates are formulated and presented in terms of confidence intervals of a species' true time of extinction.

To take a simple example, if a series of searches are made for a species at random time intervals, in some cases these searches will return a positive result and in others a negative result. Commonly a negative result (or series of negative results) will occur between two positive results, leading to the

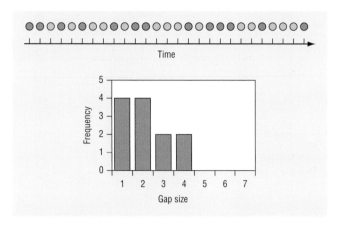

Figure 1. Conceptual diagram illustrating the principles of estimating confidence intervals for extinction times. The upper diagram represents the results of a series of searches for a hypothetical species over a defined time interval. Blue circles represent positive sightings; green circles represent no (negative) sighting. Note the "gappiness" in the distribution of positive sightings over time. This distribution can be summarized in the form of a gap-frequency histogram (lower diagram). Given assumptions of the form of the theoretical gap-frequency distribution, it is possible to estimate the area under the distribution curve to any desired level of accuracy. The 95 percent confidence interval can then be estimated as the time interval beyond the last positive sighting that encompasses 95 percent of the theoretical gaps that might occur in the species sighting distribution. Reproduced by permission of Gale, a part of Cengage Learning.

specification of a gap in the occurrence pattern (see Figure 1). Gaps formed in this way, including the gaps in time that occur as a result of the intervals between successive searches, can be summarized as a distribution of gap intervals. The species in question must have existed somewhere during the time represented by these gaps, but its presence failed to be recorded by the search that was conducted. Looked at in this way, the extinction-time problem becomes one of specifying how long beyond the last observed occurrence of the species do negative results need to be obtained before an investigator can conclude confidently (to a stated level) that the lack of observation is attributable to the lack of existence rather than to a gap in the occurrence pattern. This situation is an exact analogy to the situation in stratigraphy in which the distribution of spatial gaps in fossil species occurrence patterns are used to estimate intervals of uncertainty in the placement of extinction horizons.

Several different numerical methods are available for calculating such estimates. These differ in the assumptions they make about the character of the gap distribution. If the number of sightings is N, and the distribution of the $N-1$ sighting gaps is assumed to be random, the upper (that is, most recent) limit of the extinction time estimate $D_{ext.}$ (with 95% confidence) may be calculated as follows:

$$D_{ext.} = D_{max} + \left\{ \frac{[\ln(0.05)]}{-[(N-1)]/D_{max} - D_{min}} \right\}$$

In this expression D_{max} is the time of the species' most recent observation (see Springer and Lilje 1988).

Alternatively, if the set of observation gaps is assumed to follow an exponential distribution, the calculation of the upper

limit of the 95 percent extinction time estimate assumes the following form: $D_{ext.} = D_{max} + 4.32i$

In this expression, i represents the median gap interval time (see McFarlane 1999).

Finally, if the set of observation gaps is assumed to follow a beta (single-variable) or Dirichlet (multiple-variable) distribution, the upper limit of the 95 percent extinction time estimate may be estimated as follows:

$$D_{ext} = D_{max} + [(D_{max} - D_{min})(1 - 0.5)^{1/H-1} - 1)]$$

In this expression, H represents the number of observations used to determine the extinction time estimate (see Strauss and Sadler 1989).

These are the simplest models for obtaining probability-based estimates of extinction time based on a record of species sightings prior to the species' final disappearance. More mathematically complex—but more realistic—methods are also available that do not assume a constant and/or known distribution of observation gaps in the latter portion of a species' history (see Solow 2003, 2005; Roberts and Solow 2003; Holdaway et al. 2002a, 2002b; Buck and Bard 2007). Based on the method described by Andrew R. Solow in a 2005 article, the estimated extinction times ($\alpha = 0.05$) for Steller's sea cow and the passenger pigeon are listed in Table 3.

Although these probability-based extinction time estimates provide, in certain respects, a more realistic means of using historical data in a manner that actually bears on the question of extinction timing, and although such methods deserve to be used more widely than they are at present to test hypotheses of extinction cause, such results must, inevitably, be treated with a good deal of caution. In addition to issues arising from the reliability, accuracy, and consistency of sightings made by observers (both amateur and expert), all probabilistic methods of estimating extinction time assume that sampling efforts designed to detect the species in question are constant. The need for constant effort is bound up logically in the meaning of negative results. If a search was mounted as part of a systematic sampling program and failed to find evidence of a species, this result represents a true, negative-observation datum. If, however, any part of the interval between sightings returns a negative result because no search was conducted, the

Species name	Hydrodamalis gigas	Ectopistes migratorius
Common name	Steller's sea cow	Passenger pigeon
Last sighting	1768	1900
Extinction date estimate	1771	1902
Confidence interval on extinction date estimate (95%)	1768–1781	1900–1906
Probability	<0.001	<0.001
Observation references	Stejneger (1887); Gibson (1999)	Schorger (1955)

Table 3. Probabilistic extinction date estimates based on the method of Andrew R. Solow (2005). Reproduced by permission of Gale, a part of Cengage Learning.

Scientific name	Common name	Previous sighting	Rediscovery	Notes
Tarsius pumilus	Pygmy tarsier	1921	2000	This is a small nocturnal mammal that would require considerable effort to monitor adequately. Accidentally rediscovered when one was caught in a trap during a survey of rat populations on Mount Rore Katimbo in Indonesia's Lore Lindu National Park.
Carpomys melanurus	Greater dwarf cloud rat	1896	1960	First sighting came from specimens collected from Mount Data in the Philippines and given to a British researcher. Rediscovery was made during a survey of Mount Pulag National Park.
Cryptotis nelsoni	Nelson's small-eared shrew	1894	2009	This is a small nocturnal mammal that would require considerable effort to monitor adequately. First discovered on the slopes of San Martín Tuxtla, a volcano near Veracruz, Mexico. Rediscovery was the result of the first directed survey to the type area mounted specifically to look for the species.
Laonastes aenigmamus	Loatian rock rat	11 MYA	2005	The genus to which this modern species belongs was originally known only from the Miocene fossil record. Accidentally rediscovered in remote Laotian forests.
Solenodon cubanus	Cuban solenodon	1890	2003	A small nocturnal fossorial mammal that appears to be rare throughout its range, this species would require considerable effort to survey and/or monitor. Surveys undertaken subsequent to its rediscovery have found it to occur throughout Cuba's central and western Oriente Province.
Heosemys depressa	Arakan forest turtle	1908	1994	Always regarded as rare and never surveyed or monitored, this species was rediscovered accidentally when a few specimens were recognized in Asian feed markets. Discovery in the wild followed when individuals were identified in a Myanmar elephant sanctuary.
Burramys parvus	Mountain pygmy possum	Pleistocene	1992	An upland forest–dwelling species with low fossilization potential but previously known only from a serendipitous fossil discovery. Rediscovered accidentally when an individual was collected in a ski hut on Mount Hotham, Australia. Two viable populations have been identified thus far.
Phoboscincus bocourti	Terror skink	1876	2003	Originally described from a single specimen collected on New Caledonia. Rediscovered over a century later during a survey of New Caledonian reptiles.
Eupetaurus cinereus	Woolly flying squirrel	1924	1995	A large squirrel, but one that lives in a very remote part of the world and is quite secretive in its habits.
Gallotia gomerona	La Palma giant alizard	1500	2007	Known from zooarchaeological deposits on La Gomera. Rediscovered on two other Canary Islands.
Elephas maximus	Java elephant	1800s	2006	A dwarf island population thought to have been a victim of Western colonization of the island of Java during the Age of Exploration. Rediscovered accidentally when similar populations on Borneo turned out to have almost identical gene sequences.

Table 4. Examples of species once regarded as extinct and then later rediscovered, often as a result of accidental captures during directed surveys. Reproduced by permission of Gale, a part of Cengage Learning.

nature of the negative result is of a very different kind. The former is the outcome of a genuine lack of observation—part of the pattern of species' occurrence in nature—whereas the latter is simply the result of a human decision not to make an effort to look for the species. These are fundamentally different variables and cannot be treated as equivalents.

In addition, sampling programs must be carried out throughout the entire geographic (and stratigraphic for fossil) range(s) of the species in question in the context of the investigation's scope. If the purpose of the investigation is to estimate the extinction date for a local population (that is, the extirpation date), the sampling program must include a reasonable sample of sites from across the population's known range. By the same token, if a global extinction date estimate is needed, the sampling program must include a reasonable selection of sites from across the species' geographic range. If the only data that are available come from a single locality, probabilistic extinction time estimates can still be made, but these will be valid only for the single sampled locality. Given

the state of taxonomic information for most species, these are very stringent requirements that can be met only with logistic difficulty and, in most cases, extraordinary expense. Nonetheless, a failure to respect the logic of extinction-time estimate investigations has been responsible for many (if not most) of the somewhat embarrassing rediscoveries of perfectly viable populations of previously termed extinct species, in some cases more than a century after their "official" extinction date was declared based on historical records (see Table 4).

The historical record

For clarity, presentation of the historical extinction record is subdivided into two sections: extinctions in the sea and extinctions on the land.

Extinctions in the sea

In many ways the present-day oceans are in a state similar to that of Pliocene-Pleistocene terrestrial biotas. Humans first impacted the marine ecosystem coincident with their migration out of Africa, which in many areas took place along coastal trackways. These populations were sustained by subsistence fisheries that took advantage of the tremendous abundance and productivity of many prehistoric coastal waterways. This activity, no doubt, had some local and perhaps even regional effects. But compared to later, postindustrial times, the overall impact of preindustrial human interference with ecosystem function appears to have been slight.

Contrast this with the contemporary situation. Marine megafaunal species of all types have been decimated almost exclusively as a result of hunting (i.e., fishing) of wild stocks to feed a world population of more than 7 billion individuals, 44 percent of whom live in coastal areas. Human populations use the oceans as a ready source of animal protein, consuming an average of over 132 million tons (120 million metric tons) of fish and by-products as food per year (FAO 2012). Moreover, fish is becoming more popular with the human shopper, who has been responsible for a global per capita increase in consumption from 38.4 pounds (17.4 kg) in 2006 to 41.4 pounds (18.8 kg) in 2011. Despite the best efforts of conservation organizations to alert consumers to the biological unsustainability of many current fisheries (see below), this trend toward increasing fish consumption shows no hint of decline. Indeed, over 2010–2012 the marginal consumption rate underwent a step-change increase. Fitting a linear model to these data since 2006 indicates that the rate of annual increase in fish consumption is 21 percent.

This prodigious harvest of Earth's marine biological resources has been made possible only through the advent of technology that has allowed an increasingly smaller proportion of the human population to be occupied in the fishing industry directly but that still delivers an ever-increasing catch. The technology that has resulted in this interesting juxtaposition has been developed largely in the interval between 1800 and the present.

Surprisingly, the annual harvest of truly astounding amounts of fish from Earth's oceans, rivers, and lakes has occurred, and continues to occur, in the face of very small numbers of species extinctions. Although the long and eventful extinction history of fossil marine invertebrate and fish species is well established and tolerably well known (e.g., Stanley 1987; MacLeod, forthcoming), it may surprise many readers that only four marine invertebrates, three marine fish, and three marine mammals are currently regarded as having become extinct since 1800 (see Table 5)—at the height of the technological revolution that has wrought such profound ecological changes to marine ecosystems.

What sort of changes in marine ecosystems have occurred since 1800? Between 1990 and 2005 the average size of fish populations along the western US coast declined by 45 percent (Levin et al. 2006). This pattern of significant and sustained reduction has been a consistent feature of the marine environment for over a millennium (Jackson et al. 2001). Associated long-term trends include reductions in average age/maturity of fish stocks, changes in diversity associated with the local and regional removal of top predators, and changes to marine food webs.

As of 2012, by far the most striking extinction-related feature of marine ecosystems occurring since 1800 was the

Scientific name	Common name	Last known sighting	Cause(s) of extinction	References
Algae				
Vanvoorstia bennettiana	Bennett's seaweed	1916	Habitat loss	Millar (2003); IUCN (2006)
Invertebrates				
Lottia alveus	Atlantic eelgrass limpet	1929	Habitat loss	Carlton et al. (1991, 1999); Carlton 1993
Collisella edmitchelli	Rocky shore limpet	1861	Habitat loss	Carlton (1993); Carlton et al. (1999)
Cerithidea fuscata	Marsh horn snail	1935	Commercial exploitation	Carlton (1993); Carlton et al. (1999)
Littoraria flammea	Periwinkle	1840	Habitat loss	Carlton (1993); Carlton et al. (1999)
Fish				
Azurina eupalama	Galápagos damselfish	1982	Habitat loss, climate change	Jennings et al. (1994); Roberts and Hawkins (1999)
Anampses viridis	Mauritius green wrasse	1839	Unknown	Hawkins et al. (2000)
Prototroctes oxyrhynchus	New Zealand grayling	1923	Commercial exploitation, invasive species	Balouet and Alibert (1990); McDowell (1996)
Mammals				
Neovison macrodon	Sea mink	1860	Commercial exploitation	Anderson (1995); Carlton et al. (1999); Turvey and Risley (2006)
Zalophus japonicus	Japanese sea lion	1950s	Unknown	Campbell (1988); Youngman (1989); Carlton et al. (1999); IUCN (2006); Sealfon (2007)
Monachus tropicalis	Caribbean monk seal	1952	Commercial exploitation	Carlton et al. (1999); Wilson and Reeder (2005); IUCN (2006); McClenachan and Cooper (2008)

Table 5. Known marine species extinctions since 1800 (from Dulvy et al. 2009). Reproduced by permission of Gale, a part of Cengage Learning.

extirpation of local populations as a result of overfishing. This includes not only the removal of the species targeted directly by hunting efforts, but also inadvertent capture and killing of associated species as bycatch (e.g., turtles, elasmobranchs) and habitat destruction as a direct result of nonselective hunting strategies (e.g., bottom trawling). Global summary data (Dulvy et al. 2003) indicate that this source of environmental pressure is responsible for about 60 percent of local, regional, and global marine extinctions. The only other significant factor contributing to marine extinctions is non-hunting-related habitat loss (e.g., resulting from coastal development), which accounts for about 33 percent of local, regional, and global marine extinctions.

INVERTEBRATES On the whole, invertebrate extinction patterns between 1800 and the early 2000s have not received intensive study because of the large number of marine invertebrate species, the remote and difficult-to-access nature of their habitat, and the simple fact that most marine invertebrate species are not considered interesting commercially. What is known, however, is that, in many cases, the abundance and diversity of many marine invertebrate species are linked ecologically to the existence and behavior of top predator species, which, in many marine biomes are vertebrates.

Reef corals represent the best-understood (relative) exception to the scientific community's level of ignorance with respect to the extinction fate of most marine invertebrate species. This understanding is largely the result of the degree to which reef ecosystems provide habitat for a wide variety of vertebrate species as well as providing services (including tourism) to human communities. While reef corals are not exploited commercially to any significant extent, corals—like many marine invertebrate species—are quite sensitive to changes in the marine environment and exhibit quite narrow environmental tolerance envelopes.

No coral species became extinct between 1800 and 2012. Between the early twentieth and early twenty-first centuries, however, the coral reef biome has changed more than in the previous 220,000 years combined (Pandolfi and Jackson 2006). The example of Caribbean reefs is typical. As noted by several researchers (Jackson 1997; Jackson et al. 2001; see also Dulvy et al. 2003), Caribbean reefs had been fished extensively by local and regional inhabitants for 500 years without any noticeable effect. Even the disappearance of large vertebrates and conchs in the 1950s and 1960s had little direct effect on the structure of these reefs. However, the elimination of medium-sized predatory and grazing fish owing to continued pressure by subsistence fisheries resulted in a perturbation in the ecological structure of Caribbean reefs, mostly through an increased abundance of invertebrate grazers such as the sea urchin Diadema antillarum. Reef diversity was lowered, but structurally the Caribbean reefs remained intact. Nevertheless, this reduction in ecosystem diversity is seen retrospectively as having placed the Caribbean reefs in a finely balanced state. When this balance was destroyed as a result of a Diadema mass mortality event in 1983–1984, microalgae managed to gain a structural foothold that set up an unanticipated positive feedback loop. Subsequent investigations have shown that Diadema was the last

grazer species capable of holding the abundance of the relatively unpalatable microalgae in check. With Diadema much reduced, macroalgae were able to overgrow coral colonies across the region and flip these reefs into a new but stable structural state with low coral diversity. This process was augmented by increased use of fertilizers on the islands of the Caribbean (which stimulated algae growth), as well as storms, increased rates of sedimentation resulting from deforestation, pollution, and oil spills, all of which depressed coral growth. As of 2012, the Caribbean reefs were but a shadow of what they were as recently ago as the 1930s in terms of structure, species richness, species abundance, stability, and resilience.

Similarly Australia's Great Barrier Reef has suffered substantial and ongoing declines in coral cover since the 1960s despite the Great Barrier Reef Marine Park Act 1975, which established this biome as a protected habitat (Brodie 2012). Coral cover on the Great Barrier Reef in 2012 was approximately 20 percent, which is about half what the cover was estimated to be in the 1960s. The main causes of this decline are believed to be pollution from nearby inhabited coasts and from tourism, fishing and climate change.

In addition to these issues, which are now being played out in coral reef environments worldwide, the increased abundance of carbon dioxide in Earth's atmosphere has led to a lowering of the pH of marine surface waters—ocean acidification—as well as a concomitant rise in sea-surface water temperatures. These two factors have conspired to make it progressively harder for coral species to maintain the integrity of their calcium carbonate skeletons (as 30 percent of a coral skeleton is composed of the relatively unstable mineral aragonite). If these trends in the chemistry of marine surface waters continue, the diversity of aragonite-secreting scleractinian corals may be reduced substantially simply because the entire marine environment will have moved out of the zone of aragonite stability—a situation that last occurred some 250 million years ago.

FISH For the most part, data on fish stocks mirror those for invertebrates. For the majority of marine fish species there are very few data at hand because of the large number species, the remote habitat of most fish species, and lack of commercial interest. Excellent data are available, however, for commercial fish stocks. Although this subset of marine fish is small in terms of diversity (see Figure 2A), the numbers of fish taken from these fisheries on an annual basis are impressive (see Figure 2B). In 1950 the total marine fisheries catch was 18.5 million tons (16.8 million metric tons). In 2010 this figure stood at 85.3 million tons (77.4 million metric tons), down from a high of 95.2 million tons (86.4 million metric tons) in 1996.

Is the current rate of marine fish exploitation sustainable? The Food and Agriculture Organization of the United Nations (FAO) tracks fisheries worldwide and classifies them into three categories: "non-fully exploited," "fully exploited," and "overexploited." (Here the category "fully exploited" is formally defined as continuous fishing at the maximum sustainable yield that can reduce biomass by 60 to 70%.)

(a)

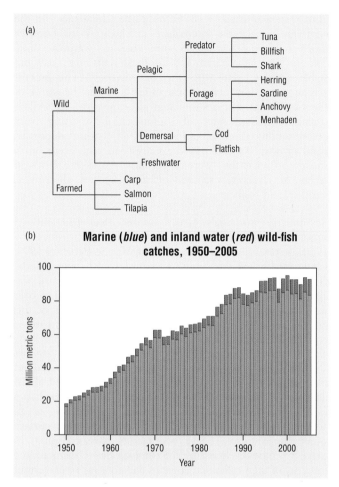

(b)

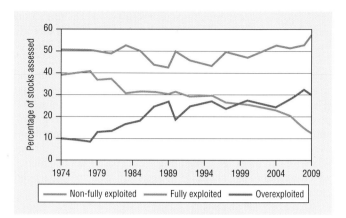

Figure 3. Time series of marine fishery risk categories between 1974 and 2009. Note the steady increase in overexploited fish stocks over time. Redrawn from FAO 2012. Reproduced by permission of Gale, a part of Cengage Learning.

Figure 2. Commercial fish catch data since 1950. A. The major marine fisheries, along with the ecological role these fish play in their local communities. B. Fish production in millions of metric tons of fish caught each year, from 1950 to 2005 (data from the Fisheries Global Information System of the Food and Agriculture Organization of the United Nations). Note the predominance of marine species in the global wild-fish catch along with the strongly (and statistically significant) positive slope of the trend over time. Reproduced by permission of Gale, a part of Cengage Learning.

Figure 3 shows the history of these categories between 1974 and 2009. Over this interval, non–fully exploited fisheries decreased from 40 percent to 10 percent of marine (commercial) fish stocks, overexploited fisheries increased from 10 percent to 30 percent, and fully exploited fisheries remained reasonably constant at about 50 percent. Based on these figures, 80 percent to 85 percent of the known marine commercial fish stocks are currently being harvested at, or over, their biological carrying capacity. Yet the clear economic trend is for consumer demand to increase monotonically for the foreseeable future. Tellingly, an independent analysis of FAO data (Mullon et al. 2005) suggested that fully 25 percent of fish stocks had collapsed between the mid-twentieth and early twenty-first centuries.

There is no better example of fish stock collapse than that of the Grand Banks cod fishery (see Figure 4). When first "discovered" by western Europeans during the 1497 voyage of

the English explorer John Cabot, cod were so abundant in the waters south and east of Newfoundland it was reputed that simply lowering a basket into the water was sufficient to land a basketful of fish. For 500 years fishermen from Canada, the United States, Russia, Germany, France, Spain, Portugal, and Greenland took cod from these waters sustainably, mainly because they relied largely on sail-based technology and hand fishing from small, open boats. This way of life for the Grand Banks fishery ended in the wake of World War II with the introduction of large steam-driven factory ships and a switch from hand fishing to trawling, which was made possible because of the greater size and power of the ships. At first the catch rose precipitously, peaking at just under 2.2 million tons (2 million metric tons) in 1968 (see Figure 4), after which the Grand Banks catch declined through the 1970s despite continued technological development (e.g., radar, electronic navigation systems, fish finding sonar, better communications technology). After a relatively stable interval through the 1980s, the fishery collapsed in 1992 when cod stocks were found to be less than 1 percent of the levels recorded in 1960. The Canadian government closed the fishing grounds in 1992, initially for two years to give fish stocks an opportunity to replenish. This action caused a collapse of the economy of Newfoundland and, to a lesser extent, in fishing industries worldwide, with over 40,000 Canadians losing their livelihoods. Unfortunately, the fish did not return in the 1990s nor have they returned as of 2013. As a result the Grand Banks cod fishery remains closed. No one has any idea when the cod will, or if they can, come back. Subsequent research has found that, not only did the factory trawling operation take too many cod over too short an interval of time, the bycatch destroyed local populations of fish species the cod fed on (e.g., capelin), and the physical disruption of the sea floor destroyed the local habitat cod require to reproduce.

In the early 2000s the Grand Banks supported fisheries in snow crab and northern shrimp. Together these were worth approximately as much as the pre-boom cod fishery. This replacement, however, represents an example of switching the target(s) of fishing in a local area from those that are no longer

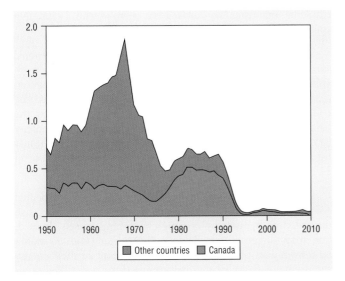

Figure 4. Fish catch data from the Grand Banks cod fishery from 1950 to 2010. Note the sudden and sustained collapse of the fishery starting in 1992. This collapse was presaged by a strong decline throughout the 1970s that cod stocks did manage to recover from in the 1980s. Data are derived from the FishStat Plus database of the Food and Agriculture Organization of the United Nations (FAO). Reproduced by permission of Gale, a part of Cengage Learning.

sustainable to those that are stable and can support a high-valued fishery, but ones that occupy (in this case) a lower trophic level in the local food web. This situation should be contrasted with the "fishing down the food chain" scenario described by Pauly et al. (1998). Here, as large predatory fish suffer the effects of overexploitation differentially because of their higher value lower generation times and high intrinsic resource requirements (Reynolds et al. 2005), they are replaced by a fishery that develops around species that occupy a lower trophic position, thus putting pressure on an ever-larger proportion of the local food web and making it even more difficult for the collapsed fishery to recover (Pauly et al. 1998; but see also Essington et al. 2006). The point, however, is neither of these fishery-development strategies address the issue of how sustainable stock management can be achieved in the face of ever-increasing consumer demand for inexpensive animal protein.

The Grand Banks story is an example, in microcosm, of patterns and trends in marine fin-fish management that have occurred throughout the twentieth and early twenty-first centuries. Various authors have provided estimates of the decline in predatory fish since the introduction of massive-scale, technology-driven fishing fleets. In the North Sea, for example, these estimates range from a low of 66 percent of all predatory fish species (Christensen et al. 2003) to a high of 99 percent of all North Sea fish within the size range from 35 to 145 pounds (16 to 66 kg; Jennings and Blanchard 2004).

REPTILES Sea turtles have been the only significant marine reptile group in the interval covered by this entry. There are seven living sea turtle species: the flatback (*Natator depressus*), green (*Chelonia mydas*), hawksbill (*Eretmochelys imbricata*), Kemp's ridley (*Lepidochelys kempi*), leatherback (*Dermochelys*

coriacea), loggerhead (*Caretta caretta*), and olive ridley (*Lepidochelys olivacea*). No sea turtle extinctions have been reported since 1800, but all seven species are considered endangered except the olive ridley (vulnerable) and flatback (data deficient). Accordingly, there is evidence that five of the seven extant sea turtle species have suffered a known or estimated 50 percent reduction in population size since 2000, along with a fragmented distribution pattern.

Sea turtles are killed routinely as bycatch victims of mechanized net-fishing and longline commercial fishing operations. In a 2004 study, Rebecca L. Lewison and colleagues estimated that as many as 260,000 loggerheads and 50,000 leatherbacks are eliminated in this manner each year. In the United States, 70 percent to 80 percent of fatal sea turtle strandings are thought to be the result of individuals becoming entangled in shrimp fishing nets (Crowder et al. 1995). Although a concerted effort has been made to encourage net-fishing operations to install turtle excluder devices on all fishing nets, compliance has been low (though increasing) and effectiveness uncertain.

In addition, coastal development often destroys or compromises sea turtle nesting sites. A 2006 summary for the IUCN (www.iucnredlist.org) found a decline of between 48 percent and 67 percent in the overall number of nesting female sea turtles. Other investigations indicate this figure may be as high as 70 percent for nesting female leatherbacks over the course of a single year (Pritchard 1982; Spotila et al. 1996).

Finally, sea turtles are harvested for food and medicine. The green sea turtle is the main ingredient in green turtle soup. This dish was developed in the Cayman Islands where it is a traditional meal, though turtle meat was also used in this region in many other ways. Between 1688 and 1730 upward of 130,000 green sea turtles were harvested each year by the Cayman Island fishery to meet the needs of the local population (Jackson 1997). By 1800 continued hunting pressure of this magnitude caused the Cayman green sea turtle fishery to collapse. In the early 2000s, the commercial need for green sea turtle meat was met through aquaculture, by raising green sea turtles in aquaria, a by-product of which is an active program to release farm-reared turtles into the wild. Sea turtle plastrons are also used in traditional Chinese medicine, and there is a small but active—though now illegal—trade in hawksbill sea turtle shells as ornamental objects and as a raw material in the creation of high-quality string instrument picks, eyeglass frames, combs, brushes, and various traditional objects (e.g., *toluk*).

BIRDS Seabirds are usually regarded as members of the Sphenisciformes and Procellariiformes along with the Pelecaniformes (except the darters), some of the Charadriiformes (skuas, gulls, terns, auks, and skimmers), and the phalaropes. Among these groups only two species are listed by the IUCN as having become extinct between 1800 and 2012: the great auk (*Pinguinus impennis*) and Pallas's cormorant (*Phalacrocorax perspicillatus*). By far the more familiar of the two, the great auk represents a classic example of island extinction in the modern era, as well as a cautionary tale for science. The great auk was a large, flightless charadriiform that lived along the coasts of Canada, the northeastern United States, Norway, Greenland, Iceland, the Faroe Islands, Ireland, Great Britain, France, and northern Spain. Cosmopolitan within this region in terms of its hunting habits, this species required rocky islands with sloping shorelines directly adjacent to the sea as nesting sites. The only known great auk rookeries were on Papa Westray in the Orkney Islands, the St. Kilda archipelago off Scotland, Grimsey Island and Eldey Island near Iceland, Funk Island near Newfoundland, and the Bird Rocks (Rochers-aux-Oiseaux) in the Gulf of St. Lawrence.

Populations of auks had been taken continuously by peoples indigenous to these areas stretching back to Paleolithic times without any sign of significant depletion. The significance of these birds to local human populations is signalled by the fact that they were depicted as religious symbols by Native American tribes. However, when great auk populations were encountered by western European sailors, initially on voyages of exploration, they were used as a convenient source of food for the voyages and as bait for catching fish. Sailors also prized the auk eggs, which were three times larger than murre eggs and had a larger yolk. Even more ominously, as a result of these sailors' visits, rats were introduced to the islands that contained great auk rookeries.

The plight of the great auk did not go unnoticed by the powers that be of the time. Nesting colonies in the eastern North Atlantic had been eliminated by 1550. The first great auk conservation order was issued in 1553. By 1794 Great Britain had made the importation of great auk feathers (used to stuff pillows) illegal. Nonetheless, by 1880 the nesting colony on Funk Island had vanished.

In a genuinely cruel twist of fate, with rarity came collectability. Private individuals and state-sponsored museums funded expeditions to the remaining great auk nesting grounds to secure skeletons, skins, and eggs for their collections, thus putting the remaining vestigial populations under even greater pressure. The last great auk in Britain was caught on Stac an Armin in the St. Kilda archipelago in 1840. Following a storm that its captors blamed the bird for calling forth, they beat it to death with sticks believing it to be a witch. When an unexpectedly large colony of fifty individuals was discovered in 1835 on Eldey Island, near Geirfuglasker off the southern coast of Iceland, museums competed with each other to collect specimens from the last remaining great auk colony. The last nesting pair was discovered on July 3, 1844, and killed by strangulation to prevent their escape on order of a merchant who wanted the specimens for his collection. The egg the pair was incubating was smashed, possibly in order to prevent it from falling into the hands of another collector.

The story of Pallas's cormorant is much the same, though less dramatic. This species was restricted to Bering Island, in the Commander Islands, Russia, and possibly the adjacent coast of the Kamchatka Peninsula (Siegel-Causey et al. 1991). Georg Wilhelm Steller, the chief naturalist on an expedition led by Vitus Bering, the Danish explorer, noted that Pallas's cormorant was common in 1741. Colonists settled on its breeding islands, however, and in 1882 the Norwegian-born American naturalist Leonhard Hess Stejneger was told by

residents that the last birds were seen in the 1850s. No specimens have come to light in the intervening years.

With regard to extirpations, the IUCN lists 131 seabird species as endangered. These range across the categories from vulnerable (67) to endangered (39) to critically endangered (25). Seabirds such as the albatross are endangered by commercial fishing because the birds are attracted to the baited hooks that are used under the longline fishing strategy to catch fish such as bluefin tuna and albacore. In 1996 only three albatross species were listed as endangered. By 2006, all 21 albatross species had been so listed.

But the modern situation is also fraught with unexpected complexity. Seabirds benefit from feeding on the fish discarded as bycatch by commercial fishing vessels. Indeed, one study suggests that Balearic shearwater (*Puffinus mauretanicus*) chicks obtain as much as 40 percent of their energy requirements from bycatch scavenging (Oro et al. 2004). This sets up the uncomfortable—almost catch-22—situation in which solving an extinction-risk problem for one set of species (in this case, fish) exacerbates the same problem for another species (in this case, some seabirds).

MAMMALS Like the great auk, Steller's sea cow (*Hydrodamalis gigas*), an extinct sirenian species, is another classic example of poor population management from the time before the basic principles of conservation were understood. A relict from the Pleistocene, Steller's sea cow was the largest sirenian species that had ever lived and had, in former times, enjoyed a distribution that encompassed the entire North Pacific. By the time Georg Wilhelm Steller encountered the animal in 1741, it was restricted, like Pallas's cormorant, to the Commander Islands, which at that time were part of Russia. Once word of this species' existence became public knowledge, the hunters descended quickly, driven by the same incentives as those of sealers and whalers. A wide variety of uses were quickly found for Steller's sea cow products, including its fur (insulation), hide (covering for boats), fat (cooking), and oil (for lamps and as a lubricant). The last known individual was killed in 1768.

Unlike Steller's sea cow, contemporary dugongs (*Dugong*) and manatees (*Trichechus*) are in reasonably good shape for large, slow-moving marine mammals. However, as all four contemporary dugong and manatee species are listed as vulnerable, there is cause for concern. For example, in a 2005 study, Helene Marsh and colleagues concluded that the seas around Queensland, Australia, had supported a population of about 72,000 dugongs in the early 1960s, but they found that this tally had declined to around 4,000 individuals by the mid-1990s. Nevertheless, compared to other marine mammal species, the global population sizes of both dugongs and manatees remain reassuringly large. All four species are at risk from hunting, habitat loss, and injuries suffered when these large, slow-moving herbivorous animals are struck by boats.

Two seal species have been acknowledged as having become extinct in the twentieth century: the Caribbean monk seal (*Monachus tropicalis*) and the Japanese sea lion (*Zalophus japonicus*). The former has the dubious honor of having been mentioned in the second voyage account of Christopher Columbus, the Genoese explorer. A party of eight men from

Columbus' ship sent to explore and secure provisions from Hispaniola killed eight of the seals (described as "sea wolves") on the beach, presumably for food. Prior to 1700 there were an estimated 250,000 Caribbean monk seals, and the species is believed to have ranged throughout the Caribbean and Gulf of Mexico, subdivided into thirteen distinct populations (McClenachan and Cooper 2008). This species, however, was hunted extensively throughout its range by Western explorers and settlers for food and for its oil, which was used at the time to fuel lamps and as a lubricant for machinery. The species' docile nature made it an easy target for the hunters and probably contributed to its selection as a commercial target over other seal species. Again, scientific research also played a role in the species' demise, with a two-day collecting trip as part of the Mexican Geographical and Exploring Survey (1886), taking no less than 42 Caribbean monk seal specimens, two of which still exist in the collections of The Natural History Museum (London) and the University Museum of Zoology, Cambridge. In addition, overfishing of the Caribbean reefs effectively stripped out the seal's fish and molluscan food sources. From 1700 to their estimated extinction year of 1952, the Caribbean monk seal's geographic range contracted progressively until it was confined to the interior of the Caribbean Sea. The last known breeding colony was located on Serranilla Bank, between Jamaica and Nicaragua, with the last sighting of the species occurring in this area. After that there were relatively frequent reports of Caribbean monk seals being sighted by local fishermen and divers around Haiti and Jamaica. Nevertheless, two scientific expeditions in the 1970s failed to find any trace of the species. While it is possible the species still exists, many local biologists suspect the sightings were more likely wandering hooded seals, which are known to occur in the area.

The Japanese sea lion was known from the northwestern Pacific, where it occurred along the coasts of Japan, Korea (North and South), and Sakhalin Island (Russia). As of 2012 the IUCN listed no confirmed sightings after the late 1950s despite several marine mammal research expeditions having been organized to search for it within its former range. The last sighting (of 50–60 individuals) came from Takeshima (a small island between Japan and South Korea, also known as Liancourt Rocks,) in 1951 (Rice 1998). Unconfirmed reports of subsequent sightings were filed, and it is possible the species is not extinct. Confusion with the California sea lion (*Zalophus californianus*) cannot be ruled out, though, especially insofar as the Japanese sea lion has been considered a subspecies of the California sea lion (*Z. c. japonicus*) by many authors. Dale W. Rice (1998) and Sylvia Brunner (2004) have argued for species status, an interpretation that was subsequently bolstered by genetic evidence (Sakahira and Niimi 2007; Wolf et al. 2007).

Worryingly, the IUCN also lists the only two other modern monk seal species, the Mediterranean (*Monachus monachus*) and the Hawaiian (*Monachus schauinslandi*), as critically endangered. Mediterranean monk seals were once distributed widely and continuously in the Mediterranean and North Atlantic. For centuries Mediterranean monk seals were exploited for human subsistence needs, commercially harvested, and killed as a competitor for marine resources. After

2000 there was increased pressure on the seals and their habitat as a result of the destruction of breeding sites (e.g., caves), fishery bycatch mortality, and aggression by fishermen even in countries and areas where the species is legally protected (Aguilar 1999). As of 2012, the estimated total population size of this species stood at 350 to 450 individuals, with 250 to 300 in the eastern Mediterranean. Another 130 seals inhabited the Cabo Blanco area (Western Sahara and Mauritania) and perhaps 20 to 23 the Desertas Island, Madeira. By 2006 the Mediterranean monk seal was thought by many to be the most endangered seal species in the world.

Similarly the Hawaiian monk seal is under pressure throughout its habitat from a number of different factors, including food limitation resulting from changes in oceanographic conditions; competition with human fisheries; competition with other predators; entanglement in fragments of net and line discarded by North Pacific fisheries; predation by sharks (especially on pre-weaned and recently weaned pups); interactions with recreational fishing gear, especially hookings and entanglements in gill nets; the possible transmission of diseases from domestic pets and livestock to seals; disturbance of seals on beaches heavily used by people; and the possible loss of terrestrial habitat stemming from sea-level increases resulting from global warming (Baker et al. 2006). In 1983 the total population of Hawaiian monk seals stood at only 1,488. By 2007 this number had declined to 935—amounting to a 37 percent reduction in twenty-four years. Monitoring programs indicate that between 1999 and 2012 the Hawaiian monk seal population declined at a rate of 4.1 percent per year. Using this estimate to project abundance forward from the 1983 estimate (to 2028), an overall reduction of 86 percent (from 1,488 to 201) is indicated after two generations. A separate investigation conducted for the IUCN (www.iucnredlist.org/details/13654/0) suggested that, given current trends, this species' population size could decline to just 37 individuals in three generations.

In 1700 the sea otter (*Enhydra lutris*) ranged around the entire northern Pacific Rim from Hokkaido, Japan, to Baja California, Mexico (Kenyon 1969). The worldwide population was estimated to be between 150,000 (Kenyon 1969) and 300,000 individuals (Johnson 1982). After Russian explorers arrived in Alaska in 1741, however, an extensive commercial harvest of sea otters over the next 150 years nearly resulted in the species' extinction. By 1911 it is likely that fewer than 2,000 animals remained, concentrated in thirteen remnant colonies in Kamchatka, Alaska, the Queen Charlotte Islands, central California, and Mexico (San Benito Islands; see Estes 1980). In 1938 biologists undertook an active program of restoration based on the Californian and Alaskan populations. By 2005 the worldwide population estimate stood at over 105,000 individuals—a truly remarkable recovery and an astounding achievement for a managed-spaces conservation program. Since 2007, however, ominous clouds have gathered over this classic conservation good-news story. In 1973 the Alaskan sea otter population was estimated at between 100,000 and 125,000 individuals. By 2006 the population had fallen to about 73,000 individuals, amounting to a drop of more than 40 percent. Most of this decline took place in the Aleutian Islands population.

The cause of this decline is unknown, although killer whale predation is suspected (Estes et al. 1998). The Prince William Sound population was also hit hard by the *Exxon Valdez* oil spill, which killed thousands of sea otters in 1989. Other sea otter populations have also declined, though none so dramatically. Sea otters were drowned in gill and trammel nets in California from the mid-1970s to the early 1980s (Riedman and Estes 1990). Between 2000 and 2012 population declines in California's sea otters were suspected to be associated with commercial fisheries insofar as mortality was elevated in the summer months when commercial fin-fish landings in the coastal live-trap fisheries are at their height (Estes et al. 2003). More mysteriously, around 10 percent of sea otters examined exhibited signs of underfeeding. Several studies found infectious disease to be an important mortality factor in California sea otter populations (Conrad et al. 2005; Miller et al. 2008; Johnson et al. 2009). The deaths of some 280 sea otters were linked to a pair of protozoan parasites—*Toxoplasma gondii* and *Sarcocystis neurona*—both of which are known to breed in cats and opossums (ibid). In Alaska, Streptococcal endocarditis, encephalitis, and/or septicemia, collectively referred to as strep syndrome, have also been found in sea otter autopsies. In addition to these factors, sea otters are preyed on by killer whales (*Orcinus orca*), great white sharks (*Carcharodon carcharias*), bald eagles (*Haliaeetus leucocephalus*), coyotes (*Canis latrans*), and brown bears (*Ursus arctos*; see Riedman and Estes 1990). Significant declines in preferred prey species populations—including the northern fur seal (*Callorhinus ursinus*), harbor seal (*Phoca vitulina*), and Steller sea lion (*Eumetopias jubatus*)—may have caused killer whales to switch prey and consume sea otters (see Estes et al. 1998).

The last marine mammal group considered in this review is the cetaceans. The Order Cetacea includes all whales and dolphins. This is an ancient group of animals that first appear in the fossil record about 54 million years ago (Eocene). In the early 2000s, eighty-eight species are assigned to the order.

Owing to their size, adult whales are largely immune from predation, though calves, ill or injured individuals, and smaller species may be taken by killer whales and large sharks. Regardless, the primary extinction threat to cetacean populations, for at least the last 20,000 years, has been humans. At first whales were used almost exclusively for food, the list of whale products has grown over the centuries to include ambergris, ambrein, baleen, blubber, cetyl alcohol, muktuk, scrimshaw, spermaceti, tabua, and whale oil. So long as whaling remained a shore-based enterprise conducted from small boats, most, if not all, whale fisheries were sustainable. But with the advent of technological advances in the late 1700s and early 1800s, starting with the creation of self-contained factory ships that could travel anywhere in the world ocean and remain "on station" for months, thereby allowing particular species to be targeted—often during particularly vulnerable phases of the life cycle—humans began to eliminate whale populations systematically.

By 1800 the pioneering voyages of the Basques, the Dutch, and the Danes had faded into history. To be sure the latter still sent whaling fleets forth, mostly to Greenland and Svalbard. But the whaling industry in Europe was small by comparison to the emerging industrial, shipbuilding, and technological might of the United States.

In 1829 the US whaling fleet consisted of over 200 ships, rising to over 600 by 1850. At first the fishery concentrated on the North Atlantic right whale (*Eubalaena glacialis*) and the humpback whale (*Megaptera novaeangliae*), though P. J. Bryant made an interesting case in a 1995 article that the original target of North Atlantic whalers was the gray whale (*Eschrichtius robustus*). Once these species had been depleted in US coastal waters, the whalers had two options: travel farther away to find their preferred species or switch to a new, more abundant species. They opted to pursue both, expanding the geographic scope of their operations (aided by technology) to hunt whales in all the world's oceans and, as resources and economics dictated, moving across the taxonomic scale. In sequence, the whale species most taken were the sperm whale (*Physeter macrocephalus*; late 1800s–1910), the blue whale (*Balaenoptera borealis*; 1920–1930), the fin whale (*Balaenoptera physalus*; 1930–1964), the sei whale (*Balaenoptera borealis*; 1968–1975), and the common minke whale (*Balaenoptera acutorostrata*; 1975–1990). Following the collapse of the fin whale fishery in the mid-1960s, the US whale catch entered a period of precipitous decline that ended in 1982, when the International Whaling Commission (IWC) banned whaling so that stocks might recover. Countries that currently refuse to abide by the ban include Japan, Russia, Norway, and Iceland. All of these are traditional whaling nations, though it must the noted that in most cases these countries have attempted to regulate their national whale catches outside the strictures of the IWC.

Subsequently, the baiji or Yangtze River dolphin (*Lipotes vexillifer*) was regarded by many as being extinct. A native of the middle and lower portions of the Yangtze River (from Yichang to the river mouth at Shanghai, a distance of some 1,050 miles [1,700 km]), the baiji had an estimated total population of 6,000 in the 1950s. The species subsequently suffered pressure from hunting for food (especially during China's so-called Great Leap Forward from 1958 to 1961), entanglement and drowning in nets deployed to catch river fish, habitat destruction caused by development along the river, collisions with ships, and pollution. The population declined to about 500 individuals by 1970, to 400 in the 1980s, and to just 13 by the late 1990s. The last Yangtze River dolphin was seen in August 2004, though there were unconfirmed reports of sightings as late as 2007.

Extinctions on land

The terrestrial extinction record since 1800 is too large to consider in as much detail as the marine record. Not only are more species known from terrestrial environments (see Vermeij and Grosberg 2010), but the extinction status of many terrestrial groups (e.g., mammals, birds) has also been studied at a greater level of detail compared to the overwhelming majority of marine groups. Accordingly, all that is attempted here is a brief overview emphasizing trends, common features, and common causes. Interested readers are encouraged to consult other sources (e.g., Baillie et al. 2004; Steadman 2006; Stork 2010). The empirical part of the brief summary that is presented here is based on the Red List of

Threatened Species database of the International Union for Conservation of Nature (IUCN) and a similar list maintained by the American Museum of Natural History's Committee on Recently Extinct Organisms (CREO).

The most striking feature of the terrestrial extinction record since 1800 is the profound distinction between extinctions on the mainlands of continental landmasses and those taking place on islands, both physical and ecological. Both the IUCN and the CREO databases show that virtually all species extinctions between 1800 and 2000 have involved insular species endemic either to physical or ecological islands (e.g., lakes, river drainages, isolated valley or montane habitats). Indeed, in the case of the best-studied groups—mammals, birds, and fish—the number of non-island extinctions that have been recorded since 1800 is so small as to be able to be ignored from a statistical point of view. This fact leads to the decidedly counterintuitive conclusion that, even though the continental landmasses of western Europe and North America have been modified extensively over this time interval as a result of industrialization, development, population pressures, the growth of agriculture, ever-increasing energy demands to fuel industrial technologies, and the concomitant climate change that meeting such demands has entailed to date, the massive species extinctions predicated by many since the early 1960s have failed to materialize as of 2013. Instead, it is the island biotas, which by comparison have remained in a relatively natural state, that have borne the brunt of extinctions induced by human activities.

As always, the terrestrial extinction record is biased in favor of species in which humans have either a commercial or recreational interest. There is no doubt that a full census of all terrestrial taxa—from the most obscure soil protist to the emblematic top predators in species-rich terrestrial ecosystems—would increase the lists of extinct and endangered species by several orders of magnitude. Nevertheless, for some of the more charismatic taxonomic groups, active monitoring programs had been in place for several decades as of 2012. No serious claim cam be made that insufficient effort has been expended to assess the extinction status of species within these groups. Yet the pattern in these data is unmistakable. As of 2013 island biotas had suffered far larger numbers of terrestrial species extinctions than continental biotas.

Between 1800 and 2012, the IUCN and CREO recorded a total of 302 species extinctions for freshwater fish, bird, and mammal species, including 67 fish, 108 birds, and 127 mammals. Based on estimates of extant species richness in these groups, these figures represent less than 1 percent of the total species in these three groups combined. Among fish, members of the Cypriniformes and Cyprinodontiformes comprise the bulk of the extinction list. These are largely fish that inhabit lakes and river systems—both of which are, in effect, ecological islands. In many cases these fish also constitute important food resources for local inhabitants and/or serve as the basis for commercial fisheries. Among birds, rails (Rallidae) and parrots (Psittacidae) are differentially represented in the lists, reflecting the fact that species from these groups are common members of island biotas. Among mammals, rodents and bats are the higher taxa that have been

most at risk, again reflecting the common occurrence of these groups on islands.

The fact that island biotas are uniquely susceptible to extinction has long been appreciated. Factors that contribute to this susceptibility include (1) small geographic range, (2) small population sizes, (3) limited resources, (4) specialized behaviors (reflected in poor ability to cope with invasive species), (5) limited resistance to introduced diseases, (6) low evolution/migration rates, and (7) low levels of competition (see Cronk 1997; Duncan and Blackburn 2004; Corlett 2010; Wright et al. 2009; Lane 2010). In a 2012 study, Craig Loehle and Willis Eschenbach reviewed the historical record of island extinctions and concluded that human hunting (including hunting for eggs in the case of birds) was the primary factor responsible for the high rate of species extinction on islands over the last 500 years, followed closely by habitat destruction, predation by or competition with invasive species (introduced by humans), and introduction of diseases via direct transfer from human populations or from invasive species.

In regard to this high proportion of human-induced island extinctions during historical times, several authors have argued that these data represent but the tip of the metaphorical iceberg. For example, in 1995 David W. Steadman published the results of a study of Holocene bird extinctions on Pacific islands, contrasting extinction rates in the interval prior to and after colonization by Polynesian peoples. In three broad regions (remote outpost islands, the Polynesian heartland, and Micronesia/Melanesia), bird faunas recovered from the zooarchaeological records were compared. In each region the number of species disappearing from the local fossil records of the islands studied increased for postcolonization samples, in most cases dramatically so. The primary victims of this group of extinctions were largely flightless birds—especially rails. Owing to their fully terrestrial nature, rails had evolved into separate species on almost every island. Moreover, being small and flightless, rails would have been relatively easy for humans and other introduced predators to catch. In Steadman's study the lowest extinction rates were recorded for seabirds that fly from island to island and hunt in the open ocean waters.

Based on these local records from a sample of islands in each region, Steadman estimated an extinction/extirpation rate of ten species or local populations on each of 800 major islands, though only anecdotal statistics were provided to support this estimate. Taken at face value these data suggest that as many as 8,000 local populations may have been lost as a result of human colonization, through a combination of direct hunting, habitat loss as a result of human occupation, the introduction of new predators or competitors that came to each island as a result of human discovery, and/or the inadvertent introduction of pathogens. Steadman then observed that, because by his estimate 25 percent of the avifauna on each island constituted flightless rails, the estimated loss of 8,000 populations suggests the extinction of as many as 2,000 rail species alone. Using alternative methods, similar estimates have been obtained by other researchers (Alcover et al. 1998; Pimm et al. 2006).

As for extinctions of species on the continents, the story is largely the same, but with a few twists. For example, the

Schomburgk's deer (*Rucervus schomburgki*). Science Source.

English zoologist Edward Blyth described Schomburgk's deer (*Rucervus schomburgki*) in 1863, naming it after Robert H. Schomburgk, the British consul in Bangkok at the time (see Figure 5). This species inhabited the swampy plains of central Thailand, particularly around the Chao Phraya River valley. Like many deer, this species' populations were organized into herds consisting of a single adult male, a group of females, and their young. When the Chao Phraya River flooded during the yearly rainy season, however, the normally separate herds were forced together on higher ground, which often turned into islands. As a result the deer herds were easy targets for hunters during this time of year. As Thai rice exports grew during the 1920s, the grasslands that Schomburgk's deer depended on for forage were progressively usurped for rice production. This put pressure on the herds, which were never abundant. A combination of habitat loss and hunting pressure finally drove the species to extinction, with the last individual dying in a zoo in 1938.

That Schomburgk's deer occurred on the Thai mainland does not negate the fact that it was known only from a few small river drainages—effectively an ecological island—that happened to be located close to a burgeoning metropolitan area. Humans did not "land" in the Chao Phraya River valley in the same way as they landed on Hawaii, Christmas Island, or Réunion Island. But the effect of humans moving into this species' habitat was largely the same. In many cases, when the biological and ecological factors surrounding the historical extinction of a continental species are examined in detail, the

characteristic outlines of an (ecological) island extinction emerge. This is also true of historical fish extinctions where it is typically the case that the draining, polluting, and/or opening of a watercourse to foreign invaders results in the decimation of the indigenous fish fauna.

Finally, it is worthwhile to consider briefly the unusual situation of Australia. A continent-sized island with an indigenous fauna of unique taxonomic character that was first populated by people with only the most rudimentary technology, Australia does not fit nicely into many of the categorizations that have been developed to study and understand the extinction pressures affecting the biotas of other parts of the world. In their 2012 study, Loehle and Eschenbach estimated that Australia suffered 23 mammal and bird extinctions on the mainland and 38 extinctions on nearby islands, both of which are unusually high values in comparison with data from other continents and islands. Moreover, almost all the Australian mammal extinctions were of marsupials.

Faced with statistics such as these, it is difficult to avoid drawing a parallel with the Great American Biotic Interchange between North America and South America, which occurred when the Isthmus of Panama was formed in the Piacenzian (about 3.5 million years ago). While humans were not involved in these extinctions, the indigenous South America mammalian fauna—which was heavy with marsupial species that were closely related to the Australian marsupials—suffered an even more intense extinction event as a result of competition with placental predators/invaders from North America. By the same token, the historical extinctions in Australia can be seen as a late-stage remnant of the Pleistocene extinctions/extirpations that took place at the end of the last glacial maximum and were driven by a combination of human hunting/environmental disruption and climate change and to which the indigenous Australian marsupial fauna was (apparently) uniquely susceptible.

Comparing the present with the past

As the debates about modern extinction rates mature it is becoming increasingly evident that, for a number of reasons, it is inappropriate to simply scale up extinction data derived from limited studies of modern species and compare the results to the great extinctions of the geological past. Attempts to do so miss the point of extinction studies in both the modern and ancient realms. As noted by Geerat J. Vermeij (2004), the extinctions seen in the modern world are essentially local in character, driven by factors such as hunting, habitat fragmentation, the introduction of predators or competitors, and the introduction of pathogens. These factors can be serious, but so far they have fallen far short of causing the elimination of substantial numbers of species over entire regions or precipitating extinction cascades. No doubt such factors played a role in ancient environments as well. But these factors are not among the causes of the great paleontological extinction events (see MacLeod, forthcoming). Rather, the great geological extinction events are characterized by the elimination of entire categories of habitat (e.g., the complete disappearance of the shallow marine floor of a large epicontinental sea as a result of a 330-foot (100-m) drop in sea level), the global refrigeration of the planet, and/or the sudden disruption in the primary productivity of the oceans. These far more powerful and intense processes operate on regional and global scales and cause the ecological collapse of major biomes (e.g., reefs) with ensuing extinction cascades, causing whole categories of biodiversity to be eliminated.

In a sense the difference between the present and the past in terms of extinction is the difference between consumers and producers. So far, modern extinctions have primarily affected ecological consumers whose loss, regrettable as it is, rarely perturbs other species living in the same habitat. Ancient extinctions—especially the great extinctions—affected the habitats of primary producers on whose existence all species

depend to a greater or lesser extent. If these players on the ecological stage go, even for a short time, their absence will not be overlooked or relegated to a few column inches in the local newspaper. The current challenge is to determine whether, and when, levels of modern species extinction are likely to become dangerous and to weigh the decisions that must be taken to prioritize economic, research, and educational resources on the most important problems facing human populations at any given time. To do this, scientists, regulators, and politicians will require the best available estimates of predicted extinction rates, as well as information on the consequences of extinctions and the strategies that can be implemented cost-effectively to avoid species loss.

Resources

Books

Aguilar, Alex. *Status of Mediterranean Monk Seal Populations*. RAC-SPA, United Nations Environment Program. Tunis, Tunisia: Aloès Editions, 1999.

Baillie, Jonathan E.M., Craig Hilton-Taylor, and Simon N. Stuart, eds. *2004 IUCN Red List of Threatened Species: A Global Species Assessment*. Gland, Switzerland: International Union for Conservation of Nature, 2004.

Balouet, Jean-Christophe, and Eric Alibert. *Extinct Species of the World*. Translated by K.J. Hollyman. English edition edited by Joan Robb. New York: Barron's, 1990.

Dulvy, Nicholas K., John K. Pinnegar, and John D. Reynolds. "Holocene Extinctions in the Sea." In *Holocene Extinctions*, edited by Samuel T. Turvey, 129–150. Oxford: Oxford University Press, 2009.

FAO (Food and Agriculture Organization of the United Nations). Fisheries and Aquaculture Department. *The State of World Fisheries and Aquaculture 2012*. Rome: Food and Agriculture Organization of the United Nations, Fisheries and Aquaculture Department, 2012. Accessed September 9, 2012. http://www.fao.org/docrep/016/i2727e/i2727e.pdf.

Gibson, James R. "*De bestiis marinis*: Steller's Sea Cow and Russian Expansion from Siberia to America, 1741–1768." In *Russkaya Amerika, 1799–1867*, edited by Nikolai N. Bolkhovitinov, 24–44. Moscow: Rossiiskaya Akademiya Nauk, Institut Vseobshchey Istorii, 1999.

Johnson, Ancel M. "Status of Alaska Sea Otter Populations and Developing Conflicts with Fisheries." In *Transactions of the Forty-Seventh North American Wildlife and Natural Resources Conference*, edited by Kenneth Sabol, 293–299. Washington, DC: Wildlife Management Institute, 1982.

MacLeod, Norman. *The Great Extinctions: What Causes Them and How They Shape Life*. London: Natural History Museum, forthcoming.

McFarlane, Donald A. "A Comparison of Methods for the Probabilistic Determination of Vertebrate Extinction Chronologies." In *Extinctions in Near Time: Causes, Contexts, and Consequences*, edited by Ross D.E. MacPhee, 95–103. New York: Kluwer Academic/Plenum, 1999.

Millar, Alan J.K. "The World's First Recorded Extinction of a Seaweed." In *Proceedings of the 17th International Seaweed Symposium, Cape Town, 2001*, edited by Anthony R.O. Chapman, Robert J. Anderson, Valerie J. Vreeland, and Ian R. Davison, 313–318. Oxford: Oxford University Press, 2003.

Rice, Dale W. *Marine Mammals of the World: Systematics and Distribution*. Lawrence, KS: Society for Marine Mammalogy, 1998.

Riedman, Marianne L., and James A. Estes. *The Sea Otter (Enhydra lutris): Behavior, Ecology, and Natural History*. Fish and Wildlife Service Biological Report 90 (14). Washington, DC: U.S. Department of the Interior, Fish and Wildlife Service, 1990. Accessed September 9, 2012. http://www.fort.usgs.gov/Products/Publications/2183/2183.pdf

Schorger, A.W. *The Passenger Pigeon: Its Natural History and Extinction*. Madison: University of Wisconsin Press, 1955.

Stanley, Steven M. *Extinction*. New York: Scientific American Library, 1987.

Steadman, David W. *Extinction and Biogeography of Tropical Pacific Birds*. Chicago: University of Chicago Press, 2006.

Turvey, Samuel T. *Holocene Extinctions*. Oxford: Oxford University Press, 2009.

Wilson, Don E., and DeeAnn M. Reeder, eds. *Mammal Species of the World: A Taxonomic and Geographic Reference*. 3rd ed. Baltimore: Johns Hopkins University Press, 2005.

Periodicals

Alcover, Josep Antoni, Xavier Campillo, Marta Macias, and Antònia Sans. "Mammal Species of the World: Additional Data on Insular Mammals." *American Museum Novitates*, no. 3248 (1998): 1–29.

Anderson, Paul K. "Competition, Predation, and the Evolution and Extinction of Steller's Sea Cow, *Hydrodamalis gigas*." *Marine Mammal Science* 11, no. 3 (1995): 391–394.

Baker, Jason D., Charles D. Littnan, and David W. Johnston. "Potential Effects of Sea Level Rise on the Terrestrial Habitats of Endangered and Endemic Megafauna in the Northwestern Hawaiian Islands." *Endangered Species Research* 4 (2006): 21–30.

Brodie, J. "Management of the Great Barrier Reef: A Success?" *Book of Abstracts*, 12th International Coral Reef Symposium. Cairns, Australia (2012): 435.

Brunner, Sylvia. "Fur Seals and Sea Lions (Otariidae): Identification of Species and Taxonomic Review." *Systematics and Biodiversity* 1, no. 3 (2004): 339–439.

Bryant, P.J. "Dating Remains of Gray Whales from the Eastern North Atlantic." *Journal of Mammalogy* 76, no. 3 (1995): 857–861.

Buck, Caitlin E., and Edouard Bard. "A Calendar Chronology for Pleistocene Mammoth and Horse Extinction in North America Based on Bayesian Radiocarbon Calibration." *Quaternary Science Reviews* 26, nos. 17–18 (2007): 2031–2035.

Campbell, R.R. "Status of the Sea Mink, *Mustela macrodon*, in Canada." *Canadian Field-Naturalist* 102, no. 2 (1988): 304–306.

Carlton, James T. "Neoextinctions of Marine Invertebrates." *American Zoologist* 33, no. 6 (1993): 499–509.

Carlton, James T., Jonathan B. Geller, Marjorie L. Reaka-Kudla, and Elliott A. Norse. "Historical Extinctions in the Sea." *Annual Review of Ecology and Systematics* 30 (1999): 515–538.

Carlton, James T., Geerat J. Vermeij, David R. Lindberg, et al. "The First Historical Extinction of a Marine Invertebrate in an Ocean Basin: The Demise of the Eelgrass Limpet *Lottia alveus*." *Biological Bulletin* 180, no. 1 (1991): 72–80.

Christensen, Villy, Sylvie Guénette, Johanna J. Heymans, et al. "Hundred-Year Decline of North Atlantic Predatory Fishes." *Fish and Fisheries* 4, no. 1 (2003): 1–24.

Conrad, P.A., M.A. Miller, C. Kreuder, et al. "Transmission of *Toxoplasma*: Clues from the Study of Sea Otters as Sentinels of *Toxoplasma gondii* Flow into the Marine Environment." *International Journal for Parasitology* 35, nos. 11–12 (2005): 1155–1168.

Corlett, Richard T. "Invasive Aliens on Tropical East Asian Islands." *Biodiversity and Conservation* 19, no. 2 (2010): 411–423.

Cronk, Q.C.B. "Islands: Stability, Diversity, Conservation." *Biodiversity and Conservation* 6, no. 3 (1997): 477–493.

Crowder, Larry B., Sally R. Hopkins-Murphy, and J. Andrew Royle. "Effects of Turtle Excluder Devices (TEDs) on Loggerhead Sea Turtle Strandings with Implications for Conservation." *Copeia* no. 4 (1995): 773–779.

Dulvy, Nicholas K., Yvonne Sadov, and John D. Reynolds. "Extinction Vulnerability in Marine Populations." *Fish and Fisheries* 4, no. 1 (2003): 25–64.

Duncan, Richard P., and Tim M. Blackburn. "Extinction and Endemism in the New Zealand Avifauna." *Global Ecology and Biogeography* 13, no. 6 (2004): 509–517.

Essington, Timothy E., Anne H. Beaudreau, and John Wiedenmann. "Fishing through Marine Food Webs." *Proceedings of the National Academy of Sciences of the United States of America* 103, no. 9 (2006): 3171–3175.

Estes, James A. "*Enhydra lutris*." *Mammalian Species*, no. 133 (1980): 1–8.

Estes, James A., Marianne L. Riedman, Michelle M. Staedler, et al. "Individual Variation in Prey Selection by Sea Otters: Patterns, Causes, and Implications." *Journal of Animal Ecology* 72, no. 1 (2003): 144–155.

Estes, James A., Martin T. Tinker, Terrie M. Williams, and Daniel F. Doak. "Killer Whale Predation on Sea Otters Linking Oceanic and Nearshore Ecosystems." *Science* 282, no. 5388 (1998): 473–476.

Gulick, J.T. "Descriptions of new species of Achetinella from the Hawaiian Islands." *Annals of the Lyceum of Natural History of New York* 6 (1858): 173–255.

Hawkins, Julie P., Callum M. Roberts, and Victoria Clark. "The Threatened Status of Restricted-Range Coral Reef Fish Species." *Animal Conservation* 3, no. 1 (2000): 81–88.

Holdaway, Richard N., Martin D. Jones, and Nancy R. Beavan Athfield. "Late Holocene Extinction of Finsch's Duck (*Chenonetta finschi*), an Endemic, Possibly Flightless, New Zealand Duck." *Journal of the Royal Society of New Zealand* 32, no. 4 (2002a): 629–651.

Holdaway, Richard N., Martin D. Jones, and Nancy R. Beavan Athfield. "Late Holocene Extinction of the New Zealand Owlet-Nightjar *Aegotheles novaezealandiae*." *Journal of the Royal Society of New Zealand* 32, no. 4 (2002b): 653–667.

Jackson, Jeremy B.C. "Reefs since Columbus." *Coral Reefs* 16, no. 5 (1997): S23–S32.

Jackson, Jeremy B.C., Michael X. Kirby, Wolfgang H. Berger, et al. "Historical Overfishing and the Recent Collapse of Coastal Ecosystems." *Science* 293, no. 5530 (2001): 629–637.

Jennings, Simon, and Julia L. Blanchard. "Fish Abundance with No Fishing: Predictions Based on Macroecological Theory." *Journal of Animal Ecology* 73, no. 4 (2004): 632–642.

Jennings, Simon, A.S. Brierley, and J.W. Walker. "The Inshore Fish Assemblages of the Galápagos Archipelago." *Biological Conservation* 70, no. 1 (1994): 49–57.

Johnson, Christine K., Martin T. Tinker, James A. Estes, et al. "Prey Choice and Habitat Use Drive Sea Otter Pathogen Exposure in a Resource-Limited Coastal System." *Proceedings of the National Academy of Sciences of the United States of America* 106, no. 7 (2009): 2242–2247.

Kenyon, Karl W. "The Sea Otter in the Eastern Pacific Ocean." *North American Fauna*, no. 68 (1969): 1–352.

Lane, David J.W. "Tropical Islands Biodiversity Crisis." *Biodiversity and Conservation* 19, no. 2 (2010): 313–316.

Levin, Phillip S., Elizabeth E. Holmes, Kevin R. Piner, and Chris J. Harvey. "Shifts in a Pacific Ocean Fish Assemblage: The Potential Influence of Exploitation." *Conservation Biology* 20, no. 4 (2006): 1181–1190.

Lewison, Rebecca L., Larry B. Crowder, Andrew J. Read, and Sloan A. Freeman. "Understanding Impacts of Fisheries Bycatch on Marine Megafauna." *Trends in Ecology and Evolution* 19, no. 11 (2004): 598–604.

Loehle, Craig, and Willis Eschenbach. "Historical Bird and Terrestrial Mammal Extinction Rates and Causes." *Diversity and Distributions* 18, no. 1 (2012): 84–91.

Marsh, Helene, Glenn De'ath, Neil Gribble, and Baden Lane. "Historical Marine Population Estimates: Triggers or Targets for Conservation? The Dugong Case Study." *Ecological Applications* 15, no. 2 (2005): 481–492.

McClenachan, Loren, and Andrew B. Cooper. "Extinction Rate, Historical Population Structure, and Ecological Role of the Caribbean Monk Seal." *Proceedings of the Royal Society* B 275, no. 1641 (2008): 1351–1358.

McDowell, Robert M. "Threatened Fishes of the World: *Prototroctes oxyrhynchus* Günther, 1870 (Prototroctidae)." *Environmental Biology of Fishes* 46, no. 1 (1996): 60.

Miller, Melissa, Patricia Conrad, E.R. James, et al. "Transplacental Toxoplasmosis in a Wild Southern Sea Otter (*Enhydra lutris nereis*)." *Veterinary Parasitology* 153, nos. 1–2 (2008): 12–18.

Mullon, Christian, Pierre Fréon, and Philippe Cury. "The Dynamics of Collapse in World Fisheries." *Fish and Fisheries* 6, no. 2 (2005): 111–120.

Oro, D., J.S. Aguilar, J.M. Igual, and M. Louzao. "Modelling Demography and Extinction Risk in the Endangered Balearic Shearwater." *Biological Conservation* 116 (2004): 93–102.

Pandolfi, John M., and Jeremy B.C. Jackson. "Ecological Persistence Interrupted in Caribbean Coral Reefs." *Ecology Letters* 9, no. 7 (2006): 818–826.

Pauly, Daniel, V. Christensen, J. Dalsgaard, R. Froese, and F. Torres Jr. "Fishing Down Marine Food Webs." *Science* 279 (1998): 860–863.

Pauly, Daniel, Jackie Alder, Elena Bennett, et al. "The Future of Fisheries." *Science* 302, no. 5649 (2003): 1359–1361.

Pimm, Stuart, Peter Raven, Alan Peterson, et al. "Human Impacts on the Rates of Recent, Present, and Future Bird Extinctions." *Proceedings of the National Academy of Sciences of the United States of America* 103, no. 29 (2006): 10941–10946.

Pritchard, Peter C.H. "Nesting of the Leatherback Turtle, *Dermochelys coriacea* in Pacific Mexico, with a New Estimate of the World Population Status." *Copeia*, 1982, no. 4, 741–747.

Reynolds, John D., Nicholas K. Dulvy, Nicholas B. Goodwin, and Jeffrey A. Hutchings. "Biology of Extinction Risk in Marine Fishes." *Proceedings of the Royal Society* B 272, no. 1579 (2005): 2337–2344.

Roberts, Callum M., and Julie P. Hawkins. "Extinction Risk in the Sea." *Trends in Ecology and Evolution* 14, no. 6 (1999): 241–246.

Roberts, David L., and Andrew R. Solow. "When Did the Dodo Become Extinct?" *Nature* 426, no. 6964 (2003): 245.

Sakahira, Fimihiro, and Michiko Niimi. "Ancient DNA Analysis of the Japanese Sea Lion (*Zalophus californianus japonicus* Peters, 1866): Preliminary Results Using Mitochondrial Control-Region Sequences." *Zoological Science* 24, no. 1 (2007): 81–85.

Sealfon, Rebecca A. "Dental Divergence Supports Species Status of the Extinct Sea Mink (Carnivora: Mustellidae: *Neovision macrodon*)." *Journal of Mammalogy* 88, no. 2 (2007): 371–383.

Siegel-Causey, Douglas, Christine Lefevre, and Arkadii B. Savinetskii. "Historical Diversity of Cormorants and Shags from Amchitka Island, Alaska." *Condor* 93, no. 4 (1991): 840–852.

Solow, Andrew R. "Estimation of Stratigraphic Ranges When Fossil Finds Are Not Randomly Distributed." *Paleobiology* 29, no. 2 (2003): 181–185.

Solow, Andrew R. "Inferring Extinction from a Sighting Record." *Mathematical Biosciences* 195, no. 1 (2005): 47–55.

Spotila, James R., Arthur E. Dunham, Alison J. Leslie, et al. "Worldwide Population Decline of *Dermochelys coriacea*: Are Leatherback Turtles Going Extinct?" *Chelonian Conservation and Biology* 2, no. 2 (1996): 209–222.

Springer, Mark, and Anneliese Lilje. "Biostratigraphy and Gap Analysis: The Expected Sequence of Biostratigraphic Events." *Journal of Geology* 96, no. 2 (1988): 228–236.

Steadman, David W. "Prehistoric Extinctions of Pacific Island Birds: Biodiversity Meets Zooarchaeology." *Science* 267, no. 5201 (1995): 1123–1131.

Stejneger, Leonhard. "How the Great Northern Sea-Cow (*Rytina*) Became Exterminated." *American Naturalist* 21, no. 12 (1887): 1047–1054. Accessed September 11, 2012. doi:10.1086/274607.

Stork, Nigel E. "Re-assessing Current Extinction Rates." *Biodiversity and Conservation* 19, no. 2 (2010): 357–371.

Strauss, David, and Peter M. Sadler. "Classical Confidence Intervals and Bayesian Probability Estimates for Ends of Local Taxon Ranges." *Mathematical Geology* 21, no. 4 (1989): 411–427.

Turvey, Samuel T., and C.L. Risley. "Modelling the Extinction of Steller's Sea Cow." *Biology Letters* 2, no. 1 (2006): 94–97. Accessed September 11, 2012. doi:10.1098/rsbl.2005.0415.

Vermeij, Geerat J. "Ecological avalanches and the two kinds of extinction." Evolutionary Ecology Research 6 (2004):315-337.

Vermeij, Geerat J., and Richard K. Grosberg. "The Great Divergence: When Did Diversity on Land Exceed That in the Sea?" *Integrative and Comparative Biology* 50, no. 4 (2010): 675–682.

Wolf, Jochen B.W., Diethard Tautz, and Fritz Trillmich. "Galápagos and California Sea Lions Are Separate Species: Genetic Analysis of the Genus *Zalophus* and Its Implications for Conservation Management." *Frontiers in Zoology* 4 (2007): 20.

Wright, Shane D., Len N. Gillman, Howard A. Ross, and D. Jeanette Keeling. "Slower Tempo of Microevolution in Island Birds: Implications for Conservation Biology." *Evolution* 63, no. 9 (2009): 2275–2287.

Other

American Museum of Natural History. Committee on Recently Extinct Organisms (CREO). "Accessing the CREO Extinctions Database." Accessed September 12, 2012. http://creo.amnh.org/pdi.html#access.

FAO (Food and Agriculture Organization of the United Nations). Fisheries and Aquaculture Department. "FishStat Plus." Accessed September 12, 2012. http://www.fao.org/fishery/statistics/software/fishstat/en.

IUCN (International Union for Conservation of Nature). IUCN Red List of Threatened Species. Version 2012.1. Accessed September 12, 2012. http://www.iucnredlist.org.

Norman MacLeod

• • • • •

Subrecent oceanic island extinctions

Oceanic islands have been more affected by recent extinctions than any other part of Earth. This is due to a number of factors that are unique to the island environment or applies more strongly on islands.

The nature of oceanic islands

A very large majority of oceanic islands are volcanic. Such volcanic islands are created whenever a submarine volcano grows tall enough to reach the surface. However, volcanic rocks are soft and erode quite quickly once eruptions have ceased, and in most cases volcanic islands have eroded down to sea level within a few million years or less. In addition, the sea bottom on which these islands stand tends to sink, partly from the weight of the volcanic pile and partly because most submarine volcanoes originate in relatively shallow water, on or near mid-ocean ridges, and the depth increases as the oceanic crust cools and is moved away from the ridge by plate tectonics.

If a volcanic island is situated in warm tropical waters, coral reefs will form along the coasts. Because corals keep growing upward toward the surface, an island can survive as an atoll long after the volcanic core has eroded and subsided below sea level. In such cases, sooner or later a period of rapid sea-level rise will outstrip the ability of the corals to "keep up," and the island will cease being an island, instead becoming a flat-topped seamount, known as a *guyot*. Although the oldest parts of the seafloor of the Pacific are more than 200 million years old, few if any atolls are older than 5 million years.

This process is famously illustrated by the islands of the Hawaiian archipelago. Hawaii, the youngest and highest island, still has active volcanoes; however, the islands to the northwest are successively older, lower, and more eroded. Beyond Nihoa, the last high island (at 892 feet [272 m]), follows a string of atolls and beyond these a long chain of seamounts—former islands that have subsided below sea level. Meanwhile, southeast of Hawaii a new undersea volcano, Loihi Seamount, has grown to within 3,000 feet (915 m) of the surface and will probably become a new island within the next several thousand years.

There are two other types of oceanic islands. Fragments of continental crust (often called microcontinents) can become detached from the main continents and isolated by plate tectonic processes. Such small continental blocks tend to subside over time, with the land on them becoming more or less completely flooded. The most prominent cases are the two continental islands of Madagascar and New Zealand, but there are also a number of smaller islands that consist of continental crust. These include the Chatham Islands, the Bounty Islands, and New Caledonia in the Pacific; the Seychelles in the Indian Ocean; and the Falklands, South Georgia, and Rockall in the Atlantic.

A third, and quite rare, type of oceanic island consists of a fragment of the deep-ocean bottom that is very occasionally thrust up high enough by tectonic forces to emerge above sea level. The few known cases are Macquarie Island, located south of New Zealand; St. Paul's Rocks, situated halfway between Brazil and West Africa; and Barbados in the West Indies. From a biogeographical point of view such islands are similar to volcanic islands, having never been in contact with a continent.

Characteristics of oceanic island biota

The composition of the fauna and flora of oceanic islands is very much influenced by the isolation of these islands. Because (with the exception of a very few islands of continental origin) the islands have never been connected to a continent, they are inhabited exclusively by taxa that have been able to colonize the islands from a continent or another, usually older, island by crossing water barriers. This means that only species that can fly, drift long distances (e.g., on floating vegetation), be blown by the wind, or hitchhike on flying animals can colonize oceanic islands.

Flying animals—that is, birds, bats, and winged insects—are of course very prominent among island biota, although it must be noted that the colonizing capabilities vary greatly within these groups. Mammals have been much less successful, and the only nonvolant mammals that have been able to reach a few oceanic islands are rodents.

Among reptiles, squamates and turtles have been quite successful colonizers, snakes and crocodiles rather less so; whereas amphibians, which are dependent on freshwater, are virtually unknown on oceanic islands. Turtles, which float well and are proverbially capable of surviving for long periods without either food or water, have been particularly successful.

As already mentioned, a few oceanic islands are actually old continental fragments. Such continental islands may in principle have fauna and flora that have evolved from species that lived on the continent of which they were once a part. These may therefore include very old lineages and forms with little or no capability for crossing water barriers. Other than the continental islands of Madagascar and New Zealand, however, the only islands where such old continental taxa actually still exist are the Seychelles and perhaps New Caledonia.

Once a species has reached an oceanic island it is usually isolated there completely, and evolution can be quite rapid under strong selection in a new and often harsh and resource-poor environment. On islands small species tend to grow larger and large species smaller. Insular mammal faunas have included both pony-sized elephants and badger-sized hedgehogs. For birds and reptiles, which were normally small to begin with, gigantism is predominant, though a few cases of small island forms do occur (e.g., among geese). Birds very often become flightless; flying makes heavy physiologic demands and may not be useful on a small and more-or-less predator-free island. Once a bird is flightless the strong

The dodo (*Raphus cucullatus*) of Mauritius is possibly the most well-known of all extinct animals ("dead as a dodo"). This picture from Strickland, H. E. & Melville, A. G. The Dodo and its kindred (1848) illustrates the "Ashmolean head," together with a foot the only existing remains of the dodos that were brought to Europe in the seventeenth century. Below it is a reconstruction of the head of a live dodo. ©SSPL/The Image Works.

constraints flying has on morphology are relaxed, and flightless birds can evolve in bizarre ways, with the dodo (*Raphus cucullatus*) of Mauritius being a particularly well-known example. Other common characteristics of taxa on oceanic islands include long generations, low breeding rates, slow movement, and lack of predator avoidance behavior, all of which tend to make island species vulnerable to predation and competition.

Natural extinctions on oceanic islands

Generally speaking, species living on oceanic islands are probably shorter lived than continental species, simply because oceanic islands are relatively short lived, in geological terms. In some cases, particularly in archipelagos such as the Hawaiian Islands, species can survive the submergence of the island where they originally evolved by colonizing a nearby island, and indeed a number of DNA studies strongly suggest that some island lineages are considerably older than the islands on which they now live (e.g., DeSalle 1995).

Natural extinctions on oceanic islands are not well documented in the fossil record. Almost nothing is known about the biota that once undoubtedly lived on the thousands of now submerged islands. A few drill cores from guyots containing terrestrial microfossils (e.g., pollen) exist and show that these sunken islands once had a flora, and presumably a fauna, but few details are known.

An important factor in natural extinctions of taxa on oceanic islands is probably glacial/interglacial sea-level cycles. During interglacials many oceanic islands are drowned by the rising seas and the area of many others shrinks drastically.

That these sea-level changes have caused extinctions seems likely, and this is supported by evidence showing that endemic species are strikingly rare on low islands. In two cases the effects of interglacial ocean highstands on oceanic island biota have been documented through paleontological studies. On Aldabra, an atoll in the Indian Ocean, a goose and a petrel (*Aldabranas cabri* and *Pterodroma kurodai*, respectively) may have gone extinct during the previous interglacial (marine isotope stage [MIS] 5e) about 130,000 years ago, when Aldabra was probably completely drowned. On Bermuda, where stratigraphic control is unusually good for an oceanic island, extinctions are known to have occurred during both MIS 11 (c. 400,000 years ago) and MIS 5e (c. 120,000 years ago), the two strongest interglacials during the last million years. During MIS 11 an albatross (very closely related, or identical, to the short-tailed albatross [*Phoebastria albatrus*] of the North Pacific) became extinct, probably because its breeding grounds were drowned. Subsequently, an endemic avifauna consisting of a duck, a flightless crane, and two flightless rails evolved, but these species became extinct during MIS 5e. Finally, a new endemic avifauna, containing among other species a new flightless rail, is known from the last glaciation. This fauna survived until Bermuda was settled in the seventeenth century but then disappeared rapidly.

As already noted, this extinction mechanism applies only to low islands. In the only case in which reasonably good fossil

Ocean	Island group/Island	Number of Islands	Number of mammalian extinctions	Number of avian extinctions
North Atlantic	Canaries	8	5	12
	Other Islands	3	0	13
Mediterranean	Corsica	1	5	4
	Sardinia	1	5	3
	Malta	1	2	5
	Other Islands	9	13	9
Caribbean	Cuba	1	35	23
	Hispaniola	1	23	4
	Jamaica	1	3	6
	Puerto Rico	1	8	10
	Other Islands	35	37	34
South Atlantic		5	2	9
Indian Ocean	Madagascar	1	26	28
	Mascarenes	3	3	39
	Other Islands	8	2	7
Pacific Ocean				
New Zealand Area	New Zealand, North Island	1	1	27
	New Zealand, South Island	1	1	32
	Other Islands	5	1	31
New Guinea and Melanesia	New Guinea	1	6	0
	New Caledonia	1	0	14
	Other Islands	15	5	25
Wallacea	Timor	1	12	0
	Other Islands	6	7	1
Northwest Pacific		7	1	9
Polynesia	Hawaiian Archipelago	8	1	117
	West Polynesia	16	3	53
	East Polynesia	23	0	75
Micronesia		7	2	18
Bering Sea		2	2	1
North America, Offshore Islands	California Channel Islands	5	7	2
	Other Islands	4	3	2
Galapagos Islands		4	9	0
Sum		186	230	613

Geographical distribution of currently (2012) known Holocene avian and mammalian extinctions on islands. Because of its size, and the large number and widely different character of islands, the Pacific has been subdivided. Note that the number of islands given in each group only includes those where extinctions are known to have occurred. The number of extinctions is larger than the number of extinct species mentioned above since some extinct species have been found on more than one island. Reproduced by permission of Gale, a part of Cengage Learning.

evidence exists going back before the last ice age (probably to MIS 11) on a high island (Oahu in the Hawaiian archipelago), no prehuman extinctions are discernible.

Holocene extinctions

Extinctions on oceanic islands during the Holocene, however, are fundamentally different. While rising sea levels may well have caused some extinctions on low islands during the Early Holocene, particularly in the Pacific where sea levels during the mid-Holocene were about 3.3 to 6.6 feet (1 to 2 m)

higher than at present, no direct evidence for this has been found.

The great wave of island extinctions during the Holocene is instead clearly related to human colonization. Whenever humans have reached oceanic islands, extinctions have followed. For islands that were colonized in prehistoric times (which in this context means before c. 1600 AD), this process must be largely reconstructed from the paleontological record, because most extinctions had already occurred before the first historical records, although the process has continued up to the present. The pattern is the same for those islands that were uninhabited before being discovered by Europeans. There is no evidence for prehuman extinctions, but extinctions invariably and rapidly followed human settlement. On some islands that were uninhabited until being discovered by European explorers, a combination of paleontological studies and a careful search of historical records allows a relatively detailed reconstruction of the extinction process and the various factors causing the extinctions. This is particularly true for the Mascarenes (Mauritius, Rodrigues, and Réunion) in the Indian Ocean, where habitat changes, introductions of exotic taxa, and extinction have been reconstructed in unique detail (Cheke and Hume 2008).

Extent

The scope and extent of extinctions on oceanic island has become known only in the last few decades through increased paleontological research, and the consequences have yet to be fully assimilated in scientific fields outside paleontology. It is still common, for example, to find ecological or biogeographic studies that assume that island biota are more or less pristine, something that is almost never true, even for very large islands.

It should be noted that extinctions on oceanic islands are no different in principle from extinctions on other kinds of islands, such as those of the West Indies, the Mediterranean, or Wallacea, or on the "almost continents" of Madagascar and New Zealand or even on the continent of Australia. In each of these cases the causes and extent of the extinctions are similar, although the process was not surprisingly rather slower on the larger landmasses.

Furthermore, there are large numbers of islands where it seems extremely likely that extinctions have occurred but where as of 2012 there were no early descriptions of the fauna or flora and no fossil sites had yet been found (or even searched for). Obvious examples are the Azores, the Cape Verde Islands, the Comoros, the Seychelles, Socotra, and several hundred islands in Wallacea and the Pacific. This hypothesis is reinforced by the illogical and highly discontinuous range of many multi-island taxa, which suggests that they have been extirpated from intervening islands.

That there are practically no oceanic islands with a reasonably good fossil record *without* known extinct species shows just how universal island extinctions have been. The few cases known are arctic islands or very small atolls, where the remains of seabirds and marine mammals have been found but where there probably have never been any native land animals.

Taxonomic scope

Extinctions on oceanic islands are often considered more-or-less synonymous with avian extinctions, but this is incorrect. Avian extinctions have certainly been severe on islands. More than 350 species on oceanic islands and a further 140 on other islands, including Madagascar and New Zealand, are known to have become extinct during the Holocene, and the true total is almost certainly several times higher. This is at least partly

attributable, however, to birds having been better studied than other groups, and the high visibility of birds also means that avian extinctions have been more often noticed in historical sources.

Because mammals have a limited capability for long-distance overwater dispersal, they only rarely occur on truly oceanic islands, and when they do they are almost invariably either bats or rodents. Where they do occur (or rather did occur) they have been at least as badly affected by recent extinctions as birds.

Thus, the Galápagos Islands, which have not yet sustained any known avian extinctions (although two species were as of 2012 critically endangered), have lost all but three of the islands' native species of rodents, of which there were at least eleven, whereas the Canaries have lost all three known endemic rodents. In all, about thirty species of rodents and bats endemic to oceanic islands have become extinct during the Holocene. On islands closer to continents, which originally had larger and more diverse mammal faunas, as of 2012, the total number of definitely known mammalian extinctions during the same interval was well over 150. The number of known avian and mammalian extinctions and their approximate geographical distribution are shown in Table 1.

Reptiles (crocodiles, squamates, and turtles) have also been badly affected by extinctions in the recent past. Fewer cases are known for snakes and amphibians, but this is mostly because these groups, like mammals, have only rarely managed to colonize oceanic islands.

Land snails, insects, and plants are also known to have suffered extensive extinction, although they have been much less well studied. On the Hawaiian Islands alone, 90 percent of the more than 750 species of native land snails are now believed to be extinct.

A particularly dramatic example involves the flightless rails. Rails, somewhat surprisingly, are very successful in colonizing oceanic islands, and being omnivorous and very eurytopic (i.e., able to live in wide variety of habitats), they can maintain viable populations even on quite small islands (as small as about 0.4 square miles [1 sq km]). Apparently, such populations rapidly become flightless and form endemic species. On virtually every oceanic island that has ever been investigated paleontologically at least one species of extinct flightless rail has been found, and on larger islands frequently there occur two or three species of different sizes. Based on the proportion of islands in the Pacific that have been investigated paleontologically, and the number of flightless rails found on these, David W. Steadman, in a 2006 study, very conservatively estimated that 440 species of flightless rails existed in the Pacific before human colonization. Of these, seven survived as of 2012 (one on New Guinea, three in the Solomons, two on New Zealand, and one on Henderson Island). New Zealand, New Guinea, and the Solomons are all very large islands with extensive areas of dense and inaccessible forest and scrub, which has undoubtedly made it possible for the flightless rails to survive there. The unique survival of the red-eyed crake (*Porzana atra*) on small Henderson Island (with an area of 14.3 square miles [37 sq km]) presumably stems from two factors: Most of the island is covered by extremely rugged karst topography that is virtually impenetrable to humans, and the Polynesian population on Henderson Island died out before AD 1600, and the island has never been resettled.

Extinction mechanisms

Holocene island extinctions almost exclusively resulted from the influence of humans on previously uninhabited islands. The immediate causes for the extinctions have been the subject of considerable debate, and many extinctions have probably stemmed from the combined effect of several factors.

The most common cause of human-induced extinctions occurring on a previously uninhabited island is probably the introduction of new predators, whether deliberate or accidental. Because mammalian predators are absent on oceanic islands, the native fauna frequently lack predator-avoidance behavior and are an easy prey for any new predators, including humans.

The most important of these new predators are probably rats (including the black rat [*Rattus rattus*], the Norway rat [*R. norvegicus*], and the Pacific rat [*R. exulans*]), but among others cats, pigs, mongooses, ferrets, stoats, snakes, toads, and even shrews have had dire effects when introduced to islands where no similar predators existed previously. Among invertebrates, introduced carnivorous snails have devastated native land snail faunas on several Pacific islands, whereas ants (which are not good at crossing water barriers and consequently are often absent from oceanic islands) have decimated arthropod faunas on many islands.

A well-documented example of the devastating effect rats can have on island biota occurred in 1962 when ship rats were accidentally introduced to Big South Cape Island, located off southern New Zealand. Within a few years these rats had exterminated five species of native birds on the island, including the last surviving populations of the bush wren (*Xenicus longipes*) and the South Island snipe (*Coenocorypha iredalei*), as well as the last population of the greater short-tailed bat (*Mystacina robusta*). Conversely, once new eradication techniques had made it possible to exterminate the rats on Campbell Island, located to the southeast of New Zealand, the local population of snipes (*Coenocorypha* cf. *aucklandica*), which had managed to survive in tiny numbers on a few rat-free skerries, almost immediately started recolonizing the main island. Rats are highly efficient predators of eggs and nestlings and are also quite capable of killing the adults of smaller bird species.

Invasive exotic species can also have a strong negative effect on the original fauna and flora without being predators, by competing for food or other resources, such as nesting sites, or, in the case of plants, simply by physically crowding out the original vegetation. Herbivores such as feral cattle, goats, and rabbits can also indirectly have large negative effects through overgrazing and consequent soil erosion.

Direct human extermination of insular species through overhunting may have been less important on oceanic islands

An example from Macquarie Island south of New Zealand of the effect of exotic herbivores on native vegetation. The Tussock grass within the fenced enclosure probably approximates the undisturbed natural vegetation. The rest of the island has literally been eaten bare by rabbits. Courtesy of Tommy Tyrberg.

than on islands in the Mediterranean or the Caribbean and has probably mostly affected the larger species such as the moas of New Zealand or the megapodes and fruit doves on Pacific islands. One group of birds that has apparently been hunted extensively, to judge from archaeological remains and modern ethnographic data, is the colonial seabirds. Yet although seabirds have been virtually exterminated on most inhabited oceanic islands, relatively few extinctions have resulted. The reason for this is that seabirds are usually vulnerable to hunting only at their breeding sites, and in most cases they have been able to survive, although in vastly reduced numbers, by breeding in inaccessible sites or on small uninhabited islands. A large proportion of the seabird extinctions that have taken place have occurred on very isolated islands where no such safe havens were available nearby (e.g., Easter Island or St. Helena). All things considered, overhunting by humans has probably been of greater importance on larger islands and continents where large and slowly breeding species of mammals and birds were clearly easy prey for the first humans to arrive.

Human influences have probably been more important in the form of the destruction of the original vegetation through deforestation and farming and very often extensive soil erosion. In many cases the original vegetation has been almost completely destroyed, as has happened on Easter Island and other relatively low islands. On high islands it is usually mainly the lowlands, particularly the drier areas, that have been denuded, with some natural vegetation surviving in rugged areas at higher altitudes.

In the Pacific, human colonization is so closely related to deforestation and soil erosion that the onsets of these, which are usually easily discerned in wetland deposits, have proved a useful proxy for dating the colonization of specific islands. At the same time, archaeological study of the earliest sites on these islands can be quite difficult, because the oldest, usually

This illustration of a moa (*Dinornis robustus*) hunt, while somewhat quaint and faulty in details (moas did not have such an upright stance for example) is still true in essence. There is no doubt that moas and many other large birds were quickly hunted to extinction once humans reached the islands where they lived. © Florilegius/Alamy.

coastal, settlements are now deeply buried by alluvial deposits that have eroded from higher ground.

One reason for extinctions that has been much discussed, but the importance of which is still disputed, is introduced pathogens. It is well established that the remaining Hawaiian endemic birds are quite sensitive to avian malaria, which has arrived in the islands with introduced birds. This seems to be the main reason for the scarcity of indigenous birds in the lowlands where malaria is most common, and it may well have been the "final straw" in the recent extinction of several native birds.

Another case in which disease seems very likely as a reason for extinction concerns the two endemic rodents—Maclear's rat (*Rattus macleari*) and the bulldog rat (*Rattus nativitatis*)—on Christmas Island. These disappeared with extreme rapidity in 1903 when black rats reached the island, even from areas that had not yet been reached by the black rat, and apparently abnormal behavior, which may have been the result of disease, was noted at the time. Disease has also been suggested as a reason in some other cases in which a formerly common species disappeared abruptly, for no obvious reason (e.g., the New Zealand quail [*Coturnix novaezelandiae*] in the 1870s). On the whole, however, it does not seem likely that introduced pathogens have been an important cause of extinctions on oceanic islands thus far.

Blitzkrieg or war of attrition?

There has been a great deal of controversy about the nature and causes of the great extinctions of large animals on the continents—specifically, whether they were caused by climatic change or by humans, and if so whether they involved an abrupt (the so-called Blitzkrieg hypothesis) or a drawn-out process. On oceanic islands, where there is essentially no doubt that virtually all extinctions have been caused by humans, directly or indirectly, and the extinctions are more recent, it is possible to determine more exactly how fast the extinctions occurred. In these cases the answer seems to be that extinctions can be both quite a fast and a drawn-out process (e.g., Cheke and Hume 2008, Steadman 2006). In general, extinctions resulting from new predators and overhunting seem to have occurred fast enough to seem virtually instantaneous in the paleontological/archaeological record, whereas extinctions caused by habitat changes have been slower, at least on larger islands.

Research history

With a few notable exceptions (New Zealand, Madagascar, and the Mascarenes), research on insular extinctions is a surprisingly recent phenomenon and has, up to now, engaged only a few scientists. Although the disappearance of a number of insular species was noticed and recorded by contemporaries in the seventeenth and eighteenth centuries (particularly in the West Indies and on the Mascarenes), it does not seem to have attracted much interest from natural historians. Indeed, for many people in the seventeenth or eighteenth century the very conception of extinction was controversial. Because God had created all creatures, to many it seemed unreasonable or even blasphemous that mere humans would be able to exterminate species.

By the early nineteenth century some naturalists even doubted the dodo had ever existed, even though several illustrations and even some scanty physical remains still existed. In addition, the excellent description of the original fauna and flora of Rodrigues Island, published in 1708 by the French explorer François Leguat, had come to be regarded as a romance in the Robinson Crusoe tradition, a view that many literary historians clung to, even long after his truthfulness had been amply confirmed by other contemporary sources and fossils.

An important milestone was the publication of the first description of a moa fossil by Richard Owen in 1840 (Owen 1840), along with the discovery of and the publication of reports on a large number of fossils in the Mare aux Songes swamp on Mauritius in the 1860s. At approximately the same time the existence of extinct, very large flightless birds on Madagascar (aepyornithids) was discovered, mostly by French explorers.

By 1900 the larger extinct animals of New Zealand, Madagascar, and a few islands in the Mediterranean were known, and the contemporary decline and disappearance of many island taxa, particularly in New Zealand and the Hawaiian Islands, were well known, but still essentially nothing was known about pre-European insular extinctions. During the twentieth century a number of more-or-less isolated finds of extinct birds occurred, particularly in the West Indies, but the great breakthrough in the study of island extinction came in the 1980s with the work of Storrs L. Olson and Helen F. James in the Hawaiian Islands (e.g., James and Olson 1991; Olson and James 1982, 1991) and David W. Steadman in Polynesia, Melanesia, and Micronesia (summarized in Steadman 2006).

Between the early 1980s and 2010, the scientific community's knowledge of insular extinctions grew quite rapidly. Nevertheless, this increase in knowledge still stemmed from the exertions of only a handful of scientists, and this, together with the logistic and economic problems of research on often distant and inaccessible islands, means that the vast majority of oceanic islands have never even been cursorily surveyed for fossils. Indeed, even the extant fauna and flora of many islands

Portrait of Sir Richard Owen with a skeleton of a giant moa. Owen was the first to describe the giant moa and did extensive research on other fossils and coined the term "dinosaur." George Bernard/Science Source.

in the Pacific and Indian Oceans are little known. There is no doubt that scientists in the early 2000s had still discovered only a small part of the insular species that existed just a few thousand years ago, and indeed many species undoubtedly went extinct without a trace.

Resources

Books

Cheke, Anthony, and Julian Hume. *Lost Land of the Dodo: An Ecological History of Mauritius, Réunion, and Rodrigues.* New Haven, CT: Yale University Press, 2008.

DeSalle, Rob. "Molecular Approaches to Biogeographic Analysis of Hawaiian Drosophilidae." In *Hawaiian Biogeography:*

Evolution on a Hot Spot Archipelago, edited by Warren N. Wagner and Vicki A. Funk. Washington. DC: Smithsonian Institution, 1995.

King, Carolyn. *Immigrant Killers: Introduced Predators and the Conservation of Birds in New Zealand.* Auckland, New Zealand: Oxford University Press, 1984.

Steadman, David W. *Extinction and Biogeography of Tropical Pacific Birds*. Chicago: University of Chicago Press, 2006.

Turvey, Samuel T., ed. *Holocene Extinctions*. Oxford: Oxford University Press, 2009.

Worthy, Trevor H., and Richard N. Holdaway. *The Lost World of the Moa: Prehistoric Life of New Zealand*. Bloomington: Indiana University Press, 2002.

Periodicals

James, Helen F., and Storrs L. Olson. "Descriptions of Thirty-two New Species of Birds from the Hawaiian Islands. Part II. Passeriformes." *Ornithological Monographs* 46 (1991): 1–88.

Olson, Storrs L., and Helen F. James. "Prodromus of the Fossil Avifauna of the Hawaiian Islands." *Smithsonian Contributions to Zoology* 365 (1981).

Olson, Storrs L., and Helen F. James. "Descriptions of Thirty-two New Species of Birds from the Hawaiian Islands. Part I. Non-passeriformes." *Ornithological Monographs* 45 (1991): 1–88.

Owen, Richard. 'On the Bone of an Unknown Struthious Bird from New Zealand." *Proceedings of the Zoological Society of London* 7, no. 83 (1840): 169–171.

Tommy Tyrberg

Tropical rain forests

Tropical rain forests cover 7 percent of Earth's surface and yet are home to about 50 percent of Earth's species. Since at least the early 1980s, logging for timber, clearing for agriculture, over hunting, and increasing human population density have been the major threats to tropical rain forests and the principle causes of the extinction of rain forest species. The extinction rate is likely to accelerate in the future with additional synergistic threats of climate change, diseases, and pests. Such anthropogenic changes potentially will alter the global carbon and hydrological cycles, cause large-scale extinctions of biodiversity, and adversely affect those people who depend on rain forests for their livelihoods as well as all other members of the human species.

What is a tropical rain forest?

Tropical rain forests are the tall, dense, and humid evergreen forests that dominate the equatorial wet parts of the landscape and are renowned both for the immense diversity of plant life that forms the forest and for the animal life that inhabits them. At the same time tropical rain forests provide crucial ecosystem services, including the storage and cycling of carbon and water. They form a terrestrial ecosystem type that occurs roughly between the Tropic of Cancer and Tropic of Capricorn (approximately 23° north and south of the equator, respectively; see Figure 1). Rain forests usually have high average temperatures and high rainfall. Typically, a rain forest receives more than 60 inches (150 cm) of rain a year, and most receive more than 100 inches (250 cm) a year. Natural, undisturbed tropical rain forest is sometimes called *primary forest*, but the term *virgin rain forest*—in other words, not touched or visited by humans—is now rarely used because anthropological and historical studies have shown that many ancient civilizations existed in many tropical rain forests. Rain forests are traditionally grouped into the Afrotropical, Australian, Indomalayan, and Neotropical rain forest realms, which include parts of Asia, Australia, Africa, South America, Central America, and Mexico as well as many islands of the Pacific, Caribbean, and Indian Oceans. Sometimes the tropical regions of Africa and Southeast Asia through to Australia are referred to as the paleotropics, contrasting with the forests of Central and South America, which are called the neotropics. This distinction reflects the ancient origins of these land masses following the breakup of

the supercontinent Gondwana (see below) and subsequent collisions of some of these land masses.

In practice, there are many definitions used for different kinds of forests, and hence comparing data on forest extent and rates of deforestation is a complex process. Within the World Wildlife Fund's biome classification, tropical rain forests are thought to be a type of tropical wet forest (or tropical moist broadleaf forest) and may also be referred to as lowland equatorial evergreen rain forest. It can, however, sometimes also include other kinds of forests such as montane rain forests, mangroves, and seasonally flooded forests. Dry tropical forests are usually excluded as the vegetation and rainfall regimes are very different.

Rain forests of different regions have their own distinctive features because of a combination of factors, including rainfall and temperature regimes, soil types, origin of the plants and animals that make up the forest, and most importantly the past history of the land mass on which they are found. Most of these land masses owe their origin to the break-up of Gondwana, a supercontinent that rifted apart between 400 and 100 million years ago. During this long geological interval climates underwent significant changes. Because high rainfall, in particular, is essential for tropical rain forests, their distributions would have contracted and expanded frequently as conditions changed. There is strong evidence that in the last few million years such expansions and contractions of rain forests occurred frequently depending on the climate of that time. At the time of the last glacial maximum around 20,000 years ago, rain forests contracted in many parts of the world as much of the freshwater was locked up in ice sheets in the upper latitudes. Some of the best evidence of glacial drying and rain forest contraction is from Australia and Africa. At the same time, though, sea levels were much lower. Hence areas such as the Malay Peninsula, including Java, Sumatra, and Borneo, were connected, and in this area of Sundaland the extent of tropical rain forest may have increased.

Seasonal and nonseasonal rain forests

Although the traditional view of tropical rain forests is that they have an equatorial climate year-round, such climates are restricted to a few tropical rain forests in the world such as the lowland forests of the Sundaland area of

Tropical Rain Forests of the World

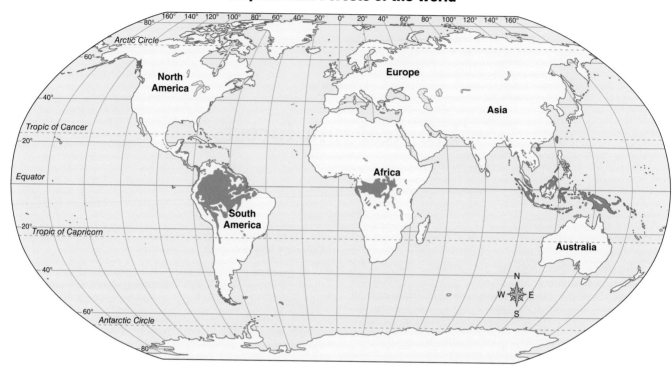

Figure 1. Map of the distribution of the tropical rain forests of the world. Reproduced by permission of Gale, a part of Cengage Learning.

Southeast Asia, central Africa, and small parts of the Amazon. In practice most rain forests show some seasonality reflecting the changing temperatures and rainfall patterns during the year, with more rainfall in some months than others. In nonseasonal rain forests many species of trees may flower throughout the year, but in more seasonal forests flowering typically occurs during the dry season. The types of rain forests are often further distinguished by characteristics of the plant communities, rainfall regimes, and soil types. Where rainfall measures below 4 inches (100 mm) a month during the dry season, such forests often lose their leaves and form tropical dry forests, as typified by the *cerrado* regions of Brazil and parts of Central America and East Africa. These forests are not discussed further here except to note that their conversion for agriculture has been at a greater rate than that of tropical rain forests, and hence they too comprise a highly endangered ecosystem.

Mangroves

Mangroves occur in the tropics along the coast where the land meets the sea and particularly in estuarine areas. Only some 110 species of trees worldwide make up mangrove ecosystems, and such species usually occur as distinct bands of one or two species from the sea edge in land within the tidal zone. Mangrove trees are adapted to living in swampy saline conditions. Most distinctive of the mangrove species are those that have large stilt roots. Although the terrestrial fauna and flora of mangroves are very low in diversity, the marine component of mangroves is rich, as the shallow waters are a crucial breeding ground for fish and invertebrates. Globally, mangroves are highly threatened as they are often cut for charcoal and cleared and drained for agriculture. In turn, the loss of mangroves impacts very significantly on breeding for many tropical estuarine and marine fish and other species. The livelihoods of coastal people dependent on such ecosystems are similarly threatened.

Flooded forests

In some lowland parts of tropical forests, particularly in some parts of the Amazon, the rainy or so-called wet season often causes temporary flooding of the forest for up to six months of the year. In these so-called *várzea* forests the invertebrate species that are normally associated with the ground layer stay submerged below ground, migrate up the tree trunks to the forest canopy, or fly to non-flooded forests. Some flooded forests, such as some of the peat swamp forests in Southeast Asia, are generally of low diversity in terms of plant species.

Montane rain forests

Many tropical forest regions include significant mountain ranges, which are often almost completely covered in montane rain forest. A number of rain forests, such as those in the Wet Tropics in northeastern Australia, the Western Ghats in India, the Usumbara Mountains in Tanzania, and the Brazilian Mata Atlântica, owe their existence to orographic rainfall, which is produced as offshore winds meet mountains along the edge of coastlines. A feature of such montane rain forests is that the trees become much shorter with increasing altitude because, although rainfall can increase considerably, temperatures also drop. Often these are called *elfin* forests. Some montane forests are shrouded in clouds for a large part of the day or for long periods of the years and rainfall can exceed 33 feet (10 m) a year. The high levels of moisture result in the branches of trees being covered in thick layers of liverworts, lichens, and epiphytes. On very high mountains rain forests are replaced at high altitudes above the tree line by shrubs, grassy vegetation, bare earth and rock, or even snow and ice. For example, Mount Kinabalu, located in the Malaysian state of Sabah, is over 13,100 feet (4,000 m) in height, and the rain forest extends from the lowlands through montane rain forest until about 10,500 feet (3,200 m). The remainder is above the tree line.

Cyclone- or hurricane-disturbed forests

Some areas of rain forests, such as in coastal areas of the Caribbean, Central America, the Pacific Islands, parts of Asia, and the northern parts of Australia, are subject to strong winds in the form of hurricanes and cyclones. Such heavily disturbed forests are often shorter, have a more uneven height canopy, and can be dominated by vines.

Biodiversity

Tropical rain forests, with a few exceptions, are the most species-rich habitats in the world. The list of the most megadiverse countries of the world (as measured by the number of bird, mammal, and plant species) is dominated by those countries that have extensive areas of tropical rain forest (see Table 1). Furthermore, even though tropical rain forest covers only around 0.01 percent of the land surface in Australia, the species from these forests contribute considerably to the overall national total. The same is true for several other countries where rain forests are not the most widespread ecosystem, such as China. Around the world researchers have identified all the species of plants above a standard 4-inch (10-cm) diameter tree (with the trunks measured at about 5 feet [1.5 m] above the ground) in plots that vary from 2.5 to 125 acres (1 to 50 ha) or more. Table 2 shows the diversity of tree species in plots of various at different latitudes; tropical rain forests far exceed other forest systems in terms of their species richness.

In a 1988 article, Norman Myers noted that the majority of the world's endemic plants were located in a few "biodiversity

Country	Mammals	Birds	Flowering plants
Mexico	450	1,026	25,000
Indonesia	436	1,531	27,500
Democratic Republic of the Congo	415	1,096	11,000
Brazil	394	1,635	55,000
China	394	1,244	30,000
Colombia	359	1,695	50,000
Peru	344	1,678	17,121
India	316	1,219	15,000
Venezuela	305	1,296	20,000
Ecuador	302	1,559	18,250
Cameroon	297	874	8,000
Malaysia	286	736	15,000
Australia	252	751	15,000
South Africa	247	790	23,000
Panama	218	926	9,000
Papua New Guinea	214	708	10,000
Vietnam	213	761	7,000
Costa Rica	205	850	11,000
Philippines	153	556	8,000
Madagascar	105	253	9,000

Table 1. World ranking of megadiversity countries as assessed by estimates of the numbers of species of mammals, birds, and flowering plants found in these countries. Reproduced by permission of Gale, a part of Cengage Learning.

CTFS plot location	Latitude	Plot size (hectares)	Number of species	Number of trees
Wind River, Washington	45.8	25.6	25	31,162
Wabikon Lake Forest, Wisconsin	45.6	25.6	36	57,388
Haliburton Forest, Canada	45.3	13.53	30	46,339
Changbaishan, China	42.4	25	52	38,902
Smithsonian Conservation Biology Institute, Virginia	38.9	25.6	65	41,031
Yosemite National Park, California	37.8	25.6	23	34,458
Gutianshan, China	29.3	24	159	140,700
Xishuangbanna, China	21.6	20	468	95,834
Huai Kha Khaeng, Thailand	15.6	50	251	72,500
Mo Singto, Thailand	14.4	30.5	262	No data
Mudumalai, India	11.6	50	72	25,500
Barro Colorado Island, Panama	9.2	50	299	208,400
Khao Chong, Thailand	7.5	24	593	121,500
Sinharaja, Sri Lanka	6.4	25	204	193,400
Korup, Cameroon	5.1	50	494	329,000
Lambir, Malaysia	4.2	52	1,182	359,600
Pasoh, Malaysia	3.0	50	814	335,400
Yasuni, Ecuador	−0.7	50	1,114	151,300

Table 2. The number of tree species and number of individual trees in surveyed plots of carious sizes from around the world, arranged by latitude. These data are for all trees on the plots that are greater than 4 inches (10 cm) in diameter (as measured on tree trunks at around 5 ft [1.5 m] above the ground). Reproduced by permission of Gale, a part of Cengage Learning.

hotspots" and that if these were protected then so too would the majority of the world's biodiversity. Today, of the 35 identified global biodiversity hotspots, 25 are areas of tropical rain forests. A number of very species-rich families of plants are pantropical in distribution. These include the legumes (formerly Leguminosae now Fabaceae), figs (Moraceae), and palms (formerly Palmae now Arecaceae). The largest family is the orchids (Orchidaceae) with around 20,000 species; 41 percent are found in the tropical Americas, 34 percent in tropical Asia and New Guinea, 15 percent in Africa, and 3 percent in Australia.

At the same time tropical rain forests arguably provide habitat for the majority of the world's vertebrate and invertebrate species. Scientists are uncertain as to how many species there are on Earth, but the best estimates suggest that there may be 5 to 10 million species of all organisms on Earth with about 15 to 20 percent in the oceans and about 50 percent in tropical rain forests. Almost 90 percent of species are arthropods and other invertebrates, most are insects, and most occur in tropical rain forests. Because only around 1.5 to 1.7 million species of organisms have been named and described, most species are yet to be discovered and the rain forest is without doubt where most of these undiscovered species reside.

A large proportion of the world's vertebrates are also found in rain forests. One of the most distinctive groups is the primates, which often comprise a large part of the vertebrate biomass, are the most abundant of the large mammals, and often play important ecological roles. Primates include lemurs, lorises, pottos and bush babies, tarsiers, New World and Old World monkeys, and apes and humans. Their diets vary and include leaves, fruits, and insects, and many species are omnivorous.

Tropical forest structure

Rain forests can be divided into several structural vertical layers. The ground layer is where leaves, fruit, and dead wood fall and decompose. Although it is usually not obvious, this layer is dominated by fungi and other microorganisms, which are crucial to the decomposition of dead plant material. More obvious are insects, which also play roles in decomposition, but are often feeding on fungi. Below ground, termites also play important roles in such processes. Above ground, the vegetation is often divided into the upper and lower canopy. The upper canopy is where most leaves and flowers are produced by trees, vines, and epiphytes. The lower canopy usually receives much less light penetration and is often darker and cooler and dominated by the tree trunks themselves and younger trees. Some tropical rain forests carry a large number of epiphytic plants (plants that grow on other plants) and vines that have roots on the ground but use other trees to reach into the upper canopy.

Although rain forests are often called evergreen, many species of trees are deciduous or semi-deciduous as they lose some or all their leaves during the year. In forests with a strong dry season leaf fall may be higher during the dry periods. Leaves usually comprise the largest component of biomass in the canopy and carry out photosynthesis. Photosynthesis is the process by which carbon dioxide from the air is converted into sugars and carbohydrates, which are used as the building blocks for plants. The organelles that carry out photosynthesis in leaves are called chloroplasts. In some species of rain forest trees, there are chloroplasts in the trunks of trees, but this is unusual. Most tropical plants produce flowers, 90 percent of which are pollinated by insects. Tree roots absorb water and nutrients, which are transported to the forest canopy and other parts of the tree through xylem cells in the trunks.

Lowland rain forests are found below altitudes of about 3,280 feet (1,000 m), and although these are usually the tallest forests, they rarely reach above 200 feet (60 m) in height. However, the lowland rain forests of Southeast Asia, from Thailand through Indonesia, are dominated by the Family Dipterocarpaceae. These trees include species such as *Koompassia excelsa* (Leguminoseae) and various *Shorea* species (Dipterocarpaceae) that rise high above the forest canopy and are called emergents or super-emergents. Such trees are known to reach over 330 feet (100 m) in height. Another distinctive feature of dipterocarps (and many of the other trees species in this forests) is their flowering, which may occur only every two to seven years during strong El Niño events. These El Niño events bring low rainfall to areas of Indonesia and Malaysia and coincide with synchronized flowering and fruiting in the dipterocarps, known as masting. The precise triggers for masting are not fully understood as of 2012.

Until the late twentieth century, the rain forest canopy was so inaccessible it was often called the last biological frontier as so little was known about it. Traditionally, researchers have used climbing rope techniques for their studies, and entomologists have hauled fogging machines up to the canopy on ropes to release knockdown insecticides. This procedure allows them to collect insects from the canopy for study. Since the early 1990s, industrial cranes have provided safer and easier access to the forest canopy in rain forests in Panama, Australia, and Malaysia. These cranes convey researchers in gondolas to the parts of the trees they wish to study.

Tropical forests as carbon and water sinks

Tropical rain forests are important global stores of carbon (see Figure 2) and contribute more than 30 percent of terrestrial carbon stocks and net primary productivity. Mapping and estimating the amounts of stored carbon in tropical rain forests is achieved by using a combination of ground-based measurements and a variety of satellite data. Analyses show that most of tropical forest carbon is stored in the forests of South and Central America. They also indicate that deforestation and forest degradation contributes 10 to 20 percent of global carbon emissions, and most of that comes from tropical forest regions.

The forest canopy in the Daintree rain forest of north-east Australia. Courtesy of Nigel Stork.

Tropical forests store large amounts of carbon in the wood and roots of their trees. When trees are cut down and decompose or are burned, the carbon is released to the atmosphere. Typically researchers use a ground-based technique to measure the height of trees from which they can estimate how much carbon they contain. To do these measurements on very large scales across whole forests is not feasible, however, particularly because forest structure and the topography of the landscapes can be varied, so data on the heights of the tops of trees are calculated using satellite information. A NASA-led team (see Figure 2) found that in the early twenty-first century, forests in the seventy-five tropical countries they studied contained 247 billion tons of carbon. To put this into perspective, about 10 billion tons of carbon are released annually to the atmosphere from combined fossil-fuel burning and land-use changes. Further, the researchers found that forests in Latin America hold 49 percent of the world's tropical forest carbon. Developing these estimates is an important first step toward establishing baselines for those countries planning to participate in the United Nations Collaborative Programme on Reducing Emissions from Deforestation and Forest Degradation in Developing Countries (REDD+) program. REDD+ is an international effort to create a financial value for the carbon stored in forests and offers financial incentives for countries to preserve their forests as a means to reduce carbon emissions and promote low-carbon development. At the same time, these maps are useful as a mechanism to monitor forest health and to see how forests contribute to the global carbon cycle and overall function of the Earth system.

Tropical forest people and cultural diversity

People have lived in and modified tropical rain forests for at least 10,000 years. There is evidence that a number of rain forest peoples' cultures have existed in the past but have since disappeared. In modern times, these forests provide drinking water, fuelwood, and animal protein to perhaps 500 million people who are culturally extremely diverse. It was only during the twentieth century that some tribes have been

An Australian canopy crane used by scientists to access the forest canopy at James Cook University's Daintree Rainforest Observatory in Australia. Courtesy of Nigel Stork.

discovered, particularly in places such as the island of New Guinea, some of the Pacific islands, Amazonia, and central Africa. The nongovernmental organization Survival International estimates that over 100 rain forest tribes still remain untouched by modern society, with New Guinea alone accounting for 44 of these. Linguistic diversity is also extremely high in tropical rain forest people, with 10 of the top 12 countries for biodiversity also being among the top 25 for linguistic diversity. There are over 850 different languages in New Guinea alone, for example. A number of studies indicate that genetic pygmies have evolved independently in several tropical rain forest regions.

Most rain forest peoples are hunter-gatherers living off wild foods collected from the forest. A wide range of animals and plants are collected for food, medicines, and building materials. Studies from many parts of the world indicate that often hundreds of species of plants and animals are used by local peoples in individual communities as part of their way of life, with the knowledge of which ones to use and how being passed down orally from one generation to another. Hunting of animals, usually larger mammals, as bushmeat is a valuable food source for some 60 million people across the tropics. Bushmeat sold at local markets also provides cash for forest people. In the Congo basin of West Africa, for example, about

30 million people regularly eat bushmeat, which can provide up to 80 percent of the protein and animal fat in their diet. Bushmeat can sometimes serve as a back-up food resource during times when other foods are in short supply.

Most of these people also practice some limited cultivation, such as planting fruit trees around frequently used sites. Some cultures have practiced more extensive farming techniques, including shifting cultivation, often called *swidden agriculture* or *slash-and-burn cultivation*. With this method, forest areas are cut down and burned to help fertilize the soil before the planting different crops. Once the nutrients are depleted in these sites or the forest is moderately regenerated, the cultivators move on and clear a new area.

The forest also provides vital materials for building houses and boats and other materials for local peoples. Climbing palms (usually many species of the genus *Calamus*) in the paleotropics are the source of rattan, which is used to build furniture, baskets, and traps to catch fish or other animals. Rattan products are now extremely important commercially. The export values of rattan products increased 250-fold in Indonesia and 75-fold in the Philippines between the early 1990s and the early 2010s. Other important commercial products include açai, Brazil nuts, and rubber.

(a)

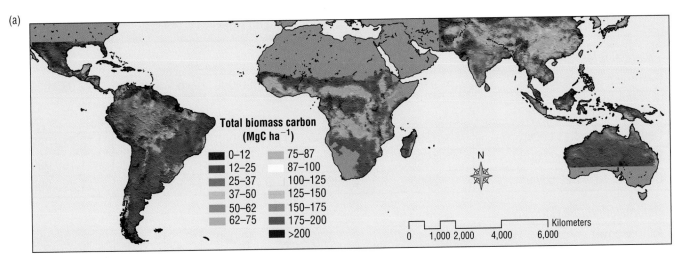

(b)

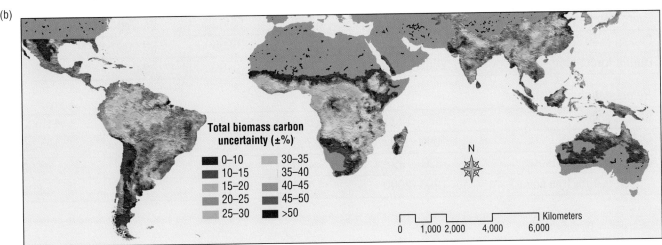

Figure 2. (a) Carbon stored in Earth's tropical forests. This is a benchmark map of carbon stored in Earth's tropical forests, covering about 6.2 million acres (2.5 million ha) of forests over more than 75 countries. (b) Percentage uncertainty of carbon estimates. Reproduced by permission of Gale, a part of Cengage Learning.

Tropical forest threats

Tropical forests are threatened by many different threats, including deforestation and climate change. Again, the realized levels of these threats vary between the major blocks of tropical forests.

Deforestation

Norman Myers first drew attention to increasing rates of deforestation in the early 1980s. Much of the deforestation at that time was from selective logging for timber extraction or the complete clearing of forests for cattle ranching in South and Central America. Myers pointed out that opening up forests for logging also provided road access for shifting cultivators, particularly in places such as Brazil and parts of Southeast Asia. In a 2010 article, S. Joseph Wright estimated that about 35 percent of tropical forests have been converted for human use and about half of the remaining area has been logged for timber or is regenerating or abandoned land (see Table 3). Governments have protected about 20 percent of tropical forests, which includes the approximately 7 percent of tropical forests that are protected strictly for conservation.

In a 2009 article, Robin L. Chazdon and colleagues contended that 50 percent of all tropical forests can be classified in a broad group of secondary forests or degraded old-growth forests, which includes forests that have been selectively logged or completely cleared for agriculture and are regenerating. How important such forests are in providing habitat for species of animals or plants that are normally found in primary forests is hotly debated and is an important issue for conservation. Some argue that conservation should focus on conserving endemic species (ones found only in that location) by retaining and preserving primary forest areas, and yet with such a high (and increasing) proportion of the world's forest now being secondary forests, it is important to understand their role as species habitat.

High population growth in many of the countries with tropical forests and high levels of poverty have meant that many people have become shifting cultivators who further

Definition of forest	Year(s)	Africa	Americas	Asia	Pantropical
Potential forest area (km²)					
Evergreen and deciduous	—	1,890,000	9,390,000	5,740,000	17,020,000
Extant forest area (km²)					
Dense (≥80%) tree cover	1997	1,720,000	7,010,000	1,990,000	10,720,000
Evergreen and seasonal	1997	1,930,000	6,530,000	2,700,000	11,160,000
Closed (>40% tree cover)	2000	2,630,000	6,790,000	1,940,000	11,350,000
Deforestation rate (km²/yr)					
All tree cover	1984–1990	4,300	43,000	18,000	65,000
All tree cover	1990–1997	3,700	43,000	26,000	73,000
Evergreen and seasonal	1990–1997	8,500	25,000	25,000	59,000
Closed (>40% tree cover)	1990–2000	8,400	39,000	20,000	68,000
Humid forest biome	2000–2005	3,000	33,000	19,000	55,000
Reforestation rate (km²/yr)					
All tree cover	1984–1990	5,600	2,800	3,500	11,900
All tree cover	1990–1997	4,300	3,700	2,600	10,600
Evergreen and seasonal	1990–1997	1,400	2,800	5,300	9,500
Closed (>40% tree cover)	1990–2000	1,100	1,300	1,100	3,500

Table 3. Potential and extant forest area and rates of deforestation and reforestation estimated from satellite imagery. Reproduced by permission of Gale, a part of Cengage Learning.

clear and burn the forest. Between 1990 and 2010, forest clearing for timber often was followed by the complete clearing of land and the planting of cash crops such as oil palm, cocoa, rubber, and soybeans. In Indonesia two-thirds of the plantations on former rain forest land are oil palm plantations, and these now cover an estimated 7.4 million acres (3 million ha). Demand for oil palm is increasingly rapidly, and oil palm plantations will likely spread to other parts of the world. Forests are important sources of fuelwood for people, and in large parts of Africa forest clearing has resulted from the gathering of firewood. Reforestation of degraded lands with industrial timber plantations occupies only a small percentage of rain forest landscapes in Africa and Central and South America, but has been more successful in Southeast Asia.

Increasing population density and development

Unlike parts of Africa or Southeast Asia, in the Amazon human population densities are low. The Amazon forested region covers some 1.4 million square miles (3.5 million km²), equivalent to approximately 80 percent of total forested area in Brazil. Much of the Amazon is accessible only by river, and hence large parts are still intact. However, billions of dollars has been available since 2000 for the creation of a

series of roads extending 4,650 miles (7,500 km) and crisscrossing this area. Many of these have since been built and have the potential to open up the Amazon for development, logging, and land clearing. Some of this development is designed to provide access for the creation of soybean plantations. Already existing development led to an increase in the human population within the region from about 2.5 million in 1960 to 20 million in 2000. William F. Laurance and colleagues, in a 2001 study, modeled the impact this infrastructure development might have on the Amazon by 2020 (see Figure 3).

One of the large remaining rain forests is on the island of New Guinea. The western half is part of Indonesia. The eastern part comprises the independent nation of Papua New Guinea. Traditionally, land is owned by local tribes, and logging has proceeded more slowly than in other parts of the region because logging operations required agreements with these traditional owners. As in many parts of the developing world, however, corruption is rife, and concessions have been given through the bribing of officials. A system of government-driven special agricultural business leases has increased the level of corruption, and scientists and conservationists have raised concerns that all accessible lowland rain forest will have been logged by around 2025.

Hunting

A further threat to the survival of rain forests is the increasing hunting for bushmeat, a development that has been exacerbated by the growing forest populations. The range of species that are now hunted has increased, and traditional methods of hunting have been discarded and replaced with the use of more modern weapons that are more likely to catch and kill animals. Many forests now appear to be almost devoid of mammal species above several pounds in weight. Such so-called empty forests no longer have some of the key species that eat fruit and disperse seeds. As well as being hunted for meat, some species are hunted and sold as living specimens for zoos and the pet trade, often through illegal trading.

Fire

Recurrent surface wildfires are now one of the most important threats to seasonally dry tropical rain forests

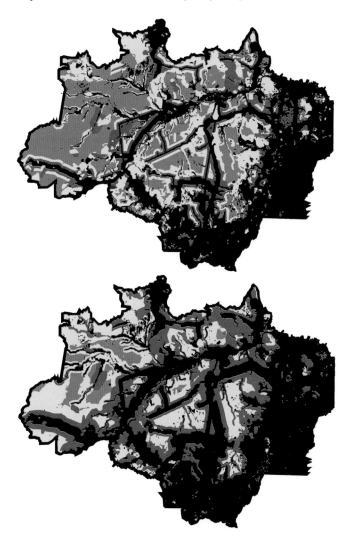

Figure 3. Optimistic (top) an non-optimistic (bottom) scenarios of deforestation of the Brazilian Amazon by 2020 if proposed road and infrastructure projects proceed. Black is deforested or heavily degraded, savannas and other nonforested areas, red is moderately degraded, yellow is lightly degraded and green is pristine (Laurance et al. 2001). Reproduced by permission of Gale, a part of Cengage Learning.

(Cochrane 2009), often acting synergistically with severe droughts. Woody plants in wet tropical forests are typically thin-barked and therefore vulnerable to even low "creeping" fires set by shifting cultivators (Barlow and Peres 2004). In a 2005 study, Mark G. L. van Nieuwstadt and Douglas Sheil found that mast-fruiting species such as Dipterocarpaceae in Southeast Asia are particularly susceptible because fires following recent recruitment events have drastic effects. In the Indonesian territory of Kalimantan, even larger-stemmed dipterocarp trees with bark thick enough to survive fire are still vulnerable to drought. Indeed, much of the plant mortality ascribed to fire stems from the severe droughts required before a forest can burn. Both larger and smaller species are vulnerable in different contexts, presumably depending on soil moisture and rooting depths. Small seedlings and nondeciduous canopy trees suffer if they cannot drop their leaves. Drought and fire together strongly affect the maintenance or recovery of forests and their role as carbon stores. In some central-eastern Amazonia regions, recurrent fires have changed the forest structure and composition so that pioneers dominate the understory (Barlow and Peres 2004). For forest vertebrates, fire poses the greatest risk to primary forest specialists, which are characterized by low mobility, poor climbing ability, poor flight capacity, small home ranges, or heavy reliance on woody shelters such as cavity nests within hollow tree trunks.

Increasing evidence suggests that different threats (e.g., fire and drought—see above) act additively or synergistically and exacerbate tropical forest species declines. The lack of observed extinctions in tropical forests between the mid-twentieth century and the early twenty-first century may belie future exponential increases. Synergistic interactions or multiplicative interactions result from simultaneous action of separate processes (extrinsic threats or intrinsic biological traits). Synergies occur between almost all anthropogenic threats. For example, hunting often acts synergistically with other threats such as fragmentation and disease to cause local extinction. Logging increases forest fires and fragmentation, thus accentuating ecoclimatic stresses, species invasions, and edge-related changes, often followed by hunting, trophic cascades, invasions of agricultural pests, and sometimes complete deforestation as colonists follow loggers into the forest frontier.

Invasive organisms, including those transported by humans or accentuated by climate change, can severely affect faunas naive to these new species or invasions. Predation by exotic species on islands has caused the extinctions of many native animals. The spread of chytrid fungus on amphibians, which has been enhanced by increasing global temperatures, is a particularly apt example of synergistic threats on a global basis.

Climate change

Tropical rain forests have become a focus for studies of the impact of climate change because of their significance for biodiversity, their roles in global carbon and water cycles, and their importance to local people whose livelihoods depend on them. The Intergovernmental Panel on Climate Change (IPCC), which through its reports offers a synthesis of all current research on climate change, has suggested that by

Illegal logging in a Sumatran rain forest in Indonesia. Courtesy of Nigel Stork.

2100 temperatures are likely to increase by 5.2°F to 9.5°F (2.9°C–5.3°C; Solomon et al. 2007). This was based on a business-as-usual scenario, and because the levels of carbon dioxide in the atmosphere increased to around 390 parts per million by 2011 and global temperatures are generally

considered to have increased by approximately 1.35°F (0.75°C) since the early 1960s, it is likely that subsequent predictions of future temperatures may be further revised upward.

The lines of evidence for climate change are based on empirical data of what has happened in the recent past, particularly focusing on the period from the mid-twentieth century to the early twenty-first century; other empirical data from the geological past, such as measures of carbon dioxide levels over millions of years in ice cores and fossil evidence; and finally modeling what might happen in the future. In a 2011 study, Richard T. Corlett reviewed the scientific literature on the potential impacts of climate change on tropical rain forests and concluded that there are a range of views, some very pessimistic and others more optimistic. The numerous global climate models vary considerably in their predictions of what will happen to rainfall patterns, and further new sources of data and research will be needed to improve these models' consistency. All these models agree, however, that temperatures will rise considerably, and this means that many species will have to move toward the poles or up mountains to cooler areas if they are to survive. Chris D. Thomas and colleagues (2004) modeled what this would mean for global biodiversity and predicted that on the basis of midrange climate-warming scenarios for 2050, 15 to 37 percent of species in a cross-section of samples from different regions (including tropical rain forest) and taxa would be "committed to extinction." Because mid-range scenarios now seem optimistic, it is likely that this prediction is an underestimate. Lapse rates of temperature with altitude suggest that a 9°F (5°C) increase in temperature might mean a species has to move roughly 0.6 miles (1,000 m) upward in altitude. Cooler-adapted rain forest species such as those found in upland forests may be near the peaks of mountains; because they will be unable to move higher, they may disappear. Others have pointed out that the temperature regimes of lowland tropical rain forests have low variability; therefore, those species that are unable to find mountainous areas to occupy will have to move very long distances to find cooler climates. Some have suggested that a 9°F (5°C) increase in temperature might see all species in the Amazon disappear. One aspect has broad support: Under increased carbon dioxide levels, plants will take up carbon dioxide more quickly and therefore lose less water.

Resources

Books

Cochrane, Mark A. *Tropical Fire Ecology: Climate Change, Land Use, and Ecosystem Dynamics.* Berlin: Springer, 2009.

Corlett, Richard T., and Richard B. Primack. *Tropical Rain Forests: An Ecological and Biogeographical Comparison.* 2nd ed. Chichester, U.K.: Wiley-Blackwell, 2011.

Ghazoul, Jaboury, and Douglas Sheil. *Tropical Rain Forest Ecology, Diversity, and Conservation.* Oxford: Oxford University Press, 2010.

Laurance, William F., and Carlos A. Peres, eds. *Emerging Threats to Tropical Forests.* Chicago: University of Chicago Press, 2006.

Myers, Norman. *The Primary Source: Tropical Forests and Our Future.* New York: Norton, 1984.

Richards, Paul W. *The Tropical Rain Forest.* 2nd ed. Cambridge, U.K.: Cambridge University Press, 1996.

Solomon, Susan, Dahe Qin, Martin Manning, et al., eds. *Climate Change 2007: The Physical Science Basis; Contribution of Working Group I to the Fourth Assessment Report of the Intergovernmental*

Panel on Climate Change. Cambridge, U.K.: Cambridge University Press, 2007. Published for the Intergovernmental Panel on Climate Change.

Periodicals

Barlow, Jos, and Carlos A. Peres. "Ecological Responses to El Niño–Induced Surface Fires in Central Brazilian Amazonia: Management Implications for Flammable Tropical Forests." *Philosophical Transactions of the Royal Society* B 359, no. 1443 (2004): 367–380.

Barlow, Jos, and Carlos A. Peres. "Fire-Mediated Dieback and Compositional Cascade in an Amazonian Forest." *Philosophical Transactions of the Royal Society* B 363, no. 1498 (2008): 1787–1794.

Chazdon, Robin L., Carlos A. Peres, Daisy Dent, et al. "The Potential for Species Conservation in Tropical Secondary Forests." *Conservation Biology* 23, no. 6 (2009): 1406–1417.

Corlett, Richard T. "Impacts of Warming on Tropical Lowland Rainforests." *Trends in Ecology and Evolution* 26, no. 11 (2011): 606–613.

Dixon, R.K., S. Brown, R.A. Houghton, et al. "Carbon Pools and Flux of Global Forest Ecosystems." *Science* 263, no. 5144 (1994): 185–190.

Field, Christopher B., Michael J. Behrenfeld, James T. Randerson, and Paul Falkowski. "Primary Production of the Biosphere: Integrating Terrestrial and Oceanic Components." *Science* 281, no. 5374 (1998): 237–240.

Laurance, William F., Mark A. Cochrane, Scott Bergen, et al. "The Future of the Brazilian Amazon." *Science* 291, no. 5503 (2001): 438–439.

Myers, Norman. "Threatened Biotas: 'Hot Spots' in Tropical Forests." *Environmentalist* 8, no. 3 (1988): 187–208.

Paine, James R. "Status, Trends, and Future Scenarios for Forest Conservation Including Protected Areas in the Asia-Pacific Region." Asia-Pacific Forestry Sector Outlook Study Working Paper 4, Food and Agriculture Organization of the United Nations (FAO), Rome, 1997.

Pounds, J. Alan, Martín R. Bustamante, Luis A. Coloma, et al. "Widespread Amphibian Extinctions from Epidemic Disease Driven by Global Warming." *Nature* 439, no. 7073 (2006): 161–167.

Thomas, Chris D., Alison Cameron, Rhys E. Green, et al. "Extinction Risk from Climate Change." *Nature* 427, no. 6970 (2004): 145–148.

Van Nieuwstadt, Mark G.L., and Douglas Sheil. "Drought, Fire, and Tree Survival in a Borneo Rain Forest, East Kalimantan, Indonesia." *Journal of Ecology* 93, no. 1 (2005): 191–201.

Wright, S. Joseph. "The Future of Tropical Forests." *Annals of the New York Academy of Sciences* 1195 (2010): 1–27.

Other

Butler, Rhett A. "Tropical Rainforests: Structure of the Tropical Rainforest." Mongabay.com. Accessed May 11, 2012. http://rainforests.mongabay.com/0201.htm.

Smithsonian Tropical Research Institute. Center for Tropical Forest Science. "Plots Summary." Accessed May 14, 2012. http://www.ctfs.si.edu/plots/summary.

United Nations Collaborative Programme on Reducing Emissions from Deforestation and Forest Degradation in Developing Countries (REDD+). "About REDD+." Accessed May 14, 2012. http://www.uncontactedtribes.org

Nigel E. Stork

Coral reef ecosystems

Coral reef ecosystems are found in all tropical ocean areas, most commonly in warm, clear, and relatively oligotrophic (i.e., low-nutrient) waters. The spatial extent of coral reef ecosystems is determined by the distribution of reef-building stony corals. Thus, to begin to understand coral reef ecosystems and its inhabitants, it is essential to first consider the biology of stony corals.

Biology and ecology of reef-building corals

Stony corals are a class of benthic cnidarians (Order Anthozoa, Class Scleractinia) that precipitate calcium carbonate skeletons under their tissue (Brusca and Brusca 2003). This is a diverse group that includes both solitary and colonial species spanning a range of growth morphologies. Stony corals can be found in a wide variety of habitats, including tropical, temperate, and polar seas from the shallow photic zone to the deep, dark seas (greater than 19,685 feet [6,000 m] deep).

Like most cnidarians, stony corals can derive nutrients and energy by consuming prey and organic material from the environment (i.e., heterotrophy). Zooplankton, including small crustaceans and other invertebrates from the water column, are a common food source for stony corals. A subset of stony corals, however, has entered into a characteristic symbiosis with photosynthetic dinoflagellates, called zooxanthellae (Muscatine and Porter 1977). Zooxanthellae live within the nutrient-rich tissue of the coral and harness sunlight to create sugar. Much of the photosynthetically derived sugar is released by the zooxanthellae directly to the coral, providing hermatypic corals (i.e., corals with symbiotic zooxanthellae) with an abundant source of energy to complement that derived from (mostly plankton-based) heterotrophy. The symbiotic relationship between corals and zooxanthellae facilitates the rapid growth with the associated accretion of calcium carbonate skeletons. As such, hermatypic corals are the most significant contributors to the growth and emergence of coral reefs. Importantly, this symbiosis between corals and zooxanthellae is supported only in warm, high-light, marine environments. As such, it is only in the shallow tropical ocean that expansive reefs of stony corals develop (Darwin 1842).

Founded on the symbiosis between corals and zooxanthellae, a biologically rich and productive community of organisms is supported. Coral colonies themselves are home to many species beyond the cnidarian host. In addition to the symbiotic zooxanthellae, there are countless fungi, algae, bacteria, and viruses that live within the coral colony and skeleton. In a healthy coral colony, these species interact in a seeming balance. Light not used by the zooxanthellae passes through the tissue and is used by the algae for photosynthesis. Some of the bacteria living on the surface of corals use energy from the coral colony to convert inorganic nitrogen into a biologically available form, a process called nitrogen fixation. Plus, in general, the bacteria and viruses living on the surface of corals help create a community dynamic that maintains the relative stability of the entire microbial consortium. As such, it has been proposed that reef-building corals be called "holobionts" rather than individual organisms (Rohwer et al. 2002), referring to the high diversity of interacting species that compose a typical colony.

Moreover, reef-building corals provide food and resources for many other organisms living within the ecosystem. The three-dimensional structure created by reef-building corals provides habitat for myriad organisms, including fish, mobile invertebrates, and seaweeds. In addition to providing shelter for organisms, the structural complexity of coral colonies also affects local water flow, leading to a thicker boundary layer and thus more mixing of passing water masses with the reef organisms. This physical oceanographic dynamic is important for dampening the energy from waves (an oft-cited benefit of coral reefs for coastal towns), but it is also biologically important for increasing the rate of delivery of pelagic zooplankton and other open-ocean food resources to the reef biota (Reidenbach et al. 2006). Beyond providing increased access to other food resources, the coral holobiont is also an important producer at the base of the food chain. A variety of fish and invertebrates consume corals directly. The guild of corallivores includes fish species with specialized mouthparts that can either scrape coral tissue or pluck individual polyps from coral colonies and invertebrate species such as the crown-of-thorns starfish (*Acanthaster planci*) that feed on corals using extracellular digestion (Rotjan and Lewis 2008).

Perhaps even more important to the reef food web, however, is the contribution that corals make through the secretion of copious amounts of mucus. Based on the

This Pacific species, *Platygyra lamellina*, is one of many species of stony corals that are suffering rapid population declines. A combination of increased rates of mortality (due to disease and thermal extremes) and decreased competitive advantages against fleshy algae (due to overfishing of herbivores and nutrient pollution) are providing novel challenges for the persistence of these species. Courtesy of Gareth J. Williams.

coral–algal symbiosis, coral colonies are very productive, and in fact there is a relative overabundance of sugar produced by the zooxanthellae. Much of this fixed carbon is converted into complex carbohydrates by the coral animal and released as mucus on the coral's surface. For some species of coral, it is estimated that 50 to 80 percent or even more of the photosynthate produced by the zooxanthellae is released as mucus. One benefit of the high rate of mucus secretion is to prevent competitive organisms from settling on or infecting the coral colony, a form of anti-foulant analogous to the sloughing paints used on the hulls of boats. This mucus, however, is very rich in lipids and proteins and is readily consumed by other reef organisms such as fish and invertebrates (Benson et al. 1978). The mucus that is not consumed is sloughed off directly, and the energy and nutrients ultimately are recycled into the reef ecosystem by microbes (Wild et al. 2004). As a source of shelter, food, and other resources, reef-building stony corals are the true foundational species of the coral reef ecosystem.

A so-called "coral garden" in the central Pacific, including stony coral species *Acropora acuminate* (branching species) and *A. cytherea* (table-top species). Historical accounts of coral reefs worldwide include descriptions of such coral-dominated habitats. Today, due to myriad stressors of human activities, only a limited number of reefs support prodigious growth of stony corals. The goal of reef management is to improve conditions to allow stony corals to grow rapidly again and into the future. Courtesy of Stuart A. Sandin.

Biodiversity of coral reefs

Because coral reef ecosystems are limited to shallow depths in the tropics, their spatial extent is geographically limited. In fact, coral reefs cover only about 0.2 percent of the world's oceans. Despite covering such a relatively small area, the biological diversity of coral reef ecosystems is impressive. For example, the vast majority of the 2,500 species of stony corals are found in coral reef ecosystems, including the more than 800 species of reef-building corals. Further, over 4,000 species of fish live on corals reefs, comprising approximately 25 percent of the world's diversity of marine fishes. Beyond these large, apparent taxa of corals and fishes, there are likely to be millions of other species that are found only on coral reefs. Because of the very high biodiversity of coral reefs, there are no reliable estimates of the total number of species. However, extrapolations from small reef areas or from other ecosystems provide estimates ranging from one million to nine million species of multicellular organisms living on coral reefs, and there are certainly additional millions of bacteria and other microbes sharing this ecosystem (Knowlton et al. 2010).

The reason that coral reefs support such an unusually high level of biodiversity is unresolved yet much discussed (Connell 1978; Bellwood and Hughes 2001; Hughes et al. 2002; Sandin, Vermeij, et al. 2008). Factors that are both extrinsic and intrinsic to coral reef organisms have been considered as putative causes for the high biodiversity. Extrinsic factors, such as patterns of environmental change (or relative stability) and geological change, may have resulted in the high levels of diversification across coral reef taxa. Further, intrinsic factors, such as ecological specialization, key innovations, and symbioses, may have acted independently or in synergy with extrinsic factors to favor the development and maintenance of new species (Kohn 1997). Regardless of the origin, coral reefs are the most biodiverse marine ecosystem per unit area on the planet.

A school of barred jacks (*Carangoides ferdau*) with one golden trevally (*Gnathanodon speciosus*); foreground. Groups of large jacks can be found on reefs worldwide, though the largest schools are most frequently found in little-fished areas, such as the Phoenix Islands Protected Area (pictured here) in the central Pacific. Courtesy of Stuart A. Sandin.

State of biodiversity on coral reefs

Despite the high level of biological diversity on coral reefs, a growing number of reef species are suffering rapid and dramatic population declines. In most cases, the reductions of reef taxa are caused by three groups of human activities and their individual and interacting effects: (1) the excessive harvesting of species is reducing the numbers of some taxa, (2) pollution and the introduction of invasive species are increasing the numbers of previously rare species, and (3) changes to climate and ocean chemistry are altering the basic conditions in which all reef organisms live (Hughes et al. 2003). Although populations of many taxa are likely to be affected by these anthropogenic activities, the best data of decline exist for fishes and large invertebrates, especially corals. Some examples of decline follow.

Impacts of overexploitation

Coral reefs are an important source of food, contributing millions of tons of fisheries yield. While this total accounts for only 2 to 5 percent of the global marine fisheries yield, the food captured from reefs feeds hundreds of millions of people worldwide. In particular, coral reefs are a critical part of the protein production for people in many tropical developing nations, where they compose over 25 percent of the national fisheries yield. Many coral reef fisheries are small scale or artisanal, with the seafood captured sold locally or eaten directly by the fishing community. Although some coral reef fisheries are managed effectively, many are not (Newton et al. 2007). The result has been the local and global depletion of some important reef fisheries targets (Jennings and Polunin 1996).

As with many marine fisheries, the predatory fish species are among the first to be targeted and are frequently overexploited (Myers and Worm 2003). Reef sharks, groupers, and other large-bodied predators have suffered dramatic population reductions on most coral reefs (Sandin et al. 2008). Direct harvesting of these species for food or for sale is the principal means of decline, although some of the species are vulnerable to incidental and nontarget capture such as by lost fishing gear. The population reductions have prompted national and international groups to declare many of these predators to be species of concern, namely as species whose populations are significantly reduced and warrant focused management to avoid local or global extinction.

Many populations of other large-bodied species associated with coral reefs are under similar threat. Some larger fish species, such as the humphead wrasse (*Cheilinus undulatus*) and bumphead parrotfish (*Bolbometopon muricatum*), are valued highly in international markets and have suffered large population reductions across the majority of their habitat (Donaldson and Sadovy 2001; Donaldson and Dulvy 2004). Populations of sea turtles that use coral reef habitats for food and shelter have also experienced significant declines as a result of harvesting principally for local consumption. Such patterns of decline have been ongoing for decades and, in some cases, centuries, leading to functional extirpations of some species from coral reefs. For example, historical accounts describe populations of large charismatic species, such as dugongs (*Dugong dugon*) and crocodiles, as common to many coral reef

A whitetip reef shark (*Triaenodon obesus*) searching for prey on a coral-covered reef. Most reef-associated sharks have status as threatened or near-threatened due to fishing activities. Reef sharks are caught for food, for their fins, and in many cases as by-catch for other reef fisheries. Courtesy of Gareth J. Williams.

areas. Today, the sightings of such species are extremely rare and limited largely to the most remote areas of the tropics.

A growing number of smaller-bodied fish species also are experiencing widespread population declines as a result of harvesting. The demand for live fish for tropical aquaria has created a productive market for a diversity of coral-reef fish species (Sadovy and Vincent 2002). Many species of angelfish, butterflyfish, and other families of attractive fish are captured and shipped worldwide for the pet trade. In heavily fished areas, particularly coveted species may be essentially extirpated from all but the most inaccessible areas (e.g., deep depths beyond the range of divers, remote or unsafe coastlines beyond the range of local boats). More recently, specialized aquaculture facilities have successfully bred a number of species common in the pet trade, thereby relaxing the global demand for wild-caught fish (Tlusty 2002). The demand for products directly from coral reefs remains high, however, and continues to affect populations of many species.

Finally, a number of non-coral invertebrate species contribute significantly to fisheries yields from coral reefs, and many of these species have experienced severe population declines globally. Giant clams (*Tridacna* spp.) are harvested actively both for their value in the aquarium trade and for consumption. Similarly, many species of echinoderms, including both sea cucumbers and sea urchins, are harvested for local consumption or international export. Other popular invertebrates that contribute to coral reef fisheries are conchs, lobsters, and some crabs. In many locations, the populations of these large-bodied and sometimes long-lived invertebrate species have been reduced dramatically and are providing little new production to local fisheries.

Although heavy exploitation is far from uncommon in nearshore tropical waters, only one case of global extinction resulting from overexploitation by humans has been well documented on coral reefs. The Caribbean monk seal (*Monachus tropicalis*) was a large predator that was distributed widely across the Caribbean Basin, feeding on reef fishes and invertebrates from coral reefs. Although active breeding colonies of Caribbean monk seals had been recorded since the time when Europeans arrived in the Americas, the hunting of the species for food and oil led to rapid depletions of the population (McClenachan and Cooper 2008). Ironically, as people began to notice the catastrophic decline of this Caribbean endemic species, a new and hurried market of collectors arrived to collect the last specimens for preservation of the skeletons in museums. The last sighting of a live Caribbean monk seal occurred in 1952.

Despite the paucity of examples of global extinction, overexploitation has led to profound reductions of many populations on coral reefs, with a growing number of cases of local or regional extirpation of some species. National and international agencies have responded to these large-scale reductions, providing conservation and management listings of threat levels for imperiled species. Further, active protection of the few remaining intact and quasi-pristine coral reefs is providing spatial refuge for highly susceptible and often-times long-lived fisheries targets. Large marine parks that span hundreds to thousands of square miles of coral reef habitat—and that are functionally analogous to large national parks on land—likely will provide the best hope to protect the most vulnerable of fisheries targets from global extinction (Spalding et al. 2010).

A school of herbivorous ember parrotfish (*Scarus rubroviolaceus*) being watched by one predatory twinspot snapper (*Lutjanus bohar*). Large schools of herbivorous fish have capacity to consume huge amounts of seaweed while maintaining conditions favoring the growth of stony corals. Courtesy of Stuart A. Sandin.

A giant clam, *Tridacna maxima*, growing within a patch of stony coral, *Porites sp.* Giant clams are coveted by people both for consumption and for trade in the aquarium community, and their populations are severely reduced in many locations. These species of bivalves are important ecologically, as they filter the seawater, removing particulate and dissolved organic matter in addition to bacteria and viruses. Courtesy of Gareth J. Williams.

Impacts of pollution and invasive species

Compounding the impact of the removal of some reef taxa through overexploitation is the introduction of chemicals or nonnative species that can compromise the persistence of many reef taxa. Pollution, both chemical and biological, can directly compromise the health of individual organisms and can indirectly affect the ecology and competitive landscape for native taxa. Although many technical solutions exist for preventing pollution and invasive species, there remain many examples worldwide of declines of coral reef species resulting from introduced chemicals and species.

A green sea turtle (*Chelonia mydas*) swimming over a coral reef. Herbivorous sea turtles use coral reefs for foraging and shelter, but also depend on neighboring habitats, like seagrass beds for foraging and beaches for nesting. This species of sea turtle, along with a number of other reef-associated species, are listed as endangered on the IUCN Red List of Threatened Species. Courtesy of Gareth J. Williams.

CHEMICAL POLLUTION Chemical pollution on coral reefs comes in many forms and most commonly is the result of release from coastal human populations (Fabricius 2005). Sewage is perhaps the most ubiquitous source of pollution. In many locations, especially in less-developed and low-income communities, human waste is treated in a limited and/or perfunctory manner, or not at all, and then released directly into the nearshore environment. Coral reefs in many cases suffer the brunt of this source of stress, being the first ecosystem exposed to the bath of particulates and nutrients. Under high exposure to sewage, reef communities suffer myriad impacts. Corals are particularly sensitive to the direct effects of sewage, with evidence linking particulates (causing abrasion and shading) and nutrients (affecting disease dynamics) to the reduced fitness and eventual mortality of entire colonies. Further, elevated nutrient loads can alter competitive dynamics among benthic taxa. Although stony corals can individually benefit from increases in nutrient concentrations, fleshy algae are disproportionately advantaged by nutrient additions. Because stony corals compete for limited space with fleshy algae and a host of other species, when any taxon receives an advantage, benthic competitive hierarchies can shift. In many circumstances, nutrient additions tip the balance from favoring the persistence and growth of stony corals toward favoring the growth and competitive advantage of fleshy algae. Especially in heavily populated areas, pollution has led to the expansion of turf and other fleshy algae with concomitant decreases in the populations of stony corals.

The effect of sewage on coral reefs is not a phenomenon limited to low-income communities. In even the most affluent societies, sewage treatment and disposal methods can result in dramatic alterations of nutrient dynamics on coral reefs (Smith et al. 1981). Sewage that has been treated to remove solid wastes and bacteria still contains high concentrations of organic nutrients, such as nitrogen and phosphorous. If this effluent is released into the coral reef environment, as through nearshore outflow pipes or coastal injection wells, the nutrient-rich waters can cause competitive transitions comparable to those of untreated sewage—with fleshy algae gaining a significant advantage to the detriment of stony corals (Dailer et al. 2010). Importantly, the deleterious effects of sewage are perhaps one of the most manageable threats to coral reef biodiversity. Through financial investment in improved sewage treatment and disposal protocols, there is potential to mitigate the negative effects of sewage on nearshore tropical ecosystems.

Organic nutrient input to coral reef ecosystems is not limited to sewage (Fabricius 2005). A variety of sources of pollution similar to those that commonly affect freshwater and other nearshore marine environments is also known to reach coral reefs. Runoff from parks and golf courses can introduce appreciable amounts of fertilizer and pesticides to nearby reef ecosystems. Similarly, agricultural fields in the tropics can affect reefs both through leaching of nutrients and through direct sloughing of top soil into the ocean. Effluent from factories and other industrial centers can bring a host of chemicals and nutrients from land to sea. Importantly, a broad collection of nonpoint source pollutants can reach coral reefs through myriad locations. Each of these forms of pollution can stress reef organisms both through direct stress (and

Bumphead parrotfish (*Bolbometopon muricatum*) is one of the species listed as vulnerable by the IUCN RedList of Threatened Species. Courtesy of Gareth J. Williams.

possible mortality) and through indirect alterations of competitive dynamics among reef organisms.

Finally, solid wastes have been shown to locally alter the structure and productivity of coral reefs. Litter from many sources can cover or abrade benthic organisms and in some cases even leach chemicals stressful to many reef species. Of prominence at many locations is the deliberate or accidental introduction of sunken ships to tropical nearshore environments. Ship groundings can have immediate and profound impacts on reef organisms, in the worst cases causing destruction of large swaths of reef habitat. Once in place in shallow tropical waters, however, there is potential for shipwrecks and other underwater structures (e.g., retired oil platforms) to serve as artificial habitats for some reef organisms. Large reef fish frequently find shelter in shipwrecks, and some benthic species can settle and grow on to the surface of the wreck. Intentional placement of retired ships in soft-bottom tropical waters has been viewed as a possible option for increasing reef habitat in areas where hard-bottom substrate is uncommon. Interestingly, this management strategy appears to be site-specific and not necessarily appropriate in all locations. Of importance are the possible leachates that can come off the shipwreck, including the metal structure of the hull itself. In some low islands, especially coral atolls, far from continental areas, shipwrecks have been linked to the local degradation of coral habitats. In some cases it appears that iron leaching from a shipwreck can fertilize fast-growing cyanobacteria or other iron-limited primary producers, leading to a dramatic competitive asymmetry, with stony corals on the losing end of the relationship (Kelly et al. 2012). This mechanism of iron fertilization is comparable to that noted in the pelagic waters of the central Pacific, with the exception that the iron source in these iron-poor waters—that is, the shipwreck—becomes a static feature of the environment and one to which organisms of the reef community can respond.

INVASIVE SPECIES Another critical form of pollution on coral reefs is the introduction of nonnative and potentially invasive species (Carlton and Geller 1993; Naylor et al. 2001). Invasive species, or so-called biological pollution, can lead to wholesale shifts of reef communities, with the most dramatic examples involving the introduction of seaweed and fish species.

An important income source for people in nearshore tropical communities of developing countries is seaweed farming. Many thick, fleshy algae produce large quantities of cell-wall constituents that are commercially important for a variety of human uses. For example, agar and carrageenan are two of the most common colloids used in commercial applications and are largely produced by seaweeds grown in shallow coastal farms. These products are part of a multimillion-dollar industry and are used as thickening agents in a variety of products such as cosmetics, foods, and beverages, as well as for medical research. A number of communities in developing countries have introduced particularly rapid-growing and profitable species of seaweed to shallow habitats, such as lagoons, to capitalize on such aquacultural opportunities. Although efforts are taken to harvest the bulk of the product, seaweed farming operations can result in the release of nonnative species of algae to adjacent coral reef habitats where they may become competitively dominant, overgrowing other benthic organisms (Williams and Smith 2007). While seaweed aquaculture can be an important source of revenue for some communities, care should be taken to ensure that associated reef communities are not harmed by these activities.

Similarly, a number of fish species have been introduced to coral reefs as a consequence of efforts to expand local fisheries. On coral reefs with naturally low biodiversity, especially of classically harvested reef fish species, fisheries programs have been employed to expand the fishery through the introduction of nonnative fish species. For example, the coral reefs of Hawaii have relatively few native, large-bodied fisheries species, with a particular paucity of groupers and snappers. In the mid-1900s fisheries managers began a program to introduce a number of harvestable fish species to the Hawaiian coral reefs, most notably the bluestripe

A school of blacktail snapper (*Lutjanus fulvus*), a popular target for reef fisheries. Courtesy of Gareth J. Williams.

snapper (*Lutjanus kasmira*) and the peacock grouper (*Cephalopholis argus*). These species rapidly became abundant across the Hawaiian Islands (Friedlander et al. 2002). While some fisheries benefits have been gained, there have been a number of ecological consequences associated with the expansion of these nonnative species, including reductions in the densities of native species as a result of predation by or competition with these new arrivals.

In addition to intentional introductions of species, an unknown number of species have been introduced to coral reefs worldwide through unintentional pathways (Carlton and Geller 1993). Oceangoing vessels are a common vector for moving species outside their native range. Fouling organisms, including algae and sessile invertebrates, inhabit the hulls of vessels and can gain safe passage to new locations. Similarly, ballast water used to stabilize larger vessels can be pumped into a ship in one port and pumped out in another port. If this water is not treated (such as through chemical or physical mechanisms), organisms of countless species can be moved broadly across the world's oceans. The taxonomy of species hitchhiking in ballast water is broad, and examples range from spores of algae to larvae of invertebrates to, in extreme cases, fully grown fish successfully moving from port to port. The impacts of these nonnative species when they arrive at locations outside their native range can vary greatly. In most cases, the few individuals that arrive in a new location will survive only briefly, without successfully reproducing. A minority of individuals, however, are known to have successfully colonized the ecosystem of their new destination, thereby adding a new species to the local ecosystem. While many of these species will maintain only small populations, a few have been known to realize wildly rapid population expansion, in some prominent cases leading to the wholesale restructuring of the local ecosystem.

A highly publicized biological invasion occurred with the introduction of the lionfish (principally the red lionfish [*Pterois volitans*]) to the coral reefs of the Caribbean in the early twenty-first century. In this case, the source of the new species was the unintentional dumping of individuals being kept in tropical aquaria. The species appears to have spread from the east coast of Florida southward throughout the majority of the Caribbean (Johnston and Purkis 2011). Lionfish are fast-growing predators that prey voraciously on juvenile fish. In locations where densities of lionfish have become particularly high, the structure of the reef fish assemblage appears to be affected strongly by these predatory newcomers. Interestingly, mounting evidence suggests that the capabilities of lionfish as an invader are reduced in areas with high densities of native predators, such as within historically little-fished areas or well-managed marine protected areas (Maljković et al. 2008). Such observations suggest that lionfish are not necessarily a dominant predator within Caribbean coral reefs but instead are opportunistic predators that can rapidly capitalize on ecological availability, specifically when competitive native predators are absent or in low numbers.

Of principal concern with the introduction of nonnative species is scientists' inability to predict which species hold potential to cause important shifts in the ecology of the local ecosystem (Williams and Smith 2007). As such, the most prudent management strategy is one of conservatism. Proposals to introduce species for aquaculture should be considered closely, weighing the direct economic advantages against the potential costs for the nearby coral reef communities if the farmed species escapes, including how such an occurrence would affect the people who depend on the reef for income from fishing or tourism. In many cases, there are native species that can be used for algal farming or fish aquaculture that perform almost as well as (and in some cases, better than) nonnative species for commercial operations, while representing no threat of nonnative introduction. For invasions resulting from long-distance shipping, a successful strategy has been to work with vessels and shipping companies to control the movement of organisms through coherent protocols of hull cleaning, ballast water management, and general attendance to biological control procedures. Finally, educational programs for aquarists can help prevent unintentional introductions resulting from the dumping of unwanted and nonnative organisms. In all, the introduction of nonnative species is a potentially dangerous ecological concern but one that is largely preventable through insightful management.

Impacts of climate change

Perhaps the most ubiquitous threats to coral reef species are those associated with climate change (Hoegh-Guldberg et al. 2007). The anthropogenic release of extreme amounts of carbon dioxide and other waste gases is changing the patterns of temperature, storm activity, and water chemistry for all of the world's oceans. Reef organisms, especially those with finely tuned symbiotic relationships such as corals, are experiencing new physiological and physical challenges that may compromise the long-term survival of these species.

Alterations of oceanographic conditions and climatological patterns introduce novel thermal conditions for coral reef organisms. Particularly sensitive to these novel thermal conditions are the stony corals (Hoegh-Guldberg 1999). Importantly, the physiology of corals appears linked tightly to local thermal conditions. Most notably, extreme thermal events lead to a breakdown of the relationship between stony corals and their symbiotic zooxanthellae. Under high and sustained temperature stress, stony corals experience so-called coral bleaching during which time the endosymbiotic algae evacuate the coral tissue, leaving the colonies with little pigmentation (thus, the pale, "bleached" appearance of the transparent coral animal on top of a white limestone skeleton; Glynn 1993). Most likely, this breakdown in symbiosis is linked to photosynthetic activity that runs rampant under thermal stress and dangerously alters the chemical and ionic conditions within the host coral's tissue. If the thermal stress is short term, the coral animal can survive through the stressful event and recover the symbionts back from the water column. If, however, the thermal event is particularly extreme in intensity or duration, the coral will not reestablish its symbiont population, and the animal will ultimately die of starvation. Through this mechanism of coral bleaching, large-scale coral mortality events have followed some of the particularly intense El Niño Southern Oscillation events of the late 1900s and early 2000s (Obura and Mangubhai 2011).

The changing thermal conditions are also changing the distribution of energy across the world's oceans. An important consequence of these shifts is an altered landscape of storm activity—the spatial patterning and likely the intensity of large storms are changing as a result of the altered climate of the globe (Knutson and Tuleya 2004). The implications for coral reefs are nontrivial. Coral reef ecosystems exposed to increased storm activity are prone to larger mortality events, especially for the reef-building corals. If the frequency of such mortality events increases, there is consequently a shorter period of time for the slow-growing corals to recover between events. Under these circumstances, some coral reefs may be expected to experience a punctuated and persistent decline in reef-building corals, thereby shifting the ecological context of the reef ecosystem. Because of the stochastic nature of storm activity, such shifts of reef communities are sometimes difficult to quantify. It is important to consider, however, that the complex suite of life, death, growth, and competition of benthic reef organisms is likely to be affected strongly by shifts in the patterns of even episodic, yet intensive, oceanographic events (Blackwood et al. 2011).

Acroporid corals (genus *Acropora*) are the most diverse groups of stony corals, made up of over 100 species. Despite being among the most rapidly group of stony corals, with reported growth rates exceeding 20 cm linear extension per year, *Acropora* species are among the most threatened. These species appear to be particularly sensitive to thermal anomalies and to diseases associated with changing environmental conditions. Courtesy of Stuart A. Sandin.

Much attention has also been paid to the rapidly changing chemical conditions of the oceans that have been observed since the early twentieth century. Of concern is the fact that the oceans serve as a buffer for industrial emissions, absorbing approximately 30 percent of all the carbon dioxide released by the combustion of fossil fuels (Doney et al. 2009). The result is that the oceans are becoming more acidic, a trend that is accelerating. Given that the hallmark of coral reefs is the biogenic structure that the stony corals and other calcifying organisms produce, and given that the rate of calcification is related to chemical conditions, there is reason to be concerned about this global pattern of ocean acidification. In particular, there is good evidence that many reef calcifiers, especially the stony corals, experience severe reductions in the rate of skeletal growth when exposed to relatively small changes in pH (Anthony et al. 2008). Further, even if calcification continues, the limestone may be of lower density under more acidic conditions, creating a much more fragile limestone framework on which the reef community depends. The linkages between ocean acidification and the fate of coral reef species into the future was only beginning to be explored in the early 2000s, but there was immediate reason to consider mitigation of continued, large-scale emissions from fossil fuels to address concerns about large-scale shifts in the composition and persistence of coral reefs.

As of 2012 much of scientists' understanding about the fate of corals and other reef organisms under a changing climate was based on extrapolations from the responses of organisms at the time. It is critical to consider, however, that climatic conditions are changing gradually and consistently and that biology is by no means static. A look at the geological record provides a reminder that the climate has changed repeatedly since the dawn of time, and species have adapted in many cases to these novel circumstances. There is no reason to imagine that the organisms inhabiting coral reefs in a century will perfectly mimic those on reefs today. There is capacity for reef species to acclimate or adapt to novel circumstances, such as through selection for more heat-tolerant genotypes or modifications of symbiotic interactions. The critical difference, however, between biological responses to climate change in the geological past relative to responses to anthropogenic climate change is the rate of change—never

before has the planet's climate changed as rapidly as it is in the early twenty-first century (Zachos et al. 2005). Despite this fact, it would be a mistake to ignore the potential of biological adaptation even in these circumstances, and a number of noteworthy studies have explored this capacity among coral reef organisms (Rowan 2004; Sampayo et al. 2007; Pandolfi et al. 2011; Donelson et al. 2012).

The fate of this ecosystem

The resultant changes to the community of coral reef organisms are many and ubiquitous. Although few species on coral reefs have been driven to extinction globally, profound regional declines have led to aggressive management actions. Many reef species are included on national and international lists identifying species of conservation concern, and many others are sure to be added in the second and third decade of the twenty-first century and beyond. As possible, these listings have or will lead to new regulations and protections to prevent further declines and ultimate extinctions of reef species. But a singular focus on just the population status of individual species ignores the more profound and unalterable changes occurring to coral reef ecosystems as a whole.

According to a 2008 estimate, upward of one-fourth of the world's coral reefs have been effectively lost (Wilkinson 2008). The definition of "effectively lost" is significant, implying that these reefs no longer have the characteristics common to a functioning coral reef. The widespread degradation of coral reefs and their sensitivity to myriad human stressors have invoked the image of the canary in the coal mine. Following this analogy, coral reef ecosystems may provide the first and clearest example of the profundity of cumulative anthropogenic impacts on the health of natural ecosystems. Overexploitation can reduce densities of key taxa, pollution can alter the ecological interactions of the remnant community, and climate change alters the most fundamental physical and chemical conditions in which the organisms live. When one considers the fate of coral reefs as an ecosystem it is clear that without a conscious commitment to effective management and conservation, it is possible that humanity will drive not just species, but this entire ecosystem, to the brink of extinction.

Resources

Books

Brusca, Richard C., and Gary J. Brusca. *Invertebrates*. 2nd ed. Sunderland, MA: Sinauer Associates, 2003.

Darwin, Charles. *The Structure and Distribution of Coral Reefs*. London: Smith, Elder, 1842.

Knowlton, Nancy, Russell E. Brainard, Rebecca Fisher, et al. "Coral Reef Biodiversity." In *Life in the World's Oceans: Diversity, Distribution, and Abundance*, edited by Alasdair D. McIntyre, 65–78. Chichester, UK: Wiley-Blackwell, 2010.

Kohn, Alan J. "Why Are Coral Reef Communities So Diverse?" In *Marine Biodiversity: Patterns and Processes*, edited by Rupert

F.G. Ormond, John D. Gage, and Martin V. Angel, 201–215. Cambridge, UK: Cambridge University Press, 1997.

Sadovy, Yvonne J., and Amanda C.J. Vincent. "Ecological Issues and the Trades in Live Reef Fishes." In *Coral Reef Fishes: Dynamics and Diversity in a Complex Ecosystem*, edited by Peter F. Sale, 391–420. San Diego, CA: Academic Press, 2002.

Spalding, Mark, Louisa Wood, Claire Fitzgerald, and Kristina Gjerde. "The 10% Target: Where Do We Stand?" In *Global Ocean Protection: Present Status and Future Possibilities*, edited by Caitlyn Toropova, Imèn Meliane, Dan Laffoley, et al., 25–40. Gland, Switzerland: International Union for Conservation of

Nature, 2010. Accessed May 26, 2012. http://data.iucn.org/dbtw-wpd/edocs/2010-053.pdf

Wilkinson, Clive, ed. *Status of Coral Reefs of the World: 2008.* Townsville, Australia: Global Coral Reef Monitoring Network, Reef and Rainforest Research Centre, 2008.

Periodicals

Anthony, K.R.N., D.I. Kline, G. Diaz-Pulido, et al. "Ocean Acidification Causes Bleaching and Productivity Loss in Coral Reef Builders." *Proceedings of the National Academy of Sciences of the United States of America* 105, no. 45 (2008): 17442–17446.

Bellwood, David R., and Terry P. Hughes. "Regional-Scale Assembly Rules and Biodiversity of Coral Reefs." *Science* 292, no. 5521 (2001): 1532–1535.

Benson, A.A., J.S. Patton, and S. Abraham. "Energy Exchange in Coral Reef Ecosystems." *Atoll Research Bulletin*, no. 220 (1978): 33–53.

Blackwood, Julie C., Alan Hastings, and Peter J. Mumby. "A Model-Based Approach to Determine the Long-Term Effects of Multiple Interacting Stressors on Coral Reefs." *Ecological Applications* 21, no. 7 (2011): 2722–2733.

Carlton, James T., and Jonathan B. Geller. "Ecological Roulette: The Global Transport of Nonindigenous Marine Organisms." *Science* 261, no. 5117 (1993): 78–82.

Connell, Joseph H. "Diversity in Tropical Rain Forests and Coral Reefs." *Science* 199, no. 4335 (1978): 1302–1310.

Dailer, Meghan L., Robin S. Knox, Jennifer E. Smith, et al. "Using δ^{15}N Values in Algal Tissue to Map Locations and Potential Sources of Anthropogenic Nutrient Inputs on the Island of Maui, Hawai'i, USA." *Marine Pollution Bulletin* 60, no. 5 (2010): 655–671.

Donaldson, Terry J., and Nicholas K. Dulvy. "Threatened Fishes of the World: *Bolbometopon muricatum* (Valenciennes 1840) (Scaridae)." *Environmental Biology of Fishes* 70, no. 4 (2004): 373.

Donaldson, Terry J., and Yvonne J. Sadovy. "Threatened Fishes of the World: *Cheilinus undulatus* Rüppell, 1835 (Labridae)." *Environmental Biology of Fishes* 62, no. 4 (2001): 428.

Donelson, J.M., P.L. Munday, M.I. McCormick, and C.R. Pitcher. "Rapid Transgenerational Acclimation of a Tropical Reef Fish to Climate Change." *Nature Climate Change* 2, no. 1 (2012): 30–32.

Doney, Scott C., Victoria J. Fabry, Richard A. Feely, and Joan A. Kleypas. "Ocean Acidification: The Other CO_2 Problem." *Annual Review of Marine Science* 1 (2009): 169–192.

Fabricius, Katharina E. "Effects of Terrestrial Runoff on the Ecology of Corals and Coral Reefs: Review and Synthesis." *Marine Pollution Bulletin* 50, no. 2 (2005): 125–146.

Friedlander, A.M., J.D. Parrish, and R.C. DeFelice. "Ecology of the Introduced Snapper *Lutjanus kasmira* (Forsskal) in the Reef Fish Assemblage of a Hawaiian Bay." *Journal of Fish Biology* 60, no. 1 (2002): 28–48.

Glynn, P.W. "Coral Reef Bleaching: Ecological Perspectives." *Coral Reefs* 12, no. 1 (1993): 1–17.

Hoegh-Guldberg, Ove. "Climate Change, Coral Bleaching, and the Future of the World's Coral Reefs." *Marine and Freshwater Research* 50, no. 8 (1999): 839–866.

Hoegh-Guldberg, Ove, Peter J. Mumby, Anthony J. Hooten, et al. "Coral Reefs under Rapid Climate Change and Ocean Acidification." *Science* 318, no. 5857 (2007): 1737–1742.

Hughes, Terry P., Andrew H. Baird, David R. Bellwood, et al. "Climate Change, Human Impacts, and the Resilience of Coral Reefs." *Science* 301, no. 5635 (2003): 929–933.

Hughes, Terry P., David R. Bellwood, and Sean R. Connolly. "Biodiversity Hotspots, Centres of Endemicity, and the Conservation of Coral Reefs." *Ecology Letters* 5, no. 6 (2002): 775–784.

Jennings, Simon, and Nicholas V.C. Polunin. "Impacts of Fishing on Tropical Reef Ecosystems." *Ambio* 25, no. 1 (1996): 44–49.

Johnston, Matthew W., and Samuel J. Purkis. "Spatial Analysis of the Invasion of Lionfish in the Western Atlantic and Caribbean." *Marine Pollution Bulletin* 62, no. 6 (2011): 1218–1226.

Kelly, Linda Wegley, Katie L. Barott, Elizabeth Dinsdale, et al. "Black Reefs: Iron-Induced Phase Shifts on Coral Reefs." *ISME Journal* 6, no. 3 (2012): 638–649.

Knutson, Thomas R., and Robert E. Tuleya. "Impact of CO_2-Induced Warming on Simulated Hurricane Intensity and Precipitation: Sensitivity to the Choice of Climate Model and Convective Parameterization." *Journal of Climate* 17, no. 18 (2004): 3477–3495.

Maljković, A., T.E. Van Leeuwen, and S.N. Cove. "Predation on the Invasive Red Lionfish, *Pterois volitans* (Pisces: Scorpaenidae), by Native Groupers in the Bahamas." *Coral Reefs* 27, no. 3 (2008): 501.

McClenachan, Loren, and Andrew B. Cooper. "Extinction Rate, Historical Population Structure, and Ecological Role of the Caribbean Monk Seal." *Proceedings of the Royal Society* B 275, no. 1641 (2008): 1351–1358.

Muscatine, L., and James W. Porter. "Reef Corals: Mutualistic Symbioses Adapted to Nutrient-Poor Environments." *BioScience* 27, no. 7 (1977): 454–460.

Myers, Ransom A., and Boris Worm. "Rapid Worldwide Depletion of Predatory Fish Communities." *Nature* 423, no. 6937 (2003): 280–283.

Naylor, Rosamond L., Susan L. Williams, and Donald R. Strong. "Aquaculture—a Gateway for Exotic Species." *Nature* 294, no. 5547 (2001): 1655–1656.

Newton, Katie, Isabelle M. Côté, Graham M. Pilling, et al. "Current and Future Sustainability of Island Coral Reef Fisheries." *Current Biology* 17, no. 7 (2007): 655–658.

Obura, D., and S. Mangubhai. "Coral Mortality Associated with Thermal Fluctuations in the Phoenix Islands, 2002–2005." *Coral Reefs* 30, no. 3 (2011): 607–619.

Pandolfi, John M., Sean R. Connolly, Dustin J. Marshall, and Anne L. Cohen. "Projecting Coral Reef Futures under Global Warming and Ocean Acidification." *Science* 333, no. 6041 (2011): 418–422.

Reidenbach, Matthew A., Stephen G. Monismith, Jeffrey R. Koseff, et al. "Boundary Layer Turbulence and Flow Structure over a Fringing Coral Reef." *Limnology and Oceanography* 51, no. 5 (2006): 1956–1968.

Rohwer, Forest, Victor Seguritan, Farooq Azam, and Nancy Knowlton. "Diversity and Distribution of Coral-Associated Bacteria." *Marine Ecology Progress Series* 243 (2002): 1–10.

Rotjan, Randi D., and Sara M. Lewis. "Impact of Coral Predators on Tropical Reefs." *Marine Ecology Progress Series* 367 (2008): 73–91.

Rowan, Rob. "Coral Bleaching: Thermal Adaptation in Reef Coral Symbionts." *Nature* 430, no. 7001 (2004): 742.

Sampayo, Eugenia M., Lorenzo Franceschinis, Ove Hoegh-Guldberg, and Sophie Dove. "Niche Partitioning of Closely Related Symbiotic Dinoflagellates." *Molecular Ecology* 16, no. 17 (2007): 3721–3733.

Sandin, Stuart A., Jennifer E. Smith, Edward E. DeMartini, et al. "Baselines and Degradation of Coral Reefs in the Northern Line Islands." *PLoS ONE* 3, no. 2 (2008): e1548.

Sandin, Stuart A., Mark J.A. Vermeij, and Allen H. Hurlbert. "Island Biogeography of Caribbean Coral Reef Fish." *Global Ecology and Biogeography* 17, no. 6 (2008): 770–777.

Smith, Stephen V., William J. Kimmerer, Edward A. Laws, et al. "Kaneohe Bay Sewage Diversion Experiment: Perspectives on Ecosystem Responses to Nutritional Perturbation." *Pacific Science* 35, no. 4 (1981): 279–395.

Tlusty, Michael. "The Benefits and Risks of Aquacultural Production for the Aquarium Trade." *Aquaculture* 205, nos. 3–4 (2002): 203–219.

Wild, Christian, Markus Huettel, Anke Klueter, et al. "Coral Mucus Functions as an Energy Carrier and Particle Trap in the Reef Ecosystem." *Nature* 428, no. 6978 (2004): 66–70.

Williams, Susan L., and Jennifer E. Smith. "A Global Review of the Distribution, Taxonomy, and Impacts of Introduced Seaweeds." *Annual Review of Ecology, Evolution, and Systematics* 38 (2007): 327–359.

Zachos, James C., Ursula Röhl, Stephen A. Schellenberg, et al. "Rapid Acidification of the Ocean during the Paleocene–Eocene Thermal Maximum." *Science* 308, no. 5728 (2005): 1611–1615.

Stuart A. Sandin

• • • • •

Extraterrestrial impact as a cause of extinction

Rare impacts of large comets and asteroids on Earth have the destructive power to be a potentially dominant cause of mass extinctions of species of life on the planet. Whether they actually caused all or most of the mass extinctions that have occurred during Earth's history is debatable. But clearly one major mass extinction was, indeed, caused, at least predominantly, by an asteroid impact: the K–Pg (Cretaceous–Paleogene) mass extinction 65.5 million years ago (previously known as the K–T [Cretaceous–Tertiary] mass extinction), which is the most recent of the major extinctions. This entry presents the astronomical context for these impacts and explores the possible roles of such impacts in changing the ecosphere so severely that a mass extinction results. It delves into the physical, chemical, meteorological, and geological effects of impacts and includes a discussion of the generalities concerning the effects on the environments within which living things live, die, and evolve. Apart from a few generalities, however, this entry does not discuss how different kinds of life and particular species might survive—or not—in the face of the environmental consequences of an extraterrestrial impact.

Asteroid and comet impacts are part of an ongoing, continuous cosmic process that is observable today and has necessarily operated during Earth's history. Impacts by such bodies are not ad hoc events but unavoidable, if rare, events that have manifestly shaped the surface of the planet. This entry also briefly examines some minority claims that (1) even the K–Pg mass extinction was *not* caused by an impact and (2) that very recent events, such as the Younger Dryas big freeze, which commenced 12,900 years ago, *were* caused by impacts.

Asteroids and comets: The astronomical context

Planet Earth, the only known abode of life, orbits the Sun with seven other major planets. In addition, numerous smaller bodies, made of ices, rocks, and metals, also orbit the Sun, mainly asteroids and comets, which are central to this entry. In addition, smaller fragments of comets and asteroids, less than 10 m (about 33 feet) in diameter, which are called *meteoroids* or *interplanetary dust particles*, orbit the Sun and exemplify, on a nightly basis, the vastly larger and rarer events that can cause extinctions. While the solar system is very large

and interplanetary space is mostly empty, asteroids and comets move very rapidly in their orbits, and, over time, if their orbits intersect, they can collide with each other and with the planets, including Earth. Inside the orbit of Jupiter, asteroids are much more numerous than comets. The largest of these, Ceres (as of 2012 designated a *dwarf planet*), is about 590 miles (950 km) across, but—like most asteroids—it circles the Sun between the orbits of Mars and Jupiter in the *main asteroid belt* and can never come close to Earth. A small fraction of asteroids can be dislodged from the asteroid belt and/or have the ellipticities of their orbits augmented so that they can cross Earth's orbit and thus have a chance of encountering Earth. (Such dislodgments are caused primarily by complex orbital resonance phenomena involving the gravity of distant planets and, to some degree, by solar thermal effects and by catastrophic collisions among the asteroids.) As these so-called near-Earth asteroids (NEAs) finally strike a planet or the Sun, which typically takes many millions of years, they are depleted. But they are also replenished gradually from the asteroid belt, maintaining a constant swarm of small bodies that can strike Earth. (The term *near-Earth object* [NEO] refers to both NEAs and to infrequent comets from the outer solar system, which occasionally enter the inner solar system. (For a review of the relevant properties of NEAs, see Chapman 2004.)

There is a whole array of small interplanetary objects, starting with the smallest ones that can be witnessed on Earth and studied today and progressing to the largest ones that can dramatically affect the ecosphere but that have done so only occasionally during the history of Earth. Among these small astronomical bodies, the smaller ones are increasingly and vastly more numerous than the larger ones. This mainly results from what is termed a *collisional cascade*: When these bodies occasionally, but repeatedly, encounter one another at cosmic speeds, they collide catastrophically, breaking into ever-smaller fragments that follow a power-law size distribution. The smaller the fragments, the more efficiently they can be moved around the inner solar system by forces related to the Sun—solar radiation, solar wind, and so on—beyond just normal gravitational forces. Tiny meteoroids, ranging from dust-sized and sand-sized particles up roughly to the size of a marble, are so numerous that they strike Earth almost continuously. A person gazing up into a clear night sky away from city lights can see an impact several times an hour as a

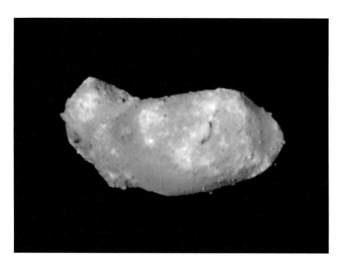

Asteroid 25143 Itokawa is a near-Earth asteroid, studied by the Japanese spacecraft Hayabusa in late 2005. It is about 1,750 feet long. An asteroid that size, if it struck Earth, would cause a regional catastrophe and there would be global atmospheric effects (e.g. possible destruction of the ozone layer). It probably would require an asteroid at least 20 times bigger to cause a major mass extinction. AP Photo/Japan Aerospace Exploration Agency, HO.

meteor or so-called shooting star. These brief darting flashes of light are caused by the disintegration of these particles in the atmosphere about 30 to 60 miles (50 to 100 km) above the ground—that is, they burn up. Larger meteoroids may produce streaks of light that are briefly even brighter than the brightest planet, Venus; they are called *fireballs*. Those that rival the brightness of the full Moon are sometimes called *bolides*; they may explode and potentially yield meteorites.

Meteoroids a couple feet (half a meter) in diameter and larger can be bright enough to be seen in daylight as they impact the upper atmosphere. These often break up into a brilliant train of pieces moving across the sky; some of the pieces can fall to the ground and be recovered as meteorites. Such meteorites, ranging in size from pebbles to monsters as big as a small automobile, can be seen in natural history museums around the world: They are pieces of asteroids, occasional pieces of the Moon or Mars and, perhaps, comets. People who are out and about during nighttime hours might see such a meteorite-producing event, within several hundred miles of their location, about once a year. Meteorites can damage houses and cars because, even though their cosmic velocities (typically 10 to 15 miles per second [15 to 25 km/sec]) are greatly slowed as they fall through the air, they can strike at terminal velocities of about 300 to 700 feet per second (100 to 200 m/sec). While at least one person has been struck by a ricocheting meteorite, there is no confirmed case of any person dying from such a strike.

Still larger cosmic bodies strike Earth even less frequently, but the larger they are the more damage they can cause. The largest impact in recent history was the impact of an NEA that exploded low in the atmosphere above Tunguska, Siberia, in 1908. It caused widespread devastation over a region of about 400 square miles (1,000 sq km; the size of a major metropolitan area). But as this area was essentially

unpopulated, there are no reliable reports of any people being killed. The Tunguska body was probably rocky and about 130 feet (40 m) in diameter. Such NEAs are likely to strike somewhere on Earth once every several hundred years or so. Near-Earth asteroids significantly larger than 300 feet (100 m) in diameter can impact the ground with most of their cosmic velocities (typically 10 to 15 miles per second [20 km/sec]) unslowed by the atmosphere and can produce craters with diameters ten or twenty times the size of the NEA. Craters such as Meteor Crater in Arizona, which is about 4,000 feet (1,200 m) in diameter and was caused by impact of a metallic body about 130 feet (40 m) across, exemplify the enormous destructive energy of such impacts. A 300-foot (100-m) NEA would explode with a force greater than the largest hydrogen bomb ever tested. A *crater* in the ocean would instantly collapse, of course, resulting in an enormous wave, or tsunami. It is a matter of some debate how rapidly such a tsunami would dissipate as it races across the ocean, but an impact by an NEA several hundred yards in diameter or larger could be very destructive near the shore around the rim of the impacted ocean.

Exceedingly rare and unlikely—but not impossible—on the timescale of human lifetimes or even human history, but very common during geological epochs, are impacts by NEAs 0.3 to 0.6 miles (500 m to 1 km) in diameter. With explosive yields equivalent to tens of thousands of megatons of TNT, such impacts would have global consequences, likely destroying the ozone layer and instigating significant climate change within weeks and months of the impact. Human civilization could probably survive such a catastrophic event, and such an event certainly would be far too puny to result in a mass extinction, although changes to habitats and ecosystems would be inevitable, especially within hundreds of miles of the impact site. Geological studies have confirmed the existence of seventy-five craters on Earth larger than 6.2 miles (10 km) in diameter, demonstrating that numerous such impacts have happened, although many times that number of craters must have been created only to be destroyed by active geological processes on Earth. (Terrestrial craters are listed in the Earth

Meteor crater, near Winslow, Arizona, is one of the best-preserved impact craters on Earth. It is about 4000 feet in diameter. The crater formed about 50,000 years ago when a small, metallic asteroid struck and exploded. © Charles & Josette Lenars/Corbis.

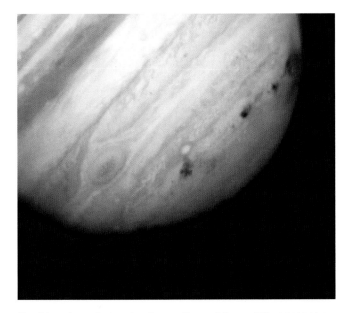

About two dozen fragments of a small comet (named Shoemaker–Levy 9) crashed into Jupiter in July 1994. This Hubble Space Telescope view of Jupiter shows black regions where four of the 1000-foot sized fragments impacted. The disturbed regions are major fractions of the size of planet Earth, illustrating the destructive power of even small asteroids and comets when they strike a planet at cosmic velocities. Science Source.

Impact Database [PASSC 2012]. Over 180 confirmed craters of all sizes are listed in this database. It is widely understood that many more likely impact craters have been identified by exploration geophysicists, but the information is considered proprietary to the oil and gas companies they work for.)

It is estimated that the impact of an NEA approaching 2 miles (3 km) in diameter, striking with the force of over a million megatons of TNT, would be so destructive on a global scale that the future of human civilization would indeed be threatened, although probably not the survival of the human species. Such an event would likely threaten the existence of some endangered species, especially those unluckily sensitive (e.g., because of their restricted location near the impact site) to the particular event. (The threshold size of an impactor capable of destroying human civilization might be even smaller than 1 mile [1.5 km] or as large as 3 miles [5 km].) This distinction between the destruction of civilization and the eradication of the human species is something to reflect on when contemplating extinctions and mass extinctions in general. What from a human perspective might seem to be an unimaginably horrendous catastrophe is actually minuscule compared to what is expected to be required to produce a mass extinction—and minuscule as well compared to what still-larger impacts have done during Earth's history. Of course, while many such impacts are known to have occurred during the Phanerozoic, scientists are at the point where only theoretical extrapolations from much smaller nuclear explosions, extrapolations to Earth from other observed cosmic impacts (e.g., the impact of the comet Shoemaker–Levy 9 on Jupiter in 1994), and studies of the actual impacts recorded in Earth's geological stratigraphy enable estimates of the actual

physical, chemical, and ultimately biological effects of such enormous events.

History of the developing awareness of impact dangers

The idea that comets might be potentially destructive, at least in some magical or superstitious way, has been around for at least centuries and probably stretches back into prehistory. As these icy bodies from the cold outer solar system evaporate and expel material in response to solar warming while they are briefly in the inner solar system, they can make a dramatic appearance in the night sky. The tails of large comets close to Earth and/or the Sun can appear like a searchlight beam stretching across the sky, and, rarely, the head of a comet can become bright enough to be seen in the daylight sky. From prehistoric times, human beings feared such unexpected appearances of comets in the sky (see, for example, Yeomans 1991). But the physical nature of comets remained largely unknown until the middle of the twentieth century when Fred L. Whipple (1906–2004), an astronomer at Harvard University, hypothesized that they were solid bodies, mixtures of rocky and icy materials, like a *dirty snowball*. Subsequent telescopic and spacecraft studies of comets have confirmed that they are solid bodies, roughly akin to asteroids, and that the larger ones could be exceedingly destructive if they collided with Earth.

The very first asteroid was discovered a little over two centuries ago. As already mentioned, Ceres is in the main asteroid belt, and no main-belt asteroid poses any known danger to Earth. The first NEA, 433 Eros, was discovered in 1898, and not until 1932 was an NEA discovered (1862 Apollo) that had an orbit actually crossing Earth's orbit. As asteroids continued to be discovered it gradually became clear, again in the middle of the twentieth century, that there is a considerable population of NEAs that come near to, or even cross, Earth's orbit. This conclusion was inescapable even though only about twenty NEAs were known as of 1970. (By 2012, 9,000 had been discovered.) From the 1950s through the 1970s, there were occasional scientific articles in obscure publications, several science fiction works, and even the report of a class study at the Massachusetts Institute of Technology recognizing the destructive potential of NEAs. But it was not until the very early 1980s that two events quickly enlightened the scientific community, and soon thereafter the general public, with regard to the dangerous consequences of an NEA strike on Earth: (1) the publication in 1980 of the hypothesis that the dinosaurs and other species went extinct 65.5 million years ago as a result of an asteroid impact (Alvarez et al. 1980) and (2) a 1981 conference, sponsored by the National Aeronautics and Space Administration (NASA) and held in Snowmass, Colorado, on the physical and human consequences of asteroid impacts.

A report published in 1980 in the widely circulated scientific journal *Science* by the Nobel Prize–winning physicist Luis W. Alvarez, his geologist son, Walter, and two colleagues proposed that the K–Pg mass extinction was caused by the impact of an asteroid with a diameter of roughly 6 miles (10 km). The idea was received with broad skepticism by most geologists and paleontologists, many of whom were influenced by the traditional philosophy of geological gradualism, which is one component of the uniformitarianism that had held sway for a century and a half (since its promulgation by the British geologist Charles Lyell [1797–1875]). A sudden bolt from the heavens as a cause of this profound change in biodiversity 65.5 million years ago seemed radical and even ad hoc. But during ensuing years, evidence mounted that the timing of the deposition around the globe of a thin layer rich in spherules, shocked grains, and the cosmically abundant element iridium (and other platinum group elements, which are rare in Earth's crust) matched the timing of what has traditionally been regarded as a sudden changeover in the proportions of fossilized species in the sedimentary records that had long been the basis for the idea that a mass extinction had occurred at that time.

The arguments of Alvarez and colleagues would have been viewed as definitive by many physical scientists, even if the crater that would necessarily have been created by such an impact had vanished as a result of subsequent reprocessing, such as subduction into Earth's mantle. But it became difficult to doubt the reality and relevance of an impact when, later in the 1980s, a crater with a diameter of about 110 miles (180 km) was discovered that indeed had formed at the required time (Hildebrand et al. 1991). The crater, named Chicxulub, is centered near the northern edge of the Yucatán Peninsula in Mexico. Such a large crater would have resulted from the impact of the hypothesized 6-mile-wide NEA (or perhaps an NEA 8 or 9 miles [12 or 15 km] wide), and mounting evidence had already been focusing the cause of the worldwide iridium-rich spherule layer toward the Caribbean and the vicinity of Chicxulub (the layer deposited at the K–Pg boundary is thicker in localities close to Chicxulub). Hence, during the subsequent quarter century, many accepted the idea that the infamous extinction of the non-avian dinosaurs and many other species was caused by an asteroid impact. (For a popular account of the development of this hypothesis, see Powell 1998.) Of course, while it is widely accepted that the formation of Chicxulub was the primary trigger for the K–Pg mass extinction, the impact naturally occurred at a specific time when other circumstances may have played roles as secondary contributors to the extent or duration of the extinction. For example, the specific state of Earth's (or a region's) climate or the robustness of one or more species as a result of unrelated reasons might have magnified or diminished the importance of the environmental consequences of the impact in causing the mass extinction.

An important point known to the Alvarez team and to astronomers is that there is nothing ad hoc about the idea of an asteroid 6 to 9 miles (10 to 15 km) in diameter striking Earth. There are several such NEAs orbiting the Sun today, and scientists' understanding of the processes that deplete and replenish the NEA population implies that there normally are several asteroids of such sizes capable of striking Earth and that such impacts probably actually occur about every 100 million to 200 million years. (Ernst J. Öpik had qualitatively explained the issues of asteroid impact in a 1952 article, and in 1973 he published an estimate that NEAs larger than 5 miles (8.5 km) in diameter would strike Earth about every 260 million years.)

In addition, although small NEAs are much more common than small comets in the inner solar system, impacts by larger comets approach the frequency of large NEA impacts. Comet Hale–Bopp, widely observed in the mid-1990s, was estimated to have a nucleus perhaps 30 miles (40 km) in diameter; the comet came inside the orbit of Earth, although the planet was more or less on the opposite side of the Sun at the time. Thus, the impact of a large NEO 65.5 million years ago is not ad hoc but rather consistent with expectations. Indeed, it is expected that several such impacts during the Phanerozoic would have happened, similar to the number of known major mass extinctions. It is statistically possible, but very unlikely, that there would have been no impacts of NEAs larger than 6 miles (10 km) in diameter since the fossil record of plants and animals began on Earth.

For this reason, astronomers found compelling a hypothesis put forward in 1991 by University of Chicago paleontologist David M. Raup, among others, that the mass extinctions were all caused by asteroid impacts. (In their 1982 study, Raup and his colleague J. John Sepkoski Jr. analyzed Sepkoski's compendium on extinctions, recognizing the five major mass extinctions.) Raup, indeed, pointed to the spectrum of large to small extinctions as roughly conforming to the size spectrum of NEAs and hypothesized that all extinctions might be the result of asteroid impacts. While this all-inclusive hypothesis is accepted by few if any extinction researchers, its firm basis in asteroid science may put the shoe on the other foot: Instead of trying to demonstrate that an ad hoc cosmic mechanism has been responsible for each extinction or mass extinction, it now might be regarded as being necessary for researchers to demonstrate that an asteroid impact was not the cause of any particular extinction and that the extinction resulted from alternative plausible and demonstrable causes. Or it is at least necessary to demonstrate that there was something unique about the K–Pg impact so that it should not be expected that other similar impacts would likely cause a mass extinction (see below for further discussion). And it somehow, eventually, needs to be demonstrated that the unavoidably enormous environmental consequences of such impacts could be survived by the kinds of species that went extinct during the various mass extinctions.

In science, the publication of a hypothesis is no guarantee that it will receive wide recognition or acceptance, especially when it argues for a changed paradigm about such a popular issue as the demise of the dinosaurs. Thus, the second event of the early 1980s, the 1981 NASA conference in Snowmass, Colorado, played a contributing role in the scientific community's gaining an understanding of how asteroid impacts might be very dangerous. The conference included a wide variety of experts in fields other than paleontology, including members of the Alvarez team, as well as experts on the properties of asteroids and comets, physicists knowledgeable about large nuclear explosions and how to simulate them with computers, space mission engineers, and astronomers who were surveying the skies for as-yet-unknown asteroids and comets. The meeting was chaired by Eugene Shoemaker, a planetary geologist partly responsible for proving in the late 1950s that Meteor Crater was formed by an asteroid impact and was not a volcano. Shoemaker had gone on to try to understand the

In early 2007, Comet McNaught became the most spectacular comet in decades, primarily as observed from the southern hemisphere. The comet came into the inner solar system from the so-called Oort cloud of comets, far beyond the planetary system. It approached the Sun closely, about half the distance from the Sun to the planet Mercury. The evaporating dust and ices created the spectacular, fan-shaped tail visible in mid-January 2007. Robert McNaught/Science Photo Library.

relationship between craters on Earth, the Moon, and other planetary bodies and the comets and asteroids responsible for their formation. Indeed, he transformed himself from a geologist into an astronomer as he founded the first organized telescopic effort to search for NEAs. The 1981 conference primarily addressed issues of how dangerous the NEO threat is today and what might be done to prevent a potentially civilization-destroying impact. But as the diverse scientists addressed the modern threat, they began to develop a better understanding of the physical and environmental consequences of such impacts, which can also be applied to Earth's history.

A book-length report of the conference findings was prepared during 1982 but was never actually published by NASA. Nevertheless, awareness of the issue spread throughout the space science community, in the United States and in other countries. For example, interest was reawakened in Russia, a country that presents a large target for an asteroid and had been struck in 1908 (the aforementioned Tunguska event) and again in 1947, by the Sikhote-Alin impact. In the mid-1980s NASA ramped up a couple telescopic observing programs and began initial planning for a spacecraft mission

to an NEA, which eventually evolved into the late 1990s NEAR Shoemaker mission to 433 Eros. Popular publications, based on the unpublished Snowmass conference volume, raised further public interest in the NEO hazard, including interest in the US Congress, which mandated NASA to study how even more NEAs might be identified and what might be done if a dangerous one were found to be on an Earth-impact course. In March 1989 a small asteroid—later designated 4581 Asclepius—with a diameter of 1,000 feet (300 m) passed 425,000 miles (684,000 km) from Earth (which is quite close by astronomical standards), making front-page news as a "near miss." It was during 1989 and 1990 that Alan R. Hildebrand (at the time a graduate student at the University of Arizona) and his associates were homing in on the Caribbean source of the K–Pg spherule layer, soon leading to the identification of Chicxulub (Hildebrand et al. 1991). The interest of the space science community in the NEO hazard has constituted a parallel and complementary avenue, along with the work of geologists and paleontologists, for developing an understanding of the nature of giant asteroid impacts (e.g., an appreciation of the environmental consequences that are a necessary precursor for understanding biological responses; see Toon et al. 1997).

Effects of cosmic impacts on the biosphere

Asteroid and comet impacts are feeble, almost insignificant events in the context of the universe or even just the solar system. As is emphasized by the output of the online impact simulator Impact: Earth! (Collins et al. 2005, 2012), which has been employed for most of the examples of environmental consequences below, even the greatest impacts have a negligible effect on the planet as a whole. An impact does not change the length of the day, the tilt of Earth's axis, or the planet's course around the Sun. (Of course, very early in solar system history, there were much larger bodies around, and it is widely thought that an impact into Earth of a Mars-sized body led to the formation of Earth's Moon. Impacts of that size would dramatically affect the geophysical character of the planet—but such bodies have not existed for 3 billion years in orbits that could come near to Earth.) Even the geology of Earth's crust is minimally affected by asteroid impacts: The giant craters that result from the occasional large impacts are relatively modest features, dwarfed by major mountain ranges, plateaus, rift valleys, and so on. And the K–Pg spherule layer around Earth is only about a centimeter thick in most places, hardly recognizable in a road-cut exposure to anyone but expert field geologists. Nevertheless, the ecosphere of Earth—the atmosphere, the oceans, and the upper few miles of the crust—is, of course, essentially at the surface of the planet, exposed to space, and it is very thin and fragile. So even if only a modest fraction of the kinetic energy of an impactor is partitioned into the ecosphere, the transformation can be profound.

The following two attributes of NEO impacts make them especially capable of transforming the biosphere: (1) the almost incomprehensible enormity of the globally destructive energy liberated in an impact and (2) the virtually instantaneous nature of the transformation and destruction.

What happens when a large NEO strikes Earth? If it is an NEA, it likely strikes at a velocity ranging between 10 and 20 miles per second (15 and 30 km/sec; Earth itself orbits the Sun moving at about 18.5 miles per second). If it is a comet, then the velocity could be two or three times greater. By way of comparison, 12 miles per second (20 km/sec) is nearly 100 times the speed of a jet airliner. Such a cosmic projectile would penetrate Earth's atmosphere in less than a few seconds and explode on contact with the surface (whether ground, ice sheet, or water). Because kinetic energy is proportional to the square of the velocity, a jumbo jet airliner crashing into a mountain cliff at such a cosmic velocity would liberate 10,000 times the destructive energy of the same airliner crashing into the cliff at ordinary airliner speeds. An asteroid with a diameter of 6 miles (10 km) is roughly 10 billion times more massive than a jumbo jet, so the destructive energy of an Earth impact explosion would be roughly 100 trillion times that of a jumbo jet crashing into a mountain.

Such numbers are impossible to comprehend. By way of comparison, the atomic bomb that the United States dropped on Hiroshima, Japan, in 1945 had a yield of about 15 thousand tons of TNT. An impact of a 6-mile-wide asteroid would explode with the energy of 100 million megatons of TNT (a megaton is a million tons). Accordingly, the destructive

energy would be 100 trillion times greater than the Hiroshima bomb. If this destructive energy were distributed equally around Earth's entire surface area, it would be equivalent to 40 simultaneous tornados present on each and every square mile of the planet. But of course the energy of an impact is not distributed evenly.

Specific kinds of damage would be wrought. First, the local damage would be continent-wide in extent. While the resulting crater would be about 100 miles (150 km) in diameter, forests would be blown down over 1,200 miles (2,000 km) away from ground zero in all directions. The fireball from the explosion might be like twenty-five suns above the horizon lasting for a half hour, resulting in forests burning over a region more than 1,000 miles across. In addition, an enormous earthquake would be felt about 600 miles (1,000 km) away, and rocks and dust would fall from the sky across the entire continent, accumulating into a layer a foot or two thick.

But these are just the *local* effects. The global effects would be equally profound and are quite well understood (as reviewed by Toon et al. 1997 and Collins et al. 2005) based on (1) extrapolations from nuclear bomb tests in the middle of the twentieth century, (2) observations of enormous craters on the Moon and other planets and satellites, and (3) modern computer hydrocode simulations of the physics. An enormous amount of material would be rocketed up and away at many miles per second in an impact plume as the explosion excavated the huge crater. Much of these ejecta would escape Earth's atmosphere and penetrate outer space; indeed some would escape Earth's gravity field entirely, but most would zoom back into the atmosphere around the entire globe during the course of the couple of hours after the strike. These returning ejecta from the crater would cause an unimaginably severe meteor storm; instead of individual bright meteors against a dark sky, the returning debris would form brilliant meteors covering the entire sky, with little or no black showing through, converting the sky into a broiler around the entire globe. Whether or not every ignitable thing exposed to the sky would burn is uncertain (the quantity and nature of soot found in the K–Pg boundary layer and calculations of whether temperatures would reach the points of ignition are subjects of debate). But the thermal pulse would be huge (e.g., about 570°F [300°C] for an hour) and would almost instantly kill all unprotected living things globally. The meteor storm would continue, although with diminished intensity, for several days. A major side effect of the meteor storm would be extreme chemical changes to Earth's atmosphere.

A long-lingering effect would be the retention in the stratosphere of dust and aerosols that would dim the sunlight worldwide for months to many years, putting Earth into a dark, deep freeze, perhaps colder on average by nearly 20°F (10°C). Photosynthesis would be shut down. (The chemistry of at least the surface waters in the oceans would also be dramatically affected as noxious chemicals, such as sulfuric acid, rained out of the atmosphere and detritus on the damaged surface of the planet washed into rivers and down into the sea, although these consequences are poorly understood.) The darkness and deep freeze—or impact winter—was the chief global environmental effect originally envisioned by the Alvarez team in 1980. It remains unclear what combination of these effects would be most destructive or fatal to specific ecosystems or species of life. Undoubtedly there are cumulative and synergistic effects that have yet to be appreciated. There might also have been countervailing effects. For example, burning forests could suddenly increase carbon dioxide in the atmosphere, possibly producing a large greenhouse effect that might moderate temperatures.

While the natures and magnitudes of each and every possible consequence of an impact are beyond the capabilities of modern science to calculate, it remains the case that the sheer destructive *energy* deposited by an asteroid impact into the large but fragile ecosphere of planet Earth would be wholly overwhelming. And the general magnitude and nature of the damage is not a matter of speculation or extrapolation: It can be directly discerned simply by looking at the cratered surface of the Moon or by understanding the enormity of effects observed on the giant planet Jupiter in 1994 when fragments of a small comet (roughly a thousandth the mass of a 6-mile-wide NEA) struck in 1994; each of the larger impacting fragments, typically 1,000 to 2,000 feet (roughly half a kilometer) in diameter, dramatically altered regions of Jupiter's atmosphere approaching the size of the whole planet Earth, as could be seen through small backyard telescopes from Earth.

The second essential attribute of an asteroid impact in transforming the ecosphere and potentially causing a mass extinction is its *suddenness*. No known or hypothesized noncosmic process (excluding all-out nuclear war) that could modify the global ecosphere or put living species in danger of

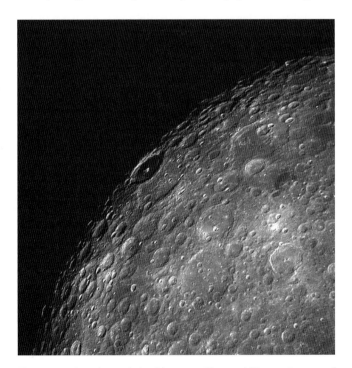

The cratered surface of the Moon testifies to billions of years of bombardment by near-Earth asteroids and comets. While most of the craters visible in this image formed around 4 billion years ago, the Moon, Earth, and other planets continue to be cratered in the modern epoch, although at a declining rate. © Corbis. Premium RF/Alamy.

extinction can happen nearly as abruptly. For example, Deccan Trap volcanism, which occurred around 65 million years ago, extended over a million years, and even spikes in the volcanism were tens of thousands of years in duration. The dramatic effect human civilization is currently having on the global environment—climate change—has a timescale of the order of a century. In contrast, some of the most devastating global consequences of a large impact take just *hours*. And the sudden climatological effects take months or a few years, at most.

Such near-instantaneous onsets of dramatic environmental changes around the globe are a unique attribute of asteroid impacts. There are other actual or at least hypothetical events on Earth that are very sudden, such as a huge earthquake or the eruption of a large volcano. But it is doubtful that the very largest such events could exceed 0.1 percent of the energy of the impact of a 6-mile-wide (10 km) NEA. The suddenness of such events can surely devastate regions (e.g., recall the ecological effects of the explosion of Mount St. Helens in the early 1980s). But sudden global devastation is out of the question because of natural caps on the magnitudes of endogenic catastrophes. Because of the power-law size-frequency character of NEA and comet populations, there is essentially no maximum limit on the size of a potential impact disaster—except that impacts by objects larger than tens of miles in diameter are so unlikely that the next one may not occur for many billions of years.

There is a qualitative difference for survivability of species between environmental changes that take place on timescales that are *short* compared with an individual's lifetime or reproductive timescale and those that are much longer (Chapman 2002). Traditionally, extinction scenarios have featured events such as sea-level changes, chemical and thermal changes in the ocean, global warming, ice ages, and hot-spot volcanic outpourings as being important to species survival. These might indeed be extreme changes, forcing members of a species into small colonies in atypical refugia, until they gradually adapt to the more general conditions. But a catastrophe that transforms the entire world on just a generational timescale or less presents a radical challenge to *every* individual and hence to the survival of its species.

There are various strategies that living things may invoke to escape changes to their ecosystem. Individuals with locomotion (mainly animals) may migrate to refugia; indeed, the survival of a species may involve rather few individuals being initially in or successfully finding some unusual hiding place (e.g., a special cave). On much longer timescales, evolution may yield adaptation to changed environments. Thus, over tens of thousands or millions of years, abundant opportunities for species survival exist in the face of large cumulative changes to their environment that are not possible in the course of just hours or months following an asteroid impact.

One of the uncertainties in assessing the effects of large, sudden asteroid impacts is that most calculations invoke attributes of Earth as it is known today. In considering mass extinctions that happened long ago, one must realize that Earth may well have been—indeed is known to have been—very different. The chemical composition of Earth's atmosphere has differed in the past, changing its susceptibility to certain kinds of perturbations (e.g., by the sudden addition of greenhouse gases). Also, the nature of plant life differed greatly in the past, perhaps changing its susceptibility to being ignited into a global firestorm, and hence whether the carbon sequestered in global forests went into the atmosphere or instead into the oceans. Beyond these uncertainties concerning the environmental consequences of an NEA impact at various past epochs, there are vital issues about the nature of individual species and of ecosystems (e.g., food chains) that may or may not explain the observed patterns of extinction, but these are beyond the scope of this entry.

Contrasting NEO impacts with alternative causes of extinctions

During the 1980s, there were lively and appropriate debates about the then-new impact-extinction hypothesis put forth in 1980 by the Alvarez team. After all, much of the evidence around the globe had yet to come to light, especially the existence of the crater Chicxulub of the appropriate size, geological age, and geographical location. For example, Alvarez and colleagues had based their hypothesis on only a few exposures of the K–Pg boundary in Europe. But the ejecta layer has now been identified at hundreds of localities around the planet. Nevertheless, minority views continue to exist about the cause of the K–Pg mass extinction. There are variations on the main hypothesis that are certainly plausible. For example, it has been suggested that other ongoing circumstances in Earth's geology or climate, or stages in the evolution of regional or global ecosystems, might have contributed to the extinction, even if the main cause was the impact. Certainly this is reasonable. Earth has been known to undergo major climatic changes, for instance, and if the impact occurred near an extreme (say, with an unusual atmospheric composition resulting from the Deccan Trap volcanism), the devastating effects of the impact might have been somewhat magnified.

Other suggestions of alternative or multiple causations are less reasonable. For example, it has been suggested that there were multiple impacts during a rather short duration that might have greatly augmented the consequences beyond those of a single impact. Despite the impact of multiple comet fragments on Jupiter during the course of one week in July 1994, well-understood dynamical processes in the solar system and the known characteristics of NEOs and their orbital evolution render multiple-impact scenarios on Earth, whether in brief succession or over hundreds of thousands of years, highly unlikely.

In a 2010 article, Peter Schulte and colleagues amply summarized the consensus (though not unanimous) view that the Chicxulub impact was coincidental with and caused the K–Pg mass extinction. They also evaluated and dismissed the remaining outlying contentions that some data are inconsistent with the Chicxulub impact causing the mass extinction and/or that other causes (e.g., Deccan Trap volcanism) were responsible. The question remains, however, about the other mass extinctions. The K–Pg extinction is the most recent and well-studied mass extinction. Necessarily, evidence becomes

spottier the farther one goes back in time. The largest mass extinction of all, the Permian–Triassic (or P–Tr) extinction 252 million years ago, has been widely researched, but no cause or group of causes has yet been advanced conclusively. It appears that the extinction may actually have been a series of extinctions, but attempts to develop a secure chronology of events from the highly incomplete stratigraphic record are highly prone to ambiguous interpretations. For example, there is only one locality (in the Karoo, South Africa) where there is an exposure of fossils of land vertebrates across the P–Tr boundary; published studies of these fossils have been contentious and not statistically robust. Evidence of the specific causal mechanism that proved that an impact triggered the K–Pg extinction, the Chicxulub crater, is unlikely to be identified for the P–Tr boundary if it existed, because 70 percent of Earth's surface area is oceanic, and ocean floors have been wholly destroyed by seafloor spreading since the end of the Permian. There have been several suggestions of evidence of a major impact that might be responsible for the P–Tr mass extinction (e.g., a possible crater in Antarctica), but the studies have not been robust and have not been widely accepted.

The kind of iridium-rich spherule layers that led to the original Alvarez team hypothesis for the K–Pg mass extinction have not been robustly identified for the P–Tr or other mass extinctions, but a comet impact might yield very little platinum-group enhancements at all as a result of the ice-rich composition of comet nuclei compared with rocky/metallic asteroids along with other differences between cometary and asteroidal impacts (e.g., the much higher impact velocities of comets, resulting in the ejection of a larger fraction of the mass of the projectile). One attribute of the K–Pg impact—the nature of the target rock—might have augmented the impact's role in causing a mass extinction compared with other similar impacts that likely occurred during the Phanerozoic. It is believed that the K–Pg asteroid impacted into rocky strata in the Chicxulub region (2 to 2.5 miles [3 to 4 km] thick) that were anomalously rich in carbonates and sulfates; this might have resulted in more potent atmospheric effects, including the production of prodigious sulfate aerosols in the stratosphere, which would have enhanced the dimming of the Sun. Conceivably it would take an appreciably larger, and less likely, asteroid impact into more common continental and ocean-floor target materials to cause a mass extinction—especially if ecosystems and species of life are more robust than scientists currently expect. Descriptions of other extinctions and possible causes are included elsewhere in this publication, and it is not the purpose of this entry to evaluate them. But the unparalleled, enormous, worldwide destructive energy and extreme abruptness of asteroid impacts should be kept in mind so that their role is not minimized as the causes of extinctions are evaluated. This contrasts with an apparently common view among paleontologists that impacts are an exceptional process for initiating mass extinctions and should be considered an explanation of last resort.

Opposite to denials that any of the largest mass extinctions were caused by impact, a more recent argument by some researchers holds that a very minor—though interesting—

extinction *was* caused by an impact. Prior to 13,000 years ago, there were numerous large animals in North America, including mammoths, mastodons, giant ground sloths, saber-toothed tigers, and short-faced bears. Rather suddenly, during a period of climatic cooling, they disappeared. This event does not qualify as a major mass extinction, but it is of considerable interest because it was very recent on a geological timescale and, indeed, affected—or was affected by—early human beings (the Clovis culture) in North America. In 2007 R.B. Firestone and colleagues proposed that a comet at least 2.5 miles (4 km) in diameter broke up over North America and impacted, causing this so-called Younger Dryas climate change and extinction. Asteroid astronomy and the physics of impacts, however, can be used to establish constraints on this hypothesis.

First, it is extraordinarily unlikely that such a large NEO would have impacted Earth so recently. Even if the impacting body were only 2 miles (3 km) in diameter, such a body impacts somewhere on Earth only once in several million years. The odds are hundreds to one against such a recent impact of that magnitude. Beyond that, although the hypothesis invokes an impact into a rocky surface to explain the nanodiamonds in the so-called black mat, there is no remnant crater of the requisite size and age. It takes a highly contrived model—one involving an impact into a thick, now-vanished ice sheet of just the appropriate thickness so that the very bottom of the crater barely penetrates into the ground below but the rest of the crater disappears when the ice melts—to explain the absence of a crater. A variation on this hypothesis, mentioned by Firestone and colleagues, suggests that the comet broke up just before entry (a very dubious proposition from the perspective of astronomical processes) and that the numerous fragments formed the hundreds of thousands of *Carolina bays*, elliptical depressions ranging southward from Delaware to Florida. These features, however, bear no resemblance to impact craters. The Younger Dryas cooling indeed happened, but there remain plausible nonimpact explanations for the Younger Dryas climate change (e.g., changes in oceanic circulation, initiated by a flood) and also for the disappearance of the megafauna (e.g., Clovis culture hunters).

In summary, extraterrestrial impacts by large asteroids and comets are inevitable events during the history of life on Earth. Because of their extreme physical and environmental consequences and the suddenness of their effects, they must be considered as potential causes of mass extinctions. One mass extinction—the K–Pg—was surely triggered by such an impact. Yet even though scientists are certain that such impacts must have occurred on geological timescales, it is also certain that impacts of kilometer-scale bodies are wholly improbable on the timescale of human history. Meteor Crater was formed about 50,000 years ago, and meteorites occasionally strike each month; but, as interesting as these recent events are in serving as reminders of Earth's cosmic environment, impacts that could cause mass extinctions are very rare, occurring no more often than the great mass extinctions recognized in Earth's geological record.

Resources

Books

Chapman, Clark R. "Impact Lethality and Risks in Today's World: Lessons for Interpreting Earth History." In *Catastrophic Events and Mass Extinctions: Impacts and Beyond*, edited by Christian Koeberl and Kenneth G. MacLeod, 7–19. Geological Society of America Special Paper 356. Boulder, CO: Geological Society of America, 2002.

Powell, James Lawrence. *Night Comes to the Cretaceous: Dinosaur Extinction and the Transformation of Modern Geology*. New York: W.H. Freeman, 1998.

Raup, David M. *Extinction: Bad Genes or Bad Luck?* New York: Norton, 1991.

Yeomans, Donald K. *Comets: A Chronological History of Observation, Science, Myth, and Folklore*. New York: Wiley, 1991.

Periodicals

Alvarez, Luis W., Walter Alvarez, Frank Asaro, and Helen V. Michel. "Extraterrestrial Cause for the Cretaceous–Tertiary Extinction." *Science* 208, no. 4448 (1980): 1095–1108.

Chapman, Clark R. "The Hazard of Near-Earth Asteroid Impacts on Earth." *Earth and Planetary Science Letters* 222, no. 1 (2004): 1–15.

Chapman, Clark R. "Meteoroids, Meteors, and the Near-Earth Object Impact Hazard." *Earth, Moon, and Planets* 102, nos. 1–4 (2008): 417–424.

Collins, Gareth S., H. Jay Melosh, and Robert A. Marcus. "Earth Impact Effects Program: A Web-Based Computer Program for Calculating the Regional Environmental Consequences of a Meteoroid Impact on Earth." *Meteoritics and Planetary Science* 40, no. 6 (2005): 817–840.

Firestone, R.B., A. West, J.P. Kennett, et al. "Evidence for an Extraterrestrial Impact 12,900 Years Ago That Contributed to the Megafaunal Extinctions and the Younger Dryas Cooling." *Proceedings of the National Academy of Sciences of the United States of America* 104, no. 41 (2007): 16016–16021.

Goldin, Tamara J., and H. Jay Melosh. "Self-Shielding of Thermal Radiation by Chicxulub Impact Ejecta: Firestorm or Fizzle?" *Geology* 37, no. 12 (2009): 1135–1138.

Hildebrand, Alan R., Glen T. Penfield, David A. Kring, et al. "Chicxulub Crater: A Possible Cretaceous/Tertiary Boundary Impact Crater on the Yucatán Peninsula, Mexico." *Geology* 19, no. 9 (1991): 867–871.

Öpik, Ernst J. "Collisions with Heavenly Bodies." *Irish Astronomical Journal* 2, no. 4 (1952): 95–98.

Öpik, Ernst J. "Our Cosmic Destiny." *Irish Astronomical Journal* 11, no. 4 (1973): 113–124.

Raup, David M., and J. John Sepkoski Jr. "Mass Extinction in the Marine Fossil Record." *Science* 215, no. 4539 (1982): 1501–1503.

Schulte, Peter, Laia Alegret, Ignacio Arenillas, et al. "The Chicxulub Asteroid Impact and Mass Extinction at the Cretaceous–Paleogene Boundary." *Science* 327, no. 5970 (2010): 1214–1218.

Toon, Owen B., Kevin Zahnle, David Morrison, et al. "Environmental Perturbations Caused by the Impacts of Asteroids and Comets." *Reviews of Geophysics* 35, no. 1 (1997): 41–78.

Other

Collins, Gareth S., H. Jay Melosh, and Robert A. Marcus. "Impact: Earth!" Accessed November 25, 2012. http://www .purdue.edu/impactearth.

PASSC (Planetary and Space Science Centre, University of New Brunswick). "Earth Impact Database." Accessed November 25, 2012. http://www.passc.net/EarthImpact Database.

Clark R. Chapman

• • • • •

Large igneous province (LIP) volcanism

Large igneous provinces (LIPs) are vast conglomerations of volcanic and intrusive igneous rocks that form in one fairly localized area on and within Earth's crust, respectively (Coffin and Eldholm 1994). Research between the late 1980s and the turn of the millennium (summarized by Courtillot and Renne 2003) has shown that LIPs usually form within a geologically brief period of time, especially the climax, or peak, of volcanic output, which is often much less than one million years in duration (reviewed by Self et al. 2006). LIPs have formed throughout much of Earth's history and are a manifestation of Earth's normal, but in this case spasmodic, internal processes. The most common hypothesis to explain the cause of LIPs, as recognized in research results between the 1990s and 2012, is that they are generated when a new mantle plume impacts Earth's lithosphere (Campbell and Griffiths 1990; Ernst and Buchan 2001; Campbell 2005). Alternative hypotheses for LIP occurrences also exist (Anderson 2005).

Large igneous provinces occur on continental crust and in the ocean basins (see Figure 1), the latter being sometimes known as oceanic plateaus. They come in two broad compositional varieties: basaltic LIPs, which are the most numerous and the only type forming oceanic plateaus, and silicic LIPs (Bryan 2007). In detail, some basaltic LIPs, which are dominated by vast piles of basalt lava flows, have silicic igneous rocks among their sequences, and vice versa for silicic LIPs (the terms *basaltic* and *silicic* are explained below). Several important LIPs were formed at the same time as major plate-rifting (splitting) events on Earth, and the remnant rock sequences of the same LIP are found on separate continents. An example is the Central Atlantic Magmatic Province (CAMP), which is possibly associated with extinction events at the Triassic–Jurassic boundary (Whiteside et al. 2007).

The process of generation and emplacement of LIPs can be thought of as unusual transient events in Earth's volcanic history when compared with magma production at mid-ocean ridges and subduction zones (see Mahoney and Coffin 1997 for a monograph on LIP characteristics). Large igneous province events result in rapid and large-volume accumulations of volcanic and intrusive igneous rock (Bryan and Ernst 2008). The volcanic and intrusive products of LIPs collectively cover areas in excess of hundreds of thousands of square miles (sometimes more than 400,000 square miles or 1 million square

kilometers) and, typically, extruded volcanic deposit and lava volumes are vast, more than 240,000 cubic miles (1 million km^3). By comparison, approximately thirty years of almost constant basaltic magma eruption at Kilauea volcano, Hawaii, from 1983 to 2012, produced about 0.4 cubic miles (1.5 km^3) of lava.

Oceanic plateaus define the upper limits of the areal and volumetric dimensions of terrestrial LIPs, with reconstruction of the Ontong Java, Hikurangi, and Manihiki Plateaus (suggested to be separated parts of the same LIP) having a pre-rift areal extent larger than the Indian subcontinent, making it the biggest recognized LIP (Kerr and Mahoney 2007). The largest subaerial, or continental, LIP in volume and area is the Siberian Traps, which is largely basaltic and formed about 250 million years ago (MYA; e.g., Reichow et al. 2005). Subaerial basaltic LIPs such as the Siberian Traps are usually known as flood basalt provinces; much of this entry is concerned with flood basalts, and the terms LIP (volcanism) and flood basalt province (volcanism) are used interchangeably. The term *trap* or *traps* is often used in the name of continental flood basalt provinces, such as the aforementioned Siberian Traps or the Deccan Traps of India. The term is of disputed origin, derived either from a Sanskrit word for south or the Swedish/Dutch word for a set of steps (stairs; *trappa*), and was first used by geologists in Victorian times (Geological Survey of India, 2012). The stepped topography of most flood basalt lava sequences is thought to be the reason for the use of this term.

This entry concentrates on basaltic LIPs as exemplified by continental flood basalt provinces because the geochronological record (dated age of occurrence) of these provinces reveals that many coincided with periods of mass extinctions and other rapid environmental changes (summarized by Courtillot 1994 and Courtillot et al. 1996; see also Saunders 2005). The growth of several oceanic plateaus also occurred during other periods of past rapid environmental change in the oceans and atmosphere (e.g., Kerr 2005). A very useful set of short summary articles on the impact that LIP volcanism may have had on Earth's environment and biotic extinctions appeared in the December 2005 (volume 1, number 5) issue of the journal *Elements*; all articles from that volume are cited herein.

A feature of LIPs is the high magma emplacement rate (e.g., Storey et al. 2007; Svensen et al. 2012) in which aggregate

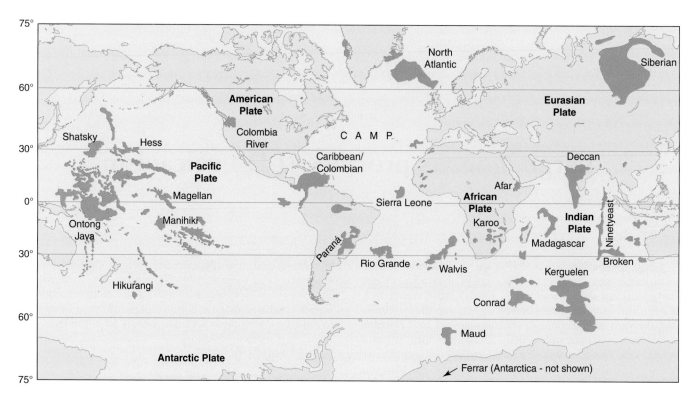

Figure 1. Large igneous provinces of the world, including continental flood basalt provinces and oceanic plateaus. CAMP is Central Atlantic Magmatic Province (after Sanders, 2005). Reproduced by permission of Gale, a part of Cengage Learning.

magma volumes of 240,000 cubic miles (1 million km^3) or more from a focused source were emplaced during periods of 1 to 5 million years. Within this period were shorter pulses (acmes) during which the output was even higher, but constraining the duration of these is difficult because of the limited precision of geochronological determinations (Kelley 2007). Unusual amounts of magma erupting in a short period of time, and the great extent of basaltic LIPs, also leads to the inference that almost all LIP eruptions had magnitudes significantly greater than those of historic eruptions, tending toward extraordinarily large-volume eruptions up to, and in excess of 240 cubic miles (1,000 km^3) or more (Tolan et al. 1989; Self et al. 2008). Consequently, it is the volume of magma emitted during these individual eruptions, the frequency of such large-volume eruptions, and the total volume of magma intruded and released during the main igneous pulses that make LIP events so exceptional in Earth's history. They have been called upon to explain environmental and climatic changes, including mass extinctions, for several reasons (summarized by Wignall 2005), as explained in later sections.

Despite the huge total cumulative erupted volumes of lava and the fact that the overall timing of LIP events is reasonably well-constrained (Coffin and Eldholm 1994; Bryan and Ernst 2008), the scientific understanding in the early 2000s of the size, duration, and frequency of individual flood basalt eruptions, which are the best exposed and studied products of LIP volcanism, is really very limited. Almost all information on the size of individual flood basaltic eruptions comes from the many studies undertaken on the Columbia River flood basalt province in the US Pacific Northwest. This is the smallest (roughly 60,000 cubic miles [250,000 km^3] of lava) and youngest example of a continental flood basalt province; most eruptions occurred between 16 and 15 million years ago, in the mid-Miocene geologic epoch (e.g., Tolan et al. 1989). Larger provinces, such as the Siberian Traps (Reichow et al. 2005; Black et al. 2012), originally contained in total ten times as much lava (480,000 to 720,000 cubic miles [2 to 3 million km^3]). It is only since 2005 that some understanding has been gained on the magnitude of flood basalt eruptions from other flood basalt provinces, mainly the Deccan Traps of India (e.g., Chenet et al. 2009).

During the growth of flood basalt provinces, hundreds to possibly thousands of eruptions produce immense lava flows, each the product of a vent or a group of vents along a fissure (a line of volcanic vents; see Self et al. 1997). Consequently, all flood basalt lava fields are dominated by basaltic lava flows, and, perhaps rather surprisingly, all flood basalt provinces examined in detail are dominated by compound pāhoehoe lavas composed of thousands of lava sheet lobes, and innumerable smaller lobes, stacked and superposed in some places and laterally arranged in others. Pāhoehoe lavas (Hawaiian, pronounced "pa·ho·e·ho·e") are the smooth, ropey-surfaced flows that are common in Hawaii; all flood basalt provinces from the very ancient geologic eras to the youngest on Earth, the Columbia River basalts, consist almost entirely of this lava type (Self et al. 1998).

It is not of vital importance to this topic that the origins of the magma erupted in LIPs are discussed in detail. What is useful to recognize is that the compositions of the basaltic magmas erupted in LIPs are similar to those being erupted by

A stack of lava flows, typical of the "trap" or stepped topography shown by flood basalt lavas, exposed along the Grande Ronde River in Washington state of the USA. The greater than 2000 feet (600 m)-thick pile of lavas seen here is part of the Columbia River Basalt LIP. Courtesy of S. Self.

some volcanoes today, such as many in Iceland, and thus a great deal can be learned about how these magmas behave and what gases are released on eruption (Self et al. 2005). For interested readers, a 2010 paper by Scott E. Bryan and colleagues provides a summary of the four most common magma-generation pathways in LIPs. Silicic LIPs have not been shown to correlate with periods of mass extinction and are thus of less importance in this discussion. Notwithstanding this fact, LIPs contain evidence of Earth's largest eruptions, and, although explosive super-eruptions of silicic magma are not restricted to LIPs (Self 2006; Bryan et al. 2010), subaerial basaltic volcanism on the scale described here have occurred only during periods of LIP (flood basalt) volcanism.

The terms *basaltic* and *silicic* (which can be used for magma, lava, or other volcanic products) refer to the amount of silica and other chemical ingredients in the magma. The compositional characteristics and source of the magmas control the physical properties when the magma erupts, giving the following broad distinction. Basaltic magmas have a lower viscosity (are runnier) and have lower gas contents, and they are thus more likely to form lava flows, with some explosive activity. The gases released as the lava is emitted include carbon dioxide (CO_2) and sulfur dioxide (SO_2), two compounds of importance to Earth's atmosphere (other sulfur gases are possibly released, plus chlorine and fluorine; Self et al. 2005; Black et al. 2012). During eruptions of silicic magma the main gas released is water, which is abundant in the atmosphere, along with some CO_2 (usually less than with basaltic magma), which is similarly quite abundant. The gases of importance to environmental and biotic changes are most likely CO_2 and sulfur, which are discussed in more detail below.

Evidence for LIP volcanism influence on mass extinctions

Biotic extinction events are important factors in the history of life on Earth, and studies beginning about 1980 radically

suggested catastrophic causes for at least some biotic mass extinctions. Two catastrophic processes have been invoked: impacts of extraterrestrial bodies such as asteroids or comets and a series of large volcanic eruptions, namely, continental flood basalt volcanism (Alvarez et al. 1980; Courtillot et al. 1988; Courtillot 1999). On the one hand, the end-Cretaceous (65.5 MYA) mass extinction (at the Cretaceous–Tertiary or K–T boundary, now sometimes referred to as the Cretaceous–Paleogene or K–Pg event) has been convincingly correlated with the impact of an asteroid with a diameter of 6.2 miles (10 km; Schulte et al. 2010), but tantalizing evidence of other impact events from extraterrestrial bodies has been found to be associated only with the times of a few other extinction events (summarized by Kelley 2007). On the other hand, the coincidence of the eruption of the Siberian flood basalt lava flow province with the even more severe end-Permian extinction (250 MYA; Renne and Basu 1991 and Black et al. 2012) and the near-coincidence of the growth of the Deccan flood basalt province and the K–T extinctions (Courtillot and Renne 2003; Keller et al. 2008) fostered further recognition that flood basalt eruptions might have contributed to a number of mass extinctions (Rampino 2010). It should be noted that when the eruptions of the Siberian Traps were first proposed as a trigger for the end-Permian mass extinction, this explanation presented a possible correlation of the emplacement of the largest flood basalt province with the greatest loss of floral and faunal diversity in Earth's history (Erwin 1994; Reichow et al. 2009).

Several workers compared the dates of extinction events of various magnitudes with the dates of flood basalt episodes and found some significant correlations (e.g., Courtillot 1994) supporting a possible cause-and-effect connection. It has also been suggested that a coincidence of both a large impact and a flood basalt eruption might be necessary to cause severe mass extinctions, and some researchers have even proposed that large impacts might in some way trigger or enhance the volcanism (e.g., Jones et al. 2002), but such hypotheses are not in the mainstream of scientific thought on this topic. Thus, it could be that extreme events of both extraterrestrial and terrestrial origin have been responsible for many of the punctuation marks of the fossil record. Both can potentially have a severe effect on biota (Keller 2003).

It should be noted that the only known convincing case of coincidence between a mass extinction and an asteroid impact revealed by study up until 2012 is the 65.5-million-year-old end-Cretaceous biotic event and the large impact at Chicxulub, Mexico (Schulte et al. 2010). This extinction event also occurred at the same time as the peak phase of volcanism of the Deccan Traps (Chenet et al. 2008; Keller et al. 2008). Because the records of LIP and extinction events are two time series that may or may not have some causal relationship, Rosalind V. White and Andrew D. Saunders in 2005 used a statistical study to examine whether there could be random coincidences in such a pair of chronologies. They concluded that up to three coincidences could be expected between the time of known LIP eruptions and mass extinctions over the last 300 million years. Other studies have shown, however, that there is a record of more numerous correlations; about six to seven out of eleven mass

Flood basalt episode	Age (MYA)	Volume (10^6 km^3)	Paleolatitude	Duration of peak pulse (MY)	Stratigraphic boundary	Age (MYA)
Columbia River	16±1	0.25	45°N	≤1 (for 90%)	Early/Mid-Miocene	16.0
Ethiopian	31±1	≤1.0	10°N	≤1	Early/Late Oligocene	30
North Atlantic (NAIP)	57±1; 55±1	>1.0	65°N	≤1	Paleocene/Eocene (Thanetian/Selandian)	54.8 (57.9)
Deccan	66±1	>2.0	20°S	≤1	Cretaceous/Tertiary (E)	65.0±0.1
Madagascar	88±1	?	45°S	≤6?	Cenomanian/Turonian (Turonian/Coniacian)	93.5±0.2 89±0.5
Rajmahal	116±1	?	50°S	≤2	Aptian/Albian	112.2±1.1
Paranà/Etendeka	132±1	>1.0	40°S	≤1 or ~5?	Jurassic/Cretaceous (Hauterivian/Valanginian)	142±2.6 (132±1.9)
Antarctica*	176±1 or 183±1	>0.5	50–60°S	≤1?	(Aalenian/Bajocian)	(176.5±4)
Karroo	183±1	>2.0	45°S	0.5–1	Early/Middle Jurassic	180.1±4
Central Atlantic (CAMP)	201±1	>1.0	~30°N	?	Triassic/Jurassic (E)	201±0.6
Siberian	249±1	>2.0	45°N?	≤1	Permian/Triassic (E)	248.2±4.8
Emeishan	259±3	0.4	25°N	1–2	End Guadalupian (E)	260.4±0.7

Figure 2. Information about continental flood basalt episodes of the last 300 million years, stratigraphic boundaries, and faunal extinction events. Bold (E) indicates a major extinction event, otherwise stratigraphic boundary coincides with sudden environmental change and/or less severe extinction. MY, million years; MYA, million years ago. NAIP, North Atlantic Igneous Province; CAMP, Central Atlantic Magmatic Province.
*Antarctica (Ferrar Basalts) considered by some to be part of Karroo LIP (after Sanders, 2005). Reproduced by permission of Gale, a part of Cengage Learning.

extinction events occurred at about the same time as LIP events (Wignall 2005). Further, as new age determinations are made, the frequency of coincidence between the two sets of events appears to be increasing. An example is a study of the sills (records of shallow intrusions where the magma was stored before feeding surface lava flows) under the remains of the Karoo province in Africa (part of an extensive LIP that is also found on Antarctica); in a 2012 study, Henrik Svensen and colleagues showed that many of the main sill rocks all have the same age: 183 million years (+/− 1.0 million years). This is synchronous with a major oceanic anoxic event and carbon cycle perturbation that occurred during the Toarcian age of the Jurassic period.

A major question regarding any possible relationship between LIP (flood basalt lava) eruptions and extinction events involves the nature and severity of the environmental effects of the eruptions and their potential impact on life. Further, as mentioned above, flood basalt episodes have been related to the inception of mantle plume activity and thus may represent one facet of a host of geological factors (e.g., changes in seafloor spreading rates, rifting events, increased tectonism [mountain building] and volcanism, sea-

level variations) that tend to be correlated and may be associated with unusual climatic and environmental fluctuations that could lead to significant faunal changes. Although the correlation between some flood basalt episodes and extinctions may implicate volcanism in the extinctions, it is also possible that other factors led to the apparent association.

To examine the hypothesis that LIP volcanism did have an effect on the environment and life on Earth, it is necessary to review the evidence for correlation of the two series of events throughout Earth's history; this evidence is abundant. The discussion below concentrates on extinction events since about 300 million years ago, roughly since the end of the Carboniferous period of geologic time (see Figure 2).

The record of LIP events and extinctions

The frequency of LIP events since the Archean eon (about 2.5 billion years ago) has been estimated at one every 20 million years (Ernst and Buchan 2001), but when the current oceanic LIP record dating back to about 270 million years is also included, the frequency increases to one per 10 million years over that part of geologic time (Coffin and Eldholm

2001; see also Prokoph et al. 2004). Thus, LIP events have been quite common over Earth's history and present a fairly frequently occurring series of events. Regarding the association between LIPs and mass extinctions, however, only those events that formed flood basalts provinces on land masses are considered. However, there is also evidence for correlation of oceanic plateau LIP formation with major environmental crises such as widespread anoxia (oxygen-poor conditions) of ocean waters (Kerr 2005). The record of mass extinctions also partly controls the major stratigraphic boundaries of the geologic record (Gradstein et al. 2004), which, being placed by appearance and disappearance of fossil species and genera, are modulated by periods of mass extinction and major environmental change. The discussion here considers LIP events as a whole, as a complete series of vast eruptions in a short period of time (often about 1 million years) and the correlation of these events with major biotic crises. Later sections examine the details and frequency of LIP eruptions that contribute to a flood basalt province in order to understand what, and (in some cases) how little, is really known about these vast eruptions.

It is both the volume of magma emitted during individual eruptions in LIP events and the total volume of magma released (up to an estimated 7.2 million cubic miles [30 million km^3] in the case of the Ontong Java Plateau) that make LIP events so exceptional in Earth's history. The largest subaerially erupted flood basalt province, the Siberian Traps, might have originally totaled 960,000 to 1.2 million cubic miles (4 to 5 million km^3) of lava (not including the volume of intrusive sills below). It is this combination of large erupted volumes and the high frequency of eruptions that led to the rapid construction of extensive lava plateaus, 0.6 to 1.9 miles (1 to 3 km) in thickness, which internally show few signs of major time breaks, erosion surfaces, and regional unconformities. The high frequency of large-magnitude eruptions also distinguishes LIP events from other tectonic settings and processes in which igneous rocks are formed. Importantly, without LIP-forming igneous events, basalt super-eruptions would not have occurred through Earth history (Bryan et al. 2010).

In some cases a more direct measure of correlation with mass extinction boundaries is possible. For example, the Deccan Traps flood basalts were previously believed to have been erupted over a period of about 10 million years, based on the large range of early (potassium–argon [K–Ar]) age determinations. Re-dating of the basalts using refined ^{40}Ar–^{39}Ar techniques, combined with paleomagnetic studies of the lavas, however, indicates that one exceedingly voluminous part of the thick stack of flows was erupted over a period of only about 500,000 to 800,000 years, bracketing the time of the Cretaceous–Tertiary boundary (Chenet et al. 2009). Also, the K–T boundary impact fallout layer has been reported to lie in sediments interbedded within the Deccan lavas, supporting the radiometric dating and paleomagnetic results that indicate that the eruptions began prior to the K–T boundary; in addition, the main series of lavas also lie on either side of a series of sediments that contain fossil evidence of the K–T die-off in marine species (Keller et al. 2008).

Studies using the ^{40}Ar–^{39}Ar geochronological method (and, to a lesser extent, uranium–lead [U–Pb] age determinations)

have shown that many LIPs erupted and grew over quite brief periods of geologic time (see Kelley 2007 for a review), especially the major, more intense pulses of lava production within a LIP event. Some LIPs show evidence of two or more peaks in output. Not every LIP may have followed this model of formation, and caution is still needed where part of a LIP is poorly known and/or very widespread and buried (a good example is the Siberian Traps; Reichow et al. 2005) or under the ocean and little explored, such as the Ontong Java Plateau. Generally, LIP volcanism appears to last for 1 to 3 million years with the main pulses being as brief as a few hundred thousand years.

There are indications of at least two periods of intense LIP activity in fairly recent Earth history, first during the Carboniferous to Permian and again in the Cretaceous geologic periods. However, the total extent, and thus importance, of older provinces is very difficult to establish because the continuous subduction (destruction) of ocean crust means that as much as four times the present area of ocean floors has been lost in the last 400 million years. Thus, it is highly likely that some older oceanic plateau LIPs may never be detected. By considering only the last 270 million years, it is possible to capture most of the recent oceanic LIPs and almost certainly all the continental LIPs (see Figures 2 and 3). In a 2003 study, Vincent E. Courtillot and Paul R. Renne reviewed the geochronology of LIPs and produced a database of the peak lava eruption times. These authors proposed a strong correlation between LIP ages and the ages of mass extinctions or periods of global oceanic anoxia. They noted, as others had before, that in particular the four most severe mass extinctions events (affecting the greatest numbers of species) in the last 270 million years—the end-Guadalupian (more recently recognized as Capitanian in age; Bond et al. 2010) at 258 million years ago, end-Permian at 251 million years ago, end-Triassic at 200 million years ago, and end-Cretaceous at 65.5 million years ago—were particularly

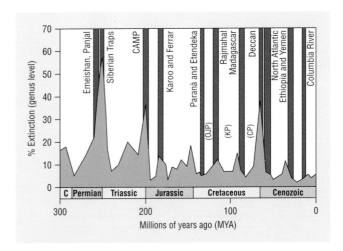

Figure 3. Extinction rate plotted against time (continuous line, blue field, based on marine genera) compared with eruption ages of continental flood basalt provinces over the past 300 million years. CAMP is Central Atlantic Magmatic Province. Three oceanic plateau LIPs are also shown in parentheses; OJP, Ontong–Java Plateau; KP, Kerguelen Plateau; CP, Caribbean Plateau. (After Saunders 2005 and work cited therein.) Reproduced by permission of Gale, a part of Cengage Learning.

closely correlated with the peak of four LIP events: the Emeishan Traps, Siberian Traps, Central Atlantic Magmatic Province, and Deccan Traps. Further, these authors held that close correlations could be found between the ages of all known LIPs and times of sudden environmental change (including mass extinction events), and they interpreted the data as indicating a causal link between LIP eruptions and mass extinctions through catastrophic climatic perturbations. They also concluded that the time sequence of LIPs shows no pattern of eruption with time, an observation confirmed in a subsequent study (Prokoph et al. 2004), but one that diverges from earlier work published in the 1980s recognizing a quasi-periodicity to LIP and mass extinction events.

A slightly different view of the relationship between LIPs and mass extinctions concluded that there is a strong relationship between some LIPs and mass extinctions, but that correlations of all LIPs with extinction events are unduly optimistic (Wignall 2001, 2005), as illustrated on Figure 4. Paul B. Wignall contended in a 2001 article that the onset of the LIP eruptions often postdated the mass extinction and that only the eruption of the Deccan Traps coincided precisely with a mass extinction (at 65.5 MYA). Later work has refined this to show that the Deccan volcanism did start before the main extinction event (Chenet et al. 2009) but that the main extinction of marine fauna coincided with one of the main pulses of Deccan volcanism (Keller et al. 2008).

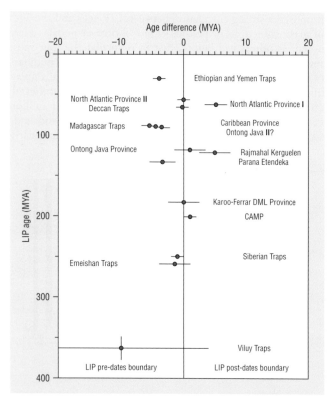

Figure 4. Age difference between a LIP and associated sudden environmental change plotted against age of a stratigraphic boundary associated with sudden environmental change. This is a different way of examining similarities in age than the types of plots shown in Figure 3, with those falling nearest the zero line having the greatest similarity in age. (From Kelley 2007.) Reproduced by permission of Gale, a part of Cengage Learning.

There appears to be no general correlation between the extinction intensity and either the eruptive volume or rapidity of the LIP episode, with the exception of the largest of the pairs of events (as noted above), although there is only about an order of magnitude difference between the largest and smallest subaerial LIPs. In particular, there is no apparent correlation between any significant global environmental event and the eruption of the extensive Paraná-Etendeka flood basalt province, which accompanied the opening of the South Atlantic Ocean, but such puzzling aspects may be explained by some of the considerations raised below.

In summary, over the past approximately 300 million years of Earth's history, there has been a strong association between extinction events and the eruption of LIPs, but proving the nature of the causal links is far from being resolved. Different studies have used different criteria to judge coincidence and reached quite disparate conclusions, from the suggestion that all LIPs correlate with times of sudden environmental change or extinction to conclusions that only about half of the two time series of events are correlated. This is almost certainly more than could be expected by random coincidence, suggesting that there is some cause-and-effect relationship between LIP volcanism and major global changes.

The associated environmental changes at times of LIPs often included global warming and the development of widespread oxygen-poor conditions in the oceans. This implicates a role for volcanic CO_2 emissions, but other perturbations of the global carbon cycle, such as the release of methane from gas hydrate reservoirs or the shutdown of photosynthesis in the oceans, may have been required for severe greenhouse warming to have occurred. The most convincing links between extinction and periods of flood basalt eruptions are seen in the interval from 300 to 150 million years ago. With the exception of the Deccan Traps eruptions (65 MYA), the emplacement of younger (post-mid-Cretaceous age, or after about 90 MYA) LIPs has been generally associated with significant environmental changes but little or no increase in extinction rates above background levels.

In his 2001 article, Wignall also noted that six out of eleven LIP provinces were associated with periods of global warming and marine anoxia or dysoxia but that the associated carbon isotope excursions could not be explained by volcanic CO_2 release alone. Carbon dioxide seems to have been the main gas that caused global warming at the times of mass extinctions and other major environmental crises. This fact raises the profile of studies on an interesting source of CO_2 associated with LIP generation, which is described in the next section.

How might LIP volcanism affect the atmosphere and environment?

It must be noted that, despite the frequent coincidence of flood basalt events with extinctions and other rapid environmental catastrophes, there is no general agreement on how LIP volcanism could have caused these changes. The evidence is still partly circumstantial, but a causal relationship between LIP eruptions and mass extinctions has been ascribed to the release of volcanic SO_2 and, more frequently, CO_2. In support of the suggestion that the associated carbon isotope excursions could not be explained by the release of volcanic CO_2 from the erupted magma alone is the consideration that the magma may not contain sufficient CO_2 to make a significant difference to the atmospheric CO_2 concentration, even with the largest individual LIP eruptions (Self et al. 2006). Previously, carbon isotope excursions (associated with the breakdown of marine and terrestrial ecosystems) thought to be related to methane clathrate release had also been associated with warming resulting from LIP events (e.g., Hesselbo et al. 2000). (Methane breaks down to form CO_2 on release to the atmosphere. Methane clathrate is an icy form of the compound found at cool temperatures and elevated pressures, for example, under the seabed in some regions.) Precisely how much CO_2 must be released over what period of time to explain the carbon isotope excursions and global warming interpreted from the geologic record remains as of 2012 a matter under scientific scrutiny.

Flood basalt eruptions have been suggested to cause several kinds of environmental effects, including climatic cooling from sulfuric acid aerosols, greenhouse warming from CO_2 (and to a much lesser extent SO_2) gases, and acid rain. Indirect environmental effects include changes in ocean chemistry, circulation, and oxygenation, especially from basaltic volcanism associated with large submarine oceanic plateaus. In this case, volcanism such as that which created the Ontong Java Plateau and the Caribbean volcanic province during the Cretaceous period could have altered ocean chemistry through the oxidation of material in hydrothermal effluents. It may also have led to the stimulation of primary productivity in the surface oceans through the release of hydrothermally derived iron. Such a submarine volcanic scenario has been proposed as a cause of global ocean anoxia at times of Cretaceous extinctions (e.g., the Cenomanian–Turonian extinction event, which occurred around 93 MYA). Before assessing these effects, however, it is worth examining the nature of flood basalt volcanism and the amounts of climatically active gases these volcanic events can release into the atmosphere.

The style of flood basalt eruptions

The eruptive style of flood basalt lavas is important in controlling the atmospheric and climatic effects during these events. In fissure-fed basaltic lava flow eruptions, the volumetric eruption rate is a critical parameter in determining the height of fire fountains over active vents, the convective rise of the volcanic gas plumes in the atmosphere, and hence the climatic impact of an eruption. Flood basalt lavas were originally envisioned to flow as turbulent sheets 33 to 330 feet (10 to 100 m) thick that covered large areas in a matter of days. By contrast, it is believed as of 2012 that flood basalts are erupted mainly as fissure-fed pāhoehoe flows that inflate to their great thicknesses during the time that they reach their great extent (Self et al. 1997). This suggests a relatively gradual emplacement of the flows over longer periods of time, most likely years to decades. Yet, the average eruption rates must still have been higher than in any historic eruption except perhaps the extremes of some Icelandic lava outpourings. For example, the time required to emplace a large (240 to 480 cubic miles [1,000 to 2,000 km³]) lava flow at the

range of peak output rates of the largest historical lava flow eruption, the Laki (Iceland) event in AD 1783, would be approximately 10 to 20 years.

The combination of large erupted volumes and apparent high frequency that led to the rapid construction of thick, areally extensive flood basalt provinces is what is exceptional about this style of volcanism. Evidence from the Columbia River LIP suggests that during the main pulse of volcanism, about 16 million years ago, huge eruptions occurred about every 4,300 years on average. A major unanswered question is whether this is frequent enough for the gases released during the sequence of eruptions to have had a cumulative effect on the environment.

Release of carbon dioxide

Greenhouse warming caused by large emissions of CO_2 from flood basalt volcanism has also been suggested as a cause of climatic change leading to mass extinctions. Results of models designed to estimate the effects of this increased CO_2 on climate and ocean chemistry suggest, however, that the total increase in atmospheric CO_2 would have been less than 200 parts per million (ppm), leading to a predicted global warming of less than 3.6°F (2°C) over the period of the eruption of the lavas. It is unlikely that such a small and gradual warming would have been an important factor in mass extinctions. Further, as mentioned above, several studies have suggested that volcanic degassing of CO_2 from the erupting magmas during flood basalt activity would alone be insufficient to generate the amounts of the gas needed to increase the atmospheric abundance indicated by geochemical studies of warming events across geologic time. Several other extinction mechanisms have been suggested as the cause of extinctions and other environmental perturbations resulting from flood basalt volcanism. In the case of the Siberian flood basalts and the Permian–Triassic extinctions, it has been suggested that the volcanism and related intrusions caused the release of vast quantities of methane from hydrates that had accumulated on adjacent high-latitude shelf areas or from carbon-rich sediments under the volcanic pile. These sedimentary deposits were intruded and heated by the sills, leading to gas release and greenhouse warming (Svensen et al. 2009). The suggested mechanism of gas release is through hydrothermal vent complexes that are now being recognized as a characteristic component of LIPs (see Svensen and Jamtveit 2010 for a review).

The geological record shows that abrupt increases in the atmospheric concentration of greenhouse gases have occurred quite a few times during the past few hundreds of millions of years. The release of several thousand gigatons of isotopically light carbon gases from sedimentary basins has been proposed to be the cause of several warm periods, including ones at the Permian–Triassic boundary (c. 251 MYA), in the Toarcian during the Early Jurassic (c. 183 MYA), and in the initial part of the Eocene epoch (c. 55.5 MYA). The resulting climate changes are extensively documented by chemical proxy data from sedimentary rocks, which commonly demonstrate a global warming of 9°F to 18°F (5°C to 10°C) lasting a few hundred thousand years, accompanied by anoxic conditions in the oceans and extinctions. As alluded to above, a series of papers by Svensen and colleagues focusing on the dynamics of LIP emplacement has helped explain how basaltic igneous activity may cause rapid climate change and the release of isotopically light carbon gases. A hypothesis proposed in 2004 by Svensen and his collaborators contends that intrusive magmatic activity (the formation of dikes and sills) into carbon-rich sedimentary basins at the end of the Paleocene epoch (early in the Tertiary geologic period) led to the formation and release of sufficiently large volumes of isotopically light carbon to cause both the global warming and the negative carbon isotope excursion observed during the Paleocene–Eocene thermal maximum. This hypothesis provides a different approach to understanding the relationship between LIPs and global environmental changes, one in which the geological characteristics of the emplacement environment is a crucial factor. Following from these ideas, it has been suggested that those flood basalt provinces where the subvolcanic sills could intrude and heat carbon-bearing sedimentary rocks are associated with extinction events, whereas those that were emplaced into other geologic sequences (barren of carbon-rich sources) were alone unable to cause a dramatic effect on faunal species (Ganino and Arndt 2009). The latter category includes the Deccan province, suggesting that the flood basalt eruptions would not have caused a mass extinction event and that the coincidental meteorite impact was necessary to do so.

The exact relationship between LIP formation and climate perturbations is poorly understood, however, and the geological processes responsible for the rapid formation and transport of greenhouse gases from the huge sedimentary reservoir to the atmosphere have been debated. The role of LIP lava degassing was played down in studies between 2005 and 2011 because of the long duration of lava emplacement and the relatively "heavy" carbon isotopic composition of magmatic CO_2 (e.g., Saunders 2005). The melting of marine gas hydrates or the heating of carbon-rich sediments have been favored climatic triggers (i.e., carbon release mechanisms) among many geoscientists between 2005 and 2011. Other proposed mechanisms bringing about environmental change at times of LIP volcanism are related to cooling resulting from CO_2 drawdown by the weathering of freshly emplaced flood basalts and changes in ocean circulation stemming from submarine LIP construction.

Possible effects of sulfur gas release

Climatic cooling at Earth's surface attributed to volcanic eruptions is primarily a result of the formation and spread of stratospheric sulfuric acid aerosols (reviewed by Robock 2000). These small droplets are formed from sulfur volatiles (largely SO_2) injected into the stratosphere by convective eruption columns and plumes rising above volcanic vents and fissures. The sulfuric acid aerosols have a residence time in the stratosphere of several years, where they backscatter incoming sunlight. Fine volcanic ash also lofted into the stratosphere largely settles out of the atmosphere in less than three months. Basaltic eruptions are commonly more effusive (lava producing) in character, last much longer, and are much less explosive than silicic eruptions of comparable volume. They are commonly considered to be much less effective in lofting volatiles to significant altitudes, where stratospheric aerosols

could be produced. Observations and model calculations suggest, however, that the high eruption rates during the AD 1783 to 17784 Laki basaltic fissure eruption in Iceland produced high fire fountaining (1,970 to 4,750 feet [600 to 1,450 m]), so that the convective plume rise above the fountains could have attained altitudes of up to 8 miles (13 km) above sea level (Thordarson et al. 1996; summarized by Self et al. 2005; see Figure 5). In support of this estimate, observations of the widespread sulfurous haze over Europe in 1783, as well as a marked sulfuric acidity peak in Greenland ice in 1783 and 1784, suggest that the eruption column reached the tropopause over Iceland (at an altitude of 5.6 miles [9 km]) and intruded into the base of the stratosphere. For the much larger volume lava flows typical of flood basalt episodes, the models of convective plume rise also indicate that sulfur-rich eruption plumes could reach the lower stratosphere (Thordarson and Self 1996).

Moreover, basaltic magmas, including those that form flood basalts, are usually rich in dissolved sulfur (commonly with sulfur concentrations of 1,500 ppm); therefore, the release of sulfur-rich gases from a large basaltic eruption can be much higher than that from an explosive silicic eruption of equal volume. For example, studies of the Laki lava flow and ash indicate that the magma originally contained about 1,700 ppm of sulfur, and the eruption could have released about 125 megatons of SO_2 (65% during the first two months of intense activity) from the vent and lava flows. (One megaton is equal to 10^{12} grams.) This could have created about 160 megatons of high-altitude sulfuric acid aerosols and about 50 megatons of local aerosols in the Icelandic troposphere. The atmospheric effects of the Laki eruption were severe and quite widespread, supporting the creation of at least some stratospheric aerosols. Haziness and dimming of sunlight were noticeable in AD 1783 and 1784 in Europe, and so-called dry fog (see Figure 5) was reported in North America and as far east as China. Climatic cooling followed, and the winter of 1783–1784 was the coldest one recorded in the eastern United States from that date forward.

The observed sulfur content of basaltic magmas and the evidence for degassing of the lava flows suggest that the sulfur release from a large flood basalt eruption (such as the Roza flow of the Columbia River basalts) can be about 10,000 megatons of SO_2, along with significant amounts of fluorine and chlorine gases. At the maximum eruption rates for Laki, the Roza eruption would have produced lava fountains more than 0.9 miles (1.5 km) in height and created a convective column rising about 9.3 miles (15 km) above the volcanic vents. If the Roza eruption had continued for 10 years, immense masses of aerosols could have been generated in the atmosphere (greater than 1,000 megatons per year), with a portion going into the lower stratosphere. Geochemical studies of the Deccan lavas that erupted about 65 million years ago also show that the venting of those basaltic magmas would have sent huge masses of sulfur gases into the atmosphere (Self et al. 2008). Amounts might have been large enough to limit the amount of solar radiation reaching Earth's surface over wide regions, and perhaps a whole hemisphere. This could occur by the conversion of SO_2 gas to sulfuric acid aerosols. Some atmospheric scientists have suggested,

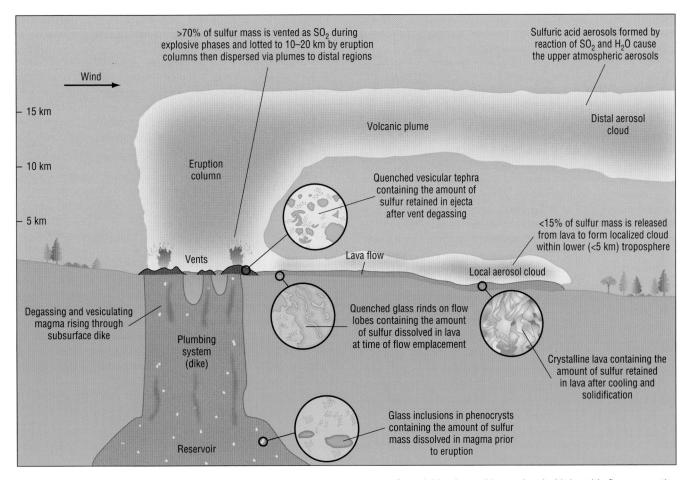

Figure 5. Illustration showing generalized eruptive style and atmospheric dispersal of gas (with minor ash) associated with basaltic fissure eruptive activity such as a flood basalt eruption. This is based on studies of smaller-scale flood lava volcanism, such as Laki 1783–1784. (Unpublished, T. Thordarson and S. Self; see Self et al. 2005 for a more detailed explanation.) Reproduced by permission of Gale, a part of Cengage Learning.

however, that physical and chemical effects in dense aerosol clouds might limit the mass of volcanic aerosols. Further modeling of dense stratospheric aerosol clouds, and their effects on atmospheric dynamics and chemistry, is needed before researchers can predict what climatic changes would happen during a flood basalt eruption. Some factors bearing on this uncertainty are discussed next.

By absorbing and reflecting incoming solar radiation, volcanic aerosols at all altitudes should cool Earth's surface, with maximum effects stemming from low-level tropospheric aerosols. During most volcanic eruptions, however, tropospheric aerosols have a very short lifetime of about a week before they are washed out of the lower atmosphere. Yet, massive gas-release from a long-lasting flood basalt eruption could generate such large amounts of volcanic aerosols that a regional aerosol cloud might be maintained despite this washout, and some of this aerosol mass would be in the stratosphere. Studies using a global climate model coupled with the expected oceanic response suggest that, if maintained for 50 years, a stratospheric loading of only about 10 megatons of aerosols would lead to a global cooling of about 9°F (5°C). Therefore, if flood basalt eruptions were continuous at an average activity level for decades, significant cooling

of climate might be possible. However, with hiatuses of perhaps thousands of years between flood basalt eruptions during the formation of a LIP province, it has not yet been shown whether such dramatic releases of sulfur and other gases would have a cumulative and lasting effect on climate and the environment.

Another possible source of greenhouse warming from flood basalt eruptions is SO_2 gas emissions. During the summer of 1783 in Europe, while the sulfurous dry fog generated by the Laki eruption hung in the air, the weather was stifling. Historical reports of acrid odor, difficult breathing, dry deposition of sulfate, and vegetation damage indicate especially high SO_2 concentrations in the lower atmosphere from mid-June to early August. The July 1783 temperatures in western Europe were up to 5.4°F (3°C) warmer than the long-term average. While it is not certain that these effects were caused by the Laki gases, the warming influence of low-elevation gas plumes downwind from basaltic eruptions is a topic that has not received much attention as of 2012.

Yet another possible environmental factor is acid rain derived not only from the SO_2 but also from chlorine and fluorine gases released from flood basalt eruptions. For the historic Laki eruption, the local atmospheric haze (an aerosol cloud) that hung over Iceland contained high concentrations of sulfuric acid, causing skin lesions in animals and humans. Grass growth was stunted, trees were killed, and 50 percent of livestock perished from fluorine poisoning. The resulting so-called haze famine led to the death of 20 percent of the Icelandic population, and acid rain effects were also reported in northern Europe. Calculations for flood basalt eruptions suggest that they can release as much fluorine and chlorine gas as sulfur (Self et al. 2008; Black et al. 2012). If spread worldwide and if deposited over the course of a year, however, the amount of acid would be quite small. The acidity of rainfall nearer to the flood basalt eruption vents would be higher (lower pH), and rainfall at distant sites would be less acidic (higher pH). Although it seems clear that the local or regional effects of acid rain from flood basalt eruptions could have been considerable, this topic has not been studied in detail as of 2012. The global effects are perhaps less likely to have an important impact on the environment. The deleterious effects of acid rain have certainly played a part in the extinction scenarios proposed for the meteorite impact hypothesis (Kring 2007; Schulte et al. 2010).

Large-scale explosive volcanism and its effects

Explosive volcanic eruptions can produce global aerosol clouds in the stratosphere, and historic explosive events such as Tambora in 1815 and more recently Pinatubo in the Philippines in 1991 have had significant short-term climatic effects on global temperatures (Robock 2000). Thus, as part of this discussion on the possible effects of huge flood-basalt eruptions, it is also worthwhile to consider the environmental effects of the largest-known explosive eruptions, so-called super-eruptions (see Self 2006 for a review), which occur from giant silicic magma bodies.

The largest and youngest explosive eruption of the last few hundred thousand years on Earth was the Toba eruption of Indonesia, which occurred 74,000 years ago and released an astounding 670 cubic miles (2,800 km^3) of silicic magma. Toba occurred during a relatively rapid period of interglacial–glacial cooling and sea-level fall, which may have played a role in triggering the eruption (Rampino and Self 1993). It has been suggested that stratospheric dust and aerosols could have accelerated the cooling by causing a "volcanic winter" with global temperature decreases of 5.4°F to 9°F (3°C to 5°C) and decreases of 18°F (10°C) during the growing season at mid- to high latitudes for a year or more, with possible longer-term climatic effects (Rampino et al. 1988). Ice core studies may indicate that the Toba aerosols remained in the stratosphere for the unusually long time of six years, but this interpretation has been called into question (Oppenheimer 2002), along with much of the perceived climatic impact of the eruption. Botanical studies suggest that very cold-sensitive tropical vegetation would be affected disastrously by abrupt low-latitude cooling. Furthermore, computer model simulations show that an 18°F (10°C) reduction in temperatures during the early growing season can kill about 50 percent of temperate to subarctic evergreen forests and severely damage the surviving trees. Temperate deciduous trees and other less cold-hardy vegetation would be expected to fare even worse.

These results suggest that a global ecological crisis, with reductions in the amounts of plants and animals, could be caused by an eruption such as Toba. However, there is no hard evidence that the Toba eruption produced huge amounts of aerosols. Evidence for such abrupt environmental changes in the aftermath of an explosive super-eruption might be detectable in high-resolution records (e.g., lake and bog sediments and ice cores), but these records have not been adequately explored for the period around the Toba event to rigorously find evidence for or against a major environmental impact.

Although no major extinctions have been discovered at the time of Toba's blast, human genetic studies may indicate that prior to about 50,000 years ago the population of humans suffered a severe bottleneck, followed by rapid population increases, technological innovations, and the spread out of Africa. The predicted environmental and ecological effects of the Toba eruption lend support to a possible connection between that volcanic event and the human population bottleneck, and other faunal and floral changes have also been predicted to have occurred. Again, such interpretations are controversial, and there is no consensus of opinion on the climatic and biotic impact of Earth's youngest and biggest volcanic outburst. When interpreting the effects of flood basalt volcanism occurring tens to hundreds of millions of years ago, considerably longer in the geologic past, it is perhaps no wonder why there is no unified view of the role that LIP volcanism played in causing extinctions and major environmental upheavals.

Summary: Weighing the evidence for LIPs as a cause of major extinction events

The geologically determined ages of the generation of large igneous provinces, especially continental flood basalt events, show a clear correlation in many instances with the ages of mass extinction events, implying that LIP volcanism is capable of causing significant environmental effects. Factors that contribute to the environmental impact of LIP volcanism include rapid temperature changes resulting from greenhouse gas emission from related magma intrusions into carbon-bearing sedimentary rocks; oceanic anoxia events; cooling stemming from CO_2 drawdown through the weathering of basalts; sea-level changes associated with LIP-related topographic adjustments; and the composition of the gases emitted, notably sulfur for short-term and CO_2 for longer-term global change. The overall size of the LIP may not be a dominant factor, although the largest subaerial flood basalt province is correlated with the greatest mass extinction. A key factor is the geological setting—the type of rocks—with which the magma interacted. There is potential for significant volatile release when the intrusive component (sills and dykes) interacts with sedimentary rocks such as evaporites, coal horizons, and, indirectly, methane hydrates (clathrates), which may melt as a result of global warming (a feedback mechanism). An important method of gas release is through hydrothermal vent complexes that are being recognized as of 2012 as a characteristic component of LIPs.

Despite continued debate between 2000 and 2011 on the causal relationships and pace of extinctions during the past 300 million years on Earth, and indeed the fact that mass extinctions may have multiple causes (MacLeod 2005), much work has centered on LIP volcanism and its intrusive components as a major causal mechanism. In the best-known mass extinction, at the end of the Cretaceous period, both a large LIP and a large meteorite impact have been advocated as mechanisms. This, the Cretaceous–Paleogene (K–Pg or K–T) boundary event, remains the only time in Earth's history, according to knowledge in the early 2000s, at which a large asteroid impact, a LIP (the Deccan Traps, formed by a series of large continental flood basalt eruptions), and a well-documented mass extinction coincide.

It is clear that over the last 300 million years or so, up to six of 15 flood-basalt LIPs have peak eruption ages coinciding to within 1 to 2 million years of the dates for sudden environmental change and extinction events. This repeated coincidence has been confirmed by improvement in the precision and accuracy of geochronologically determined age dates on LIP lava sequences, but it has also been demonstrated that there have been large LIP-eruptions for which no marked environmental effects have yet been detected. There is a good case, based on the geochronologic record of the two time series, for a link between LIP volcanism and sudden environmental change (including mass extinctions), although the precise mechanism remains unclear. Direct evidence of causality is extremely difficult to demonstrate, and there are continuing debates on the cause or causes of many of the known mass extinction events.

Resources

Books

Coffin, Millard F., and Olav Eldholm. "Large Igneous Provinces: Progenitors of Some Ophiolites?" In *Mantle Plumes: Their Identification through Time*, edited by Richard E. Ernst and Kenneth L. Buchan, 59–70. Geological Society of America Special Paper 352. Boulder CO: Geological Society of America, 2001.

Courtillot, Vincent E. *Evolutionary Catastrophes: The Science of Mass Extinctions*. Translated by Joe McClinton. New York: Cambridge University Press, 1999.

Courtillot, Vincent E., J.-J. Jaeger, Z. Yang, et al. "The Influence of Continental Flood Basalts on Mass Extinctions: Where Do We Stand?" In *The Cretaceous–Tertiary Event and Other Catastrophes in Earth History*, edited by Graham Ryder, David Fastovsky, and Stefan Gartner, 513–525. Geological Society of America Special Paper 307. Boulder, CO: Geological Society of America, 1996.

Ernst, Richard E., and Kenneth L. Buchan. "Large Mafic Magmatic Events through Time and Links to Mantle-Plume Heads." In *Mantle Plumes: Their Identification through Time*, edited by Richard E. Ernst and Kenneth L. Buchan, 483–575. Geological Society of America Special Paper 352. Boulder, CO: Geological Society of America, 2001.

Gradstein, Felix M., James G. Ogg, and Alan G. Smith, eds. *A Geological Time Scale, 2004*. Cambridge, UK: Cambridge University Press, 2004.

Mahoney, John J., and Millard F. Coffin, eds. *Large Igneous Provinces: Continental, Oceanic, and Planetary Flood Volcanism*. Geophysical Monograph Series 100. Washington, DC: American Geophysical Union, 1997.

Rampino, Michael R., and Stephen Self. "Volcanism and Biotic Extinctions." In *Encyclopedia of Volcanoes*, edited by Haraldur Sigurdsson, 263–269. San Diego, CA: Academic Press, 2000.

Self, Stephen, Thorvaldur Thordarson, and Laszlo P. Keszthelyi. "Emplacement of Continental Flood Basalt Lava Flows." In *Large Igneous Provinces: Continental, Oceanic, and Planetary Flood Volcanism*, edited by John J. Mahoney and Millard F. Coffin, 381–410. Geophysical Monograph Series 100. Washington, DC: American Geophysical Union, 1997.

Tolan, Terry L., Stephen P. Reidel, Marvin H. Beeson, et al. "Revisions to the Estimates of the Areal Extent and Volume of the Columbia River Basalt Group." In *Volcanism and Tectonism in the Columbia River Flood-Basalt Province*, edited by Stephen P. Reidel and Peter R. Hooper, 1–20. Geological Society of America Special Paper 239. Boulder, CO: Geological Society of America, 1989.

Periodicals

Alvarez, Luis W., Walter Alvarez, Frank Asaro, and Helen V. Michel. "Extraterrestrial Cause for the Cretaceous–Tertiary Extinction: Experimental Results and Theoretical Interpretation." *Science* 208, no. 4448 (1980): 1095–1108.

Anderson, Don L. "Large Igneous Provinces, Delamination, and Fertile Mantle." *Elements* 1, no. 5 (2005): 271–275.

Black, Benjamin A., Linda T. Elkins-Tanton, Michael C. Rowe, and Ingrid Ukstins Peate. "Magnitude and Consequences of Volatile Release from the Siberian Traps." *Earth and Planetary Science Letters* 317–318 (2012): 363–373.

Bond, David P.G., Jason Hilton, Paul B. Wignall, et al. "The Middle Permian (Capitanian) Mass Extinction on Land and in the Oceans." *Earth-Science Reviews* 102, nos. 1–2 (2010): 100–116.

Bryan, Scott E. "Silicic Large Igneous Provinces." *Episodes* 30, no. 1 (2007): 20–31. Accessed May 11, 2012. http://www.episodes.co.in/www/backissues/301/20-31%20Bryan.pdf.

Bryan, Scott E., and Richard E. Ernst. "Revised Definition of Large Igneous Provinces (LIPs)." *Earth-Science Reviews* 86, nos. 1–4 (2008): 175–202.

Bryan, Scott E., Ingrid Ukstins Peate, David W. Peate, et al. "The Largest Volcanic Eruptions on Earth." *Earth-Science Reviews* 102, nos. 3–4 (2010): 207–229.

Campbell, Ian H. "Large Igneous Provinces and the Mantle Plume Hypothesis." *Elements* 1, no. 5 (2005): 265–269.

Campbell, Ian H., and Ross W. Griffiths. "Implications of Mantle Plume Structure for the Evolution of Flood Basalts." *Earth and Planetary Science Letters* 99, nos. 1–2 (1990): 79–93.

Chenet, Anne-Lise, Frédéric Fluteau, Vincent E. Courtillot, et al. "Determination of Rapid Deccan Eruptions across the Cretaceous–Tertiary Boundary Using Paleomagnetic Secular Variation: Results from a 1200-m-Thick Section in the

Mahabaleshwar Escarpment." *Journal of Geophysical Research: Solid Earth* 113 (2008): B04101.

Chenet, Anne-Lise, Frédéric Fluteau, Vincent E. Courtillot, et al. "Determination of Rapid Deccan Eruptions across the Cretaceous–Tertiary Boundary Using Paleomagnetic Secular Variation: 2. Constraints from Analysis of Eight New Sections and Synthesis for a 3500-m-Thick Composite Section." *Journal of Geophysical Research: Solid Earth* 114 (2009): B06103.

Coffin, Millard F., and Olav Eldholm. "Large Igneous Provinces: Crustal Structure, Dimensions, and External Consequences." *Review of Geophysics* 32, no. 1 (1994): 1–36.

Courtillot, Vincent E. "Mass Extinctions in the Last 300 Million Years: One Impact and Seven Flood Basalts?" *Israeli Journal of Earth Sciences* 43, nos. 3–4 (1994): 255–266.

Courtillot, Vincent E., G. Féraud, H. Maluski, et al. "Deccan Flood Basalts and the Cretaceous/Tertiary Boundary." *Nature* 333, no. 6176 (1988): 843–846.

Courtillot, Vincent E., and Paul R. Renne. "On the Ages of Flood Basalt Events." *Comptes Rendus Geoscience* 335, no. 1 (2003): 113–140.

Erwin, Douglas H. "The Permo-Triassic Extinction." *Nature* 367, no. 6460 (1994): 231–236.

Ganino, Clément, and Nicholas T. Arndt. "Climate Changes Caused by Degassing of Sediments during the Emplacement of Large Igneous Provinces." *Geology* 37, no. 4 (2009): 323–326.

Hesselbo, Stephen P., Darren R. Gröcke, Hugh C. Jenkyns, et al. "Massive Dissociation of Gas Hydrate during a Jurassic Oceanic Anoxic Event." *Nature* 406, no. 6794 (2000): 392–395.

Jones, Adrian P., G. David Price, Neville J. Price, et al. "Impact Induced Melting and the Development of Large Igneous Provinces." *Earth and Planetary Science Letters* 202, nos. 3–4 (2002): 551–561.

Keller, Gerta. "Biotic Effects of Impact and Volcanism." *Earth and Planetary Science Letters* 215, nos. 1–2 (2003): 249–264.

Keller, Gerta, T. Adatte, S. Gardin, et al. "Main Deccan Volcanism Phase Ends Near the K–T Boundary: Evidence from the Krishna–Godavari Basin, SE India." *Earth and Planetary Science Letters* 268, nos. 3–4 (2008): 293–311.

Kelley, Simon. "The Geochronology of Large Igneous Provinces, Terrestrial Impact Craters, and Their Relationship to Mass Extinctions on Earth." *Journal of the Geological Society* 164, no. 5 (2007): 923–936.

Kerr, Andrew C. "Oceanic LIPs: The Kiss of Death." *Elements* 1, no. 5 (2005): 289–292.

Kerr, Andrew C., and John J. Mahoney. "Oceanic Plateaus: Problematic Plumes, Potential Paradigms." *Chemical Geology* 241, nos. 3–4 (2007): 332–353.

Kring, David. A. "The Chicxulub Impact Event and Its Environmental Consequences at the Cretaceous–Tertiary Boundary." *Palaeogeography, Palaeoclimatology, Palaeoecology* 255, nos. 1–2 (2007): 4–21.

MacLeod, Norman. "Mass Extinction Causality: Statistical Assessment of Multiple-Cause Scenarios." *Russian Geology and Geophysics* 46, no. 9 (2005): 979–987.

Oppenheimer, Clive. "Limited Global Change Due to the Largest Known Quaternary Eruption, Toba ~ 74 kyr BP?" *Quaternary Science Reviews* 21, nos. 14–15 (2002): 1593–1609.

Prokoph, Andreas, Richard E. Ernst, and Kenneth L. Buchan. "Time-Series Analysis of Large Igneous Provinces: 3500 Ma to Present." *Journal of Geology* 112, no. 1 (2004): 1–22.

Rampino, Michael R. "Mass Extinctions of Life and Catastrophic Flood Basalt Volcanism." *Proceedings of the National Academy of Sciences of the United States of America* 107, no. 15 (2010): 6555–6556.

Rampino, Michael R., and Stephen Self. "Climate–Volcanic Feedback and the Toba Eruption of ~74,000 Years Ago." *Quaternary Research* 40, no. 3 (1993): 269–280.

Rampino, Michael R., Stephen Self, and Richard B. Stothers. "Volcanic Winters." *Annual Review of Earth and Planetary Sciences* 16 (1988): 73–99.

Rampino, Michael R., and Richard B. Stothers. "Flood Basalt Volcanism during the Past 250 Million Years." *Science* 241, no. 4866 (1988): 663–668.

Reichow, Marc K., M.S. Pringle, A.I. Al'Mukhamedov, et al. "The Timing and Extent of the Eruption of the Siberian Traps Large Igneous Province: Implications for the End-Permian Environmental Crisis." *Earth and Planetary Science Letters* 277, nos. 1–2 (2009): 9–20.

Reichow, Marc K., Andrew D. Saunders, Rosalind V. White, et al. "Geochemistry and Petrogenesis of Basalts from the West Siberian Basin: An Extension of the Permo-Triassic Siberian Traps, Russia." *Lithos* 79, nos. 3–4 (2005): 425–452.

Renne, Paul R., and Asish R. Basu. "Rapid Eruption of the Siberian Traps Flood Basalts at the Permo-Triassic Boundary." *Science* 253, no. 5016 (1991): 176–179.

Robock, Alan. "Volcanic Eruptions and Climate." *Reviews of Geophysics* 38, no. 2 (2000): 191–219.

Saunders, Andrew D. "Large Igneous Provinces: Origin and Environmental Consequences." *Elements* 1, no. 5 (2005): 259–263.

Schulte, Peter, Laia Alegret, Ignacio Arenillas, et al. "The Chicxulub Asteroid Impact and Mass Extinction at the Cretaceous–Paleogene Boundary." *Science* 327, no. 5970 (2010): 1214–1218.

Self, Stephen. "The Effects and Consequences of Very Large Explosive Volcanic Eruptions." *Philosophical Transactions of the Royal Society* A 364, no. 1845 (2006): 2073–2097.

Self, Stephen, Stephen Blake, Kirti Sharma, et al. "Sulfur and Chlorine in Late Cretaceous Deccan Magmas and Eruptive Gas Release." *Science* 319, no. 5870 (2008): 1654–1657.

Self, Stephen, Anne E. Jay, Mike Widdowson, and Laszlo P. Keszthelyi. "Correlation of the Deccan and Rajahmundry Trap Lavas: Are These the Longest and Largest Lava Flows on Earth?" *Journal of Volcanology and Geothermal Research* 172, nos. 1–2 (2008): 3–19.

Self, Stephen, Laszlo P. Keszthelyi, and Thorvaldur Thordarson. "The Importance of Pāhoehoe." *Annual Review of Earth and Planetary Sciences* 26 (1998): 81–110.

Self, Stephen, Thorvaldur Thordarson, and Mike Widdowson. "Gas Fluxes from Flood Basalt Eruptions." *Elements* 1, no. 5 (2005): 283–287.

Self, Stephen, Mike Widdowson, Thorvaldur Thordarson, and Anne E. Jay. "Volatile Fluxes during Flood Basalt Eruptions and Potential Effects on the Global Environment: A Deccan Perspective." *Earth and Planetary Science Letters* 248, nos. 1–2 (2006): 518–531.

Storey, Michael, Robert A. Duncan, and Christian Tegner. "Timing and Duration of Volcanism in the North Atlantic Igneous Province: Implications for Geodynamics and Links to the Iceland Hotspot." *Chemical Geology* 241, nos. 3–4 (2007): 264–281.

Svensen, Henrik, Fernando Corfu, Stéphane Polteau, et al. "Rapid Magma Emplacement in the Karoo Large Igneous Province." *Earth and Planetary Science Letters* 325–326 (2012): 1–9.

Svensen, Henrik, and Bjørn Jamtveit. "Metamorphic Fluids and Global Environmental Changes." *Elements* 6, no. 3 (2010): 179–182.

Svensen, Henrik, Sverre Planke, Anders Malthe-Sørenssen, et al. "Release of Methane from a Volcanic Basin as a Mechanism for Initial Eocene Global Warming." *Nature* 429, no. 6991 (2004): 542–545.

Svensen, Henrik, Sverre Planke, Alexander G. Polozov, et al. "Siberian Gas Venting and the End-Permian Environmental Crisis." *Earth and Planetary Science Letters* 277, nos. 3–4 (2009): 490–500.

Thordarson, Thorvaldur, and Stephen Self. "Sulfur, Chlorine, and Fluorine Degassing and Atmospheric Loading by the Roza Eruption, Columbia River Basalt Group, Washington, USA." *Journal of Volcanology and Geothermal Research* 74, nos. 1–2 (1996): 49–73.

Thordarson, Thorvaldur, Stephen Self, Níels óskarsson, and Thomas Hulsebosch. "Sulfur, Chlorine, and Fluorine Degassing and Atmospheric Loading by the 1783–1784 AD Laki (Skaftár Fires) Eruption in Iceland." *Bulletin of Volcanology* 58, nos. 2–3 (1996): 205–225.

White, Rosalind V., and Andrew D. Saunders. "Volcanism, Impact, and Mass Extinctions: Incredible or Credible Coincidences?" *Lithos* 79, nos. 3–4 (2005): 299–316.

Whiteside, Jessica H., Paul E. Olsen, Dennis V. Kent, et al. "Synchrony between the Central Atlantic Magmatic Province and the Triassic–Jurassic Mass-Extinction Event?" *Palaeogeography, Palaeoclimatology, Palaeoecology* 244, nos. 1–4 (2007): 345–367.

Wignall, Paul B. "Large Igneous Provinces and Mass Extinctions." *Earth-Science Reviews* 53, nos. 1–2 (2001): 1–33.

Wignall, Paul B. "The Link between Large Igneous Province Eruptions and Mass Extinctions." *Elements* 1, no. 5 (2005): 293–297.

Other

International Association of Volcanology and Chemistry of the Earth's Interior. "Large Igneous Provinces Commission." Accessed May 12, 2012. http://www.largeigneousprovinces. org.

Saunders, Andrew D., and Marc K. Reichow. "Flood Basalts and Mass Extinctions." The Siberian Traps. Last modified March 6, 2009. http://www.le.ac.uk/gl/ads/SiberianTraps/FBandME. html.

Geological Survey of India, Deccan Basalt volcanism. Accessed May 23 2012. www.portal.gsi.gov.in/pls/portal/url/page/ .../GSI_STAT_DECCAN.

Stephen Self

Sea-level change

Sea-level change has been postulated as a significant possible cause of extinction at a variety of spatiotemporal scales. Sea-level rise is one of the major direct changes expected from the current increase in global mean annual temperatures, but this will not result in uniform, synchronous changes in water depth around the globe. In the recent past (20,000 years ago) sea levels were over 330 feet (100 m), lower at the Last Glacial Maximum as glaciers have resulting in the draining of the continental shelves and previously isolated landmasses becoming single areas. The reversal of this trend resulted in once connected landmasses, such as Australia and Tasmania or Ireland and Great Britain being separated by water once again. These are only the latest manifestations of a process that has gone on for the whole of the history of life. While images of the changes brought about by a "drowned world" are easy to visualize, the effects of sea-level change are multifaceted and can operate through many indirect mechanisms to bring about extirpation and extinction in both the terrestrial and marine realms.

The daily tides offer an accessible means of thinking about the physical and biological effects of sea-level change, as they capture many phenomena that are useful analogies. Tidal rise and fall is measured relative to sea level, which is not uniform across the globe. Some areas are inundated by the tide more quickly than others, with the variability depending on the topography of the coastline. Some areas are cut off by tides, leaving them as islands part of the time. Extremely high, or spring, tides can flood areas with salt water, with severe consequences. This drowned-world analogy, however, operates at a much smaller spatiotemporal scale. Sea-level change is a longer-term process with more facets. The first part of this entry focuses on the earth-systems science of sea-level change and how the geological record provides direct evidence about sea-level rise and fall. The second part of the entry focuses on how these changes can drive extinction.

Elements of sea-level change

When dealing with sea-level change it is important to separate the various elements and be aware that sea-level change has a mixture of local, regional, and global components. Also important are the different reference frames used to discuss sea-level change (see Figure 1).

Eustatic sea-level change

Eustatic sea-level change is measured relative to a fixed point, often the center of Earth, and is often related to the growth and shrinking of glaciers that are on the land surface; thus, the phrase "glacio-eustatic sea-level change" is often used. The other major influence on eustatic sea level is change in the volume of the ocean basins, which is controlled by the production and destruction of ocean crust (tectono-eustasy). The amalgamation of continental plates to form supercontinents such as Pangaea results in deeper ocean basins. Numerous attempts have been made to track eustasy through time, and these sea-level curves have been an important factor in extinction studies (see Figure 2).

Relative sea-level change

The datum for relative sea-level change measurement is a fixed point in a sedimentary sequence and gives the distance from the sea surface to that fixed point. Whereas eustasy influences relative sea level, a number of other factors can influence relative sea level at regional and local scales, including local uplift and subsidence within a sedimentary basin, and it is thus more useful for discussing sea-level change in individual depositional basins.

Water depth

Water depth is not the same as relative sea level, as water depth is measured from the sea surface to the seabed. Water depth can be affected by factors such as sediment input, even if a basin is not subsiding and eustatic change is zero. More sediment input means the basin will fill up, reducing water depth but not relative sea level. Conversely, if basin subsidence matches sediment input, the relative sea level increases, but the water depth remains constant. The balance between the amount of sediment input and the generation of space within the basin to hold the sediment is called *accommodation space*.

The observable effects of sea-level change

In the early twenty-first century, perhaps the most dramatic and well-publicized element of sea-level change is the real threat to a number of Pacific island nations. Already

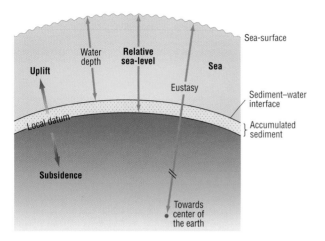

Figure 1. Diagram showing the scope of different terms and concepts used in the discussion of sea level change. Reproduced by permission of Gale, a part of Cengage Learning.

two uninhabited islands in the central Pacific have sunk beneath the waves ("Are Arctic Sea Ice Melts Causing" 2008). The threats to Tuvalu (an island nation lying between Hawaii and Australia) from rising sea level, and the threat of legal action against other states, have generated considerable media coverage. Tuvalu has an average elevation of 3.3 feet (2 m) above sea level and could be mostly overwhelmed in 50 to 100 years. As well as the loss of nonhuman biodiversity in such situations, the loss of cultural and linguistic diversity in such island communities should not be overlooked. The dispersal of people among the islands of the Pacific provides a fine study of the effects of the dispersal of small founder groups among islands and the plight of Tuvalu's inhabitants is

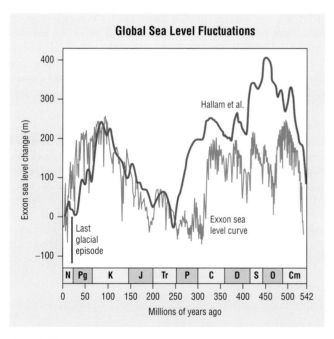

Figure 2. Two curves showing the trajectory of sea level over Phanerozoic time. The brown curve shows less detail, while the blue curve shows more short-term variation, giving it a saw-tooth appearance. Reproduced by permission of Gale, a part of Cengage Learning.

captured by the following famous quote from Hugh MacDiarmid's 1932 poem "On a Raised Beach":

What happens to us
Is irrelevant to the world's geology
But what happens to the world's geology
Is not irrelevant to us.
We must reconcile ourselves to the stones,
Not the stones to us. (MacDiarmid 1932. p. 69)

Tuvalu consists of nine islands, six of which are atolls. The atolls are a product of the dynamic process whereby volcanic islands subside and eventually become subsea mountains (seamounts), which is a necessary step in the pathway by which fringing coral reefs emerge. In the case of atolls, the volcano top is submerged completely, and only the reef remains as a land surface.

Corals, and other photosynthetic organisms, perform a range of ecosystem services, but a key, limiting factor to the persistence of these communities is light penetration. Sea-level change can influence light penetration in the following ways:

1. The water column itself attenuates the penetration of sunlight, which is a vital part of photosynthesis. If the sea level rise outstrips the growth rate of reefs, then photosynthesis will begin to shut down on the reefs, which can affect many other organisms that inhabit the reef.

2. If changes in sea level result in increased sediment or nutrient fluxes, then corals can become threatened. Increased sediment loads in the water reduce the depth to which light penetrates, while increased nutrient fluxes can result in eutrophication, blooms in the phytoplankton population, which thrive on the extra nutrients. However, this positive outcome for the phytoplankton results in the depletion of oxygen in the ecosystem, which can result in mass deaths of other organisms such as fish and clams. Reef-building corals tend to favor locations with low nutrient inputs, and part of their evolutionary success stems from their ability to recycle nutrients many times within the ecosystem, a property they share with tropical rain forests.

While corals are an exemplar of the effects of the direct and indirect influence of sea-level change, many other organisms are vulnerable to other changes in their environment that are driven by sea level variation. While it may be difficult to understand how sea-level change affects terrestrial organisms initially—other than by direct drowning of the land—a range of other physiochemical processes that can threaten terrestrial organisms are mediated by sea level.

Water table changes

The water table is the height at which water is encountered in the subsurface. It can be thought of as a plane that intersects the topography. On land, the situations where the water table intersects with the topographic surface can result in springs, lakes, and other bodies of standing water. Sea-level plays an important role in determining the height of the water table in

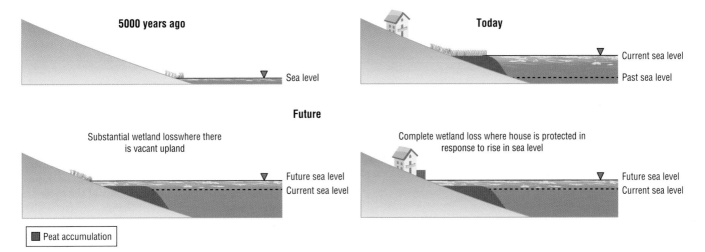

Figure 3. A diagram showing the problems that can occur when sea level change occurs at a rate greater than that at which the wetland ecosystem can adapt, resulting in both the loss of the wetland community and threats to human settlement. Reproduced by permission of Gale, a part of Cengage Learning.

coastal areas. This relationship can influence the extent and quality of certain habitat types profoundly. Many wetland areas provide vital habitats for both permanent residents and migratory species such as birds. Wetlands are not inevitably destroyed by sea-level change. In a 1990 article, James G. Titus presented a scenario in which wetlands can keep pace with relatively slow sea-level change. But if the rate of change becomes too great, then the wetlands will be drowned, not only immediately affecting taxa that cannot disperse to other sites but also influencing other taxa more indirectly (see Figure 3). Geophysical systems, like species, need space and time to evolve, and if the rate of change is too great, they cannot adapt with sufficient speed.

Salinity changes

The zone where oceans and freshwater come into contact is clearly expressed at the surface in the form of river estuaries and delta systems, but there is also an interface in the subsurface. One physical consequence of a rise in sea level can be that the amount of marine, saline water in the estuary will increase, while the input of freshwater from rivers remains fixed. The most extreme outcome of this process is for a thin upper layer of less dense freshwater to develop on top of a much larger body of more saline waters, an outcome that results from density differences. This condition can result in a decrease in the amount of oxygen reaching the lower parts of the water column, eventually leading to an almost complete lack of oxygen at the seabed. The Baltic Sea is an example of such a system, and almost no bottom-dwelling organisms are known from the Baltic.

The ingress of higher salinity waters can allow marine species to invade estuaries, and species with a wide tolerance for salinity can enter communities where they may outcompete incumbent taxa by direct (e.g., space, food) or indirect mechanisms (e.g., higher reproduction rate in stressed conditions). Many species that are tolerant of a wide variety of conditions (i.e., generalists) are often successful invaders,

especially into disturbed ecosystems. Another important influence of generalists, however, is that they can prevent further adaptation and evolution of specialist populations at the point where the two taxa come into contact; if adaptation is constrained, then the extirpation or extinction of the specialist populations becomes more likely.

The potential consequences of changes in the position of the contact zone between fresh and saline waters are not so obvious in the subsurface, as the effects are much less visible. To return to Tuvalu, the islanders are already finding that their freshwater wells are being infiltrated by saline water, making the well water undrinkable (see Figure 4). Other terrestrial animals and plants have far less ability to overcome such changes through engineering or movement to more favorable sites. Terrestrial vegetation, which is the foundation of most terrestrial ecosystems, is often the first element of such ecosystems to be threatened. Even if the increased salinity is not apparent at the surface, plants with root systems can easily be threatened by the uptake of additional salt, which can alter the composition of communities and threaten the specialist host plants on which other species rely, especially in the tropics.

Alteration of river profiles

Change in sea level can result in changes in the profile of rivers. Many rivers have experienced shifts in their *knickpoints*—places in the profile of the river where there are marked steps, often expressed as waterfalls or gorges, in an otherwise smooth, cross-sectional curve of the river from the hills to the sea. Sea-level rise changes the profile by making the gradient shallower, which leads to less energy in the river system, whereas sea-level fall increases the energy of the river system by causing the river gradient to steepen, which causes the current knickpoints to be breached and moves the new knickpoints upstream. These changes in the dynamics of the river, which are also influenced by the rock types the river cuts through, can influence the relative quality, variety, and

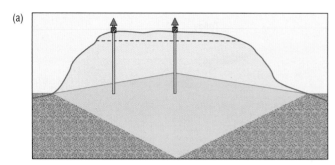

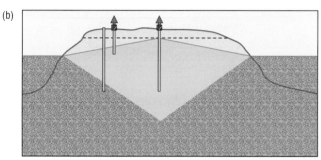

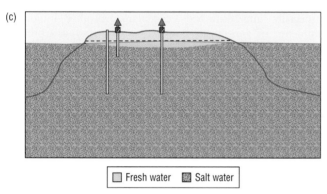

☐ Fresh water   ▨ Salt water

Figure 4. A representation of the effects of sea level rise on the availabiity of fresh water. As sea level rises, the zone of freshwater decreases. Wells have to be abandoned as they become tainted and finally only a small lens of fresh water is left. While this focuses on human use of the water, plants can also be badly affected as the pore waters change salinity. Reproduced by permission of Gale, a part of Cengage Learning.

distribution of habitats within a river system, such as the proportion of rocky to sandy/muddy bottom and the flow rate in different parts of the river. In the worst-case scenario, these changes can cut off populations in unfavorable habitats and deny them the chance to migrate to other rivers.

Impacts on biogeochemistry mediated by sea-level change

The final impact of sea-level change to be detailed is perhaps the least obvious, as it is seldom visible, but it can also be the first part of the system to be altered by sea-level change. Biogeochemistry studies the impacts on the environment of the interaction of geological, physical, and biological processes. The most important effects for life can be found in the various cycles of elements (e.g., global carbon cycle) and certain trace elements, such as iron, which can have profound

impacts on organisms. All of the cycles can be considered to have various sources, sinks, and reservoirs for the elements under consideration, and shifts in the amounts of the element in different parts of the system can have profound effects on organisms, leading to extinction.

Global carbon cycle

Life on Earth is based on carbon, and this is why carbon is of such importance in biogeochemistry and why the global carbon cycle is often regarded as the central cycle in biogeochemistry (Schlesinger 1991; see Figure 5). The possible effects of the burning of fossil fuels on global mean average temperature have been discussed widely, but other elements of the cycle are capable of being altered by sea-level change. In particular, falling sea levels can promote the erosion of soils, which sequester carbon from dead vegetation on land, as rivers strive to regain their equilibrium profile. In addition to the increased sediment flux and decrease in light penetration, additional carbon can be carried to the ocean reservoir. This action can result in the depletion of oxygen in shelf waters to the point at which oxygen-breathing organisms can no longer survive in an area for a range of biological reasons.

The other significant impact of increases in the amount of carbon in the atmospheric pool is that the excess carbon gets absorbed into the ocean. This change can result in the lowering of the pH of ocean waters, which can in turn make it harder for organisms that build shells or structures of calcium carbonate (e.g., corals, clams, cuttlefish) to manufacture these parts.

One important mechanism for mass extinctions that could be triggered, in part, by sea-level change is the release of massive amounts of methane gas (CH_4) held on the continental shelves in gas hydrates or clathrates. A *clathrate* is a structure in which one molecule traps a second molecule inside a framework. In hydrates, water molecules have trapped gas molecules, with the result that changes in the physical state of the water molecules in response to changes in temperature and pressure can release the trapped gas. One mechanism by which sea-level change might directly cause major extinction events is by a sufficient drop in sea level, which would reduce the pressure on methane clathrates. This effect would act to both reduce the confining pressures and increase temperature to the point at which water changes phase from solid to liquid, releasing the methane, which has a tremendous potential to increase global temperatures. A smaller sea-level change might alter these factors enough to allow other factors, such as the failure of sediments in submarine landslides or earthquakes, to release the remaining confining pressure and complete the degassing process. The effects of methane release would depend, in part, on the prevailing atmospheric conditions. Clathrates certainly exist in vulnerable positions on the continental shelf and have formed the basis of the so-called clathrate gun hypothesis (Ryskin 2003).

Global nitrogen and phosphorus

Nitrogen and phosphorus are critical for life, as both elements are required for the manufacture of molecules vital for life (including DNA, RNA, enzymes to help catalyze reactions at temperatures suitable for life, and ATP for energy). As such, the availability of nitrogen and phosphorus in forms that can be used

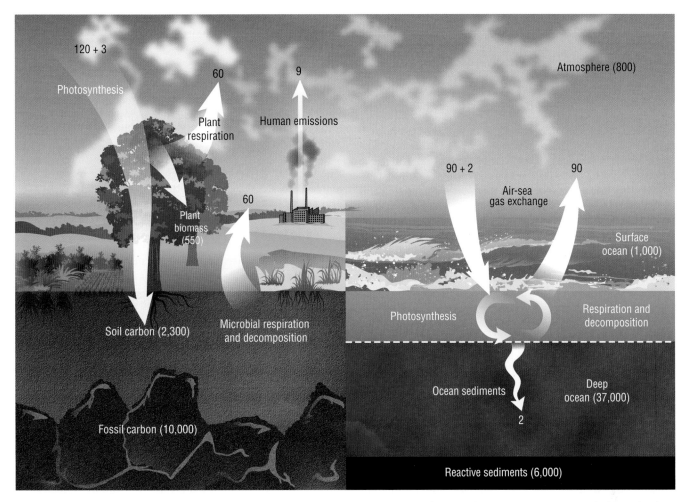

Figure 5. The global carbon cycle, one of a number of biogeochemical cycles. The arrows pointing upwards represent sources, or emissions, of carbon. The arrows pointing downwards go towards sinks, or storage mechanisms, where the carbon is "locked up." Reproduced by permission of Gale, a part of Cengage Learning.

by organisms is a key limiting factor on life. An important element in the global phosphorus cycle is the need for marine sedimentary rocks containing phosphorus from the weathering of continental rocks to be uplifted to allow them to be weathered again. Intervals of high sea level can reduce the supply of phosphorus to the terrestrial realm, and it is possible for this to happen on more local scales.

Global sulfur

The global sulfur cycle is an instance of a system that is controlled by the actions of sulfur-reducing bacteria in a relatively small area on Earth. Most of the sulfate reduction by bacteria takes place in wetlands and shallow intertidal sites (Cárdenas and Harries 2010). The discussion above about the impact of sea-level change on wetlands should make it clear that these biogeochemical factories are very vulnerable to shifts in sea level.

Habitats most vulnerable to sea-level change

While the preceding sections detail some of the broader impacts on Earth systems, including the biosphere, in the

shorter term, sea-level changes are most likely to affect the littoral zone and the zones immediately adjacent seaward and landward. The coastline is a dynamic system, and human engineering can often have unintended consequences on the coastal environment. Beach starvation can result from sea defenses, which can strip sediment supplies to other parts of the coast, destroying habitats. As sea-level rise becomes more pressing, the question of whether to defend human habitations and infrastructure becomes loaded with choices. One concept to mitigate sea-level rise is to allow managed retreat. As discussed above, however, rising sea levels can threaten terrestrial biodiversity in a number of ways, so even environmentally friendly measures are not without extinction risks.

Wetland environments are often among the most threatened, as their existence is dependent on a number of dynamic factors. The 1971 Ramsar Convention (officially known as the Convention on Wetlands of International Importance), along with its amendments, requires signatories to protect wetlands, and if an area needs to be altered in such a way that it will cease to be a wetland, then an equivalent new area of habitat is supposed to be created. While these are laudable aims, this convention does not take account of site-specific and site-fidelity driven processes. For instance, migrating birds will often return to the same estuary time and time again. Wetland environments can stretch for considerable distances on coasts with a low slope and can have an important buffering role both in terms of the physical effects of storms and biogeochemically (namely, the removal of pollutants).

In a 2005 paper, Marianne R. Fish and coworkers used detailed analysis of the topography of beaches that host sea-turtle nesting grounds in the Caribbean to show that the beaches could be lost with a sea level increase of only around 1.6 feet (0.5 m). This eventuality represents a potential loss of a vital link in the reproductive cycle—a loss that would commit a species or population to extinction, but without the obvious signs of mass deaths among adults. Such threats, by their indirect and less visible nature, are harder to monitor and predict than obvious cases in which adults are dying in large numbers, although the establishment of marine protected areas has often focused on spawning or nursery areas.

Extinction in the fossil record

Sea level has fluctuated by 10 to 1,300 feet (3 to 400 m) during the history of life. As noted above, one must be cautious about invoking the notion of global mean sea level change, as this will be mediated by a range of other factors to produce local changes in water depth, which often have a more direct impact on life.

Instances of direct effects

The flooding of epicontinental seas can have major regional effects on connectivity among continental areas. The formation of the Strait of Dover is one good example (Gibbard 1995). For much of the Pleistocene there was a chalk barrier between the English Channel and the southern North Sea, forming a land bridge. An ice-dammed lake formed when waters were not able to flow into the Atlantic. This barrier failed and the lake drained catastrophically, and major rivers were diverted into the area. The breach would have destroyed the terrestrial ecosystems present, and the evidence of the drowned fauna has emerged in trawler nets in the form of mammoth teeth and other vertebrate remains.

Perched basins

Perched, or silled, basins comprise another instance of the clear-cut effects of sea level change in the geological past. When sea level is sufficiently high, the basin is part of the ocean, albeit with a shallow bar that can act as a barrier to the spread of some taxa. As sea level drops, a situation develops in which the basin may receive some recharge but the salinity may increase as a result of evaporation (see Figure 6). Once the marine connection is severed, the basin is then doomed to eventually dry out. A good example of this occurred during the Triassic in the form of the Muschelkalk Basin of central Europe, which saw different phases of faunal immigration from the Tethys Ocean via a set of gateways. Extensive salt deposits from the Middle Muschelkalk provide evidence of the withdrawal of the sea from much of the basin, dooming its marine inhabitants. Even when marine conditions prevailed,

the environments were probably stressed, as shown by studies of body size and community structure (e.g., McGowan et al. 2009).

Low oxygen

Sea-level highs often occur when continental mountains have been worn down by weathering, which limits the removal of carbon dioxide from the atmosphere by weathering. This results in increasing temperatures, and much of the continental shelf becomes flooded and many more shallow seas appear. Although this might appear to be increasing the area of ocean available for marine organisms, these oceans can in fact be stratified. This means that, unlike the present ocean system, there is only a thin surface layer that is oxygenated, whereas the rest of the water column and the sediments of the seafloor are in conditions that lack oxygen. This condition will tend to make the area unsuitable for many organisms. Records of such events are routinely found in the geological record in black mudrocks, which are fine-grained sediments that contain no fossils of bottom-dwelling organisms and are often filled with minerals such as pyrite that are produced only in anoxic conditions.

Sea level and the species–area effect

The most general driving mechanism for extinction in the geological record by sea-level change is through the species–area effect. This effect, first noticed by and developed in theoretical form by Robert H. MacArthur and Edward O. Wilson (authors of *The Theory of Island Biogeography* [1967]), has been an extremely important way of thinking about the relationship between area and extinction. Simply put, the larger the area sampled, the greater the number of species, or taxa of interest at a particular taxonomic level, found (see Figure 7). Although the process(es) behind this pattern are still debated, it is one of the few relationships in ecology that comes close to the status of a law. All units of area, however, are not equally favorable for all species. So in addition to simply considering area, one must consider the habitable area for a species. The quality of a habitat can be degraded, so that the number of individuals in the population may be below what it would be if the habitat was optimal, and shifts in abundance can often be the first event in the chain leading to local or global extinction.

Related to this, then, is the minimum contiguous area required for a particular species to maintain a population in an area. Considerably less is known about this in the marine realm than is known about the terrestrial realm, and even so there is a strong bias in scientists' understanding toward birds and mammals, which tend to have very different life histories and ecologies than most species.

To return to island biogeography and the terrestrial realm, islands often act as natural laboratories for ecological and evolutionary experiments. Archipelagoes such as Indonesia, and their equivalents in the geological past, exhibit other features. Island biogeography postulated a mixture of speciation, extinction, and immigration to explain the community assemblages and levels of diversity on different islands. It also considered different arrangements of islands in space. For

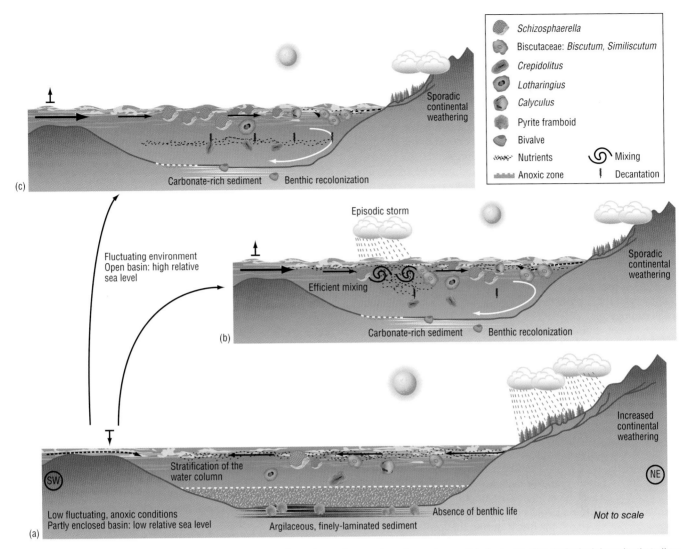

Figure 6. The development of a perched basin and how this affects the environments and organisms within the basin. The geological deposits that allow the inference of changing environmental conditions are shown forming on the sea floor. Reproduced by permission of Gale, a part of Cengage Learning.

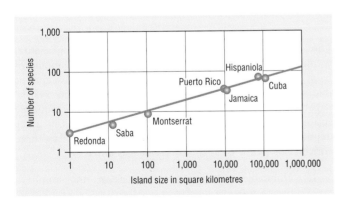

Figure 7. An example of the species–area relationship. The close fit of the line to these data points indicates that the area of each island is a very good predictor of the number of species to be found on the island. This relationship is so widely found it is often described as one of the few "laws" in ecology. Sea level change can be thought of moving the habitable area back and forth along the line, resulting in predictable increase or decrease in species richness. Reproduced by permission of Gale, a part of Cengage Learning.

instance, large islands close to a mainland coast are likely to receive many more immigrants and potential invaders than islands at the end of a chain. Long, linear chains of islands will exhibit declines in richness along the chain, whereas islands in a more "gridded" pattern will tend to show a less marked drop-off in richness.

The major fluctuations in sea level have provided an obvious means of scaling this effect up to continental and global levels in the geological past. Sea-level fluctuations leave clear traces in the sedimentary rock record. One of the elementary rules of understanding vertical sequences of sedimentary rocks is called Walther's law, which states that an idealized vertical sequence of rocks records a succession of adjacent lateral environments. One of the best examples of this are the glacio-eustatic cycles of the Carboniferous period that record the repeated land–sea transitions driven by glacio-eustatic sea-level change. Sea-level change leaves physical traces in the same record that contains the fossils that are the primary evidence of biodiversity changes, so it is no surprise that many authors have advocated sea-level as a primary control on biodiversity and have invoked it to explain mass extinctions (Simberloff 1974; Hallam 1992; Rosenzweig 1995).

The case is not so clear cut, however. Many geologists now think of sedimentary rock units as being the product of cycles bounded by intervals of nondeposition that represent the beginning and end of cycles (Emery and Myers 1996). During the different systems tracts in a basin, different amounts of rock, which represent different habitats, are available. During a low-stand tract, many rivers cut directly down to the sea, and the deeper water facies are not well represented. As relative sea level starts to rise, then the influx and increase of marine areas is accompanied by a substantial increase in marine biodiversity, both in single clades and across communities. A high-stand tract often witnesses a decrease in diversity, despite the possible climax community situation. The return to a low-stand tract can often involve down-cutting, which may be a source of bias in the systems as previous deposits of fossils are eroded and destroyed. In a 1998 article, Carlton E. Brett provides a background to these effects along with a succinct summary of the interdependence of sequence stratigraphy and paleoecology as well as an excellent illustration of the succession of different faunas as the sea level changes in a cyclical fashion (see Figure 8).

This framework has been an influential model for studying the relationships among the sequence stratigraphic cycles, taxonomic richness, and community composition. Computer simulations by Steven M. Holland (1995) and Holland and Mark E. Patzkowsky (1999) have demonstrated that patterns observed in the field, where originations were clustered at the base of cycles and extinctions at the top, could be the result of control by the rock record, rather than real events. Thus, the possibility is raised that the evidence for extinction in the rocks could, in fact, be evidence of sampling biases driven by changes in the different rocks types or the limited amount of rock available to search for fossils during the low-stand tracts.

Sea level and bias/common cause

A topic that bears clear relation to sea-level change and extinction is the debate over the common cause hypotheses of Shanan E. Peters (2005). Peters divided the possible mechanisms driving the observed correlation between the amount of rock available to search for fossils (rock availability) and the observed taxonomic richness into the following three possible scenarios:

1. Fluctuations in rock availability drive short-term fluctuations in biodiversity by increasing or decreasing the amount of rock available to sample as sea level rises and falls (the position of Andrew B. Smith [2001]).

2. Biodiversity really does increase, via the species–area effect.

3. The correlation between rock availability and biodiversity is driven by a third variable, sea level, as revealed by the correlation.

In a 2008 article, Peters made a strong case that a species–area effect was at work throughout the Phanerozoic. It is not always the case, however, that rising sea levels will increase the amount of habitat area and in turn boost taxonomic richness. In a 1995 paper, A.R. Wyatt demonstrated that during the

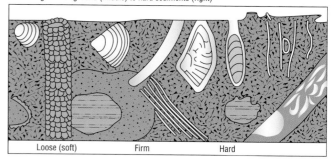

(a) Spectrum of types of burrows and traces ranging those typical of soft-sediments (left) through firmer ground (middle) to hard sediments (right)

Loose (soft)   Firm   Hard

(b) Facies change

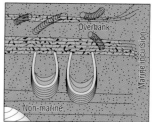

(c) Salinity change

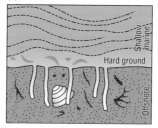

(d) Omission surface

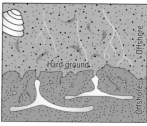

(e) Omission surface

Figure 8. Using fossil shells and burrows to track changes in the environment: a) an array of fossils found in soft sediments on the left to hard sediments on the right; b) the transition recorded here is a shift from shallower water at the bottom of the panel to deeper water at the top; c) a sort interval of marine conditions is recorded by U-shaped burrows in the middle of the panel; d) a change from deeper water to shallow, marked by line of shells; e) no change in depth but sedimentation pauses. Reproduced by permission of Gale, a part of Cengage Learning.

Late Ordovician the fall in sea levels caused by glaciation would have led to an increase, not a decrease, in the amount of habitable shelf area. Thus, there are likely to be exceptions that require further study. Steven M. Holland, in a 2012 paper, took his work in the fossil record on sequence stratigraphy and used it to forecast future changes in water depth and the likely implications for marine biotas. His three conclusions below echo the thrust of this entry.

1. Habitable area in shallow seas is not simply a function of sea level.

2. The same amount of change in sea level will produce very different effects on different coasts.

3. Even if a species–area relationship controls species richness, different parts of the world will follow different trajectories in terms of species gains and losses.

An important aspect of sea-level change in a regional context is the relative proportion of different elements of the sea. Much of the biodiversity that humanity can sample and exploit lies in the continental shelf seas, sometimes referred to as epicontinental seas, as the waters have inundated the continental shelf. This is the zone in which interactions between the sea and land are most pronounced.

Sea-level change and changes in provinciality

At the largest scale, biomes or provinces have underpinned scientists' ideas of major increases and decreases in taxonomic richness (Valentine 1971). The relative position of the continents and the amount of connectivity among ocean bodies are important factors in these models. Under these models, events that bring together continents can place previously isolated faunas and floras into environments with increased competition, which can cause extinction through direct competition, although other factors such as disease may also be involved. Within regions it is possible to have similar events at smaller scales.

A good example of the effects of isolation is the British Isles during the Quaternary, as detailed in a 1995 book edited by Richard C. Preece. Individual chapters of the book detail the exclusion of some taxa when sea levels finally rose but also make the point that the geography of the British Isles as an archipelago means that proportionately more extinction has occurred than would be expected on a single landmass of equivalent area. Other effects include the reduction in body size among red deer (*Cervus elaphus*) and the reduced diversity of the British Isles versus continental Europe and even Great Britain versus Ireland. These examples provide further evidence that it is not simply the amount of available habitat that influences extinction but how the habitat is connected and fragmented. Rising sea levels, while increasing marine habitats, can act to decrease connectivity and habitable area in the terrestrial realm.

While a biblical flood may be the mental picture that many people have, the effects of sea-level change are wider ranging than this and can be as subtle, yet as devastating, as shifts in the biogeochemical cycles that lead to limited supplies of vital chemical elements and can be as consequential as more obvious events such as the loss of habitats. The potential for anything from extirpations to full-scale extinction is very spatially dependent, and it can be difficult to predict the outcome of a given amount of sea-level change because the responses of the physical habitat and the various species that inhabit it are dynamic processes that are not yet understood fully, which makes the job of conserving littoral habitats and their species one of the most challenging tasks facing conservation biologists. Across geological time, the headline trend is for biodiversity to be positively correlated with eustatic sea-level change, but it is often harder to predict the outcomes of eustatic sea-level changes in a given region or for a particular taxon.

Resources

Books

Coe, Angela L., ed. *The Sedimentary Record of Sea-Level Change.* Cambridge, U.K.: Cambridge University Press, 2003.

Emery, Dominic, and Keith J. Myers. *Sequence Stratigraphy.* Oxford: Blackwell Science, 1996.

Gibbard, P.L. "The Formation of the Strait of Dover." In *Island Britain: A Quaternary Perspective*, edited by Richard C. Preece, 15–26. Geological Society Special Publication 96. London: Geological Society, 1995.

Hallam, Anthony. *Phanerozoic Sea-Level Changes.* New York: Columbia University Press, 1992.

MacArthur, Robert H., and Edward O. Wilson. *The Theory of Island Biogeography.* Princeton, NJ: Princeton University Press, 1967.

MacDiarmid, Hugh. "On a Raised Beach." *Stony Limits and other Poems.* London: Victor Gollancz, 1934.

Preece, Richard C., ed. *Island Britain: A Quaternary Perspective.* Geological Society Special Publication 96. London: Geological Society, 1995.

Rosenzweig, Michael L. *Species Diversity in Space and Time.* Cambridge, U.K.: Cambridge University Press, 1995.

Schlesinger, William H. *Biogeochemistry: An Analysis of Global Change.* San Diego, CA: Academic Press, 1991.

Periodicals

"Are Arctic Sea Ice Melts Causing Sea Levels to Rise?" *Scientific American*, June 13, 2008. Accessed August 7, 2012. http:// www.scientificamerican.com/article.cfm?id=arctic-ice-melts-cause-rising-sea.

Brett, Carlton E. "Sequence Stratigraphy, Paleoecology, and Evolution: Biotic Clues and Responses to Sea-Level Fluctuations." *Palaios* 13, no. 3 (1998): 241–262.

Cárdenas, Andrés L., and Peter J. Harries. "Effect of Nutrient Availability on Marine Origination Rates throughout the Phanerozoic Eon." *Nature Geoscience* 3, no. 6 (2010): 430–434.

Fish, Marianne R., Isabelle M. Côté, Jennifer A. Gill, et al. "Predicting the Impact of Sea-Level Rise on Caribbean Sea Turtle Nesting Habitat." *Conservation Biology* 19, no. 2 (2005): 482–491.

Haq, Bilal U., Jan Hardenbol, and Peter R. Vail. "Chronology of Fluctuating Sea Levels since the Triassic." *Science* 235, no. 4793 (1987): 1156–1167.

Holland, Steven M. "The Stratigraphic Distribution of Fossils." *Paleobiology* 21, no. 1 (1995): 92–109.

Holland, Steven M. "Sea Level Change and the Area of Shallow-Marine Habitat: Implications for Marine Biodiversity." *Paleobiology* 38, no. 2 (2012): 205–217.

Holland, Steven M., and Mark E. Patzkowsky. "Models for Simulating the Fossil Record." *Geology* 27, no. 6 (1999): 491–494.

Lessa, Enrique P., and Richard A. Fariña. "Reassessment of Extinction Patterns among the Late Pleistocene Mammals of South America." *Palaeontology* 39 (1996): 651–662.

McGowan, Alistair J., Andrew.B. Smith, and Paul.D. Taylor. "Faunal Diversity, Heterogeneity, and Body Size in the Early Triassic: Testing Post-extinction Paradigms in the Virgin Limestone of Utah, USA." *Australian Journal of Earth Sciences* 56, no. 6 (2009): 859–872.

Peters, Shanan E. "Geologic Constraints on the Macroevolutionary History of Marine Animals." *Proceedings of the National Academy of Sciences of the United States of America* 102, no. 35 (2005): 12326–12331.

Peters, Shanan E. "Environmental Determinants of Extinction Selectivity in the Fossil Record." *Nature* 454, no. 7204 (2008): 626–629.

Ryskin, Gregory. "Methane-Driven Oceanic Eruptions and Mass Extinctions." *Geology* 31, no. 9 (2003): 741–744.

Simberloff, Daniel S. "Permo-Triassic Extinctions: Effects of Area on Biotic Equilibrium." *Journal of Geology* 82, no. 2 (1974): 267–274.

Smith, Andrew B. "Large-Scale Heterogeneity of the Fossil Record: Implications for Phanerozoic Biodiversity Studies." *Philosophical Transactions of the Royal Society* B 356, no. 1407 (2001): 351–367.

Titus, James G. "Greenhouse Effect, Sea Level Rise, and Land Use." *Land Use Policy* 7, no. 2 (1990): 138–153.

Valentine, James W. "Plate Tectonics and Shallow Marine Diversity and Endemism, an Actualistic Model." *Systematic Zoology* 20, no. 3 (1971): 253–264.

Wyatt, A.R. "Late Ordovician Extinctions and Sea-Level Change." *Journal of the Geological Society* 152, no. 6 (1995): 899–902.

Other

Ramsar Secretariat. "About the Ramsar Convention." Accessed August 7, 2012. http://www.ramsar.org/cda/en/ramsar-about/main/ramsar/1-36_4000_0.

Alistair J. McGowan

• • • • •

Anoxia

Fortunately for the survival of marine life, today's oceans are very well ventilated, and low oxygen stress is rarely encountered. This benign situation has prevailed since the start of the Cambrian, 542 million years ago (Li et al. 2010). But there have been exceptional, short-lived episodes called oceanic anoxic events. These are times when large parts of the seas and oceans either lost their oxygen entirely (true anoxia) or were characterized by very low concentrations of oxygen that are variously termed hypoxic, dysoxic, or dysaerobic. In this entry, the term *anoxia* is used in a loose sense to denote either conditions that truly lack oxygen or those situations in which oxygen levels are so low that they inhibit the ability of most species to survive. In modern oceans once oxygen concentrations fall below 1.0 milliliter of dissolved oxygen per liter of water only a few, hardy animals can survive, and once values drop below 0.1 milliliter per liter then all but microbial life is extinguished (Rhoads and Morse 1971). Anoxic deposition favors the preservation of organic matter, because organic decay is slower in the absence of oxygen. The high organic content imparts a black color to the resultant rocks, and, thus, black shales are the typical examples of anoxic deposition.

Mechanisms for generating ocean anoxia

Although the modern ocean is well ventilated, there are small regions where anoxia is found. These offer potential analogues for past times when such conditions were much more widespread. In essence, anoxia develops when either the supply of oxygen is insufficient or the demand for oxygen is too great. The "supply" comes from diffusion across the ocean surface, and the degree of oxygenation is then dependent on how effectively the water is circulated. The "demand" for dissolved oxygen primarily comes from the decay of organic matter by bacteria. Anoxia develops today in regions where either the supply is poor because of sluggish circulation or the demand is high because of high primary productivity (Wignall 1994; Arthur and Sageman 1994). These are two very different sets of circumstances, with the result being that modern anoxic settings are the product of either a failure of oxygen supply or an excess of oxygen demand, but not both. In other words, when studying anoxic deposits in the geological record, there are two competing models to choose from (Demaison and Moore 1980).

Poor supply conditions occur when the bottom waters are isolated from well-oxygenated surface waters by a density interface known as the pycnocline. A strong pycnocline makes it difficult for waves to mix the water column and move oxygenated surface waters to greater depths, resulting in sub-pycnocline waters becoming oxygen starved. Density contrasts can be driven by salinity, temperature, or a combination of both. When this occurs and the lateral mixing of bottom waters is restricted by topographic highs (known as sills), then bottom-water anoxia can result.

The largest example of a silled basin today is the Black Sea, a virtually landlocked basin that has a tenuous connection to the world's oceans via the shallow seaway of the Bosporus. The large amount of freshwater runoff going into the Black Sea ensures that the upper water column is brackish and sits on slowly circulating, salty deeper water. More water leaves the basin than enters it (because of all the rivers that drain into the Black Sea), with the result being that it is said to have a positive water balance. This contrasts, for example, with the Mediterranean where the warm and arid climate causes evaporation that is replaced by waters that flow in through the Straits of Gibraltar. The sub-pycnocline waters of the Black Sea are isolated from the oxygenating affects of the surface waters and receive only a small amount of salty Mediterranean water as an underflow. As a result they have become intensely anoxic to the point that hydrogen sulfide is present in solution. This is known as a euxinic condition—*euxinic* being a term that derives from Pontus Euxinus, the old Greek name for the Black Sea. The Black Sea is a very deep basin (about 0.9 miles [1.5 km] in its deepest parts), but silled basins can also develop in much shallower settings, such as some fiords and the modern Baltic Sea, where anoxia develops in depths of less than 330 feet (100 m; Arthur and Sageman 1994). All anoxic silled basins, however, share the common attribute that anoxia is encountered in the deepest parts of the basin beneath a pycnocline.

Productivity in the Black Sea is not especially high because surface-water nutrients are "lost" as descending and decaying organic matter sinks into the euxinic lower-water column and is not recirculated. This contrasts with upwelling zones where deep water rises to the surface bringing nutrients, such as phosphorus and nitrates, which become available for phytoplankton to utilize. Upwelling is a phenomenon of

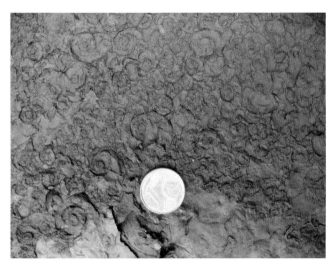

Evidence for anoxia at the Permian–Triassic boundary. A mountain-top location in central Spitsbergen in the high Arctic. The geologists are walking on black shales that accumulated in anoxic conditions at the start of the Triassic whilst the steep cliffs just below are formed of Permian sandstones that formed in well oxygenated, shallow marine conditions. Courtesy of Paul Wignall.

western-facing continental margins where the prevailing offshore-blowing winds drive surface waters oceanward, causing deeper waters to upwell in their place. The resultant supply of nutrients to the surface stimulates high productivity and incidentally accounts for some of the world's most valuable fisheries, such as off the shore of Chile. The abundant rain of decaying organic matter in upwelling zones causes oxygen levels to decline with depth, resulting in the development of an oxygen minimum zone. Beneath oxygen minimum zones, oxygen levels rise once again because little organic matter survives to this depth, and so the bacterial demand for dissolved oxygen becomes less intense. Where oxygen minimum zones intersect with the seafloor, oxygen-poor deposition

Abundant fossil ammonoids in a Carboniferous black shale from western Ireland. Although fossils are often abundant in black shales they are often consist solely of organisms that lived in the water column (like these ammonoids) whilst bottom-dwelling organisms are absent because of the anoxic conditions. Courtesy of Paul Wignall.

occurs and organic-rich sediment results. Black shales formed beneath upwelling zones form at intermediate water depth and pass downslope into better-oxygenated sediments. By contrast, black shales formed in silled basins are found in the deepest water locations.

During oceanic anoxic events, marine anoxia is extremely widespread, and it is instructive to consider what factors might cause this situation. Examination of the geological record shows that the spread of black shale deposition frequently goes hand-in-hand with sea-level rise. Thus, many black shales are described as being "transgressive" because they formed at times when the sea-level was rising and the coastline was retreating. Several types of transgressive black shale have been identified (Wignall and Newton 2001). In some cases the increased area of black shale deposition simply reflects an increase in the deepwater depositional area beneath a pycnocline. Some transgressive black shales, however, apparently record the development of shallow-water anoxia during sea-level rise, although the reason for this is not clear. It has been suggested that a major sea-level rise may allow ocean currents to penetrate into shelf seas and cause upwelling (Piper and Calvert 2009), although it is difficult to envisage how the deepwater mass, with its abundant nutrients, is not used by plankton during its long-distance transport through extensive shelf seas. Nutrients are used at the point they become available in the photic zone; there is no delay as they are transported through shallow-shelf seas.

Global warming is the third partner in the black shales–transgression link for reasons that are relatively straightforward. Transgression can be caused by the melting of the continental ice caps, which can be caused by a rise in temperatures. Even without deglaciation, warming causes the thermal expansion of water, with the result being that a modest sea-level rise is driven by global warming. There is also a further link between climate, warming, and anoxia. Warmer conditions are associated with a more humid climate and consequently increased runoff from land. It is the amount of river-derived nutrients that are the ultimate control on oceanic productivity, and therefore a more humid climate favors higher productivity with a consequent increase in total oxygen demand in the water column. Thus, it is no surprise that many of the examples of anoxia and extinction described below show the same hallmarks of ocean–climate interactions, with a chain of reactions as follows: warming, sea-level rise, increased nutrient runoff, increased productivity, increased anoxia (e.g., Johnson et al. 1985). Although somewhat beyond the scope of this entry, the final link in this cause-and-effect cascade (in fact, the first link in the chain) is flood-basalt volcanism. Many mass extinctions coincide with the voluminous eruptions of giant basalt flows, and the associated release of carbon dioxide provides a possible trigger for the coincidental warming trends (Wignall 2001).

Anoxia and marine extinctions

A multitude of models have been proposed to explain mass extinctions, many of which focus on the consequences of either giant volcanic eruptions or a giant meteorite impact (Hallam and Wignall 1997). It is unlikely that every extinction

Upwelling zone (developed on westernly-facing continental margin)

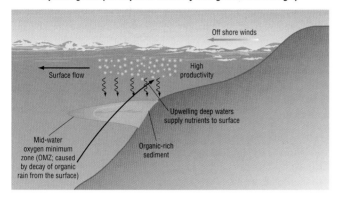

Silled basin (with positive water balance)

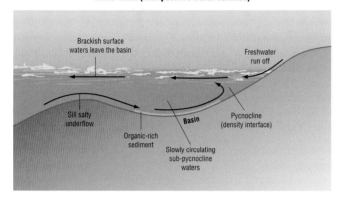

Models illustrating areas of oxygen-poor deposition in modern seas. Silled (restricted) basins such as the Black Sea are characterized by modest/low organic productivity but good preservation due to anoxic conditions in the lower water column. Upwelling zones have very high productivity which causes a zone of oxygen-poor water to develop in the mid-water column due to the decay of sinking organic matter. Where this zone intersects the sea bed organic-rich sediments accumulate. Reproduced by permission of Gale, a part of Cengage Learning.

crisis has exactly the same cause, but, as the following sections demonstrate, anoxia is second to none when it comes to marine extinction mechanisms. The principal evidence for all these examples is the close correspondence between the development of anoxic sedimentary rocks (typically black shales) and extinction levels. A more indirect connection comes from the observation that those animals capable of surviving hypoxia sometimes show better survival rates across extinction horizons. Yet, implicit in this connection between extinction selectivity and anoxia is that anoxia causes the direct death of organisms. The link may in fact be more subtle than this. The consequence of the spread of anoxic waters over the seafloor is that the available marine habitat area decreases. This places a premium on those marine species with good dispersal mechanisms that enable them to seek out regions devoid of anoxia where they can survive. In this scenario it is not hypoxia tolerance that is important to surviving anoxic events but dispersal ability, and those organisms with a long-lived, planktonic larval stage are generally the best dispersed.

It has been argued that, so long as the atmosphere is well oxygenated—which it no doubt has been throughout the Phanerozoic (the current geologic eon)—then shallow-water habitats should always have remained well oxygenated because of wind-generated mixing of surface waters (Bambach 2006). It is therefore especially difficult to envisage how anoxia can kill shallow-water ecosystems, such as reefs, despite the observation that these ecosystems are frequently the most vulnerable ones during mass extinction events (Wood 1999). This is a valid argument that is frequently deployed by critics of anoxia as an extinction mechanism. But proponents counter by arguing that anoxia frequently expands into very shallow waters during mass extinction episodes (Wignall and Twitchett 1996; Bond et al. 2004). There is no modern natural analogue for such a scenario, although intensely polluted shelf seas often develop transient nearshore anoxia in the summer months that leads to the mass death of shellfish. Whether such a situation is sustainable over longer, geological time frames and is capable of causing massive species losses is another matter.

The late Early Cambrian extinctions

The first mass extinction of the fossil record shows many of the hallmarks associated with the later extinction events. It has been called the Sinsk event, and it saw a major transgression and an associated widespread deposition of black shale (Zhuravlev and Wood 1996). The most notable victims of this crisis were the inhabitants of first reef ecosystems, which were constructed by a group of spongelike organisms called archaeocyathids, alongside trilobites. Other than these observations there has so far been little study of this crisis.

The end-Ordovician extinctions

The Sinsk event can be regarded as the first of a repeating pattern of marine mass extinctions. But the next generally recognized mass extinction, the end-Ordovician crisis, is clearly not part of this pattern. This event coincided with an intense phase of glaciation of the southern hemisphere continents of Gondwana (Brenchley et al. 1995) and also a major sea-level fall caused by the trapping of water in continental ice sheets (Hallam and Wignall 1999). The relationship between faunal diversity and oxygenation in this interval is intriguing but difficult to study because no oceanic sedimentary records are known from this time. However, deep-water sediments in the Southern Uplands of Scotland record oxygenation trends on a continental slope that is probably as close as researchers can get to studying oceanic conditions (Armstrong and Coe 1997).

These sediments reveal a change from anoxic to oxic conditions (black shales to pale gray mudstones), coincident with the onset of glaciations, and then a return back to anoxia during deglaciation at the start of the Silurian. The well-ventilated glacial interval corresponds to the Hirnantian stage, and it is characterized by an extraordinarily widespread but low-diversity assemblage of brachiopods and trilobites known as the Hirnantian fauna. This group of animals is thought to have been adapted to cool-water conditions. The first, and most severe, phase of extinction occurred at the cessation of anoxia, at the start of the Hirnantian, and offers a unique example of a crisis triggered by the onset of oxygenated deposition. Obviously, oxygenated conditions are not, of themselves, stressful, but the answer to this paradox probably

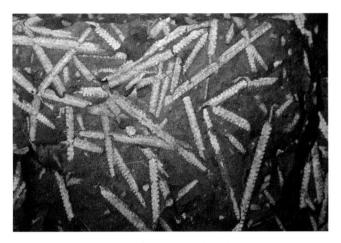

Fossil graptolites *Diplograptus multidens* with two branches back-to-back, Llanvirn and Ordovician, from Conwy, Wales. © Sabena Jane Blackbird/Alamy.

lies in the change in environmental conditions from a warm world, with poorly ventilated deeper ocean waters, to a colder world with well-circulated and oxygenated waters. One notable victim of the extinction was the graptolites, a group of colonial organisms that constructed small, branching colonies and floated far and wide in the world's oceans. It is thought that many of the graptolites were adapted to the high-nutrient conditions found at the edges of oxygen minimum zones. It is likely that the improved oceanic ventilation and weakening of the oxygen minimum zone in the cold, glacial Hirnantian world resulted in this habitat shrinking or disappearing entirely, with unfortunate consequences for the graptolites (Berry et al. 1990). Many of the other losses at this time were among animals with planktonic larval stages, and these too may have seen their larval habitats in the water column disappear in the glacial oceans of the Hirnantian.

The second phase of the end-Ordovician extinction wiped out the Hirnantian fauna and coincided with transgression and black shale development. Ostensibly, this could be regarded as a more typical anoxia-and-extinction crisis. Many of the extinction losses, however, are among shallow-water species, and it is unclear whether anoxic waters were ever developed at these depths (Brenchley et al. 1995). The combination of warming and the elimination of cold-adapted species is an equally plausible cause of these later extinctions. The end-Ordovician mass extinction and anoxia link, therefore, remains an intriguing one.

Late and end-Devonian extinctions

Several extinction-with-anoxia crises are known from the later part of the Devonian marine record, of which the most serious were at the Frasnian–Famennian stage boundary and at the end of the Devonian (Joachimski and Buggisch 1993; Hallam and Wignall 1997; House 2002). These are the Kellwasser and Hangenberg crises, respectively, which are named after distinctive horizons in Germany where they were first studied. The Kellwasser horizons are two closely spaced, organic-rich limestone beds seen in deepwater offshore locations. They record the spread of anoxic deposition during

Late Frasnian transgressions (Buggisch 1991; Bond and Wignall 2008), but it is only the upper horizon that is associated with mass extinction. This probably relates to the relative extent of anoxia: The Upper Kellwasser Horizon is widespread and manifest in sections throughout Europe, North Africa, and North America, while the Lower Kellwasser Horizon has a more sporadic regional development (Bond et al. 2004). Thus, marine habitat loss at the time of the Upper Kellwasser event was much more significant.

It is interesting that the area of anoxic deposition, while widespread in shelf seas, does not appear to have extended into the oceanic realm, in contrast to the situation during other anoxic events (see below). No true oceanic record is known from the Late Devonian, but shelf sections that faced onto the oceans provide the best clue to conditions in the open ocean. These contain sediments that are among the best oxygenated of this time interval (Bond et al. 2004). This observation may also explain why the Kellwasser event failed to leave evidence of oceanic anoxia in the geochemical records (e.g., sulfur isotopes; see John et al. 2010). Nevertheless, even if the world's oceans generally remained well oxygenated in the Late Devonian, this was of little consolation to the denizens of shallow-shelf seas that saw their habitat areas contract with the spread of anoxia during the Upper Kellwasser event.

Extinction losses during the Kellwasser crisis are best seen among many benthic or bottom-living groups such as brachiopods and trilobites, as well as among reef inhabitants. Heavily calcified sponges, called stromatoporoids, produced these reefs, which were the largest the world has ever seen. These spectacular reefs are found today in widely dispersed locations such as the Rocky Mountains of British Columbia, the Holy Cross Mountains of Poland, and the Canning Basin of Western Australia. Devonian reefs are typically capped by an erosion surface, caused by a sea-level fall (Becker and House 1997) that probably happened prior to the Kellwasser crisis (Bond and Wignall 2008). Only a few, small microbial reefs were reestablished when the sea level rose again following the Famennian age. This sedimentary gap at the top of Frasnian stromatoporoid reefs makes it difficult to determine what caused their permanent demise and the extinction of their constituent biotas. The extinction event may be related to the sea-level fall and habitat loss, but the regressions are frequent in the geological record, and reefs usually have little problem reestablishing themselves once flooding begins again. The sedimentary gap also makes it difficult to evaluate whether anoxia played a large role in reef extinction, although many have argued against a hypothesis centering on anoxia killing the reefs because they think it unlikely that oxygen restriction could have spread into waters shallow enough to affect reefs (e.g., Sandberg et al. 1988; House 2002). The Holy Cross Mountains, however, provide an example of a Frasnian reef that is directly overlain by a thin bed that records anoxic deposition. Therefore, locally at least, anoxia may indeed have played a role in the Late Devonian reef extinction (Bond et al. 2004).

The important role of anoxia in the extinction of seafloor-dwelling species at the end of the Frasnian is well established. But there were also many extinction losses among the pelagic (water-column) fauna that also coincide with anoxia. These losses include the total extinction of the homoctenids, an enigmatic group consisting of tiny, cone-shaped shells that were prolifically abundant in pelagic sediments (Bond 2006), and the entomozoaceans, a pelagic group of ostracods that also suffered substantial losses (Olempska 2002).

The homoctenid extinction occurred within the Upper Kellwasser Horizon, and so was later than the loss of bottom dwellers, which succumbed at the start of the Upper Kellwasser event, while the entomozoaceans disappeared in a series of steps during the Kellwasser event. Both these groups were comprised of surface-water dwellers, and it is unlikely that persistent anoxia ever developed in their habitat. So can their losses be ascribed to anoxia? The answer may be an indirect yes because the development of anoxia involves changes in the structure of the entire water column. For example, more intense water-column stratification can lead to a decreased oxygen supply to the seabed. Planktonic species are finely tuned to the different conditions encountered in surface waters because factors such as light intensity and food availability change rapidly with depth below the surface. Any change in the overall structure of the water column will, therefore, change the surface-water conditions—a habitat change that seems to have had serious consequences for the Devonian pelagic groups.

Whereas the Kellwasser crisis wreaked its damage from seafloor to sea surface, the subsequent Hangenberg crisis, at the end of the Devonian, was mainly a crisis for organisms within the water column, with many groups of fish and ammonoids disappearing from the fossil record at this time (Hallam and Wignall 1997). The most spectacular group to go was the placoderms, which were armored fish that included large predatory types with scissorlike jaws. These losses coincided with the spread of black shales, known as the Hangenberg Shales in Germany. Unsurprisingly, anoxia is thought to be closely linked to this crisis. Yet, unlike the anoxia/warming/transgression link seen for other crises reviewed here, this anoxic event has been attributed to global cooling prior to the onset of a glacial episode a few million years later (Caplan and Bustin 1999). Cooling and a change to a more vigorously circulated ocean are argued to have strengthened upwelling and ocean productivity and thereby to have caused the spread of anoxic waters. What is not apparent in this model is why there was also a major sea-level transgression at this time.

The Middle Permian (Guadalupian) extinctions

Following the Hangenberg crisis, the next major mass extinction in the history of life did not occur for another 100 million years, a remarkably long gap, as it took place in the middle of the Permian during the Capitanian stage. Most of the extinctions at this time were focused in shallow-marine tropical habitats and included many brachiopods, corals, and foraminifers (Bond et al. 2010). The extinction losses have been clearly linked to the eruption of a flood basalt province (Wignall et al. 2009), but the event has been included here because some researchers have suggested that it coincided with the onset of an oceanic anoxic event that persisted until the end of the Permian (Isozaki 1997; Clapham and Bottjer

2007). This claim was based on evidence from deep-ocean sediments from Japan that were thought to record a decline in ocean-floor oxygenation at this time. Improved dating of these sediments suggests, however, that this change occurred in the latest Permian and is more clearly related to the next mass extinction. There is little field evidence for anoxia at the time of the Guadalupian extinctions, either in shelf seas or oceans (Wignall et al. 2010).

The end-Permian extinctions

Following the discovery that anoxic deposits became widespread in both shallow-shelf and deep-ocean locations at the time of the end-Permian mass extinction (Wignall and Hallam 1992; Isozaki 1997), anoxia has been established as one of the main contenders for the cause of this great crisis. The extent of anoxic deposition at this time is truly astounding. Only a few locations in the end-Permian marine record remained free of anoxia, and these are concentrated in areas that were found along the southern margin of the Tethyan Ocean (present-day Oman, Pakistan, and southern Tibet; Wignall and Newton 2003). Evidence for anoxia derives from the nature of the sediments, which are typically laminated, with well-preserved organic matter, and the chemical signatures they contain. The chemistry of anoxic waters is fundamentally different from that of oxygenated waters, and this has a profound effect on sediment geochemistry; for example, many trace metals present in solution in seawater are precipitated out of anoxic waters. End-Permian sediments are typically enriched in these trace metals. Evidence for end-Permian anoxia extends into very shallow waters (Wignall et al. 1998), and a distinctive organic molecule, called isorenieratene, has been found in the sediments of this age. This is a marker for bacteriochlorophylls, which belong to green sulfur bacteria and require both sunlight and anoxic conditions, and indicates that euxinia extended into the surface waters at this time (Grice et al. 2005).

Despite the clear evidence for global anoxia, the notion that much of world's oceans were anoxic in the Permian–Triassic boundary interval is difficult to reconcile by simple comparison with modern analogues alone. Equally extraordinary is the fundamental change in the site of organic carbon burial in the Early Triassic. Usually this is split 50–50 between terrestrial locations (mostly peat swamps) and shallow-marine sediments where it occurs as dispersed particles. In the Early Triassic nearly all organic carbon was accumulating in the deep ocean where prevailing anoxic conditions greatly aided its preservation. Productivity in the surface waters of the oceans at this time must have been considerable, and this suggests (indeed it requires) a huge nutrient flux from land to support the ocean's productivity. That this was the case is supported by substantial evidence for a major increase of runoff and sediment accumulation rates in the Early Triassic (Algeo and Twitchett 2010).

Earth-system modelers have attempted to work out what the conditions responsible for global anoxia at the end of the Permian were like, but so far they have not succeeded (e.g., Zhang et al. 2001). Partial success was achieved in 2008 by Katja M. Meyer and Lee R. Kump using a model of intermediate complexity. In this model they increased the land-derived nutrient flux to a value ten times that of today (as noted above, this is not an unreasonable scenario). The result was elevated productivity and anoxia in the Tethyan Ocean, but the Panthalassa Ocean remained stubbornly well ventilated despite the clear field evidence for anoxia in this huge ocean at the time. In a 2005 study, Jeffrey T. Kiehl and Christine A. Shields used a linked ocean–atmosphere general circulation model to predict a sluggish, poorly circulating ocean when atmospheric carbon dioxide concentrations were raised to high values (3,550 parts per million by volume) and the world became hot. Although these authors did not model oxygen concentrations, this situation is likely to have led to ocean anoxia.

In contrast, one of the most sophisticated attempts to model the end-Permian world has shown that ocean mixing does not simply decrease with global warming in a predictable way but rather it changes and may in fact increase (Montenegro et al. 2011). Today, ocean mixing is primarily driven by the formation of cold, dense waters at high latitudes that become the mid-latitude and low-latitude bottom waters. But in a warmer world this system could be replaced by bottom waters that are warm and densely saline and formed in low latitudes. Only the modern Mediterranean has a circulation regime with such warm, saline bottom waters, and, significantly, it is very well ventilated at all water depths (Demaison and Moore 1980). Excellent oxygenation levels are also seen in end-Permian ocean models with the same style of circulation (Montenegro et al. 2011). If warm, saline bottom-water regimes are not intrinsically anoxic, it appears that circulation changes alone cannot be responsible for end-Permian whole-ocean anoxia. Perhaps extremely large values of nutrient runoff from land are the key factor.

The end-Triassic extinctions

The end-Triassic mass extinction was one of the so-called Big Five crises in life's history. Like many other extinctions discussed here, it coincides with eruption of a flood basalt province (Pálfy et al. 2002). There is no consensus as to the cause of this extinction, although ocean acidification is a popular culprit (Hautmann 2004). Marine anoxia has featured in some extinction models (Ward et al. 2001), and contemporaneous black shales are well developed in western Canada and western Europe (Wignall et al. 2007). There is no evidence, however, for an intensification or increase in the extent of black shales at the extinction level, and the oceans remained well ventilated throughout the crisis (Wignall et al. 2010). Anoxia is therefore unlikely to be a major player in this extinction event.

The early Jurassic (Toarcian) extinctions

One of the most famous black shales of the geological record occurs in the Lower Jurassic of northwestern Europe. This has been the focus of numerous geochemical studies and has various names in the different countries in which it occurs (the Jet Rock of England, the Schistes Carton of the Paris Basin, and the Posidonienschiefer of Germany), although it originally formed a continuous extent of black shale deposition in the basins of the region (Jenkyns 1998). Further black

Burgess Shale Quarry in British Columbia. Courtesy of Mark A. Wilson.

shale horizons from southern Europe predate the more northern examples, but there is a slight overlap in the age of their deposition early in the Toarcian. Perhaps not surprisingly, this moment of peak extent of black shales coincides with an extinction of many bottom-living species in the region (Wignall et al. 2005). However, the global extent of Toarcian anoxia is unclear. The oceanic record shows an anoxic event at this time (Wignall et al. 2010), and a similar black shale/ extinction story is seen in northern Siberia (Suan et al. 2011). Anoxia is not clearly developed, however, in all regions. For example, it is weakly manifest in Spanish marine successions, and an alternative extinction model involving warming has been proposed (García Joral, Gómez, and Goy 2011).

Cretaceous oceanic anoxic events

In the 1970s the international scientific community embarked on an ocean drilling program that remains active under the auspices of the Integrated Ocean Drilling Program. One of the first discoveries was the presence of several thin black shale horizons in deep-ocean sediments of Cretaceous age. In stark contrast to the well-oxygenated deep-ocean conditions of the twenty-first century, these shales clearly record episodes of anoxic deposition on the ocean floor. Initially two of these oceanic anoxic events were identified (Schlanger and Jenkyns 1976)—one around 94 million years ago, at the Cenomanian–Turonian boundary, and one in the early Aptian stage, 26 million years earlier. Several more Cretaceous black-shale horizons have been discovered since then, but these two were clearly the most widespread (Jenkyns 2010).

Only the Cenomanian-Turonian event is associated with an elevation of extinction rates whilst the Aptian event did not coincide with a mass extinction. However, even for the Cenomanian-Turonian anoxic event the link with extinctions is not clear because many of the losses occurred throughout the Cenomanian Stage before the anoxia happened (Harries and Little 1999). Even the extinction losses that seem to be synchronous with the oceanic anoxic event may be linked with other phenomena. In detail the extinctions coincide with a phase of cooling, called the Plenus cold event, late in the anoxic episode (Pearce et al. 2009). It is therefore a mystery as to why this and other Cretaceous oceanic anoxic events are not associated with elevated extinction rates. The answer may lie in the nature of their development. They are primarily deepwater phenomena, best known from the Atlantic Ocean, while many shelf seas and much of the Pacific Ocean remained well oxygenated at this time (Takashima et al. 2011). Thus, a well-oxygenated shallow-shelf sea area may have been more than sufficient to support normal levels of marine diversity during these times.

Indirectly, the Cretaceous oceanic anoxic events may offer a crucial, but subtle, insight into ongoing efforts to understand the relationship between anoxia and extinction: It is the spread of anoxia into shallow waters that causes extinctions, not the spread of deepwater anoxia. This insight is driven home by the youngest example of deep-sea anoxia, one that occurred around 56 million years ago at the Paleocene–Eocene boundary. This was not a full-blown example of an oceanic anoxic event, such as the Cretaceous examples, because oxygen-poor conditions are well known only from the Atlantic Ocean floor. Global extinction rates remained low at this time, and it is no surprise that the most noteworthy losses are seen among groups endemic to the Atlantic Ocean floor.

Synthesis

Marine anoxia typically coincides with both global warming and sea-level rise. The phenomenon also frequently coincides with mass extinction but only when the following two criteria are met: (1) Anoxia is developed in shallow waters, and (2) Anoxia is extensive in shelf seas (oceanic anoxic events appear to exert a relatively unimportant influence on marine life). The end-Permian mass extinction coincided with probably the most extensive development of shallow-water anoxia of the past 600 million years and was also incredibly widespread in oceanic settings. The lesser crisis of the Early Jurassic (Toarcian) also coincided with both these criteria, but it is significant that the extent of anoxia was not as great at this time. The Cretaceous anoxic events were primarily oceanic phenomena, and contemporaneous extinction losses were only modest or minor.

For seafloor-dwelling species it is probably the loss of habitat caused by the development of uninhabitable anoxic bottom waters that causes extinction stresses. Yet, anoxic events frequently also coincide with the loss of surface-dwelling, pelagic groups (e.g., the Late Devonian and end-Permian crises). In this case it may be the fundamental and consequential changes in water-column structure (e.g., nutrient availability, density structure) associated with the development of anoxia in the lower water column that causes widespread extinctions.

The future

It takes a lot to change the ambient conditions in the world's oceans, and even with predicted future global warming it is unlikely that their overall ventilation state will change much in the twenty-first century. Nonetheless, modeling suggests that the ocean's oxygen minimum zones will start to intensify in the next few decades. Within a few centuries researchers may start to see significant areas of the seabed becoming uninhabitable (Falkowski et al. 2011). Humanity must hope it is not responsible for recreating an episode like the end-Permian anoxia-led mass extinction.

Resources

Books

Berry, William B.N., Pat Wilde, and Mary S. Quinby-Hunt. "Late Ordovician Graptolite Mass Mortality and Subsequent Early Silurian Re-radiation." In *Extinction Events in Earth History*, edited by Eric G. Kauffman and Otto H. Walliser, 115–123. Berlin: Springer-Verlag, 1990.

Hallam, Anthony, and Paul B. Wignall. *Mass Extinctions and Their Aftermath*. Oxford: Oxford University Press, 1997.

Pálfy, József, Paul L. Smith, and James K. Mortensen. "Dating the End-Triassic and Early Jurassic Mass Extinctions, Correlative Large Igneous Provinces, and Isotopic Events." In *Catastrophic Events and Mass Extinctions: Impacts and Beyond*, edited by Christian Koeberl and Kenneth G. MacLeod, 523–532. Geological Society of America Special Paper 356. Boulder, CO: Geological Society of America, 2002.

Wignall, Paul B. *Black Shales*. Oxford: Clarendon Press, 1994.

Wood, Rachel. *Reef Evolution*. Oxford: Oxford University Press, 1999.

Periodicals

Algeo, Thomas J., and Richard J. Twitchett. "Anomalous Early Triassic Sediment Fluxes Due to Elevated Weathering Rates and Their Biological Consequences." *Geology* 38, no. 11 (2010): 1023–1026.

Armstrong, Howard A., and Angela A. Coe. "Deep-Sea Sediments Record the Geophysiology of the Late Ordovician Glaciation." *Journal of the Geological Society of London* 154, no. 6 (1997): 929–934.

Arthur, Michael A., and Bradley B. Sageman. "Marine Black Shales: Depositional Mechanisms and Environments of Ancient Deposits." *Annual Review of Earth and Planetary Sciences* 22 (1994): 499–551.

Bambach, Richard K. "Phanerozoic Biodiversity Mass Extinctions." *Annual Review of Earth and Planetary Sciences* 34 (2006): 127–155.

Becker, Ralph Thomas, and Michael R. House. "Sea-Level Changes in the Upper Devonian of the Canning Basin,

Western Australia." *Courier Forschungsinstitut Senckenberg* 199 (1997): 129–146.

Bond, David P.G. "The Fate of the Homoctenids (Tentaculitoidea) during the Frasnian–Famennian Mass Extinction (Late Devonian)." *Geobiology* 4, no. 3 (2006): 167–177.

Bond, David P.G., Jason Hilton, Paul B. Wignall, et al. "The Middle Permian (Capitanian) Mass Extinction on Land and in the Oceans." *Earth-Science Reviews* 102, nos. 1–2 (2010): 100–116.

Bond, David P.G., and Paul B. Wignall. "The Role of Sea-Level Change and Marine Anoxia in the Frasnian–Famennian (Late Devonian) Mass Extinction." *Palaeogeography, Palaeoclimatology, Palaeoecology* 263, nos. 3–4 (2008): 107–118.

Bond, David P.G., Paul B. Wignall, and Grzegorz Racki. "Extent and Duration of Marine Anoxia during the Frasnian–Famennian (Late Devonian) Mass Extinction in Poland, Germany, Austria, and France." *Geological Magazine* 141, no. 2 (2004): 173–193.

Brenchley, Patrick John, Graham A.F. Carden, and James D. Marshall. "Environmental Changes Associated with the 'First Strike' of the Late Ordovician Mass Extinction." *Modern Geology* 20 (1995): 69–82.

Buggisch, Werner. "The Global Frasnian–Famennian 'Kellwasser Event.'" *Geologische Rundschau* 80, no. 1 (1991): 49–72.

Caplan, Mark L., and R. Mark Bustin. "Devonian–Carboniferous Hangenberg Mass Extinction Event, Widespread Organic-Rich Mudrock, and Anoxia: Causes and Consequences." *Palaeogeography, Palaeoclimatology, Palaeoecology* 148, no. 4 (1999): 187–207.

Clapham, Matthew E., and David J. Bottjer. "Prolonged Permian–Triassic Ecological Crisis Recorded by Molluscan Dominance in Late Permian Offshore Assemblages." *Proceedings of the National Academy of Sciences of the United States of America* 104, no. 32 (2007): 12971–12975.

Demaison, G.J., and G.T. Moore. "Anoxic Environments and Oil Source Bed Genesis." *Organic Geochemistry* 2, no. 1 (1980): 9–31.

Falkowski, Paul G., Thomas Algeo, Lou Codispoti, et al. "Ocean Deoxygenation: Past, Present, and Future." *Eos* 92, no. 46 (2011): 409–410.

García Joral, Fernando, Juan J. Gómez, and Antonio Goy. "Mass Extinction and Recovery of the Early Toarcian (Early Jurassic) Brachiopods Linked to Climate Change in Northern and Central Spain." *Palaeogeography, Palaeoclimatology, Palaeoecology* 302, nos. 3–4 (2011): 367–380.

Grice, Kliti, Changqun Cao, Gordon D. Love, et al. "Photic Zone Euxinia during the Permian–Triassic Superanoxic Event." *Science* 307, no. 5710 (2005): 706–709.

Hallam, Anthony, and Paul B. Wignall. "Mass Extinctions and Sea-Level Changes." *Earth-Science Reviews* 48, no. 4 (1999): 217–250.

Harries, Peter J., and Crispin T.S. Little. "The Early Toarcian (Early Jurassic) and the Cenomanian–Turonian (Late Cretaceous) Mass Extinctions: Similarities and Contrasts." *Palaeogeography, Palaeogeography, Palaeoecology* 154, nos. 1–2 (1999): 39–66.

Hautmann, Michael. "Effect of End-Triassic CO_2 Maximum on Carbonate Sedimentation and Marine Mass Extinction." *Facies* 50, no. 2 (2004): 257–261.

House, Michael R. "Strength, Timing, Setting, and Cause of Mid-Palaeozoic Extinctions." *Palaeogeography, Palaeoclimatology, Palaeoecology* 181, nos. 1–3 (2002): 5–25.

Isozaki, Yukio. "Permo-Triassic Boundary Superanoxia and Stratified Superocean: Records from Lost Deep Sea." *Science* 276, no. 5310 (1997): 235–238.

Jenkyns, Hugh C. "The Early Toarcian (Jurassic) Anoxic Event: Stratigraphic, Sedimentary, and Geochemical Evidence." *American Journal of Science* 288, no. 2 (1998): 101–151.

Jenkyns, Hugh C. "Geochemistry of Oceanic Anoxic Events." *Geochemistry, Geophysics, Geosystems* 11 (2010): Q03004.

Joachimski, Michael M., and Werner Buggisch. "Anoxic Events in the Late Frasnian: Causes of the Frasnian–Famennian Faunal Crisis?" *Geology* 21, no. 8 (1993): 675–678.

John, Eleanor H., Paul B. Wignall, Robert J. Newton, and Simon H. Bottrell. "$\delta^{34}S_{CAS}$ and $\delta^{18}O_{CAS}$ Records during the Frasnian–Famennian (Late Devonian) Transition and Their Bearing on Mass Extinction Models." Chemical Geology 275, nos. 3–4 (2010): 221–234.

Johnson, J.G., G. Klapper, and C.A. Sandberg. "Devonian Eustatic Fluctuations in Euramerica." *Geological Society of America Bulletin* 96, no. 5 (1985): 567–587.

Kiehl, Jeffrey T., and Christine A. Shields. "Climate Simulation of the Latest Permian: Implications for Mass Extinction." *Geology* 33, no. 9 (2005): 757–760.

Li, Chao, Gordon D. Love, Timothy W. Lyons, et al. "A Stratified Redox Model for the Ediacaran Ocean." *Science* 328, no. 5974 (2010): 80–83.

Meyer, Katja M., and Lee R. Kump. "Oceanic Euxinia in Earth History: Causes and Consequences." *Annual Review of Earth and Planetary Sciences* 36 (2008): 251–288.

Montenegro, A., P. Spence, K.J. Meissner, et al. "Climate Simulations of the Permian–Triassic Boundary: Ocean Acidification and the Extinction Event." *Paleoceanography* 26, no. 3 (2011): PA3207.

Olempska, Ewa. "The Late Devonian Upper Kellwasser Event and Entomozoacean Ostracods in the Holy Cross Mountains, Poland." *Acta Palaeontologica Polonica* 47, no. 2 (2002): 247–266.

Pearce, Martin A., Ian Jarvis, and Bruce A. Tocher. "The Cenomanian–Turonian Boundary Event, OAE2, and Palaeoenvironmental Change in Epicontinental Seas: New Insights from the Dinocyst and Geochemical Records." *Palaeogeography, Palaeoclimatology, Palaeoecology* 280, nos. 1–2 (2009): 207–234.

Piper, D.Z., and S.E. Calvert. "A Marine Biogeochemical Perspective on Black Shale Deposition." *Earth-Science Reviews* 95, nos. 1–2 (2009): 63–96.

Rhoads, Donald C., and John W. Morse. "Evolutionary and Ecologic Significance of Oxygen-Deficient Marine Basins." *Lethaia* 4, no. 4 (1971): 413–428.

Sandberg, Charles A., Willi Ziegler, Roland Dreesen, and Jamie L. Butler. "Late Frasnian Mass Extinction: Conodont Event Stratigraphy, Global Changes, and Possible Causes." *Courier Forschungsinstitut Senckenberg* 102 (1988): 263–307.

Schlanger, Seymour O., and Hugh C. Jenkyns. "Cretaceous Oceanic Anoxic Events: Causes and Consequences." *Geologie en Mijnbouw* 55, nos. 3–4 (1976): 179–184.

Suan, Guillaume, Boris L. Nikitenko, Mikhail A. Rogou, et al. "Polar Record of Early Jurassic Massive Carbon Injection." *Earth and Planetary Science Letters* 312, nos. 1–2 (2011): 102–113.

Takashima, Reishi, Hiroshi Nishi, Toshiro Yamanaka, et al. "Prevailing Oxic Environments in the Pacific Ocean during the Mid-Cretaceous Oceanic Anoxic Event 2." *Nature Communications* 2 (2011), article no. 234

Ward, P.D., J.W. Haggart, E.S. Carter, et al. "Sudden Productivity Collapse Associated with the Triassic–Jurassic Boundary Mass Extinction." *Science* 292, no. 5519 (2001): 1148–1151.

Wignall, Paul B. "Large Igneous Provinces and Mass Extinctions." *Earth-Science Reviews* 53, nos. 1–2 (2001): 1–33.

Wignall, Paul B., David P.G. Bond, Kiyoko Kuwahara, et al. "An 80 Million Year Oceanic Redox History from Permian to Jurassic Pelagic Sediments of the Mino-Tamba Terrane, SW Japan, and the Origin of Four Mass Extinctions." *Global and Planetary Change* 71, nos. 1–2 (2010): 109–123.

Wignall, Paul B., and Anthony Hallam. "Anoxia as a Cause of the Permian/Triassic Extinction: Facies Evidence from Northern Italy and the Western United States." *Palaeogeography, Palaeoclimatology, Palaeoecology* 93, nos. 1–2 (1992): 21–46.

Wignall, Paul B., Ric Morante, and Robert J. Newton. "The Permo-Triassic Transition in Spitsbergen: $\delta^{13}C_{org}$ Chemostratigraphy, Fe and S Geochemistry, Facies, Fauna, and Trace Fossils." *Geological Magazine* 135, no. 1 (1998): 47–62.

Wignall, Paul B., and Robert J. Newton. "Black Shales on the Basin Margin: A Model Based on Examples from the Upper Jurassic of the Boulonnais, Northern France." *Sedimentary Geology* 144, nos. 3–4 (2001): 335–356.

Wignall, Paul B., and Robert J. Newton. "Contrasting Deep-Water Records from the Upper Permian and Lower Triassic of South Tibet and British Columbia: Evidence for a Diachronous Mass Extinction." *Palaios* 18, no. 2 (2003): 153–167.

Wignall, Paul B., Robert J. Newton, and Crispin T.S. Little. "The Timing of Paleoenvironmental Change and Cause-and-Effect Relationships during the Early Jurassic Mass Extinction in Europe." *American Journal of Science* 305, no. 10 (2005): 1014–1032.

Wignall, Paul B., Sun Yadong., Bond, David P.G., Izon, Gareth, Newton, Robert J., Védrine, Stéphanie, Widdowson, Mike, Ali, Jason R., Lai Xulong, Jiang Haishui, Cope, Helen & Bottrell, Simon H. 2009. Volcanism, mass extinction and carbon isotope fluctuations in the Middle Permian of China. *Science*, 324, 1179-1182.

Wignall, Paul B., and Richard J. Twitchett. "Oceanic Anoxia and the End Permian Mass Extinction." *Science* 272, no. 5265 (1996): 1155–1158.

Wignall, Paul B., John-Paul Zonneveld, Robert J. Newton, et al. "The End Triassic Mass Extinction Record of Williston Lake, British Columbia." *Palaeogeography, Palaeoclimatology, Palaeoecology* 253, nos. 3–4 (2007): 385–406

Zhang, R., M.J. Follows, J.P. Grotzinger, and J. Marshall. "Could the Late Permian Deep Ocean Have Been Anoxic?" *Paleoceanography* 16, no. 3 (2001): 317–329.

Zhuravlev, Andrey Yu., and Rachel A. Wood. "Anoxia as the Cause of the Mid-Early Cambrian (Botomian) Extinction Event." *Geology* 24, no. 4 (1996): 311–314.

Paul B. Wignall

Climate change

The average temperature of Earth is increasing, driven primarily by human activities. The question therefore arises as to whether such climate disruption poses a new extinction threat. Answering that question involves understanding what is causing current climate change, how the warming experienced over the last century and predicted for the next one compare in rate and magnitude to previous climate changes, what factors other than temperature increase are involved in climate change, and what the potentials and limits of species response to climate change are.

Causes of current climate change

It is now widely recognized that a primary driver of ongoing climate change is the release of greenhouse gases into the atmosphere by human activities, mainly (but not exclusively) from burning fossil fuels. Greenhouse gases are so named because, even though their molecules compose only a small fraction of the total atmosphere, they prevent a portion of the sun's heat from radiating back into space after it reaches Earth, much as the glass panels of a greenhouse trap heat and raise the temperature inside.

Anthropogenic (human-caused) emission of greenhouse gases accelerated dramatically as the Industrial Revolution went into full swing, beginning about 1750—times prior to this are called preindustrial. Carbon dioxide (CO_2) is a key greenhouse gas. In the absence of significant anthropogenic additions, its concentration in the atmosphere was about 280 parts per million (ppm) for most of the last 11,700 years, which is typical for times when the planet is not gripped in an ice age. [Note: Ice ages, which geologists call *glacial stages*, came and went several times over the past 2.6 million years. Each ice age was separated by an interglacial stage when the glaciers, which at their maximum advance covered some 30 percent of Earth's land, shrunk back to approximately the condition that typified most of the last 11,700 years. During the glacial stages, CO_2 concentrations were much lower than the 280 ppm that characterizes interglacial stages.] Emissions then accelerated as the human population began to triple relative to the population in 1950. As of January 2012 the atmospheric concentration of CO_2 stood at 394 ppm, more than one-third above the preindustrial norm. The extra CO_2 comes mostly from the burning of coal, oil, and natural gas, which accounts for about 91 percent of the excess.

Land-use changes, such as deforestation, produce most of the remainder.

Two other important greenhouse gases that are added to the atmosphere by anthropogenic activities are methane (CH_4) and nitrous oxide (NO), both of which also have increased dramatically since the Industrial Revolution. Methane today is measured at around 1,870 ppm and 1,750 ppm in the northern and southern hemispheres, respectively, up from a preindustrial value of about 700 ppm. Nitrous oxide has gone up to 323 ppm, from a preindustrial norm of 270 ppm.

Human activities also contribute to the rise in global temperature by producing ozone-forming chemicals such as carbon monoxide, hydrocarbons, and halocarbons. The warming that results from adding greenhouse gases and ozone-forming chemicals to the atmosphere is further enhanced by landscape changes that affect the amount of heat absorbed by Earth's surface—for example, anthropogenic deposition of black carbon airborne particles on what would otherwise be reflective snow. Absorption rather than reflection of heat is then exacerbated as the reflective snow and ice fields melt and are replaced with dark-colored vegetation that absorbs sunlight.

Using physics it is possible to calculate how much warming should result from any given increase in greenhouse gases. The observed rise in the global average since the year 1900, approximately 1.4°F (0.8°C), conforms to expectations of these calculations. The observed temperature rise is based on data compiled from weather stations worldwide. In addition, computer simulations that also take into account the amount of nonanthropogenic climate variation that would be expected indicate that most of the historically observed temperature rise is attributable to human activities.

By estimating the addition of greenhouse gases that are likely to be emitted under various scenarios of human population growth, fossil fuel use, economic growth, global cooperation, and other relevant factors, it is possible to model the expected global temperature rise over coming decades and centuries (see Figure 1). The best estimates for a range of possible scenarios suggest that the average global temperature will increase 3.2 to 7.2°F (1.8 to 4°C) in the century from 2000 to 2100, on top of the 1.4°F (0.8°C) that temperature rose in the century from 1900 to 2000. These estimates

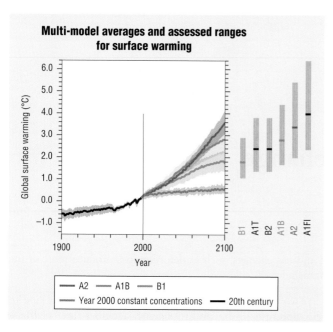

Multi-model averages and assessed ranges for surface warming

Figure 1. Range of forecasts for global temperature increase as determined by the Intergovernmental Panel on Climate Change. Black line based on measured temperatures. Colored lines illustrate multi-model global average surface warming under different emissions scenarios. Purple line holds emissions constant at the values they were at in the year 2000; in actuality emissions have increased considerably since then. Bars at right indicate best estimate (colored line) and likely range (gray bar) for each scenario. "Likely" means there is more than a 66% chance that the actual temperature increase will be within this range (based on Solomon et al., 2007). Reproduced by permission of Gale, a part of Cengage Learning.

represent the consensus agreement of the International Panel on Climate Change (IPCC), which is composed of hundreds of climate scientists working across national boundaries to evaluate and interpret evidence about how the global climate is changing.

Is current and projected climate change unusual?

Various lines of geological and paleontological evidence are available to track how the global climate has changed over hundreds, thousands, and millions of years. This evidence, as summarized below, demonstrates that the climate change in progress now is unusual with respect to Earth's previous climate changes in two ways. First, it is more rapid than past global climate changes. Second, the amount of temperature increase is very high and is moving Earth toward a higher average temperature than many species living today have experienced in their evolutionary history.

One important line of evidence that puts the speed of today's climate change into context with respect to the past several hundred thousand years comes from the chemical composition of ancient air trapped in glacier ice. During each year's snowfall, air is trapped within tiny spaces between snowflakes. These spaces are then sealed off, and the air is encapsulated in tiny bubbles, as the snow compresses and

eventually turns into glacial ice. By drilling cores through the glaciers, it is possible to retrieve and sample these tiny air bubbles, which provide a more-or-less continuous record of what the atmosphere was like from the present back through time. In the case of the Vostock ice core from Antarctica, the record goes back about 420,000 years. Cores from Greenland extend back more than 100,000 years, and those from other ice fields throughout mountainous parts of the world provide local records that go back thousands of years. From measuring the CO_2 and other greenhouse gases present in the ice-core bubbles, it is possible to directly compare past greenhouse gas levels with those of today and to calculate how regional temperatures in the southern and northern high latitudes have changed through time.

Such analyses show that the most rapid and highest-magnitude temperature changes known in Earth history occur at the transitions between glacial and interglacial episodes, the last of which began about 14,300 years ago and ended about 11,000 years ago, with most of the warming occurring from 12,900 to 11,300 years ago. At such natural global warming events (driven by Earth's orbital variations), the mean global temperature increases about 9°F (5°C) from the coldest part of the glacial to the warmest part of the interglacial over a minimum of about 1,600 years, or an average rate of about 0.0056°F/year (0.0031°C/year).

In comparison, the worst-case likely IPCC scenario suggests a 13°F (7.2°C) rise from 1900 to 2100—a warming rate (0.065°F/year [0.036°C/year]) that is about twelve times faster than occurs during glacial–interglacial transitions. The best estimate for the best-case scenario (8.6°F/200 years [4.8°C/200 years]) yields a rate about eight times faster than the usual glacial–interglacial rate. The lowest rate estimated to be likely for the best-case scenario (3.4°F/200 years [1.9°C/200 years]) is still about three times greater than glacial–interglacial transitions. These elevated rates remain evident even after they are adjusted to account for the fact that rates measured over shorter intervals can appear faster than those measured over long intervals.

The warming likely to occur by 2100 also appears abnormally fast and high with respect to the climate changes that have occurred over the longer course of geological time, reaching back 600 million years. For example, a major global warming event apparently triggered by the sudden release of CO_2 from the ocean floor occurred about 54.9 million years ago, at which time the global temperature increased a maximum of 14.4°F (8°C) within as little as 10,000 years. This event, known as the Paleocene–Eocene thermal maximum, is the most rapid, high-magnitude climate change known outside of glacial–interglacial transitions, but its rate of change (maximum 0.0014°F/year [0.0008°C/year]) also was lower than that predicted for the years 1900 to 2000.

Paleoclimatic information for these longer time spans comes from the chemical composition of fossil shells from marine organisms called foraminifera, or *forams* for short. Forams are one-celled protists (simple one-celled organisms) that are extremely common in the oceans. Represented by many different species, almost all of which are microscopic in size, forams consist of a single cell of protoplasm contained within a tiny shell of calcium carbonate. When a foram dies, its shell sinks to the ocean floor, where it is incorporated into the mud and fossilized. Over time, the layers of mud accumulate, such that the oldest layers are on the bottom and the youngest ones on the top. By pushing a coring device into the ocean floor, extracting the cylinder of mud that is retrieved, taking slices of the mud at regular intervals, separating out the fossilized foram shells from each slice, and then analyzing the chemistry of the shells, it becomes possible to track how ocean temperatures have changed through long periods of time, in some cases millions of years.

The paleotemperature signature in foram shells is revealed by measuring the ratio of oxygen isotopes, specifically, oxygen-16 (^{16}O) to oxygen-18 (^{18}O). In general, at increasingly colder ocean temperatures, the foram shells contain more ^{18}O. After accounting for other factors that influence how much ^{16}O is in the oceans (particularly the effect of locking up ^{16}O in glacial ice during cold times), the ratio of ^{16}O to ^{18}O in the foram shells can be used as a paleothermometer of sorts that records how the temperature near the ocean floor (the benthic part of the ocean) changed at the place the core was recovered, through the entire time the core samples. By analyzing the forams from many temporally overlapping cores from many parts of the oceans, determining the temperature changes at each location through time, and then averaging them in appropriate ways, it becomes possible to estimate how the average temperature of Earth varied through time. This procedure has resulted in a global oxygen–isotope curve, which provides a relatively detailed record of temperature change for the past 70 million years. Combined with oxygen–isotope data derived from organisms other than forams, other chemical signatures, and modeling techniques, even longer-term temperature records have been reconstructed, reaching back to around 600 million years.

Such paleoclimate proxy information indicates that the mean ocean temperature near the ocean floor has varied less than about 18°F (10°C) over 600 million years (see Figure 2). More detail on ocean temperature emerges from the oxygen–isotope curve for the last 70 million years (see Figure 3). From about 70 to 38 million years ago (the Paleocene epoch and most of the Eocene), ocean-bottom temperatures varied within a 14.4°F (8°C) range.

It is important to realize that most of the paleoclimate records (especially those that are older than a few tens of thousands of years, back to millions of years) are for ocean-bottom temperatures, whereas current climate records and future projections are for the mean global temperature at Earth's surface. Therefore, comparing past climate changes with current projections requires understanding how ocean-bottom temperatures correlate with mean global surface temperature. This process is not straightforward. The general assumption, however, is that the gradient prevailing today can be used as a rough guide. For the period from 1901 to 2000, the average bottom-water temperature was about 35.6°F (2°C) whereas the average sea-surface temperature was about 61°F (16.1°C), the average land temperature was about 47.3°F (8.5°C), and the average global temperature at the surface (combining land and sea temperature) was about 57°F (13.9°C). These observations suggest that the average global

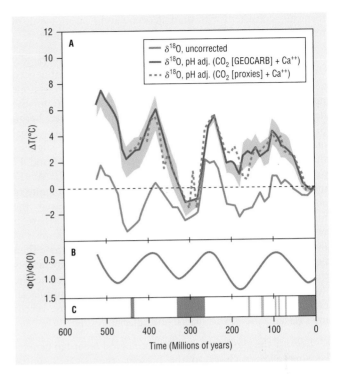

Figure 2. Estimated temperature variations for the past 600 million years. Estimated from oxygen–isotope ratios and other Paleoclimate proxy data as determined from a variety of fossil organisms and geologic information (based on Royer et al., 2004). Reproduced by permission of Gale, a part of Cengage Learning.

surface temperature is about 21.6°F (12°C) above the average temperature near the ocean floor.

Given that the estimates from oxygen–isotope data suggest a mean ocean-floor temperature of about 48.2°F (9°C) between 70 and 38 million years ago, the mean global surface temperature then is thought to be in the neighborhood of 70°F (21°C), which would mean much of the world would be

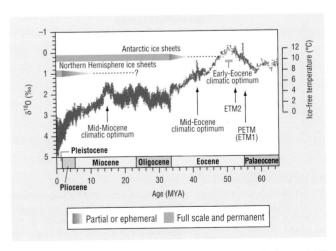

Figure 3. Relative global ocean-floor temperature through the past 70 million years as reconstructed from oxygen-isotope data derived from benthic foraminifera (based on Zachos et al. 2008). Reproduced by permission of Gale, a part of Cengage Learning.

tropical—consistent with the Paleocene and Eocene fossils of tropical plants and animals found even at high latitudes.

Around 33 million years ago, the global climate rapidly cooled as a result of plate-tectonic processes that influenced ocean currents. The relative change in the oxygen–isotope ratios of benthic forams, combined with adjustments for the depletion of ^{16}O in the oceans as continental ice sheets began to form and with a variety of information from terrestrial plant and animal fossils and paleoclimate modeling, suggests that the mean global temperature between about 33 and 17 million years ago fluctuated around 59°F (15°C). At 17 million years ago, a warming event that progressed at a rate of about 0.0000036°F/year (0.000002°C/year) increased the mean global temperature to about 65°F (18.5°C), where it remained until about 14 million years ago—a time known as the Mid-Miocene climatic optimum. The global climate then began a long, gradual cooling toward the global temperatures that eventually led into the *icehouse* world that became the norm about 2.6 million years ago, when glacial–interglacial cycling began. The present day is part of the most recent interglacial time of this glacial–interglacial cycling.

Because the geological and paleontological record in general becomes clearer closer to the present, the oxygen–isotope record of benthic foraminifera becomes more detailed for the last 5 million years (see Figure 4). That detailed record indicates that the current interglacial time and the previous interglacial, which occurred about 130,000 years ago, are the warmest times the planet has experienced in the last 2 million years.

For interpreting paleoclimate of the past few thousand years, a wide variety of paleoclimate proxy data is available, including lake-sediment cores that yield fossil pollen and various geochemical data; tree-ring records that provide information on annual climatic variation over thousands of years; lake-level fluctuations; and fossils of vegetation and animals collected and preserved in wood rat middens. For the past 1,000 or so years, historical records and temperature data

are available. The warmest time in the past several millennia was the Medieval Warm Period, which lasted from AD 950 to 1250; average global temperature as of 2012 is already higher than it was then.

These observations provide a context for understanding how the magnitude of warming that is plausible in just the next century compares with the range of climate changes that Earth's species have experienced in the past. A 11.5°F (6.4°C) rise (considered to have at least a 66 percent chance of occurring under the *business-as-usual* scenario by the IPCC) would elevate the mean global temperature to about 70°F (21°C). The last time it was that hot was between 38 and 70 million years ago, when subtropical forests grew in central North America (for example, in Wyoming) and as far north as Greenland. A 7.2°F (4°C) increase by 2100, the best estimate under the business-as-usual scenario, would equate to an average global temperature of 65°F (18.2°C), making the planet hotter than it has been in the past 14 million years. If the temperature rises 3.2°F (1.8°C) by 2100 (the IPCC best estimate for a reduced-emission scenario), the mean global temperature of 61°F (16°C) would be hotter it has been in the past 3 million years. In any of these scenarios, the planet will be hotter by 2070 than it has been since *Homo sapiens* emerged as a species. More recent studies suggest an even faster temperature rise is likely, with the mean global average temperature reaching 60 to 63°F (15.4 to 17°C) by 2050. These considerations suggest that Earth's temperature is rising to higher magnitudes, and at faster rates, than most species alive today have experienced in their evolutionary histories, especially when taking into account that the flora and fauna alive on Earth today are the legacy of a biota adapted to the Pleistocene icehouse world.

Climate change involves more than temperature

As the mean global temperature changes, other aspects of the climate system inevitably are forced to change as well. Even temperature change itself will not be uniform across the globe: Hidden in an *average global temperature* rise is the fact that not all local areas experience the same degree of warming. In general, high-latitude areas warm more than areas in lower latitudes: Arctic temperatures have already risen about twice as fast as temperatures are rising in the rest of the world. As the average global temperature increases, the Northern Hemisphere is expected to get a little warmer than the Southern Hemisphere, and continental interiors in general will experience more change in temperature than areas near the coast. Such differences are pronounced. IPCC projections indicate that by 2100, the temperature increase in northern Canada and northern Eurasia will be as much as 14.4°F (8°C), twice as much as the high-end estimate near the equator.

Going along with locally differing temperature changes will be changes in the amount of rain and snow that falls in a given area. In general, low-latitude areas—including the present-day tropics and subtropics—are expected to become drier, and higher-latitude areas are expected to become wetter. Drying already is evident in many tropical, subtropical, and midlatitude temperate regions, which have been

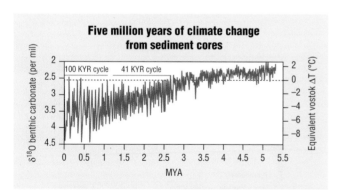

Figure 4. Relative global ocean-floor temperature through the past 5-million years based on oxygen–istope ratios from benthic foraminifera, compared to Antarctic relative temperature variations as interpreted from CO_2 in the Vostock ice core. The increased amplitude of the variations beginning 2.6 million years ago signals the onset of glacial-interglacial cycling, with values towards the top of the graph indicating interglacial times, and those towards the bottom indicating glacials. The dotted line shows the modern value for dO18 (left-side scale) and relative temperature (right-side scale) (based on Lisiecki 2005). Reproduced by permission of Gale, a part of Cengage Learning.

experiencing longer and more frequent periods of drought since the 1970s. At the same time, the nature of the climate system implies that, with global warming, the intensity and frequency of extreme weather events will increase. This means, for example, that the amount of rain that does fall becomes more concentrated in fewer, but more intense, storms, and that heat waves (several days in a row when temperatures are higher than normal for a given place) become more common, a development potentially important to the survival of species in some regions.

A warmer mean global temperature also implies that both the length of seasons and differences between the seasons will change in different ways in different areas. For example, in the northern Rocky Mountains of the United States, summer temperatures have not increased much, but winter temperatures have, accounting for most of the approximately 2.9°F (1.6°C) temperature increase in that area over the last century. One effect is that rainfall has increased at the expense of snowfall, which means the disappearance of snow in the spring. Similar trends hold throughout much of the Rocky Mountain region, where spring as defined by temperature is now arriving ten to thirty days earlier than was the case during the 1950s, contributing to a longer warm season of hot, dry weather.

Similar kinds of data indicate that these seasonal changes are underway worldwide. In England warm temperatures have been arriving six days earlier in the spring and lasting two days later into autumn—for each decade beginning in the late twentieth century. This means that in 2012 the English growing season was almost a month longer than it was in 1970.

Climate change also affects marine systems in diverse ways. For example, warming surface waters and more persistent coastal winds have influenced ocean currents along the coast of the US Pacific Northwest, ultimately depleting the water of oxygen and having caused dead zones to intermittently form during the early twenty-first century. A key concern is that through the interaction of atmospheric CO_2 with ocean water the ocean chemistry changes such that the water becomes more acidic. Between 1990 and 2010, the pH value of the ocean decreased by 0.05. Continued acidification as more CO_2 is added to the atmosphere is expected to disrupt the balance of ocean carbonate chemistry over the next century in ways that have no analog over the past 300 million years. Ongoing climate change is also causing the sea level to rise, as the ocean thermally expands and as glaciers melt, which influences species that live in the nearshore environment.

Biological responses to climate change

All these rapid changes—increasing temperature, changing patterns of precipitation, alterations in the length of seasons, extreme weather events, and changing ocean chemistry and currents—pose challenges for the continued survival of many species in the places where they now live. They face these challenges because most species can live only within a certain range of climatic conditions to which they have become adapted through the course of their evolutionary history. If a given climatic factor changes to be outside that livable range, the species has to respond, or else face extinction. If they are

to avoid extinction, species must respond to climatic changes either by moving as the climate changes (habitat tracking) or by adapting to the new climatic conditions. Examples of both kinds of response are evident for past climatic changes, although those changes were in general slower and of lower magnitude than the climatic changes projected over the next century.

Habitat tracking

The most typical response of species to climate change is simply to move, tracking their preferred climate space as it shifts across the surface of the globe. This response has been widely documented for species that survived the last natural global warming event, which was the transition from the last ice age into the present interglacial. For instance, in North America, species such as bog lemmings (*Synaptomys* spp.), which require cool, moist tundra-like conditions, followed their retreating habitat for some 930 miles (1,500 km) as continental glaciers, and the distinctive climate zone that existed just south of the ice front, receded northward. Likewise, in Eurasia, the geographic ranges of now-extinct species, such as Irish elk (*Megaloceros giganteus*) and woolly mammoth (*Mammuthus primigenius*), and extant species, such as musk oxen (*Ovibos moschatus*), reindeer (*Rangifer tarandus*), horses (*Equus ferus* and *Equus caballus*), and bison (*Bison priscus* and *Bison bison*), all shifted in coincidence with documented climate changes over the past 130,000 years, especially the last 50,000 years. Similar tracking of shifting climate space has been documented for many tree species in Europe and the United States. In general, moving to track climate change is easier for species that live in mountainous areas versus flatlands. Species that live in the mountains have to move only short distances upslope or downslope to stay in their required climate zone, whereas flatland species have to move farther to keep up with latitudinal shifts of their climate zones. This difference has been shown both by empirical observations of how mammals shifted their geographic ranges in response to the last glacial–interglacial transition, as well as through modeling efforts that examine how fast plants would have to shift their geographic ranges to keep up with the predicted rates and directions of climate change.

Ecophenotypic response

If a species cannot move to track its climate space, either because the pace of climate change outpaces its ability to disperse or barriers to dispersal are in the way, then it may respond by adapting to the new conditions. The first, and biologically easiest, step in adaptation is called ecophenotypic response and results from the inherent property of most organisms to exhibit phenotypic plasticity in growth and development. An example of phenotypic plasticity is well known for humans: Children fed nutritious diets grow up to be bigger adults than genetically similar counterparts that experience poor nutrition in childhood. Phenotypic plasticity is also influenced through climatic triggers. For example, warm-blooded animals in warmer places tend to be smaller than their counterparts of the same species in colder places. This relationship, known as Bergmann's rule, relates to physiological constraints that ultimately mean that larger animals lose proportionately less heat through the surface area of their body. Therefore, as the climate warms in a given area, an expectable ecophenotypic response is for individuals within many mammal species to get smaller. Fluctuations in body size that concord with Bergmann's rule are well documented in the fossil record as a response to past climate changes, with examples ranging from rodent species (for instance, the wood rat *Neotoma cinerea*) to bison (*Bison bison*).

Temperature, however, is not the only control for phenotypic response. For example, some species, such as the California ground squirrel (*Spermophilus beechyi*), exhibit a reverse Bergmann's response, indicating that the major control on their body size is precipitation rather than temperature. Where there is greater precipitation, the ground squirrels are bigger, probably because of increasing nutrition related to more abundant vegetation. Likewise, changes in the length of seasons can affect the nutritional quality of plants, which in turn affects the growth and development of the animals that eat them. For instance, experiments have shown that feeding bighorn sheep (*Ovis canadensis*) the highly nutritious spring green-up vegetation for two weeks more than that vegetation lasts under today's climate increases both their adult body size and horn size substantially. These experimental data are consistent with observations that many large-bodied herbivores, such as wild sheep, deer, and bison, were larger near the end of the last ice age when the spring season was longer because the position of Earth in relation to the sun was somewhat different then.

Other well-known ecological rules that highlight potential ecophenotypic responses to climate change are Allen's rule, which predicts that individuals within warm-blooded species will have relatively longer limbs (arms and legs) in warmer climates, and Gloger's rule, which predicts that individuals of species in warmer, wetter climates will generally have darker coloration than their counterparts in cooler, drier climates. Allen's rule, like Bergmann's rule, is based on physiological constraints that influence heat loss and retention, whereas Gloger's rule probably arises from the fact that warmer, wetter regions have darker soils and more vegetation, thus conferring a selective advantage on individuals with darker coloration.

The extent to which ecophenotypic response can allow an animal to adapt to climate change is limited, however, by its underlying genetic composition, or genotype. There is only so much variation a given individual can express. For instance, the offspring of two very short parents may grow up to be somewhat taller than the parents if its childhood nutrition is better, but its underlying genetics will never allow it to grow as tall as most offspring from two very tall parents (all other factors being equal).

Evolutionary response

When climate change pushes a species past its ecophenotypic response limit, evolutionary change is the only option if it is going to avoid extinction. Evolutionary change requires changing the underlying genome. The fast pace and consequently strong selective forces that future patterns of climate change are expected to exert on many species means that genetic change will need to happen very fast if a species is going to effectively adapt by evolution. The speed at which

evolutionary change can proceed is constrained, however, by several inherent limits. These limits are imposed by the rate at which genetic mutations occur (the rate of changes in the molecular structure of the genes), by the genetic variation contained in the species and the many different populations that compose it (in essence, the size of the gene pool), by the population size (the number of individuals that compose the species and each individual population within the species), and by the strength of the selective force (the pressure that climate change is exerting on the species to change). Given these constraints, only certain kinds of species are capable of evolutionary change that can keep pace with the predicted rates of climate change.

The chief constraint is mutation rates. Mutation rates, measured in generations, are generally slow—perhaps about one mutation in about 500 generations. Most mutations are neutral, meaning they do not confer any advantage or disadvantage to the organism. Therefore, all things being equal, the more generations produced in a given time, and the more individuals produced with each breeding, the more mutations in the gene pool, and the higher the chance that one of those mutations will cause a change in the organism that is advantageous and that can be passed on to its offspring. Thus, only in animals that produce new generations very quickly (within days or a few weeks) and that produce many individuals in each generation can mutation rates keep up with the rapidity of today's climate changes. Species in this category include some marine plankton, which have been shown to be capable of evolving fast enough to keep pace with ocean acidification; many insects (notable examples include fruit flies and possibly mosquitoes); and some rodents. Red squirrels (*Tamiasciurus hudsonicus*) in Canada's Yukon, for example, have exhibited a genetic response to the earlier arrival of warm spring temperatures, which requires an earlier breeding season. Between 1989 and 1998 the squirrels adapted by advancing their breeding season about six days per generation. Most of that adaptation was simply through behavioral plasticity, but about 0.8 days per generation seems to have been a result of changes in their genome.

However, the evolutionary speed limit imposed by the interaction of mutation rate, generation time, and number of offspring produced each generation implies that most vertebrate animals (mammals, birds, reptiles, amphibians, and fish) and many invertebrate animals (clams, snails, starfish, and so on) have little chance of adapting to ongoing climate change through genetic change induced by mutations. For them, it typically takes thousands to hundreds of thousands of years for mutations to accumulate into adaptively significant genetic change.

Much faster, but ultimately more limited adaptive responses are possible when the selection pressures caused by climate change act on existing genetic variation. This, in fact, is the process of natural selection that the English naturalist Charles Darwin (1809–1882) wrote about in his famous 1859 treatise *On the Origin of Species*, in which he laid out his ideas about evolution. *Natural selection* is the process in which less adapted individuals are culled from populations, leaving only the ones most fit to produce the next generation. The more genetic variation that exists within a species' gene pool, the greater is the potential adaptive response that can be expected over the short term. Generally, genetic variation increases with increasing numbers of individuals and increasing numbers of populations that compose a species, so the species that are most likely to adapt in this way will be those that have many individuals. Without mutation, however, there is only a fixed amount of genetic variation even in the largest populations. Therefore, adaptation driven only by culling the least-fit variants will proceed rapidly at first and then come to a sudden stop when most of the genetic variation has been culled by strong, unidirectional, and persistent selection forces. Extinction can occur if a rapidly changing climate produces intense selective pressures throughout a species' geographic range, such that existing genetic variation is essentially used up faster than mutations can produce new, suitable variants. There are some indications that the current, anthropogenically driven climate change is locally intensifying selection pressures to this extent, though species-wide effects are not yet documented.

Studies of the northern pocket gopher (*Thomomys talpoides*) and the Uinta ground squirrel (*Spermophilus armatus*) show that a climate-induced decline in genetic variation occurred in these two species around 1,000 years ago in Yellowstone National Park (located primarily in Wyoming). Climate change over the last century seems to have caused a loss of genetic diversity in living alpine chipmunks (*Tamias alpinus*) in Yosemite National Park (located in California).

Extinction from climate change

Climate change can lead to extinction by direct selection pressures, such as when individuals of a species find themselves unable to cope physiologically with warmer conditions or when their development is adversely affected. Extinction-causing pressures can also be indirect, as when the timing of the seasons gets out of synch with the genetically programmed seasonal rhythms of a species, or when effects on one species percolate to influence others. Finally, the potential for novel climates to emerge, as well as the synergistic interaction of biological changes induced by climate changes with those of other human impacts (primarily human population growth, energy use, habitat fragmentation, overharvesting, and invasive species), can cause biological tipping points that manifest as the extinction of a wide range of species.

Physiology

At high risk of extinction with continued warming are species whose physiological mechanisms are inadequate to cool them sufficiently in hot weather. Such species today are restricted to cool environments, usually of relatively small geographic extent and found at high elevations or high latitudes. A classic example is the pika (genus *Ochotona*), a small mammal related to rabbits (see Figure 5). Pikas die of heat stress if they are even briefly exposed to temperatures exceeding 78 to 85°F (25.5 to 29.4°C). As a result, they live only at high elevations, basically on mountaintops. Information from fossils shows that as conditions became warmer when the last glacial interval gave way to the present interglacial, pikas disappeared from many mountaintops. Between the 1950s and the early twenty-first century, as global warming increased temperatures in the mountains where

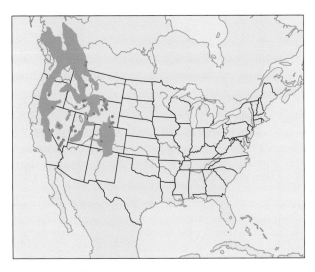

Figure 5. Pikas (*Ochotona princeps*) cannot tolerate high temperatures and are becoming locally extinct. The orange shading shows the geographic range of pikas in the late twentieth century. The blue dots show where pikas are known as fossils from the last ice age. The twentieth century range is smaller because pikas disappeared from the areas that warmed too much (based on data from FUANMAP). Reproduced by permission of Gale, a part of Cengage Learning.

populations survived the glacial–interglacial transition, pikas became locally extinct from about 30 percent of the mountaintops that had been studied in the Great Basin of the United States. The pikas died out primarily on the warmest and driest mountaintops. Chinese pikas (*Ochotona iliensis*) seem to be showing similar trends, although they are less well studied.

Musk oxen (*Ovibos moschatus*) also are at risk of extinction from warm temperatures. These bovids have several adaptations to efficiently retain body heat, which allows them to thrive in the high-latitude arctic areas in which they evolved.

Adult American pika (*Ochotona princeps*), Alberta, Canada. © All Canada Photos/Alamy.

For example, their fur is double-layered—structured for maximum insulation qualities—and they sweat only through their feet. Thus, when summer temperature persists above 50 to 54°F (10 to 12°C), they simply cannot lose the heat their metabolism produces fast enough. Besides direct death from overheating, musk oxen also illustrate the indirect mechanisms by which approaching maximum survival temperatures can cause extinction. Populations subjected to an abnormally warm late summer and fall in Norway exhibited high death rates, but not from overheating per se. Instead, the heat stress made them more susceptible to disease pathogens that caused a form of fatal pneumonia.

Besides potential extinction of entire species such as musk oxen from heat-related stresses, there is also a high chance of widespread extinction of populations within species as certain parts of the species' range experience lethal temperatures. In this situation the species may survive climate change itself, but with a much-reduced gene pool that makes the species susceptible to extinction from other direct impacts. This is particularly the case for various species of cold-water fish in freshwater habitats. As the rivers and lakes they live in warm, trout and salmon are expected to disappear from 18 percent to 30 percent of their current habitats by the year 2090, with some species seeing up to 17 percent loss by 2030, 34 percent by 2060, and 42 percent by 2090.

A warming climate can also cause cold-blooded animals to exceed physiological thresholds. Lizards in Mexico have seen 12 percent of their local populations go extinct coincident with warming trends, probably because the hotter temperatures cause them to spend too much time hiding in cool crevices to avoid overheating, which, in turn, means they do not spend enough time eating. This is a fatal trade-off, because if they hunt in the sun, they overheat and die, and if they hide from the heat, the loss of foraging time means they cannot sustain their needs. Such mechanisms, and the observed losses in Mexico since 1975, are consistent with model predictions that 39 percent of lizard species may experience local extinctions, and 20 percent may be lost worldwide, by the year 2080.

Likewise, rising ocean temperatures have already been shown to exceed the physiological limits of important species of marine protozoa called zooxanthellae. These species are *symbionts*, which means that they live in a mutually beneficial relationship with the coral species that form coral reefs. When ocean temperatures rise too high, the zooxanthellae die, which in turn kills the corals. This effect was evident in 1998, when a rise in the summer temperature of 0.9 to 1.8°F (0.5 to 1°C) above normal in the Indian Ocean bleached 80 percent of the corals, a sign of dying zooxanthellae, after which 20 percent of the coral reefs died.

Development

Development refers to the sequence of biological events that unfold to turn an embryo into an adult. The effects of climate changes on the development of wild animals have not yet been studied in detail; the work that has been done reveals that developmental impacts are important enough to decrease the chances of many species' survival. This is especially

evident in the oceans, where elevated atmospheric CO_2 is not only contributing to increasing water temperature but also spurring chemical reactions in the water that have the net effect of making the oceans more acidic. Compared to preindustrial values, surface ocean waters today have seen a decrease of about 0.1 pH units (the lower the pH, the higher the acidity), and predictions (based on climate models) call for a further increase in acidity to levels unprecedented in the past 300 million years. These rising ocean temperatures and acidity already are disrupting the developmental biology of some marine species, such *Crassostrea gigas*, an oyster raised on farms in the US Pacific Northwest. There, the warmer, more acidic waters cause the oyster eggs to die after a few days of apparently normal development. Experimental work on another oyster species, *Pinctada fucata*, has documented that the pH and temperature values predicted to characterize the oceans within the next few decades influence the genes that regulate the formation of their shells (that is, calcification mechanisms). This problem is likely to be widespread in a variety of marine species whose calcification pathways for shell formation depend on existing ocean chemistry.

Experiments also show the detrimental effects of predicted pH and temperature on certain fish. Organ damage and death result in very young fish of the species *Menidia beryllina* (commonly known as the inland silverside and found in estuaries of North America). In waters that have acidity and temperature predicted for the oceans in coming decades and centuries, Atlantic cod (*Gadus morhua*) experience severe damage to their liver, pancreas, kidney, eye, and gut within a month of hatching from their eggs. Sea bass (*Atractoscion nobilis*) grow enlarged otoliths—structures in their ears that are important in navigation and mobility. The sense of smell does not develop normally in orange clown fish (*Amphiprion percula*), which impacts their feeding. Damselfish (*Pomacentrus amboinensis*) raised in waters mimicking future ocean chemistry and temperature highlight a particularly interesting developmental effect: Their learning ability is impeded, to the extent that they fail to recognize one of their key predators. Such experiments suggest that researchers still have much to learn about potential climate-induced developmental relations.

Seasonal mismatches

Many species have a life cycle that depends on seasonal rhythms of reproduction, migration, and so on, timed to coincide with certain climatically controlled events. The study of the relation between a species' life cycle and interannual variations in climate is called phenology. Phenological data have revealed that the biological trigger for the seasonal rhythm often is not climate itself, but instead is hardwired (in a sense) into the organisms' genetic makeup to coincide with climatically independent aspects of the seasons, such as day length. This sets up the possibility for the seasonal life cycle to be disrupted with climate change, which, particularly in cases in which a species occurs only in a small geographic area, can lead to its extinction.

That ongoing climate change is altering the relationship between the life cycle and seasonal climate cycle in many species has been widely documented. More than 1,000 species of birds, mammals, frogs, butterflies, and plants now breed, nest, migrate, wake up from hibernation, bud, or flower several days earlier in the spring than they did in the mid-twentieth century. On average, such events have been happening earlier and earlier each spring, at the rate of about five days earlier each decade, with some species showing particularly dramatic phenological shifts. For instance, the North American common murre (*Uria aalge*) advanced its breeding season two months between 1975 and 2000, while the yellow-bellied marmot (*Marmota flaviventris*) advanced its emergence from hibernation by over a month in the same time period.

Such changes have the potential to trigger extinctions if the cue for the biological response puts the species in the wrong place at the wrong time with respect to climate. For example, certain populations of a bird that ranges through much of Europe, the blackcap (*Sylvia atricapilla*) overwinters in Spain, but migrates to Germany to breed, lay eggs, and raise young. Day length is the migration trigger to leave Spain. Through much of the bird's evolutionary history, day length in Spain has coincided with climatic conditions in Germany that promote hatching caterpillars. Thus, when the birds arrived in Germany and laid their eggs, the chicks emerged when the caterpillars they needed to eat were abundant. Since the early 1980s, however, spring has been arriving earlier and earlier in Germany, stimulating earlier hatching of caterpillars there, so when the blackcap chicks are born, food is scarce and many of them starve.

Paleontological evidence also indicates that seasonal mismatches can lead to the extinction of geographically isolated populations. For example, a shortening of the spring green-up season, combined with vegetation changes initiated by the last glacial–interglacial climate shift, apparently caused all populations of the Irish elk (*Megaloceros giganteus*) to go extinct on the island of Ireland some 3,000 years before their final extinction on the Eurasian mainland. This phenological mismatch stemmed from the genetic hard-wiring of the Irish elk to shed and regrow huge antlers each spring. Essentially, this demanded rapidly depositing some 100 pounds (45 kg) of calcium phosphate on their head, a physiologically taxing feat that required ingesting high-quality plants through the spring, when antlers grow and vegetation is most nutritious. Small antlers and small bodies of the elk near the time of their extinction document that a nutritional imbalance occurred when the nutritious spring-time vegetation became less abundant.

So far, fatal phenological mismatches have been observed primarily in particular populations rather than species-wide. Moreover, in some cases, what was interpreted as a detrimental mismatch initially was later shown not to be. For example, yellow-bellied marmots were at first suspected to suffer increased mortality with an earlier emergence from hibernation because there was less food available when they exited their burrows (vegetation was still under snow because spring snowstorms had also become more common). Continued study, however, demonstrated that adult survival actually increased, probably as a result of an ultimately longer spring and summer over which the animals could accumulate fat before the next season's hibernation. These caveats indicate that much work remains before scientists understand fully the extent to which

species extinctions can be caused by mismatches between emerging climate patterns and species life cycles.

Species interactions

Most of the examples of seasonal mismatches noted above and the dependence of corals on zooxanthellae and vice versa (mentioned in the section on physiology) also illustrate that species do not exist in isolation—they interact in complicated ways that promote at least one of the interacting species' survival. Little known so far are the ways in which climate change may alter species interactions and thereby amplify extinction. There is, however, a growing body of work on this topic. Local studies have found that disruptions of species interactions by climate change often lead to the local extinction (extirpation) of many species.

For example, climatically induced depletions of sardines in parts of the Pacific Ocean also cause reductions of the predators that rely on them. In the Sonoran Desert of the southwestern United States, following increased annual precipitation from 1977 to 1992 that changed vegetation, at least two species of rodents (the banner-tailed kangaroo rat [*Dipodomys spectabilis*] and the silky pocket mouse [*Perognathus flavus*]) and two species of seed-harvesting ants (*Pogonomyrmex rugosus* and *Pogonomyrmex desertorum*) went locally extinct in a study area near Portal, Arizona. As a result, other species in the area saw dramatic reductions in their populations, including horned lizard species (*Phrynosoma cornutum* and *Phrynosoma modestum*), which eat ants, and the burrowing owl (*Athene cunicularia*) and the Mojave rattlesnake (*Crotalus scutulatus*), both of which eat rodents. In the fossil record of Alaska, an apparent increase in moisture led to a vegetational change that correlated with the local extinction of a horse species related to the living onager (*Equus hemionus*) from Asia. These effects at local levels indicate that the extinction of a single species through climate-driven pressures can precipitate the extinction of dependent species.

This collateral-loss effect is most clear for affiliate species—that is, those that require the presence of another species in order to survive. Examples include butterflies and their larval host plants, and parasites and their hosts. One study estimated that accounting for affiliate species would add at least 6,300 co-endangered species to the list of host species currently considered endangered. Likewise, the loss of species that play critical roles in a particular ecosystem's food web would also be expected to amplify extinctions in the ecosystems in which they live.

Habitat loss

Habitat loss from human impacts, such as the conversion of natural landscapes to agriculture, deforestation, and trawling in the oceans, already is known to constitute a major threat for many species. Additional habitat loss is also predicted to result from climate change. This is because climate models show that, by the year 2100, particular combinations of climate characteristics important in determining a species' habitat—temperature, precipitation, extremes of hot or cold, and so on—will no longer be found where they are today. That the loss of suitable habitat space for many species is already underway is illustrated by the iconic species for the impact of global warming, polar bears (*Ursus maritimus*), which are experiencing both a significantly smaller geographic extent of the pack ice on which they rely and a shorter season of pack ice.

Theoretical models that take into account how much habitat will be lost as the global climate changes and then relate the number of species living in particular habitats today to how much the habitat of these species will shrink and move predict substantial species losses from the changing climate. An often-cited 2004 study by Chris D. Thomas and colleagues suggests that of the 1,000-plus species of mammals, birds, frogs, reptiles, butterflies, other invertebrates, and plants whose habitats the researchers defined by climatic parameters, 15 percent to 37 percent would probably experience extinction from climate change, possibly as early as 2050. A later study by Fangliang He and Stephen P. Hubbell published in 2011 suggested that other equations are more appropriate, which if applied would reduce the potential losses to as low as 6 percent to 11 percent of studied species, which is still a relatively high proportion.

It is also becoming apparent through modeling efforts that climatically defined habitats will not simply shift, shrink, or expand. Instead, by the year 2100 it is likely that many habitats

defined by certain combinations of climatic parameters will be lost entirely, so much so that between 10 percent and 48 percent of the planet will no longer have the climate that contemporary species evolved to exist in (these are called *disappearing climates*). Conversely, new habitats are expected to emerge as new combinations of climate parameters (so-called *novel climates*) become widespread, such that 12 percent to 39 percent of Earth's surface will be characterized by a climate that modern species have never experienced. Most of these kinds of changes will take place in tropical and subtropical areas that now harbor the majority of Earth's biodiversity (see figures 6 and 7). Therefore, the potential for climate-caused extinctions in these biodiversity hot spots is high, particularly given that species in these areas evolved under a climate that has changed relatively little through time and are adapted to only slight seasonal differences, relative to species at higher latitudes.

Synergies and tipping points

The extinction threats posed by human-caused global climate change are magnified through interactions with other human-induced extinction pressures: growing human populations (which now appropriate some 40 percent of Earth's primary productivity at the expense of other species), habitat

Novel climates

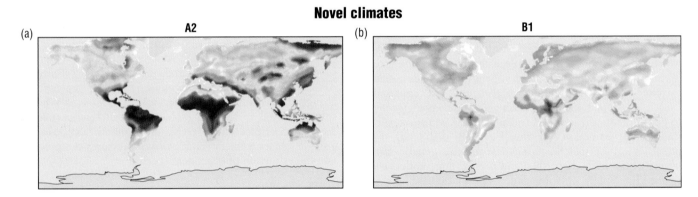

Disappearing climates

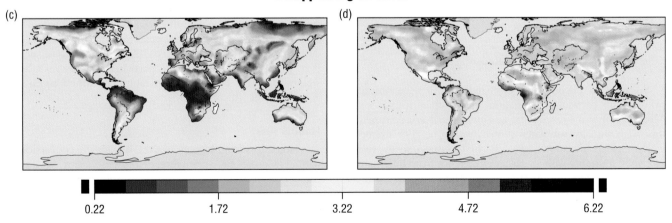

Figure 6. Areas in the world where novel climates (a and b) and disappearing climates (c and d) are projected to occur by the 2100 under two different models of climate change (A2 and B1). Warmer colors indicate increasing dissimilarity of projected climate with respect to existing climate. The B1 scenario assumes actions will be implemented to slow climate change; the A2 scenario assumes less slowing of climate change (see Figure 1). Note that the areas that will experience the most change in terms of novel and disappearing climates broadly overlap with biodiversity hotspots illustrated in Figure 7 (based on Williams 2007). Reproduced by permission of Gale, a part of Cengage Learning.

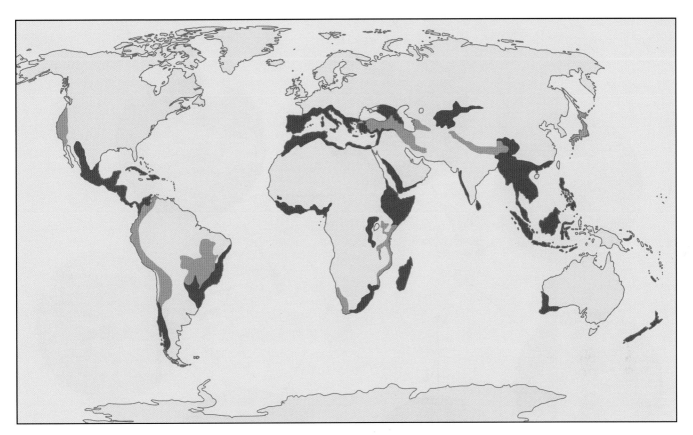

Figure 7. Biodiversity hotspots (dark tan and brown). Notice broad overlap with areas of disappearing and novel climates as shown in Figure 6 (based on data from DedicaGroup). Reproduced by permission of Gale, a part of Cengage Learning.

destruction and fragmentation, overharvesting, and replacement of native species by human-dispersed ones. For example, the rapidity of today's climate change means that in order to track their preferred habitat species would have to shift their current distributions even faster than they did during the transition from the last glacial to the present interglacial times. But more than 40 percent of Earth's lands are now dominated by farms, ranches, cities, roads, and other entirely human-constructed habitats, which means that even if species could move fast enough to keep up with climate change, they face impenetrable barriers as a result of severe habitat fragmentation. Many of the last refuges for biodiversity—national parks and the like—are so widely separated by human-constructed landscapes that a species losing its preferred climate in one protected area has no chance of dispersing into another area where its habitat may still exist. This interaction effect means that there will be a much higher loss of species than if the climate changed in the absence of habitat fragmentation. For example, the bay checkerspot butterfly (*Euphydryas editha bayensis*), after having most of its habitat destroyed in California's San Francisco Bay area, had two surviving populations left at Stanford University's Jasper Ridge Biological Preserve. Those populations went extinct in 1991 and 1998 as a result of unusual precipitation events that disrupted the phenological overlap of the butterfly larvae and their host plants.

At a larger scale, interactions between multiple extinction drivers can result in wholesale planetary transformations (see Figure 8)—so-called tipping points—which the geological record demonstrates can result in mass extinctions. Mass extinctions are times when 75 percent to 95 percent of species go extinct in a geologically short time. These highest-intensity extinction events have happened only five times in the past 540 million years. Each of these so-called big five mass extinctions was coincident with unusually fast changes in atmospheric and ocean chemistry, notably in CO_2; such changes are also evident today.

Minimizing species loss from climate change

So far, the time scale over which anthropogenically caused climate change has been influencing species is short, but biotic responses are already evident. Although it may be possible to slow the rate of climate change by concerted, international cooperation to transition from fossil fuels to renewable energy resources, by more efficient use of fossil fuels during the transition period, and by reducing global per capita energy consumption, even the best-case scenarios mean that species will likely continue reacting to climate change in all of the ways noted above. Present and anticipated greenhouse gas emissions are, in fact, much closer to the worst-case scenarios.

Therefore, climate change in the next few decades has the potential to cause significant numbers of extinctions. Evolutionary adaptation on the time scale needed will not be an option for most species; the primary and easiest response of animals will be limited to simply tracking their needed climate as it shifts across Earth's surface.

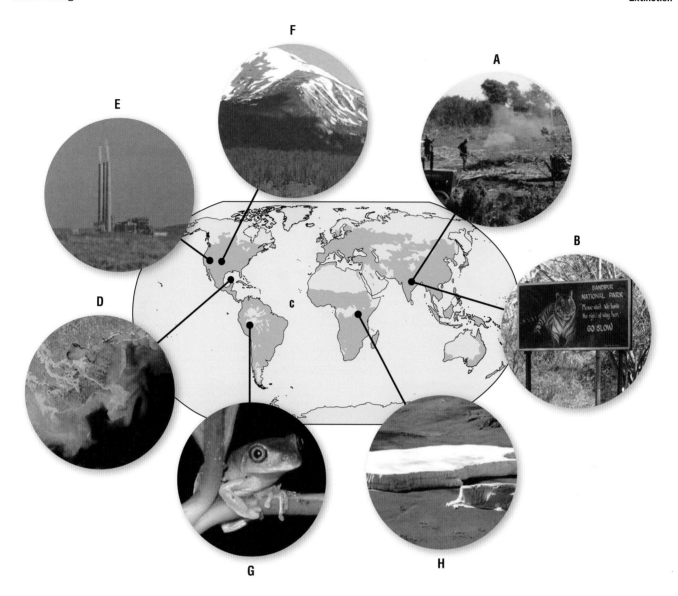

Figure 8. Interaction of individual causes can cause global pressures that significantly magnify extinction potential. Today's pressures emerge as humans transform local habitats (a), causing effects on nearby undisturbed areas (b). Local distrubances that now have turned more than 40% of Earth's surface has been turned into human-constructed habitats, primarily agricultural lands (c, agricultural lands in green). Habitat transformation precipitates changes in distant areas, such as dead zones in oceans (d). Energy required to sustain the ever-growing human population produces unusually high emissions of greenhouse gases (e), which change climate and ocean chemistry, exerting planetary-scale forces for biological change even in remote areas: forests of dead conifers as warm winter temperatures promote bark-beetle infestations (f); extinctions (g); and melting glaciers (h) (based on Barnosky 2012). Reproduced by permission of Gale, a part of Cengage Learning.

In view of this fact, a key strategy to minimize extinction under climate change is to facilitate connections between the areas where a threatened species is now found, and where it will have to go in the future to follow its needed climate. The ideal is connecting areas of relatively undisturbed habitat with migration corridors through which species can move naturally, but this will not always be feasible. Indeed, in more cases than not, it is likely that already-existing human use of land and seascapes will prevent the creation of dispersal corridors. A solution that is now being explored is the feasibility of human intervention to relocate imperiled species and populations to climatically more suitable areas. This has already been done with the marbled white butterfly (*Melanargia galathea*) in England, as summarized by in a 2009 article by Carl Zimmer in Environment360 and the Florida torreya (*Torreya taxifolia*) in the southeastern United States (see a 2007 article in *Conservation Biology* by Jason McLachlan and coauthors). Obvious concerns arise in moving species anthropogenically, in light of the potential for disruption to the ecosystem in which the species is inserted; the risks and benefits must be weighed on a case-by-case basis. This practice also highlights the need for an evolving philosophy of nature conservation, whereby the concept of saving species becomes separated from the concept of saving wilderness and maintaining nonanthropogenic ecological processes. The former is accomplished by managed reloca-tion of species, or even through zoos, but the latter is not, as by definition wild places are those ecosystems unmanaged (or at least only lightly managed) by people. The species conservation community is currently grappling with such issues.

Resources

Books

Barnosky, Anthony D. *Heatstroke: Nature in an Age of Global Warming*. Washington, DC: Island Press/Shearwater Books, 2009.

Intergovernmental Panel on Climate Change (IPCC (). "Summary for Policymakers." In *Climate Change 2007: The Physical Science Basis; Contribution of Working Group I to the Fourth Assessment Report of the Intergovernmental Panel on Climate Change*, edited by Susan Solomon, Dahe Qin, Martin Manning, et al., 1–18. Cambridge, U.K.: Cambridge University Press, 2007. Published for the Intergovernmental Panel on Climate Change.

Periodicals

Barnosky, Anthony D., Nicholas Matzke, Susumu Tomiya, et al. "Has the Earth's Sixth Mass Extinction Already Arrived?" *Nature* 471, no. 7336 (2011): 51–57.

Blois, Jessica L., and Elizabeth A. Hadly. "Mammalian Response to Cenozoic Climate Change." *Annual Review of Earth and Planetary Sciences* 37 (2009): 181–208.

Brown, James H., Thomas J. Valone, and Charles G. Curtin. "Reorganization of an Arid Ecosystem in Response to Recent Climate Change." *Proceedings of the National Academy of Sciences of the United States of America* 94, no. 18 (1997): 9729–9733.

Dawson, Terence P., Stephen T. Jackson, Joanna I. House, et al. "Beyond Predictions: Biodiversity Conservation in a Chang-ing Climate." *Science* 332, no. 6025 (2011): 53–58.

Doney, Scott C. "The Growing Human Footprint on Coastal and Open-Ocean Biogeochemistry." *Science* 328, no. 5985 (2010): 1512–1516.

Grayson, Donald K. "A Brief History of Great Basin Pikas." *Journal of Biogeography* 32, no. 12 (2005): 2103–2111.

He, Fangliang, and Stephen P. Hubbell. "Species–Area Rela-tionships Always Overestimate Extinction Rates from Habitat Loss." *Nature* 473, no. 7347 (2011): 368–371.

Hönisch, Bärbel, Andy Ridgwell, Daniela N. Schmidt, et al. "The Geological Record of Ocean Acidification." *Science* 335, no. 6072 (2012): 1058–1063.

Koh, Lian Pin, Robert R. Dunn, Navjot S. Sodhi, et al. "Species Coextinctions and the Biodiversity Crisis." *Science* 305, no. 5690 (2004): 1632–1634.

Lisiecki, Lorraine E., and Maureen E. Raymo. "A Pliocene-Pleistocene Stack of 57 Globally Distributed Benthic δ^{18}O Records." *Paleoceanography* 20, no. 1 (2005): PA1003.

Loarie, Scott R., Philip B. Duffy, Healy Hamilton, et al. "The Velocity of Climate Change." *Nature* 462, no. 7276 (2009): 1052–1055.

Lorenzen, Eline D., David Nogués-Bravo, Ludovic Orlando, et al. "Species-Specific Responses of Late Quaternary Megafauna to Climate and Humans." *Nature* 479, no. 7373 (2011): 359–364.

McInerney, Francesca A., and Scott L. Wing. "The Paleocene–Eocene Thermal Maximum: A Perturbation of Carbon Cycle, Climate, and Biosphere with Implications for the Future." *Annual Review of Earth and Planetary Sciences* 39 (2011): 489–516.

McLachlan, Jason F, Jessica J. Hellmann, and Mark W. Schwartz. "A Framework for Debate of Assisted Migration in an Era of Climate Change." *Conservation Biology* 21 (2007): 297–302.

McLaughlin, John F., Jessica J. Hellmann, Carol L. Boggs, and Paul R. Ehrlich. "Climate Change Hastens Population Extinctions." *Proceedings of the National Academy of Sciences of the United States of America* 99, no. 9 (2002): 6070–6074.

Miller-Rushing, Abraham J., Toke Thomas Høye, David W. Inouye, and Eric Post. "The Effects of Phenological Mismatches on Demography." *Philosophical Transactions of the Royal Society* B 365, no. 1555 (2010): 3177–3186.

Ozgul, Arpat, Dylan Z. Childs, Madan K. Oli, et al. "Coupled Dynamics of Body Mass and Population Growth in Response to Environmental Change." *Nature* 466, no. 7305 (2010): 482–485.

Parmesan, Camille. "Ecological and Evolutionary Responses to Recent Climate Change." *Annual Review of Ecology, Evolution, and Systematics* 37 (2006): 637–639.

Pimm, Stuart L. "Climate Disruption and Biodiversity." *Current Biology* 19, no. 14 (2009): R595–R601.

Royer, Dana L., Robert A. Berner, Isabel A. Montañez, et al. "CO_2 as a Primary Driver of Phanerozoic Climate." *GSA Today* 14, no. 3 (2004): 4–10.

Sinervo, Barry, Fausto Méndez-de-la-Cruz, Donald B. Miles, et al. "Erosion of Lizard Diversity by Climate Change and Altered Thermal Niches." *Science* 328, no. 5980 (2010): 894–899.

Thomas, Chris D., Alison Cameron, Rhys E. Green, et al. "Extinction Risk from Climate Change." *Nature* 427, no. 6970 (2004): 145–148.

Williams, John W., Stephen T. Jackson, and John E. Kutzbach. "Projected Distributions of Novel and Disappearing Climates by 2100 AD." *Proceedings of the National Academy of Sciences of the United States of America* 104, no. 14 (2007): 5738–5742.

Ytrehus, Bjørnar, Tord Bretten, Bjarne Bergsjø, and Ketil Isaksen. "Fatal Pneumonia Epizootic in Musk Ox (*Ovibos moschatus*) in a Period of Extraordinary Weather Conditions." *EcoHealth* 5, no. 2 (2008): 213–223.

Zachos, James C., Gerald R. Dickens, and Richard E. Zeebe. "An Early Cenozoic Perspective on Greenhouse Warming and Carbon-Cycle Dynamics." *Nature* 451, no. 7176 (2008): 279–283.

Anthony D. Barnosky

Habitat loss

Most basically, *habitat* is defined as the resources and conditions necessary to allow survival and reproduction of a given organism (Hall et al. 1997). Each species performs best near an optimum value of each resource and cannot survive when this value diverges beyond its tolerance (Schwerdtfeger 1977). Habitat is best thought of as individualistic and multidimensional (Hutchinson 1957). It is individualistic because each species has a unique ecological niche that differs from that of all other species. It is multidimensional because there are typically several to many important environmental variables that define each species' habitat. It is also multiscale because each of these environmental variables is likely to be related to space or other resource use at different spatial scales (Grand et al. 2004; Wasserman et al. 2012). For example, in a 2002 study, Craig M. Thompson and Kevin McGarigal found that bald eagle habitat selection was strongly driven by a number of environmental variables but that each kind of habitat selection (such as for foraging or roosting locations) was driven by different variables at different scales.

Most studies of habitat loss and fragmentation have used a simplified model of habitat known as the *island biogeographic perspective*. This perspective assumes that one can represent the environmental characteristics that are important to an organism as a single map class, defined as habitat, and that all locations that are nonhabitat are equivalent and can be treated in a uniform manner (Turner et al. 2001). This simplistic view of habitat, however, is usually insufficient to reflect the multidimensional nature of habitat structure in most species (e.g., Cushman et al. 2010). Thus, most studies of the effects of habitat loss and fragmentation begin with untested assumptions related to how a habitat is being defined. This is not merely a semantic issue; imprecise definition of the term *habitat* has contributed greatly to confusion and disagreements with regard to the effects of habitat loss and fragmentation on populations. Reliable assessment of the effects that habitat loss and fragmentation have on a population depends on defining habitat in a way that is biologically meaningful and represents the environmental features at the scales that determine suitability for the species.

Habitat loss and fragmentation defined

When a habitat is defined as a single cover type, it is easy to define habitat loss as the reduction in the extent of the cover type and habitat fragmentation as the breaking of habitat into pieces. When there is no focal habitat type, and the landscape instead is represented as a mosaic of different patch types, it is substantially harder to define habitat loss and fragmentation. When habitat is represented as a series of gradients of varying quality, there are no patches at all (e.g., Cushman et al. 2010). In such situations should habitat loss be defined as a reduction in the total aggregate quality of habitat across the study area, or a reduction in the proportion of area meeting a certain threshold of quality, or something else? Also, in a gradient representation of habitat, how is fragmentation defined? In the absence of patches, it is not possible to measure the "breaking apart" of habitat. Instead, one might quantify changes in the surface pattern (Cushman et al. 2010) of habitat quality.

There has been considerable debate since the late 1990s about the relative impacts of habitat loss and fragmentation. Is habitat loss always important? Are there thresholds at which habitat loss begins to reduce population viability? Does the fragmentation of habitat, independent of habitat loss, have consistent negative impacts? How much habitat and in what configuration is required to maintain viable populations of target species? Distinguishing the effects of habitat fragmentation from those resulting from habitat loss has important implications for conservation biology. As Lenore Fahrig noted in a 1997 article:

> If fragmentation is important, then within some limits it should be possible to mitigate effects of habitat loss by ensuring that remaining habitat is not fragmented. On the other hand, if the effects of fragmentation are trivial in comparison to the effects of [habitat] loss, then the assumption that loss can be mitigated by reduced fragmentation… has potentially dangerous consequences for conservation. (p. 603)

Despite the challenges of defining habitat and quantifying the effects of changes in habitat area and configuration on population responses, a clear and compelling message emerges from the technical literature. It is akin to scientists measuring the topology of a crater left by an asteroid impact. Methods improve, perspectives change, and scientists debate which measurement approaches are best. But nobody argues whether or not the impact has occurred.

The relationship between habitat and population

Population size is the total number of individuals of a species inhabiting a defined region. Populations change as a result of birth, death, immigration, and emigration. The science of population viability analysis (PVA; Boyce 1992) focuses on predicting the long-term viability of populations as a function of rates of births, deaths, immigration, and emigration. Reliably predicting the probability of population extinction over any time period is challenging because PVA depends on a number of parameters, including age-specific survival and fecundity rates, dispersal probabilities, initial population size, and spatial distribution (Beissinger and McCullough 2002). Typically, these parameters are difficult to obtain, and any error in their estimation can bias estimates of extinction probability (Reed et al. 2002). In addition, the processes governing birth, death, immigration, and emigration are influenced by a number of factors, including the extent, quality, and connectivity of habitat (Andrén 1994), as well as interspecific interactions (Kareiva 1987). Thus, reliable PVA requires extensive, high-quality data, usually drawn from intensive, long-term studies, which are rare (Flather et al. 2011).

Despite the challenges of formal PVA, there are several general principals that apply. Generally speaking, a given extent and quality of habitat can support a particular number of individuals, known as the *carrying capacity* (Verhulst 1838; Pearl 1927). Given an initial population at the carrying capacity, the loss or fragmentation of habitat could have several effects on the population. First, the disturbance leading to the loss of habitat may kill individuals directly, increasing the mortality rate and decreasing the population size. Second, if individuals survive the disturbance, they will be left with a lesser extent, lower quality, or more fragmented distribution of habitat that will likely reduce the carrying capacity of the landscape. This produces an excess of individuals in a population temporarily larger than what can be sustained. These individuals can either emigrate in search of other habitat or face reduced fitness in the current landscape. This reduced fitness translates into reduced survival of adults, reduced fecundity rates, and/or reduced survival of offspring (Pearl 1927). Over time, the population would be expected to decline to the new carrying capacity that would be supported by the new reduced extent or higher fragmentation of habitat (Verhulst 1838).

This is an idealized characterization of processes that are, in reality, much more complex. For example, the spatial configuration of habitat may play a substantial role in affecting persistence (With and King 1999). Long-term persistence often requires a sufficient number, size, and spatial configuration of habitat fragments (Hanski and Ovaskainen 2002). The *metapopulation capacity* is the likelihood of long-term population viability given a particular extent, configuration, and quality of habitat. Habitat loss and fragmentation reduce the metapopulation capacity of a landscape and make extinction more likely. Thus, in addition to knowing the extent and quality of the remaining habitat, identifying the habitat's spatial configuration and connectivity is essential to determining the effects on population size (Ovaskainen 2002).

Aerial view of destroyed rain forest in Brazil near Amazon River. © Chad Ehlers/Alamy.

Variation in demographic processes and interspecific interactions can also influence the impacts of habitat loss and fragmentation on a population. Variable demographic rates can often lead to short-term population fluctuations that may not reflect the effects of a reduction in habitat carrying capacity (Schaffer and Kot 1986). In addition, habitat loss and fragmentation can have strong influences on predator-prey and competitive interactions (Ives 1995), changing the net effect from what might be predicted by habitat loss and fragmentation alone (Kareiva 1987; Kotlar and Holt 1989). Under some conditions, habitat subdivision and isolation can allow for predator-prey coexistence, by providing refuges for prey and reducing the ability of predator populations to track prey abundance (Huffaker 1958; Holyoak 2000). In contrast, when habitat subdivision leads to extensive refuges and the greatly reduced ability of predators to track prey populations, the prey population may erupt (Oksanen et al. 1992). In other situations, habitat loss and fragmentation may result in the elimination or depression of the prey species by destabilizing predatory-prey population dynamics to a degree that periodic predator population spikes lead to prey extinction (May and Robinson 1985).

Habitat subdivision and isolation, in some circumstances, can moderate competition and allow coexistence, for example, when it enables species with partially overlapping niches to utilize different resources (Holyoak and Lawler 1996) or when temporal patterns of disturbance result in an equilibrium between dispersal rate and competitive ability (Fahrig 2003). For example, in a 1981 study, W. D. Atkinson and B. Shorrocks found that the coexistence of two competing species could be prolonged by dividing the habitat into smaller patches. In contrast, habitat loss and subdivision may increase the intensity of competition and lead to competitive exclusion in cases in which resource displacement is not possible and the competitors are forced to compete for more limited resources in smaller habitat remnants. In this regard, Brent J. Danielson predicted in a 1991 article that interactions between competitors are very sensitive to the structure of the landscape and that as fragmentation increases, competition may increase in severity until one of the species is eliminated.

As a result of demographic fluctuations and interspecific action, patterns of occupancy may not always reflect habitat suitability. Bea Van Horne argued in a 1983 article that it is highly questionable to infer habitat quality based on patterns of occupancy because habitat quality should reflect fitness benefits accrued to individuals by selecting the habitat resources in a particular place. Furthermore, occupancy can often be decoupled from actual habitat quality because of attractive sink habitats (Delibes et al. 2001), dispersals from nearby high-fitness landscapes that are above carrying capacity and thus produce an excess of emigrants, environmental fluctuations, or other processes (e.g., Cushman 2010).

These factors make predicting the effects of habitat loss or fragmentation challenging. In a 2008 study, Samuel A. Cushman and colleagues evaluated the sufficiency of habitat characteristics for predicting species occurrence patterns. These authors found that more than half of the variance in species abundance patterns was not predictable by patterns of habitat. This finding suggests that while existence of sufficient amounts of suitable habitat is a necessary condition for occupancy, it is not a sufficient condition, as there are many reasons why suitable habitat may not be occupied or be occupied at densities less than carrying capacity, related to historical factors of past population fluctuations and source-sink dynamics. Cushman and his colleagues also found that definitions of habitat differed substantially in how well they predicted species abundance patterns. Species responded to multiple environmental variables across a range of spatial scales, from vegetation patterns in the immediate vicinity of sampling locations, to patch size and shape, to the composition and configuration of the landscape in terms of a patch mosaic of forest cover types and successional stages. Importantly, they found that the area of forest cover types in the landscape was a poor predictor of species abundance, showing that a mismatch between the definition of habitat and species responses can lead to incorrect or equivocal conclusions.

The challenges of defining habitat quality and predicting how changes in extent and connectivity of habitat will affect birth, death, immigration, and emigration often make it difficult to produce reliable, detailed predictions of how a given landscape change will affect a population. Fortunately, several major relationships have been found to have general validity and provide reliable guidance as to the direction and nature of expected effects. The adverse effects of habitat loss and fragmentation on biodiversity can be divided into two dominant categories. First, as habitat is lost from the landscape, at some point there will be insufficient area of habitat to support a population, and the species will be extirpated from the landscape (Flather and Bevers 2002). This is referred to as the *area effect*. Second, as habitat is lost and fragmented, individual habitat patches become more isolated from one another. As populations become subdivided, the movement of individuals among habitat patches (e.g., dispersal) may decrease or cease altogether, which may affect critical metapopulation processes such as gene flow, demographic rescue, and recolonization following local extinction (Fahrig and Merriam 1994). This is referred to as the *isolation effect*.

Area effects

Reduction in the extent or quality of habitat in a landscape reduces the potential of the landscape to support the species that depend on that habitat by reducing the ability of individual organisms to acquire the resources needed to survive and reproduce (Wiens et al. 1993). When habitat is correctly defined relative to the requirements of a species, a loss of habitat has consistently negative effects (Fahrig 2003), including reductions in species richness (Steffan-Dewenter et al. 2002), population declines, changes in distribution (Donovan and Flather 2002), and extinction (Pimm and Raven 2000).

Habitat loss generally has two kinds of effects on population size, depending on the severity of habitat loss in relation to the area requirements of the species. First, as habitat is lost from a system with extensive habitat, there generally is an initial pattern of linear decrease in population size as a function of habitat loss. Each unit area of habitat can support a certain number of individuals, and a reduction in habitat area can be expected to reduce the population in proportion to the reduction in total carrying capacity that results from the reduced habitat extent. Most species, however, require at least a minimum area of habitat in order to meet all of their life requirements (e.g., Robbins et al. 1989). Thus, as habitat is progressively lost, a point will be reached at which the habitat can no longer support a viable population. This point is known as the *extinction threshold* (Fahrig 2003). The amount of habitat required for species persistence depends on species' behavior and life history (Fahrig 2001). The effects of habitat loss on each species will depend on the interaction of its ecological requirements and capabilities with the degree of habitat loss in the surrounding landscape (Fahrig 2003). For example, many Neotropical migratory bird species display a marked area-sensitivity whereby their probability of occurrence in a forest patch increases nonlinearly with patch size (Trzcinski et al. 1999). Also, large-bodied, high-trophic-level species appear to be particularly vulnerable to local extinction due to habitat loss (Gibbs and Stanton 2001). As the habitat is lost, the most area-sensitive species will be lost first. As the habitat is further fragmented, other species will drop out according to their minimum area requirements (e.g., Flather and Bevers 2002). Thus, smaller patches generally contain fewer species than larger patches (Debinski and Holt 2000), and the set of species remaining in small patches is often a predictable subset of those found in large patches in the same region (Fahrig 2003).

There is often a time lag between a change in habitat area and the resulting population response, as it takes time for a reduction in habitat extent or quality to be expressed as reduced population size. Thus, a population may persist in a landscape for some time after its extinction threshold of habitat loss has been exceeded. These pending extinctions are sometimes referred to as an *extinction debt* (Tilman et al. 1994).

There has been considerable research to evaluate the extinction debt concept. In a 1994 study, David Tilman and colleagues modeled species persistence in a patchy environment, finding that even moderate habitat destruction caused a time-delayed but deterministic extinction of the dominant competitor in remnant patches. Additional species were predicted to become extinct as habitat destruction increases, in order from the best to the poorest competitors. The expected debt will be especially large in communities in which many species are close to their extinction threshold following habitat loss, and landscapes that have recently experienced substantial habitat loss and fragmentation are expected to show a transient excess of rare species (Hanski and Ovaskainen 2002). Several researchers have documented evidence of ongoing extinctions that are consistent with the extinction debt. In a 2002 study, Ilkka Hanski and Otso Ovaskainen found that the number of regionally extinct old-growth forest beetles is much greater in areas where human impact on forests has been lengthy than in areas where extensive habitat loss is recent. The time required for an extinction debt to reach equilibrium is highly dependent on the rates of population extinction and colonization expected given the species' mobility and demographic attributes. For example, in a 2006 study, Mark Vellend and colleagues found that, more than a century after forest fragmentation reached its current level, an extinction debt persists for species with low rates of population turnover.

Isolation effects

As habitat is lost and/or fragmented, the extent of connected habitat is reduced. In this process large, extensive regions of habitat may be broken into separate fragments. The subdivision and isolation of populations caused by fragmentation can lead to reduced dispersal success and patch colonization rates, which may reduce persistence of local populations and enhance the probability of regional extinction for the entire metapopulation (e.g., Hanski and Ovaskainen 2002). Specifically, increased population isolation increases extinction risk by reducing demographic and genetic input from immigrants and reducing the chance of recolonization after local extinction (Kareiva and Wennergren 1995). The degree to which landscape fragmentation affects connectivity is a threshold phenomenon depending on the mobility of the species and its sensitivity to the type of changes occurring in the landscape. The landscape change may not affect connectivity if movement among patches is largely unimpeded by intervening habitats and connectivity across the landscape is maintained. In a 2011 study, Ruth A. Short Bull and colleagues showed that the population connectivity of the American black bear (*Ursus americanus*) was unaffected by roads or loss of forest cover when the forest mosaic remained highly connected. In contrast, gene flow became highly affected by landscape structure when loss of forest cover and road development exceeded particular thresholds. Similarly, in another 2011 study, Cushman and colleagues found that American marten (*Martes americana*) foraging behavior was not strongly influenced by nonforest areas or roads when these features were rare in the landscape and did not limit foraging in high-quality habitats. These researchers found, however, that American marten movements became highly constrained by roads and areas of recent timber harvest when landscape change exceeded a threshold, with longer, more tortuous paths that were frequently forced through suboptimal habitats. Such movements through suboptimal habitats often require higher energetic expenditures (Mahan and

American marten (*Martes americana*) movements became highly constrained by recent timber harvest when the landscape change exceeded a threshold, with longer, more tortuous paths that were frequently forced through suboptimal habitats. © Steve Brigman/ShutterStock.com.

Yahner 2000) and expose individuals to higher rates of predation (Bergin et al. 2000) and reduced breeding (Kurki et al. 2000) and dispersal success (With and King 1999).

In a 2012 study, Cushman, Shirk, and Landguth used simulation modeling to explore connectivity thresholds as functions of the degree of landscape fragmentation. They found that when habitat area was very extensive, the gene flow within a population was not limited by nonhabitat patches, regardless of the relative cost of moving through the nonhabitat space or its configuration in the landscape. Conversely, when habitat was at intermediate to low extents and was fragmented into a number of separate patches, landscape fragmentation was found to always significantly limit gene flow, regardless of the degree of contrast in movement cost between habitat and nonhabitat.

These examples indicate that when habitat loss or fragmentation sufficiently impedes movement, local populations in remnant habitat patches may become functionally isolated (Tischendorf et al. 2003). However, the degree of population disruption following habitat loss and fragmentation will depend on how the species perceives and interacts with landscape patterns (With et al. 1997). It is often proposed that less mobile species with restrictive habitat requirements and limited gap-crossing ability will be most sensitive to isolation effects (e.g., Rothermel and Semlitsch 2002). In contrast, Cushman, in a 2006 review of several hundred empirical papers on the effects of fragmentation on amphibian populations, found that the most mobile species typically suffered the most rapid and large impacts resulting from habitat loss and fragmentation. In a 2010 study, Cushman, Bradley W. Compton, and Kevin McGarigal evaluated possible causes of this counterintuitive result by simulating landscape connectivity across a large combination of dispersal abilities, population sizes, and sensitivities to landscape change. These researchers found that all species were negatively affected by habitat loss and road building, experiencing reductions in total population density and extent of occupied habitat. Species with intermediate population sizes and large dispersal abilities, however, showed the largest reductions in habitat connectivity and total expected population density. Populations of highly mobile species should be impacted more quickly by landscape change than less mobile species because their high mobility leads them to interact more with changes in the landscape, leading to elevated mortality rates and changes in the pattern of movement and habitat use. This does not mean, however, that less mobile species are less affected. Cushman and his colleagues found that habitat loss and fragmentation rapidly led to the complete isolation of populations of low mobility species. Over longer time periods, this complete isolation would eliminate gene flow and demographic rescue effects, increasing the probability of local extinction and eliminating recolonization. Thus, highly mobile species often decline quickly following habitat loss and fragmentation, whereas less mobile species may decline slowly, but more certainly, in an extinction debt (Vellend et al. 2006).

Climate change, habitat loss, and extinction

The impacts of climate change on the distribution and connectivity of populations has emerged as one of the largest conservation issues of the twenty-first century. Climate change will affect populations by changing the distribution of suitable ecological conditions (niche migration), the ability of populations to migrate from current to future suitable habitat conditions, and the ability to maintain connectivity within populations as these changes in the landscape occur. In a 2004 study, Chris D. Thomas and colleagues evaluated extinction risk across 20 percent of Earth's terrestrial surface using changes in predicted habitat suitability from species distribution models. They found that mid-range climate-warming scenarios for 2050 committed between 15 percent and 37 percent of species to extinction. John Harte and colleagues argued in a 2004 article that these results underestimated the likely extinction load resulting from current climate change because of local adaptation of subpopulations to regional climates and the limited ability of populations to migrate from current to future suitable habitats. In a 2012 study, Mark C. Urban, Josh J. Tewksbury, and Kimberly S. Sheldon modeled interactions of competition and dispersal ability on extinction risk due to climate change. They found that competition slowed the advance of colonists into newly suitable habitats, implying that climate change may result in dramatically more extinctions than Thomas and colleagues had predicted, especially among species that have narrow

niches (e.g., tropics), vary in dispersal (most communities), and compete strongly.

In a 2011 study, Cushman and colleagues modeled the effects of a range of dispersal abilities and ecological associations on landscape connectivity in the northern Rocky Mountains of the United States and found that species associated with low-elevation forests and high-elevation habitats were the species most limited by the extent and connectivity of habitat in the current landscape. These species are also the ones that are most at risk with respect to the effects of climate change. Species associated with low-elevation forests are at risk because climate change will lead to the upward migration of the lower tree-line in this landscape, reducing the extent and increasing the fragmentation of low-elevation forests. Species associated with high-elevation habitats are at risk because as the climate warms, suitable conditions migrate upslope into smaller and increasingly isolated mountain islands, with some species perhaps "going off the map" as warming passes levels at which even the highest habitat islands become unsuitable.

Empirical landscape genetic modeling has enabled researchers to rigorously predict the potential effects of climate change on two species associated with high elevations (the American marten [*Martes americana*] and the wolverine [*Gulo gulo*]). For example, in a 2009 study, Michael K. Schwartz and colleagues demonstrated that wolverine gene flow is restricted in areas of the landscape without spring snow and facilitated along corridors and stepping stones of high-elevation habitat. In a 2011 study, Kevin S. McKelvey and colleagues projected the extent and pattern of likely future spring snowpacks and modeled changes in the extent of suitable wolverine habitat and population connectivity. They found that expected climate change by 2080 would dramatically reduce the extent of suitable habitat and population connectivity for the wolverine across the species' range in the lower 48 states. In a 2010 study, Tzeidle N. Wasserman and colleagues modeled American marten gene flow as a function of a large number of landscape variables, finding that dispersal is facilitated by the extent and connectivity of cool, moist high-elevation habitats. A 2012 study headed by Wasserman used individual-based, landscape genetic simulation modeling to evaluate the potential impacts of these changes on gene flow, population differentiation, and genetic diversity. This was extended by a forthcoming Wasserman-led study covering the full range of American marten territory in the northern Rocky Mountains of the United States. The results indicate that the extent of habitat in the landscape was highly related to genetic diversity and that when habitat area was reduced and fragmented by climatic warming, gene flow was attenuated across the landscape; genetic diversity was greatly reduced, particularly in areas of relatively low elevation; and many local populations were predicted to become fully isolated or to be eliminated altogether.

When do habitat loss and fragmentation imperil species?

The answer to the question of when do habitat loss and fragmentation imperil species is a simple and resounding one: every day and all across the world. Many species are being

Expected climate change by 2080 would dramatically reduce the extent of suitable habitat and population connectivity for the wolverine (*Gulo gulo*) across the species' range in the lower 48 states. © Geoffrey Kuchera/ShutterStock.com.

The Hoover Dam is credited with the decline and endangerment of many species of fish. © Matej Hudovernik/ShutterStock.com.

progressively eliminated directly by habitat loss. The very rapid loss of natural ecosystems, particularly in the tropics, results in the elimination of many populations and, eventually, the extinction of species. For example, every year thousands of species are driven to extinction by tropical deforestation (Wilson 1999). Many other species have had their populations depressed by habitat loss and fragmentation to the extent that their extinction thresholds have been passed. Some of these species have already gone extinct, while others are in an extinction debt phase of uncertain duration but clear destination. Many other species are threatened by habitat loss that is approaching extinction thresholds.

The explosive loss of biodiversity is well documented, as is the dominant role played by habitat loss and fragmentation. Yet, despite the widespread recognition of the importance of habitat area and connectivity, scientists have a limited ability to predict accurately when and to what degree individuals, populations, and communities will be eliminated. Much of the apparent lack of predictive power stems from a failure to

properly consider habitat loss and fragmentation from the perspective of the organism. Specifically, human activities often lead to changes in the spatial extent and configuration of land cover types, but whether these changes result in the loss and fragmentation of suitable habitat for a particular organism depends on the scale and nature of those changes in relation to how that organism perceives and interacts with landscape patterns (With et al. 1997). For example, changes in the size and isolation of mature forest patches at a particular scale may have little or no detectable impact on species that perceive and respond to landscape patterns at a different scale or that select habitat on the basis of other environmental variables (e.g., shrub cover, litter depth) or that use a broad range of habitats (i.e., generalist or multihabitat species).

The current biodiversity crisis requires rapid decisions about population risk and conservation responses to this risk (Flather et al. 2011). This has led to efforts to identify robust, general guidelines to define minimum viable populations (Sanderson 2006). The oft-cited *50/500 rule* proposes that an effective population size (N_e) of 50 is required to prevent deleterious inbreeding, while an N_e of 500 is required to ensure overall genetic variability (e.g., Soulé 1987). In a 2002 book, Richard Frankham, Jonathan D. Ballou, and David A. Briscoe estimated that an N_e of at least 50 is required to avoid inbreeding depression, an N_e of 500 to 5,000 to retain evolutionary potential, and an N_e of 12 to 1,000 to avoid the accumulation of deleterious mutations. Given that the average ratio of effective population size to census population size is roughly 0.10, these rules of thumb translate to census sizes of 500 to 50,000 individuals. The 50/500 rule has been criticized frequently, however, as ad hoc and not rigorously supported by scientific observation and analysis (Brook et al. 2006). A number of scientists have proposed subsequently that evolutionary and demographic processes require an effective population size of at least 5,000 individuals for long-term viability (e.g., Clabby 2010). Curtis H. Flather and colleagues' 2011 review suggested that there is no "magic number" for population viability, noting that the minimum population sizes required for viability depend on consumer–resource relationships, the use of energy within ecosystems, and the relative roles of these factors.

Given the urgency of the biodiversity crisis, and the absence of a "magic number" for population viability that would apply across all species, it would be wise to adopt a conservative approach in which habitat conditions are maintained well within their ranges of natural variability across broad landscapes (Swanson et al. 1990) and to err on the side of caution with regard to habitat area and connectivity. To paraphrase Aldo Leopold from his work titled *A Sand County Almanac* (1949), the first rule of intelligent management in the face of uncertainty is to maintain flexibility. Land management often imparts a legacy of habitat loss and fragmentation that can take centuries to erase (Wallin et al. 1994). Thus, adaptive natural resources management ideally would use treatments that are reversible in terms of their impacts on habitat areas and configurations within reasonably short planning horizons. While in an ideal world great care should be taken to ensure that current and future management actions do not impair the sustainability of ecological systems, the reality is that in many regions habitat destruction is proceeding at such a tremendous pace that there is no practical way to protect intact ecosystems within their dynamic ranges of variability. Instead, in many ecosystems the task has devolved to protecting remnant habitats and threatened populations and trying to establish and protect connectivity between them.

The human population, having taken millions of years to reach one billion, tripled between the early twentieth century and 2012, the year the seven billion mark was reached. The human ecological footprint, a measure of the total impact of humanity on the biosphere, has more than doubled since the mid-twentieth century. There is hardly a location on the planet that has not been markedly influenced by humans, and the majority of the world's surface (both terrestrial and marine) is heavily impacted and exploited. Every year more than 54,050 square miles (14 million ha) of natural forest are converted to other land uses (FAO 2011), and many other ecosystems, such as temperate grasslands, are mere remnants of their original area. More than 30 percent of Earth's annual primary productivity is directly exploited by humans (Vitousek et al. 1986). Between one-third and one-half of Earth's land surface has been transformed by human action; the carbon dioxide concentration in the atmosphere has increased by nearly 30 percent since the beginning of the Industrial Revolution; more atmospheric nitrogen is fixed by humanity than by all natural terrestrial sources combined; more than half of all accessible surface freshwater is put to use by humanity; and about one-quarter of the bird species on Earth have been driven to extinction (Vitousek et al. 1997). Regardless of how one defines habitat, whether in an island biogeographic, patch mosaic, or gradient perspective, there can be little doubt that human impacts have dramatically reduced the quality, extent, and connectivity of natural ecosystems and have reduced the populations of a very large proportion of plant and animal species. As a result, many scientists estimate

An aerial view of land used for farming in Afghanistan. © Picture Contact BV/Alamy.

that between 33 percent and 50 percent of all species may become extinct by the end of the twenty-first century (Leakey and Lewin 1996; Wilson 2002).

As habitat loss and fragmentation are the dominant drivers of biodiversity loss, efforts should be redoubled to protect the remaining intact and undegraded ecosystems, protect areas of high species richness, and maintain and promote connectivity among these expanded networks of protected areas. The predominant importance of habitat area and the typically lesser importance of habitat fragmentation for the short-term conservation of most species (Fahrig 2003) suggest that management targeted at maintaining or enhancing biodiversity should first focus on preserving and expanding core protected habitat areas or increasing the number of such areas. These strategies have well-documented positive benefits in all cases, if habitat is adequately defined with respect to the species of concern and is adequately protected after designation. Habitat fragmentation and population connectivity are also important, however. The long-term viability of most species depends on maintaining demographic and genetic exchange among multiple dispersed subpopulations (Hanski and Ovaskainen 2002). In addition, global climate change will likely reorganize ecological systems in ways that require species to change their ranges in terms of latitude and elevation. Static island habitat reserves are unlikely to remain effective refuges for many of the species currently residing in them. Thus, a combined approach that focuses first on saving the remnants of natural ecosystems and areas of high biodiversity, but is directly coupled with efforts to maintain and enhance functional connectivity among reserves networks, will be essential (Noss 1983). The focus should remain firmly on protecting remaining habitats, expanding core habitats, and restoring degraded habitat areas, but attention must also be given to intervening areas that provide buffers and maintain population connectivity (Fahrig 2003).

Resources

Books

Beissinger, Steven R., and Dale R. McCullough, eds. *Population Viability Analysis*. Chicago: University of Chicago Press, 2002.

Cushman, Samuel A. "Space and Time in Ecology: Noise or Fundamental Driver?" In *Spatial Complexity, Informatics, and Wildlife Conservation*, edited by Samuel A. Cushman and Falk Huettmann, 19–42. Tokyo: Springer, 2010.

Cushman, Samuel A., Bradley W. Compton, and Kevin McGarigal. "Habitat Fragmentation Effects Depend on Complex Interactions between Population Size and Dispersal Ability: Modeling Influences of Roads, Agriculture, and Residential Development across a Range of Life-History Characteristics." In *Spatial Complexity, Informatics, and Wildlife Conservation*, 369–86.

Cushman, Samuel A., Kevin Gutzwiller, Jeffrey S. Evans, and Kevin McGarigal. "The Gradient Paradigm: A Conceptual and Analytical Framework for Landscape Ecology." In *Spatial Complexity, Informatics, and Wildlife Conservation*, 83–110.

Frankham, Richard, Jonathan D. Ballou, and David A. Briscoe. *Introduction to Conservation Genetics*. 2nd ed. Cambridge, U.K.: Cambridge University Press, 2010.

Leakey, Richard, and Roger Lewin. *The Sixth Extinction: Patterns of Life and the Future of Humankind*. New York: Doubleday, 1995.

Leopold, Aldo. *A Sand County Almanac*. New York: Oxford University Press, 1949.

Schwerdtfeger, Fritz. *Ökologie der Tiere* [Animal ecology]. Hamburg, Germany: Verlag Paul Parey, 1977.

Soulé, Michael E., ed. *Viable Populations for Conservation*. Cambridge, U.K.: Cambridge University Press, 1987.

Swanson, Frederick J., Jerry F. Franklin, and James R. Sedell. "Landscape Patterns, Disturbance, and Management in the Pacific Northwest, USA." In *Changing Landscapes: An Ecological Perspective*, edited by Izaak S. Zonneveld and Richard T. T. Forman. New York: Springer-Verlag, 1990.

Turner, Monica G., Robert H. Gardner, and Robert V. O'Neill. *Landscape Ecology in Theory and Practice: Pattern and Process*. New York: Springer, 2001.

Wilson, Edward O. *The Diversity of Life*. Rev. ed. New York: Norton, 1999.

Wilson, Edward O. *The Future of Life*. New York: Knopf, 2002.

Periodicals

Andrén, Henrik. "Effects of Habitat Fragmentation on Birds and Mammals in Landscapes with Different Proportions of Suitable Habitat: A Review." *Oikos* 71, no. 3 (1994): 355–366.

Atkinson, W. D., and B. Shorrocks. "Competition on a Divided and Ephemeral Resource: A Simulation Model." *Journal of Animal Ecology* 50, no. 2 (1981): 461–471.

Balmford, Andrew, Rhys E. Green, and Martin Jenkins. "Measuring the Changing State of Nature." *Trends in Ecology and Evolution* 18, no. 7 (2003): 326–330.

Bergin, Timothy M., Louis B. Best, Kathryn E. Freemark, and Kenneth J. Koehler. "Effects of Landscape Structure on Nest Predation in Roadsides of a Midwestern Agroecosystem: A Multiscale Analysis." *Landscape Ecology* 15, no. 2 (2000): 131–143.

Boyce, Mark S. "Population Viability Analysis." *Annual Review of Ecology and Systematics* 23 (1992): 481–506.

Brook, Barry W., Lochran W. Traill, and Corey J.A. Bradshaw. "Minimum Viable Population Sizes and Global Extinction Risk Are Unrelated." *Ecology Letters* 9, no. 4 (2006): 375–382.

Clabby, Catherine. "A Magic Number? An Australian Team Says It Has Figured Out the Minimum Viable Population for Mammals, Reptiles, Birds, Plants, and the Rest." *American Scientist* 98, no. 1 (2010): 24–25.

Cushman, Samuel A. "Effects of Habitat Loss and Fragmentation on Amphibians: A Review and Prospectus." *Biological Conservation* 128, no. 2 (2006): 231–240.

Cushman, Samuel A., Kevin S. McKelvey, Curtis H. Flather, and Kevin McGarigal. "Do Forest Community Types Provide a Sufficient Basis to Evaluate Biological Diversity?" *Frontiers in Ecology and the Environment* 6, no. 1 (2008): 13–17.

Cushman, Samuel A., M. G. Raphael, L. F. Ruggiero et al. "Limiting Factors and Landscape Connectivity: The American Marten in the Rocky Mountains." *Landscape Ecology* 26, no. 8 (2011): 1137–1149.

Cushman, Samuel A., Andrew Shirk, and Erin L. Landguth. "Separating the Effects of Habitat Area, Fragmentation, and Matrix Resistance on Genetic Differentiation in Complex Landscapes." *Landscape Ecology* 27, no. 3 (2012): 369–380.

Danielson, Brent J. "Communities in a Landscape: The Influence of Habitat Heterogeneity on the Interactions between Species." *American Naturalist* 138, no. 5 (1991): 1105–1120.

Debinski, Diane M., and Robert D. Holt. "A Survey and Overview of Habitat Fragmentation Experiments." *Conservation Biology* 14, no. 2 (2000): 342–355.

Delibes, Miguel, Pilar Gaona, and Pablo Ferras. "Effects of an Attractive Sink Leading into Maladaptive Habitat Selection." *American Naturalist* 158, no. 3 (2001): 277–285.

Donovan, Therese M., and Curtis H. Flather. "Relationships among North American Songbird Trends, Habitat Fragmentation, and Landscape Occupancy." *Ecological Applications* 12, no. 2 (2002): 364–374.

Fahrig, Lenore. "Relative Effects of Habitat Loss and Fragmentation on Population Extinction." *Journal of Wildlife Management* 61, no. 3 (1997): 603–610.

Fahrig, Lenore. "How Much Habitat Is Enough?" *Biological Conservation* 100, no. 1 (2001): 65–74.

Fahrig, Lenore. "Effects of Habitat Fragmentation on Biodiversity." *Annual Review of Ecology, Evolution, and Systematics* 34 (2003): 487–515.

Fahrig, Lenore, and Gray Merriam. "Conservation of Fragmented Populations." *Conservation Biology* 8, no. 1 (1994): 50–59.

Flather, Curtis H., and Michael Bevers. "Patchy Reaction-Diffusion and Population Abundance: The Relative Importance of Habitat Amount and Arrangement." *American Naturalist* 159, no. 1 (2002): 40–56.

Flather, Curtis H., Gregory D. Hayward, Steven R. Beissinger, and Philip A. Stephens. "Minimum Viable Populations: Is

There a 'Magic Number' for Conservation Practitioners?" *Trends in Ecology and Evolution* 26, no. 6 (2011): 307–316.

Gibbs, James P., and Edward J. Stanton. "Habitat Fragmentation and Arthropod Community Change: Carrion Beetles, Phoretic Mites, and Flies." *Ecological Applications* 11, no. 1 (2001): 79–85.

Grand, Joanna, John Buonaccorsi, Samuel A. Cushman et al. "A Multiscale Landscape Approach to Predicting Bird and Moth Rarity Hotspots in a Threatened Pitch Pine–Scrub Oak Community." *Conservation Biology* 18, no. 4 (2004): 1063–1077.

Hall, Linnea S., Paul R. Krausman, and Michael L. Morrison. "The Habitat Concept and a Plea for Standard Terminology." *Wildlife Society Bulletin* 25, no. 1 (1997): 173–182.

Hanski, Ilkka, and Otso Ovaskainen. "Extinction Debt at Extinction Threshold." *Conservation Biology* 16, no. 3 (2002): 666–673.

Harte, John, Annette Ostling, Jessica L. Green, and Ann Kinzig. "Biodiversity Conservation: Climate Change and Extinction Risk." *Nature* 430, no. 6995 (2004): 145–148.

Holyoak, Marcel. "Habitat Patch Arrangement and Metapopulation Persistence of Predators and Prey." *American Naturalist* 156, no. 4 (2000): 378–389.

Holyoak, Marcel, and Sharon P. Lawler. "The Role of Dispersal in Predator–Prey Metapopulation Dynamics." *Journal of Animal Ecology* 65, no. 5 (1996): 640–652.

Huffaker, C. B. "Experimental Studies on Predation: Dispersion Factors and PredatorPrey Oscillations." *Hilgardia* 27, no. 14 (1958): 343–383.

Hutchinson, G. Evelyn. "Concluding Remarks." *Cold Spring Harbor Symposia on Quantitative Biology* 22 (1957): 415–427.

Ives, Anthony R. "Measuring Competition in a Spatially Heterogeneous Environment." *American Naturalist* 146, no. 6 (1995): 911–936.

Kareiva, Peter. "Habitat Fragmentation and the Stability of Predator–Prey Interactions." *Nature* 326, no. 6111 (1987): 388–390.

Kareiva, Peter, and Uno Wennergren. "Connecting Landscape Patterns to Ecosystem and Population Processes." *Nature* 373, no. 6512 (1995): 299–302.

Kotlar, Burt P., and Robert D. Holt. "Predation and Competition: The Interaction of Two Types of Species Interaction." *Oikos* 54, no. 2 (1989): 256–260.

Kurki, Sami, Ari Nikula, Pekka Helle, and Harto Linden. "Landscape Fragmentation and Forest Composition Effects on Grouse Breeding Success in Boreal Forests." *Ecology* 81, no. 7 (2000): 1985–1997.

Mahan, Carolyn G., and Richard H. Yahner. "Effects of Forest Fragmentation on Behaviour Patterns in the Eastern Chipmunk (*Tamias striatus*)." *Canadian Journal of Zoology* 77, no. 12 (2000): 1991–1997.

May, Robert M., and Scott K. Robinson. "Population Dynamics of Avian Brood Parasitism." *American Naturalist* 126, no. 4 (1985): 475–495.

McKelvey, Kevin S., Jeffrey P. Copeland, Michael K. Schwartz et al. "Climate Change Predicted to Shift Wolverine Distributions, Connectivity, and Dispersal Corridors." *Ecological Applications* 21, no. 8 (2011): 2882–2897.

Myers, Norman, and Andrew H. Knoll. "The Biotic Crisis and the Future of Evolution." *Proceedings of the National Academy of Sciences of the United States of America* 98, no. 10 (2001): 5389–5392.

Noss, Reed F. "A Regional Landscape Approach to Maintain Diversity." *BioScience* 33, no. 11 (1983): 700–706.

Oksanen, Lauri, Jon Moen, and Peter A. Lundberg. "The Time-Scale Problem in Exploiter—Victim Models: Does the Solution Lie in Ratio-Dependent Exploitation?" *American Naturalist* 140, no. 6 (1992): 938–960.

Ovaskainen, Otso. "The Effective Size of a Metapopulation Living in a Heterogeneous Patch Network." *American Naturalist* 160, no. 5 (2002): 612–628.

Pearl, Raymond. "The Growth of Populations." *Quarterly Review of Biology* 2, no. 4 (1927): 532–548.

Pimm, Stuart L., and Peter Raven. "Biodiversity: Extinction by Numbers." *Nature* 403, no. 6772 (2000): 843–845.

Reed, J. Michael, L. Scott Mills, John B. Dunning Jr. et al. "Emerging Issues in Population Viability Analysis." *Conservation Biology* 16, no. 1 (2002): 7–19.

Robbins, Chandler S., Deanna K. Dawson, and Barbara A. Dowell. "Habitat Area Requirements of Breeding Forest Birds of the Middle Atlantic States." *Wildlife Monographs* 103 (1989): 3–34.

Rothermel, Betsie B., and Raymond D. Semlitsch. "An Experimental Investigation of Landscape Resistance of Forest versus Old-Field Habitats to Emigrating Juvenile Amphibians." *Conservation Biology* 16, no. 5 (2002): 1324–1332.

Sanderson, Eric W. "How Many Animals Do We Want to Save? The Many Ways of Setting Population Target Levels for Conservation." *BioScience* 56, no 11 (2006): 911–922.

Saunders, Denis A., Richard J. Hobbs, and Chris R. Margules. "Biological Consequences of Ecosystem Fragmentation: A Review." *Conservation Biology* 5, no. 1 (1991): 18–32.

Schaffer, W. M., and M. Kot. "Chaos in Ecological Systems: The Coals that Newcastle Forgot." *Trends in Ecology and Evolution* 1, no. 3 (1986): 58–63.

Schwartz, Michael K., Jeffrey P. Copeland, Neil J. Anderson et al. "Wolverine Gene Flow Across a Narrow Climatic Niche." *Ecology* 90, no. 11 (2009): 3222–3232.

Short Bull, Ruth A., Samuel A. Cushman, Richard Mace et al. "Why Replication Is Important in Landscape Genetics: American Black Bear in the Rocky Mountains." *Molecular Ecology* 20, no. 6 (2011): 1092–1107.

Steffan-Dewenter, Ingolf, Ute Münzenberg, Christof Bürger et al. "Scale-Dependent Effects of Landscape Context on Three Pollinator Guilds." *Ecology* 83, no. 5 (2002): 1421–1432.

Thomas, Chris D., Alison Cameron, Rhys E. Green et al. "Extinction Risk from Climate Change." *Nature* 427, no. 6970 (2004): 145–148.

Thompson, Craig M., and Kevin McGarigal. "The Influence of Research Scale on Bald Eagle Habitat Selection Along the Lower Hudson River, New York (USA)." *Landscape Ecology* 17, no. 6 (2002): 569–586.

Tilman, David, Robert M. May, Clarence L. Lehman, and Martin A. Nowak. "Habitat Destruction and the Extinction Debt." *Nature* 371, no. 6492 (1994): 65–66.

Tischendorf, Lutz, Darren J. Bender, and Lenore Fahrig. "Evaluation of Patch Isolation Metrics in Mosaic Landscapes for Specialist vs. Generalist Dispersers." *Landscape Ecology* 18, no. 1 (2003): 41–50.

Trzcinski, M. Kurtis, Lenore Fahrig, and Gray Merriam. "Independent Effects of Forest Cover and Fragmentation on the Distribution of Forest Breeding Birds." *Ecological Applications* 9, no. 2 (1999): 586–593.

Urban, Mark C., Josh J. Tewksbury, and Kimberly S. Sheldon. "On a Collision Course: Competition and Dispersal Differences Create No-Analogue Communities and Cause Extinctions during Climate Change." *Proceedings of the Royal Society Series B*. Published electronically January 4, 2012.

Van Horne, Bea. "Density as a Misleading Indicator of Habitat Quality." *Journal of Wildlife Management* 47, no. 4 (1983): 893–01.

Vellend, Mark, Kris Verheyen, Hans Jacquemyn et al. "Extinction Debt of Forest Plants Persists for More Than a Century following Habitat Fragmentation." *Ecology* 87, no. 3 (2006): 542–548.

Verhulst, Pierre-François. "Notice sur la loi que la population suit dans son accroissement." *Correspondance mathématique et physique* 10 (1838): 113–121.

Vitousek, Peter M., Paul R. Ehrlich, Anne H. Ehrlich, and Pamela A. Matson. "Human Appropriation of the Products of Photosynthesis." *BioScience* 36, no. 6 (1986): 386–373.

Vitousek, Peter M., Harold A. Mooney, Jane Lubchenco, and Jerry M. Melillo. "Human Domination of Earth's Ecosystems." *Science* 277, no. 5325 (1997): 494–499.

Wallin, David O., Frederick J. Swanson, and Barbara Marks. "Landscape Pattern Response to Changes in Pattern Generation Rules: Land-Use Legacies in Forestry." *Ecological Applications* 4, no. 3 (1994): 569–580.

Wasserman, Tzeidle N., Samuel A. Cushman, Jeremy S. Littell et al. "Climate Change, Population Connectivity, and Genetic Diversity of American Marten (*Martes americana*) in the United States Northern Rocky Mountains." *Conservation Genetics* (forthcoming).

Wasserman, Tzeidle N., Samuel A. Cushman, Michael K. Schwartz, and David O. Wallin. "Spatial Scaling and Multimodel Inference in Landscape Genetics: *Martes americana* in Northern Idaho." *Landscape Ecology* 25, no. 10 (2010): 1601–1612.

Wasserman, Tzeidle N., Samuel A. Cushman, Andrew S. Shirk et al. "Simulating the Effects of Climate Change on Population Connectivity of American Marten (*Martes americana*) in the Northern Rocky Mountains, USA." *Landscape Ecology* 27, no. 2 (2012): 211–225.

Wiens, John A., Nils Chr. Stenseth, Beatrice Van Horne, and Rolf Anker Ims. "Ecological Mechanisms and Landscape Ecology." *Oikos* 66, no. 3 (1993): 369–380.

With, Kimberly A., Robert H. Gardner, and Monica G. Turner. "Landscape Connectivity and Population Distributions in Heterogeneous Environments." *Oikos* 78, no. 1 (1997): 151–169.

With, Kimberly A., and Anthony W. King. "Extinction Thresholds for Species in Fractal Landscapes." *Conservation Biology* 13, no. 2 (1999): 314–326.

Other

Food and Agriculture Organization of the United Nations (FAO). "Global Forest Resources Assessment 2010: The Global Remote Sensing Survey." Last modified December 16, 2011. http://www.fao.org/forestry/fra/remotesensingsurvey/en.

Samuel A. Cushman

• • • • •

Agriculture

Sedentary agriculture began on Earth approximately 11,000 years ago, stimulated by rapid warming and plentiful water resources in areas such as the Indus River Valley and Mesopotamia. The terms *agriculture* and *sedentary agriculture* are used interchangeably here, both functionally defined as the pursuit of crop growing or livestock tending within the context of fixed settlements that endure at least for one season; agriculture obviates the need for continual hunting and gathering, with attendant, almost constant movement of dwelling locations. By 5,000 years before the present, a wide arc of human agriculture existed, including temperate parts of Europe and portions of China and North Africa. Although the earliest agricultural uses may have been in concert with the natural environment due to their small scale, the rapid explosion of the human population underway by the mid-Holocene began to exact a much greater toll in terms of habitat conversion for agricultural use and habitat fragmentation; this toll included significant extinction pressures on flora and fauna almost everywhere on the globe. Present rates of extinction have been variously estimated as many times the rate of background extinction (on the order of 25,000 times as great). About 25 to 30 percent of pre-Holocene species may have been lost as of 2000 (Whitmore et al. 1992). Some have estimated that more species are presently becoming extinct each year than new species can be described.

For example, in the United States at least one-third of all threatened species are linked to pressures of agriculture, but this percentage is likely understated due to the multiplicity of agricultural impacts and the difficulty of tracing secondary effects (e.g., fertilizer additions causing plant diversity losses, which in turn alter herbivore composition). This estimate for the fraction of agriculturally driven extinctions in the United States can be assumed to be a reasonable proxy for the rest of the world; in fact, the rate of agriculturally caused extinctions in developing countries can be expected to exceed that of Western countries in upcoming decades due to the forecast of massive deforestation caused by agriculture in developing nations.

In 1993 E.O. Wilson estimated the rate of species loss to be on the order of 30,000 species per year, a major percentage of which is attributable to agriculture as practiced since the Industrial Revolution. Loss of even a single species from an ecosystem has been demonstrated to have a potentially major impact upon stability of the entire ecosystem as well as the outcome of species richness, given the cascading impacts upon other species within the ecosystem (Symsted et al. 1998).

Transition from hunting to agriculture

Considerable attention has been directed to the loss of megafaunal species (e.g., *Elephas namadicus*, *Mammuthus* ssp., *Megalocnus* ssp.) during the period 18,000 to 9,000 years before present. Although *Homo sapiens* and their refined use of hunting tools caused these extinctions, they are small in number compared to the massive number of species lost that began in the age of sedentary agriculture. These megafaunal extinctions deserve mention here to demarcate the periods of hunting and gathering from agriculture. In addition, although these extinctions are low in number compared to the mass extinction that began in the Holocene, they are an important precursor to destabilizing ecosystems by eliminating considerable numbers of high-level carnivore predators and cumulatively high-biomass-content herbivores. Thus, the ecological damage wrought by human ancestors in the Late Pleistocene is significant beyond the number of species lost in setting the stage for the Holocene mass extinction, the chief driver of which is sedentary agriculture.

Furthermore, Paul S. Martin (1989) effectively argued that the wave of Late Pleistocene large mammal extinctions could not have been driven fundamentally by climate change, because the plant palette was becoming more lush in the Late Pleistocene, and the most restrictive environmental period (with respect to climate extremes) to large mammal survival had passed. In addition, the ability of large mammals to migrate militates against adverse climate effects of the Late Pleistocene.

As in the case of other drivers of extinction, agriculture does not necessarily deal the final blow to a given species. Just as in the case of pollution, disease, overexploitation, or any type of habitat destruction, the extinction driver reduces a given species population to a level below its minimum viable population size, to a point within the stochastic extinction vortex. The chief difference is that overexploitation normally diminishes when the numbers of the subject population under extinction pressure have dwindled to a level below the useful

An endangered cheetah (*Acinonyx jubatus*), a species whose chief threat is from livestock farming. Courtesy of C. Michael Hogan.

economic return; in the case of agriculture, the unrelenting pressure of removing habitat from its natural form and applying extraneous chemicals continues indefinitely because the motivation of food production is not reduced by the diminishing population of the endangered species.

Prehistoric agriculture

Prehistoric agriculture consists of an early origins phase, when there were small, scattered settlement units engaging in sedentary agriculture, and a phase marking the appearance of large cities as centers of agriculture.

In the Early Holocene, the human population grew from 1 million to 10 million. Settled agriculture became established rapidly on a widespread basis from 11,000 to 7,000 years before present throughout southern and southwestern Asia, diffusing to Mesoamerica and northern and central Africa in the succeeding millennia. Archaeological recovery shows practically no evidence of domesticated sheep until 11,000 years before present, but soon after, domestication becomes broadly practiced and evidence of moving livestock from native ranges is abundant. For example, in much of southwest Asia, clear evidence exists of pigs and cattle being regularly domesticated by 9,000 to 7,000 years before present (Gupta 2004). Rice cultivation in the Yangtze River Valley is imputed to the era 13,900 to 13,000 years before present, then went into a decline attributed to climatic factors but was revived by 10,000 years before present (Richardson et al. 2001). Other grains appeared to have been cultivated and selected for productivity as early as 10,000 years before present; the cultivation process included trial and error as well as crossbreeding of cultivated with wild species.

The Indus River Valley of India is likely one of the earliest societies transitioning to agriculture, with evidence in the humid monsoonal conversion of the Early Holocene of wheat, barley, and sheep as early as 11,000 years before present. Shanidar Cave in Iraq is important for showing that Neanderthals had plant knowledge 60,000 to 80,000 years ago, even though no evidence of settled agriculture presented at that early time. However, Shanidar also evinced settled agriculture for *Homo sapiens* at 11,000 years before present with possible early domestication of sheep at that time (Gupta 2004). In the Fertile Crescent, from the mouths of the Tigris and Euphrates rivers north to Turkey and west to coastal Israel and Lebanon, cultural conversion to sedentary agriculture occurred as early as 11,000 to 9,000 years before present, with cultivation of a number of grains and other plant species.

In the Orkney Islands of Scotland, about 5,300 to 4,900 years before present, early sedentary agriculture involved oats, barley, cattle, and sheep. In particular, at the site of Skara Brae, 12 homes are evident in the extant ruins; the human population of the Orkney Islands in this era has been estimated at about 20,000. Considerable deforestation in the Orkneys facilitated this agricultural revolution (Darvill 2009).

Ireland offers valuable insights into prehistoric agriculture because of extant remains of several distinct eras of agricultural settlement. The earliest transition of Mesolithic hunter-gatherer culture to sedentary agriculture in Ireland is placed at about 7,000 years before present. Daniel Webster Hollis (2001) points out that the Irish Neolithic was much more rural than the Mesopotamian Neolithic. In particular, these early Irish settlements were widely scattered. Slavery practices in the Middle East may have been a possible determinant for higher densities in Uruk and other Mesopotamian locales, including Egypt. The period about 7,000 years before present has evidence of small settlements where Neolithic herders kept their stock and posed rather slight impacts to the environment.

In the period 6,000 to 3,000 years before present, important large population centers arose, based upon permanent agricultural settlements, which appeared in a number of disparate regions, including the British Isles, Mesopotamia, the Iberian Peninsula, Minoan Crete, Mauritania, southern Peru, and the Indus Valley. The scale of these large settlements, for the first time, was sufficient to begin to destroy sizable natural habitats such that numerous species began their descent into an extinction vortex; this connection to extinction can be viewed as a matter of attaining an agricultural scale whereby habitats of sufficient size to span a limited range of endemic species began to be eliminated.

By 5,700 to 5,300 years before present, Ceide Fields in the west of Ireland was a large Neolithic farming settlement of more than 2,471 acres (1,000 ha) where woodland was cleared for cropping and livestock grazing (Cooney 2000). This city is thought to be Western Europe's largest Neolithic settlement and would have occurred in a warmer, more hospitable climate than today.

In Ireland's Bronze Age, agricultural development changed more toward river valleys and showed movement from

Contrast of natural habitat (far side of ravine) with historically grazed land (near side of ravine) that had a similar rich habitat value. Courtesy of C. Michael Hogan.

the coastal woodland clearing. At the end of the Iron Age a farming collapse occurred in Ireland, as noted by regeneration of woodland based upon pollen records (Barry 2000). This setback in Irish agriculture is mirrored in some other Western European locales and has been suggested to have been linked to a worsening climate.

The ancient Mesopotamian city of Uruk may have had as many as 50,000 people at about 4,900 years before present, making Uruk arguably the largest world city of that era (Adams 2005). Uruk began as a settled agricultural center about 6,000 years before present at the southwest of the course of the ancient Euphrates River at that time. At Mohenjo Daro in the Indus River basin about 4,600 years before present a population of about 40,000 was present and built a platform under its citadel requiring an estimated 300,000 to 400,000 man-days of labor. These data imply sedentary agriculture created substantial discretionary time for *Homo sapiens* (McIntosh 2008). Total land at Mohenjo Daro (including agricultural use) was approximately 150,000 acres (60,703 ha) (Martell 2001). Settlements such as Uruk, Mohenjo Daro, and Ceide Fields were likely the earliest centers whose ecological impacts began to have a clear impact on species local extirpation.

Transition to the medieval era

The Iron Age in Ireland was witness to about 3,000 scattered *souterrains*, underground passages or chambers, 40 percent of which had associated surface structures (Barry 2000). A typical settlement population of an average ring fort or *souterrain* may have been about 100 people, implying an Irish human population about AD 700 of approximately 300,000. This was an age of hill forts and ring forts and gradually increasing population around AD 400 to 900, at a time when the world population was little more than 250 million.

Ireland is a microcosm of developed countries, illustrating the progression of agriculture and human population expansion.

The population of Ireland was 800,000 in AD 1500; 1 million in 1600 (when the world population was almost 500 million); 2 million by the year 1687; 2.4 million in 1750; and 5 million by 1800 (Houston 1995). One can envision that in this example of population growth, the Iron Age population may have begun to make a very significant impact upon endemic species extinctions, based upon the agricultural landscape footprint. Thus, in the 1,100 years subsequent to 700, the effective agricultural footprint grew by a factor of 16.7, greatly magnifying the number of species exposed to critical habitat loss.

Horizontal watermills of Ireland represent the largest assemblage of first millennium watermills in the world. Most of these Irish mills date accurately from the period AD 581 to 843, dates well correlated with crannogs and ring forts of this early medieval period of Ireland. Evidence in Ireland shows that a diet derived from dairy farming and oats production in the early Christian era may explain the considerable growth in the Irish population of this time period (Barry 2000).

During the Bronze and Chalcolithic eras in Spain, cities of considerable size were founded based upon organized farming of the surrounding landscape. Even in severe terrain such as the Spanish Sierra Nevada, terraced farming was developed since at least as early as ancient Roman times. Los Silillos and Los Millares are examples of the larger sites that involved thousands of human occupants in the Andalusian region of Spain. For instance, land clearing for agricultural conversion in the massive Bronze Age settlement of Los Silillos and its environs caused a massive loss of natural habitat and eventual collapse of the local agricultural economy (Hogan 2007).

The ancient Anasazi archaeological site of Chaco Canyon in the southwestern United States presents a classic example of resource overexploitation, as proved from plant macrofossils recovered from pack rat middens (National Research Council 1995). Nearly a millennium later, the plant community remains depauperate based upon its over-intensive use from the height of the Anasazi culture about AD 1050. For example, no pinyon pines remain, even though they represented an earlier dominant species. Not only did the Anasazi lower the water table, thus decimating native species populations, but they traveled nearly 47 miles (75 km) from Chaco to clear land for agriculture as well as firewood and timber.

The transition from a hunter-gatherer culture to sedentary agriculture was driven largely by labor savings realized from producing food in a given locale; this phenomenon was accentuated by drafting children into the food production labor force at earlier years than in the hunter-gatherer lifestyle. Robert McCormick Adams (2005) estimates that in ancient Mesopotamia the labor requirement of early sedentary agriculture consumed from 125 to 249 man-days per year of labor per family. Such a small yearly labor input would have created substantial leisure time and thus considerable time for communal works.

Industrial era agriculture

According to the US Congress Office of Technology Assessment (1995) and others, modern agriculture is the chief

Ancient terraced farming in the southern Sierra Nevada, Spain, where hand made stone walls were created millennia earlier to allow crop production on steep terrain. Courtesy of C. Michael Hogan.

driver of species extinctions, including loss of mammal and amphibian taxa. For example, in the United States the landmass of roughly half of the contiguous 48 states has been converted to agriculture; other developed nations such as France and Spain have even higher percentages of agricultural conversion.

In the past 300 years, land converted to agriculture has risen by a factor of five (Matson et al. 1997), enabled by modern irrigation techniques, advancements in farm machinery technology, and chemical additives to soil. The underlying driver of this expansion has been almost certainly the unrelenting explosion of the human population from 1700 onward.

Roads, canals, and other agricultural infrastructure

Although all conversion of natural habitat to agriculture contributes to habitat fragmentation, certain linear structures contribute additionally to this fragmentation phenomenon. Especially beginning in the industrial era, agriculture assumed a scale not previously encountered; such scale demanded a higher level of infrastructure, including road networks, canals, and dams. Roads were developed with some systematics at least as early as AD 500. For example, drovers roads used throughout the British Isles for driving livestock to market served as some of the early constructs supporting agriculture. In some cases these roads were built through boggy areas such as the Causey Mounth of northeast Scotland. Some of these early medieval roads severed bog habitat (bisected bogs) and began the assault upon species populations, particularly those species depending upon continuous wetland or bog habitat

where the road structures were necessarily elevated to avoid the boggy conditions.

Earthen canals were used by the Mesopotamians and other early cultures, but only in modern times have massive concrete-lined structures been employed to deliver substantial irrigation flow to agricultural uses. One of the largest such projects is the California State Water Project, one aim of which is delivery of water to more arid parts of California's Central Valley. These canals impede terrestrial animals from normal migration and dispersal patterns and serve to fragment habitats. Additionally, the size and intensity of these canals, including return flow functionality, aggravate the buildup of selenium and other salts in previously fertile soil. Thus, despite the short-term agricultural benefit, these projects generate a long-term soil productivity loss and a decided impact upon primary native plant productivity that in turn adversely affect herbivores and higher trophic levels.

Agricultural chemicals: pesticides and herbicides

Since ancient times humans have employed certain inorganic chemicals to control weeds from agricultural plots. However, in the 1940s an explosion of organic compounds emerged targeting unwanted insects and weedy plant materials. To give a regional example, by the late 1980s 137,788 tons (125,000 metric tons) of herbicide (active ingredients) were being applied to the Mississippi River basin every year. As of 2012, pesticide use exceeds 220,462,262 pounds (1 billion kg) worldwide. The excessive amount of pesticide toxins applied to the environment is illustrated by stating that the amount of herbicide used to kill one honeybee, approximately .01 microgram, represents enough toxin to kill the world honeybee population many times over.

Widespread herbicide use beginning in the 1940s is responsible for numerous species extinctions, including birds, amphibians, fish, and arthropods (Lannoo 2005). In many cases, herbicide use contributes, along with habitat destruction, to species endangerment. Many herbicides have persistent effects in the environment, retaining their toxicity as they remain in soils for decades in some cases; furthermore, some pesticides are highly soluble, so they may enter aquatic systems where nonselective lethal effects can occur (Jekel and Reemtsma 2006). Often the herbicides undergo chemical change after release into the environment; in some cases, the altered chemicals may have entirely different toxicity effects upon plants and animals from the original chemicals applied.

A typical instance is the decline of northern cricket frog (*Acris crepitans*) populations in North America, where strong correlations have been noted between metamorphism decline and agricultural herbicide applications (Lannoo 2005). In many cases the species impacts are through food-chain impacts; for example, decline of the grey partridge (*Perdix perdix*) in Britain has been attributed to reduction in chick prey of sawflies, which in turn have had their populations decimated by herbicide kills of the native flora (considered weeds by the farmers) needed to support the sawfly populations (Potts and Aebischer 1991). This example is important because the direct avian toxicity to the herbicides used was quite low, allowing farmers to declare the chemicals safe for birds.

Arthropod extinctions are particularly vulnerable to pesticide use because these chemicals are specifically aimed at insects. A notable example is the use of pyrethroids, which are not only lethal to the target pest species but also exhibit similar toxicity to a broad range of nontarget crustaceans and aquatic insects (Shukla et al. 1999). The pyrethroid class of pesticides exhibits extremely high toxicity to a broad range of nontarget arthropods, including numerous freshwater isopods, crayfish, and amphipods, as well as many species of

Expanse of oil palm monoculture that has replaced native lowland dipterocarp forests, and produced loss of countless species. Kinabatangan Basin, Sabah, Borneo. Courtesy of C. Michael Hogan.

marine shrimp and lobster. Acute toxicity to the microcrustacean *Daphnia magna* is in the range of LC50 values in the parts per billion with three to ninety-six hours of exposure (depending upon precise pyrethroid molecule utilized). (The LC50 value is the concentration of a toxin that can be expected to kill one half of a population of organisms.)

Agricultural chemicals: fertilizers

Application of massive quantities of chemical fertilizers has also altered terrestrial and aquatic ecosystems. Trevor Rowley (2006) has stated that agricultural nitrogen fertilizer may be the greatest ongoing threat to biodiversity because excess soil nitrogen favors dominance of more vigorous flora, such as nettles, thistles, bracken, and cow parsley, which end up crowding out great numbers of other native vegetative species. For example, in one controlled study, cumulative application of 352 to 666 pounds (160 to 303 kg) nitrogen per 2.47 acres (1 ha) resulted in loss of about one-fourth of the native flora in prairie-like acid grasslands. Much of the industrialized world has nitrogen deposition rates of 11 to 44 pounds (5 to 20 kg) per 2.47 acre (1 ha) per year, and most of the lesser-developed nations can be expected to follow this rate of application in the next 50 years (Clark 2007). Plant species loss has been shown to be particularly high where nitrogen addition has been monitored in temperate forests and savanna; species loss has been documented to be as high as 50 percent of native flora (Carpenter 2005). Note that nitrogen addition not only includes fertilizer runoff but also deposition of agricultural nitrogen in rainfall reentry. Flora lost via nitrogen addition were disproportionately rare flora species. Recovery of lost species was extremely low even after 13 years of cessation of nitrogen fertilizer addition.

Excess nutrient loading has also been shown to have an adverse effect on species richness in aquatic systems that may be recipients of nitrate and phosphate runoff. To be sure, nitrate loading to the environment arises from activities other than agriculture. Nevertheless, sedentary agriculture accounts for about one-half of the total loading to natural systems. Significant species losses continue to occur not only from outright toxicity of agricultural chemicals, but also through encouragement of dominant forbs that are tolerant of nitrates and phosphates, thereby diminishing flora biodiversity.

Besides eutrophication effects that may deplete oxygen and harm aquatic faunal species, controlled tests have shown that aquatic flora diversity is depressed with excess nitrate deposition to freshwater systems (Barker et al. 2008). Direct metabolic interference has been demonstrated to many species of fish and amphibians when nitrate or nitrite levels are elevated from agricultural runoff. Decline of amphibian species is of particular concern because they represent an important trophic link, consuming insects below them in the food chain and serving as a key food source to higher level predators, including snakes and raptors.

There are inverse cases of benefit from phosphate runoff; for example, in the Neva Estuary of the Baltic Sea, such runoff mitigates the formation of persistent algal mats so that a wider variety of aquatic species is able to thrive in this phosphate-enhanced milieu (Berezina et al. 2008).

Pesticides are viewed widely as the chief chemical set involved in species extinctions because of their high direct toxicity to a myriad of organisms, notably amphibians and arthropods. Moreover, massive doses of herbicides are responsible for widespread killing of nontarget flora. Less publicized as major drivers in species loss are the effects of the tonnages of chemical fertilizers upon which modern agriculture has become dependent. The highly touted green revolution, increasing fertilizer use by sevenfold from 1975 to 2010 to double the world's food production, has been responsible for a large toll in biodiversity loss. Nitrogen fertilizers are known for their acute faunal toxicity in the environment. By creating nitrogen content alteration in soils, in almost every world region nitrogen fertilizers are responsible for vast depletions of native plant species by producing competitive imbalances in the flora associations. A study spanning 20 years, conducted by Carly Stevens, found that doubling the nitrogen content in soils from fertilizer use caused a loss of over 20 percent of native plants; moreover, higher fertilizer use commonly led to species losses of more than 20 percent.

Species loss in the oceans has also been linked to excessive fertilizer use. Not only is this effect seen in eutrophic areas such as the Caspian and Baltic seas, but nitrogen additions are suggested to broaden and perpetuate marine dead zones in the Gulf of Mexico, Chesapeake Bay, and Pacific Ocean off the Oregon coast, even though the original dead zone was initiated by other drivers. For example, the agronomically valuable German pasturelands (Lolio-Cynosuretum) have diminished their flora native species by 50 percent since 1945 (Tscharntke et al. 2005). Correspondingly, cereal yield growth in Europe since 1945 is shown to be responsible for the decline of about 30 percent of native bird species.

Genetic engineering factors

Late twentieth-century agriculture added the dimension of genetic engineering to enable higher yield crops to be grown, but this has further accentuated the trend toward monoculture and simplification of species diversity in modern agriculture. Genetic engineering began in the 1980s with molecular biology assisting in guiding new crop types. However, subsequent decades added gene splicing and irradiation to develop new DNA material in the quest for improved crop yields and insect resistance. Not only is the genetic stock of the crop itself simplified by development of super-crop varieties, the diversity of native plant associates in the crop environment is reduced; the outcome is a drastically reduced content of coexisting native species in a genetically modified crop landscape. For example, where a superior-yielding corn taxon exists, the agricultural landscape will move toward a more homogeneous type of plant association of species that aggregates (in the understory or fringe of the crop area) with the preferred high-yielding taxon.

Government intervention

Especially beginning in the twentieth century, many national and provincial governments have sought to intervene in agriculture, chiefly through economic incentives. The goals of these programs have been variously derived from well-meaning largess, social engineering experiments, or political cronyism; however, the outcomes have often been to drive agricultural practices to ever more intensive uses, with concomitant consequences to species loss.

Two chief defects in these intervention programs are paying farmers not to grow crops (unrelated to fallow cycles) and leasing grazing lands at below-market rates. In the first case, land may be taken out of habitat or kept from habitat restoration in order for the farmer to earn an agricultural welfare payment. Below-market lease programs, rampant in the western United States, induce overgrazing of marginal lands and effectively reduce the diversity of native flora over large portions of public lands. Although the goals of both types of intervention are superficially laudable, the practical outcome of each is to reduce biodiversity.

Additional risks exist in centrally planned economies because bureaucrats far from the farming regions may make inappropriate decisions affecting massive areas of marginal farmland. For example, large tracts of the Asian steppe were converted to farmland during the Soviet era, which led to the decline of many species of wildlife on the Great Steppe such as the Sociable Lapwing and Pallid Harrier (Donald et al. 2010). Similar ecological disasters have occurred on the North China Plain, where central planning has demanded mining of groundwater, resulting in large-scale ecological disturbance with net impacts of diminishing grain yields.

Species redundancy issues

Those focusing on the functional aspects of a robust ecosystem tend to cite the phenomenon of species redundancy or the well-known effect that an ecosystem can lose many species before overt deterioration of ecosystem services appears. However, this argument is shortsighted, especially in the context of dynamic agricultural landscapes, because the presence of redundant species is pivotal in the resilience of an ecosystem to rebound from disturbance regimes (e.g., overgrazing, fire, intensive deep ploughing).

Tragedy of the commons

A phenomenon arising whereby certain lands are vulnerable to common usage with the effective absence of property rights, tragedy of the commons occurs most often in developing countries where natural resource management capabilities are spread thin and the tradition of property rights is not well established. Extreme cases of this phenomenon have been seen in such places as the central highlands of Madagascar, where the power vacuum of French colonial withdrawal and the exploding human population of native peoples beginning in the 1970s created an atmosphere of desperate exploitation of rain forest. The soils of this region were low in nutrients to start with, and the slash-and-burn tactics of undernourished people have left indelible marks on a vast portion of central Madagascar, not only losing the native forest habitat and countless species, but also losing massive tonnages of soil that will take millennia to repair by natural processes.

Even more widespread is agricultural exploitation of lands in nominal protection. In many developing countries throughout the world, local farmers commonly appropriate national park or other public lands for agricultural usage. These practices persist in remote areas of dozens of countries such as Panama, Brazil, Namibia, South Africa, China, Morocco, and Romania. In such areas there are simply not enough enforcement personnel, and in some cases national will, to protect lands from agricultural appropriation. In some cases the issue is chiefly the need to feed hungry people where the land's carrying capacity for human food has been exhausted; in other cases, such as parts of China, public corruption permits the agricultural conversion of protected habitat.

Extinction half lives

Upon disturbance or fragmentation of an ecosystem, species extinctions are sometimes measured in the time interval it takes for 50 percent of the original species to become locally extinct. For example, Kenyan forest remnants, heavily fragmented by agriculture and other human disturbance, denominated in sizes of 247 to 2471 acres (100 to 1,000 ha), have been predicted to have extinction half-lives of 23 to 80 years (Carpenter 2005).

Vulnerability classes of biota to extinction risk

C. Barry Cox and Peter D. Moore (2010) summarized species characteristics most vulnerable to extinction risk:

1. Species having a restricted range

2. Epiphytes

3. Low breeding success animals

4. Plants pollinated by mammals, such as bats

5. Species lacking connectivity between metapopulations

Specific adverse agricultural practices

A number of specific agricultural practices have a strong impact upon species diversity. Some of the chief issues that generally have an adverse effect upon biodiversity include the following (Adams et al. 2004):

1. Fragmenting surrounding habitat

2. Filling ponds

3. Removing or reducing hedgerows and other edge boundaries such as riparian zones

4. Practicing monoculture versus multiple crop production

5. Ceasing to undersow grasses and clovers

6. Abandoning crop rotation

7. Over-relying on agricultural chemicals

8. Overgrazing by livestock

Most of the above practices have been addressed in this entry except for pond filling and hedgerow loss. Filling or draining of ponded areas and wetlands is often carried out in order to expand arable land area. In some cases ponds or wetlands are simply filled by silt runoff from poorly managed fields. Significant losses of amphibians may ensue due to reduction of breeding habitat; furthermore, declines in aquatic insects as well as loss of aquatic plants usually result.

For centuries, hedgerows have been employed to separate fields in order to organize land ownership and provide windbreaks and drainage infrastructure. With pressures to produce more food in an overpopulated world, many farmers have narrowed the width of these habitat resources, putting additional pressure on the species seeking relict habitats in agricultural areas that have replaced previous expansive natural habitat. These residual hedgerows or other field perimeters have provided valuable refuge for numerous flora and fauna that would otherwise have no haven from the crops and pastures checkering the landscape. Reduction of hedgerows causes particular impacts to small mammals and to birdlife who rely upon berry production from many hedgerow species. A different type of impact arises when farmers select alien species such as eucalyptus, which form more effective windbreaks but supplant native shrubs.

Although a variety of claims of positive impacts to biodiversity from agriculture have been touted by agricultural interests and other commercial enterprises, these assertions appear unfounded. A more accurate form of these claims, which began to appear in the mid-twentieth century, is that prudently managed agriculture practices militate against biodiversity loss. Examples of such practices include hedgerow protection, pond preservation, use of organic farming methods, and crop rotation. Use of organic farming methods minimizes application of pesticides and chemical fertilizers, which practices are particularly adverse to soil microbes, earthworms, avifauna, butterflies, and spiders. The prime benefits to agriculture going forward would be the avoidance of monoculture and the integration of native flora into cropping areas, one of example of which has been accomplished on a small scale with certain Mesoamerican coffee and cacao plantations.

Outlook

Sedentary agriculture is the prime driver to the present Holocene mass extinction, which began approximately 10,000 years ago, coinciding with widespread conversion of *Homo sapiens* lifestyle from early hunter-gatherer cultures. Initial species extinctions were driven by habitat loss and fragmentation as large areas of natural habitat were converted to cropping and grazing lands.

Temperate forests were decimated in Europe in the past 4,000 years, but rainforest destruction in Asia, Africa, and South America commenced in earnest in the last four to six decades, chiefly led by slash and burn and logging. Thus, the bulk of species loss from 2000 to 2040 will arise in closed tropical forests. Although forecasting future species loss is difficult, even more complex is distinguishing agriculturally caused biodiversity loss from other causes of biodiversity loss. Nevertheless, the following global estimates are useful in

Massive loss of rainforest cover for the Madagascar Central Highlands, leaving land unfit for agriculture and almost devoid of native species; this outcome is the result of widespread slash-and-burn farming by indigenous peoples. Courtesy of C. Michael Hogan.

visualizing the magnitude of species loss that can be expected to occur by the year 2040 relative to a base year of 1990. Under a middle loss habitat scenario (namely world total of

24,710,538 acres [10 million ha] of native habitat loss per year), the following percentages of additional species decline are expected by region: Africa, 6 to 13 percent; Latin America, 8 to 18 percent; Asia, 12 to 26 percent (Cox and Moore 2010).

The trend of ongoing extinctions due to agricultural expansion and intensification is not without hope of reversal. Available technologies support polycultures and minimum use of deleterious fertilizer and pesticide substances. Examples of successful application of sustaining native plant assemblages compatibly with a valuable crop include retention of native rainforest with coffee plantations in Mesoamerica and sowing of native wildflowers amidst vine rows in California's Sonoma County wine country. However, the economics of such practices are not always favorable, unless benefits of long-term sustainability and ecosystem services are taken into account and monetized, or unless consumers appreciate the added value of such ecologically favorable practices.

Other methods of mitigation of reversing species losses due to agriculture include forsaking governmental subsidies that encourage livestock grazing of marginal rangelands; ensuring the retention of biological corridors to connect fragmented habitats in agricultural areas; and educating ranchers to promote compatibility of endangered apex predators with livestock operations.

Resources

Books

Adams, Robert McCormick. *The Evolution of Urban Society: Early Mesopotamia and Prehispanic Mexico*. New Brunswick, NJ: Aldine Transaction, 2005.

Adds, John, Erica Larkcom, and Ruth Miller. *Genetics, Evolution and Biodiversity*. Cheltenham, U.K.: Nelson Thornes, 2004.

Barry, Terry. *A History of Settlement in Ireland*. London: Psychology Press, 2000.

Carpenter, Stephen R. *Millennium Ecosystem Assessment Program, Vol. 2: Ecosystems and Human Well-Being*. Washington, D.C.: Island Press, 2005.

Clark, Christopher Michael. *Long Term Effects of Elevated Nitrogen Inputs on Plant Community Dynamics and Biogeochemistry: Patterns and Process of Community Recovery*. Ann Arbor, MI: ProQuest, 2007.

Cooney, Gabriel. *Landscapes of Neolithic Ireland*. London: Psychology Press, 2000.

Cox, C. Barry, and Peter D. Moore. *Biogeography: An Ecological and Evolutionary Approach*. 8th ed. Hoboken, NJ: Wiley, 2010.

Darvill, Timothy C. *Prehistoric Britain*. London: Taylor & Francis, 2009.

Donald, Paul, Nigel Collar, Stuart Marsden, and Debbie Pain. *Facing Extinction: The World's Rarest Birds and the Race to Save Them*. London: A & C Black, 2010.

Hollis, Daniel Webster. *The History of Ireland*. Westport, CT: Greenwood, 2001.

Houston, R.A. *The Population History of Britain and Ireland 1500–1750*. Cambridge, U.K.: Cambridge University Press, 1995.

Thorston Reemtsma and Jose Benito Quintana. *Analytical Methods for Polar Pollutants*. in Jekel, Martin, and Thorsten Reemtsma, eds. *Organic Pollutants in the Water Cycle: Properties, Occurrence, Analysis and Environmental Relevance of Polar Compounds*. Weinheim, Germany: Wiley-VCH.

Lannoo, Michael J., ed. *Amphibian Declines: The Conservation Status of United States Species*. Berkeley: University of California Press, 2005.

Martell, Hazel Mary. *The Kingfisher Book of the Ancient World: From the Ice Age to the Fall of Rome*. Oxford, U.K.: Oxford University Press, 2001.

Martin, Paul S. *Prehistoric Overkill: The Global Model*. In Martin, Paul S., and Richard G. Klein, eds. *Quaternary Extinctions: A Prehistoric Revolution*. Tucson: University of Arizona Press, 1989.

McIntosh, Jane R. *The Ancient Indus Valley: New Perspectives*. Santa Barbara, CA: ABC-CLIO, 2008.

National Research Council. *Science and the Endangered Species Act*. Washington, D.C.: National Academy Press, 1995.

Potts, G.R. "The Grey Partridge." In *Bird Conservation and Farming Policy in the European Union*, edited by D. Pain and J. Dixon. London: Academic Press, 2000.

Potts, G.R., and N.J. Aebischer. "Modeling the Population Dynamics of the Grey Partridge: Conservation and Management." In *Bird Population Studies: Relevance to Conservation and Management*, edited by C.M. Perrins, J.D. Lebreton, and G.J. M. Hirons, 373–390. Oxford, U.K.: Oxford University Press, 1991.

Rowley, Trevor. *The English Landscape in the Twentieth Century*. London: Hambledon Continuum, 2006.

Shukla, O.P., Omkar, and A.K. Kulshretha. *Pesticides, Man and Biosphere*. New Delhi: Ashish, 1999.

Simmons, Ian Gordon. *Changing the Face of the Earth*. Oxford, U. K.: Blackwell, 1989.

Stevens, Carly. *Nitrogen Impacts on Plant Communities*. PhD diss. Milton Keynes Open University, 2004.

U.S. Congress, Office of Technology Assessment. *Agriculture, Trade and Environment: Achieving Complementary Policies*. OTA-ENV-617. Washington, D.C.: Author, 1995.

Whitmore, Timothy Charles, and Jeffrey Sayer, eds. *Tropical Deforestation and Species Extinction*. IUCN Forest Conservation Programme. London: Chapman & Hall, 1992.

Periodicals

Adams, William M., Ros Aveling, Dan Brockington, et al. "Biodiversity Conservation and the Eradication of Poverty." *Science* 306 (2004): 1146–1149.

Barker, Tom, Keith Hatton, Mike O'Connor, et al. "Effects of Nitrate Load on Submerged Plant Biomass and Species Richness: Results of a Mesocosm Experiment." *Fundamental and Applied Limnology* 173 (2008): 89–100.

Berezina, N.A., S.M. Golubkov, and Yu Gubelit. "Structure of Littoral Zoocenoses in the Macroalgae Zones of the Neva River Estuary." *Inland Water Biology* 2 (2009): 340–347.

Clark, Christopher Michael, and David Tilman. "Loss of Plant Species after Chronic Low-Level Nitrogen Deposition to Prairie Grasslands." *Nature* 451 (2008): 712–715.

Gupta, Anil K. "Origin of Agriculture and Domestication of Plants and Animals Linked to Early Holocene Climate Amelioration." *Current Science* 87 (2004): 54–59.

Matson, P.A., W.J. Parton, A.G. Power, and M.J. Swift. "Agricultural Intensification and Ecosystem Properties." *Science* 277 (1997): 504–509.

Richardson, Peter J., Robert Boyd, and Robert L. Bettinger. "Was Agriculture Impossible during the Pleistocene but Mandatory during the Holocene? A Climate Change Hypothesis." *American Antiquity* 66 (2001): 387–411.

Symsted, Amy J., David Tilman, John Wilson, and Johannes M. H. Knops. "Species Loss and Ecosystem Functioning: Effects of Species and Community Composition." *OIKOS* 81 (1998): 389–397.

Tscharntke, Teja, Alexandra M. Klein, Andreas Kreuss, et al. "Landscape Perspectives on Agricultural Intensification and Biodiversity—Ecosystem Service Management." *Ecology Letters* 8 (2005): 857–874.

Other

Dani, A.H. "Critical Assessment of Recent Evidence on Mohenjodaro. *Second International Symposium on Mohenjodaro*, February 1992.

Hogan, C. Michael. "Los Sililos—Ancient Village or Settlement in Spain in Andalucía." The Megalithic Portal. Accessed April 9, 2012. http://www.megalithic.co.uk/article.php?sid=17974.

C. Michael Hogan

Development

Economic growth is an overarching driver of species declines and extinctions (Woodward 2009). It leads to development of land, in the form of towns, cities, roads, and farms. Land development creates increasing human pressure on natural resources and limits the resources that species need to survive. Extinctions can occur in countries with fewer financial resources for reasons besides development; for example, there may be little or no money to protect habitat and species, and a need may exist to exploit animals and plants for food, clothing, and trade. However, poor countries, such as Costa Rica, can also have incentive to protect habitats for ecotourism (Woodward 2009).

History of urban development and sprawl

Urban development represents a fairly new condition for people and has occurred mainly since 1800. The transition from hunter-gatherer societies without permanent homes to those with more-or-less permanent dwellings in rural areas took place around 10,000 years ago. Then, with the development of agriculture, including the domestication of animals and plants, cities and economies began to emerge. The first urban settlement appeared in Mesopotamia around 3,500 years ago. After that, settlements arose in the Nile Valley (Egypt), the Indus Valley (Pakistan), and the Huang Ho Valley (China). Urban settlements later developed independently in Mexico and Peru. This early type of urban development, along with a sizable portion of the population living in rural areas, has lasted over 5,000 years.

A large part of the world population, including more than half the US population, came to reside in cities during the Industrial Revolution, roughly dated from 1750 to 1850. Machine-based manufacturing and trade that expanded as a result of improved roads and railways allowed the building of large, highly dense cities. These early cities were walkable and surrounded by open lands devoted to agriculture, but they also contained wilderness areas. In the period just after the industrial era, cities remained compact because of limitations in their transportation infrastructure, which consisted mainly of trains and trolleys.

Compact urban development continued in the United States until the early twentieth century, when technological

innovations and several policy incentives offered opportunities for workers to spread out from urban areas, resulting in what is known as *sprawl*. Sprawl is the spreading outward of a city and its suburbs to the city's outskirts with low-density development that encourages car dependency. It is often thought of as poorly planned, land-consumptive development of large-lot, single-family homes, office campuses, and strip malls. Developments may be disconnected, jump over large tracts of undeveloped land, or may jump beyond established settlements onto farmland or forestland. Over time, the areas between the urban core and the satellite developments begin to fill in with buildings, parking lots, and lawns, creating a dense suburb (Daniels 1999). Sprawl is also found in remote wilderness settings, around national parks and other recreational areas, and where people make second homes in once undeveloped areas that become accessible by roads.

In the early 1990s, buildings were estimated to occupy more than 5 percent of Earth's land surface, and 10 percent of that area consisted of impervious surfaces (Meyer and Turner 1992). In the early 2000s, urban areas covered approximately 5 percent of the United States, which tallied more than the combined area of the national parks, state parks, and Nature Conservancy preserves in the country (Blair 2004). Roads were estimated to affect 15 percent to 20 percent of the land area in the United States. People drove increasing distances in their daily work commutes, and new roads were being added to suburban development all the time (Forman and Deblinger 2000).

Economic incentives for urban development and sprawl

Urban development and sprawl have been driven largely by economic incentives. Over the last century, and continuing in the early 2000s, new job opportunities in cities have drawn people from rural areas. As a result, the global urban population has risen from 13 percent of the total (or 220 million people) in 1900 to 49 percent (or over 3 billion) in 2005. The portion of the global population living in cities is projected to rise to 60 percent, or almost 5 billion people, by 2030 (United Nations 2005). The US population has changed from 25 percent urban and 75 percent rural in 1870 to about 80 percent urban and 20 percent rural as of 2012 (Kareiva

Early development in ancient Mesopotamia. Ancient Assyrian royal palace at Ninevah on the Tigris River destroyed in 612 BC. © North Wind Picture Archives/Alamy.

2012). This population growth is a major factor contributing to urban development.

Nevertheless, population growth is thought to be unassociated with sprawl at the local or regional level. For example, between 1982 and 1997, the rate of land development in the United States was almost twice the rate of population growth. Alternatively, sprawling development in the United States is thought to be a consequence of economic incentives for

Aerial residential urban sprawl west of Phoenix, Arizona. © Stock Connection Blue/Alamy.

various parties, motivated by the invention of the automobile and changes in public policies that govern housing, transportation, taxation, public investment, and neighborhood zoning (Johnson and Klemens 2005).

From the 1930s to the 1960s a number of housing policies in the United States created economic incentives for people to relocate from urban areas to the suburbs. To prevent foreclosures as a result of the Great Depression, the Home Owners' Loan Corporation was formed in 1933 to help refinance mortgages. This action led, however, to the devaluation of buildings in urban areas because of their higher risk for mortgages. The National Housing Act of 1934 created the Federal Housing Authority (FHA), with the goals of increasing employment in the homebuilding industry and making it easier for many Americans to obtain mortgages and own houses. Yet the FHAs housing appraisal system resulted in the further devaluation of urban areas, producing several negative social consequences. Namely, the new system made it harder for urbanites, especially African Americans, to obtain mortgages; it promoted whites-only subdivisions; and it led to many African Americans moving into public housing in cities, which exacerbated problems. The Servicemen's Readjustment Act of 1944 (commonly called the GI Bill) intensified these trends by making it less expensive to build a new home in the suburbs than in the urban areas (Johnson and Klemens 2005).

On top of these housing policies, taxation policies further encouraged the buying of homes in the suburbs. These included the homeowners' mortgage interest deduction and the exclusion of taxes on capital gains on sales up to a certain level of profit. This arrangement resulted in the reduction in the cost of home ownership and provided incentives to purchase larger and more expensive homes in the suburbs. Additionally, zoning led to reduced urban density and encouraged sprawl. Zoning decisions are made by local governments to protect the health, welfare, and safety of citizens, and they are a factor in determining where developers can build. Zoning results in single-use zones for separate uses, such as residential or commercial (Knaap et al. 2001).

Finally, transportation policies brought about the creation of roads and highways. Roads are built to lead to new opportunities for economic gain and initiate the process of sprawl. The Federal-Aid Highway Act of 1956 set in motion the construction of tens of thousands of miles of interstate highways and roads, and the decision was made to disinvest in public-transit infrastructure. This change marked the transition from public-transit infrastructure, such as trolleys, streetcars, and commuter rail lines, to a car-centered interstate highway system. Additionally, the effects of roads and highways have been amplified by the US government's subsidization of gasoline, making prices lower there than their true value compared to that in other countries.

Some of the damaging effects of these older policies of development were curtailed (e.g., civil rights advocates brought down discriminatory housing policies in the 1960s and 1970s, people moved back into urban cores, and governments reinvested in public-transit infrastructure). Nevertheless, the economic incentives and momentum generated by the older policies still produce sprawl. In the early 2000s, a consensus formed regarding the need for new policies for development to reverse some of the damaging ecological effects of urban development and sprawl, to prepare for an expected surge in gas prices in subsequent decades, and to adjust for the needs of changing population demographics.

Ecological impacts of urban development and sprawl

Urban development and sprawl are the main anthropogenic causes of habitat loss, habitat fragmentation, and habitat degradation. In fact, habitat loss, fragmentation, and degradation are the main causes of species extinction worldwide. Urban development and sprawl endanger more species in the United States than any other human activity except for agriculture. According to the US Fish and Wildlife Service, urbanization endangers upward of 300 different American species (Czech et al. 2000). The effects of urban development have led to the endangerment of 8 percent of vertebrate species, and that number is expected to rise in the future (McDonald et al. 2008). While habitat fragmentation and degradation are aspects of habitat loss, they do not necessarily involve blatant habitat loss.

Much of what is known about the impacts of urban development and sprawl on the diversity of life comes from studies of the effects of urbanization on species and ecosystems. In general,

urban areas tend to have fewer numbers of species and poorer ecosystem quality, and these areas support, on average, less than half the diversity found in rural areas or more natural habitats. Urban areas support plant and animal species that are adaptable to environmental stressors, such as pollution, and that are considered less valuable to people. By contrast, undeveloped rural areas support a range of natural communities and a larger diversity of plants and animals with more specialized life history requirements, including species that are considered more valuable (McKinney 2002).

Habitat fragmentation and degradation work together to contribute to the endangerment and extinction of species by affecting critical factors necessary for their survival. Survival factors for individual species include the following: abundance and quality of food and shelter; presence of mates; presence of predators, parasites, and diseases; quality of soil; availability of nutrients; availability of seed dispersers; quality of the landscape, including the amount of connectivity between habitat patches; and the amount of light, shade, moisture, dryness, and windiness (Johnson and Klemens 2005).

Habitat fragmentation

Habitat fragmentation occurs when large, continuous areas are broken into isolated fragments or patches by natural or human processes. Road construction initiates the fragmentation process for urban development and sprawl. After roads are built, land is cleared for the construction of subdivisions or vacation homes. Then, the remaining natural habitat patches are built on and broken up even further until they are completely and permanently replaced with pavement and structures.

Habitat fragmentation resulting from urban development and sprawl affects species survival in two important ways. First, it reduces the amount of quality habitat available because the resulting fragments or patches are smaller. They also have more edge habitat, which is the transition between two habitats. The edges in human-dominated landscapes tend to be abrupt and intolerable to many sensitive species; by contrast, the transitions between habitats are more gradual in natural landscapes. Thus, over time, the smaller patches support fewer and fewer species.

Second, fragmentation alters species' movement patterns among patches. When patches are disconnected, individuals have more difficulty finding food and mates. When plant and animal populations are separated or unable to spread out into a habitat, there is a higher risk of genetic inbreeding and local extinction. Furthermore, because plant species are often dispersed by animals, reduced animal movement affects them as well.

On top of these indirect effects of fragmentation stemming from urban development and sprawl, roads contribute to species endangerment and extinction directly through vehicle mortalities. For example, vehicles kill about one million vertebrates in the United States every day (Johnson and Klemens 2005). Other environmental factors thought to determine the severity of the effects of fragmentation on wildlife, especially on birds, include the pattern of natural disturbances (e.g., fire, wind,

A fragmented landscape showing habitat patches. Courtesy of Google.

insects, disease), the similarity of the human-dominated and natural landscapes, and the persistence of the human-induced change (Marzluff and Ewing 2001).

Most species are sensitive to fragmented landscapes, although some more opportunistic varieties can adapt to them. Species sensitivity to fragmentation is determined by the following factors:

1. having particular habitat requirements (e.g., wetlands, old-growth forests, or a combination of habitat types, such as for amphibians and dragonflies);

2. having particular site requirements, such as for nesting or migration;

3. being a species that may be collectible by people, such as large spiders;

4. being highly mobile, which includes migrant songbirds and raptors;

5. being wide ranging, which includes carnivores and large predators, such as the black bear and gray wolf;

6. being a plant with limited dispersal, such as an orchid; and

7. having small home ranges, among which are small mammals (Johnson and Klemens 2005).

Habitat degradation

Habitat degradation is the alteration of a species' habitat, such that it reduces the habitat's ability to meet that species' needs. Pollution and invasive species are two of the main introduced threats that contribute to habitat degradation.

Habitat degradation affects species' survival in one major way: by altering or degrading environmental conditions to the point that species have difficulty tolerating them. The quality of the habitat is a main determinant of population persistence. As the environmental quality in isolated patches degrades, there is a reduction in the quality of ecological processes there, including the water cycle, carbon cycle, photosynthesis,

pollination, and soil formation and decomposition. In turn, species numbers are reduced (Forman 1995).

Habitat degradation in urban environments has led to water and air that is polluted, soils that are compacted and that impede root growth, decreased natural lighting and increased artificial lighting, and increased disturbance by humans and vehicles. The buildings and impervious surfaces in urban areas tend to alter the movement of water through the city, increase runoff and channelize stream flows, and affect climate (Adams 1994). The heat island effect is a climate-altering phenomenon that makes cities warmer than surrounding areas because pavement absorbs and radiates solar heat.

Pollution in urban areas comes from point (i.e., single, identifiable) and non-point (i.e., diffuse) sources. Non-point sources of pollution are considered the greatest threat to water quality. The main source is storm-water runoff from roads and lawns that carries pesticides, fertilizers, nutrients, and sediments into rivers, lakes, and aquifers. The chemicals ending up in water bodies can:

1. be toxic to organisms at certain concentrations;

2. be bio-accumulative, or have the potential to build up in the tissues of animals at each trophic level in the food chain;

3. be endocrine disrupting, or have the potential to mimic the effects of hormones and interfere with reproduction; or

4. negatively affect critical ecosystem processes.

Point sources of pollution, such as wastewater treatment plants, septic systems, and polluting industries, also reduce water quality by contributing to nutrient loading, which can lead to effects such as phytoplankton blooms and depleted oxygen levels, which reduce populations of fish and other animals.

Migratory birds are examples of species that are particularly sensitive to the effects of habitat degradation or lowered environmental quality in urban areas. In one study on the Acadian flycatcher (*Empidonax virescens*), researchers found that the number of young produced per nesting season averaged half as many in urban sites as in rural sites (one young compared to two, respectively). This study took into account predation in urban areas. It found that compared to rural areas, urban areas attracted so-called lower-quality birds that arrived later in the spring, left earlier in the fall, made fewer attempts to nest, and were much less likely to return to the same nesting sites (Rodewald and Shustack 2008).

Development efforts that promote conservation

Between the 1970s and the early 2000s, public awareness grew as scientific studies recognized the negative impacts of older policies of development. New development began to respond to this awareness and to the needs and desires of shifting population demographics in the United States—from

Point-source pollution contributes to habitat degradation. © Marvin Dembinsky Photo Associates/Alamy.

may include protecting natural spaces, ensuring affordable housing, and preserving buildings of historic value. The techniques used for growth management include imposing impact fees for new development and enacting zoning laws to reduce the area affected by urbanization. In the 1970s, seven US states adopted some form of growth management (DeGrove and Miness 1992). Additionally, urban growth boundaries were placed around some municipalities, such as Portland, Oregon, and Boulder, Colorado. The latter mandated that the area inside the boundaries be used for higher density development and the area outside be used for conservation of open space and wildlands, in order to protect natural resources and species. The boundaries were drawn to allow enough developable land to accommodate projected population growth. Growth management plans went into effect in several locations in the early 1970s, and habitat and species protection are among the listed goals of many state and city growth management strategies; however few, if any, retrospective studies of their effects on species had been undertaken as of 2012.

CASE STUDY ON SAGE GROUSE A proposal to protect the endangered sage grouse in Idaho is an example of using growth management to prevent further habitat loss and the extinction of a species. This example of managing development would occur at a smaller scale, in the sage grouse's habitat, than many other examples of growth management, which occur at the urban scale. The sage grouse (*Centrocercus urophasianus*) is a popular game bird as well as a species selected by conservationists to provide protection for other species in its shrub-steppe ecosystem, most notably, sagebrush and a large variety of wildflowers. Sage grouse depend wholly on sagebrush habitats for their diet and cover. Unfortunately for the sage grouse, over half of its historical habitat was lost by the early twenty-first century as a result of urban and suburban sprawl, ranching, and energy development. As a result, the sage grouse's population in eleven states declined from 16 million in 1900 to between 200,000 and 500,000 by 2010. Though the sage grouse is believed to be facing extinction, it has not yet been listed under the Endangered Species Act. It is argued that the grouse is a lower priority for protection than other species.

In 2012 a legal settlement required the US Fish and Wildlife Service to decide on the sage grouse's status for protection. In Idaho, the governor devised a strategy to persuade the agency not to list the grouse, in an effort to balance protecting the bird with protecting the state's economy by allowing development by ranchers and energy producers. The main mechanism for protecting the bird would be to place restrictions on new development (e.g., residences, transmission lines, wind and solar energy plants, and ranching) within its habitat. For example, in its core habitat of 6 million acres, which contains most of the grouse's known mating sites, projects would be limited to those that cannot increase in size by more than 50 percent. Any other new projects beyond those would be prohibited. In other areas of its habitat, however, new projects would be limited but not prohibited. This proposal presents one way that could protect the sage grouse from extinction, but it is not clear whether it would be effective if implemented or whether it will ever get implemented as it must first pass legal scrutiny.

predominantly traditional families who wanted to live in large houses in the suburbs to increasingly higher numbers of nontraditional families (e.g., single parents, childless couples, and singles) who want to live in urban areas to be closer to their jobs and amenities. Additionally, the population was expected to increase by 60 million by the early 2030s, which will require the building of appropriate new housing.

New development movements have formed that are aimed at reversing the trend toward sprawl, habitat destruction, and species losses and possible extinctions. These movements aim to create more environmentally sustainable development practices that conserve habitats and species, that are more economically efficient and socially responsible, and that improve the quality of life for people. These movements include: growth management, smart growth, new urbanism, sustainable development, urban conservation, and restoration ecology/urban planning.

Growth management

Growth management is concerned with ensuring that, as populations grow, public services will meet the demand. It also

The sage grouse (*Centrocercus urophasianus*) dancing on a lek, an area where birds gather during the breeding season for community courtship displays to attract mates. © franzfoto.com/Alamy.

Smart growth

Smart growth is an urban planning theory that advocates for compact urban centers that are transit-oriented, walkable, and bicycle-friendly, and that contain mixed-use development and a range of housing types. It also aims to see that development promotes environmental conservation by protecting green spaces. According to the nonprofit organization and nationwide coalition Smart Growth America, this type of growth "protects farmland and open space, revitalizes neighborhoods, makes housing more affordable and provides more transportation choices" (Smart Growth America 2012). Some studies have found that communities with smart growth characteristics, such as compact development with extensive park systems, have greater bird species abundance and diversity compared with communities with lower-density development that contain the same number of buildings spread over the same area (US Environmental Protection Agency 2001).

New urbanism

New urbanism is an urban planning theory that advocates for "the restoration of existing urban centers and towns within coherent metropolitan regions, the reconfiguration of sprawling suburbs into communities of real neighborhoods and diverse districts, the conservation of natural environments, and the preservation of our built legacy" (Congress for the New Urbanism 2012). New urbanist–style communities are becoming a popular response to the increasing demand for attractive and convenient high-density development.

Sustainable development

Sustainable development is development that meets the needs of the present without compromising the ability of future generations to meet their own needs (Institute for Sustainable Development 2012). Sustainable development often incorporates green building with Leadership in Energy and Environmental Design (LEED) certification. LEED promotes "a whole building approach to sustainability by recognizing performance in sustainable site development, water savings, energy efficiency, materials selection, and indoor environmental quality" (US Green Building Council 2012).

Urban conservation

Urban conservation is a movement centered on restoration efforts in cities that inspire young urbanites to care about conservation. Opportunities include: (1) restoration of rivers and streams to bring back fish species and protect other species upstream or downstream; (2) planting city forests to provide cooling of an urban heat island effect; and (3) restoration of nearshore habitats for fish and shoreline protection. All these efforts improve the quality of life for people and for flora and fauna. Peter Kareiva of the Nature Conservancy has heralded urban conservation and uses the example of the Three Rivers in Pittsburgh, Pennsylvania, and of how they "were once (and not that long ago) so polluted that they were a dead zone; their waters corroded boats and drove people away. Now, after 40 years of cleanup and restoration, fish are abundant, and the riverfront is a draw for wildlife and people" (Kareiva 2012).

CASE STUDY ON THE THREE RIVERS The restoration of the Three Rivers in Pittsburgh is an example of urban conservation of an urban ecosystem degraded as a result of development. Pittsburgh's Three Rivers, consisting of the Monongahela and Allegheny Rivers which join to create the Ohio River, comprise a vast watershed of creeks and river systems that extend for tens of thousands of miles into New York and West Virginia. The Allegheny River still has wild and scenic stretches and provides Pittsburgh's water supply, whereas the Monongahela River traditionally was and still is a "working river" and supports barge traffic.

When Pittsburgh became a center of iron and steel manufacturing during the Industrial Revolution, the rivers were used to transport products to other cities around the world. Their economic value was of paramount importance, leaving their value as ecosystems largely disregarded. For more than a century the rivers were polluted by industrial waste and municipal sewage, which was released to the river untreated. The rivers carried diseases such as cholera, typhoid, and dysentery. Excessive boat traffic drove temperatures to 130°F. Nearly every fish species became regionally extinct. People were warned to avoid the rivers as a matter of public health and safety; the rivers were not enjoyed for recreation.

The turning point came in 1972, when the enactment that year of the Clean Water Act led to strict controls on substances discharged into the rivers. Over the next few decades increased government and public attention on the watershed environment led to further improvements in water quality and additional protection of the rivers. By the early twenty-first century, the rivers were much cleaner and accommodating to fish species, which were returning in healthy numbers. Largemouth and smallmouth bass, walleyes, and saugers were common catches for recreational fishers. Mayflies, insects known to survive only in clean water, were

In this photo elephants exit Africa's first dedicated elephant underpass near the slopes of Mt. Kenya. Conservationists say the tunnel connects two elephant habitats that had been cut off from each other for years by human development. © AP Photo/Jason Straziuso.

also returning each spring. Grant-funded restoration projects had restored 440 miles of previously dead streams. People had returned to the rivers to live along their shores; to work and play at the new parks, stadiums, and malls built there; and to enjoy recreation activities on and near them, including boating, kayaking, rafting, fishing, birding, and camping. Still, the rivers have some problems, such as a legacy of contaminated sediment and sewage overflows following heavy rains. Nevertheless, the renewed awareness and appreciation of the rivers in the quality of life of the region, as a result of their environmental transformation, are likely to support their continued restoration.

Restoration ecology/urban planning

Restoration ecology and urban planning can be used together to maintain native species in fragmented landscapes. For example, to protect native birds, actions can be taken such as maintaining native vegetation; managing the landscape surrounding a fragment (the matrix) and making it more like the native habitat fragments; designing buffers that reduce passage of unwanted species or environmental threats from the matrix; actively managing mammals in fragments; anticipating urbanization; and seeking creative ways to increase native habitat (Marzluff and Ewing 2001). Another example of how to maintain native species in fragmented environments is the effort to protect mammals (e.g., large game, bears, and panthers in North America; koalas in Australia; elephants in Kenya) where busy highways cross

migration pathways or bisect their habitats. Restoration ecologists have added highway underpasses, creating corridors or strips of landscape, to add connectivity to other patches. These efforts reduce animal mortalities, as well as vehicle damage and human fatalities.

CASE STUDY ON BUMBLEBEES The reintroduction of the regionally extinct short-haired bumblebee into the United Kingdom is an example of restoration ecology and planning to maintain native species in landscapes that were fragmented by development. The short-haired bumblebee (*Bombus subterraneus*) was last observed in the United Kingdom in 1988 and was declared regionally extinct several years later. The bumblebee's dramatic decline across Europe has been attributed to habitat loss, fragmentation, and degradation as a result of intensive farming and pesticide use after World War II. These practices were supported and encouraged by government grants and subsidies aimed at increasing self-sufficiency. Over the next 60 years, virtually all of the wildflower meadows and farmland borders that were the bumblebees' home and source of food (pollen and nectar), were lost. Increasing habitat destruction led to increasing isolation and increasing vulnerability of the bees to inbreeding and extinction.

Because of the significance of bumblebees as pollinators of agricultural crops and flowers, a renewed interest in reintroducing them developed. In 2012 up to 100 queen bumblebees were collected from the southern region of Sweden and released to the English countryside, in the hope that they would recolonize meadows and farmlands. The species' survival in Sweden is attributed to less intensive farming, because of a smaller human population, and the preservation of critical wildflower meadows. To attract the bumblebees, farmers in the United Kingdom are working to re-create native habitats by planting wildflowers around the edges of their farmlands. These wildflower borders are predicted to enhance bumblebee populations, not only by providing food but also by maintaining natural habitat that would not be disrupted every season.

The creation of habitat for bumblebees across the countryside should restore a flower-rich ecosystem that has largely been lost in the United Kingdom and that has additional benefits for a range of other species. This ecosystem can provide pollen and nectar for insects, such as moths, butterflies, and flies, which provide food for bats and birds; nesting habitat for birds and rodents; and protection for species sensitive to grazing, such as the marshmallow plant and the streamside canopy cover–dependent water vole.

Conclusion

Economic growth is an overarching driver of species declines and extinctions because it leads to the development of land, in the form of towns, cities, roads, and farms, which limits the resources needed by species to survive. People have a long history of generating urban development and sprawl, which has resulted in a sizable area of Earth's land surface being covered with urban areas, buildings, and pavement. The major mechanism driving urban development and sprawl has been incentives for economic gain, which have been enabled and furthered by government policies. Urban development and sprawl have contributed to extinctions and to the endangerment of modern diversity by causing habitat fragmentation and degradation. In the United States urban development is responsible for the endangerment of hundreds of species. Habitat fragmentation and degradation work together to negatively affect a species' survival by reducing the amount of available quality habitat; by reducing movement patterns among fragments or patches, thus making it difficult for the members of the species to find food or mates; and by altering and degrading environmental conditions, so that the habitat's ability to meet the needs of the species is reduced. Because of heightened awareness of the negative impacts of existing policies of urban development and sprawl, new development movements are beginning to reverse this trend. There is hope that if new development is implemented in ways that make an effort to consider and ensure the survival needs of species, it will work to protect the diversity of life, thus preventing further species extinction and endangerment.

Resources

Books

Adams, Lowell W. *Urban Wildlife Habitats: A Landscape Perspective.* Minneapolis: University of Minnesota Press, 1994.

Bennett, Andrew F. *Linkages in the Landscape: The Role of Corridors and Connectivity in Wildlife Conservation.* 2nd ed. Gland, Switzerland: International Union for Conservation of Nature, 2003.

Daniels, Tom. *When City and Country Collide: Managing Growth in the Metropolitan Fringe.* Washington, DC: Island Press, 1999.

DeGrove, John Melvin, and Deborah A. Miness. *Planning and Growth Management in the States.* Cambridge, MA: Lincoln Institute of Land Policy, 1992.

Forman, Richard T.T. *Land Mosaics: The Ecology of Landscapes and Regions.* Cambridge, U.K.: Cambridge University Press, 1995.

Jacobs, Jane. *The Death and Life of Great American Cities.* New York: Random House, 1961.

Jacobs, Jane. *The Economy of Cities.* New York: Random House, 1969.

Johnson, Elizabeth A., and Michael W. Klemens, eds. *Nature in Fragments: The Legacy of Sprawl.* New York: Columbia University Press, 2005.

Marzluff, John M. "Worldwide Urbanization and Its Effects on Birds." In *Avian Ecology and Conservation in an Urbanizing World,* edited by John M. Marzluff, Reed Bowman, and Roarke Donnelly, 19–47. Boston: Kluwer Academic, 2001.

Mumford, Lewis. *The City in History: Its Origins, Its Transformations, and Its Prospects.* New York: Harcourt, Brace and World, 1961.

U.S. Environmental Protection Agency. *Our Built and Natural Environments: A Technical Review of the Interactions between Land Use, Transportation, and Environmental Quality.* Washington, DC: U.S. Environmental Protection Agency, Development, Community, and Environment Division, 2001.

Periodicals

Blair, Robert B. "The Effects of Urban Sprawl on Birds at Multiple Levels of Biological Organization." *Ecology and Society* 9, no. 5 (2004): 2.

Blair, Robert B., and Alan E. Launer. "Butterfly Diversity and Human Land Use: Species Assemblages along an Urban Gradient." *Biological Conservation* 80, no. 1 (1997): 113–125.

Czech, Brian, Paul R. Krausman, and Patrick K. Devers. "Economic Associations among Causes of Species Endangerment in the United States." *BioScience* 50, no. 7 (2000): 593–601.

Day, Simon. "Extinct Short-Haired Bumblebee to Be Reintroduced in England." *Guardian* (London), April 26, 2012. Accessed August 17, 2012. http://www.guardian.co.uk/environment/2012/apr/26/extinct-bumblebee-uk-release.

Forman, Richard T.T., and Robert D. Deblinger. "The Ecological Road-Effect Zone of a Massachusetts (U.S.A.) Suburban Highway." *Conservation Biology* 14, no. 1 (2000): 36–46.

Marzluff, John M., and Kern Ewing. "Restoration of Fragmented Landscapes for the Conservation of Birds: A General Framework and Specific Recommendations for Urbanizing Landscapes." *Restoration Ecology* 9, no. 3 (2001): 280–292.

McDonald, Robert I., Peter Kareiva, and Richard T.T. Forman. "The Implications of Current and Future Urbanization for

Global Protected Areas and Biodiversity Conservation." *Biological Conservation* 141, no. 6 (2008): 1695–1703.

McKinney, Michael L. "Urbanization, Biodiversity, and Conservation." *BioScience* 52, no. 10 (2002): 883–890.

Meyer, William B., and B.L. Turner II. "Human Population Growth and Global Land-Use/Cover Change." *Annual Review of Ecology and Systematics* 23 (1992): 39–61.

Rodewald, Amanda D., and Daniel P. Shustack. "Urban Flight: Understanding Individual and Population-Level Responses of Nearctic–Neotropical Migratory Birds to Urbanization." *Journal of Animal Ecology* 77, no. 1 (2008): 83–91.

Woodward, Tali. "The Nature of the Fiscal World." *Conservation Magazine* 10, no. 1 (2009).

Other

Congress for the New Urbanism. "Charter of the New Urbanism." Accessed August 17, 2012. http://www.cnu.org/charter.

Dawes, John. "America's Three Rivers." Pittsburgh Green Story. Accessed August 17, 2012. http://www.pittsburghgreenstory.org/html/3_rivers.html.

Gammans, Nikki. "The Short-Haired Bumblebee Reintroduction Project Report, 2009–2011." Hymettus. Accessed August 17, 2012. http://hymettus.org.uk/downloads/B_subterraneus_Project_report_2011.pdf.

Institute for Sustainable Development. "What Is Sustainable Development?" Accessed August 17, 2012. http://www.iisd.org/sd.

Kareiva, Peter. "Urban Conservation: Conservation Should Be a Walk in the Park, Not Just the Woods." *Nature Conservancy Magazine*. Accessed August 17, 2012. http://magazine.nature.org/features/think-about-it-urban-conservation.xml.

Knaap, Gerrit, Emily Talen, Robert Olshansky, and Clyde Forrest. "Government Policy and Urban Sprawl." Urbana: University of Illinois, Department of Urban and Regional Planning, 2001. Accessed August 17, 2012. http://dnr.state.il.us/orep/pfc/balancedgrowth/pdfs/government.pdf.

Smart Growth America. "Americans Want Smarter Growth: Here's How to Get There." Accessed August 17, 2012. http://www.smartgrowthamerica.org/SGBOOK.pdf.

United Nations. Department of Economic and Social Affairs, Population Division. "World Urbanization Prospects: The 2005 Revision." Accessed August 17, 2012. http://www.un.org/esa/population/publications/WUP2005/2005wup.htm.

U.S. Green Building Council. Accessed August 17, 2012. http://www.usgbc.org/News/PressReleaseDetails.aspx?ID=3743.

Amanda P. Rehr

・・・・・

Invasive species

Invasive species have been described as the second-greatest extinction threat in the world today, behind only habitat loss (Wilcove et al. 1998). Is this true? Are invasive species a major cause of animal extinctions, or has the extinction threat of invasive species been exaggerated? By what mechanisms have invasive species driven animal species to extinction? Are certain animal groups more threatened by invasive species than others? Do certain environments increase the vulnerability of animal species to invasive species? Before these questions can be answered, it is necessary to define what is meant by the term *invasive species*.

Definition of invasive species

In the 1980s most ecologists used the term *invader* to describe any species that colonized a territory or ecosystem in which it had never occurred before (Mack 1985; Mooney and Drake 1989). In the latter 1990s ecologists and policymakers began to distinguish between nonnative species that did and did not cause harm, with the term *invasive* being reserved for only those nonnative species that cause harm. For example, in former president Bill Clinton's 1999 executive order on invasive species, invasive species were defined as nonnative species whose introduction causes, or is likely to cause, harm to the economy, the environment, or human health. Since about 2000, this has been the most common usage of the term invasive species, both in the fields of ecology and conservation and in most national and international doctrines and policies addressing problems caused by nonnative species. In this entry, the term invasive species refers to nonnative species that have been deemed harmful by humans.

Are invasive species a major cause of animal extinctions?

Invasive species are known to have caused many animal extinctions. The brown tree snake (*Boiga irregularis*) was accidentally introduced into Guam following World War II (1939–1945). Because the native animals of Guam lacked predator defenses against snakes, they were easy prey for this new predator. Within several decades, these snakes had caused the extirpation (localized extinction) of 12 of the 22 native bird species. For similar reasons, introduced rats and cats have also caused many island bird and small mammal species to go extinct. Often experiencing little predation themselves, the rat and feral cat populations can grow mostly unchecked following their introduction, resulting in large numbers of novel predators that can drive island prey species to extinction in decades, or even years. Particularly vulnerable to rat and cat predation are nestlings of oceanic birds such as puffins, shearwaters, and petrels that live entirely in the open ocean, except when they come to shore to breed. These birds typically breed in large dense colonies, with nesting pairs often numbering into the hundreds of thousands or even millions. Adult birds, although susceptible to predation while brooding the eggs or chicks, are not nearly as vulnerable to predation by the rats and cats as are their flightless nestlings, which are essentially defenseless. Perhaps responding innately to the desperate behavior of thousands of defenseless prey, whose cries and futile efforts to escape inundate the predators' senses, the cats and rats often kill far more chicks than they can possibly eat. As a result, even modest numbers of rats and cats can decimate entire breeding colonies.

The Nile perch (*Lates niloticus*), a large freshwater fish (individuals can exceed 6.5 feet [2 m] in length and weigh more than 440 pounds [200 kg]) that is native to many of the large African rivers, was introduced into Lake Victoria in the early 1950s to enhance the local fisheries. Prior to the introduction, Lake Victoria was home to hundreds of species of fish, many of them found nowhere else in the world. These included more than 300 species in the family Cichlidae. While the introduction succeeded in substantially boosting Lake Victoria's commercial fishing industry, the large introduced predator is believed to have caused the extinction of more than 100 of the lake's endemic cichlids. In several Russian lakes a number of native amphipod species (small crustaceans) are believed to have been extirpated, replaced by an introduced amphipod from Lake Baikal, *Gmelinoides fasciatus*, which had been introduced intentionally into many lakes to enhance fish production. It is thought that the most likely cause of the extirpations of the native amphipods has been predation by *G. fasciatus* on the juveniles of the native species.

Introduced predatory snails, such as *Euglandina rosea*, have driven many native land snails to extinction on Pacific islands.

The brown tree snake, *Boiga irregularis*, drove most of the native bird species of Guam to extinction on the island following the snake's introduction in the middle of the twentieth century. Photo By Martin Cohen Wild About Australia/Lonely Planet Images/Getty Images.

Ironically, *E. rosea*, native to the southeastern United States, was introduced to Hawaii as a biological control agent in the 1950s in an effort to reduce the abundance of another invasive snail, *Achatina fulica*, an African herbivorous snail

that had become a serious crop pest. Introduced flatworms are also thought to have caused the extinction of some land snails. For example, *Platydemus manokwari*, a flatworm native to New Guinea has been introduced, both intentionally and unintentionally, to many Pacific islands where they have fed on endemic snails and are believed to be the primary cause of extinction for some of these species. Like *E. rosea*, *P. manokwari* was sometimes introduced to control the invasive African snail *A. fulica* but ended up becoming invasive itself.

Introduced diseases are another major cause of animal extinctions. Avian malaria and avian pox virus, along with their introduced mosquito vectors, are believed to have been the primary causes of extinctions of many Hawaiian native bird species. The pathogen currently threatening the most species with extinction is likely *Batrachochytrium dendrobatidis*, a chytrid fungus that is lethal to many amphibians. This fungus is believed to have originated in South Africa and to have been transported around the world during the twentieth century via the international trade in the African clawed frog (*Xenopus laevis*), a frog species commonly used for research purposes in developmental biology laboratories. Now found on all continents except Antarctica, this chytrid fungus has

The Nile perch (*Lates niloticus*), a voracious predator, is believed to have caused the extinction of more than 100 species of cichlid fish in Lake Victoria following its twentieth century introduction into this African lake. © Tom McHugh/Photo Researchers, Inc.

The predatory snail, *Euglandina rosea*, has driven many native land snails to extinction on Pacific islands. It is shown here attacking a native Hawaiian snail, *Anchatinella vulpina*. © Photo Resource Hawaii/Alamy.

already caused the extinction of many frog species, and it is thought that this single pathogen may be one of the primary causes of the ongoing worldwide decline in amphibians.

The chytrid fungus, *Batrachochytrium dendrobatidis*, is believed be one of the primary causes of the ongoing worldwide decline in amphibians. Shown is a dead wood frog (*Rana sylvatica*) in early spring, a possible victim of chytrid fungus. © John Cancalosi/Alamy.

A different fungus, *Geomyces destructans*, is currently devastating bat populations in the northeastern United States and adjacent Canadian provinces. Infecting the skin of the bats and causing a white growth around their noses (which is the basis for the disease's name: white-nose syndrome), this fungus has killed more than one million bats since it was first identified in bats from a cave in New York state in 2006. The origin of this disease is still not definitively known, but most research has suggested a possible European origin. The fungus is found in Europe but does not have the lethal effect there it is having in North America. This suggests European bat species have evolved some immunity to this particular pathogen. The fungus is believed to disrupt the bats' winter roosting—the time when they enter a state of torpor to reduce energetic demands. By waking the bats repeatedly over the winter, the fungus causes the bats to use up all their stored energy so they end up starving to death before the insects emerge in the spring. Although no bat species has yet gone extinct as a result of the fungus, populations have been extirpated. Given the virulence of the fungus and its apparent ability to be spread widely and quickly, there is concern that at least regional extinction of some species could be possible in upcoming decades.

A little brown bat (*Myotis lucifugus*) in Greeley Mine, Vermont, showing symptoms of white-nose syndrome (WNS). WNS is caused by a fungus, *Geomyces destructans*, which is thought to have killed more than one million bats since it was first identified in a cave in New York state in 2006. Courtesy of U.S. Fish and Wildlife Service.

Do certain environments increase the vulnerability of animal species to extinctions?

As the reader may have noticed, most of the examples of extinctions and extirpations caused by invasive species that have been presented involve the introduction of new species to islands or freshwater systems. With few exceptions, it is difficult to find examples of invasive species that have driven native species to extinction on continents or in marine systems. Thus, most documented extinctions caused by invasive species have occurred in isolated environments.

While introduced enemies may be able to reduce the size of local populations of continental and marine species greatly, and sometimes even cause extirpations, it is rare for introduced enemies to drive continental or marine species to extinction. The native continental and marine species generally are able to escape total eradication by persisting in parts of their range that are unoccupied by the introduced enemies.

Although in certain instances the last remaining individual of an island or lake species may meet its demise at the hands (or jaws) of the introduced enemy, it is likely that the last individuals probably die for other reason(s). Introduced enemies can cause extinctions of native species without having to kill every last individual. Once the invaders have driven population sizes to very low levels, other factors come into play that increase the probability of extinction. This is because small populations are at much greater risk to various random processes. For example, genetic diversity can be lost due to chance when populations are very small, increasing the

vulnerability of the species to environmental change. Simply because of chance, small populations can experience a significant skew in the ratio of males to females, which can seriously reduce subsequent reproductive output. Small populations are also more vulnerable than large populations to natural catastrophes and extreme weather events.

The combined effect of these different processes is to create a positive feedback loop that forces the population into an extinction vortex. In a phenomenon that ecologists call the Allee effect, a species in the extinction vortex exhibits a negative growth rate, meaning that the death rate exceeds the birth rate. Under these conditions, and without sufficient immigration to compensate for the low birth rate, the population is doomed. It is only a matter of time until it goes extinct. Because of the isolation of their environments, populations inhabiting islands and lakes are often smaller to begin with, compared with their continental and marine counterparts. This means that it is more likely that population declines on islands and in lakes will be susceptible to the Allee effect, and hence populations resident in these environments will be more likely to go extinct.

Pathogens are the one type of introduced species that do seem to have the capability to cause extinctions on continents. Fungal infections in particular have demonstrated this potential, as exhibited by the devastating effects of the white-nose fungus on North American bats and the chytrid fungus on frogs worldwide. Both of these fungi infect only the skin, but they damage the skin's structural integrity and disrupt various vital physiological processes, eventually causing the death of the bat or frog. An important aspect of the biology of these two fungal pathogens is that they do not require a host to persist in an infected region. Most pathogens become less abundant as the density of their hosts decline, thereby representing less of an infection risk when host numbers are low. When not infecting frogs or bats, however, the chytrid fungus lives in water and the white-nose fungus lives in the soil, respectively. This means that even after they have killed large numbers of frogs or bats in an area, these fungi are able to persist and continue to infect remaining individuals and/or immigrants. This may explain why both fungi have been able to drive continental populations of their animal hosts to extinction so quickly.

By what mechanisms do invasive species cause extinctions?

Predation and disease have been the primary causes of animal extinction by invasive species. This indicates that disease and top-down effects (effects coming from a higher trophic level, that is, from predators) are stronger extinction forces, and threats, than other processes such as competition and bottom-up effects (effects stemming from changes in food type and abundance). For example, although changes in vegetation can cause local declines and even the disappearance of particular herbivores because of a diminished food supply (such as during secondary succession or when an introduced plant species displaces preferred native food plants), there are few examples of animals actually being driven to extinction by invasive plants. The primary exception to the general absence of extinctions by invasive species based on bottom-up causes involves species that are feeding specialists or host specialists. Obviously, the extinction of a particular plant or animal species on which one or more other species are dependent for their own survival (such as specialist herbivores or host-specific parasites) will necessarily result in the extinction of these other species as well.

Prior to colonization by humans, many islands and lakes lacked predators or diseases that were present on continents or marine systems. This often meant that animal species that had lived for long periods of time on islands or in lakes had not evolved effective defenses against these new enemies. Ecologists have argued that prey naïveté among island animals probably has contributed to their extinctions by introduced predators. Long-term isolation from certain predatory archetypes (e.g., snakes and ground mammals) is believed to be the cause of prey naïveté for many of these island species. Continental terrestrial prey are generally not as likely to exhibit naïveté to an introduced predator, because it is unlikely that any new predator would represent a new predatory archetype. This is not the case, however, for continental aquatic systems, in which the isolation of many freshwater systems is believed to have similar effects as the isolation of oceanic islands. For example, the introductions of the European brown trout (*Salmo trutta*) into South America and New Zealand and the eastern mosquitofish (*Gambusia holbrooki*) into Australia have caused major reductions in native fish and amphibians. Prey naïveté is believed to have played a role in these reductions (Hamer et al. 2002). Although a number of freshwater extinctions that resulted from the introduction of a predator have been documented, there are few examples of recent extinctions of marine species caused by an introduced predator, a finding that is consistent with the hypothesis of increased prey naïveté in freshwater systems.

The type of naïveté just described is evolutionary naïveté, in which the species has not evolved recognition abilities for certain predator types, as opposed to ontogenetic naïveté, which refers to the lack of individual exposure to a particular predator type during the prey's lifetime. In species where learning plays a large role in predator defense, animals can lose effective predator defenses rather quickly. Tammar wallabies (*Macropus eugenii*), which had been introduced in the late 1800s onto Kawau Island, New Zealand, which was free of large wallaby predators, have been reported to have lost some of their predator-recognition abilities. In a 2001 study, Joel Berger, Jon E. Swenson, and Inga-Lill Persson found that native North American moose that have lived for multiple generations in the absence of predators, such as wolves and grizzly bears, exhibited prey naïveté when these predators were reintroduced. Berger and his colleagues also found, however, that predator recognition and avoidance behavior in the moose developed quite quickly through learning, leading the researchers to conclude that it was highly unlikely that the moose would experience a predation "blitzkrieg" because of these predator introductions. Given the life span of the moose, as well as the rapidity with which they regained their predator-avoidance behavior, the change almost certainly resulted from individual moose learning through experience. In other instances, though, the

acquisition of antipredator responses to a novel predator may involve natural selection and genetic changes. For example, the red-legged frog (*Rana aurora*), an endangered California species, is reported to have developed recognition abilities (chemical cues) and antipredator responses to the introduced American bullfrog (*Rana catesbeiana*)—changes that are believed to have a genetic component to them (Kiesecker and Blaustein 1997).

While these findings provide some hope for prey species threatened by extinction from introduced predators, prey need time to develop defenses against a new predator archetype. As shown by some of the examples of island extinctions, some novel predators are simply too effective and the prey are extinguished before they have time to develop or evolve effective defenses. Even if a new predator does not represent a new predatory archetype—and hence the prey does not suffer from naïveté—this does not mean the new predator cannot drastically reduce the size of the prey population, or even cause its extinction. If the predator is highly efficient, prey populations can be substantially reduced even if the prey recognizes the new species as a predator and tries to take evasive action. Examples of this phenomenon include the very heavy predation on the European water vole (*Arvicola terrestris*) by the introduced American mink (*Mustela vison*; Macdonald and Harrington 2003), the predatory impact of the red fox (*Vulpes vulpes*) on eastern gray kangaroos (*Macropus giganteus*; Banks, Newsome, and Dickman 2000), and the Nile perch on cichlid species in Lake Victoria (as well as human hunters using modern technology on just about any species).

In species where learning plays a large role in predator defense, animals can lose effective predator defenses rather quickly. Tammar wallabies (*Macropus eugenii*), which had been introduced in the late 1800s onto Kawau Island, New Zealand—which was free of large wallaby predators—have been reported to have lost some of their predator-recognition abilities. © Jose Gil/ShutterStock.com.

Although one often hears claims that invasive species threaten to drive native species to extinction by outcompeting them, there are very few documented examples of extinctions caused by competition (Davis 2003). The belief that competition from invasive species represents a major extinction threat is grounded in traditional niche theory, which holds that resident species have partitioned up the environment so that each species uses a unique set of resources, thereby minimizing competition among species. The notion that communities could be saturated with species is implied by this niche-based argument. In a species-saturated environment, species would have partitioned the resources in the environment to the maximum extent possible, with any more partitioning resulting in insufficient resources to support a species. If many communities are species saturated, then either a species introduction must fail because the new species cannot gain access to resources already monopolized by the residents, or if the species successfully establishes, it must be a better competitor than one or more of the resident species. Because the community is species saturated, this means that the establishment of the new species must be accompanied by the extirpation of one or more of the native species through a process known as competitive exclusion. The introductions of species throughout the world have provided a test of this niche-based perspective of how communities are maintained. This natural experiment has consistently shown that communities have not been species saturated and that, more often than not, communities are able to accommodate new species without any accompanying extinctions or extirpations of native species (Davis 2009).

If competition is a relatively weak threat, extinctions caused by competition should take longer than those caused by predation and habitat loss. This raises the possibility that so few competition-driven extinctions have been documented because not enough time has passed for competitive exclusion to occur. If this is the case, it has been suggested that more competition-driven extinctions may be observed in the future. Yet, the increased time needed for these extinctions to occur also provides more time for other factors to disrupt the competitive asymmetry between the new and long-term resident species, thereby reducing the likelihood that such extinctions would ever occur. These possible factors include events and processes that would reduce the abundance of the new species, such as disturbances, disease, environmental fluctuations, or even a new introduced species. For example, in a 1999 study, Michael P. Marchetti concluded that although the Sacramento perch (*Archoplites interruptus*) is threatened by the aggressive dominance of an introduced bluegill (*Lepomis macrochirus*), competitive exclusion of the perch may never occur because of fluctuating environmental conditions.

A longer time frame also means that the resident species may have time to adapt to the new competition pressure in its environment and thereby reduce the intensity of competition to a level that permits coexistence. For example, the introduction of more than 250 new fish species into the Mediterranean Sea following the completion of the Suez Canal has resulted in only a single extinction (Por 1978). This has been attributed to the ability of the long-term residents to respond to

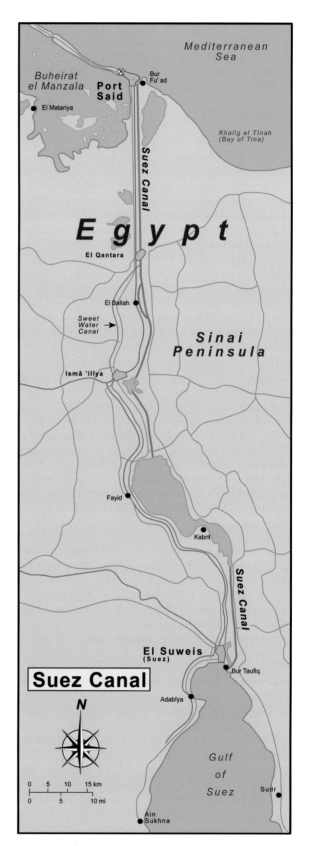

During the years following the completion of the Suez Canal in 1869, more than 250 new fish species dispersed into the Mediterranean Sea. Despite this large number of species introductions, the new fish are believed to have resulted in only a single extinction of a native Mediterranean fish species. © MAPS.com/Corbis.

the competitive interactions with the Red Sea species by adjusting their foraging depths. This niche adjustment enabled the long-term residents, which prefer to feed in the warmer surface waters of the Mediterranean, to accommodate the introductions.

Are certain animal groups more threatened by invasive species than others?

Birds, mammals, amphibians, reptiles, fish, and invertebrates (e.g., mollusks) have all been driven to extinction by invasive species. Thus, there does not seem to be any particular taxonomic group of animals that are inherently more vulnerable than other groups to extinctions caused by invasive species. Any generalizations from data that can be made are more likely geographic than taxonomic. Specifically, as described above, animal species living in isolated environments (e.g., actual or ecological islands) are far more vulnerable to extinction than are species living on continents or in marine environments. Beyond this geographic generalization, when it comes to causing animal extinctions, invasive predator and pathogen species seem to be an equal-opportunity destroyer.

Have extinction threats by invasive species been overstated?

In a 1998 study, David S. Wilcove and colleagues concluded that invasive species are the second-greatest extinction threat to species in peril. This conclusion has been cited more than 1,600 times since the article's publication, as well as in countless research proposals, management documents, and university classes throughout the world. By the first years of the twenty-first century, it had become common boilerplate for invasion literature, the conclusion often presented as fact without any reference at all. However, there are serious limitations and some biases in the information that Wilcove and his colleagues used to come to their conclusion. First, little of the information used to declare nonnative species the second-greatest threat to species survival was based on actual data at all, as the authors were careful to make very clear:

> We emphasize at the outset that there are some important limitations to the data we used. The attribution of a specific threat to a species is usually based on the judgment of an expert source, such as a USFWS [US Fish and Wildlife Service] employee who prepares a listing notice or a state Fish and Game employee who monitors endangered species in a given region. Their evaluation of the threats facing that species may not be based on experimental evidence or even on quantitative data. Indeed, such data often do not exist. With respect to species listed under the ESA [Endangered Species Act], Easter-Pilcher (1996) has shown that many listing notices lack important biological information, including data on past and possible future impacts of habitat destruction, pesticides, and alien species. Depending on the species in question, the absence of information may reflect a lack of data, an oversight, or a determination by USFWS that a particular threat is not

harming the species. The extent to which such limitations on the data influence our results is unknown. (Wilcove et al. 1998, 608–609)

Second, the article dealt with species only in the United States, as its title made very clear: "Quantifying Threats to Imperiled Species in the United States." Thus, it has never been justifiable to cite this article when making claims about global extinction threats. Third, the findings are dramatically affected by the inclusion of Hawaii, which, while of course part of the United States, has a dramatically different invasion history than does the continental, and substantially majority, portion of the country. A similar review of extinction threats in Canada found introduced species to be the *least* important of the six categories analyzed (habitat loss, overexploitation, pollution, native species interactions, introduced species, and natural causes, the latter including stochastic events such as storms and factors inherent to the species, such as limited dispersal ability; Venter et al. 2006). When the Hawaiian species were excluded from Wilcove and colleagues' data, the United States and Canada did not differ with respect to the threats posed by introduced species (Venter et al. 2006), meaning that nonnative species would have ranked very low on the list of threats to the survival of species in the United States. Other studies that have examined species threats over a much larger global area have come to similar conclusions. For example, an analysis of the causes of species depletions and extinctions in estuaries and coastal marine waters concluded that the threat of nonnative species was negligible compared to that of exploitation and habitat destruction (Lotze et al. 2006).

Biodiversity impacts of invasive species

As mentioned earlier, there is abundant evidence that introduced predators and pathogens can cause extinctions, mainly on islands and in freshwater systems. It does not necessarily follow, however, that biodiversity is reduced in these regions because of species introductions. Species richness in a region will decline only if the number of species that have gone extinct exceeds the number of new species that have been introduced. This is not the case in most regions of the world, where species introductions have typically exceeded extinctions, often by a great margin.

For example, although more than 80 nonnative marine species are believed to have established themselves in the North Sea since the early nineteenth century, with respect to species richness, their impact has been primarily additive, with little evidence that they have driven any native species to extinction (Reise et al. 2002). This may be the case with inland seas as well. Although more than 100 nonnative species are believed to have been introduced into the Baltic Sea since the early nineteenth century, at least seventy of which have become established, no extinctions of native species had been recorded as of 2002 (Leppäkoski et al. 2002), and this was still the case at the end of 2007 (personal communication with Erkki Leppäkoski). Also, in their characterization of the fauna in the Caspian Sea in a 2002 study, Nikolai V. Aladin, Igor S. Plotnikov, and Andrei A. Filippov concluded that, while some of the introduced species produced some undesirable effects, they primarily contributed to the Caspian Sea's rich biodiversity.

In a 2006 study of the impacts of nonnative species on coastal marine environments, Karsten Reise, Stephan Gollasch, and Wim J. Wolff reported that there was no indication that nonnative species were causing a decline in biodiversity. On the contrary, they concluded that, more often than not, the new species expand ecosystem functioning by adding new ecological traits, intensifying existing ones, and increasing functional redundancy.

The opening of the Suez Canal in 1869 enabled many residents of the Red Sea and the Indo-Pacific to move into the Mediterranean Sea, a phenomenon often referred to as the Lessepsian migration, named after the French engineer who supervised the construction of the canal, Ferdinand de Lesseps (1805–1894). Although there have been some local extinctions of some native species, the primary biodiversity impact on a regional scale has been a substantial increase in species richness. Likewise, the species richness of European aquatic coastal communities has been enhanced by the introductions of nonnative species, particularly in the historically biodiversity-poor estuaries. Reise and his colleagues concluded in their 2006 study that in coastal aquatic ecosystems, there is no support for the idea that if new species come in, others have to go extinct.

Although animal species on islands typically have been much more vulnerable to extinction from invasive species than mainland species, island faunas have also usually exhibited the most dramatic increases in species richness resulting from species introductions. This has often been because island fauna has lacked entire groups of animals. For example, Hawaii, which had no terrestrial amphibian or reptile species and only one terrestrial mammal species (a bat) before the arrival of humans, now has a diverse terrestrial fauna of amphibians, reptiles, and mammals, all introduced except for the endemic bat.

While it is true that introductions of animals have increased the animal diversity in most regions of the world, it is also true that these introductions have caused a reduction in the number of species at the global level. At the global level, the rate of animal extinctions caused by invasive species far exceeds animal speciation rates. Also, even when animal introductions increase regional species diversity, they also usually homogenize regional faunas. Homogenization is the combined result of introductions of nonnative species and the extirpation of native species. In the United States, the similarity in the fish faunas of the 50 states has increased dramatically since European settlement, a finding that was determined to have been caused primarily by widespread introductions of game fish, with extinctions of native species having less of an impact. Frank J. Rahel reported in 2000 that 89 pairs of states in the United States that had no species in common prior to European settlement shared, by the end of the twentieth century, an average overlap of 25 species.

Documented cases of animal extinctions and extirpations caused by invasive species are numerous. In most instances, these events have taken place in isolated environments, particularly islands and freshwater systems. Comparatively few animal extinctions that can be primarily attributed to invasive species have occurred on continents or in marine systems. Introduced predators and pathogens have been the primary agents of animal extinction caused by invasive species during the past few hundred years. In contrast, competition-driven extinctions have been rare.

That invasive species seldom drive continental or marine animals to extinction does not mean that invasive species have little effect on these animals or their communities. Although a species may not be eliminated by an invasive species totally, its numbers may be so reduced that it becomes ecologically extinct. Ecological extinction occurs when a species has been reduced to such an extent that it has little effect on other species or ecosystem processes. Although the species technically is still present, any role that it played in its environment has essentially vanished.

The extinction threat posed by invasive species is real. On continents and in marine environments, however, animals face far more serious extinction threats than introduced species. Habitat loss, overharvesting, and pollution are the primary causes of animal extinction in these environments, and it is these causes, along with climate change, that will continue to be the primary threats for the foreseeable future.

Resources

Books

Aladin, Nikolai V., Igor S. Plotnikov, and Andrei A. Filippov. "Invaders in the Caspian Sea." In *Invasive Aquatic Species of Europe: Distribution, Impacts, and Management*, edited by Erkki Leppäkoski, Stephan Gollasch, and Sergej Olenin. Dordrecht, Netherlands: Kluwer Academic Publishers, 2002.

Davis, Mark A. *Invasion Biology*. Oxford: Oxford University Press, 2009.

Lafferty, Kevin D., Katharine F. Smith, Mark E. Torchin, et al. "The Role of Infectious Diseases in Natural Communities: What Introduced Species Tell Us." In *Species Invasions: Insights into Ecology, Evolution, and Biogeography*, edited by Dov F. Sax, John J. Stachowicz, and Steven D. Gaines. Sunderland, MA: Sinauer, 2005.

Leppäkoski, Erkki, Sergej Olenin, and Stephan Gollasch. "The Baltic Sea: A Field Laboratory for Invasion Biology." In *Invasive Aquatic Species of Europe: Distribution, Impacts, and Management*, edited by Erkki Leppäkoski, Stephan Gollasch, and Sergej Olenin. Dordrecht, Netherlands: Kluwer Academic Publishers, 2002.

Mack, Richard N. "Invading Plants: Their Potential Contribution to Population Biology." In *Studies on Plant Demography*, edited by James White. London: Academic Press, 1985.

Mooney, Harold A., and James A. Drake. "Biological Invasions: A SCOPE Program Overview." In *Biological Invasions: A Global Perspective*, edited by James A. Drake, Harold A. Mooney, Francesco di Castri, et al. Chichester, UK: Wiley, 1989.

Por, Francis Dov. *Lessepsian Migration: The Influx of Red Sea Biota into the Mediterranean by Way of the Suez Canal.* Berlin: Springer-Verlag, 1978.

Reise, Karsten, Stephan Gollasch, and Wim J. Wolff. "Introduced Marine Species of the North Sea Coasts." In *Invasive Aquatic Species of Europe: Distribution, Impacts, and Management*, edited by Erkki Leppäkoski, Stephan Gollasch, and Sergej Olenin. Dordrecht, Netherlands: Kluwer Academic Publishers, 2002.

Periodicals

Banks, Peter B., Alan E. Newsome, and Chris R. Dickman. "Predation by Red Foxes Limits Recruitment in Populations of Eastern Grey Kangaroos." *Austral Ecology* 25, no. 3 (2000): 283–291.

Berger, Joel, Jon E. Swenson, and Inga-Lill Persson. "Recolonizing Carnivores and Naïve Prey: Conservation Lessons from Pleistocene Extinctions." *Science* 291, no. 5506 (2001): 1036–1039.

Davis, Mark A. "Biotic Globalization: Does Competition from Introduced Species Threaten Biodiversity?" *BioScience* 53, no. 5 (2003): 481–489.

Easter-Pilcher, Andrea. "Implementing the Endangered Species Act." *BioScience* 46, no. 5 (1996): 355–363.

Hamer, A.J., S.J. Land, and M.J. Mahony. "The Role of Introduced Mosquitofish (*Gambusi holbrooki*) in Excluding the Native Green and Golden Bell Frog (*Litoria aurea*) from Original Habitats in South-Eastern Australia." *Oecologia* 132, no. 3 (2002): 445–452.

Kiesecker, Joseph M., and Andrew R. Blaustein. "Population Differences in Responses of Red-Legged Frogs (*Rana aurora*) to Introduced Bullfrogs." *Ecology* 78, no. 6 (1997): 1752–1760.

Lotze, Heike K., Hunter S. Lenihan, Bruce J. Bourque, et al. "Depletion, Degradation, and Recovery Potential of Estuaries and Coastal Seas." *Science* 312, no. 5781 (2006): 1806–1809.

Macdonald, David W., and Lauren A. Harrington. "The American Mink: The Triumph and Tragedy of Adaptation Out of Context." *New Zealand Journal of Zoology* 30, no. 4 (2003): 421–441.

Marchetti, Michael P. "An Experimental Study of Competition between the Native Sacramento Perch (*Archoplites interruptus*) and Introduced Bluegill (*Lepomis macrochirus*)." *Biological Invasions* 1, no. 1 (1999): 55–65.

Rahel, Frank J. "Homogenization of Fish Faunas across the United States." *Science* 288, no. 5467 (2000): 854–856.

Venter, Oscar, Nathalie N. Brodeur, Leah Nemiroff, et al. "Threats to Endangered Species in Canada." *BioScience* 56, no. 11 (2006): 903–910.

Wilcove, David S., David Rothstein, Jason Dubow, et al. "Quantifying Threats to Imperiled Species in the United States." *BioScience* 48, no. 8 (1998): 607–615.

Mark A. Davis

· · · · ·

Fishing

Global fish consumption is at a record high. Global population growth has increased the demand for fish products. Per capita fish consumption has also increased, by an average of 3.2 percent per year between 1961 and 2009, to reach an estimated 41 pounds (18.6 kg) per capita per year in 2010 (FAO 2012). At the same time, global fishing capacity is greater than ever before. The world's fishing fleets have expanded their range to areas and depths of the ocean that were previously inaccessible. This expansion has stimulated a vigorous and influential debate about the sustainability of fisheries and the effects of fishing on freshwater and marine ecosystems. Within this context, this entry provides a review of the potential of fishing to contribute to extinctions and biodiversity loss in both freshwater and marine systems.

Development of modern fishing capacity

Fishing is an ancient practice. Indications of fishing have been found in Late Paleolithic archeological sites from some 50,000 years ago. Fishing developed on all inhabited continents and continues to provide an important food source, trade base, and source of cultural value in many coastal and inland communities in modern times.

Preindustrial fishing primarily targeted accessible resources in inland or shallow coastal waters, although Basque fishermen crossed the Atlantic from Europe to exploit cod on the fishing banks off North America as early as the sixteenth century. Although total fishing power was substantially less than today and more spatially restricted, a challenge in assessing the impacts of fishing on biodiversity is that significant biodiversity loss probably occurred before the development of modern records (Jackson et al. 2001).

The Industrial Revolution provided the basis for the development of modern fisheries. Development of steam power in the late nineteenth century initiated the transition from sail to steam along with the development of larger and heavier gear and new technologies, such as trawls and dredges. The industrialization of fishing was further catalyzed by World War II (1939–1945). The number, size, and power of fishing vessels increased as the costs of shipbuilding were driven down and diesel engines became cheaper and more widely available. There were also substantial improvements

in efficiency with the development of new gear and equipment, such as synthetic nets, modern navigation systems, and fish-finding devices. Processing technologies, such as canning and marine refrigeration, enabled vessels to stay at sea longer. Postwar global population growth increased the demand for fish products, while the development of modern transport systems and new processing techniques enabled fisheries to supply more distant markets. In many countries governments invested heavily in building fishing capacity as part of national economic development plans, especially following the establishment of exclusive economic zones in the 1970s.

Through this process, fishing fleets not only expanded in size and power, they also increased their fishing range substantially. Modern fisheries have now spread over most of the world's oceans, and fishing gear is capable of reaching depths formerly inaccessible. This expansion has reduced the natural refugia previously available to some fish stocks and brought some inherently vulnerable deepwater species and fragile habitats into the domain of fisheries.

Current state of the world's fisheries

Fishing is defined here as the capture and harvest of wild fish, crustaceans, mollusks (including bivalves, gastropods, and cephalopods), echinoderms (e.g., sea cucumbers), and similar species, in freshwater and marine systems. Globally, the largest fisheries by weight are those for herrings, sardines, and anchovies; cods, hakes, and haddocks; tunas, bonitos, and billfish; squids, cuttlefish, and octopuses; and shrimps and prawns. Fisheries for crabs and sea spiders; salmons, trouts, and smelts; lobsters and spiny rock lobsters; marine crustaceans; and scallops have high value but are lower in volume. The targeted hunting of marine mammals, seabirds, and sea turtles is not included in this definition of fishing, but the effects of fishing on these species are considered here.

The Food and Agriculture Organization of the United Nations (FAO) compiles data on fisheries around the world and every two years publishes a review titled *The State of World Fisheries and Aquaculture* (FAO 2012). Overall, global capture-fisheries production increased steadily starting in the

Fishing boats at dock and at moorings in San Felipe, Yucatan, Mexico. Courtesy of Mark D. Callanan.

1950s, before stabilizing at about 99.2 million tons (90 million metric tons) in the late 1980s. Most capture-fisheries production (approximately 87.5%) comes from marine fisheries. Marine fisheries production increased from 18.5 million tons (16.8 million metric tons) in 1950 before leveling off at about 88.2 million tons (80 million metric tons). Marine capture fisheries have probably reached their maximum potential, given limits to primary production and the natural resource base. Further growth will be limited and based mainly on the rebuilding of depleted stocks. Inland capture fisheries are still growing, reaching an estimated 12.3 million tons (11.2 million metric tons) in 2010, but the scope for further expansion is limited. Future increases in global fish production to meet the demands of the world's growing population are therefore likely to depend on rebuilding capture fisheries and on aquaculture. As of 2010, aquaculture provided nearly half of global fish production for human consumption, following a nearly twelvefold increase in volume between 1980 and 2010.

Note on fisheries sectors

Fisheries are conventionally segregated into industrial, small-scale, and recreational sectors. Small-scale fisheries are diverse but generally less capital intensive than industrial fisheries. Small-scale fisheries may use traditional fishing techniques (often described as artisanal fishing) or modern technologies and may be primarily for subsistence or fully commercial.

The effects of fishing on biodiversity are not sector specific but rather depend on the target species, the area and season, the fishing gear and techniques, and the amount of fishing effort. There is substantial overlap in the species targeted by industrial, small-scale, and recreational fisheries. In general, however, small-scale and recreational fisheries are more likely to operate inland or nearshore than in the open ocean. Many coastal and inland habitats, such as coral reefs, coastal lagoons, tidal flats, estuaries, rivers, lakes, and wetlands, are inaccessible to large vessels and thus are exploited mainly by small-scale and recreational fisheries. Recreational fishing is the dominant use of freshwater fisheries resources in many industrial countries.

The amount of fishing effort is a function of the size or power of each fishing vessel or unit, the number of fishing units, and the duration of fishing. When small-scale fisheries involve large numbers of fishing vessels, aggregated fishing effort may be similar to a much smaller number of more powerful industrial fishing vessels. Often, the distinction between sectors is unclear. The Bristol Bay salmon fishery in Alaska, for example, is characterized by a large number of small vessels (up to 32 feet in length), often run as family enterprises. Most of these vessels make use of powerful engines, mechanized winches, refrigerated seawater systems, global positioning systems, and fish finders. Increasingly, recreational fishers also use modern equipment and technologies that can make them equivalent to commercial fishers in terms of fishing capability. The total annual catch by recreational fishers was estimated at about 12 percent of the total world catch in 2004 (FAO 2012). For some species, such as salmon, marlins, sailfish, and swordfish, recreational fisheries contribute an even greater share of the total catch. The effects of fishing on biodiversity reviewed in this entry may be attributable to any of these fisheries sectors.

Hawksbill turtle (*Eretmochelys imbricata*) hooked on long line. Cocos Island, Costa Rica–Pacific Ocean. © Jeff Rotman/Alamy.

Extinctions in freshwater and marine systems attributable to fishing

Freshwater and marine systems both support high species diversity. Marine habitats cover approximately 70 percent of Earth's surface and hold about 97 percent of the world's water. While the full extent of species diversity has yet to be documented in any biome, marine systems host approximately 15 percent of all documented species (May 1994). In contrast, freshwater systems comprise approximately 1 percent of Earth's surface and hold only 0.01 percent of the world's water, yet they host approximately 6 percent of described species, including about one-fourth to one-third of the world's vertebrates and 40 percent of the world's fishes (Dudgeon et al. 2006). Thus, marine systems have greater absolute species diversity, but freshwater systems support high concentrations of species diversity, packed into a small proportion of Earth's surface or habitable volume.

Freshwater and estuarine systems have been heavily affected by human development around the world. Species in freshwater and estuarine systems are subject to a wide range of threats, including habitat degradation, removal and diversion of water, flow modification, siltation, eutrophication, pollution, invasive species, and projected global changes in temperature and precipitation patterns (Dudgeon et al. 2006). These threats often act synergistically, so that it can be challenging to distinguish the effects of fishing from other factors. In contrast, the dominant current threat to marine species is fishing (Dulvy et al. 2003). Historically, coastal and nearshore systems have been more heavily affected than offshore systems. With the expansion of fishing effort since the mid-twentieth century, however, and the global nature of emerging threats such as climate change and ocean acidification, there are growing concerns about threats to offshore systems.

The Red List of Threatened Species of the International Union for Conservation of Species (IUCN) provides a standardized, peer-reviewed assessment of the global threat status of the world's species. Mammals, birds, amphibians, sea turtles, and various families of fishes have been comprehensively assessed for the IUCN Red List, but this represents only a subset of freshwater and marine taxa. Based on these assessments, species in this entry are identified as extinct (EX), extinct in the wild (EW), critically endangered (CR), endangered (EN), vulnerable (VU), near threatened (NT), least concern (LC), data deficient (DD), or not evaluated (NE). Species listed as critically endangered, endangered, or vulnerable are considered globally threatened. Summaries of species' current status and major threats have been drawn where possible directly from information compiled as part of the IUCN assessment process and available at the IUCN Red List website (IUCN 2012).

As of 2012, the IUCN Red List listed eleven freshwater and marine species as extinct as a result of fishing and the harvest of aquatic resources. Only two of these eleven species were fish, and in only one case was overfishing the primary cause of extinction. This highlights the importance of looking at the full range of freshwater and marine species when assessing the contribution of fishing to extinctions and biodiversity loss and when investigating the various ways in which fishing can contribute to extinctions. In addition to the overfishing of target species, fishing may lead to incidental mortality and injury of nontarget species, persecution of species viewed as competitors, reductions in food availability as a result of competition with fisheries, and the introduction of nonnative species for the purposes of fishing.

Five freshwater species were listed as extinct on the IUCN Red List in 2012 as a result of fishing and the harvest of aquatic resources. The blackfin cisco (*Coregonus nigripinnis*, EX), a freshwater salmonid, used to occur in the Great Lakes of North America. The primary cause of extinction was overfishing. For the gocke baligi (*Alburnus akili*, EX), a freshwater cyprinid endemic to Beyşehir Lake in Turkey, the primary cause of extinction was the introduction of pike perch. Overfishing may also have contributed to its decline. The Atitlán grebe (*Podilymbus gigas*, EX) was endemic to Lake Atitlán in Guatemala. The initial population decline was probably caused by competition and predation with introduced largemouth bass, but habitat degradation also played a role, possibly along with drowning in gillnets. The Alaotra grebe (*Tachybaptus rufolavatus*, EX) was endemic to Lake Alaotra and surrounding lakes in Madagascar. Introduction of exotic fish species and direct harvesting were probably the major causes of extinction. Drowning in gillnets may also have contributed. The Yunnan lake newt (*Cynops wolterstorffi*, EX) was restricted to Kunming Lake and the surrounding areas in Yunnan, China. Several factors are implicated in its extinction, including the introduction of nonnative fish species. In addition, the baiji (*Lipotes vexillifer*, CR), endemic to the Yangtze River in China, has been listed as critically endangered but may now be extinct. As with many freshwater species, a combination of factors is implicated. Major threats include entanglement in fishing gear and electric fishing.

Six marine species were listed as extinct in 2012 on the IUCN Red List as a result of fishing and the harvest of aquatic resources. Hunting was the primary cause of the extinction of the Caribbean monk seal (*Monachus tropicalis*, EX) and the Japanese sea lion (*Zalophus japonicus*, EX), but both species were also persecuted by fishers. Steller's sea cow (*Hydrodamalis gigas*, EX) was probably driven to extinction by hunting alone. Overharvesting of eggs and hunting were probably the main factors in the extinction of the Labrador duck (*Camptorhynchus labradorius*, EX). Reduced food availability as a result of competition with shellfish fisheries has been identified as the likely main cause of extinction of the Canary Islands oystercatcher (*Haematopus meadewaldoi*, EX), and may also have contributed to the extinction of the Labrador duck. Bennett's seaweed (*Vanvoorstia bennettiana*, EX) is known from only two sites, both in Sydney Harbor in Australia. It was subject to a wide range of threats, such that it is difficult to distinguish the impact of fishing.

Extinctions may be difficult to detect or confirm in both freshwater and marine systems. Several species listed as critically endangered may be extinct, and it is likely that some extinctions in lesser-known taxa remain undetected.

Major threats to species diversity associated with fishing

As of 2012, few fish species had been driven extinct through overfishing. The question was whether that pattern would continue. This section reviews major direct threats to species or populations associated with fishing, specifically overfishing of target species, incidental mortality and injury of nontarget species (bycatch), and persecution and culling. The ecosystem effects of fishing, including the effects of reduced food availability as a result of competition with fisheries and the effects of invasive species introduced for the purposes of fishing, are reviewed in a later section.

Overfishing

The potential for overfishing was famously debated by the English biologist Thomas Huxley (1825–1895) and English zoologist Ray Lankester (1847–1929) in the 1880s. Huxley believed "that the cod fishery, the herring fishery, the pilchard fishery, the mackerel fishery, and probably all the great sea fisheries, are inexhaustible; that is to say, that nothing we do seriously affects the number of the fish. And any attempt to regulate these fisheries seems consequently . . . to be useless" (Huxley 1884, 16). Lankester counterargued that the many thousands of young produced by fish are not superfluous: "It is a mistake to suppose that the whole ocean is practically one vast store-house, and that the place of the fish removed on a particular fishing-ground is immediately taken by some of the grand total of fish, which are so numerous in comparison with man's depredations as to make his operations in this respect insignificant" (Lankester 1884, 413). Huxley's assertion that fisheries have a limited impact on fish populations was disproved by the so-called Great Fishing Experiment brought about by the forced reduction of fishing activity during World War I (1914–1918). At the end of the war, there were more large fish and catch rates were higher, demonstrating that fishing can reduce fish biomass and that stocks may recover when pressure is reduced.

The fundamental concepts that underpin modern fisheries science, including density-dependent population growth, surplus production, and maximum sustainable yields, were established by the mid-1950s, based on the work of pioneers such as Johan Hjort (1869–1948), Michael Graham (1898–1972), William E. "Bill" Ricker (1908–2001), Milner B. Schaefer (1912–1970), Raymond J.H. "Ray" Beverton (1922–1995), and Sidney J. Holt (b. 1926) (see Smith for a review). Populations rarely grow without limit. Sooner or later, all the available habitat is occupied or other essential resources are used. As a population approaches some finite carrying capacity, the net production of new individuals per current individual is expected to decrease, so that the population growth rate declines toward zero. Unfished stocks tend to have high biomass levels, but the number of new recruits per spawning fish (net recruitment rate) is low, and hence the number of new fish or biomass added to the stock each year (surplus production) is also low. At the other end of this spectrum, when fish stocks are low and habitat and other resources are plentiful, the net recruitment rate may be high, but surplus production is low because there are few spawners. Surplus production is maximized at some intermediate level of biomass, often at 20 percent to 40 percent of unfished biomass in commercial fisheries. In theory, if this surplus production is harvested each year, a fishery can be maintained at this biomass, producing maximum sustainable yields (MSY) in perpetuity. For many decades, maintaining fish stocks at the level of biomass that produces maximum sustainable yields (B_{MSY}) has been the holy grail of fisheries management.

In practice, however, fisheries managers and scientists have to deal with a wide range of real-world complications not captured in the simplest population models. Most fish stocks are also notoriously difficult to count. Environmental conditions in freshwater and marine ecosystems vary at interannual, multidecadal, and even longer scales. Observed stock-recruitment relationships are highly variable for most commercial fish stocks. It can be extremely difficult to disentangle the effects of fishing and the environment, and there is continued debate about the relative contribution of fishing and changes in environmental conditions to declines and variability in many fish stocks. Uncertainty about actual stock size and environmental variation both imply that managers should take a precautionary approach and aim to maintain biomass at a level considerably higher than B_{MSY}.

From an economic perspective, the optimal level of effort occurs where the difference between the costs of fishing and the revenue from fishing is greatest. As catching more fish usually requires more effort, total profits will generally be maximized at harvest rates below the maximum and at levels of biomass greater than B_{MSY}. This provides an economic argument for maintaining stocks greater than B_{MSY}. Nevertheless, it may still be in the interests of individual fishers to continue to fish until profits have been driven to zero, and no more profit can be extracted from the fishery.

Furthermore, removing large amounts of fish biomass may have a wide range of indirect effects on aquatic ecosystems. Removing a large share of available forage fish, for example, may have negative effects on predators, including large piscivorous fish, seabirds, and marine mammals, that may be of commercial interest or conservation concern. An ecosystem approach to fisheries management may therefore require levels of biomass well above B_{MSY} in order to maintain the broad range of services provided by aquatic ecosystems.

For all these reasons, many fisheries scientists and managers now advocate that B_{MSY} should be seen as a limit reference point—the minimum allowable biomass—below which stocks are considered depleted (Hilborn 2007). Even so, for fisheries managers, stocks that are fully exploited and maintained at target biomass levels in the region of 40 percent to 50 percent of unfished biomass may be regarded as well managed. In contrast, ecologists might regard a 50 percent reduction in stock biomass as evidence of severe depletion (e.g., Worm et al. 2006). This difference in perspective is fundamental to the current debate about whether there is a crisis in world fisheries.

The FAO conducts a biennial assessment of the status of marine fish stocks, using MSY as a benchmark. According to the 2012 assessment (FAO 2012), most of the world's marine capture fisheries (57%) are now fully exploited, meaning that they produce catches very close to the MSY with no room for further expansion. This proportion has remained fairly stable since 1974 when the first FAO assessment was completed. About 30 percent of stocks are overexploited and are in need of rebuilding before yields can increase. The percentage of overexploited stocks has increased since 1974, although the rate of increase has slowed since the early 1990s. The remaining 13 percent of stocks are not fully exploited and have some potential for increased yields. This proportion has decreased gradually since the mid-1970s. It is important to note that this assessment is based on fisheries for which adequate data are available, and these fisheries may include a disproportionate share of well-managed fisheries. There are many regions of the world where adequate data and controls on exploitation rates are lacking.

As is discussed further below, the inherent vulnerability of species to overfishing depends in large part on their productivity. Species with low productivity are at greater risk of extinction through overfishing, and will take longer to recover from severe depletion. Examples of species listed as globally threatened as a result of overfishing include several species of sturgeons and paddlefish and of sawfishes, both of which are groups characterized by low productivity.

There are 27 species of sturgeons and paddlefish (Acipenseriformes), of which 23 are listed as globally threatened on the IUCN Red List. Overfishing is identified as a major threat to 19 of these 23 species. The Chinese paddlefish (*Psephurus gladius*) is endemic to the Yangtze River in China. This species is listed as critically endangered (possibly extinct). Declines have been attributed to historical overfishing and habitat degradation, including the construction of dams that have blocked migration routes. Only two adult specimens have been recorded since 2002, and it is likely that there are less than fifty mature individuals left in the wild (Qiwei 2010).

The sawfishes (Pristidae) are a family of rays. All seven species of sawfishes are listed as critically endangered on the IUCN Red List, with fishing implicated in all cases. The knifetooth sawfish (*Anoxypristis cuspidata*, CR) was historically fairly common in inshore estuaries through much of the Indo-West Pacific. The major threat is fishing, both as a target species and as bycatch. Fishing has led to substantial reductions in population size and the fragmentation of the remaining populations. The species has virtually disappeared from commercial catches in regions where it was once considered common (Compagno et al. 2006).

The totoaba (*Totoaba macdonaldi*, CR) is a large fish found in the shallow coastal waters of the Gulf of California, Mexico. Once abundant and the target of both commercial and recreational fisheries, the population has been severely depleted through a combination of overfishing and the degradation of spawning habitat. Recovery is further impeded by the continued bycatch of juveniles in shrimp trawl fisheries (Findley 2010).

Bycatch

Bycatch, which refers to incidental mortality or injury of nontarget species through fishing activities, is a major threat to marine mammals, seabirds, marine turtles, and some species of fish. In a 2005 study, the FAO estimated global discards at 8 percent of landings by weight, or on the order of 8 million tons (7.3 million metric tons) per year (Kelleher 2005). The FAO defines discards as "the proportion of catch that is returned to the sea for whatever reason," so these figures include discards of target species as well as bycatch of nontarget species.

Unintentional catch of nontarget species occurs in almost all fisheries, but some fishing techniques are more selective than others. In the 2005 FAO study, shrimp trawl fisheries accounted for approximately 39 percent of total estimated discards and demersal finfish trawls for approximately 35.6 percent. Discard rates were highest in shrimp trawl fisheries, with a weighted average discard rate of 62.3 percent, followed by longline fisheries for tuna and highly migratory species with a discard rate of 28.5 percent. Fisheries with low or negligible discard rates included midwater trawl fisheries for small pelagics, fixed trap and pot fisheries, and diver and collection fisheries. Fishing practices, in particular the location and timing of fishing, can also influence bycatch rates. Gear modifications, such as turtle excluder devices, can improve selectivity. Understanding the effects of fishing on biodiversity requires a basic understanding of the main fishing gear and techniques. Industrial fisheries mainly use trawls, dredges, gillnets, purse seines, longlines, and pots or traps.

TRAWLING Trawls are generally large, funnel-shaped nets towed through the water with one end held open and the other closed. Bottom trawls target benthic or demersal species (species associated with the seabed), such as sand lance, cod, haddock, halibut, sole, plaice, shrimp, and, more recently, deepwater species that aggregate at seamounts and other benthic features, such as orange roughy and hoki. Demersal habitats often support a high diversity of species, including many juveniles, and bottom trawls are associated with high rates of bycatch. Gear modifications, such as turtle excluder devices, have been used to reduce bycatch in some bottom-trawl fisheries. In contrast, midwater trawls target pelagic species that occur in large shoals or extended aggregations in the water column, such as capelin, herring, mackerel, sardine, and pollack, and are generally more selective.

DREDGING Modern fishing dredges usually comprise heavy scoop nets that are towed along the seabed and used to excavate benthic species such as clams, oysters, and scallops. Like bottom trawls, dredges are associated with high rates of bycatch, but they represent a much smaller share of global landings.

GILLNETTING Gillnets are designed to form an invisible mesh wall suspended vertically through the water column. Fish that try to swim through the net are entangled by their gills. Drift gillnets are generally long nets that drift astern of a fishing vessel. Gillnet fisheries are often associated with high levels of bycatch, including marine mammals and seabirds, and high-seas drift gillnets have been banned. Salmon fisheries provide a good example of how fishing practices influence bycatch rates. In many salmon fisheries, gillnets are set close to the entrance of river mouths to capture salmon returning to spawn. Hauls are dominated by the target species, resulting in low bycatch rates.

PURSE SEINING Purse seines are large nets designed to catch dense shoals or schools of fish. Like gillnets, the net forms a long wall suspended vertically through the water column. In purse seines, one end is fixed to the primary vessel, while a second vessel tows the other end around a shoal of fish and then rejoins the primary vessel. The bottom of the net is then closed by pulling a purse line that acts as a drawstring at the base of the net, trapping the fish from below. Target species include tunas and small pelagic species, such as anchovy, sardine, capelin, and herring.

As purse seines generally target dense shoals of a single species, bycatch is often limited. But purse seining for tuna schooling with dolphin has led to high rates of dolphin bycatch in the past. Following government regulation and consumer pressure, dolphin mortalities have been substantially reduced through changes in fishing practices, including procedures that enable dolphins to escape by leaping over the back of the net.

Many purse seine fisheries make use of fish aggregating devices (FADs). These are usually human-made floating objects, such as bamboo rafts. Small fish shelter under the FADs and attract larger predatory fish, such as tunas and billfish, which are then targeted by purse seiners. There are concerns that fishing the multispecies aggregations associated with FADs leads to higher rates of bycatch than fishing on free-swimming schools.

LONGLINING Longlines are long fishing lines (up to 93 miles [150 km] in length) set with large numbers of baited hooks. Pelagic longlines target pelagic fish, such as tunas and billfish, while demersal longlines lie along the seabed and target demersal species, such as cod, haddock, hake, and halibut. Bycatch in longline fisheries is a major concern, especially of sharks, turtles, and seabirds. Various methods to reduce bycatch, such as modifications of hook design and baiting systems, have been developed and tested.

POTS AND TRAPS Pots and traps are designed to target specific species, such as crabs and lobsters. The location of a pot on the seabed is usually marked by a buoy at the surface tied to the pot by a long fishing line. Large cetaceans or sea turtles may get entangled in these lines. Bycatch in pots may be limited through design features that allow small individuals and nontarget species to escape.

SMALL-SCALE FISHING TECHNIQUES Small-scale fisheries make use of a diverse range of fishing techniques. In many fisheries, small-scale fishers use gear similar to that of industrial fisheries, including trawls, gillnets, and seines. Small-scale fisheries, however, generally have lower discard rates than industrial fisheries because nontarget species are often used rather than thrown back. Traditional or artisanal fishing techniques include hand gathering, hooks, nets, baskets, traps, spears, poisons, and harpoons. Several of the fishing techniques used in small-scale fisheries, such as hand gathering and spearfishing, are highly selective, whereas

Lobster pots on deck of boat Hobart Tasmania Australia. © Rob Walls/ Alamy.

others, such as blast fishing, are associated with very high rates of bycatch.

RECREATIONAL FISHING TECHNIQUES Recreational fishing techniques include angling, gathering, trapping, spearing, and the use of nets. Many recreational fisheries are highly selective, often targeting larger individuals of specific species.

GHOST FISHING Ghost fishing refers to lost or abandoned gear that continues to catch fish after it has been lost. Ghost fishing is mostly associated with passive fishing gear, such as set gillnets as well as traps and pots, that are left unattended. Limited data are available on gear loss rates, the persistence of lost gear, and the effects of ghost fishing on biodiversity. The few studies available suggest that gear loss rates may be low (Brown and MacFadyen 2007). Fishing gear is costly, and fishers may make considerable effort to avoid losing or to recover lost gear. Nevertheless, the total amount of lost gear depends on the scale of the fishery, and so it may be substantial in large fisheries despite low loss rates. In most European fisheries investigated, the loss of commercial species as a result of ghost fishing has been found to be small compared to commercial landings (Brown and MacFadyen 2007).

Species threatened by bycatch

As with overfishing, species characterized by low productivity are less able to sustain high rates of bycatch. Species that are listed as globally threatened as a consequence of bycatch include several species of marine mammals, seabirds, sea turtles, and sharks and rays (Elasmobranchii).

MAMMALS Estimates of marine mammal (cetacean and pinniped) bycatch in US fisheries were on the order of 6,000 individuals annually in the 1990s (Read et al. 2006). This is likely an underestimate resulting from underreporting. In the United States most observed bycatch of marine mammals occurs in gillnet fisheries. Even large cetaceans can be entangled and drowned in gillnets. Large cetaceans may also be taken in other types of fisheries, especially those that use vertical lines to connect gear on or near the seabed to the surface, as in pot and demersal trawl fisheries. The global population of the North Atlantic right whale (*Eubalaena glacialis*, EN) was severely depleted through whaling. The western population is currently estimated at 300 to 350 individuals, and sightings in the historical range of the eastern population are extremely rare. This small population size renders the species vulnerable to the loss of any individuals, especially adult females. The major current threats to the species are entanglement in gillnets and ship strikes (Reilly et al. 2012).

BIRDS The major current threats to seabirds at sea are incidental mortality in fishing gear, reduced food availability, and pollution. In a review of the conservation status of seabirds, 97 of 346 species were identified as globally threatened based on the 2010 IUCN Red List. Of the 97 threatened species, 40 (41%) were identified as threatened by bycatch (Croxall et al. 2012). Seabirds often forage behind longline vessels for bait and fish waste and then are caught on baited hooks or entangled in lines. Estimates of the number of seabirds killed in longline fisheries range from 160,000 to more than 320,000 annually (Anderson et al. 2011). Seabirds are also killed in trawl fisheries. They collide with the cables that run between fishing vessels and the net or are entangled in the net as it is deployed or hauled in. Birds may also be entangled and drowned in gillnets in both freshwater and marine systems. Drowning in gillnets may have contributed to the extinction of both the Atitlán grebe and the Alaotra grebe.

The species most frequently caught in longline fisheries are albatrosses, petrels, and shearwaters (Procellariiformes). According to the IUCN Red List, a total of 17 of the 22 species of albatross (Diomedeidae) are globally threatened, and the remaining five are near threatened. Bycatch is identified as a major threat to all seventeen threatened species. Many of these species are characterized by small global populations and are long lived with low maximum population growth rates—factors that increase their inherent vulnerability to extinction and the sensitivity of population growth rates to adult mortality. The breeding population of the Tristan albatross (*Diomedea dabbenena*, CR) is essentially restricted to Gough Island, but the birds range across the southern Atlantic Ocean while foraging. The Tristan albatross has an estimated annual breeding population of fewer than 2,000 pairs and a total population of about 7,100 individuals. The population declined by about 28 percent between the 1960s and 2012.

Predation by the introduced house mouse (*Mus musculus*, LC) causes very low breeding success. The main threat to population persistence is incidental mortality in longline fisheries, with an estimated 500 individuals killed each year (BirdLife International 2012).

TURTLES Six of the seven species of sea turtles are listed as globally threatened on the IUCN Red List. The major threats are the harvesting of eggs and incidental mortality through bycatch in gillnet, longline, and trawl fisheries. A comprehensive database of bycatch of marine turtles based on observer records for 1990 to 2008 indicates that a total of 85,000 individuals were caught and recorded as bycatch during that period (Wallace et al. 2010). Only a small proportion of global fishing effort is observed, however, so these records likely represent only 1 percent to 5 percent of the total global bycatch. The number of adult female leatherback turtles (*Dermochelys coriacea*, CR) at many nesting beaches in the Pacific has fallen by more than 80 percent (Sarti Martinez 2000). In the year 2000, an estimated 1.4 billion longline hooks were set worldwide, catching an estimated 50,000 to 60,000 leatherbacks (Lewison et al. 2004b). The mortality rate of leatherbacks hooked on longlines may be only 4 percent to 27 percent, but the total bycatch is still substantial because of the scale of the fishery (Lewison and Crowder 2007).

SHARKS AND RAYS Sharks and rays are targeted in some fisheries but are also captured incidentally in fisheries where they are not the target species. Following the rapid increase in the value of shark fins, many individuals taken as bycatch were retained. Sharks and rays occupy a wide range of coastal and open-ocean habitats. Pelagic sharks, such as the shortfin mako (*Isurus oxyrinchus*, VU) are taken in pelagic longline, purse seine, gillnet, and midwater trawl fisheries, whereas demersal species such as the angel shark (*Squatina squatina*, CR) are taken in benthic trawls and demersal longlines.

Sharks and rays cover a wide range of life-history types. The species that are most at risk of extinction through bycatch are those with relatively low productivity compared to the target species with which they are caught. The blue skate (*Dipturus batis*, CR) was once abundant in the continental shelf areas of the northeastern Atlantic and Mediterranean, but it is taken as bycatch in multispecies trawl fisheries and has now disappeared from much of its former range (Dulvy et al. 2006). The barndoor skate (*D. laevis*, EN), which occurs in the northwestern Atlantic off Canada and the United States, was never directly targeted but is taken as bycatch in multispecies trawl fisheries and longlines. The species is now very rare in the shallow-shelf portion of its range, but bycatch data indicate that the species' range extends farther north and into deeper waters than previously considered (Dulvy et al. 2003) This deepwater portion of its range may act as a natural refuge from fishing, but there is concern about increasing fishing effort in these areas.

Culling and persecution

Often overlooked, the deliberate culling and persecution of species viewed as competing predators of targeted fish species is a major threat contributing to extinction risk in some species. The extinct Caribbean monk seal and Japanese sea lion were both directly targeted for their skins and oil, but persecution by fishers is also thought to have been a major threat that led to their extinction. Persecution remains a threat to several marine mammal species, with the persecution often increasing when individuals learn to steal fish from gear, a behavior known as depredation. The Mediterranean monk seal (*Monachus monachus*, CR) is the most endangered pinniped species in the world, with an estimated total population size of 350 to 450 animals. The deliberate killing of monk seals, even in countries and areas where the species is legally protected, is considered the single most important source of mortality (Aguilar and Lowry 2008).

Mechanisms of extinction biodiversity loss through fishing

Increased mortality associated with fishing may lead to extinctions through reductions in population size, fragmentation of populations, and changes in the population characteristics of fished species. All three mechanisms may involve the loss of unique genetic diversity even if the species remains extant. Loss of genetic diversity is a major concern in the context of global change, as reduced genetic diversity may limit species' scope to adapt to global change.

Reduction in population sizes

The primary direct effect of increased mortality associated with fishing is reductions in population sizes. Fishing inevitably leads to a reduction in the population size or biomass of targeted species. It is important to distinguish the effects of sustainable fishing at target biomass levels from the effects of the initial fishing-down phase and the effects of continued overfishing and stock depletion. In well-managed single-species fisheries, managers may intentionally allow the biomass to be reduced to a level at or above B_{MSY} in order to increase yields. These population reductions may be controlled and may be reversible.

But in multispecies fisheries that target a mix of high- and low-productivity species, fisheries management strategies designed to maximize multispecies yields may lead to population reductions of some target species below B_{MSY} (Worm et al. 2009). The less productive species may be severely depleted, while populations of more productive target species remain robust. Similarly, among species caught as bycatch, low-productivity species are more vulnerable, especially when caught in fisheries that target higher-productivity species. Longline fisheries directed toward tunas and billfish (Scombridae, Istiophoridae, and Xiphiidae) also capture blue sharks (*Prionace glauca*, NT) as incidental catch. Blue sharks have slow population growth rates compared to yellowfin tuna (*Thunnus albacares*, NT). Simple population models indicate that blue shark populations are very sensitive to fishing pressure, while yellowfin tuna populations are more robust (Schindler et al. 2002). In fisheries that are poorly managed or barely managed at all, economic incentives may push fishers to deplete stocks well below B_{MSY}.

Fragmentation of populations

Because the global populations of many species are composed of distinct local and regional subpopulations, the process of extinction may involve the sequential extirpation of these subpopulations. Some of these may be at much greater risk of extinction than the global population or species. For example, the global population of the sockeye salmon (*Oncorhynchus nerka*) is considered least concern, but five subpopulations have gone extinct since the early twentieth century with four listed as critically endangered, twelve as endangered, and three as vulnerable (Rand et al. 2012). The extirpation of subpopulations is a conservation concern because it may be a step in the process of global extinction.

For many species, the global population comprises a metapopulation of subpopulations, linked through the dispersal and exchange of individuals. In this context, the persistence of the global population depends on that of subpopulations and the potential for individuals from healthy subpopulations to sustain, rescue, or recolonize others. Population fragmentation through the extirpation of some subpopulations and the isolation of others undermines these metapopulation processes and potentially the long-term viability of the global population.

Variation among subpopulations may be essential to the resilience of the global population in the face of environmental variation. The Bristol Bay sockeye salmon fishery in Alaska is based on the exploitation of a large number of distinct subpopulations that return to the various streams and rivers that empty into the bay. These subpopulations are characterized by variation in a range of ecological attributes, including run timing and the amount of time that young fish spend rearing in freshwater and that adults spend at sea before returning. The diverse subpopulations respond differently to variation in environmental conditions, such that variation at the scale of Bristol Bay is substantially less than variation at the subpopulation scale (Schindler et al. 2010). Overfishing of individual stocks could undermine this portfolio effect.

In many species, some subpopulations are at least partially reproductively isolated. Isolated subpopulations may have genetically based local adaptations that are critical to the survival of individuals in that region. The potential for the recolonization of the ecological niches occupied by genetically distinct subpopulations may be limited. In cases in which subpopulations are genetically distinct, the extirpation of subpopulations through fishing may lead to the loss of genetic diversity. For example, in a 2009 review of the status of the loggerhead turtle (*Caretta caretta*, EN) under the U.S. Endangered Species Act, the global population of loggerheads was found to comprise at least nine distinct population segments. The review team argued that each population segment is genetically unique and that the loss of any one population segment would represent a significant loss of genetic diversity (Conant et al. 2009).

Changes in productivity

Much of the theory of sustainable fisheries management depends on the concept of density-dependent population growth, as discussed in the section on overfishing. Recruitment rates are generally expected to be highest at low stock sizes because of factors such as reduced competition for resources. Some species, however, may exhibit reduced population growth rates at low population densities, as a result of reduced ability to find mates (a phenomenon referred to as the Allee effect), increased per-capita predation rates, or other factors (Gascoigne and Lipcius 2004). In fisheries, the general term for these dynamics is *depensation*. Species that combine low productivity and depensation are especially vulnerable to overexploitation. For these species, the fishing rate that would be expected to trigger population collapse may be very close to the fishing rate expected to support maximum sustainable yields (Punt 2000). Once stocks are depleted, they may not be able to recover. Unfortunately, it is hard to predict how populations will respond to low densities.

While there is limited evidence for depensation in commercial fish species (Myers et al. 1995), some sessile or semisessile broadcast spawners have suffered repeated recruitment failure following severe depletion through fishing. For broadcast spawners, fertilization success is expected to decline with distance between males and females. In many broadcast-spawning species, adults aggregate to spawn, but some marine invertebrates are highly sedentary. Current densities of the white abalone (*Haliotis sorenseni*, NE), for example, are too low to allow adequate fertilization to support population recovery. Even in the absence of fishing pressure, the species is not expected to recover without intervention (Hobday & Tegner 2000). For some nonbroadcast spawners, such as the queen conch (*Strombus gigas*, NE), a heavily exploited large Caribbean gastropod, reproductive activity also appears to decline at low densities, probably as a result of the reduced probabilities of encountering a mate (Gascoigne & Lipcius 2004).

Other changes in population characteristics

Fishing may lead to other changes in the population characteristics of fished species, with these changes having implications for reproduction and resilience to overfishing. Most fishing is selective and can lead directly to changes in the size structure, age structure, and sex structure of fished populations.

Fecundity tends to increase with body size so that populations with a higher proportion of larger fish have greater reproductive potential. For some species, larger females are more fecund and produce higher-quality eggs or larvae with higher survival rates. This is the case for several species of Pacific rockfish (*Sebastes* spp.). For these species, fishing that targets larger individuals removes older, larger females from the stock and may lead to a disproportionate reduction in population growth rates. A similar process may occur when larger males are targeted, as in many lobster and crab fisheries.

The epinepheline groupers (Epinephelinae) are commercially valuable tropical marine fish. Most epinepheline groupers are protogynous hermaphrodites, with individuals first maturing as females and only some large adults becoming males. Species with this type of life-history strategy may be at greater risk of extinction because fishing that targets larger individuals may alter the sex structure of the stock (Morris et al. 2000). The dusky grouper (*Epinephelus marginatus*, EN), which occurs in the eastern Atlantic and western Indian Ocean, is a protogynous hermaphrodite. The primary threat is overfishing. The sex ratio is heavily skewed toward females

Dusky grouper (*Epinephelus marginatus*) in open water in the Atlantic Ocean. Norbert Probst/imagebroker/Corbis.

and may become more skewed if fishing is targeted toward the largest individuals (males). Heavily skewed sex ratios may lead to reduced reproductive rates (Cornish and Harmelin-Vivien 2004).

Even when fishing is not highly selective, an increase in mortality is expected to lead to changes in life-history traits, in particular age at maturity. There is evidence for long-term trends toward reduced size or age at maturity in many commercially exploited fish stocks, including stocks that are considered well managed and stable, such as the Australian spiny lobster (*Panulirus cygnus*, LC), as well as stocks that have seen severe declines, such as the Atlantic cod (*Gadus morhua*, VU) in the northwestern Atlantic (Allendorf et al. 2008).

In cases in which life-history traits are partially heritable, changes in population characteristics may be phenotypic or genetic. It can be challenging to distinguish between the two, but when changes do have a genetic basis, they will be extremely difficult to reverse.

Vulnerability to extinction through fishing

Some species are inherently more vulnerable than others to population reductions, population fragmentation, reduced productivity, and other changes in population characteristics associated with fishing. Species that are especially vulnerable include those with low productivity, restricted ranges, and low dispersal rates, as well as those whose catchability remains high even as populations decline. Many species at high risk of extinction associated with fishing combine several of these traits.

Low productivity

The maximum rate at which a population can grow is one of the key factors determining a species' vulnerability to extinction through overexploitation. Species with low maximum population growth rates are at greater risk, especially if growth rates decline at low population sizes. A species' maximum population growth rate depends on several life-history traits, in particular age at maturity, annual reproductive rates, and annual survival. These traits are often correlated, and species may be described as having fast or slow life-history strategies (see Musick 1999).

Species with fast life-history strategies typically are small, are fast growing, reach sexual maturity quickly, and die young. These species tend to have high maximum rates of population growth and may be more resilient to overfishing as they have the potential to recover more rapidly. Several species that have recovered from a severe depletion fall into this category, such as two of the clupeids—the Peruvian anchoveta (*Engraulis ringens*, LC) and the Pacific sardine (*Sardinops sagax*, NE). This does not imply, however, that it is safe to overexploit species with fast life-history strategies. Populations of species with fast life-history strategies may grow and decline rapidly in response to changes in environmental conditions. Fisheries for both the Peruvian anchoveta and the Pacific sardine have collapsed in the past as a result of a combination of overfishing and unfavorable environmental conditions. Unfavorable environmental conditions may impede recovery for many years, even after fishing pressure has been reduced (Beverton 1990).

Species with slow life-history strategies typically have slow growth, have large maximum body size, reach sexual maturity relatively late, have low annual reproductive rates, and are often long lived. Many freshwater and marine fish species produce large numbers of eggs during spawning. This high fecundity, however, does not necessarily translate into high rates of population growth. The key question is how many of these eggs produce offspring that survive to produce offspring themselves. For highly fecund species, survival rates in early life stages are generally very low. Studies of commercial fish species, which are often the most productive, indicate that adult spawners of most species generally produce one to seven replacement spawners per year, a figure comparable to terrestrial vertebrates (Myers et al. 1999).

Species with slow life-history strategies tend to have low maximum rates of population growth. They cannot sustain heavy fishing pressure and may require a long time to recover from depletion, even if fishing pressure is reduced. In particular, late age-at-maturity implies a delay in population response. Many freshwater and marine fishes threatened by overfishing have slow life-history strategies; these include the sturgeons and paddlefish, rockfishes and rockcods (Sebastidae), epinepheline groupers, and sharks, rays, and skates. These characteristics are also shared by many marine mammals (especially large cetaceans and sirenians), seabirds (especially albatrosses and shearwaters), and sea turtles.

Sturgeons and paddlefish are characterized by long life histories, including late age-at-maturity, and are extremely sensitive to overfishing. The Atlantic sturgeon (*Acipenser sturio*, CR) is the source of one of the most expensive forms of caviar. This species was once common in rivers throughout Europe but is now found only in a single river in France. A population assessment in 2005 estimated the population at

Shrimp trawl catch, shrimp trawling results in tremendous bycatch and waste, up to 12 times bycatch for one pound of shrimp, Gulf of Mexico, Texas. © Alamy.

2,000 individuals, but more recent estimates indicate that only 20 to 750 native wild fish remain. Bycatch is the major threat, along with habitat degradation and loss of spawning sites (Gesner et al. 2010).

Large-bodied species are also often identified as at risk of extinction as a result of fishing (Olden et al. 2007). Such species often have slow life-history strategies, as body size is linked to growth, which is correlated with age at maturity. Large species may also be specifically targeted because of their body size and in some cases, such as the blue skate, may be vulnerable to bycatch well before reaching sexual maturity. The Atlantic halibut (*Hippoglossus hippoglossus*, EN) is the largest flatfish species in the northwestern Atlantic Ocean. It is also long lived with a late age-at-maturity. The species supported important commercial fisheries from the early 1800s, but stocks collapsed by the mid-twentieth century as a result of overfishing and have not yet recovered (Brodziak and Col 2006). The Mekong giant catfish (*Pangasianodon gigas*, CR), endemic to the Mekong River basin, is one of the world's largest freshwater fish. This species has a slow life-history strategy with an age at maturity of ten years or more. While fishing effort for this species has been fairly constant or increasing, annual catch rates indicated a population decline of over 80 percent over a thirteen-year span. (Hogan 2011).

Many deepwater species, such as the orange roughy (*Hoplostethus atlanticus*, NE) and the Patagonian toothfish (*Dissostichus eleginoides*, NE), are slow growing, late maturing, and long lived, with low reproductive rates. Deepwater sharks and rays generally have a higher age at maturity and are longer lived than shallow-water species and are therefore inherently more vulnerable to fishing pressure (Garcia et al. 2008).

For species with low annual reproductive rates, high survival rates are required for population growth. For species that rarely encounter conditions that favor high rates of larval survival, strong recruitment events may occur only episodically, and the longevity of mature individuals is key to population persistence. The rockfishes and rockcods compose a family of long-lived marine fishes that are targeted by commercial and recreational fisheries. Within this family, the bocaccio (*Sebastes paucispinis*, CR) occurs in the California Current System in the northeastern Pacific. Steep declines in bocaccio populations since the mid-1970s have been attributed to a combination of changes in environmental conditions and overfishing. Females are highly fecund, producing some 20,000 to 2 million eggs per year, but on average only about seven larvae survive for every 10 million eggs. The long-term population growth rate depends on the frequency of favorable oceanographic conditions and strong recruitment events. When fishing pressure is high, strong recruitment events need to occur more often in order to offset fishing mortality (Tolimieri and Levin 2005).

Species that have evolved in ecosystems characterized by high variability, such as the major upwelling systems, have developed mechanisms to cope with that variability, including long life spans in the case of many species with slow life-history strategies (e.g., Pacific rockfish), high maximum population growth rates in species with fast life-history strategies (e.g., Pacific sardine), and biocomplexity (as in Pacific salmon). Yet anthropogenic factors such as fishing can undermine the resilience of populations to environmental variability, for example by selectively removing larger, more fecund fish from the population, by truncating the age structure of species that depend on long life spans, or by extirpating distinct subpopulations in species that depend on biocomplexity.

Restricted-range species

Species with restricted ranges may be at greater risk from overfishing and other threats associated with fishing. For restricted-range species, even localized threats may impact their entire global range. The vaquita (*Phocoena sinus*, CR), a harbor porpoise with an estimated global population of less than 600 individuals, is restricted to the northern Gulf of California, Mexico. Its entire global range is subject to intense fishing pressure. Incidental bycatch in gillnets is the major threat to the species, with annual bycatch rates estimated at 7 percent to 15 percent of the total population size (Rojas-Bracho et al. 2008).

Many species of reef fish have restricted ranges. One study found that 9.2 percent of a sample of 1,677 coral reef fish species had geographic ranges of less than 1,930 square miles (50,000 sq km) (Hawkins et al. 2000). For reef-associated fish, the actual area occupied is likely to be substantially smaller than the geographic range. Many of these restricted-range reef fish species are small and can occur at high densities, but heavy exploitation may threaten extinction. For example, the Banggai cardinalfish (*Pterapogon kauderni*, EN) is a small reef fish endemic to the Banggai Archipelago off Sulawesi, Indonesia. It has a geographic range of about 2,125 square miles (5,500 sq km) and an area of occupancy of only 13 square miles (34 sq km). It is highly valued in the aquarium trade and is heavily exploited. One subpopulation of about 50,000 individuals was completely extirpated in the early twenty-first century (Allen and Donaldson 2007).

Some more broadly distributed species share the vulnerability of restricted-range species because they are habitat specialists or depend on a few locations or limited habitats for specific stages in their life cycles, such as species dependent on particular spawning locations or nursery habitat. Habitat specialists that depend on highly threatened habitats, such as coral reefs, estuaries, mangroves, and wetlands, are especially vulnerable to habitat degradation. The sawfishes generally have large geographic ranges but occur mostly in coastal bays and estuaries. They are late maturing and have low annual reproductive rates, which makes them vulnerable to overfishing. Their sawlike rostrum makes them especially vulnerable to entanglement and capture in net fisheries. When captured incidentally, they are often retained for the high value of their meat, fins, and rostra. The smalltooth sawfish (*Pristis pectinata*, CR) is broadly distributed in coastal regions of the Mediterranean, Atlantic, and Gulf of Mexico but has been extirpated from much of its former range by fishing and habitat modification. The remaining subpopulations are now small and fragmented. The major threat to these subpopulations is fishing, along with habitat degradation. Stocks are now so severely depleted that commercial targeted fishing is no longer economically viable in most regions, but individuals are still taken incidentally in a wide range of fisheries targeting other species (Adams et al. 2006).

Limited dispersal

Species that have low rates of exchange between subpopulations are especially vulnerable to local extirpation. Limited dispersal implies that subpopulations subject to elevated threats may not be sustained or recolonized by individuals from other subpopulations.

For many marine fishes, dispersal occurs mainly during the larval stage. There has been substantial research on larval dispersal. Dispersal distances depend on circulation patterns, the length of time that larvae spend in the plankton, and whether they are passive or active swimmers. In Pacific rockfishes, for example, dispersal is assumed to occur mainly during the larval and pelagic juvenile phases, as adults are fairly sedentary. Even though pelagic stages may last many months, coastal currents and eddies may serve to retain rather than disperse larvae and juveniles during pelagic stages. Behavioral changes during the late-pelagic stage may also contribute to retention. Overall, populations range along a spectrum from fully open to fully closed, but some fish species that were thought to have fairly open populations are now recognized as having more limited dispersal and consequently greater vulnerability to local extirpation (Cowen and Sponaugle 2009).

For many species of marine mammals, seabirds, and sea turtles, juvenile dispersal is also important. After weaning, fledging, or hatching, juveniles may disperse widely and may select a breeding site other than their natal site. In many of these species, breeding adults are mostly philopatric, meaning that once they have selected a rookery, colony, or nesting beach, they will return to breed at the same site for the rest of their adult lives. Given the sensitivity of the population growth of these long-lived species to adult mortality, localized increases in adult mortality can lead to the extirpation of subpopulations and population fragmentation.

Bangaii cardinalfish (*Pterapogon kauderni*) brooding eggs in its mouth. © Norbert Wu/Science Faction/Corbis.

Catchability

For some species, catchability remains high even while populations decline, undermining the myth that fishing cannot cause extinctions because commercial extinction will occur long before biological extinction. Several schooling and shoaling species respond to population reduction by reducing their occupied habitat so that catchability remains relatively stable in the face of population declines. This behavior provides a partial explanation for the potential of fishing to lead to the collapse of highly productive small pelagic fish, such as anchovies, sardines, and herring (Clupeiformes). The collapse of the northwestern stocks of Atlantic cod has also been attributed in part to the aggregative behavior of fish stocks combined with technological improvements that enabled fishers to continue catching cod even when stocks were severely depleted.

Species that migrate through narrow bottlenecks, such as anadromous fish, can still be caught efficiently even when numbers are low. Examples include the sturgeons and paddlefish and many salmonids (Salmonidae). Many anadromous species face a combination of heavy fishing pressure, dams that block their migration routes, and other changes to freshwater systems.

Species that aggregate in large numbers at locations that are predictable in time and space can also be caught efficiently when

numbers are low, leading to local extirpation (Sadovy and Domeier 2005). Many species of epinepheline groupers form large spawning aggregations at specific sites. Heavy fishing of these aggregations is thought to be a major reason for the severe decline of species such as the Nassau grouper (*Epinephelus striatus*, EN). The Nassau grouper is distributed from Bermuda and Florida throughout the Bahamas and the Caribbean Sea, but spawning aggregations have collapsed in many countries, leading to gaps in the distribution (Cornish and Eklund 2003). The bumphead parrotfish (*Bolbometopon muricatum*, VU) is a large-bodied, long-lived parrotfish. This species was once common throughout much of its range across the Indian and Pacific Oceans, but it is now globally rare and has probably been extirpated from some parts of its range. Gregarious as adults, bumphead parrotfish are always found in small shoals or aggregations numbering many tens of individuals, and they spawn in densely packed schools of about 100 individuals at specific sites. This aggregative behavior renders the species highly vulnerable to spearfishing (Chan et al. 2012). The orange roughy also forms large spawning aggregations and forages in dense shoals at predictable locations close to seamounts and other benthic features that foster concentrations of prey, where it is targeted by deepwater trawl fisheries (Lack et al. 2003).

In a few high-value species, catchability declines as stocks decline, but value increases with scarcity, which more than offsets the additional fishing effort required so that the species continues to be targeted. A comprehensive status assessment of tunas and billfish (Scombridae, Istiophoridae, and Xiphiidae) for the IUCN Red List found that most of the long-lived economically valuable species are globally threatened (Collette et al. 2011). All three species of bluefin tunas (*Thunnus maccoyii*, CR; *T. thynnus*, EN; and *T. orientalis*, LC) are relatively late maturing and rely on geographically restricted spawning sites, with both of these factors making them vulnerable to overfishing. Bluefin tunas also have very high economic value, which has increased with scarcity. Similarly, the high value of sturgeon caviar implies that fishing pressure on sturgeons, including high rates of illegal poaching, continues despite dwindling populations. For sawfishes, the high value of a single large rostrum provides extremely strong incentives to resource-poor fishers to target the few remaining individuals in some regions (Raloff 2009).

Indirect effects of fishing on biodiversity

Beyond the direct effects of fishing on target and nontarget species, such as reductions in the population sizes, fragmentation of populations, and changes in population characteristics, fishing may have broader indirect effects on the structure and function of ecosystems. In particular, fishing may lead to changes in the composition of species assemblages and the relative abundance of functional groups, with potential impacts on species interactions.

Species interactions

Since the early 1990s, there has been vigorous debate about the reduction in the mean trophic level of landings observed in many regions and about the processes underpinning those shifts. Under the "fishing down" hypothesis (Pauly et al. 1998), fisheries development involves an unsustainable progression in target species from large-bodied upper-trophic-level predators to smaller lower-trophic-level species. Many predatory fish are characterized by slow life histories that make them vulnerable to fishing. Heavy exploitation of large predators reduces their abundance and may release populations of their prey from predation pressure, thus transforming the species composition of aquatic ecosystems. Alternatively, under the "fishing through" hypothesis (Essington et al. 2006), shifts in the mean trophic level of catches may be explained by the sequential addition of fisheries for lower-trophic-level species. Over time, fisheries transition from being more selective to less selective as a broader range of trophic levels are exploited. Many lower-trophic-level groups, such as shellfish and invertebrates, support relatively high-value, low-volume fisheries. From an economic perspective, fisheries development may be best characterized as driven by profits, targeting the most valuable and accessible species initially and gradually adding less desirable species (Sethi et al. 2010).

In practice, shifts in the composition of landings are highly context dependent. They may reflect changes in fishing strategies driven by economic factors as well as changes in the availability of target species attributable to fishing pressure or changes in environmental conditions. Off the coast of Argentina and Uruguay in the early 1990s, landings were characterized by catches of relatively large slow-growing and late-maturing species, such as the Argentine hake (*Merluccius hubbsi*, NE). In the late 1990s and early twenty-first century, during a period of growth and diversification of fishing effort, landings shifted in favor of crustaceans, mollusks, and medium-sized fishes (Jaureguizar and Milessi 2008). In contrast, off southern Brazil, the mean trophic level of landings has tended to increase, following the collapse of the sardine fishery and a shift in fishing effort to target upper-trophic-level sharks and tunas farther offshore (Vasconcellos and Gasalla 2001).

The focus here is on how fishing affects the composition of species assemblages and on the consequences for biodiversity. As with landings, the effects of fishing on species composition are context dependent, depending especially on the selectivity of fishing pressure. Fishing that selectively targets particular species or functional guilds is more likely to alter the composition of species assemblages. Comparative analysis of reef fish communities in remote, lightly fished areas of the northwestern Hawaiian islands and more densely populated, heavily fished areas of the main Hawaiian islands showed that fish biomass was dominated by large apex predators in the northwestern Hawaiian islands but by herbivores and lower-trophic-level predators in the main Hawaiian islands. Target species at all trophic levels were generally larger in the remote northwestern Hawaiian islands (Friedlander and DeMartini 2002).

Where fishing is more uniform, the effects on species composition may be more limited. Studies at two sites in the Philippines, both with fairly uniform fishing pressure, showed that large predatory species with slow life histories, such as epinepheline groupers, declined in abundance under intensive fishing pressure and recovered slowly when fishing ceased (Russ and Alcala 1998). Smaller lower-trophic-level species with faster life-history strategies, such as fusiliers (Caesionidae), were also subjected to intensive fishing pressure. This group also declined in densities, but recovered more rapidly when fishing ceased. Overall, the relative abundance of the major families and trophic groups was not altered, except during a period of destructive fishing. These findings were attributed to the relatively nonselective nature of fishing at the two sites, as well as the openness of the sites to recruitment from elsewhere and the diversity of species assemblages.

The consequences of changes in the relative abundance of species and functional groups depends on the scale of the changes in species composition, the functional role of the species directly affected, their dominance in the system, and whether other species in the system have the potential to fill a similar role (functional redundancy). Most evidence is drawn from less diverse systems, such as freshwater lake systems and temperate marine systems. In less diverse systems, the main functional groups often comprise fewer species with less functional redundancy. A few keystone species may dominate a functional guild, such that changes in the abundance of these species can have significant effects. Some evidence, however, is also available from more diverse systems in which a single species plays a dominant role or an entire functional guild is removed.

REMOVAL OF PREDATORS There is growing global concern about the depletion of predatory fish communities and the indirect effects of their removal on aquatic ecosystems (Myers and Worm 2003). Apex predators, such as tunas and billfish, are often among the first species targeted by fisheries, and their life-history strategies may make them especially vulnerable to fishing and slow to recover. The removal of top predators may have pronounced effects on species assemblages through the release of prey species from predation pressure.

In general, the removal of piscivorous fish from tropical reefs has not led to an increase in the abundance of their prey, apparently because piscivorous fish belong to large and diverse functional groups, so that removing individual predator species may have a limited impact. Removing entire functional groups of species, however, may have more pronounced effects. In tropical reef systems, fisheries often target fish families, such as emperors (Lethrinidae) and triggerfish (Balistidae), that play a key role in regulating sea urchin populations (Kaiser and Jennings 2002). Urchin populations have increased following the reduction in the biomass of their predators. Removal of piscivorous fish through fishing has had greater effects in less diverse systems, including several freshwater and temperate marine systems. In the northwestern Atlantic, fishing of predatory fish, such as cod, haddock, and pollack, led to a system dominated by lower-trophic-level fish species and to an increased abundance of macroinvertebrates, including crabs, lobsters, and urchins (Frank et al. 2005).

REMOVAL OF COMPETITORS Fishing may also have indirect effects through changes in the relative abundance of competitors. The selective removal of one species in a functional group may lead to the release and expansion of competing species. For example, skates tend to be generalist bottom feeders, and there is considerable dietary overlap among species. In the northeastern Atlantic, the removal of larger skates may have led to an increase in smaller skates through increased food availability (Dulvy et al. 2000). When the removal of upper-trophic-level predators leads to mesopredator release, prey populations may decline. In the northwestern Atlantic, overfishing of the entire guild of large sharks led to the release of their prey, including the cownose ray (*Rhinoptera bonasus*, NT). Elevated populations of the cownose ray led to the depletion of their prey, including bay scallops and possibly other bivalves (Myers et al. 2007).

REMOVAL OF PREY There are also growing concerns about the effects of fishing on the availability of food for upper-trophic-level predators, such as marine mammals, seabirds, and large predatory fish. Reductions in food availability may be most severe in regions where large commercial fisheries target stocks of forage fish, such as anchovies, sardines, capelin, and sand eels, that form the main prey of upper-trophic-level predators. The decline of the Canary Islands oystercatcher, endemic to the Canary Islands but now extinct, was probably a result of overharvesting of intertidal invertebrates, although other factors have also been implicated.

To some extent, upper-trophic-level predators can compensate for reduced prey availability through increased foraging effort and other adaptive mechanisms. But species that have limited flexibility in one or more dimensions may be more vulnerable. Pinnipeds and seabirds have limited spatial flexibility during the breeding season as they must return to colonies to protect and feed their young. Reductions in prey availability lead first to reduced breeding success and ultimately to increased adult mortality. As seabirds and pinnipeds are mostly long lived, population growth rates are sensitive to increased adult mortality. In the North Sea, industrial fisheries target the lesser sand eel (*Ammodytes marinus*, NE), which is an important prey species for pinnipeds, cetaceans, and seabirds. Substantial reductions in breeding success of several seabird species in the Shetland Islands coincided with a marked decline in landings in the industrial sand eel fishery off the Shetland coast, although it is unclear whether the decline in sand eel abundance was attributable primarily to the fishery or environmental variation (Furness 2003). The causes of the substantial decline in western populations of the Steller sea lion (*Eumetopias jubatus*, NT) in the North Pacific remain unclear, but it is likely that reductions in the biomass of prey species has played a role. As elsewhere, these reductions have been attributed to a combination of fishing and changes in environmental conditions (Gelatt and Lowry 2012).

Seabirds that forage close to the surface, including many tropical species, lack flexibility in the vertical dimension. Tropical seabirds are often dependent on subsurface predators, such as tunas and billfish and cetaceans, to chase prey to the surface where it is accessible. Fishing may thus reduce the availability of prey to surface-foraging seabirds by reducing populations of subsurface predators (Balance and Pitman 1999).

Species that lack dietary flexibility may also be vulnerable to reduced prey availability. The Peruvian tern (*Sterna lorata*, EN), endemic to Peru and northern Chile, is heavily dependent on Peruvian anchoveta. Populations of Peruvian tern were severely impacted by the 1972 collapse of anchoveta, which has been attributed to a combination of fishing pressure and environmental change (BirdLife International 2012b). The southern resident stock of the killer whale (*Orcinus orca*, DD) off the west coast of the United States and Canada is listed as endangered under the U.S. Endangered Species Act. Several factors may explain the failure of this population to recover following a series of live captures in the 1960s and 1970s to supply marine parks. One factor is the dietary dependence of this population on Fraser River chinook salmon (*Oncorhynchus tshawytscha*, NE), which has been heavily exploited (Ayres et al. 2012).

Conversely, in systems in which the harvesting of predatory fish has led to an increase in populations of forage fish, such as sand eels and capelin, that are an important food source for upper-trophic-level predators, fishing may have increased the availability of prey.

REMOVAL OF SPECIES THAT FORM OR MAINTAIN HABITAT Habitat diversity provides a foundation for species diversity by providing refuges for prey from their predators and enabling the coexistence of competitors. Fishing can degrade habitat and reduce habitat diversity indirectly through changes in the relative abundance of species responsible for habitat development (e.g., reef-forming species), maintenance (e.g., grazers), and degradation.

Reef-forming organisms, such as oysters in temperate estuarine systems, add structural complexity to the seabed and increase habitat and species diversity. Oyster beds, for example, provide habitat for fishes, invertebrates, and algae. Oysters and other filter feeders may also play an important role in maintaining water quality through their filtration activities. Eastern oysters (*Crassostrea virginica*, NE) have been heavily exploited, and over 50 percent of the area of oyster reef habitat in Chesapeake Bay on the east coast of the United States has disappeared as a result of destructive fishing techniques and habitat degradation (Lenihan and Peterson 1998).

In coral reef systems, grazing fishes play an important role in maintaining the composition of coral-dominated versus algal-dominated reefs. In Jamaica, heavy fishing pressure reduced the densities of large herbivorous fish. Reef habitat apparently remained healthy for several decades as the expansion of macroalgae was limited by abundant populations of the herbivorous sea urchin *Diadema antillarum* (NE). But the reduction in functional redundancy undermined the resilience of the system. When a pathogen triggered a mass mortality of this sea urchin, the depleted functional guild of herbivorous species was no longer able to control the abundance of algae, leading to a broad and persistent algal bloom, which transformed the reef system (Hughes 1994).

INTRODUCTION OF INVASIVE SPECIES In terms of extinctions and threats to freshwater biodiversity, one of the most significant threats associated with fishing is the introduction of nonnative species. A total of 32 of the 248 freshwater species listed as extinct or extinct in the wild on the IUCN Red List, including 10 ray-finned fishes, 12 amphibians, 8 birds, 1 crustacean, and 1 insect, have invasive nonnative or alien species identified as a major threat. For a further 1,137 freshwater species that are listed as globally threatened, invasive nonnative or alien species have been identified as a major threat. Many of the invasive species involved were introduced for the purposes of fishing. Invasive nonnative species threaten native species through predation, competition, hybridization, and other effects.

Lake Victoria in eastern Africa is a center of diversity for haplochromine cichlid fish (*Haplochromis* spp.), with approximately 350 endemic species. The introduction of the piscivorous Nile perch (*Lates niloticus*, LC) for the purposes of fishing devastated this assemblage (Kitchell et al. 1997). A total of sixty haplochromine cichlids are currently listed as globally threatened, with invasive species as a major threat, although some are showing signs of recovery following the reduction of Nile perch populations through heavy fishing pressure.

Many high alpine lakes in the western United States are naturally fishless but have been artificially stocked with various species of salmon and trout (e.g., *Oncorhynchus* spp. and the brown trout [*Salmo trutta*]) to create recreational fisheries. The introduced fish prey on a range of native species

(Knapp et al. 2001). The southern mountain yellow-legged frog (*Rana muscosa*, EN), for example, occurs in the montane regions of California. As a result of predation by introduced trout on its larvae, this frog has suffered precipitous declines and been extirpated from much of its range. The remaining global population is small and fragmented (Hammerson 2008).

CHANGES IN NUTRIENT FLOWS More broadly, changes in species composition associated with fishing can lead to changes in the flows of nutrients and energy in aquatic ecosystems. Many stocks of Pacific salmon (*Oncorhynchus* spp.) have been reduced substantially through a combination of fishing and habitat degradation. Salmon returning from the ocean to spawn provide an influx of marine-derived nutrients to freshwater systems and riparian habitats. They may contribute a large portion of available phosphorous and nitrogen that are essential for the productivity of freshwater systems and their communities, including juvenile salmon. Returning salmon also provide an important food source to scavengers and predators, such as the brown bear (*Ursus arctos*, LC) and the bald eagle (*Haliaeetus leucocephalus*, LC). Reductions in salmon populations have led to concern about the effects of reduced nutrient inputs to these systems (Gende et al. 2002).

Discards from fisheries can also provide subsidies to scavengers at the surface or on the seabed (Kaiser and Hiddink 2007). Increases in some populations of seabirds that scavenge on fisheries discards have been detected. Seabirds are highly mobile and are able to search for fishing vessels and follow them over large areas. The effects on benthic scavengers, including crustaceans and fishes, are more challenging to detect and may be more limited as these species are less mobile than seabirds.

CASCADING EFFECTS Fisheries-induced changes at one ecosystem level may have cascading effects through the system and may reduce the resilience of aquatic communities to fishing and a range of other threats (Pinnegar et al. 2000). Trophic cascades involve changes in three or more trophic levels of an ecosystem, connected by predation. Evidence for trophic cascades is stronger for freshwater systems and those marine systems that have relatively low diversity at all trophic levels, such that interactions between species at different levels may be tightly coupled.

In the Black Sea, for example, heavy fishing pressure on upper-trophic-level predators released the lower trophic level of zooplanktivores and led to a reduction in zooplankton and an increase in phytoplankton abundance (Daskalov et al. 2007). Conversely, several studies have focused on the effects of increasing densities of large piscivorous fish through the artificial stocking of lakes to support recreational fishing, along with associated reductions in mid-level species, including zooplanktivores. Experimental removals of zooplanktivorous fish have led to increases in densities of larger zooplankton, which consume phytoplankton, and consequently reductions in phytoplankton biomass (Carpenter et al. 1985).

There is limited evidence of trophic cascades in marine ecosystems characterized by high diversity of generalist species at each level. Trophic cascades may occur, however, when an entire functional group is removed or the species removed are specialized and have few substitutes. In tropical reef and temperate kelp systems, a few species may play a keystone role in regulating sea urchin (Echinoidea) populations. Changes in urchin abundance can have a profound influence on the structure and function of coral reef and kelp ecosystems through their role as grazers on macroalgae. In the Galápagos off Ecuador, coral and macroalgal habitats have been transformed to heavily grazed reefs and urchin barrens. The severe El Niño of 1982–1983 was a major driver of this transformation, but the shift may have been exacerbated by the reduction of populations of large lobsters and predatory fish through fishing, leading to reduced predation pressure on urchins. This shift represents a threat to the biodiversity that depends on these habitats and may have contributed to the possible extinction of the restricted-range Galápagos damsel (*Azurina eupalama*, CR) (Edgar et al. 2010)

Habitat degradation through destructive fishing techniques

Fishing may also impact habitat through the direct effects of gear on habitat structure. This type of effect is mostly associated with destructive fishing techniques, such as trawling, dredging, and blast fishing (Jennings and Kaiser 1998). Habitat degradation is especially threatening to restricted-range species as their entire area of occupancy may be affected. Even for highly mobile and more widespread species, degradation of habitat that is essential for the completion of critical stages of species' life cycle, such as spawning and nursery habitat, represents a threat.

The scale of impacts of destructive fishing techniques depends on the scale of the area affected. In general, the impacts of mobile fishing gear, such as trawls and dredges, are greater than those of static gear, as mobile gear has the potential to cover greater areas. Even for mobile gear, however, fishing effort is usually patchy. In the case of trawling, for example, some areas are trawled repeatedly and may be maintained in a permanently altered state, whereas others are unsuitable for trawling and rarely impacted.

Bottom trawling and dredging can have destructive effects on benthic habitats (Thrush and Dayton 2002). Some bottom trawls are designed to maintain contact with the seabed, while others are designed to be towed just above the bottom. Trawls designed to operate on hard substrates, such as rocks and corals, often carry heavy gear such as chains at the base of the net to prevent snags and damage to the net. Some dredges have metal teeth on the base of the net that act like a rake to dig up clams and other organisms buried in the surface sediment. Bottom trawls and dredges may degrade hard and soft habitats by destroying hard bottom features, disturbing and resuspending soft sediments, simplifying topographically complex habitats, and destroying sea-grass beds and other biogenic structures, such as corals, sponges, and oyster reefs. The effects depend on the specific gear, how frequently the area is trawled, and how adapted the habitat is to natural disturbance processes, such as storms, strong tidal currents, and wave action. Deepwater habitats, such as seamounts, are generally less exposed to disturbance and are therefore likely to be more fragile and less resilient.

A reduction in habitat diversity may lead to reductions in the diversity of associated species assemblages. On the North West Shelf of Australia, fishes such as snappers (*Lutjanus* spp.) and emperors (*Lethrinus* spp.) prefer habitats with large epibenthic organisms, such as sponges and soft-coral communities. Demersal trawling leads directly to the loss of suitable habitat for snappers, emperors, and other species associated with large epibenthic organisms and reductions in their abundance. The recovery of large epibenthic organisms in areas closed to trawling appears to be slow, but it eventually leads to the recovery of associated species (Sainsbury et al. 1992).

Blast fishing, which involves the use of explosives to stun fish to facilitate harvesting, is a destructive form of fishing used on coral reefs. The impacts of blast fishing are localized but may be severe, especially given the restricted ranges of many coral reef species. In addition to the direct loss of coral cover, blast fishing reduces the growth potential of scleractinian corals by creating exposed and unstable conditions. The potential for coral reefs to recover from blast fishing depends on the scale of blasting. After a single blast, coral cover may recover within five years, but reefs that have been extensively blasted may not recover significantly over the same period (Fox and Caldwell 2006).

Regime shifts and fishing-induced phase shifts

Regime shifts are fairly abrupt transformations of ecosystems from one relatively stable state to another at broad spatial scales attributable to changes in climate or oceanographic conditions (Scheffer et al. 2001). Climate is a major forcing factor that may induce basin-scale shifts that alter energy pathways at all levels, from physical processes through primary productivity and mid-trophic-level species to upper-trophic-level predators. Since the late twentieth century, there has been growing recognition of the profound effects on aquatic ecosystems of climate variability at multiple scales, including interannual variability, such as El Niño events; interdecadal variability, such as the Pacific Decadal Oscillation and the North Atlantic Oscillation; and even intercentennial variability. Regime shifts may be detected by synchronous changes in the physical properties of water masses (e.g., sea-surface temperature) and in the relative abundance and distribution of ecosystem components at multiple levels. In upwelling systems, for example, periodic decadal-scale shifts between anchovy-dominated and sardine-dominated systems have been linked to changes in major current patterns, water masses, and phytoplankton and zooplankton communities (Chavez et al. 2003). These shifts provoke changes in populations of upper-trophic-level predators, such as large predatory fish, seabirds, and pinnipeds, that feed on anchovy and sardine. Regime shifts are distinguished from short-term variability by their relative persistence.

Interactions between regime shifts and the effects of fishing on aquatic ecosystems are difficult to predict. Fishing pressure may reduce the resilience of target species and ecological communities to climate-driven regime shifts, exacerbating their effects. Once a regime shift has occurred, it can be challenging to distinguish the roles of climate forces and fishing in transforming ecosystems. Collapses in various commercial fish stocks, including those of the Peruvian anchoveta and the Pacific sardine, have been attributed to a combination of environmental variation and fishing.

Fishing may also lead to changes in the structure and function of ecosystems, through changes in species assemblages and interactions and habitat degradation. Here, fishing-induced phase shifts are defined as broad-scale shifts in ecosystems from one relatively stable state to another that are attributable primarily to fishing rather than climate forces. For example, the overfishing of grazing fish, together with the mass mortality of urchins, led to a phase shift from coral-dominated to algal-dominated reef systems in the Caribbean (Hughes 1994), as outlined above.

A key question is whether fishing-induced phase shifts are reversible through reduced fishing pressure. The recovery of coral reefs from extensive blasting is inhibited as the unstable and exposed conditions are unsuitable for the recruitment and growth of scleractinian corals. In tropical reef systems and temperate kelp forests, fishing may lead to a reduction in grazing fish and the proliferation of urchins. Once sea urchins are established, it may be difficult for grazing fish to compete, even if fishing pressure is reduced. In the northwestern Atlantic, predation on urchins by adult cod may be essential for the maintenance of suitable nursery habitat for juvenile cod (Steneck et al. 2002). Fishing may also create opportunities for invasive species to become established, leading to a phase shift in ecological communities. In the Black Sea, following the depletion of upper-trophic-level predators, a shift in fishing effort led to a reduction in planktivorous fish that enabled an outbreak of populations of an invasive nonnative comb jelly, *Mnemiopsis leidyi* (Daskalov et al. 2007). Once established, invasive species may be extremely difficult to eliminate.

Addressing the effects of fishing on biodiversity

Given the continued global demand for wild-caught fish, how can the negative effects of fishing on freshwater and marine biodiversity be reduced?

Avoiding overfishing

Many industrial fisheries are now managed through target and limit reference points that are designed to prevent overfishing and enable the rebuilding of overexploited fish stocks (Worm et al. 2009). For several decades, the level of biomass that produces maximum sustainable yields (B_{MSY}) has been a common target reference point. As discussed above, however, many fisheries scientists and managers now advocate that B_{MSY} be considered a limit reference point, below which fish stocks are considered depleted and fishing effort should be reduced to allow stocks to rebuild. A wide range of management tools are available to manage fishing effort in line with these reference points, especially in high-value fisheries that can support the costs of fisheries management. Management tools include gear restrictions, seasonal or area closures (e.g., marine reserves), limited access, and rights-based approaches (e.g., catch shares). In regions where industrial fisheries management is failing, weak governance is often the major constraint rather than lack of fisheries management tools. In particular, the governance of high-seas fisheries that fall beyond national boundaries needs strengthening, and the illegal, unreported, and unregulated fishing that is prevalent in many regions and represents a major threat to both fisheries and biodiversity needs to be combated.

In the case of small-scale fisheries, considerable progress has also been made in developing effective management systems to avoid overfishing, especially for specific high-value sedentary resources such as lobsters and Chilean 'abalone' (*Concholepas concholepas*, NE) (Salas et al. 2007). Greater attention needs to be paid to developing cost-effective solutions for the management challenges of small-scale fisheries that target low-value species, especially those characterized by large numbers of small vessels or fishing units, such that the costs of conventional management are high relative to the value of the catch.

Reducing bycatch

Efforts to reduce bycatch of nontarget species have focused on technical solutions, such as gear modifications, and bycatch quota systems that encourage fishers to avoid areas with high bycatch and develop and adopt a range of other practices to reduce bycatch (Hall and Mainprize 2005). Bycatch quotas require onboard observers and so are costly to implement, especially in fisheries with a large number of vessels. But they have been effective in several industrial fisheries. Yellowfin tuna often school together with dolphins. Purse seining of tuna has led to high mortalities of several species, including the pantropical spotted dolphin (*Stenella attenuata*, LC) and the eastern spinner dolphin (*S. longirostris*, DD). The US government required US vessels to adopt several measures to reduce this bycatch and avoid harm to dolphins. These regulations together with 100 percent observer coverage and strong consumer pressure led to the widespread adoption of backing-down procedures and other practices designed to reduce dolphin mortality (Perrin et al. 2002). The short-tailed albatross (*Phoebastria albatrus*, VU) is listed as endangered under the U.S. Endangered Species Act. Observed bycatch of a few individuals could lead to the disruption or closure of Alaska longline fisheries. Research scientists have worked closely with the fishing industry to identify and test proposed solutions. This collaborative research initiative has led to the identification of several low-cost, practical solutions, such as streamers and weighted lines (Melvin et al. 2001).

Addressing the ecosystem effects of overfishing

Reserves have been proposed as a key tool for reducing the effects of fishing on biodiversity and ecosystems (Lubchenco et al. 2003). Reserves can be an effective tool for safeguarding some species and populations, especially restricted-range species and relatively closed populations. For highly mobile species and species that occur at low densities over large global ranges, reserves may play a more limited role in protecting populations directly but can help to safeguard habitats that are critical for specific life stages or functions. Nevertheless, freshwater and marine ecosystems are relatively open, and reserves alone may not address the full range of threats. In the context of fishing, reserves are likely to displace rather than reduce fishing effort, so the effects of fishing outside reserves still need to be addressed.

Until recently, fisheries management and science has largely focused on single species. But the growing concern about the ecosystem effects of fishing is leading to a reorientation from single species to a broader ecosystem approach to fisheries (Pikitch et al. 2004). The FAO's "Code of Conduct for Responsible Fisheries" provides a framework of principles for the responsible management of freshwater and marine fisheries based on an ecosystem approach (FAO 2012b). This code has been adopted by many fisheries management agencies at both the national and international level. But challenges remain in designing and implementing ecosystem approaches to fisheries. The Convention on the Conservation of Antarctic Marine Living Resources represents one of the very few long-standing examples of an ecosystem approach to fisheries in practice (Constable 2011).

An ecosystem approach to fisheries requires a shift in management objectives from maximization of yields toward objectives that recognize the value of safeguarding the structure and function of ecosystems and maintaining a broad suite of ecosystem services. In an ecosystem approach, target and limit reference points may need to be adjusted to take into account the ecosystem effects of fishing, but the question of how to set these reference points represents a major challenge. The spatial and temporal dynamics of aquatic ecosystems are complex, and species interactions are difficult to predict. Implementing management experiments on scales that are relevant to fisheries management is rarely possible. Fisheries scientists are building multispecies models and ecosystem models to inform ecosystem approaches to fisheries, but the lack of data and understanding of fundamental processes is a major constraint.

An ecosystem approach will therefore require investments in new research on the effects of fishing and fisheries management in an ecosystem context. Further research is also needed on the incentives faced by fishers and how various management actions can realign incentive patterns consistent with an ecosystem approach to fisheries. Substantial investments will also be required to strengthen governance. Stakeholders, including fishing communities and industry, will need to be engaged in the process. Above all, political commitment and leadership is required at multiple levels of society.

Resources

Books

Boyd, Charlotte. "Fisheries Sector." In *Importance of Biodiversity and Ecosystems in Economic Growth and Equity in Latin America and the Caribbean: An Economic Valuation of Ecosystems*, edited by Andrew Bovarnick, Francisco Alpizar, and Charles Schnell, 83–126. New York: United Nations Development Programme, 2010.

Dayton, Paul K., Simon Thrush, and Felicia C. Coleman. *The Ecological Effects of Fishing in Marine Ecosystems of the United States*. Arlington, VA: Pew Oceans Commission, 2002.

Dulvy, Nicholas K., John K. Pinnegar, and John D. Reynolds. "Holocene Extinctions in the Sea." In *Holocene Extinctions*, edited by Samuel T. Turvey, 129–150. Oxford: Oxford University Press, 2009.

FAO Fisheries and Aquaculture Department. *The State of World Fisheries and Aquaculture, 2012*. Rome: Food and Agriculture Organization of the United Nations, 2012.

Hilborn, R., and C.J. Walters. *Quantitative Fisheries Stock Assessment: Choice, Dynamics, and Uncertainty*. New York: Chapman and Hall, 1992

Johannes, R.E. *Words of the Lagoon: Fishing and Marine Lore in the Palau District of Micronesia*. Berkeley: University of California Press, 1981.

Kaiser, M.J., and S. Jennings. "Ecosystem Effects of Fishing." In *Handbook of Fish Biology and Fisheries*, edited by Paul J.B. Hart and John D. Reynolds, vol. 2, *Fisheries*, 342–366. Malden, MA: Blackwell, 2002.

Kurlansky, Mark. *Cod: A Biography of the Fish That Changed the World*. New York: Walker, 1997.

Misund, Ole Arve, Jeppe Kolding, and Pierre Fréon. "Fish Capture Devices in Industrial and Artisanal Fisheries and Their Influence on Management." In *Handbook of Fish Biology and Fisheries*, edited by Paul J.B. Hart and John D. Reynolds, 13–36. Malden, MA: Blackwell, 2002.

Musick, John A., ed. *Life in the Slow Lane: Ecology and Conservation of Long-Lived Marine Animals*. American Fisheries Society Symposium 23. Bethesda, MD: American Fisheries Society, 1999.

National Research Council. Committee on Ecosystem Effects of Fishing. *Dynamic Changes in Marine Ecosystems: Fishing, Food Webs, and Future Options*. Washington, DC: National Academies Press, 2006.

Pauly, Daniel, and Jay Maclean. *In a Perfect Ocean: The State of Fisheries and Ecosystems in the North Atlantic Ocean*. Washington, DC: Island Press, 2003.

Perrin, William F., Bernd Wursig, and J.G.M. Thewissen, eds. *Encyclopedia of Marine Mammals*. San Diego, CA: Academic Press, 2002.

Reynolds, John D., Nicholas K. Dulvy, and Callum M. Roberts. "Exploitation and Other Threats to Fish Conservation." In *Handbook of Fish Biology and Fisheries*, edited by Paul J.B. Hart and John D. Reynolds, vol. 2, *Fisheries*, 319–341. Malden, MA: Blackwell, 2002.

Sainsbury, K.J., R.A. Campbell, and A.W. Whitelaw. "Effects of Trawling on the Marine Habitat on the North West Shelf of Australia and Implications for Sustainable Fisheries Management." In *Sustainable Fisheries through Sustaining Fish Habitat*, edited by D.A. Hancock, 137–145. Canberra: Australian Government Publishing Service, 1993.

Smith, Tim D. *Scaling Fisheries*. Cambridge: Cambridge University Press, 1994.

Periodicals

Allan, J. David, Robin Abell, Zeb Hogan, et al. "Overfishing of Inland Waters." *BioScience* 55, no. 12 (2005): 1041–1051.

Allan, J. David, and Alexander S. Flecker. "Biodiversity Conservation in Running Waters." *BioScience* 43, no. 1 (1993): 32–43.

Allendorf, Fred W., Phillip R. England, Gordon Luikart, et al. "Genetic Effects of Harvest on Wild Animal Populations." *Trends in Ecology and Evolution* 23, no. 6 (2008): 327–337.

Anderson, Orea R.J., Cleo J. Small, John P. Croxall, et al. "Global Seabird Bycatch in Longline Fisheries." *Endangered Species Research* 14, no. 2 (2011): 91–106.

Ayres, K.L., Booth, R.K., Hempelmann, J.A., Koski, K.L., et al. (2012) "Distinguishing the Impacts of Inadequate Prey and Vessel Traffic on an Endangered Killer Whale (Orcinus orca) Population." *PLoS ONE* 7, no. 6 (2012).

Bascompte, Jordi, Carlos J. Melián, and Enric Sala. "Interaction Strength Combinations and the Overfishing of a Marine Food Web." *Proceedings of the National Academy of Sciences of the United States of America* 102, no. 15 (2005): 5443–5447.

Beverton, Raymond J.H. "Small Marine Pelagic Fish and the Threat of Fishing: Are They Endangered?" *Journal of Fish Biology* 37, supp. sA (1990): 5–16.

Birstein, Vadim J., William E. Bemis, and John R. Waldman. "The Threatened Status of Acipenseriform Species: A Summary." *Environmental Biology of Fishes* 48, nos. 1–4 (1997): 427–435.

Boehlert, George W. "Biodiversity and the Sustainability of Marine Fisheries." *Oceanography* 9, no. 1 (1996): 28–35.

Botsford, Louis W., Juan Carlos Castilla, and Charles H. Peterson. "The Management of Fisheries and Marine Ecosystems." *Science* 277, no. 5325 (1997): 509–515.

Branch, Trevor A. "How Do Individual Transferable Quotas Affect Marine Ecosystems?" *Fish and Fisheries* 10, no. 1 (2009): 39–57.

Browman, Howard I., and Konstantinos I. Stergiou. Introduction to "Perspectives on Ecosystem-Based Approaches to the Management of Marine Resources." *Marine Ecology Progress Series* 274 (2004): 269–270.

Brown, James, and Graeme Macfadyen. "Ghost Fishing in European Waters: Impacts and Management Responses." *Marine Policy* 31, no. 4 (2007): 488–504.

Caddy, J.F., and K.L. Cochrane. "A Review of Fisheries Management Past and Present and Some Future Perspectives for the Third Millennium." *Ocean and Coastal Management* 44, nos. 9–10 (2001): 653–682.

Carpenter, Stephen R., James F. Kitchell, and James R. Hodgson. "Cascading Trophic Interactions and Lake Productivity." *BioScience* 35, no. 10 (1985): 634–639.

Carr, Mark H., Joseph E. Neigel, James A. Estes, et al. "Comparing Marine and Terrestrial Ecosystems: Implications for the Design of Coastal Marine Reserves." *Ecological Applications* 13, supp. (2003): S90–S107.

Chavez, Francisco P., John Ryan, Salvador E. Lluch-Cota, and Miguel Ñiquen C. "From Anchovies to Sardines and Back: Multidecadal Change in the Pacific Ocean." *Science* 299, no. 5604 (2003): 217–221.

Coleman, Felicia C., and Susan L. Williams. "Overexploiting Marine Ecosystem Engineers: Potential Consequences for Biodiversity." *Trends in Ecology and Evolution* 17, no. 1 (2002): 40–44.

Collette, B.B., K.E. Carpenter, B.A. Polidoro, et al. "High Value and Long Life—Double Jeopardy for Tunas and Billfishes." *Science* 333, no. 6040 (2011): 291–292.

Constable, Andrew J. "Lessons from CCAMLR on the Implementation of the Ecosystem Approach to Managing Fisheries." *Fish and Fisheries* 12, no. 2 (2011): 138–151.

Cooke, Steven J., and Ian G. Cowx. "The Role of Recreational Fishing in Global Fish Crises." *BioScience* 54, no. 9 (2004): 857–859.

Cowen, Robert K., Kamazima M.M. Lwiza, Su Sponaugle, et al. "Connectivity of Marine Populations: Open or Closed?" *Science* 287, no. 5454 (2000): 857–859.

Cowen, Robert K., and Su Sponaugle. "Larval Dispersal and Marine Population Connectivity." *Annual Review of Marine Science* 1 (2009): 443–466.

Crowder, Larry B., Elliott L. Hazen, Naomi Avissar, et al. "The Impacts of Fisheries on Marine Ecosystems and the Transition to Ecosystem-Based Management." *Annual Review of Ecology, Evolution, and Systematics* 39 (2008): 259–278.

Crowder, Larry B., and Steven A. Murawski. "Fisheries Bycatch: Implications for Management." *Fisheries* 23, no. 6 (1998): 8–17.

Croxall, John P., Stuart H.M. Butchart, Ben Lascelles, et al. "Seabird Conservation Status, Threats, and Priority Actions: A Global Assessment." *Bird Conservation International* 22, no. 1 (2012): 1–34.

Cucherousset, Julien, and Julian D. Olden. "Ecological Impacts of Nonnative Freshwater Fishes." *Fisheries* 36, no. 5 (2011): 215–230.

Dagorn, Laurent, Kim N. Holland, Victor Restrepo, and Gala Moreno. "Is It Good or Bad to Fish with FADs? What Are the Real Impacts of the Use of Drifting FADs on Pelagic Marine Ecosystems?" *Fish and Fisheries*. Published electronically May 16, 2012.

Daskalov, Georgi M., Alexander N. Grishin, Sergei Rodionov, and Vesselina Mihneva. "Trophic Cascades Triggered by Overfishing Reveal Possible Mechanisms of Ecosystem Regime Shifts." *Proceedings of the National Academy of Sciences of the United States of America* 104, no. 25 (2007): 10518–10523.

Dudgeon, David, Angela H. Arthington, Mark O. Gessner, et al. "Freshwater Biodiversity: Importance, Threats, Status, and Conservation Challenges." *Biological Reviews* 81, no. 2 (2006): 163–182.

Dulvy, Nicholas K., Julia K. Baum, Shelley Clarke, et al. "You Can Swim but You Can't Hide: The Global Status and Conservation of Oceanic Pelagic Sharks and Rays." *Aquatic Conservation: Marine and Freshwater Ecosystems* 18, no. 5 (2008): 459–482.

Dulvy, Nicholas K., Julian D. Metcalfe, Jamie Glanville, et al. "Fishery Stability, Local Extinctions, and Shifts in Community Structure in Skates." *Conservation Biology* 14, no. 1 (2000): 283–293.

Dulvy, Nicholas K., Yvonne Sadovy, and John D. Reynolds. "Extinction Vulnerability in Marine Populations." *Fish and Fisheries* 4, no. 1 (2003): 25–64.

Duncan, Jeffrey R., and Julie L. Lockwood. "Extinction in a Field of Bullets: A Search for Causes in the Decline of the World's

Freshwater Fishes." *Biological Conservation* 102, no. 1 (2001): 97–105.

Edgar, Graham J., Stuart A. Banks, Margarita Brandt, et al. "El Niño, Grazers, and Fisheries Interact to Greatly Elevate Extinction Risk for Galapagos Marine Species." *Global Change Biology* 16, no. 10 (2010): 2876–2890.

Essington, Timothy, Anne E. Beaudreau, and John Wiedenmann. "Fishing through Marine Food Webs." *Proceedings of the National Academy of Sciences of the United States of America* 103, no. 9 (2006): 3171–3175.

Fox, Helen E., and Roy L. Caldwell. "Recovery from Blast Fishing on Coral Reefs: A Tale of Two Scales." *Ecological Applications* 16, no. 5 (2006): 1631–1635.

Frank, Kenneth T., Brian Petrie, Jae S. Choi, and William C. Leggett. "Trophic Cascades in a Formerly Cod-Dominated Ecosystem." *Science* 308, no. 5728 (2005): 1621–1623.

Friedlander, Alan M., and Edward E. DeMartini. "Contrasts in Density, Size, and Biomass of Reef Fishes between the Northwestern and the Main Hawaiian Islands: The Effects of Fishing Down Apex Predators." *Marine Ecology Progress Series* 230 (2002): 253–264.

Furness, Robert W. "Impacts of Fisheries on Seabird Communities." *Scientia Marina* 67, supp. 2 (2003): 33–45.

García, Verónica B., Luis O. Lucifora, and Ransom A. Myers. "The Importance of Habitat and Life History to Extinction Risk in Sharks, Skates, Rays, and Chimaeras." *Proceedings of the Royal Society* B 275, no. 1630 (2008): 83–89.

Gascoigne, Joanna, and Romuald N. Lipcius. "Allee Effects in Marine Systems." *Marine Ecology Progress Series* 269 (2004): 49–59.

Gende, Scott M., Richard T. Edwards, Mary F. Willson, and Mark S. Wipfli. "Pacific Salmon in Aquatic and Terrestrial Ecosystems." *BioScience* 52, no. 10 (2002): 917–928.

Gislason, Henrik, Michael Sinclair, Keith Sainsbury, and Robert O'Boyle. "Symposium Overview: Incorporating Ecosystem Objectives within Fisheries Management." *ICES Journal of Marine Science* 57, no. 3 (2000): 468–475.

Hall, Stephen J., and Brooke M. Mainprize. "Managing By-Catch and Discards: How Much Progress Are We Making and How Can We Do Better?" *Fish and Fisheries* 6, no. 2 (2005): 134–155.

Harrison, Susan. "Local Extinction in a Metapopulation Context: An Empirical Evaluation." *Biological Journal of the Linnean Society* 42, nos. 1–2 (1991): 73–88.

Hawkins, Julie P., Callum M. Roberts, and Victoria Clark. "The Threatened Status of Restricted-Range Coral Reef Fish Species." *Animal Conservation* 3, no. 1 (2000): 81–88.

Hilborn, Ray. "Defining Success in Fisheries and Conflicts in Objectives." *Marine Policy* 31, no. 2 (2007): 153–158.

Hilborn, Ray, Trevor A. Branch, Billy Ernst, et al. "State of the World's Fisheries." *Annual Review of Environment and Resources* 28 (2003): 359–399.

Hilborn, Ray, and Martin Liermann. "Standing on the Shoulders of Giants: Learning from Experience in Fisheries." *Reviews in Fish Biology and Fisheries* 8, no. 3 (1998): 273–283.

Hughes, Terence P. "Catastrophes, Phase Shifts, and Large-Scale Degradation of a Caribbean Coral Reef." *Science* 265, no. 5178 (1994): 1547–1551.

Hutchings, Jeffrey A. "Conservation Biology of Marine Fishes: Perceptions and Caveats Regarding Assignment of Extinction Risk." *Canadian Journal of Fisheries and Aquatic Sciences* 58, no. 1 (2001): 108–121.

Hutchings, Jeffrey A., and John D. Reynolds. "Marine Fish Population Collapses: Consequences for Recovery and Extinction Risk." *BioScience* 54, no. 4 (2004): 297–309.

Huxley, Thomas H. "Inaugural Address." *Fisheries Exhibition Literature* 4 (1884): 1–22.

Jackson, Jeremy B.C., Michael X. Kirby, Wolfgang H. Berger, et al. "Historical Overfishing and the Recent Collapse of Coastal Ecosystems." *Science* 293, no. 5530 (2001): 629–637.

Jennings, Simon. "Indicators to Support an Ecosystem Approach to Fisheries." *Fish and Fisheries* 6, no. 3 (2005): 212–232.

Jennings, Simon, and Michel Kaiser. "The Effects of Fishing on Marine Ecosystems." *Advances in Marine Biology* 34 (1998): 201–212, 212a, 213–266, 266a, 268–352.

Jennings, Simon, and N.V.C. Polunin. "Impacts of Predator Depletion by Fishing on the Biomass and Diversity of Non-target Reef Fish Communities." *Coral Reefs* 16, no. 2 (1997): 71–82.

Jaureguizar, A., and A. Milessi. "Assessing the Sources Of The Fishing Down Marine Food Web Process in the Argentinean-Uruguayan Common Fishing Zone." *Scientia Marina* 72, no. 1 (2008): 25–36.

Jiao, Yan. "Regime Shift in Marine Ecosystems and Implications for Fisheries Management, a Review." *Reviews in Fish Biology and Fisheries* 19, no. 2 (2009): 177–191.

Kaiser, M.J., and J.G. Hiddink. "Food Subsidies from Fisheries to Continental Shelf Benthic Scavengers." *Marine Ecology Progress. Ser.* 350 (2007): 267–276.

Kitchell, James F., Daniel E. Schindler, Richard Ogutu-Ohwayo, and Peter N. Reinthal. "The Nile Perch in Lake Victoria: Interactions between Predation and Fisheries." *Ecological Applications* 7, no. 2 (1997): 653–664.

Knapp, Roland A., Kathleen R. Matthews, and Orlando Sarnelle. "Resistance and Resilience of Alpine Lake Fauna to Fish Introductions." *Ecological Monographs* 71, no. 3 (2001): 401–421.

Lankester, E. Ray "The Scientific Results of the Exhibition." *Fisheries Exhibition Literature* 4 (1884): 405–442.

Law, Richard. "Fishing, Selection, and Phenotypic Evolution." *ICES Journal of Marine Science* 57, no. 3 (2000): 659–668.

Law, Richard. "Fisheries-Induced Evolution: Present Status and Future Directions." *Marine Ecology Progress Series* 335 (2007): 271–277.

Lenihan, Hunter S., and Charles H. Peterson. "How Habitat Degradation through Fishery Disturbance Enhances Impacts of Hypoxia on Oyster Reefs." *Ecological Applications* 8, no. 1 (1998): 128–140.

Lewin, Wolf-Christian, Robert Arlinghaus, and Thomas Mehner. "Documented and Potential Biological Impacts of Recreational Fishing: Insights for Management and Conservation." *Reviews in Fisheries Science* 14, no. 4 (2006): 305–367.

Lewison Rebecca L., and Larry B. Crowder. "Putting Longline Bycatch of Sea Turtles into Perspective." *Conservation Biology* 21 (2007): 79–86.

Lewison, Rebecca L., Larry B. Crowder, Andrew J. Read, and Sloan A. Freeman. "Understanding Impacts of Fisheries Bycatch on Marine Megafauna." *Trends in Ecology and Evolution* 19, no. 11 (2004a): 598–604.

Lewison, Rebecca L., Sloan A. Freeman, and Larry B. Crowder "Quantifying the Effects of Fisheries on Threatened Species: The Impact of Pelagic Longlines on Loggerhead and Leatherback Sea Turtles." *Ecology Letters* 7 (2004b): 221–231.

Lubchenco, Jane, Stephen R. Palumbi, Steven D. Gaines, and Sandy Andelman. "Plugging a Hole in the Ocean: The Emerging Science of Marine Reserves." *Ecological Applications* 13, supp. (2003): S3–S7.

Mangel, Marc, and Phillip S. Levin. "Regime, Phase, and Paradigm Shifts: Making Community Ecology the Basic Science for Fisheries." *Philosophical Transactions of the Royal Society B* 360, no. 1453 (2005): 95–105.

Mantua, Nathan J., Steven R. Hare, Yuan Zhang, et al. "A Pacific Interdecadal Climate Oscillation with Impacts on Salmon Production." *Bulletin of the American Meteorological Society* 78, no. 6 (1997): 1069–1079.

May, Robert M. "Biological Diversity: Differences between Land and Sea." *Philosophical Transactions of the Royal Society of London B* 343, no. 1303 (1994): 105–111.

McClanahan, T.R., N.A. Muthiga, A.T. Kamukuru, et al. "The Effects of Marine Parks and Fishing on Coral Reefs of Northern Tanzania." *Biological Conservation* 89, no. 2 (1999): 161–182.

McManus, John W., Rodolfo B. Reyes Jr., and Cleto L. Nañola Jr. "Effects of Some Destructive Fishing Methods on Coral Cover and Potential Rates of Recovery." *Environmental Management* 21, no. 1 (1997): 69–78.

Morris, Annalie V., Callum M. Roberts, and Julie P. Hawkins. "The Threatened Status of Groupers (Epinephelinae)." *Biodiversity and Conservation* 9, no. 7 (2000): 919–942.

Mumby, Peter J., Craig P. Dahlgren, Alastair R. Harborne, et al. "Fishing, Trophic Cascades, and the Process of Grazing on Coral Reefs." *Science* 311, no. 5757 (2006): 98–101.

Murawski, Steven A. "Definitions of Overfishing from an Ecosystem Perspective." *ICES Journal of Marine Science* 57, no. 3 (2000): 649–658.

Musick, John A. "Criteria to Define Extinction Risk in Marine Fishes: The American Fisheries Society Initiative." *Fisheries* 24, no. 12 (1999): 6–14.

Myers, Ransom A., Nicholas J. Barrowman, Jeffrey A. Hutchings, and Andrew A. Rosenberg. "Population Dynamics of Exploited Fish Stocks at Low Population Levels." *Science* 269, no. 5227 (1995): 1106–1108.

Myers, Ransom A., Julia K. Baum, Travis D. Shepherd, et al. "Cascading Effects of the Loss of Apex Predatory Sharks from a Coastal Ocean." *Science* 315, no. 5820 (2007): 1846–1850.

Myers, Ransom A., Keith G. Bowen, and Nicholas J. Barrowman. "Maximum Reproductive Rate of Fish at Low Population Sizes." *Canadian Journal of Fisheries and Aquatic Sciences* 56, no. 12 (1999): 2404–2419.

Myers, Ransom A., and Boris Worm. "Rapid Worldwide Depletion of Predatory Fish Communities." *Nature* 423, no. 6937 (2003): 280–283.

Olden, Julian D., Zeb S. Hogan, and M. Jake Vander Zanden. "Small Fish, Big Fish, Red Fish, Blue Fish: Size-Biased Extinction Risk of the World's Freshwater and Marine Fishes." *Global Ecology and Biogeography* 16, no. 6 (2007): 694–701.

Orensanz, J.M. (Lobo), Ana M. Parma, Gabriel Jerez, et al. "What Are the Key Elements for the Sustainability of 'S-Fisheries'? Insights from South America." *Bulletin of Marine Science* 76, no. 2 (2005): 527–556.

Pauly, Daniel. "Anecdotes and the Shifting Baseline Syndrome of Fisheries." *Trends in Ecology and Evolution* 10, no. 10 (1995): 430.

Pauly, Daniel, Villy Christensen, Johanne Dalsgaard, et al. "Fishing Down Marine Food Webs." *Science* 279, no. 5352 (1998): 860–863.

Pauly, Daniel, Villy Christensen, Sylvie Guénette, et al. "Towards Sustainability in World Fisheries." *Nature* 418, no. 6898 (2002): 689–695.

Pikitch, E.K., C. Santora, E.A. Babcock, et al. "Ecosystem-Based Fishery Management." *Science* 305, no. 5682 (2004): 346–347.

Pinnegar, J.K., N.V.C. Polunin, P. Francour, et al. "Trophic Cascades in Benthic Marine Ecosystems: Lessons for Fisheries and Protected-Area Management." *Environmental Conservation* 27, no. 2 (2000): 179–200.

Pinsky, Malin L., Olaf P. Jensen, Daniel Ricard, and Stephen R. Palumbi. "Unexpected Patterns of Fisheries Collapse in the World's Oceans." *Proceedings of the National Academy of Sciences of the United States of America* 108, no. 20 (2011): 8317–8322.

Powles, Howard, Michael J. Bradford, R.G. Bradford, et al. "Assessing and Protecting Endangered Marine Species." *ICES Journal of Marine Science* 57, no. 3 (2000): 669–676.

Punt, A.E. "Extinction of Marine Renewable Resources: A Demographic Analysis." *Population Ecology* 42, no. 1 (2000): 19–27.

Raloff, Janet. "Hammered Saws: Shark Relatives with Threatening Snouts Win Global Protection." *Science News* 172, no. 6 (2007): 90–92.

Rand, Peter S., Matthew Goslin, Mart R. Gross, et al. "Global Assessment of Extinction Risk to Populations of Sockeye Salmon *Oncorhynchus nerka*." *PLoS ONE* 7, no. 4 (2012): e34065.

Read, Andrew J. "The Looming Crisis: Interactions between Marine Mammals and Fisheries." *Journal of Mammalogy* 89, no. 3 (2008): 541–548.

Read, Andrews J., Phebe Drinker, and Simon Northridge. "Bycatch of Marine Mammals in U.S. and Global Fisheries." *Conservation Biology* 20, no. 1 (2006): 163–169.

Reynolds, John D., Nicholas K. Dulvy, Nicholas B. Goodwin, and Jeffrey A. Hutchings. "Biology of Extinction Risk in Marine Fishes." *Proceedings of the Royal Society B* 272, no. 1579 (2005): 2337–2344.

Reynolds, John D., Thomas J. Webb, and Lorraine A. Hawkins. "Life History and Ecological Correlates of Extinction Risk in European Freshwater Fishes." *Canadian Journal of Fisheries and Aquatic Sciences* 62, no. 4 (2005): 854–862.

Roberts, Callum M., and Julie P. Hawkins. "Extinction Risk in the Sea." *Trends in Ecology and Evolution* 14, no. 6 (1999): 241–246.

Rose, G.A. "Reconciling Overfishing and Climate Change with Stock Dynamics of Atlantic Cod (*Gadus morhua*) over 500 Years." *Canadian Journal of Fisheries and Aquatic Sciences* 61, no. 9 (2004): 1553–1557.

Russ, G.R., and A.C. Alcala. "Natural Fishing Experiments in Marine Reserves, 1983–1993: Roles of Life History and Fishing Intensity in Family Responses." *Coral Reefs* 17, no. 4 (1998): 399–416.

Sadovy, Yvonne, and Michael Domeier. "Are Aggregation-Fisheries Sustainable? Reef Fish Fisheries as a Case Study." *Coral Reefs* 24, no. 2 (2005): 254–262.

Salas, Silvia, Ratana Chuenpagdee, Juan Carlos Seijo, and Anthony Charles. "Challenges in the Assessment and Management of Small-Scale Fisheries in Latin America and the Caribbean." *Fisheries Research* 87, no. 1 (2007): 5–16.

Scheffer, Marten, Stephen R. Carpenter, Jonathan A. Foley, et al. "Catastrophic Shifts in Ecosystems." *Nature* 413, no. 6856 (2001): 591–596.

Schindler, Daniel E., Timothy E. Essington, James F. Kitchell, et al. "Sharks and Tunas: Fisheries Impacts on Predators with Contrasting Life Histories." *Ecological Applications* 12, no. 3 (2002): 735–748.

Schindler, Daniel E., Ray Hilborn, Brandon Chasco, et al. "Population Diversity and the Portfolio Effect in an Exploited Species." *Nature* 465, no. 7298 (2010): 609–612.

Sethi, Suresh A., Trevor A. Branch, and Reg Watson. "Global Fishery Development Patterns Are Driven by Profit but Not Trophic Level." *Proceedings of the National Academy of Sciences of the United States of America* 107, no. 27 (2010): 12163–12167.

Shears, Nick T., and Russell C. Babcock. "Marine Reserves Demonstrate Top-Down Control of Community Structure on Temperate Reefs." *Oecologia* 132, no. 1 (2002): 131–142.

Sibert, John, John Hampton, Pierre Kleiber, and Mark Maunder. "Biomass, Size, and Trophic Status of Top Predators in the Pacific Ocean." *Science* 314, no. 5806 (2006): 1773–1776.

Steneck, Robert S., Michael H. Graham, Bruce J. Bourque, et al. "Kelp Forest Ecosystems: Biodiversity, Stability, Resilience, and Future." *Environmental Conservation* 29, no. 4 (2002): 436–459.

Stevens, John D., Ramón Bonfil, Nicholas K. Dulvy, and Paddy A. Walker. "The Effects of Fishing on Sharks, Rays, and Chimaeras (Chondrichthyans), and the Implications for Marine Ecosystems." *ICES Journal of Marine Science* 57, no. 3 (2000): 476–494.

Strong, Donald R. "Are Trophic Cascades All Wet? Differentiation and Donor-Control in Speciose Ecosystems." *Ecology* 73, no. 3 (1992): 747–754.

Tasker, Mark L., C.J. Camphuysen, John Cooper, et al. "The Impacts of Fishing on Marine Birds." *ICES Journal of Marine Science* 57, no. 3 (2000): 531–547.

Thrush, Simon F., and Paul K. Dayton. "Disturbance to Marine Benthic Habitats by Trawling and Dredging: Implications for Marine Biodiversity." *Annual Review of Ecology and Systematics* 33 (2002): 449–473.

Tolimieri, Nick, and Phillip S. Levin. "The Roles of Fishing and Climate in the Population Dynamics of Bocaccio Rockfish." *Ecological Applications* 15, no. 2 (2005): 458–468.

Vasconcelos, M., and M.A. Gasalla. "Fisheries Catches and the Carrying Capacity of Marine Ecosystems in Southern Brazil." *Fisheries Research* 50 (2001): 279–295.

Wallace, Bryan P., Rebecca L. Lewison, Sara L. McDonald, et al. "Global Patterns of Marine Turtle Bycatch." *Conservation Letters* 3, no. 3 (2010): 131–142.

Ward, Peter, and Ransom A. Myers. "Shifts in Open-Ocean Fish Communities Coinciding with the Commencement of Commercial Fishing." *Ecology* 86, no. 4 (2005): 835–847.

Waugh, Susan M., Dominique P. Filippi, David S. Kirby, et al. "Ecological Risk Assessment for Seabird Interactions in Western and Central Pacific Longline Fisheries." *Marine Policy* 36, no. 4 (2012): 933–946.

Worm, Boris, Edward B. Barbier, Nicola Beaumont, et al. "Impacts of Biodiversity Loss on Ocean Ecosystem Services." *Science* 314, no. 5800 (2006): 787–790.

Worm, Boris, Ray Hilborn, Julia K. Baum, et al. "Rebuilding Global Fisheries." *Science* 325, no. 5940 (2009): 578–585.

Other

Adams, W.F., S.L. Fowler, P. Charvet-Almeida, et al. "Pristis pectinata" (2006). In IUCN 2012. IUCN Red List of Threatened Species. Version 2012.2. Accessed December 9, 2012. www.iucnredlist.org.

Aguilar, A., and L. Lowry. IUCN SSC Pinniped Specialist Group 2008: "Monachus monachus." In IUCN 2012. IUCN Red List of Threatened Species. Version 2012.2. Accessed December 9, 2012. www.iucnredlist.org.

Allen, G.R and Donaldson, T.J. 2007. Pterapogon kauderni. In: IUCN 2012. IUCN Red List of Threatened Species. Version 2012.2. Accessed December 9, 2012. www.iucnredlist.org.

Ballance, Lisa Taylor, and Robert Lee Pitman. "Foraging Ecology of Tropical Seabirds." Paper presented at the 22nd International Ornithological Congress, Durban, South Africa, August 16–22, 1998.

BirdLife International 2012. "Diomedea dabbenena" (2012). In IUCN 2012. IUCN Red List of Threatened Species. Version 2012.2. Accessed December 9, 2012. www.iucnredlist.org.

BirdLife International 2012b. "Sterna lorata" (2012b). In IUCN 2012. IUCN Red List of Threatened Species. Version 2012.2. Accessed December 9, 2012. www.iucnredlist.org.

Brodziak, J., and L. Col. "Status of fishery Resources off the Northeastern US—Atlantic Halibut." Northeastern Fisheries Science Center: Resource Evaluation and Assessment Division. Rev. December 2006. Accessed December 9, 2012.

Chan, T., Y. Sadovy, and T.J. Donaldson. "Bolbometopon muricatum" (2012). In IUCN 2012. IUCN Red List of Threatened Species. Version 2012.2. Accessed December 9, 2012. www.iucnredlist.org.

CCAMLR (Commission for the Conservation of Antarctic Marine Living Resources). "CAMLR Convention." Last modified July 16, 2012. http://www.ccamlr.org/en/organisation/camlr-convention.

Compagno, L.J.V., S.F. Cook, M.I. Oetinger, and S.L. Fowler. "Anoxypristis cuspidate" (2006). In IUCN 2012. IUCN Red List of Threatened Species. Version 2012.2. Accessed December 9, 2012. www.iucnredlist.org.

Conant, Therese A., Peter H. Dutton, Tomoharu Eguchi, et al. "Loggerhead Sea Turtle (*Caretta caretta*) 2009 Status Review under the U.S. Endangered Species Act." Report of the Loggerhead Biological Review Team to the National Marine Fisheries Service, August 2009. Accessed December 3, 2012. http://www.nmfs.noaa.gov/pr/pdfs/statusreviews/loggerhead-turtle2009.pdf.

Cornish, A., and A.-M. Eklund. "Epinephelus striatus" (2003). In IUCN 2012. IUCN Red List of Threatened Species. Version 2012.2. Accessed December 9, 2012. www.iucnredlist.org.

Cornish, A., and M. Harmelin-Vivien. "Grouper and Wrasse Specialist Group: Epinephelus marginatus" (2004). In IUCN 2012. IUCN Red List of Threatened Species. Version 2012.2. Accessed December 9, 2012. www.iucnredlist.org.

Drake, Jonathan S., Ewann A. Berntson, Jason M. Cope, et al. "Status Review of Five Rockfish Species in Puget Sound, Washington: Bocaccio (*Sebastes paucispinis*), Canary Rockfish (*S. pinniger*), Yelloweye Rockfish (*S. ruberrimus*), Greenstriped Rockfish (*S. elongatus*), and Redstripe Rockfish (*S. proriger*)." U.S. Department of Commerce, NOAA Technical Memorandum NMFS-NWFSC-108, 2010. Accessed December 3, 2012. http://www.nmfs.noaa.gov/pr/pdfs/statusreviews/rock-fish.pdf.

Dulvy, N.K. "Dipturus laevis" (2003). In IUCN 2012. IUCN Red List of Threatened Species. Version 2012.2. Accessed December 9, 2012. www.iucnredlist.org.

Dulvy, N.K., G. Notarbartolo di Sciara, F. Serena, et al. "Dipturus batis" (2006). In IUCN 2012. IUCN Red List of Threatened Species. Version 2012.2. Accessed December 9, 2012. www.iucnredlist.org. Downloaded on December 9, 2012.

FAO. *The State of World Fisheries and Aquaculture 2012*. Rome: FAO Fisheries and Aquaculture Department, 2012.

Findley, L. "Totoaba macdonaldi" (2010). In IUCN 2012. IUCN Red List of Threatened Species. Version 2012.2. Accessed December 9, 2012. www.iucnredlist.org.

Food and Agriculture Organization of the United Nations. "Code of Conduct for Responsible Fisheries." Accessed December 3, 2012. http://www.fao.org/docrep/005/v9878e/v9878e00.HTM.

Gelatt, T., and L. Lowry. "Eumetopias jubatus" (2012). In IUCN 2012. IUCN Red List of Threatened Species. Version 2012.2. Accessed December 9, 2012. www.iucnredlist.org.

Gesner, J., P. Williot, E. Rochard, J. Freyhof, and M. Kottelat. "Acipenser sturio" (2010). In IUCN 2012. IUCN Red List of Threatened Species. Version 2012.2. Accessed December 9, 2012. www.iucnredlist.org.

Hammerson, G. "Rana muscosa" (2008.) In IUCN 2012. IUCN Red List of Threatened Species. Version 2012.2. Accessed December 9, 2012. www.iucnredlist.org.

Hobday, Alistair J., and Mia J. Tegner. "Status Review of White Abalone Haliotis Sorensoni Throughout Its Range in California and Mexico." NOAA Technical Memorandum NMFS, Southwest Region Office-035.

Hogan, Z. "Pangasianodon gigas" (2011). In IUCN 2012. IUCN Red List of Threatened Species. Version 2012.2. Accessed December 9, 2012. www.iucnredlist.org.

IUCN (International Union for Conservation of Nature). IUCN Red List of Threatened Species. Version 2012.2. Accessed December 3, 2012. http://www.iucnredlist.org.

Kelleher, Kieran. "Discards in the World's Marine Fisheries: An Update." Food and Agriculture Organization of the United Nations (FAO). FAO Fisheries Technical Paper 470, 2005. Accessed December 3, 2012. ftp://ftp.fao.org/docrep/fao/008/y5936e/y5936e00.pdf.

Lack, M., K. Short, and A. Willock. "Managing Risk and Uncertainty in Deep-Sea Fisheries: Lessons from Orange Roughy." TRAFFIC Oceania and WWF Endangered Seas Programme, 2003. Accessed December 3, 2012. http://awsassets.panda.org/downloads/oranger0.pdf.

Melvin, E.F., J.K. Parrish, K.S. Dietrich, and O.S. Hamel. "Solutions to Seabird Bycatch in Alaska's Demersal Longline Fisheries" (2001). Washington Sea Grant Program. Project A/FP-7.

Qiwei, W. "Psephurus gladius" (2010). In IUCN 2012. IUCN Red List of Threatened Species. Version 2012.2. Accessed December 9, 2012. www.iucnredlist.org.

Reilly, S.B., J.L.Bannister, P.B. Best, et al. "Eubalaena glacialis" (2012). In IUCN 2012. IUCN Red List of Threatened Species. Version 2012.2. Accessed December 9, 2012. www.iucnredlist.org.

Rojas-Bracho, L., R.R. Reeves, A., Jaramillo-Legorreta, and B.L. Taylor. "Phocoena sinus" (2008). In IUCN 2012. IUCN Red List of Threatened Species. Version 2012.2. Accessed December 9, 2012. www.iucnredlist.org.

Charlotte Boyd

Hunting/food preferences

The loss of species and ecosystems in the modern era can be attributed to the accelerating transformation of Earth by a growing human population. Humans appropriate roughly half of the world's net primary productivity, most of the available freshwater, and virtually all of the available productivity of the oceans. Largely as a consequence of overhunting by growing numbers of people, defaunation (the extinction of large vertebrates as a result of human action) of the world's remaining tropical forests is considered a major cause of biodiversity loss, in some cases more important than deforestation (Redford 1992). Since the mid-twentieth century exploitation of bushmeat—meat derived from wild animals in the forest, which is also called the bush—has risen dramatically. This increase has resulted from the growing human population, greater access to undisturbed forests, changes in hunting technology, a scarcity of alternative protein sources, and the fact that bushmeat is often a preferred food. Mammals hunted for subsistence or commercial purposes are particularly affected (Robinson and Bodmer 1999; Robinson and Bennett 2000). Uncontrolled exploitation threatens to bring about marked wildlife population declines and eventually the extinction of a number of hunted species. This exploitation, coupled with threats from habitat loss and deforestation, could lead to global extinctions of the most sensitive species, such as primates, through an accumulation of local disappearances. This trend may result in long-term changes in tropical forest dynamics through the loss of seed dispersers and habitat landscapers such as large forest mammals, especially elephants. Elephants directly influence forest composition and density and can alter the broader landscape. In tropical forests, elephants create clearings and gaps in the canopy that encourage tree regeneration.

This rapid acceleration in losses of tropical forest species owing to unsustainable hunting occurred first in Asian forests. For example, between the early 1960s and the first years of the twenty-first century, 12 large vertebrate species were extirpated in Vietnam largely because of hunting (Bennett and Rao 2002). By the beginning of the twenty-first century, the problem had become most acute in the bushmeat heartlands of West and Central Africa, and losses owing to hunting seemed to be occurring in even the remotest parts of Latin America (Peres 2001). This pattern follows the major impacts of development and forest loss on the three continents linked to dramatic human population growth: There are 522 people per square kilometer of remaining forest in South Asia, 99 per square kilometer in West and Central Africa, and 46 per square kilometer in Latin America (Fa and Peres 2001).

The Asian situation appears different from that on the other continents because large-scale wildlife trade in Asia involves a network of long-distance, international commodity links. As a result of rapid economic growth, the demand for natural resources, such as land, timber, and nontimber forest resources, has exploded across Asia. Moreover, the East Asian and Southeast Asian regions comprise a center for the consumption of wildlife derivatives, ranging from tiger bone medicines to shark fin cuisine. This region is also a key supplier to the international wildlife market, both legal and illegal. In Africa, bushmeat is sold for public consumption either fresh or smoked. Bushmeat may provide the primary source of animal protein for the majority of forest families, as well as for many urban families (particularly the many people without access to refrigeration), and can also constitute a significant source of revenue.

In tropical areas worldwide the meat of wild animals has long been a part of the staple diet of forest-dwelling peoples. Contrary to a popular view, this cultural preference is not a matter of lack of awareness or entrepreneurship. It relates ultimately to the low productivity of domestic livestock in tropical forest conditions and the high risks and investment costs associated with such husbandry. Given the low purchasing power of most forest dwellers and the dynamics of the agricultural economy in forest-rich (and therefore labor-constrained) environments, extensive livestock husbandry is rarely a feasible option. Most livestock are kept as a form of reserve banking and are used to satisfy particular cultural needs. The wild harvest is often the most accessible and sustainable source of protein.

During the late twentieth and early twenty-first century, there was an important transition in the scale of commercial hunting and trading of wildlife because of accelerating population growth, the modernization of hunting techniques, and greater accessibility to remote forest areas. Researchers have compiled much evidence showing that these trends are seriously threatening the stock. The elevated demand for bushmeat and the lucrative trade associated with it are the main reasons for the very large number of animals extracted

A bushmeat hunter with smoked gorilla carcasses in southeast Cameroon. © Photoshot Holdings Ltd/Alamy.

seasonality of rainfall, latitude, altitude, and edaphic (soil) conditions. M. J. Coe, D. H. M. Cumming, and J. Phillipson, in a 1976 study, and R. F. W. Barnes and S. A. Lahm, in a 1997 study, used annual rainfall as a simple index of ecosystem type to explore variation in standing mammalian biomass. Because plant biomass decreases steadily as rainfall declines in the tropics, under conditions of low seasonality and at low altitudes, "wet forests" can be supported in areas with rainfall above 4,000 millimeters, "moist forests" between 2,000 and 4,000 millimeters, and "dry forests" between 1,000 and 2,000 millimeters. Between 100 millimeters and 1,000 millimeters of rainfall, savannahs, scrub, and even dry woodlands thrive, but little plant biomass is possible under arid conditions of less than 100 millimeters of annual rainfall (see Figure 1).

Mammalian biomass increases with rainfall, but falls as forest canopy covers over habitat suitable for herbivorous ungulates. In wet and moist forests, much of the plant biomass is in the form of inedible tree trunks, and most of the leaves are heavily defended by plant secondary compounds such as lignins and toxins and are thus inedible to most mammals (McKey et al. 1981; Waterman et al. 1988; Waterman and McKey 1989). Moreover, a high proportion of the primary productivity is in the canopy and available only to relatively small mammals such as primates, sloths, and rodents; the amount of food available for large ungulates in tropical forests is low (e.g., Glanz 1982; Hart 2000). Neither supply nor demand is uniform across the tropics. Tropical landscapes are heterogeneous, featuring different wildlife communities and dynamics and different human pressures.

Intercontinental differences in prey availability

The availability of mammalian prey biomass varies substantially between sites and also between continents. A 2001 study

every year in many West and Central African countries (Fa and Peres 2001). There is considerable evidence that commercial hunting has been growing in importance for some time (see Hart 2000), and thus increasing numbers of hunters have been supplementing their incomes with the sale of meat. Such commerce increases the amount of hunting and reduces the sustainability of numerous wildlife species largely because it enlarges the effective human population density of consumers eating meat from an area of forest (Bennett and Robinson 2000).

Prey availability in tropical environments

Mammals are the main source of bushmeat protein throughout the tropics worldwide. The three most important taxa for human consumption—large-bodied ungulates (animals with hooves), primates, and rodents (see below)—occur at different relative and absolute densities in different ecosystems, with ungulates predominating in open habitats and primates occurring most commonly in forested ones. Ecosystem types are generally predictable from total rainfall,

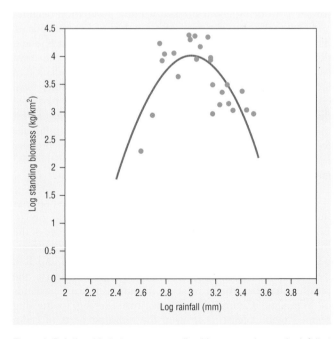

Figure 1. Relationship between mammalian biomass and annual rainfall at tropical sites. Reproduced by permission of Gale, a part of Cengage Learning.

A hunter with two goliath frogs (*Conraua goliath*) in Cameroon. Goliath frogs are the largest amphibian in the world, and consumed as bushmeat where they occur. Courtesy of Renaud Fulconis/Awely.

by John E. Fa and Carlos A. Peres, comparing the differences in nonflying mammalian biomass in a number of representative African and Neotropical moist forest sites, showed that an average of around 6,600 pounds per square mile (around 3,000 kg/sq km) was typical in Africa, a figure that far exceeded estimates for Neotropical sites (around 2,200 lb/sq mi [1,000 kg/sq km]).

Total biomass can differ substantially among sites, even within a single geographical area. For example, in Gabon's Lopé Reserve (White 1994) and the Virungas mountains of East Africa (Plumptre 1991), estimates of mammalian biomass have ranged between 2,2000 and 13,200 pounds per square mile (1,000–6,000 kg/sq km). This enormous range in productivity can be attributed largely to differences in ungulate and elephant densities (Barnes et al. 1993). In some areas, however, duikers (small African antelope) are known to attain a higher biomass than elephants (Dubost 1978, 1979), and in several other sites primates dominate (Oates et al. 1990). The latter is probably typical of the mammal communities in most tropical rain forests, where a large proportion of the primary biomass is made up of leaf-eating primates (namely, colobines [*Colobus* and *Procolobus* spp.] in mainland Africa and howler monkeys [*Alouatta* spp.] in South America).

More larger-bodied mammals (averaging 82.7 pounds [37.5 kg] in body mass) are typical of African forests in comparison with those in Neotropical forests (which average 10.6 pounds [4.8 kg]). This means that African forest species, although more profitable for hunters to take (see below), are relatively more vulnerable to extinction. There are 110 mid-sized to large-bodied mammal species whose average body mass is equal to or exceeds 2 pounds (1 kg) in African forests (with a mean body mass of 213 pounds [96 kg]), representing 70 percent of the 157 species recorded in African game harvest profiles. In contrast, only 73 of the Neotropical forest mammal species have a body mass equal to or greater than 2 pounds (1 kg) (with a mean body mass of 27 lb [12.2 kg]), including 50 species that represent 94 percent of all species

known to be hunted. Moreover, a total of 39 very large species (with a body mass equal to or greater than 22 lb [10 kg]) can be found in African forests (with a mean body mass of 582.2 lb [264.1 kg]) in contrast with only 13 species in Neotropical forests (with a mean body mass of 110.2 lb [50.9 kg]). (*Mean body mass* here refers to the average size of the prey.)

The accessibility of prey to hunters can be estimated by determining whether forest mammalian biomass is largely arboreal or terrestrial. Significant contrasts exist between African and Neotropical forests, essentially related to the observed differences in body mass. African forests are dominated by terrestrial species, whereas arboreal taxa dominate Neotropical forests. Arboreal species account for no more than 20 percent of the mammalian biomass in the few African forests that have been surveyed, whereas this figure is typically between 50 percent and 90 percent in Neotropical forests. The structure and distribution of plant production in these forests may explain, to some extent, the spread of mammalian consumers.

Prey size and accessibility to hunters may thus explain the wider range of species hunted in African moist forests. Of all species known to be hunted, 55 percent of a total of 284 African forest mammals have been reported as game species, compared to only 28 percent of 192 species in South American (Amazonian) forests (Fa and Peres 2001). The prominence of terrestrial large-bodied mammals in African forests can explain their greater vulnerability to indirect hunting techniques such as traps, nets, and snares (Hladik et al. 1990; Wilkie and Curran 1991; Noss 1998).

Disappearance of large-bodied species from overhunted areas

A 2005 study by Fa, Sarah F. Ryan, and Diana J. Bell showed that, as expected, hunters are most likely to be active in the areas with the most game. Heavily hunted areas had, on average, smaller prey and showed evidence of the depletion of larger-sized animals. This has also been shown for Amazonian forests (Jerozolimski and Peres 2003). As a corollary, a higher proportion of ungulates have been found to be extracted in lightly hunted areas, but more rodents have been found to be taken in the more hunted sites (see Fa 2000; Cowlishaw et al. 2005).

In most continental sites in Africa, rodents are important prey items only in disturbed areas (Eves and Ruggiero 2000). Therefore, increases in rodents hunted seem to be linked to reductions in the availability of more favored bushmeat species. In a 2000 study, Peres showed that large-bodied species rapidly decline with the increased presence of hunters. Poorer reproduction rates coupled with lower population densities renders these large species more vulnerable to overexploitation.

Biological consequences of overhunting in moist forests

It was only in the 1990s that researchers began documenting the impact of hunting on mammal communities in tropical forests (Robinson and Redford 1991; Robinson and Bennett

2000; Peres 2000). Some studies suggest that the impact of hunting on mammals such as primates is greater than that of moderate habitat disturbance such as logging (Oates 1996; Peres 1999; Wilkie et al. 2000).

The available information indicates that if hunting pressure is not too great and large neighboring tracts of undisturbed forest can buffer and replenish hunted areas, game populations can readily bounce back after exploitation. Peres's 2000 study of vertebrate assemblages at 25 Amazonian forest sites found that higher levels of hunting pressure were correlated with significantly lower levels of vertebrate density and biomass. For these sites, game vertebrate biomass totaled 1,543 pounds per square mile [700 kg/sq km) in nonhunted sites but only about 440 pounds per square mile (200 kg/ sq km) in heavily hunted areas. Data from 20 forest sites in Africa clearly demonstrate similar decreases in primate numbers as hunting pressure becomes increasingly heavier, but no quantitative data are available (Oates 1996).

A decline in vertebrate densities is also expected in overharvested areas. Yet, overall game densities in hunted and nonhunted Amazonian forest sites do not differ significantly (Peres 2000). This is because in areas where larger-bodied species have been significantly depressed, the abundance of small and mid-sized species remains largely unaffected or even increases. In the case of Amazonian primates, Peres and Paul M. Dolman argued in a 2000 article that this is evidence of density compensation (or undercompensation) of the residual assemblage of nonhunted mid-sized species.

Data for Africa are limited in comparison to what is available for the Neotropics. The depletion of larger-bodied species has been demonstrated—for example in Sally A. Lahm's 1993 study of mammal abundance in nonhunted and hunted sites in Makokou, Gabon. Lahm found that body mass and population density were negatively correlated with impact on the species. Nevertheless, the question of whether density compensation occurs in African hunted sites, as has been observed in the Neotropics, remains to be answered.

Observations of rapid faunal declines

The rapid decline of fauna after intensive periods of hunting in Africa has been demonstrated in studies in Equatorial Guinea. Counts of the numbers of animal carcasses arriving at Malabo market, Bioko Island, Equatorial Guinea, made during two eight-month study periods by Fa, Juan E. García Yuste, and Ramon Castelo (2000) showed that the number of species and carcasses in 1996 was significantly greater than the number in 1991. In biomass terms, the increase was much less, only 12.5 percent, compared to the almost 60 percent increase in carcasses entering the market between 1991 and 1996. A larger number of carcasses of the smaller-bodied species, such as rodents and the blue duiker, were recorded in the latter study period. Concurrently, the researchers observed a dramatic reduction in the larger-bodied species, including the Ogilby's duiker (*Cephalophus ogilbyi*) and seven diurnal primates. In another study in Equatorial Guinea, this one focusing on continental forests, records of hunting indicated that in less than five months the

Ogilby's duiker (*Cephalophus ogilbyi*) foraging in the forest. © Michael Gore/Minden Pictures.

numbers of animals from all taxonomic groups fell to less than one-fourth of what was hunted in the first month (Fa and García Yuste 2001). In a 2005 study in Takoradi, Ghana, using market profiles and hunter reports, Cowlishaw, Mendelson, and Rowcliffe demonstrated that the market was sustainable as the result of a series of nonrandom extinctions from historical hunting. Vulnerable taxa (slow reproducers) had been depleted heavily in the past, so that only robust taxa (fast reproducers), such as rodents and small antelope, were being traded. These robust taxa were supplied from a predominantly agricultural landscape around the city. The productivity of agricultural landscapes for many bushmeat species indicates that these areas may play an important role in supporting a sustainable bushmeat trade.

These findings suggest that the mere existence of a thriving bushmeat trade does not necessarily imply that all the species involved in it will be at risk. If settlements are long-established, they have generally gone through an *extinction filter* where the fauna is reduced to the most resilient species, allowing hunters to bag rodents, small antelopes, and small primates. Perversely, in areas where it has been passed, the trade could well become sustainable. By contrast, in areas where it has not yet been passed, high extraction may lead to local extinctions. In their 2005 article, Cowlishaw, Mendelson, and Rowcliffe suggest that bushmeat management policy might be improved by adopting a two-pronged approach in which vulnerable species are protected from hunting but robust species are allowed to supply a sustainable trade.

Consequences of the loss of seed dispersers

Studies of how hunting might affect the population dynamics of bushmeat species are limited for both Africa and the Neotropics. However, some data are available on the impact of hunting on animals from different age classes. For example, in separate studies, Gérard Dubost (1978, 1980) and François Feer (1988) concluded that the hunting and trapping of chevrotains and duikers by Gabonese villagers most severely affected young adults, the age class with the greatest reproductive potential. However, population age structures and demographics for hunted versus nonhunted sites are rarely available (see Bodmer et al. 1997; Hart 2000).

In many areas, the heavy presence of hunters, deforestation, and habitat fragmentation disrupt the spatial dynamics of animal populations source-sink (Novaro et al. 2005), leading to potential overexploitation. Large-bodied species are more vulnerable to overexploitation (see above), and these often play a key role in the structuring and functioning of the forest ecosystem (Fa and Peres 2001). Most mammals in tropical forests are frugivores (including frugivore-granivores, frugivore-herbivores, and frugivore-omnivores), making them important in seed dispersal and predation (e.g., Wright et al. 2000; Roldán and Simonetti 2001). Overexploitation of wildlife is expected to alter forest composition, architecture, and biomass, as well as alter ecosystem dynamics, such as regrowth and succession patterns, deposition, of soil nutrients and carbon sequestration (Apaza et al. 2002). Because of the intricate association between bushmeat species in moist forests in West and Central Africa and the habitat itself, the alteration and especially the fragmentation of forests have important negative impacts on bushmeat productivity. Hunters, it should be noted, do not deliberately destroy habitat to obtain bushmeat species.

Effects on species in protected areas

Several studies (Woodroffe and Ginsberg 1998; Robinson and Bennett 2000; Fa et al. 2002; Laurance et al. 2006) have suggested that bushmeat hunting is the single most geographically widespread form of resource extraction in tropical forests, affecting even remote protected areas. Evidence of penetration of protected areas by hunters is unavailable, although this is known to occur throughout African moist forests. With few exceptions, protected areas in these regions are not secured, and they have questionable levels of support among the resident human populations.

In a 2006 study, John E. Fa and colleagues collected data on the numbers of animals traded as bushmeat at about 100 sites in the Cross–Sanaga region in Nigeria and Cameroon. This study showed that the numbers of animals traded declined dramatically the farther away from the Korup (Cameroon) and Cross River (Nigeria) National Parks the trading was conducted. This effect has also been confirmed by a study focusing on the relationship between bushmeat prices and the distance of trading points to protected areas (Macdonald et al. 2012). It is not that case that protected areas are sources for animals bagged outside the parks but rather that the parks themselves are being penetrated. This fact is further supported by the observation that the endangered species (e.g., red colobus, chimpanzee, drill) that appear in markets outside the parks can only have originated from the protected areas because the species are almost exclusively found within these sanctuaries.

Congo Basin off-take estimates

Estimates of bushmeat off-takes in tropical forests range from global appraisals of what proportion wild animal protein contributes to people's diets (Prescott-Allen and Prescott-Allen 1982) to extrapolations of the numbers and biomass consumed within particular regions. For the Congo Basin, two studies (Fa and Peres 2001; Fa et al. 2003) estimated bushmeat extraction by compiling data from hunting profiles from forest sites in a number of countries (see above). From these studies, extraction rates were calculated for 57 reported mammalian taxa for a rural human population of 24 million within a forest area of 695,000 square miles (1.8 million sq km).

These calculations showed that a total of about 579 million animals were being consumed in the Congo Basin annually, amounting to around 4 million metric tons of dressed bushmeat (Fa et al. 2003). This figure contrasts somewhat with the findings of David S. Wilkie and Julia F. Carpenter's 1999 study, which produced an estimate of only 1 million metric tons. The latter figure is based on extrapolations of actual meat consumed (see below). Despite their magnitude, these figures are likely to be significant underestimates because they do not account for more than a fraction of the rural and even urban households that also consume large amounts of bushmeat. Despite this caveat, the amount of bushmeat extracted and consumed per unit area in the Congo Basin is still orders of magnitude higher than in the Amazon. In terms of actual yields of edible meat (given that muscle mass and edible viscera account on average for 55 percent of body mass), around 2 million metric tons was being consumed in Africa, in contrast to the 62,808 metric tons being consumed in the Amazon, according to Carlos A. Peres's 2000 study. More specifically, estimated hunting rates for Amazonian and Congo Basin species, shown in a graph of production versus extraction (see Figure 2), indicate that most are exploited unsustainably in the Congo, whereas most hunted Amazonian taxa are still within the sustainable part of the graph. In particular, Congo Basin primates appear more heavily hunted than other species; twelve of the represented 17 species (71%) fall above the 20 percent line.

These differences in species exploitation between the two continents are largely a result of larger human population sizes within a smaller forest area in the Congo, as well as the fact that a large proportion of the hunters' bag is sold in towns and villages for profit. Thus, per capita harvest rates in relation to number of consumers show a lower variation for South American settlements, whereas for Africa they decline significantly from an average of about 1,100 pounds (500 kg) per person per year in smaller settlements to 2 pounds (1 kg) per person per year in the largest settlements. This does not necessarily reflect differential consumption rates of bushmeat per person but rather dependence on commercialization of meat. The hunting of wild species does not just provide meat for household consumption. It is also an important source of income for many hunters. This distinction between subsistence and commercial hunting is often blurred because many subsistence hunters will sell a portion of their bag under certain circumstances. In terms of marketing of game, many authors agree that there are significant differences between

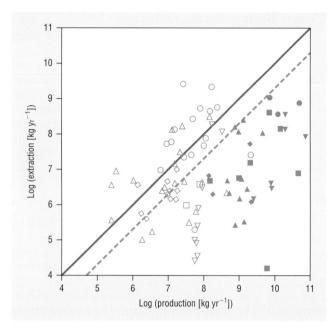

Figure 2. Hunting rates are unsustainably high across large tracts of tropical forests as seen in the relationship between extraction and total production of bushmeat throughout the Amazon and Congo Basin (solid and open symbols, respectively, by mammalian taxa). The solid line indicates where extraction equals production; the dashed line indicates exploitation at 20 percent of production, considered to be sustainable for long-lived taxa. Taxon symbols are as follows: ungulates (squares), primates (triangles), carnivores (circles), rodents (inverse triangles), and other taxa (diamonds). Reproduced by permission of Gale, a part of Cengage Learning.

Africa and South America, both in emphasis and volume of wild meat sold.

There are reports of around 100 metric tons of game meat marketed in a year in Iquitos, Peru (Fa and Peres 2001), but this seems extremely atypical of the situation in South and Central America (Redford 1993), where traded levels are low. By contrast, most towns and cities in African rain-forest regions operate markets that regularly produce well in excess of 100 metric tons monthly (Colyn et al. 1987; Steel 1994; Juste et al. 1995). Such production also constitutes an important, though often underestimated, part of the economy of many African countries (Butynski and von Richter 1974; Feer 1993).

Bushmeat and its role in human livelihoods

According to one estimate (Scherr et al. 2004), it is mainly in the tropics that human populations are highly dependent on forests. A gloomy view of bushmeat from the perspective of biological sustainability must be contrasted with the very high value of bushmeat in human livelihoods. The stark contrast between these two raises important issues for development and conservation policy. When assessing the value of bushmeat in human livelihoods, consideration must be given not only to absolute levels of production but also to bushmeat's importance relative to other components of the

livelihoods profile and to its role in risk-mitigation strategies. The starting point in any discussion of the institutional context for bushmeat trade must be its positive characteristics, as these are easily overlooked. Such characteristics help to account for the wide acceptance of such practices in the producer societies, despite heavy campaigning by anti-bushmeat groups, as well as the low levels of support for (indeed, the frequent hostility to) many conservation activities. This wide acceptance of bushmeat exists despite the severe conservation threats in many areas, as documented above.

In relative terms, the bushmeat trade is marked by high social inclusivity. There are generally few barriers to entry, and markets are often heavily penetrated by the poor. While information on the structure of bushmeat markets is limited (in part because heavy conservation pressures have tended to drive the trade underground, inhibiting research efforts), such evidence as does exist does not generally suggest a high degree of centralized control, beyond some degree of *affermage* in the more capital-intensive side of the hunt (*affermage* being when rich patrons subcontract the hunt to hunters who cannot afford rifles, shotguns, bullets, or cartridges). Important aspects of the hunting economy are nearly impossible to centralize or control, with cable snaring being an obvious example. A fair proportion of the value of the product is retained by the primary producer (the hunter)—much more so indeed than has historically been the case with other forest products such as timber and beverage crops (see Ntiamoa-Baidu 1997; Mendelson et al. 2003; Brown and Williams 2003).

Unlike the case of domestic animal husbandry, the labor inputs that hunting requires are often discontinuous and easily reconciled with the agricultural cycle. Except in cases in which full-time hunting is feasible (as on Bioko Island in the late 1990s [Fa et al. 2000]), bushmeat tends to be the product of a system of farm and forest management that collectively offers high returns from a range of activities. For the risk-averse small farmers to whom labor is the major constraint, all this has much to commend it. The trade is, likewise, low risk and flexible, with minimal capital costs and thus particularly attractive to the poor. Extractive technology is generally low level and accessible. Gender aspects are also remarkably

The smoking of bushmeat, such as these monkeys, is a way of preserving the meat. Courtesy of Renaud Fulconis/Awely.

positive. In most situations men do the hunting, and women take charge of all the downstream processing and commerce, including the point of sale in the scores of so-called chop bars and restaurants that are a familiar feature of the urban scene in Africa. Bushmeat has excellent storage qualities, in a manner compatible with the storage of agricultural produce. It is also easily transportable and has a high value/weight ratio. Accordingly, bushmeat hunting fits well with the realities of rural life in the tropics in other respects, particularly for the poor. Whether it can continue to do so at the off-take levels that prevailed in the early twenty-first century is, in many areas, seriously in doubt.

Though detailed analyses are limited, there is substantial evidence that bushmeat figures strongly in rural economies not only as a traded item but also as a pillar of livelihood security, including food security. Consideration has to be given not only to the volume of bushmeat traded and consumed but also to the incidence of hunting and other bushmeat-related matters relative to the overall array of livelihood activities. Livelihood strategies for the poor tend to be wide ranging and low risk (see Chambers 1987; Carney 1998). Their value tends to lie not only in the gains relative to minimizing the inputs (hence the level of risk), but also in the balance between activities, maintaining complementarities between different elements and ensuring a basic level of guaranteed subsistence and income, even at the expense of more lucrative—but higher-risk—options (see McSweeney 2004, 2005; de Merode et al. 2003, 2004). Thus, it would be mistaken to argue that because bushmeat forms only one component of livelihood activities and income, its benefits can therefore be easily foregone. The reverse is more likely to be the case.

Estimates of bushmeat volume and protein consumed

Bushmeat is eaten as fresh or smoked meat in soups and stews and also, occasionally, roasted or fried. A 1997 study undertaken in Ghana by Yaa Ntiamoa-Baidu (1997) found that most of the people interviewed (96%) ate bushmeat in soups. The majority of people cooked the meat in their homes, but chop bars were also important (for around 20% of interviewees). In their 2005 article, Guy Cowlishaw, Samantha Mendelson, and J. Marcus Rowcliffe stressed the importance of chop bars in the bushmeat commodity chain in Ghana, while Tasmyn East and colleagues, in a 2005 study, found the same thing to be true in continental Equatorial Guinea. The literature assessing the relative and absolute contribution of bushmeat to household economies is as sparse as that evaluating the ecological impacts of hunting. This makes it difficult to design mitigation approaches, given that the role of bushmeat in the diet as its contribution to household income is not well understood.

Most studies of bushmeat consumption report the frequency (in days in a week) with which bushmeat is consumed (e.g., Ntiamoa-Baidu 1997; East et al. 2005) or, less commonly, actual quantities of bushmeat eaten from weighed amounts of meat consumed in households (e.g., Koppert and Hladik 1990; Koppert et al. 1993). Other studies have calculated

Person eating a meal containing tree pangolin (*Manis tricuspis*) meat. This species is highly priced because of its taste. Courtesy of Renaud Fulconis/Awely.

game meat eaten from 24-hour recalls in which households are interviewed and asked to name what meats and quantities were eaten the day before (Starkey 2004; Albrechtsen et al. 2005). Comparison across studies is problematic because it is often unclear what sample sizes have been employed in estimating amounts consumed or whether the measurements are based on whole-carcass, dressed, or boned-out weights.

Allowing for these caveats, estimates of game meat consumed have been published. The estimates are wide ranging, from 1.8 ounces (50 g) per person per day to 10 ounces (280 g) per person per day (Chardonnet et al. 1995; Ntiamoa-Baidu 1997). These varying estimates could reflect differences in the study population's dependence on game meat versus fish (see below), but they also could reflect differences in the time of year during which the studies were undertaken. Often, there is insufficient information reported to assess these potential sources of error.

A 1993 study by Georgius J. A. Koppert and colleagues is probably one of the most extensive investigations of the diet of a number of human populations in equatorial Africa, including forest-dwelling families. This study concentrated on a number of ethnic groups in Cameroon where food intake was weighed in large samples of households and in different seasons. In all populations studied, the staple food (e.g., cassava) is the main source of energy, but fish and meat are the main sources of proteins. Diets based on roots and tubers, especially cassava, are known to be very low in proteins and other nutrients. This paucity is met by an important intake of animal proteins from fish and bushmeat. Thus, agricultural crops provide most of the calories to human populations, while animal meat, including bushmeat, is the main source of protein. The study indicated that game meat accounted for between 70 percent and 88 percent of the protein in the diets of the various ethnic groups, but the source of protein varied according to the population's proximity to the coast. As a result, the Yassa, who live on the Atlantic coast, fish at sea and grow cassava, while for the Kola pygmies, living in a climax forest, the main protein source is game meat. One conclusion to be drawn from this study is that unless families have access to true substitutes for bushmeat, any attempt to curtail bushmeat production may result in children suffering the consequences of protein deficiency—that is, slowed growth and learning delays. As of 2012, fish and domestic animals are the only plausible substitutes for bushmeat as a source of protein.

The firm conclusion to be drawn from this review is that there is a disjunction between the importance of bushmeat in human livelihoods and the likely sustainability of the source populations on which these livelihoods depend. Loss of bushmeat as a food resource or even as a source of income would have a range of effects on the welfare of the poor, including—for the majority of the human populations—very negative effects on the available protein supply. Bushmeat is an important commodity in the economies of producer states and an important livelihood asset for the poor, particularly (though not only) in West and Central Africa. It is a primary source of protein for many rural and urban families. It is also a major source of income for forest dwellers. Its value lies both in the absolute contribution that it makes to household welfare and in its role in household safety nets, food security, and risk mitigation.

At present off-take levels, however, there are major threats to the conservation of biodiversity. There is strong evidence that, within the range states, bushmeat is being depleted on an unprecedented scale. There has been a major transition in the scale of off-take from forest areas in the last century or so, with West and Central Africa being the most affected region. Extraction rates in West and Central Africa are orders of magnitude higher than those in, say, the Amazon, and much less likely to be sustainable.

The evidence suggests that this drawdown is likely to have negative consequences for future generations, in terms of both biodiversity and livelihoods effects. The effects are particularly evident in relation to the primates and slow-breeding, large-bodied mammals, which have major roles in forest ecology. While endangered primates comprise only a small percentage of the bushmeat trade, there are nevertheless grounds

for concern, in relation both to local extinctions and their effects on the genetic stock.

That much of the off-take consists of fast-reproducing "vermin" species does present the possibility of a sustainable trade (albeit passively and by default). Nevertheless, the predominance of such species in the trade can be misleading. The operation of an extinction filter needs to be considered—implying a threshold for the extinction of vulnerable species beyond which the trade is limited to non-threatened species and hence becomes sustainable. Where this threshold has already been passed there may be no problem in the sense that the damage has already been done. Where it has not yet been reached, however, then the dominance of non-threatened species in the trade cannot be taken to imply the lack of a conservation threat.

In regard to geographical considerations, all tropical regions have been affected by unsustainable hunting, though the problem has become most acute in the bushmeat heartlands of West and Central Africa. Latin America may present somewhat similar problems in the future (perhaps by the year 2030), although more optimistic scenarios are conceivable there. Asia presents a contrasting picture, to some degree, because of the reliance on large-scale wildlife trade involving long-distance, international commodity chains.

Given the characteristics of the threatened species, there are strong reasons to argue that the core mitigation response should be targeted on protected areas. Protected area management is problematic, however, particularly in the bushmeat heartlands. It cannot be claimed that this management has been very successful in securing the protected areas. Support is often weakest at the local level, among the populations whose buy-in is most important—indeed, essential—for success. Whatever the merits of protected areas as a conservation strategy, there remain many questions about their effectiveness in dealing with the issues under review.

Evidence also suggests that the long-term effects on the welfare of the poor could be severe, through the denial of an important protein source. It is difficult to predict this in a vacuum because (1) the future development of the societies in question is subject to many other influences and (2) the effects are likely to be of a "slow drip" type, which means that, where other protein sources are available (an important caveat), populations would have time to switch to new protein sources.

Resources

Books

Bennett, Elizabeth L., and John G. Robinson. *Hunting of Wildlife in Tropical Forests: Implications for Biodiversity and Forest Peoples.* Washington, DC: World Bank, Environment Department, 2000.

Bennett, Elizabeth L., and Madhu Rao. "Wild Meat Consumption in Asian Tropical Forest Countries: Is This a Glimpse of the Future for Africa?" In *Links between Biodiversity, Conservation, Livelihoods, and Food Security: The Sustainable Use of Wild Species for Meat*, edited by Sue Mainka and Mandar Trivedi,

39–44. Gland, Switzerland: International Union for Conservation of Nature, 2002.

Carney, Diana, ed. *Sustainable Rural Livelihoods: What Contribution Can We Make?* London: Department for International Development, 1998.

Chardonnet, Philippe, H. Fritz, N. Zorzi, and E. Féron. "Current Importance of Traditional Hunting and Major Contrasts in Wild Meat Consumption in Sub-Saharan Africa." In *Integrating People and Wildlife for a Sustainable Future*, edited by

John A. Bissonette and Paul R. Krausman, 304–307. Bethesda, MD: Wildlife Society, 1995.

Colyn, M. M., Akaibe Dudu, and Mankoto ma Mbaele. "Data on Small and Medium Scale Game Utilisation in the Rain Forest of Zaire." In *Wildlife Management in Sub-Saharan Africa: Sustainable Economic Benefits and Contribution towards Rural Development*. Paris: Fondation Internationale pour la Sauvegarde du Gibier, 1987.

Eves, Heather E., and Richard G. Ruggiero. "Socioeconomics and the Sustainability of Hunting in the Forests of Northern Congo (Brazzaville)." In *Hunting for Sustainability in Tropical Forests*, 427–454.

Fa, John E. "Hunted Animals in Bioko Island, West Africa: Sustainability and Future." In *Hunting for Sustainability in Tropical Forests*, 168–198.

Fa, John E., and Carlos A. Peres. "Game Vertebrate Extraction in African and Neotropical Forests: An Intercontinental Comparison." In *Conservation of Exploited Species*, edited by John D. Reynolds, Georgina M. Mace, Kent H. Redford, and John G. Robinson, 203–242. Cambridge, U.K.: Cambridge University Press, 2001.

Feer, François. "The Potential for Sustainable Hunting and Rearing of Game in Tropical Forests." In *Tropical Forests, People, and Food: Biocultural Interactions and Applications to Development*, 691–708.

Glanz, William E. "The Terrestrial Mammal Fauna of Barro Colorado Island: Censuses and Long-Term Changes." In *The Ecology of a Tropical Forest: Seasonal Rhythms and Long-Term Changes*, edited by Egbert G. Leigh Jr., A. Stanley Rand, and Donald M. Windsor. Washington, DC: Smithsonian Institution Press, 1982.

Hart, John A. "Impact and Sustainability of Indigenous Hunting in the Ituri Forest, Congo-Zaire: A Comparison of Unhunted and Hunted Duiker Populations." In *Hunting for Sustainability in Tropical Forests*, 106–153.

Hladik, Claude Marcel, Serge Bahuchet, and Igor de Garine, eds. *Food and Nutrition in the African Rain Forest*. Paris: UNESCO, 1990.

Hladik, Claude Marcel, Olga F. Linares, Hélène Pagezy, et al., eds. *Tropical Forests, People, and Food: Biocultural Interactions and Applications to Development*. Paris: Parthenon, 1993.

Koppert, Georgius J. A., Edmond Dounias, Alain Froment, and Patrick Pasquet. "Food Consumption in Three Forest Populations of the Southern Coastal Area of Cameroon: Yassa, Mvae, Bakola." In *Tropical Forests, People, and Food: Biocultural Interactions and Applications to Development*, 295–310.

Koppert, Georgius J. A., and Claude Marcel Hladik. "Measuring Food Consumption." In *Food and Nutrition in the African Rain Forest*.

Ntiamoa-Baidu, Yaa. *Wildlife and Food Security in Africa*. FAO Conservation Guide 33. Rome: Food and Agriculture Organization of the United Nations, 1997.

Peres, Carlos A. "Effects of Subsistence Hunting and Forest Types on Amazonian Primate Communities." In *Primate Communities*, edited by John G. Fleagle, Charles H. Janson, and Kaye E. Reed, 268–283. Cambridge, U.K.: Cambridge University Press, 1999.

Prescott-Allen, Robert, and Christine Prescott-Allen. *What's Wildlife Worth? Economic Contributions of Wild Plants and Animals to Developing Countries*. London: Earthscan, 1982.

Redford, Kent H. "Hunting in Neotropical Forests: A Subsidy from Nature." In *Tropical Forests, People, and Food: Biocultural Interactions and Applications to Development*, 227–248.

Robinson, John G., and Elizabeth L. Bennett, eds. *Hunting for Sustainability in Tropical Forests*. New York: Columbia University Press, 2000.

Robinson, John G., and Kent H. Redford. "Sustainable Harvest of Neotropical Forest Mammals." In *Neotropical Wildlife Use and Conservation*, edited by John G. Robinson and Kent H. Redford, 415–429. Chicago: Chicago University Press, 1991.

Scherr, Sara J., Andy White, and David Kaimowitz. *A New Agenda for Forest Conservation and Poverty Reduction: Making Forest Markets Work for Low-Income Producers*. Washington, DC: Forest Trends, 2004.

Steel, Elisabeth A. *Study of the Value and Volume of Bushmeat Commerce in Gabon*. Libreville, Gabon: World Wildlife Fund, 1994.

Waterman, Peter G., and Doyle B. McKey. "Herbivory and Secondary Compounds in Rain-Forest Plants." In *Tropical Rain Forest Ecosystems: Biogeographical and Ecological Studies*, edited by Helmut Lieth and Marinus J. A. Werger. Amsterdam: Elsevier, 1989.

Periodicals

Albrechtsen, Lise, John E. Fa, Brigid Barry, and David W. Macdonald. "Contrasts in Availability and Consumption of Animal Protein in Bioko Island, West Africa: The Role of Bushmeat." *Environmental Conservation* 32, no. 4 (2005): 340–348.

Apaza, Lilian, David S. Wilkie, Elizabeth Byron, et al. "Meat Prices Influence the Consumption of Wildlife by the Tsimane' Amerindians of Bolivia." *Oryx* 36, no. 4 (2002): 382–388.

Barnes, Richard, M. Agnagna, M. P. T. Alers, et al. "Elephants and Ivory Poaching in the Forests of Equatorial Africa: Results of a Field Reconnaissance." *Oryx* 27, no. 1 (1993): 27–34.

Barnes, Richard, and Sally Lahm. "An Ecological Perspective on Human Densities in the Central African Forests." *Journal of Applied Ecology* 34, no. 1 (1997): 245–260.

Bodmer, Richard E., John F. Eisenberg, and Kent H. Redford. "Hunting and the Likelihood of Extinction of Amazonian Mammals." *Conservation Biology* 11, no. 2 (1997): 460–466.

Brashares, Justin S., Peter Arcese, Moses K. Sam, et al. "Bushmeat Hunting, Wildlife Declines, and Fish Supply in West Africa." *Science* 306, no. 5699 (2004): 1180–1183.

Brown, David, and Andrew Williams. "The Case for Bushmeat as a Component of Development Policy: Issues and Challenges." *International Forestry Review* 5, no. 2 (2003): 148–155. Republished in *The Earthscan Reader in Forestry and Development*, edited by Jeffrey Sayer. London: Earthscan, 2005.

Butynski, Thomas M., and Wolfgang von Richter. "In Botswana Most of the Meat Is Wild." *UnaSylva* 26, no. 106 (1974): 24–29.

Chambers, Robert. "Sustainable Livelihoods, Environment, and Development: Putting Poor Rural People First." IDS Discussion Paper 240, Institute of Development Studies, University of Sussex, Brighton, UK, 1987.

Coe, M. J., David H. M. Cumming, and J. Phillipson. "Biomass and Production of Large African Herbivores in Relation to Rainfall and Primary Production." *Oecologia* 22, no. 4 (1976): 341–354.

Cowlishaw, Guy, Samantha Mendelson, and J. Marcus Rowcliffe. "Evidence for Post-depletion Sustainability in a Mature Bushmeat Market." *Journal of Applied Ecology* 42, no. 3 (2005): 460–468.

De Merode, Emmanuel, Katherine Homewood, and Guy Cowlishaw. "Wild Resources and Livelihoods of Poor Households in Democratic Republic of Congo." ODI Wildlife Policy Briefing 1, Overseas Development Institute, London, 2003.

De Merode, Emmanuel, Katherine Homewood, and Guy Cowlishaw. "The Value of Bushmeat and Other Wild Foods to Rural Households Living in Extreme Poverty in Democratic Republic of Congo." *Biological Conservation* 118, no. 5 (2004): 573–581.

Dubost, Gérard. "Un aperçu sur l'écologie du chevrotain africain *Hyemoschus aquaticus* Ogilby, Artiodactyle Tragulidé." *Mammalia* 42, no. 1 (1978): 1–62.

Dubost, Gérard. "The Size of African Forest Artiodactyls as Determined by the Vegetation Structure." *African Journal of Ecology* 17, no. 1 (1979): 1–17.

Dubost, Gérard. "L'écologie et la vie sociale du céphalophe bleu (*Cephalophus monticola* Thunberg), petit ruminant forestier africain." *Zeitschrift für Tierpsychologie* 54, no. 3 (1980): 205–266.

East, Tasmyn, Noëlle F. Kümpel, Eleanor Jane Milner-Gulland, and J. Marcus Rowcliffe. "Determinants of Urban Bushmeat Consumption in Río Muni, Equatorial Guinea." *Biological Conservation* 126, no. 2 (2005): 206–215.

Fa, John E., Dominic Currie, and Jessica Meeuwig. "Bushmeat and Food Security in the Congo Basin: Linkages between Wildlife and People's Future." *Environmental Conservation* 30, no. 1 (2003): 71–78.

Fa, John E., Javier Juste, Jaime Pérez del Val, and Javier Castroviejo. "Impact of Market Hunting on Mammal Species in Equatorial Guinea." *Conservation Biology* 9, no. 5 (1995): 1107–1115.

Fa, John E., Carlos A. Peres, and Jessica Meeuwig. "Bushmeat Exploitation in Tropical Forests: An Intercontinental Comparison." *Conservation Biology* 16, no. 1 (2002): 232–237.

Fa, John E., Sarah F. Ryan, and Diana J. Bell. "Hunting Vulnerability, Ecological Characteristics, and Harvest Rates of Bushmeat Species in Afrotropical Forests." *Biological Conservation* 121, no. 2 (2005): 167–176.

Fa, John E., Sarah Seymour, Jef Dupain, et al. "Getting to Grips with the Magnitude of Exploitation: Bushmeat in the Cross–Sanaga Rivers Region, Nigeria and Cameroon." *Biological Conservation* 129, no. 4 (2006): 497–510.

Fa, John E., and Juan E. García Yuste. "Commercial Bushmeat Hunting in the Monte Mitra Forests, Equatorial Guinea: Extent and Impact." *Animal Biodiversity and Conservation* 24, no. 1 (2001): 31–52.

Fa, John E., Juan E. García Yuste, and Ramon Castelo. "Bushmeat Markets on Bioko Island as a Measure of Hunting Pressure." *Conservation Biology* 14, no. 6 (2000): 1602–1613.

Feer, François. "Stratégies écologiques de deux espéces de Bovidés sympatriques de la forêt sempervirente africaine (*Cephalophus callipygus* et *C. dorsalis*): Influence du rythme d'activité." PhD diss., Université Pierre et Marie Curie, 1988.

Jerozolimski, Adriano, and Carlos A. Peres. "Bringing Home the Biggest Bacon: A Cross-Site Analysis of the Structure of Hunter-Kill Profiles in Neotropical Forests." *Biological Conservation* 111, no. 3 (2003): 415–425.

Juste, Javier, John E. Fa, Jaime Pérez del Val, and Javier Castroviejo. "Market Dynamics of Bushmeat Species in Equatorial Guinea." *Journal of Applied Ecology* 32, no. 3 (1995): 454–467.

Lahm, Sally A. "Ecology and Economics of Human/Wildlife Interaction in Northeastern Gabon." PhD diss., New York University, 1993.

Laurance, William F., Barbara M. Croes, Landry Tchignoumba, et al. "Impacts of Roads and Hunting on Central African Rainforest Mammals." *Conservation Biology* 20, no. 4 (2006): 1251–1261.

Macdonald, David W., Paul J. Johnson, Lise Albrechtsen, et al. "Bushmeat Trade in the Cross–Sanaga Rivers Region: Evidence for the Importance of Protected Areas." *Biological Conservation*. Published electronically January 11, 2012.

McKey, Doyle B., J. Stephen Gartlan, Peter G. Waterman, and Gillian M. Choo. "Food Selection by Black Colobus Monkeys (*Colobus satanas*) in Relation to Plant Chemistry." *Biological Journal of the Linnean Society* 16, no. 2 (1981): 115–146.

McSweeney, Kendra. "Forest Product Sale as Natural Insurance: The Effects of Household Characteristics and the Nature of Shock in Eastern Honduras." *Society and Natural Resources* 17, no. 1 (2004): 39–56.

McSweeney, Kendra. "Forest Product Sale as Financial Insurance: Evidence from Honduran Smallholders." ODI Wildlife Policy Briefing 10, Overseas Development Institute, London, 2005. Accessed March 15, 2012. http://www.odi.org.uk/resources/docs/3292.pdf.

Mendelson, Samantha, Guy Cowlishaw, and J. Marcus Rowcliffe. "Anatomy of a Bushmeat Commodity Chain in Takoradi, Ghana." *Journal of Peasant Studies* 31, no. 1 (2003): 73–100.

Noss, Andrew J. "The Impacts of Cable Snare Hunting on Wildlife Populations in the Forests of the Central African Republic." *Conservation Biology* 12, no. 2 (1998): 390–398.

Novaro, Andrés J., Martín C. Funes, and R. Susan Walker. "An Empirical Test of Source–Sink Dynamics Induced by Hunting." *Journal of Applied Ecology* 42, no. 5 (2005): 910–920.

Oates, John F. "Habitat Alteration, Hunting, and the Conservation of Folivorous Primates in African Forests." *Australian Journal of Ecology* 21, no. 1 (1996): 1–9.

Oates, John F., George H. Whitesides, A. Glyn Davies, et al. "Determinants of Variation in Tropical Forest Primate Biomass: New Evidence from West Africa." *Ecology* 71, no. 1 (1990): 328–343.

Peres, Carlos A. "Effects of Subsistence Hunting on Vertebrate Community Structure in Amazonian Forests." *Conservation Biology* 14, no. 1 (2000): 240–253.

Peres, Carlos A. "Synergistic Effects of Subsistence Hunting and Habitat Fragmentation on Amazonian Forest Vertebrates." *Conservation Biology* 15, no. 6 (2001): 1490–1505.

Peres, Carlos A., and Paul M. Dolman. "Density Compensation in Neotropical Primate Communities: Evidence from 56 Hunted and Nonhunted Amazonian Forests of Varying Productivity." *Oecologia* 122, no. 2 (2000): 175–189.

Plumptre, Andrew John. "Plant-Herbivore Dynamics in the Birungas." PhD diss., University of Bristol, 1991.

Redford, Kent H. "The Empty Forest." *BioScience* 42, no. 6 (1992): 412–422.

Robinson, John G., and Elizabeth L. Bennett. "Having Your Wildlife and Eating It Too: An Analysis of Hunting Sustainability across Tropical Ecosystems." *Animal Conservation* 7, no. 4 (2005): 397–408.

Robinson, John G., and Richard E. Bodmer. "Towards Wildlife Management in Tropical Forests." *Journal of Wildlife Management* 63, no. 1 (1999): 1–13.

Roldán, Alejandra I., and Javier A. Simonetti. "Plant-Mammal Interactions in Tropical Bolivian Forests with Different Hunting Pressures." *Conservation Biology* 15, no. 3 (2001): 617–623.

Starkey, Malcolm. "Commerce and Subsistence: The Hunting, Sale, and Consumption of Bushmeat in Gabon." PhD diss., Cambridge University, 2004.

Waterman, Peter G., Jane A. M. Ross, Elizabeth L. Bennett, and A. Glyn Davies. "A Comparison of the Floristics and Leaf Chemistry of the Tree Flora in Two Malaysian Rain Forests and the Influence of Leaf Chemistry on Populations of Colobine Monkeys in the Old World." *Biological Journal of the Linnean Society* 34, no. 1 (1988): 1–32.

White, Lee J. T. "Biomass of Rain Forest Mammals in the Lopé Reserve, Gabon." *Journal of Animal Ecology* 63, no. 3 (1994): 499–512.

Wilkie, David S., and Julia F. Carpenter. "Bushmeat Hunting in the Congo Basin: An Assessment of Impacts and Options for Mitigation." *Biodiversity and Conservation* 8, no. 7 (1999): 927–955.

Wilkie, David S., and Bryan Curran. "Why Do Mbuti Hunters Use Nets? Ungulate Hunting Efficiency of Archers and Net-Hunters in the Ituri Rain Forest." *American Anthropologist*, 93, no. 3 (1991): 680–689.

Wilkie, David S., Ellen Shaw, Fiona Rotberg, et al. "Roads, Development, and Conservation in the Congo Basin." *Conservation Biology*, 14, no. 6 (2000): 1614–1622.

Woodroffe, Rosie, and Joshua R. Ginsberg. "Edge Effects and the Extinction of Populations inside Protected Areas. *Science* 280, no. 5372 (1998): 2126–2128.

Wright, S. Joseph, Horacio Zeballos, Iván Domínguez, et al. "Poachers Alter Mammal Abundance, Seed Dispersal, and Seed Predation in a Neotropical Forest." *Conservation Biology* 14, no. 1 (2000): 227–239.

John E. Fa

Ecological economics

Ecological economics arose in the final decades of the twentieth century out of concerns for environmental protection and economic sustainability. It was largely a response to a real or perceived lack of physical and biological underpinnings in neoclassical economics. It was also intended to infuse economics with a moral philosophy, in contrast with the amoral implications of models portraying humans as rational, utility-maximizing automatons.

Ecological economics is a transdisciplinary endeavor, incorporating and synthesizing concepts and findings from an array of natural and social sciences. Of particular importance are the laws of thermodynamics and basic principles of ecology. Limits to economic growth are understood thoroughly only via the first two laws of thermodynamics. The first law establishes that there is a limit to the inputs required for economic production, and the second establishes that there are limits to the efficiency with which those inputs may be transformed into goods and services. Similarly, ecological concepts such as trophic levels, niche breadth, and competitive exclusion are required for a thorough understanding of the relationship between the human economy and the diversity of nonhuman species, or the so-called economy of nature.

Given its roots in the natural sciences and moral philosophy, the major themes of ecological economics are scale, distribution, and allocation. Scale refers to the size of the human economy relative to its containing and sustaining ecosystem. Because the scale is limited—that is, there is a limit to economic growth—the distribution of wealth is a topic that must be addressed, with public policy if necessary, if poverty is to be alleviated. Prioritizing scale and distribution distinguishes ecological economics from neoclassical economics, which posits unlimited economic growth and therefore implies that a "rising tide lifts all boats." In neoclassical economics, efficient allocation of resources among producers is the primary concern. Efficient allocation of labor and capital, especially, is thought to help maximize production and boost rates of economic growth. In ecological economics, efficient allocation is also recognized as an important objective, but the importance of land and natural resources as factors of production is emphasized. Natural resources are found to be only partially substitutable by labor and manufactured capital. In ecological economics, individual natural resources are

scrutinized to determine whether they have the properties necessary for being allocated efficiently in the market. Many natural resources and services provided by ecosystems (e.g., pollination, climate regulation, water purification) are often found to be lacking such properties and are therefore overused or ignored unless protected by forces outside the market.

Based on its themes and findings, the application of ecological economics produces a number of distinctive policy implications. Some new policies are recommended and many existing policies must be reformed if the goals of sustainable scale, fair distribution, and efficient allocation of resources are to be met. For sustainable scale, the vast array of fiscal, monetary, and trade policies that are designed to stimulate economic growth may be readjusted gradually to make them conducive to a steady-state economy with stabilized production and consumption of goods and services in the aggregate. Additional policies such as caps on the extraction of resources and pollution may be necessary for assuring sustainable scale and more closely approximating optimal scale.

Facing limits to growth, societies are likewise faced with challenging choices with regard to ongoing efforts to deal with poverty. Progressive taxes are a traditional method for doing so. Caps on income and wealth, minimum income, and the distribution of returns from natural resources are additional options proffered in ecological economics.

For efficient allocation of resources, many of the policy recommendations stemming from environmental economics, or neoclassical economics as applied to environmental issues, are supported by ecological economics as well. These policies are focused on correcting for market imperfections in the management of natural resources where it is feasible to do so. The contribution of ecological economics to the use of these corrective policies is primarily in gaining a deeper understanding of the components, structures, and functions of ecosystems that need to be evaluated in order to identify a corrective course of action. This understanding is usually procured through the collaboration of economists with ecologists or by the cross-training of individuals in ecology and economics, and it is often used in estimating values of natural capital and ecosystem services in monetary terms.

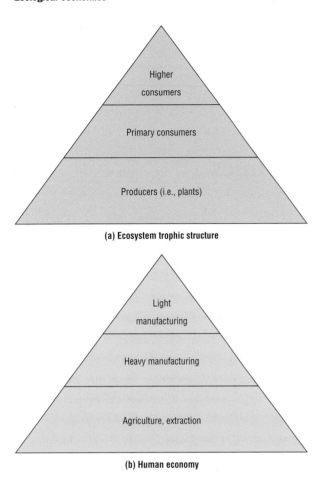

(a) Ecosystem trophic structure

Higher
consumers

Primary consumers

Producers (i.e., plants)

(b) Human economy

Light
manufacturing

Heavy manufacturing

Agriculture, extraction

Trophic levels diagram from: Czech, B. 2008. Prospects for reconciling the conflict between economic growth and biodiversity conservation with technological progress. Conservation Biology 22(6):1389-1398. Reproduced by permission of Gale, a part of Cengage Learning.

With such estimates, markets can be designed or modified to allocate the available resources more efficiently and equably. In ecological economics, however, the need for nonmarket mechanisms to control the allocation or conservation of some natural resources and ecosystem services is readily recognized, and regulations are viewed as efficient policy tools in many such cases. This contrasts with the neoclassical view in which faith in the market tends to dissuade the polity from adopting conservation regulations.

Ecological economics will be one of the most important human endeavors of the twenty-first century as nations and the world population approach, breach, and adjust to supply shocks such as peak oil and environmental crises such as climate change. For numerous reasons, including the vast reach of neoclassical economists in academia, commerce, and government, ecological economics will be challenged to avoid a preoccupation with natural capital valuation exercises at the expense of its distinguishing emphasis on sustainable scale. Ecological economics has come along none too soon, given that the steady-state economy, as a macroeconomic policy goal, must also be reconciled with legitimate social calls for economic degrowth.

Historical development of ecological economics

Ecological economics arose in response to the mounting environmental problems that were witnessed by the public and documented by authors in books such as Rachel Carson's *Silent Spring* (1962), Barry Commoner's *The Closing Circle* (1971), and *The Limits to Growth* (1972), by Donella H. Meadows and the Club of Rome. Many observers were disappointed with the approach of conventional or neoclassical economics to environmental degradation, exemplified by Howard J. Barnett and Chandler Morse (*Scarcity and Growth* [1963]), who believed that prices in a well-functioning market would prevent crippling resource shortages. Neoclassical economists and business professors such as Julian Simon (1932–1998) invariably prescribed economic growth as the solution to virtually all social problems, even environmental problems and especially pollution. According to them, conflicts between economic growth and environmental protection could be solved via technological progress.

One of the first well-trained economists to part ways with the neoclassical school on environmental grounds was Herman E. Daly, whose *Steady-State Economics* (1977) provided an alternative vision for a sustainable, equitable economy in balance with the environment. Daly was a professor of economics at Louisiana State University when he wrote *Steady-State Economics*, and he served as a senior economist at the World Bank from 1988 to 1994. His professional leadership and writing talents attracted many other economists, as well as ecologists concerned with environmental protection. Ecologists found in *Steady-State Economics* a refreshing familiarity with the natural sciences as well as economic principles. Daly, a protégé of Nicholas Georgescu-Roegen (*The Entropy Law and the Economic Process* [1971]), was particularly adept with the laws of thermodynamics and the implications of thermodynamics for economic growth. Other prominent and productive figures with similar emphases and outlooks are Kenneth E. Boulding, Robert Ayres, and E.F. Schumacher.

Key figures in the development of ecological economics assembled during the 1980s, most notably in Stockholm in 1982 (organized by AnnMari Jansson) and Barcelona in 1987 (organized by Joan Martinez-Alier). These meetings helped the participants identify common ground, complementary skills, and major challenges to developing a more ecologically sound theory and practice of economics. Many of the attendees later became prominent contributors to the ecological economics literature and related institutions. One of them was Robert Costanza, who took the lead in establishing the International Society for Ecological Economics (ISEE) in 1988. A student of the systems ecologist Howard T. Odum (1924–2002), Costanza brought his own mastery of thermodynamics with additional ecological and economic applications. Costanza served as the editor of *Ecological Economics* from its inception in 1989 until 2002 and has been one of the most prolific authors in the ecological economics literature at large.

The first ISEE conference was held in 1990, with biannual conferences held thereafter. By 2007 there were nine ISEE-affiliated regional societies representing Africa,

Herman Daly and Donella "Dana" Meadows, founders of ecological economics. Daly and colleagues clarified the relationship between the economy and Earth with a diagram which was simple but powerful for illustrating limits to growth. AP Photo/Eric Roxfelt.

Argentina and Uruguay, Australia and New Zealand, Brazil, Canada, Europe, India, Russia, and the United States. (There was also a nonaffiliated Chinese Ecological Economics Society and an Iberian and Latin American Network of Ecological Economics.)

With regard to the broader sweep of history, one of the more noteworthy roots of ecological economics was the work of François Quesnay (1694–1774) and the physiocrats of late-eighteenth-century France. Quesnay was brought into the king's court as a physician and became a general adviser. He developed a strong interest in agriculture and, with his medical background, viewed the French economy as a circulatory system of goods and services, as described in his *Tableau économique* (1758; *Economic Table*). The most important point of the *Tableau* was Quesnay's designation of agriculture as the sole source of economic production, with all other economic activities deriving from that production.

The Scottish economist Adam Smith (1723–1790) met Quesnay and studied the *Tableau* prior to writing *The Wealth*

of Nations (1776). Although he disagreed with Quesnay's categorization of agriculture as the sole source of production, Smith nevertheless described how agricultural surplus was necessary for the division of labor. There was no argument about the primacy of agricultural surplus among the classical economists, even in the midst of the Industrial Revolution, but as their studies of political economy splintered into neoclassical economics and political science at the dawn of the twentieth century, microeconomics eclipsed the broader, integrated vision of the economy. Future economists would not be as familiar with the interrelationships among economic sectors, much less with the natural sciences or agricultural practices. Meanwhile, much of the vacuum in political economy was occupied by Marxists and followers of the American economist Henry George (1839–1897), the latter calling for a singular and substantial tax on land rents in *Progress and Poverty* (1879).

When George followed up on *Progress and Poverty* with political activism and attained broad support from populist followers, land barons teamed with hand-picked economists to play down the role of land in economic production in order to refocus tax policy on wages. Many economics departments in the United States were in their formative stages, and the anti-George backlash manifested itself in the development of neoclassical economics. By the time macroeconomics was borne of the Keynesian revolution in the second quarter of the twentieth century, agricultural economics was consigned to its own corridors. Among the broader economics community, land was generally overlooked as a factor of production while economists focused on labor and capital. Wartime economics were especially focused on capital mobilization, whereas the Great Depression prompted a focus on labor and employment. Furthermore, the developed countries were urbanizing at a rapid rate, with citizens increasingly removed from the land. These developments in the social and political spheres help to explain the growing propensity of twentieth-century neoclassical economists to underestimate the magnitude and implications of natural resource scarcity and environmental deterioration. Conversely, in ecological economics, the fundamental requirement of agricultural surplus for a fully developed economy—and an increasing surplus for a *growing* economy—is a cornerstone in the theoretical foundation.

One classical economist with exceptional relevance to ecological economics is the English philosopher and economist John Stuart Mill (1806–1873). In *Principles of Political Economy* (1848), Mill synthesized the state of the art in economics to that time. He was also perhaps the first economist to advance with hope the notion of the *stationary state* as opposed to warning of it as had Thomas Malthus and David Ricardo, who pointed gloomily to the collision of population growth and agricultural capacity, prompting observers to refer to economics as the "dismal science." Mill believed that an informed human citizenry could come to control its population, achieve a comfortable standard of living, and then turn its attention to matters of social justice. The stationary state—a nongrowing, nondeclining economy—is synonymous for practical purposes with the steady-state economy of ecological economics.

Scenes from the Industrial Revolution, such as "Coalbrookdale at Night" accompanied the "dismal science" of Thomas Malthus and other classical economists. © SSPL/Science Museum/The Image Works.

John Stuart Mill called for the "stationary state;" not in cultural affairs but in levels of economic activity. © Pictorial Press Ltd/Alamy.

The role of Marxist thought in the development of ecological economics is not entirely clear. The founders of ecological economics recognized the preoccupation with growth in capitalist (and other) economies as a major threat to the environment and society, so "green" Marxists were natural allies. Marx himself, however, appeared to have substantial faith in technology to obviate limits to growth; his critique of capitalism stemmed more from his thoughts on the concentration of power and the maldistribution of wealth. One of the legacies of Marxist versus capitalist ideology was an arms race between the United States and the Soviet Union, a Cold War in which the score was kept in terms of economic production. The preoccupation of these powers with economic growth was one factor in speeding the human race into environmental deterioration—and into the study of ecological economics.

Approach and philosophy of ecological economics

The general approach and philosophy of any endeavor are interrelated, so they are treated here in the same section. Ecological economics has an approach and philosophy that distinguishes it from neoclassical economics and from most heterodox economics traditions such as the Austrian school, Keynesian economics, and Marxism. The approach and philosophy of ecological economics may be concisely described as transdisciplinary and normative, respectively.

Transdisciplinarity

Ecological economics is sometimes referred to as a *transdisciplinary* endeavor to distinguish it from a long line of *interdisciplinary* studies that arose in academia during the latter decades of the twentieth century. The movement toward integration and synthesis of disciplinary studies in some corners of academia resulted from a concern that the policy implications stemming from reductionist science were impractical or misguided. Even numerous efforts at interdisciplinary studies were criticized for merely coupling reductionist disciplines, however, and the transdisciplinary approach was advanced as cooperative problem-solving with dynamic integration of philosophical perspectives and scientific findings.

The concern with disciplinary reduction was especially warranted with regard to the ecological aspects of economic systems, because many national economies had grown to an extent that pushed the limits of sustainability, and global environmental problems related to economic production such as depletion of the ozone layer, biodiversity loss, and climate change were becoming evident. Most ecologists knew little about the economic processes giving rise to environmental problems, and most economists knew little about the severity or economic implications of ecological degradation. Many ecologists and economists knew little about the political and sociological influences on their studies and their occasional policy recommendations. It was in this context that Daly, Costanza, Richard B. Norgaard, and others advanced the concept of transdisciplinarity, which may itself be considered a theme or an emphasis of ecological economics. Nevertheless, a transdisciplinary approach assumes there is something to apply it to, and ecological economics applies it to three primary themes: scale, distribution, and allocation.

Ends, means, and a normative stance

Perspectives on human nature and civil rights strongly influence how economic theory is developed, interpreted, and applied. Although there is no consensus in ecological economics about the spiritual origins or ethical nature of humans, there is a general consensus that economics is irreducibly a normative endeavor, in study and in practice. This distinguishes ecological economics from neoclassical economics, in which humans are modeled as *Homo economicus*, self-interested, utility-maximizing automatons, with utility indicated by the consumption of goods and services. In ecological economics, humans are viewed as having multifarious motives that derive not only from economic exigencies but also from evolutionary, cultural, and spiritual factors deeply embedded in the human psyche. Although the consumption behavior of humans may be modeled as an academic exercise, such modeling exercises produce few practical or dependable policy implications.

Given a broader view of human nature, a spectrum of ends and means helps to place the academic terrain in context. Sciences that reduce the sphere of observation to physical and biological minutia provide insights to the *means* by which various human goals and objectives may be pursued. However, the meaning of life and the corresponding *ends* are beyond science to ascertain and are often manifested in or interpreted through religion. Social sciences, interdisciplinary studies, and transdisciplinary approaches help to bridge the gap from reductionist science to meaningful lives—that is, from means to ends. For example, physics is a study of ultimate means; theology is a study of ultimate ends; and social sciences, including economics, are studies of intermediate means (e.g., economic institutions) and ends (e.g., economic welfare).

Ecological economics explicitly and consciously encompasses a longer portion of the ends–means spectrum than neoclassical economics does. As ecological economics has arisen out of environmental concerns, the ecological expertise of its practitioners has been coupled with a closer analysis of all natural sciences of particular relevance to economic affairs, such as the laws of thermodynamics. In other words, ecological economics is concerned with ultimate means, virtually by definition, and how those ultimate means affect human economic prospects. Meanwhile, the normative stance of ecological economics requires a consideration of ultimate ends, including religious callings and needs. This is an ironic aspect of ecological economics to the extent that ecologists are often characterized as atheistic scholars with a purely evolutionary view of *Homo sapiens*. There are logical and faith-based reasons, however, for linking ultimate means and ultimate ends in economic affairs, as revealed in the section below on the distribution of wealth.

Themes and emphases in ecological economics

In conventional economics textbooks, economics is defined as the allocation of scarce resources among competing end uses. Neoclassical economics tends to be focused on the issue of efficiency—that is, the efficient allocation of resources. Neoclassical economists acknowledge the scarcity of resources at any given point in time—it is because of scarcity that efficient allocation is called for—but do not often acknowledge long-run scarcity of resources. Neoclassical economists usually posit that innovation and new technology continuously push back the limits to production and consumption that are temporarily imposed by scarcity.

Ecological economics, by contrast, emphasizes the scarce resources that must be allocated. Long-run limits are recognized as well as short-term limits, giving rise to the scale issue. This acknowledgment of long-run limits to growth also leads to a strong concern about the distribution of wealth (as is shown below). The scale issue and the distribution of wealth provide the context within which allocative efficiency is assessed.

The scale issue

As noted in the historical background section above, the importance of land as a factor of production has been unrecognized or played down in neoclassical economics. In economics and business textbooks the economy is often portrayed as a circular flow of money between firms and households. In the basic circular flow model, households provide labor for firms, while firms possess the capital that, combined with labor, is required for the production of goods and services. Money passes from firms to households in the form of wages, which are eventually spent on the goods and services produced by firms.

In more detailed models of the circular flow, other aspects of the economy are introduced, either as *leakages* from the flow (e.g., savings) and *injections* into the flow (e.g., investment) or as other entities that occupy the circle. For example, a Keynesian version of the circular flow includes the government, which taxes firms and households, pays wages to some individuals, and purchases goods and services from firms. Most circular flow models, however, do not illuminate the extraction of natural resources used as inputs to the production process, much less the environment as the context in which the economy functions.

In ecological economics, the circular exchange between firms and households is acknowledged, but graphical models of the economy emphasize the context within which this exchange transpires. The economy with all its firms, individuals, and government sectors is shown to exist within its containing, sustaining ecosystem. The ecosystem is shown to provide the energy (primarily the solar energy required for photosynthesis and therefore agriculture) and the natural resources (such as water, timber, and minerals) that are required for the production of consumer goods and services and for the manufacturing of capital and infrastructure. As important, the ecosystem is shown to absorb the wastes and pollutants of the economic production process. This graphical image of the economy creates an awareness of the primacy of land as a source of economic inputs and the importance of the environment as a sink for pollutants.

This image also helps one to recognize and appreciate the issue of scale, which in ecological economics refers to the size of the economy relative to its containing, sustaining ecosystem. The concept of scale raises numerous analytical questions with increasingly important policy implications. The most pressing questions are the following: (1) What is the

maximum sustainable scale? (2) What is the optimum scale? Both of these questions may be asked at local, national, regional, or global levels.

Considerations of maximum sustainable scale are enlightened by the ecological literature pertaining to carrying capacity, which refers to the number of animals an ecosystem may support. For any species, carrying capacity is determined by a mix of welfare and decimating factors. For wildlife species, the welfare factors are those that comprise a species' habitat: food, water, cover, and space. Decimating factors include predators, diseases, and severe weather. Carrying capacity for the typical species of wildlife is expressed in terms of the number of individuals the ecosystem may support.

In ecological economics, the relevance of carrying capacity to *Homo sapiens* is highlighted. Humans, however, differ from other animal species with regard to the use or consumption of habitat components per individual. In fact, per capita consumption among humans varies by orders of magnitude. Therefore, a better metric (than numbers of humans) for expressing human carrying capacity is gross domestic product (GDP), which is an indicator of human population and per capita consumption. In other words, GDP is a reasonably good indicator of the size of the human economy—that is, the level of production and consumption of goods and services in the aggregate. As such, it is also a good starting point for determining the scale of the economy (i.e., the size of the economy relative to the size of the ecosystem).

The fact that GDP is expressed in value units should not lead to the false conclusion that it is not a physical measure with ecological implications. GDP is a value-based aggregate of physical goods and services. A dollar's worth of X is a physical quantity thereof, and GDP is an aggregate index of physical quantities. The accuracy and precision with which GDP represents physical activity and throughput is an issue requiring further attention and research (see the "Ecological implications of money volumes and flows" section below).

The scale issue encompasses all aspects of the economy/ecosystem relationship—pollution, crowding, climate stability, and so on—and one aspect that has received substantial attention is biodiversity conservation. Conservation biologists have contributed to ecological economics by describing the principles of ecology that are most relevant to the human/biodiversity relationship, such as niche breadth and competitive exclusion. For example, they have described how, because of the tremendous breadth of the human niche, which expands via new technology, the human economy grows at the competitive exclusion of nonhuman species in the aggregate. These and related principles have led professional, scientific societies such as the Wildlife Society and the American Society of Mammalogists to take policy positions on the *fundamental conflict* between economic growth and biodiversity conservation. The word *fundamental*, in this context, indicates that the conflict is based on principles of physics and ecology, and not mere observation.

Meanwhile, ecologists and economists have teamed up to describe, quantify, and estimate the economic value of ecological services provided by nonhuman species and other features of the natural environment. For example, many

Agricultural, extractive, and information sectors. Farming Alfalfa in Saudi Arabia. Circle pivot irrigation, fed by wells pumping fossil water from aquifer. © George Steinmetz/Corbis.

species are beneficial to the human economy because, during the courses of their life cycles, they incidentally pollinate wild and domestic plants that are valued by humans for food and fiber. When the fundamental conflict between economic growth and biodiversity conservation is recognized in tandem with the value of the ecological services provided by nonhuman species, then economic growth is recognized as a threat not only to biodiversity but also to the continued functioning of the human economy.

Maximum sustainable scale, then, cannot be estimated without an understanding of the following:

1. the natural resource stocks and ecological services provided by nature, collectively referred to as *natural capital*;

2. how natural capital stocks and services are converted or used up in the process of economic growth;

3. to what extent natural capital is substitutable by human technology; and

4. the prospects for human technology to progress in a manner and at a rate sufficient for finding, and putting into production, substitutes for natural capital.

All four of these topics are highly complex, and ecological economists do not presume that humans will develop a thorough and accurate synthesis, espousing instead the *precautionary principle* in environmental and economic management. Nevertheless, meaningful approaches to the assessment of scale have been developed to inform citizens and policymakers. An example is the ecological footprint concept, pioneered by Mathis Wackernagel and William E. Rees, authors of *Our Ecological Footprint* (1996), which is used to demonstrate the amount of area required to support human economies. Ecological footprinting and related analyses have centered on inventories of natural capital and the natural capital requirements of economic activity.

The ecological footprint literature has indicated that many national economies, as well as the global economy, are already operating beyond their maximum sustainable scale. For example, some estimates suggest that, if all humans on Earth consumed at the same level per capita as Americans, it would take twenty-three Earth equivalents to sustain them. Globally, other estimates suggest that the level of per capita consumption at the dawn of the twenty-first century would require four Earth equivalents to be sustainable. According to these types of studies, humans have been able to consume at levels higher than those sustainable in the long run only because they have been using natural capital at a rate too fast for its replenishment or replacement; that is, humans have been liquidating natural capital such as fossil fuels (especially petroleum).

Ecologists, too, have long recognized that species may exist for periods of time at population sizes that exceed their long-run carrying capacity. In many such cases, however, the result is a long-term reduction in carrying capacity, such as when an ungulate species decimates its food source and damages the

soils, leading to erosion and the development of a different type of ecosystem that is not as supportive of the ungulate species.

Many neoclassical economists and others sometimes referred to as *technological optimists* have opined that the concept of carrying capacity does not apply to humans because, unlike other animals, humans are able to manipulate their environment and develop increasingly efficient modes and methods of production. This belief has spawned a long-running argument about the limits to growth. In ecological economics, which has some of its roots in the work of Georgescu-Roegen, the first two laws of thermodynamics are invoked to refute the notion of perpetual growth. The first law and its derivatives establish that neither energy nor matter may be created nor destroyed. From the second, or entropy, law comes the implication that it is impossible to achieve (much less exceed) 100 percent efficiency in an economic production process. Taken together, these two laws of thermodynamics imply that there is an absolute limit to economic growth. A related conclusion is that economic growth may not be continuously reconciled with environmental protection via technological progress and that apparent, intermediate reconciliation is often overestimated or nonexistent when all of the environmental impacts are entered into account.

For all of the complexities in determining maximum scale and maximum sustainable scale, limits to growth are becoming more evident, pervasive, and politically accepted. As a result, many ecological economists have turned their analytical focus to optimum scale. The starting point for assessing optimum scale is the equimarginal principle of maximization, a staple concept in microeconomics that tells the firm to stop producing as the costs of production rise to match the revenue obtained from each new (or marginal) unit produced. Extending this logic to the economy at large, ecological economists recognize that the costs (in a broad sense) of growing an economy eventually exceed the benefits when the interests of society at large are considered. For the sake of maximizing social welfare, then, an economy should cease growing when the marginal disutility of growth has risen to the level of the marginal utility of growth. Economic growth beyond that point no longer serves as a net benefit to society; therefore, some ecological economists refer to it as *uneconomic growth*. (There is some debate among ecological economists about the merits of using the term *uneconomic growth*, especially in the policy arena, because *economic growth* is used in the vernacular and by conventional economists to refer to increasing production and consumption of goods and services, regardless of the net benefits to society.)

In ecological economics, most scholars concur that GDP is a reasonably good indicator of the size of an economy, but all are united in noting that GDP is not a good indicator of social welfare—thus the concept of uneconomic growth. Therefore, other metrics have been developed to provide clues about optimum scale. For example, in the 1980s Herman E. Daly and John B. Cobb Jr. developed an Index of Sustainable Economic Welfare (ISEW) that accounts not only for the monetary value of final goods and services (as with GDP) but also for the natural capital that has been liquidated in the economic process. Stemming from Daly and Cobb's work, the public policy think tank, Redefining Progress, based in Oakland, California, developed the Genuine Progress Indicator (GPI) in the 1990s. The GPI incorporates social factors of human welfare beyond purely economic factors. For numerous nations with the appropriate available data, when the GPI (or ISEW) is plotted against GDP, GPI has been stagnant or declining since the 1970s, even as GDP has continued to grow, suggesting growth beyond the optimum. Many other indicators in various stages of development may be used in similar assessments, such as the long-established Human Development Index or the more recent gross national happiness. The term *gross national happiness* was coined by the king of Bhutan in 1972, and the concept received substantial theoretical and empirical analysis between the mid-1990s and 2012.

Distribution of wealth

A useful way to view the distribution of wealth as an issue in ecological economics is to consider briefly how distribution is viewed in neoclassical economics, in which the belief in unlimited economic growth leads to the attitude or philosophy that "a rising tide lifts all boats." In other words, because the scale issue does not exist in neoclassical economics, neoclassical economists assume that poverty may be solved by growing the economy ever larger so that financial wealth may trickle down via employment opportunities or philanthropy. State-sponsored welfare programs are generally not condoned in neoclassical economics because they interfere with the free functioning, and therefore the efficiency, of market economies.

Conversely, heightened concern about the distribution of wealth follows naturally from the scale issue in ecological economics because limits to economic growth are emphasized. To borrow from the "rising tide" metaphor, the tide can rise only so far, and only a certain number of boats may be lifted. Given this knowledge, which is rooted in an understanding of ultimate means (matter and energy), the normative stance of ecological economics becomes paramount. Although evolution and natural selection are acknowledged and to some extent analyzed, in ecological economics, ecological economists do not believe in leaving the prospects of the poor to the market. They recognize a broader suite of market failures than do neoclassical economists (see the section on allocation of resources below) and recognize that some factors of production are not conducive to efficient allocation via the market. Inefficient allocation may lead to or exacerbate distribution problems. Ecological economists agree with neoclassical economists about many other market failures and corrections that can be pursued with public policy, but they also believe in policies and programs to redistribute wealth if necessary for the purposes of distributive justice and the common good.

This leads to difficult policy decisions, because there are no scientific or mathematical formulae available to ascertain a just distribution of wealth. By definition, ethics are required even to acknowledge the concept of justice. For most people, religion provides moral authority or at least guidance in constructing an ethical framework. Ecological economics includes some analysis of religious teachings with regard to economic justice. All major religions teach moderation and generosity in economic affairs and warn of the spiritual perils of greed, riches, and luxury. Even simple reminders of such teachings help the public and policymakers with economic decision-making.

With regard to methods for analyzing the distribution of wealth, many of the concepts and tools used in the international development community are likewise used in ecological economics. For example, the Gini coefficient is used to determine the equality with which wealth is distributed among the populace, providing implications for public policy and international diplomacy. By contrast, Pareto optimality (discussed further in the section on allocation of resources) is not deemed relevant to economic justice as it is in neoclassical economics.

International trade is one of the featured issues in ecological economics pertaining to the distribution of wealth. In neoclassical economics, free trade among nations is generally considered a desirable feature of the global economy. Free trade is considered a mechanism for the efficient allocation of

The New York Stock Exchange power centers where GDP trends are closely monitored. © Justin Guariglia/Corbis.

resources and for facilitating global economic growth. This viewpoint is rooted in the principle of comparative advantage, developed by the English economist David Ricardo (1772–1823). A nation has a comparative advantage in the production of a good if its opportunity costs (as opposed to absolute costs) of producing the good are less than those in other nations. The implication for economic growth is that a higher level of global production may occur as nations specialize in the production of goods for which they have a comparative advantage.

In ecological economics, the principle of comparative advantage is neither denied nor deemed applicable to the modern era entirely, because one of the assumptions underlying the principle is that the factors of production do not move across national boundaries. By the latter decades of the twentieth century, capital had become highly mobile and numerous mass international movements of labor had occurred, so the principle applies to a lesser degree than it did during the classical era of the nineteenth century. More important, however, with economic growth threatening to breach global ecological capacity, any phenomenon facilitating further growth is not necessarily deemed desirable in ecological economics, regardless of its conduciveness to allocative efficiency. Finally, voluminous and fast-paced international trade is seen to cause disruptions in the social fabric of nations and facilitates the concentration of wealth in nations with multiple comparative advantages. Advantageous terms of trade for nations with an industrial and institutional head start suggest that a laissez-faire approach to free trade will produce an increasingly skewed macroeconomic distribution of wealth.

Allocation of resources

The term *allocation* in economics refers primarily to the use of resources in a general sense, meaning the factors of production. Allocation of factors may be analyzed or described macroeconomically, as in the proportions of land, labor, and capital that are used to produce the goods and services of a nation, but the primary concern in neoclassical economics is how the factors of production are allocated among firms and thenceforth commodities. Indeed, a common definition of economics is the study of the allocation of resources among competing end uses, and many scholars would say that neoclassical economics is synonymous with microeconomics and that macroeconomics should be identified with another tradition, such as Keynesian economics. In any event, the primary measure of success in neoclassical economics (and among many Keynesians as well) is the efficient allocation of resources.

The phrase *competing end uses* requires some elaboration. With a focus on production, competing end uses connotes the productive activities of firms. For example, one firm may use oak logs to produce flooring, another to produce furniture. Flooring and furniture are end uses in the economic production process. With a focus on consumption, competing end uses emphasizes the choices of consumers. For example, one consumer may desire oak flooring whereas another desires oak furniture. It would not be efficient if most consumers wanted the supply of oak logs to take the form of

furniture, while firms used most of the logs to produce flooring. The quintessential finding and focus of neoclassical economics is that prices, as they evolve in a freely functioning market, dynamically lead to an optimum allocation of oak logs (and all other resources, for a given distribution of wealth and income among consumers).

Efficient allocation is important in ecological economics too, but the ecological foundation and ethical framework of ecological economics result in a different philosophy of allocation and different implications for allocative policy. First, the primacy of land among the factors of production is emphasized. Manufactured capital is not deemed substitutable for land, as it is to a large degree in neoclassical economics, but rather is recognized as deriving from the land, with energy expended by labor. Once it is manufactured, capital typically becomes complementary, not substitutable, to land in the production process. Second, efficiency is assessed in material or energetic terms more so than in financial terms. For example, it may be efficient financially for a firm to employ a particular ratio of capital to labor without being efficient in the use of material or energy. Third, because of the emphasis on scale, the macroeconomic aspect of efficiency is emphasized. A financially efficient mixture of inputs for the firm, or even a fiscally efficient mixture of government spending, may not be efficient or even feasible for the economy at large in material or energy terms.

All else being equal—*ceteris paribus* in economics jargon—it is not efficient to employ a particular resource to produce a good or a service if another, more plentiful resource may be used instead. Prices help to provide information about the scarcity of resources; more plentiful resources are likely to go into production because the prices of those resources should be lower. For neoclassical economists, prices assure that no resources become so scarce as to cripple the attentive and competitive firm or economy, because new technologies and institutions are developed as prices send signals to firms and governments.

The faith in prices to obviate problematic resource shortages is less fundamental in ecological economics. All economists agree that various market imperfections result in faulty prices, but in addition, ecological economists emphasize that many natural resources and ecological services cannot be substituted for by capital or synthetic products, regardless of how high prices might rise. Also, prices reflect conditions today, with little consideration of future generations. For example, oil prices during the twentieth century did not reflect the energy shortages that would arise in the twenty-first century, much less the costs of global warming that ensued largely as a function of low oil (and other fossil fuel) prices.

Furthermore, certain goods and services (collectively referred to as *goods* in this section) do not have the characteristics required for efficient allocation in the market. These characteristics include rivalness and excludability. Rivalness is a natural property. A good is rival if one's consumption of it prevents its consumption by anyone else. For example, food is a rival good. Excludability is also a natural property, but, unlike rivalness, can manifest only in the context of a legal institution. A good is excludable if others may be prevented from using it. Excludability is required for property rights to be assigned, and some goods are more excludable than others. For example, nonmigratory fisheries are more excludable than migratory fisheries. Regardless of how excludable a good may be, its *exclusion* must be enforced.

Rival goods are usually excludable to some degree. Many excludable goods and especially services are not rival, however. For example, singing in a concert hall may be enjoyed by one or many people. The singer, hall owner, and/ or producer negotiate the level of exclusion and the price of tickets based on the supply of singers (and concert halls) and the demand for singing.

Rivalness and excludability are required for prices to function as indicators of scarcity. Prices are particularly good indicators of scarcity for goods that are rival and excludable such as food, clothing, and housing. Conversely, prices are not particularly good indicators of scarcity for goods that are excludable but nonrival such as information and entertainment services. For goods that are not even excludable, prices can be neither demanded nor taken. Therefore, a free market in which prices arise as a function of supply and demand can result in the efficient allocation of rival resources; a less efficient allocation of nonrival but excludable resources; and no allocation of nonexcludable resources. In other words, nonexcludable resources must be provided, or maintained, through some other mechanism than the market.

The requirements of rivalness and excludability for efficient market allocation are particularly relevant in ecological economics because many natural resources are nonexcludable or have low levels of excludability. Oceanic fisheries, large forests and rangelands, and remote mineral deposits are examples of natural resources that are rival but nonexcludable or not readily excludable. They are susceptible to overuse by extractors who accrue the benefits without absorbing the full costs of overuse, resulting in prices that are too low for efficient allocation. This "tragedy of the commons," as described by Garrett Hardin in his classic 1968 paper with that title, was one of the conceptual foundations of ecological economics (although the word *commons* was somewhat of a misnomer because traditional commons were often protected from overuse by complex social contracts and customs). The allocation of nonexcludable resources that are even nonrival, such as the ozone layer, is even less sufficient. Protection of the ozone layer, crucial for human health and survival, required laws and international agreements to overrule the market forces that favored the use of chlorofluorocarbons as refrigerants.

Neoclassical economists have also acknowledged the tragedy of open-access, nonexcludable resources and have emphasized that the cost to society of overexploitation was an *externality* of the market, an externality that could be corrected by using various institutional arrangements. In ecological economics, it is agreed that some market externalities may be *internalized* to a degree with taxes, user fees, and so on. In ecological economics, however, such efforts are viewed as a somewhat Pyrrhic victory for the market, because they amount to the regulatory contrivances loathed in free market ideology. More important, however, the very term *externality* has symbolized to many ecological economists the problematic paradigm of neoclassical economics—that is, that something falling outside the market system is tangential to the focus of economics, which is the functioning of the market to allocate resources efficiently.

Policy implications of ecological economics

Given the normative stance of ecological economics, public policy is viewed as an intermediate means along the ends–means spectrum (see the section on ends and means above). Given the themes and emphases of ecological economics, the policies of central concern are those that affect scale, distribution, and allocation. New policies are needed for sustainable scale, fair distribution, and efficient allocation, and reforms are needed for many existing policies that are unsustainable, unfair, and inefficient.

There are numerous theories, traditions, or schools of thought in public policy studies that are, to various degrees, applied in public policy. Most of these traditions have some type of economic basis or propensity. Public choice theory, for example, is essentially the application of neoclassical economics, whereby the will of the public is freely and efficiently expressed through the choices individuals make in the market. In this tradition, public policy is designed to keep the market operating efficiently and, if necessary, to preclude government intervention. Critical theory, with Marxist roots, is focused instead on the oppressive nature of political and economic powers. It calls for policy reforms as the needs for them are inevitably unveiled, and often these reforms interpose on market forces.

Policy design theory is a more recent effort, led by the political scientists Anne Larason Schneider and Helen Ingram (authors of *Policy Design for Democracy* [1997]), to meld the best traits of other public policy traditions. In policy design theory, a public policy is also judged by its adherence to and nurturing of democracy. As an integrating, synthesizing endeavor with a penultimate end of democracy, policy design theory is perhaps the tradition of public policy most conducive to the goals of ecological economics.

In their groundbreaking textbook, *Ecological Economics: Principles and Applications* (first published in 2004 and revised in 2010), Herman E. Daly and Joshua Farley presented a set of six policy design principles. These include some standard, general-purpose principles, including several that reflect the approach and philosophy of ecological economics. These principles are as follows:

1. Economic policy always has more than one goal.

2. Policies should strive to attain the necessary degree of macro-control with the minimum sacrifice of micro-level freedom and variability.

3. Policies should leave a margin of error when dealing with the biophysical environment.

4. Policies must recognize that one must always start from historically given initial conditions.

5. Policies must be able to adapt to changed conditions.

6. The domain of the policymaking unit must be congruent with the domain of the causes and effects of the problem with which the policy deals.

Each of these principles has its unique level of importance or prominence in the pursuit of sustainable scale, equitable distribution, and efficient allocation.

Sustainable scale

Ecological economics is often looked to for creative policy solutions to the problems of unsustainable scale and uneconomic growth (i.e., growth beyond optimal scale). Certainly there have been some original policy tools proposed in the ecological economics literature. However, the first and perhaps most important terrain in the policy arena, as it pertains to sustainable scale, is the myriad of already existing policies that lead to economic growth. These may be generally categorized as fiscal, monetary, and trade policies. Fiscal and trade policies generate the most attention in ecological economics. Not much is said in the ecological economics literature about reforming particular monetary policies, such as money supplies and interest rates, presumably because the proper reforms for sustainable scale are too obvious and also because the challenge is quite daunting politically. Monetary authorities are expected to cut interest rates and increase money supplies to stimulate so-called sluggish economies. Nonetheless, if an economy has grown beyond optimal scale, and especially beyond maximum sustainable scale, monetary policy to stimulate economic growth does more societal harm than good. In this context, higher interest rates and tighter money supplies are appropriate. As of the first decade of the twenty-first century, however, ecological economics was not known widely enough in public and policymaking circles to precipitate a serious dialogue about monetary policy toward a steady-state economy. Monetary authorities typically favor higher interest rates and restrictive money supplies only when inflation threatens.

Economists (of all persuasions) also understand that monetary policy has limited effects. For example, when an economy is operating at full capacity, lowering interest rates and flooding the economy with money will result only in inflation. Similarly, as economic capacity diminishes as a result of the liquidation of natural capital (i.e., as limits to growth are reached), the economy must become sluggish and will almost certainly be forced to contract in the wake of major and global supply shocks such as peak oil (i.e., the peak on global per capita oil production). No amount of monetary tinkering can change this biophysical reality.

If a polity is determined to have economic growth, however, the monetary authority may "pull out all the stops," and, more importantly, fiscal policy too will be designed for growth. Taxes are likely to be lessened, with the hope that consumers will then spend more and stimulate the economy. Budgets will be reallocated in a manner also designed to stimulate the economy. An early-twenty-first-century example is the subsidizing of corn farming to increase the production of ethanol, a hoped-for alternative to petroleum as a primary energy source for economic growth.

These standard, traditional, expected responses of fiscal and monetary policy authorities are not consistent with ecological economics and the six policy design principles presented above. Most ecological economists believe that the global economy and many national economies are beyond maximum sustainable scale and probably far beyond optimum scale. Therefore, the ecologically economic approach would be to readjust fiscal and monetary levers downward. As of 2012, however, this type of policy reform was not politically feasible, which may explain to a large extent why the ecological economics literature is not replete with such policy recommendations. This also points to the primacy of identifying the appropriate policy *goal*, in contrast to particular policies. To some degree this is an issue of semantics because the formal acknowledgement of a policy goal may itself be deemed a policy. Such is the case, for example, with the U.S. Full Employment Act. When it was originally passed in 1946, full employment was the goal, for which the Full Employment Act established a programmatic approach—that is, a policy or set of policy tools for achieving full employment. But full employment itself is a policy of the United States.

This example is especially relevant to sustainable scale, because the Full Employment Act is also perhaps the most codified manifestation of the general policy goal of economic growth in the United States. Although the number one goal of the Full Employment Act is full employment, it was and is assumed that the US population will grow. Therefore, all else being equal, a policy of full employment is equivalent to a policy of economic growth.

In the United States and most other nations, however, the expectation is that each generation will have a higher quality of life, especially in material terms. This is not codified, as is full employment in the United States, but it permeates public policy. Monetary authorities, for example, speak as often about their efforts to promote economic growth as about preventing unemployment or inflation. In addition, in the United States, economic growth is part of the mission for at least four federal agencies beyond the Federal Reserve (the monetary authority). With all of the formal and less formal government policies and programs for economic growth, one may surmise that economic growth is the number one domestic policy goal of the United States and many other nations, and perhaps of the international governing community too (e.g., Organisation for Economic Co-operation and Development [OECD], signatories of the North American Free Trade Agreement (NAFTA), World Bank, International Monetary Fund).

The linkage with full employment points to another crucial political issue for achieving sustainable scale: population size. Although an economy may grow beyond optimal and maximum sustainable scale either via population or per capita production and consumption, in one important sense overpopulation is the most fundamental problem to solve. Humans require a minimum amount of consumption to exist, such that continuous population growth must eventually lead to the breaching of carrying capacity. Conversely, per capita consumption may increase, at least in theory, without an inevitable increase in the global ecological footprint, if population is decreasing simultaneously. Neither population nor per capita consumption can increase perpetually, and to the extent that full employment remains a high priority, population is the more crucial parameter to stabilize. With a growing population, movement from economic growth to a steady-state economy will entail some level of unemployment. The politics of advancing a steady-state economy in this context are exceptionally daunting.

Summarizing to this point, sustainable scale entails replacing national and international goals of economic growth with the goal of a steady-state economy at the optimal scale. This means that the policy of economic growth must be replaced with the policy of a steady-state economy, and the policy complex designed to facilitate economic growth must be reformed to facilitate a steady-state economy.

Reforming existing policies is a necessary but probably insufficient condition for establishing a steady-state economy. New policies will almost surely be required, including policies designed to help with stabilizing population, per capita production, per capita consumption, throughput, and natural capital stocks. The most draconian approach in all cases is direct regulation, whereby the state imposes behavioral and commercial limits. Direct regulation is socially unpalatable and politically infeasible in most nations; otherwise, it could be highly effective.

In addition to direct regulation, Pigouvian taxes and subsidies may be designed to contribute to sustainable scale. What distinguishes Pigouvian policies (named after the English economist Arthur C. Pigou [1877–1959]) from other taxes and subsidies is their focus on correcting for market failures. Thus they are palatable to most economists and useful for social justice as well. Particular Pigouvian instruments may contribute substantially to sustainable scale, too. For example, if polluters are taxed the full social cost of the pollution, the rate of the pollution will decrease.

The most distinctive form of policy with regard to sustainable scale is the cap-and-trade policy, which also combines several of the policy design principles listed above. A cap-and-trade policy (or policy mechanism) may be effectively applied to most stocks of natural capital and many pollutants. The relevance to scale is exhibited by the word *cap*. When the use of a material or energy source that is integral to the economy is capped, the cap puts up a de facto sideboard to economic growth. The clearest example is with greenhouse gas emissions, especially from the combustion of fossil fuels. The global economy is primarily fossil fueled, with petroleum constituting the primary transport fuel and coal a significant electric power fuel. Capping greenhouse gas emissions in this context is tantamount to capping economic growth.

The primary objective of policymakers in capping, or attempting to cap, greenhouse gas emissions is not to stop economic growth but rather to protect the atmosphere and prevent catastrophic levels of global warming. Yet the predictable economic dampening effects of a strict cap on greenhouse gas emissions has prevented some of the wealthiest nations from participating in international agreements to lower greenhouse gas emissions. This experience demonstrates the primacy of macroeconomic policy goals in the policy arena.

Theoretically, starting from the perspective of ecological economics, one could prescribe a cap on greenhouse gas emissions explicitly for scale-limiting purposes. This would be feasible, however, only if the international community agreed, consistent with the tenets and findings of ecological economics, that global economic growth was no longer an appropriate goal. Then the level at which the greenhouse gas emissions (or other ecologically relevant) cap would be set would be informed by policy design principles 3 through 5 (above). Pursuant to principle 3, a precautionary approach is called for, so that any benefit of the doubt would be applied to environmental protection and future generations. Pursuant to principle 4, however, the cap would be applied gradually to avoid shocking the economic system. In an age of environmental crisis, the principle or necessity of gradualism indicates the need to accept the steady-state economy as a policy goal quickly enough, while (or assuming) there is still time for gradual policy adjustments. Pursuant to principle 5, policies to cap throughput must be designed with flexibility such that, as limits to growth and optimum scale become more apparent, caps can be readjusted. Because many throughput issues are global in nature, international policy entities and instruments are required pursuant to principle 6. For example, the Kyoto Protocol of 1997 was an early attempt at capping the rates of greenhouse gas emissions via global convention.

Principle 2 calls for the minimum sacrifice of micro-level freedom. Cap-and-trade policies meet this principle better than direct regulations because firms are free to trade throughput permits within the limits established by the cap. Markets are established, permits are allocated among firms, and thenceforth some of the allocative advantages of laissez-faire markets are engaged. Other advantages are not, however. For example, whereas a laissez-faire market requires no government interference and expenditure, a cap-and-trade system is essentially a government-established system, with rules enforced by the government. This reflects the fact that natural capital is typically not wholly or readily rival and excludable. Nevertheless, the throughput permits are entirely rival and excludable and therefore tend to be allocated efficiently among the firms.

A cap-and-trade policy, then, is a legitimate compromise between laissez faire and central planning, and most if not all public policy traditions will appreciate various aspects of it. The cap-and-trade policy is emblematic of policy design principle 2, as it is designed to attain the necessary degree of macro-control with the minimum sacrifice of micro-level freedom.

Fair distribution

In ecological economics, the goal of fair distribution cannot be effectively pursued unless the goal of sustainable scale is already achieved or is being achieved. If sustainable scale is not a policy goal, and economic growth is occurring beyond maximum sustainable scale and remains a goal of the state, then efforts toward fair distribution are certain to fail. As limits to growth are breached, history shows that conflict invariably ensues and the victors claim the natural resources, including the land itself. Peaceful and equitable coexistence requires a social contract in which citizens agree to live sustainably, as a society, and to share, within reason, natural resources and other wealth. In ecological economics, this social contract would be manifest in caps on income and wealth, minimum income, and the distribution of returns from the factors of production, especially natural capital.

Ecological economists tend to be more aligned with so-called Georgists (modern-day followers of the nineteenth-century American economist Henry George) than are neoclassical economists. George and other classical economists made the compelling point that, unlike labor or capital stocks, the land cannot grow. While labor and capital stocks proliferate and become more prominent relative to land, the value of land increases. In other words, the landowner becomes wealthier by virtue of others' toil. Georgists and many ecological economists believe that the unearned rents of landowners should enter into the commonwealth instead.

Socialists may even advocate the state holding all the land. Ecological economists tend to advocate a balance of public and private lands, so long as firms are not subsidized to extract natural capital from public lands and landowners are taxed unearned rents. Taxing unearned rents is advanced as a highly practical approach to fair distribution because systems of land taxing already exist. It is the rationale and formulae that would be transformed by ecological economics, more than the institutional arrangements of taxing.

The rationale for capping income and wealth flows directly from the scale issue to the distribution issue. If the global economy is at maximum sustainable scale, the acquisition of more income or wealth by an individual entails the breaching of long-term carrying capacity, unless an equivalent amount of income or wealth is taken away from someone else. Especially if that individual is already very wealthy, it would run against the ethical stance of ecological economics for he or she to jeopardize the environment, fellow citizens, and future generations by consuming at an even higher and unsustainable level.

On the other side of the same ethical coin is the logic for capping or redistributing income and wealth. Given a global economy exceeding its maximum sustainable scale, the only ethical and ecologically economic approach to alleviating poverty while moving closer to sustainable scale is the capping of income and wealth, with preexisting excess used to alleviate poverty. Precisely at what level to cap income or wealth is a matter to be determined, ideally in a democratic manner (according to policy design theory), whereby the majority of citizens understand the need for caps on throughput, and therefore caps on income, and support the goals of sustainability and social justice. Presumably a gradualist approach would entail formal but voluntary capping, followed if necessary by imposed capping. Data pertaining to the existing scale of the economy, a range of optimal scales, and the ecological footprints associated with different levels of income and wealth would be necessary for determining appropriate capping levels.

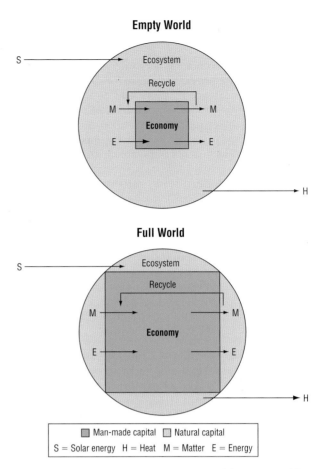

Empty versus full. Reproduced by permission of Gale, a part of Cengage Learning.

At the other end of the distributional policy spectrum is minimum income. A minimum income policy has logical and ethical foundations pertaining to scale and distribution. Logically, impoverished individuals are highly unlikely to prioritize environmental considerations, which is crucial for establishing sustainable scale. For example, landless, unemployed peasants may resort to poaching timber from public lands. Poor people typically have been victims of circumstance rather than lazy, and the ethical response is to help them without jeopardizing the environment and future generations. In other words, in a full world economy, the logical and ethical approach is to distribute a minimum income to the needy, procured from the overcapped excesses of the wealthy. This approach combines a steady-state economy with a fairer distribution of wealth.

Efficient allocation

Efficient allocation, the top goal of neoclassical economics, is deprioritized in ecological economics, but only relative to the urgent, first-order needs of sustainable scale and equitable distribution. Yet efficient allocation is important in ecological economics from the normative perspective of reducing waste and from the macroeconomic perspective that efficiency allows for higher sustainable scale. In addition, a key distinction between conventional economics and ecological economics is that, in ecological economics, the prospects for *technical* efficiency are recognized as limited by the second law of thermodynamics.

Because ecological economists acknowledge that the market is often a reasonably efficient mechanism for allocating private (rival and excludable) goods, they tend to focus on the estimation of economic values of nonmarket or public natural capital and ecosystem services. Such estimation exercises help to educate the public and policymakers about the opportunity costs of private goods production and consumption that are incurred by society as natural capital and ecosystem services are eroded. In some cases the estimated values can also be used in the development of Pigouvian taxes and subsidies (see above). They also assist decision makers in cost–benefit exercises. A widely cited example is the decision of New York metropolitan officials to acquire and protect portions of the Catskill Mountains. In 1997 the city of New York had the choice of installing a water filtration plant at a cost of $4 billion to $8 billion, plus $250 million to $300 million per year in operating costs, or to invest approximately $1.5 billion in the natural capital of the Catskills, maintaining the already existing ecosystem service of water filtration. New York opted for the latter.

Basic methods for estimating the values of natural capital—including biodiversity—and ecosystem services (including those flowing from stocks of biodiversity) include the following:

1. Market Price Method. Many ecosystem goods or services are bought and sold in commercial markets. Meat, fur, and ivory are well-known examples of marketed goods from the mammalian sector in the economy of nature. The most prominent example of a marketed service attributable to wild mammals is ecotourism. Hunting and nonhunting trips and safaris are a major source of income in areas with charismatic megafauna such as grizzly bears (*Ursus arctos horribilis*), giraffes (*Giraffa camelopardalis*), or gorillas (*Gorilla* spp.). Although externalities exist, market prices provide a starting point in estimating the value of related natural capital and ecosystem services.

2. Productivity Method. Economic values may be estimated for intermediate ecosystem goods or services that contribute to the production of commercially marketed final goods. For example, red squirrels (*Sciurus vulgaris*) have an estimable value in the production of American marten (*Martes americana*), the furs of which are a marketable good.

3. Hedonic Pricing Method. Economic values may be estimated for ecosystem goods or services that directly affect prices of other marketed goods or services. For example, land prices in the Rocky Mountains of Colorado may be higher, ceteris paribus, as a result of the presence of elk (*Cervus canadensis*), mule deer (*Odocoileus hemionus*), or pronghorn (*Antilocapra americana*), species that many people like to view. Hedonic pricing may be used to estimate the value added to the land by the presence of such mammals.

4. Travel Cost Method. Based on the assumption that the value of a recreational site is reflected in how much people are willing to pay to visit the site, economic values are associated with ecosystems or parcels of land that are used for recreation. For example, when a hunter purchases a license to hunt gemsbok (*Oryx gazella*) in South Africa and spends an additional $3,000 in travel expenses to carry out the hunt, one may conclude that an individual gemsbok (one envisioned by the hunter as reasonably representative of the herd in the hunting area) is *worth* a substantial proportion of the $3,000 of value sacrificed by the hunter in addition to the value suggested by other monetary exchanges associated with the hunt (such as the acquisition of the license and the hiring of a guide). This method is complicated by the realization that travel expenditures indicate the value of other aspects of the hunt, including (in this example) the general South African experience. Nevertheless, travel costs are too significant to ignore in attempts to value ecosystem goods and services.

5. Damage Cost Avoided, Replacement Cost, and Substitute Cost Methods. When an ecosystem is protected from economic or other disruptive activities, damage costs are avoided, as are the costs of replacing ecosystem goods and services or of providing substitute goods and services.

6. Contingent Valuation Method. Economic values for virtually anything may be estimated contingent upon certain hypothetical scenarios. For example, individuals from Canada to Panama may be asked how much they are willing to pay for the protection of the polar bear (*Ursus maritimus*), and responses in the aggregate provide an estimate of the polar bear's value (or, more precisely, of a subset of the polar bear's value related to its mere existence).

7. Benefit Transfer Method. In this method, economic values are estimated by *transferring* (or extrapolating) existing estimates obtained from studies already completed in other areas. This method is often used in estimating the value of ecosystem services. For example, if the chiropterophily (pollination by bats) of mangos in the Mexican state of Sinaloa has been calculated at $1.2 million per year, and the acreage of mango orchards in the neighboring state of Nayarit is half that of Sinaloa's, the value of chiropterophily in Nayarit may be estimated at $600,000 per year, ceteris paribus.

None of the above methods is unique to ecological economics; all have been described in environmental economics or in the application of neoclassical economics to environmental issues. The contribution of ecological economics to the use of these methods is primarily in the deeper understanding of the components, structures, and functions of ecosystems that need to be evaluated in economic accounting and decision-making. Mammalian ecologists have a key role to play in providing the insights necessary to estimate the economic value of mammals in providing ecosystem goods and services. From a macroeconomic perspective, however, ecologists also have a role to play in reminding economists and policymakers that no market activity is ecologically benign. Because of the trophic structure of the human economy, generating money to spend on the likes of land, safaris, or mangos (see examples above) entails the use of natural capital and ecosystem services elsewhere. This points to a *trophic conundrum* whereby it takes the liquidation of natural capital to spend money on *conserving* natural capital elsewhere (Czech, forthcoming). In other words, a society cannot buy its way to biodiversity conservation, which helps explain why, in ecological economics, the steady-state economy is seen as a prerequisite for biodiversity conservation and general sustainability.

Future directions and challenges

In the early twenty-first century, ecological economics is somewhat established in academia, but it remains a mere infant in policy circles. Because it embraced a diversity of perspectives and methods from the beginning, its emphases and tendencies have always been subject to challenges and have varied from one region to another. For example, the European tradition of ecological economics has emphasized the distribution of wealth to a relatively greater degree than the American tradition, which conversely has emphasized sustainable scale and, especially since the 1990s, efficient allocation of natural capital. However, the first decade of the twenty-first century ushered in dramatic social, political, ecological, and economic changes. These ongoing developments simultaneously empower and challenge ecological economics and will affect the course it takes for the remainder of the twenty-first century.

Reinforcing the primacy of sustainable scale

One challenge for ecological economics is to revisit and reinforce the primacy of sustainable scale as the most distinctive and original aspect of the field. Prioritizing sustainable scale constitutes the so-called Dalyist tradition of ecological economics (see the historical development section above). Although sustainable scale is often listed as the highest priority in ecological economics textbooks or overviews, the Dalyist tradition has been overshadowed in academia and in practice by exercises in which the value of natural capital and ecosystem services are estimated in monetary terms, often in great econometric detail, sometimes with little ecological grounding, and almost always with little macroeconomic context. This is evident in the literature at large and even in the flagship journal *Ecological Economics*.

The emphasis on natural capital valuation has resulted from at least three phenomena. First, sustainable scale will clearly entail macroeconomic policy reform, including the introduction of new policy tools and the adjustment of existing policies, and such policy reform is a daunting challenge. While widespread agreement exists on the importance of "getting the prices right" with ecologically informed microeconomics, replacing the macroeconomic goal of growth with the goal of a steady-state economy entails a veritable paradigm shift on the part of conventional economists,

policymakers, and the society at large. Fiscal, monetary, and trade policies are crafted at high levels of government and are affected by powerful corporate interest groups, or what is sometimes called the *corporatocracy* to indicate the concerted nature of corporate influence in economic policymaking. Using a concept from political science, some observers refer to an *iron triangle* of corporations, politicians endeared to corporations, and influential economists who are hired by corporations or appointed by politicians. This tripartite network surrounds and pervades the macroeconomic policy arena, making it extremely difficult to access or succeed in. Monetary policy in the United States, for example, is developed and implemented by a central bank, the Federal Reserve System. The Federal Reserve's board members, who typically come from and return to prestigious posts in academic bastions of neoclassical economics, are appointed by the president of the United States and serve fourteen-year terms. In other words, the conventional, neoclassical approach to monetary policy will be difficult and time consuming to supplant, and many ecological economists view the prospect of reform as impractical to undertake at this point in history.

Second, as environmental concerns intensify, more neoclassical economists are focusing on environmental issues and even joining ecological economics organizations such as the International Society for Ecological Economics. Given that there are far more neoclassical economists than ecological economists, the ratio of neoclassical economists to ecological economists in the ecological economics community has been increasing. Their education and training have prepared them to ascertain and analyze prices and to publish papers thereon, but not usually to ascertain and analyze the ecological limits to economic growth.

Third, the funders of research tend to be more interested in natural capital valuation than in sustainable scale. This is partly a function of the earlier observation that macroeconomic policy reform is too daunting to attract participants, including funders who often want to see clear and relatively quick results from their grants. There is the additional reason that much macroeconomic policy—especially monetary policy—is handled almost exclusively at national levels, by relatively few officials. Fiscal policy, by contrast, is handled by numerous officials in local, regional, and national authorities, but fiscal policy is often microeconomic in nature (for example, taxing a type of good), although it can be macroeconomic (for example, establishing rates for income taxes). While the primary goals of conventional monetary policy are stimulating growth and preventing inflation, the goals of fiscal policy are much more diverse and often entail pricing mechanisms or adjustments. For nonmarket goods and services, such as ecosystem services, economic values must first be estimated to enable the use of pricing mechanisms. The volume and diversity of issues requiring valuation exercises has led to a condition in which neoclassically (or microeconomically) trained economists outnumber the ecologically (and macroeconomically) trained economists, contributing to the prominence of valuation exercises in ecological economics.

Given these reasons for the emphasis on natural capital valuation in ecological economics, the ecological economics

The Federal Reserve, power centers where GDP trends are closely monitored. © Philip Scalia/Alamy.

community faces the following question: Is such an emphasis a problem? If so, how may the problem be solved? Concerns are often expressed within the ecological economics community about the emphasis on natural capital valuation, so there must be a real or perceived problem. Perhaps the most common concern is that, by focusing on natural capital valuation, ecological economics becomes little more than environmental economics—that is, an extension of neoclassical economics. Given that ecological economics arose from the realization that neoclassical economics was inadequate for illuminating sustainability challenges and helping to solve them, a merger of neoclassical and ecological economics may be considered a weakening compromise. This concern has led some of the early participants in ecological economics to distance themselves from and/or establish other, mostly informal alternatives to neoclassical economics. For example, some distinguish their research as biophysical economics to indicate the prominence of natural sciences in their work.

Natural capital valuation has, however, helped the ecological economics community to become more immediately relevant to the conventional economics community and to policymakers faced with difficult decisions about allocating natural resources. The relative ease of natural capital valuation exercises and the political and economic support for such studies have also resulted in a plethora of opportunities for graduate students to

engage in ecological economics, and presumably many of these students will graduate further into ecological macroeconomics and issues of economic justice. Valuation studies have been reported in numerous journals, helping to familiarize diverse scholars and professionals with at least the allocation aspect of ecological economics.

To summarize without casting judgment on the merits to date of natural capital valuation relative to sustainable scale and distributional investigations, clearly the emphasis on natural capital valuation has been at least somewhat problematic for the ecological economics community. One way to lessen this problem is by providing more detail on the macroeconomic context of valuation studies. In journal articles the basic concepts of limits to growth and sustainable scale are usually highly relevant to the contexts of valuation scenarios and may be described in introductions and conclusions or discussion sections. For example, the economic value of biodiversity has become a research topic because biodiversity has been lost as a function of economic growth. Instead of delving immediately into descriptive details of particular species and ecosystems, and then presenting valuation methodologies, authors can briefly describe the aggregate (macroeconomic) pressures that led to the scarcity of the species or ecosystems in the first place. Similarly, in the conclusions of such articles, authors may duly note that getting the prices right is indeed helpful for efficiently allocating biodiversity but that ultimately, if biodiversity is to be conserved, a steady-state economy will be required.

A reemphasis on sustainable scale, as well as more attention to fair distribution of wealth, may also be instituted in academia via program development, curriculum development, faculty qualifications, graduate student examinations, and community service. Government agencies and nongovernmental organizations with conservation interests may also contribute to these emphases via program development, staff qualifications and training, and public education campaigns.

Ecological implications of money volumes and flows

As with any academic endeavor, ecological economics raises as many questions as it answers, and it is beyond the current scope to list many such questions. Yet one question stands out as exceptionally relevant and important to answer soon, given the scenario(s) of ecological unsustainability developing concurrently with financial crises. In particularly ecological terms, or in terms that are most relevant to sustainable scale as well as financial solvency, the question is: What is the nature of money?

It has been posited by some that the volumes or flows of real (inflation-adjusted) money are reliable indicators of scale. If this is so, then real GDP, for example, could be used by a society to gauge its sustainability. In other words, actual GDP would serve as a surrogate for the ecological footprint, and estimates of maximum and optimum scale could also be expressed in GDP terms, greatly simplifying the application of ecological economics to macroeconomic policymaking.

Nevertheless, there is considerable disagreement among ecological economists about the ecological nature of money. Some think that money cannot serve as an indicator of scale because prices are determined by demand as well as supply, and demand is a psychic function as opposed to an ecological function. Also, because different goods and services enter the market as new technologies are developed, the ratio of throughput to money may change over time, preventing policymakers from equating money volumes and flows from environmental impact.

The possibility of using money volumes and flows as indicators of scale warrants a concerted investigative effort in ecological economics. A clear and convincing demonstration that standard measurements of money volumes or flows may be used to assess scale could become one of the most important academic accomplishments of the twenty-first century. It would help guide the formulation of macroeconomic policy goals, the administration of monetary and banking policies, and the expectations of international financial institutions and capital markets.

Conceivable need for degrowth

With an emphasis on sustainable scale, scholars of the Dalyist tradition suggested for decades that societies and polities undertake conscious planning for steady-state economies so that the ravages of *overshoot* could have been avoided. Total avoidance no longer seems feasible. Peak oil, climate

change, the ecological footprinting literature, and financial crises suggest that the global economy has already caused grave ecological and economic damage and is substantially beyond long-term sustainable scale. In the context of the large preexisting ecological footprints of wealthy nations such as the United States, rapid economic growth in twenty-first-century China and India appears to assure that vast regional economies and the global economy will suffer a protracted and deep recession. However, to the extent that economic growth may be consciously slowed by determined polities (including citizens as conscientious consumers in addition to policymakers working toward economic policy reform), overshoot damages may be lessened. It is in this context that some scholars and activists have begun to advocate for immediate and long-lasting economic *degrowth*. An example is the movement for La Décroissance (The Decline) in Western Europe.

Some of the more ardent advocates of economic degrowth have gone as far as critiquing the goal of a steady-state economy as already anachronistic and insufficient for ecological and economic sustainability in the twenty-first century. This critique has its logical merits, as briefly indicated in the preceding paragraph. In the long run, however, a degrowing economy is no more sustainable than a growing one.

The challenge for ecological economics, then, is to incorporate degrowth research and policy implications without losing sight of the long-term goal of a steady-state economy. Questions for researchers to explore include the following:

1. How far beyond carrying capacity is the economy? The economy in this context may be the global economy or an economy at any geographic scale, such as a state or province. (For less-than-global economies, scale may be analyzed with respect to the respective endowments of natural resources.)

2. What is the long-run maximum sustainable scale of the economy?

3. What is the optimum scale?

4. When maximum sustainable scale is breached, how much is carrying capacity compromised and how quickly must an economy recede to avoid further compromising the carrying capacity?

5. With or without breaching, how will maximum and optimum scales change over time as a result of natural and anthropogenic forces?

6. What types of policies and institutions are required for degrowth and for the maintenance of steady-state economies?

Ecological economics for a sustainable future

In the context of climate change, peak oil, financial meltdowns, resource conflicts, and other indications of environmental and economic crisis, economics is at a crossroads. Citizens, economists, and policymakers have numerous choices among the economic pathways of thought. The conventional choice is neoclassical economics with its focus on the efficient allocation of resources. Yet historical perspective and scientific analysis strongly suggest that the path paved by neoclassical economics, regardless of how efficiently traversed, does not lead from crisis to sustainability.

Ecological economics was developed partly as a response to the real and perceived shortcomings of neoclassical economics. In ecological economics, limits to economic growth are recognized as stemming directly from the laws of thermodynamics and principles of ecology. A key concept is that efficiency is itself limited, so that perpetually increasing efficiency is not an alternative and may not overcome the limits to economic growth.

The existence of limits to growth suggests that the theory and practice of economics need to be expanded to include the issue of scale, or the size of the economy relative to its containing, sustaining ecosystem. This leads, in turn, to the need to address the distribution of wealth. If the tide of the global economy can rise only so far, then only a limited fleet may be accommodated. In ecological economics, economic justice is not about trying to defy the laws of physics by raising the tide past the realm of possibility, but rather ensuring that tiny, law-abiding boats are not capsized in the wakes of hulking luxury liners.

Ecological economics faces numerous challenges stemming primarily from the political difficulties entailed by a critical analysis of economic growth as a policy goal. As with any endeavor that develops in academia prior to manifesting in society, there are also numerous theoretical and methodological issues to be developed, and the list of such issues is likely to lengthen as the body of research expands. To the extent that ecological economics research is conducted, transmitted, and understood by publics and polities, it is likely to have major effects on consumer behavior, economic policy, and international diplomacy.

Resources

Books

Barnett, Harold J., and Chandler Morse. *Scarcity and Growth: The Economics of Natural Resource Availability*. Baltimore: Johns Hopkins University Press, 1963. Published for Resources for the Future.

Boulding, Kenneth E. *The Structure of a Modern Economy: The United States, 1929–89*. New York: New York University Press, 1993.

Carson, Rachel. *Silent Spring*. Boston: Houghton Mifflin, 1962.

Commoner, Barry. *The Closing Circle: Nature, Man, and Technology*. New York: Knopf, 1971.

Costanza, Robert, John Cumberland, Herman E. Daly, et al. *An Introduction to Ecological Economics*. Boca Raton, FL: St. Lucie Press, 1997.

Czech, Brian. *Shoveling Fuel for a Runaway Train: Errant Economists, Shameful Spenders, and a Plan to Stop Them All*. Berkeley: University of California Press, 2000.

Czech, Brian. *Supply Shock: Economic Growth at the Crossroads and the Steady State Solution*. Gabriola Island, B.C.: New Society, forthcoming.

Daly, Herman E. *Steady-State Economics: The Economics of Biophysical Equilibrium and Moral Growth*. San Francisco: W.H. Freeman, 1977.

Daly, Herman E., ed. *Toward a Steady-State Economy*. San Francisco: W.H. Freeman, 1973.

Daly, Herman E., and John B. Cobb Jr. *For the Common Good: Redirecting the Economy toward Community, the Environment, and a Sustainable Future*. 2nd ed. Boston: Beacon Press, 1994.

Daly, Herman E., and Joshua Farley. *Ecological Economics: Principles and Applications*. 2nd ed. Washington, DC: Island Press, 2010.

Daly, Herman E., and Kenneth N. Townsend, eds. *Valuing the Earth: Economics, Ecology, Ethics*. Cambridge, MA: MIT Press, 1993.

Gaffney, Mason, and Fred Harrison. *The Corruption of Economics*. London: Shepheard-Walwyn, 1994. Published in association with the Centre for Incentive Taxation.

George, Henry. *Progress and Poverty: An Inquiry into the Cause of Industrial Depressions and of Increase of Want with Increase of Wealth—The Remedy*. San Francisco: W.M. Hinton, 1879.

Georgescu-Roegen, Nicholas. *The Entropy Law and the Economic Process*. Cambridge, MA: Harvard University Press, 1971.

Kingdon, Jonathan. *Self-Made Man: Human Evolution from Eden to Extinction?* New York: Wiley, 1993.

Krishnan, Rajaram, Jonathan M. Harris, and Neva R. Goodwin, eds. *A Survey of Ecological Economics*. Washington, DC: Island Press, 1995.

Meadows, Donella H., and the Club of Rome. *The Limits to Growth: A Report for the Club of Rome's Project on the Predicament of Mankind*. New York: Universe Books, 1972.

Mill, John Stuart. *Principles of Political Economy, with Some of Their Applications to Social Philosophy*. Rev. ed. 2 vols. New York: Colonial Press, 1899. First published 1848 by John W. Parker.

Odum, Howard T., and Elisabeth C. Odum. *The Prosperous Way Down: Principles and Policies*. Boulder: University Press of Colorado, 2001.

Quesnay, François. *The Economic Table*. New York: Bergman, 1968. Reprint of the 1766 translation of *Tableau économique*.

Schneider, Anne Larason, and Helen Ingram. *Policy Design for Democracy*. Lawrence: University Press of Kansas, 1997.

Schumacher, E.F. *Small Is Beautiful: Economics as If People Mattered*. New York: Harper and Row, 1973.

Smith, Adam. *An Inquiry into the Nature and Causes of the Wealth of Nations*. Edited by R.H. Campbell and A.S. Skinner. 2 vols. Oxford: Clarendon Press, 1976. First published 1776 by W. Strahan and T. Cadell.

Wackernagel, Mathis, and William E. Rees. *Our Ecological Footprint: Reducing Human Impact on the Earth*. Gabriola Island, B.C.: New Society, 1996.

Periodicals

Costanza, Robert, Ralph d'Arge, Rudolf de Groot, et al. "The Value of the World's Ecosystem Services and Natural Capital." *Nature* 387, no. 6630 (1997): 253–260.

Czech, Brian. "Prospects for Reconciling the Conflict between Economic Growth and Biodiversity Conservation with Technological Progress." *Conservation Biology* 22, no. 6 (2008): 1389–1398.

Czech, Brian, and Herman E. Daly. "The Steady State Economy: What It Is, Entails, and Connotes." *Wildlife Society Bulletin* 32, no. 2 (2004): 598–605.

Daly, Herman E. "The Economics of the Steady State." *American Economic Review* 64, no. 2 (1974): 15–21.

Hardin, Garrett. "The Tragedy of the Commons." *Science* 162, no. 3859 (1968): 1243–1248.

Norgaard, Richard B. "Economic Indicators of Resource Scarcity: A Critical Essay." *Journal of Environmental Economics and Management* 19, no. 1 (1990): 19–25.

Brian Czech

Ecosystem services

When thinking about the environment, many people conceptualize it as being something other, something that is not a part of human civilization. From this perspective, a landscape is a black-and-white checkerboard with people, cities, towns, and industries in one category and nature, its other species, and wild areas in another. But the reality is, of course, significantly more nuanced. There are no sharp distinctions between natural and not natural. The environment includes humanity. Humans impact it even in the most remote parts of the globe and are dependent on it for all of the things that make life possible and worth living. Far from black and white, coupled human and natural systems present every shade of gray with varying degrees of natural arrayed across the landscape. Indeed, the only hues missing are pure white and deep black—solely natural and solely human simply do not exist. To succeed, modern conservation efforts to prevent extinction or recover imperiled species must embrace these shades of gray.

Arising implicitly from the false dichotomy between humans and nature is a debate that has underpinned almost all conservation efforts: preservation versus conservation or, stated more broadly, nature for nature's sake versus nature for the benefit of humans. Undergraduate environmental studies courses throughout the United States teach what is often framed as an epic battle between John Muir and Gifford Pinchot over the Hetch Hetchy Valley in Yosemite National Park. In 1906 the city of San Francisco proposed a dam to provide drinking water for the city. The combatants were represented as seeing the world in two fundamentally different ways. In one camp were the adherents to Muir's idea of preservation—setting nature aside to protect it from people and their projects. In the other were Pinchot's soldiers of conservation—wise use of nature to ensure sustainability into the future. After a seven-year struggle, Muir and his partisans lost. The US Congress approved the flooding of the valley in 1913, and the dam was built in 1923. Water from the dam now provides over two million Californians with drinking water and electricity. But the urge to set aside nature has lived on, as has the debate between conservation and preservation.

For many years, practitioners of what is now (confusingly) called *conservation* have focused on preservation, using protected areas as the primary tool for conservation. Moreover, the conservation community has used biodiversity and its evil twin—the threat of extinction—as central themes to inspire people to act. This strategy has had some success. Indeed, by some estimates 13 percent of world landmasses are protected—an area larger than all of South America. (The story is quite different for the 71% of the surface of the globe covered by oceans, where significantly less area [~1%] is set aside in marine protected areas [IUCN 2010].)

But by some accounts, the aims and methods of this sort of traditional conservation are neither politically compelling nor felt deeply. In opinion polls, only about 1 percent of Americans say that environmental issues are the "most important problems facing the country today" (Gallup.com 2012). And conservation or the environment remains a niche concern of a narrow demographic. For example, the average age of members in the Nature Conservancy, the world's largest conservation organization, is 65. According to Peter Kareiva, the Nature Conservancy's lead scientist, only 5 percent of members are under 40 (P. Kareiva, pers. comm.).

Particularly troubling is the apparent inefficiency of preservation. Despite setting aside large swaths of protected areas, species are becoming endangered or going extinct at an alarming rate. The International Union for Conservation of Nature (IUCN) Red List Index (an analysis of trends of movements of species through the categories on the IUCNs' Red List of Threatened Species) shows that all species groups with known trends are deteriorating in status as more species move toward extinction than away from it. As well, protection in name does not ensure protection in deed. Without proper enforcement, some preserves are *paper parks* that do not lead to protection of species on the ground. Moreover, though often meant to confer protection in perpetuity, protected areas are vulnerable to population trends. As people are added to the planet, they will of necessity get resources from what is at hand.

In a world with a growing population of seven billion people, preservation is certainly necessary. Humans and other species need regions of the globe where human impacts are minimized. But to succeed in slowing down extinction rates and learn to sustainably coexist within the natural systems on which people depend, a larger effort is needed that reclaims Pinchot's version of conservation as including people in the equation. Conservation has to be made relevant in new ways.

Hetch Hetchy Reservoir in Yosemite National Park, California. © Kip Evans/Alamy.

One promising solution is to make clear the connections between people and nature—to see people as inextricably embedded in nature, to recognize that every square mile of Earth's surface is on a continuum between the nonexistent poles of entirely human-dominated and without human influence, to recognize the potential for a landscape to support both people and a wide range of biota, and to name, appreciate, and research the rich and varied ways that ecosystems sustain and fulfill human life. If humanity—and its institutions—can recognize and demonstrate the benefits that ecosystems provide to people, it will be possible to simultaneously enhance investments in conservation (Pinchot's version) and foster human well-being.

Countryside biogeography and conservation

Gretchen C. Daily, a thought leader in conservation biology and early proponent of specifying nature's benefits as a pathway to mainstreaming conservation, has helped establish the field of countryside biogeography. She tells the story of visitors to her remote research station in Costa Rica.

Whenever friends visit me, they're dying to see the amazing birds, butterflies, and other brightly colored creatures emblematic of tropical rainforest. They're always surprised when I suggest we walk around farms, rather than in the forest. Unless there's a canopy tower, most people are much more likely to be dazzled by biodiversity out on low- to intermediate-intensity farms, where the "canopy" is lower and a spectacular variety of organisms can be seen close-up. (Daily pers. comm.)

Proponents of countryside biogeography and conservation argue that because human impacts on the biosphere continue to intensify, the future of Earth's biodiversity depends to a large extent on the conservation value of countryside, the growing fraction of Earth's unbuilt land surface where human impacts are nevertheless still dominant. Like that binary checkerboard, the prevailing view has often been that most biodiversity is in remnants of native habitat—white islands in a sea of black. The logic of this prevailing perspective is that

most plants, animals, and microorganisms are highly adapted to their native habitats and that few would be able to exploit those areas used for agriculture or human settlement. In general, those few—typically undesirable—species would not require or merit protection.

Yet since the late 1990s, conservation scientists have come to realize that the simple checkerboard with human-dominated areas hostile to biodiversity is, in reality, full of shades of gray. These gray lands are multiple use in the broadest sense. They are managed to provide food, fiber, and fuel, but if managed carefully, they also can provide a range of other benefits such as water purification, soil retention, and habitat for species important to humans. A broad range of native species live in human-dominated landscapes. In studies of the neotropics, for example, well over half, and in some places over 90 percent, of the native biota in various taxonomic groups is found in countryside habitats (Daily et al 2003, Horner-Devine et al. 2003, Mayfield and Daily 2005, Ricketts et al. 2001).

Some worry that countryside is the inevitable end result of lessened attention to preservation—that serious consideration of and investment in these human-dominated landscapes might foster complacency and inaction across all of conservation. This argument overlooks three critical issues. First, although protected areas are an important building block of conservation strategies, protected areas alone are unlikely to ensure the survival over the long term of more than a tiny fraction of Earth's biodiversity. They are too small, too few, too isolated, and too subject to change. Second, countryside is not monolithic. It ranges in conservation value from very low (supporting less than 10% of the native biota; e.g., extensive monocultures of annual crops) to very high (supporting more than 90% of the native biota; e.g., diverse landscapes with significant native vegetation cover and little hunting). With growing population comes a growing threat of intensification. Thus, efforts to protect human-dominated areas with significant complements of native biota conservation can help achieve biodiversity conservation goals. Third, many of the ecosystem service benefits (see below) that biodiversity confers to people are supplied on scales from local to regional (e.g., pollination, pest control, renewal of soil fertility, flood control, water purification). Therefore, people must coexist on the landscape with countryside biota to receive these benefits.

Countryside biogeography and conservation are not meant to supplant the preservation of extensive native habitats. Instead, they serve to expand the purview of conservation to include human-dominated systems, thereby magnifying its effects.

The concept of ecosystem services

While countryside biogeography and conservation seek to recognize that human-dominated systems can still provide for biota, the larger and function-oriented framework of ecosystem services seeks to recognize that biota and their environments together provide for people all along the spectrum from lightly to heavily managed systems. Nature's benefits make life on Earth possible and worthwhile. Humans commonly take the things they get from nature for granted and thus do not consider their value when engaged in decision-making. This tendency often leads to decisions that imperil natural systems and their ability to provide life-support systems for humans in the long term. Better accounting for nature's benefits and explicitly considering nature's many values in decision-making will lead to better outcomes for societies and the natural systems on which they depend.

Put simply, ecosystem services are the benefits that people obtain from ecosystems. They sustain and fulfill human life and flow from many states and processes of ecosystems and the species inherent in them. The Millennium Ecosystem Assessment (MEA) of 2005 was a global effort involving more than 1,300 experts from 95 countries to assess the consequences of ecosystem change for human well-being. The MEA categorized ecosystem services into the following four classes:

1. *provisioning services*, which include the production of goods such as food, water, timber, and fiber

2. *regulating services*, which keep processes operating within comfortable bounds (e.g., stabilizing climate, moderating risk of flooding and disease, protecting or enhancing water quality)

3. *cultural services*, which provide recreational, aesthetic, educational, community, and spiritual opportunities

4. *supporting services*, which underlie the provision of the other three classes of benefits through soil formation, photosynthesis, nutrient cycling, and other foundational processes (see Table 1)

The MEA highlighted the diversity of services provided by the planet, the range of natural processes and natural-human interactions that underlie them, and the fundamental dependence of people on the flow of ecosystem services. A major, and decidedly alarming, finding of the MEA was the worldwide decline and degradation of ecosystem services. The MEA categorized approximately 60 percent of ecosystem services as degraded or used unsustainably. Imperiled services include capture fisheries, air and water purification, and the regulation of regional and local climate, natural hazards, and pests.

The MEA was instrumental in establishing the relevance of the concept of ecosystem services to policy and governance. Ongoing efforts to use the concept in understanding linked human and natural systems, setting up new markets, and designing sustainable management schemes has opened a new function—that of providing practical guidance for decision-making.

Biodiversity and ecosystem services

Biodiversity underpins the ability of ecosystems to provide the full range of services on which people depend. It is, therefore, intimately and causally connected to human well-being. The broad spectrum of biological diversity—from large

Ecosystem Service Category	Examples
Provisioning Services	
Food production	Food from agriculture, aquaculture, wild fisheries, hunting and gathering
Fiber production	Wood, cotton, hemp
Biomass fuel production	Wood, peat, algae
Production of biochemicals	Taxol from Pacific Ewe, antiviral drugs from sponges, food additives and cosmetics from seaweed
Regulating Services	
Climate regulation	Carbon storage and sequestration of tropical forests, ocean regulation of global temperatures and CO_2 cycles
Water regulation	Stormwater management by watersheds and floodplains
Disease and pest regulation	Natural enemies keep crop pests in check, fish graze coral reefs and prevent algal overgrowth
Natural hazard regulation	Nearshore habitats attenuate waves and decrease the likelihood and severity of erosion and flooding on shore
Pollination	23 percent of US agricultural production is pollinator-dependent
Cultural Services	
Provision of conditions that support or enhance ethical or existence values	Spiritual fulfillment from nature, belief that all species are worth protecting—no matter their direct value to humans
Provision of recreation and ecotourism opportunities	Hiking, mountain biking, climbing, scuba diving, kayaking, beachcombing, bird watching, fishing, whale-watching, clamming, gathering berries
Supporting Services	
Nutrient and water cycling	Forests, oceans and other biomes are critical to the cycling of water, Nitrogen, Oxygen, and Carbon
Soil formation	Decomposition returns carcasses to soil, vegetation traps soils
Primary production	Plant and algae transform solar energy into energy that fuels life

Table 1. The Millennium Ecosystem Assessment categories of ecosystem services and some examples of each. Reproduced by permission of Gale, a part of Cengage Learning.

charismatic mammals to the tiniest microbe—provides a wide range of ecosystem services. For example, birds disperse seeds, control pests, and provide bird-watching opportunities and inspiration; trees provide fiber, stabilize soil, regulate climate, and provide habitat for other critical species; bees pollinate crops and provide honey; decomposers return carcasses to soil. As extinctions occur—as they continue to do at up to 1,000 times the background rate—the ability of ecosystems to continue to provide these services is at risk.

In 1983 Paul R. Ehrlich and Harold A. Mooney argued cogently that it was in humanity's best interest to minimize anthropogenic extinctions in order to maintain ecosystem services. In that paper, they discuss the role of some species—or guilds of species—serving as *controllers* that determine the structure of the ecosystem and are critical to the flows of energy and materials through the system. The loss of these controllers—particularly producers and decomposers—can lead to the collapse of the system.

In the field of ecology, this relationship between biodiversity and ecosystem functioning has been the subject of intense

Makah Indian women dress salmon steaks for a community potlatch, Neah Bay, Washington. © B. Anthony Stewart/Getty Images.

production of ecosystem services also enhance biodiversity conservation. In a 2008 analysis of ecosystem services and conservation priorities, Robin Naidoo and colleagues found that while areas selected for (theoretical) conservation to maximize biodiversity did not produce more ecosystem services than randomly selected areas at the global scale, there were win-win regions of the globe where areas were important to both biodiversity and the provisioning of ecosystem services. There is much research to do here in understanding potential tradeoffs and synergies between approaches that seek to maximize the conservation of biodiversity and those that emphasize the provisioning of ecosystem services.

It is also worth considering that an ecosystem services approach is essential for biodiversity conservation. Mirroring the discussion of countryside biogeography and conservation above, the protection of biodiversity for biodiversity's sake can only go so far; Earth is dominated by people, and a focus on protected areas alone cannot work. In addition, the protection of biodiversity for biodiversity's sake can be considered an ecosystem service (with people benefiting from the intrinsic value of biodiversity) and thus brought under the larger theoretical umbrella.

Adding to the arsenal of those concerned with preventing extinctions, a focus on ecosystem services can bring new supporters into the cause of conservation. When decisions regarding natural resources are on the table, a focus on ecosystem services might provide incentives for decision makers to choose conservation and therefore benefit people and the other species with whom people share the planet.

research. Researchers have used observational, theoretical, and experimental approaches to try to understand this complex and wide-ranging link. Though significant controversy remains about some aspects of this relationship—particularly the mechanisms driving it—there is general agreement that a wide range of species with their inherently diverse traits matters to ecosystem functioning and thus the delivery of ecosystem services. It is also clear that diversity matters more under changing environmental conditions than it does under relatively constant ones.

Ecosystem services and the prevention of extinction

The framework of ecosystem services is inherently anthropocentric. At its core, it focuses on what ecosystems do for humans. To some, this is heretical. Indeed, opponents of this kind of thinking worry that an emphasis on ecosystem services will hinder the achievement of the goals of biodiversity conservation. Others suggest that the two can complement one another. For example, in a 2009 article Erik Nelson and colleagues showed that land-use scenarios that enhance the

Services provided by different ecosystems

On land, various different types of habitat provide a wide range of overlapping and non-overlapping ecosystem services. For example, vegetation in forests, grasslands and rangelands, and croplands can protect soils from erosion and improve soil fertility by retaining moisture and storing and recycling nutrients. Vegetation and soils together regulate the quantity, quality, and timing of water flows, thus moderating floods and droughts and providing a cleaner, more reliable water supply. Forests are critically important in regulating water and carbon cycles, influence the climate on local regional and global scales, and provide natural products for human use, including timber, firewood, mushrooms, fruits and seeds, medicinal plants, rubber, cork, and game. Forest and woodland habitats also harbor species that provide pollination and pest control for commercial or subsistence crops. Grassland and other dryland systems play many of these roles and also support vast livestock populations. Croplands and pasturelands, which encompass about 38 percent (12.1 billion acres [4.91 billion ha]) of Earth's land that is not covered in ice, are dedicated to the production of food, fodder, and fuel. Each of these systems, however natural or managed, can provide habitat for biodiversity and opportunities for recreational activities, spiritual experiences, and creative, cultural expression.

Freshwater ecosystems provide a suite of highly visible services to people. Freshwater that is regulated by terrestrial systems and the atmosphere is used for drinking, hydropower production, irrigation, household activities (e.g., washing), industrial purposes (e.g., cooling, manufacturing), and cultural experiences. People also gain large revenues and important nutrition from freshwater fisheries and aquaculture. Less appreciated is the value of sediment transport and deposition in rivers that supply important habitats within rivers and downstream beaches with important sand and gravel resources. Wetlands occupy a small fraction of Earth's surface, but provide a wide array of water quality, flood mitigation, coastal protection, and biogeochemical services. Finally, freshwater systems serve as pathways for human transportation and recreational or cultural activities.

The oceans encompass 71 percent of Earth's surface and provide a diverse array of ecosystem services. They hold over 95 percent of Earth's water and are thus the center of the global water cycle and a primary driver of the atmosphere's temperature, moisture content, and stability. Oceans are key players in the global cycles of carbon, oxygen, nitrogen,

Newfoundland marshs and bogs with spruce and fir trees. © Aurora Photos/Alamy.

phosphorus, and other major elements. They also transform, detoxify, and sequester wastes. Home to unimaginably vast populations of phytoplankton, they are responsible for approximately 40 percent of global net primary productivity. Marine fisheries and aquaculture provide nutrition, feed for animals, livelihoods, and important recreational and cultural opportunities. Harvests of other species for pharmaceuticals, food additives, and cosmetics also support health, nutrition, and human adornment. Marine habitats such as coral reefs, oyster reefs, and marshes regulate natural hazards, including storm surges, and may play an ever more critical role in helping coastal communities adapt to rising sea levels. Finally, coastal communities reap many benefits from coastal tourism and often define their very identities in relation to the sea and all it brings—materially and metaphorically.

Natural capital

If ecosystem services are the flows of benefits to people from coupled human and natural systems, then it must be that these flows come from reservoirs of *natural capital*—a term that is puzzling to many. Generally, economists recognize five forms of capital: financial, built, human, social, and natural. People are broadly aware of the first four but tend not to make natural capital an explicit category.

Financial capital provides a familiar metaphor for thinking about *natural capital*. When ecosystems are degraded, they are bankrupt, providing little in the way of returns. When ecosystems are healthy and diverse, they provide significant returns that are buffered from stresses. Dividends can be spent/harvested without impacting the principal, leading to sustainability into the future. Oyster reefs are a salient example of natural capital (see Figure 1). They provide the foundation for oyster fisheries, which not only deliver food but also help support livelihoods and define the character of thriving local communities. Destructive harvesting of reefs leads to a one-time gain but results in long-term losses.

Importantly, a natural capital perspective incorporates a systems perspective by encouraging the examination of a wide

Cork oak forms a thick, rugged bark containing high levels of suberin. Over time the cork cambium layer of bark can develop considerable thickness and can be harvested every 9 to 12 years to produce cork. The harvesting of cork does not harm the tree, in fact, no trees are cut down during the harvesting process. © Danita Delimont/Getty Images.

Figure 1. Oyster reefs provide many services. Oysters provide food and livelihoods. Reefs provide habitat for many species, some of which support additional fisheries (e.g., finfish and crab). Oysters also filter sediments and nutrients, supporting healthy vegetation (e.g., sea grass). Sea grass supports additional species important for fisheries and recreation. Finally, oyster reefs attenuate waves and can decrease erosion and flooding (compare the dashed and solid water levels and the dashed and solid shoreline profiles). Reproduced by permission of Gale, a part of Cengage Learning.

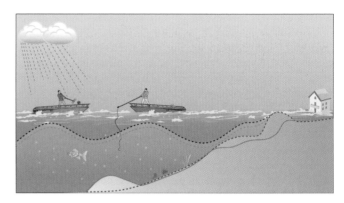

Figure 2. Diminished oyster populations cannot provide the full range of services. Sparse oysters decrease oyster harvests, are not habitat for other species (decreasing yields in other fisheries), and have diminished filtration and wave attenuation capacities. Reproduced by permission of Gale, a part of Cengage Learning.

range of benefits. To extend the oyster reef example, these reefs provide habitat for a number of commercially important fished species, thus providing support for additional fisheries. Moreover, oysters filter suspended sediments and excess nutrients from the water column. In the 1800s populations of native oysters in Chesapeake Bay, which lies along the US Atlantic coast, filtered a water volume equivalent to the entire bay in 3.3 days (Newell 1988). Improvements in water quality lead to improvements in local habitats (e.g., submerged aquatic vegetation), which provide further habitat for critical species and improve and diversify fishery yields and recreational opportunities. Finally, well-developed oyster reefs attenuate waves during big storms, decreasing the likelihood and severity of erosion and flooding onshore.

Despite the diverse flows of services provided by the natural capital of oyster reefs, Michael W. Beck and colleagues, in a 2011 study, showed that destructive harvesting, declines in habitat, and other pressures have degraded oyster reefs worldwide, with 85 percent of native reefs lost. Degraded reefs may still perform some services, and oysters without reefs can still be harvested, but neither provides the full range of services of healthy—and hence well-populated—reefs (see Figure 2). The current reduced oyster populations in the Chesapeake Bay filter the equivalent of its volume in 325 days (Grabowski and Peterson 2007), submerged aquatic vegetation is at historically low levels, and fisheries are in trouble. Still, oyster restoration efforts in the United States generally use a *put-and-take* system that focuses on restoration for short-term oyster fishery yields rather than on rebuilding reefs. A natural capital perspective argues for restoring oyster reefs to harvest oysters (the interest on the principal, to revisit the vernacular of financial capital), while leaving the principal in place to support fin-fish fisheries, improve water quality, and protect shorelines from coastal hazards. Indeed, in a 2003 study, Charles H. Peterson and colleagues showed that if only the harvest of fin fish that use oyster reefs as habitat in the Chesapeake Bay are considered, the dollar values of oyster reefs are greater than those generated

from the destructive harvesting of the reef. When taking a systems perspective, the restoration of natural capital makes economic sense and leads to better outcomes for both participants in coupled human-natural systems.

Valuing ecosystem services

Assessing the value of ecosystem services can be useful in many contexts—be they more complete (or less limited) cost-benefit analyses of alternate decisions or simply articulating the often implicit things that matter to people. Ecosystem service assessment tools can generate conceptual depictions of the ways in which human activities depend on and affect ecosystems, or they can measure the monetary value of particular services. Overall, the goal of such assessments is to link management actions, the development of new markets, and other activities directly to changes in ecosystem conditions and to gain an understanding of how those changes may affect the benefits that various individuals and groups derive from ecosystems. Critically, valuation does not have to be in monetary terms. Ecosystem services can be valued using monetary metrics, but they can also be assessed qualitatively by groups of people or by using social metrics (e.g., the number of people displaced by a flood).

One key method for valuing ecosystem services in monetary terms is to be very careful about what is being valued and for whom. For example, the concept of ecosystem services has people at its core—ecosystem services must be defined and assessed in terms of their direct impact on people. When estimating total ecosystem value, supporting services such as nutrient cycling often should not be valued in their own right because their value is captured in the total value of the final service, such as clean drinking water and food. This careful accounting avoids double counting. In other contexts, however, such as monitoring the effectiveness of management practices, it can be important to measure and value supporting services (e.g., using changes in water quality against targets in legislation such as total maximum daily loads).

A wide range of methodologies have been developed and applied to ecosystem service valuation. Each of these has strengths and weaknesses and needs to be matched carefully to the questions being asked, the data available, and other details about the application (see Marc N. Conte [forthcoming] Shuang Liu st al. [2010] for a review of ecosystem services valuation). Some services have explicit prices or are traded in open markets. In these cases, value may be directly obtained from what people pay for a good (e.g., timber, crops, or fish). Many services, however, are not traded in markets and represent public goods (e.g., climate regulation, coastal protection). A suite of non-market valuation techniques can be used to assign monetary value in these cases. Some approaches estimate value based on the costs people incur to enjoy a particular service (e.g., the travel-cost method uses estimates of what people spend to enable their enjoyment of a service), whereas others estimate value from what people are willing to pay for the service through purchases in related markets (e.g., purchases of properties near natural amenities such as lakes and wilderness areas). Some approaches that are used to represent the value of a service involve surveying people to ask them what they would be willing to pay for that service, whereas others use the cost of replacing a lost service (e.g., water purification by water treatment plants instead of forested lands and wetlands or coastal protection by seawalls instead of coastal wetlands) or the estimate of the cost of damage that would have occurred without the service (e.g., avoided flood damage). Finally, one can use the value, type, number, and level of services calculated in one place to estimate those attributes in new contexts that lack data. For example, one can take the value of a wetland calculated in one region and apply that value, on a per-unit-area basis, to wetlands in other regions or across the globe (a technique known as *benefits transfer*). Done carefully, through matching the conditions and services at the sites with and without local data, this approach can serve to provide reasonable estimates where data are lacking. Done haphazardly, this approach can yield nonsense (see below).

The economic approach to measuring benefits has many limitations. Some arise from the paucity of data necessary to apply otherwise useful approaches; others arise from the inappropriate use of particular approaches. For example, in a 2009 study, Mark L. Plummer offered a story of the perils of the benefits transfer approach. A small wetland in Louisiana was estimated to have a high ecosystem service value because of its adjacency to a potato chip factory. To comply with clean water legislation, the factory used the water-purification capacity of the wetland instead of incurring wastewater treatment costs, and the value of the adjacent wetland specifically reflected this cost savings to the factory owners. Given that few forested wetlands are adjacent to potato chip factories, this particular value is not transferrable to other wetlands. The careful matching of the characteristics and contexts of the studied and unstudied sites is imperative.

When primary site-specific data on ecosystem service provision and valuation are not available (or even when they are but future conditions are of interest), approaches that explore how changes in ecosystems lead to changes in their functions and services are less sensitive to the problem of matching sites because they generalize mechanistic relationships rather than estimated values. So-called production

function models, which show the relationship between inputs (e.g., fertilizer, labor) and outputs (e.g., crop production), have been used extensively in agriculture, manufacturing, and other sectors of the economy. Similar relationships exist between natural inputs (e.g., quality of coral reefs) and natural capital or ecosystem services (e.g., tourism and recreation). These models can provide estimates of changes in ecosystem structure, function, and service provision that can then be valued in various currencies using the techniques described above.

Various types of data required to understand the values of services in a particular place or to specific groups (e.g., local beneficiaries versus society as a whole) are often lacking. For example, the effects of a particular action on the benefits to private entities from the tangible production of goods (e.g., timber, fish) are generally accessible through the observation of prices in markets. Data and information about less tangible services such as the provisioning of recreational opportunities, spiritual value, food security, water purification, and climate regulation that might accrue to a wide range of local and global beneficiaries are significantly harder to come by.

Also, many services—including spiritual and cultural values—cannot easily (or, in many cases, ever) be quantified in monetary terms. Disputes among different stakeholders will often require extensive dialogue and explicit discussion of trade-offs. Such disputes are often multifaceted, and their languages and values cannot be reduced to a common currency. For example, commercial entities, which readily deal with monetary values, and indigenous groups, which do not, need to measure services in a wide range of currencies to provide transparent and fruitful discussion. Non-monetary indicators of ecosystem benefits are better suited to address services that cannot or should not be valued in monetary terms, including spiritual, cultural, or aesthetic values. Interviews, surveys, and other analyses can yield insights about deeply held beliefs of individuals and groups and the benefits they derive from ecosystems.

Estimates of the monetary value of ecosystem services are useful in some contexts and difficult or impossible to establish in others. Irrespective of the currencies used, the framework of ecosystem service is useful for conveying people's links to and dependence on functioning ecosystems and to informing environmental decision-making.

Ecosystem services in practice

An increasing recognition by scientific research, management, and governance sectors of the considerable and diverse values to society of ecosystem services is beginning to change the management of natural resources in a fundamental way. Impelled by new management and policy appetites for rigorous and flexible valuations of ecosystem services, scientific advances in conceptual frameworks and measurement tools have been developed and have begun to be translated into decision-making processes.

An ecosystem services perspective can be used to move the management of natural resources from simplistic, single-sector

In this August 26, 2009 photo birds are reflected in the calm water in the marshes of the J.D. Murphree wildlife management area near Port Arthur, Texas one year after Hurricane Ike slammed into the Texas coast. © AP Photo/Pat Sullivan.

approaches that can lead to unintended negative impacts on other sectors toward more holistic and pluralistic schemes. Analyses that focus on a wide range of values rather than only those that are easy to express in economic terms can guide decision makers toward actions that yield improved delivery of a broad range of ecosystem services to society. For example, analyses of the benefits of intact mangroves versus mangrove areas cleared for shrimp aquaculture have shown that the net value of shrimp production (market value minus subsidies and social costs) pales in comparison to the social and ecological benefits that standing mangroves provide. In addition, an ecosystem services perspective has the capacity to inform large-scale spatial planning that looks across traditional management silos and strives to maximize benefits to society from management actions. Along the west coast of Vancouver Island, modeling of ecosystem services from alternative management plans is informing the development of a marine spatial plan that balances aquaculture, recreation, and other uses of the marine environment (Guerry et al. 2012).

In many cases, investing in natural capital is more efficient than using built capital to provide desired services. The case of the provisioning of New York City's drinking water from the Catskill Mountains is a classic case in point. In the 1990s US law required water suppliers to filter surface water unless they could demonstrate alternative ways of keeping water clean. Managers explored building a filtration plant (at a cost of $6 billion to $8 billion, excluding annual operating and maintenance costs) or implementing a variety of watershed protection measures (acquisition of land, reduction of contamination, etc., at a cost of $660 million). Adding the additional benefits of watershed protection (e.g., recreational opportunities) only further tipped the scales in favor of investing in natural over built capital.

The ecosystem services paradigm is also being used to generate new funding streams for conservation. So-called payment for ecosystem service schemes are being set up around the world. Examples are emerging from the United States, Costa Rica, Australia, Colombia, and elsewhere. The Nature Conservancy, with many partners, has helped to set up over a dozen water funds in the northern Andes. In these programs, water users voluntarily invest in a trust fund that subsidizes conservation projects, improves water quality, and avoids significant water treatment costs. A public-private partnership consisting of water users and other stakeholders makes decisions about using the fund to finance conservation activities in the watershed. Such programs are protecting rivers and watersheds, maintaining the livelihoods of farmers and ranchers upstream, and helping provide people with irrigation, hydropower, and clean drinking water. Together, these promise positive results in social, financial, and conservation metrics.

Finally, the conceptual and organizing power of ecosystem services has begun to inform policy and institutions on a global scale. As described above, the Millennium Ecosystem Assessment was the first major effort to establish the utility of the ecosystem services framework in the international policy arena. Inspired by the MEAs compelling findings with respect to the concrete value of ecosystem services and the cost of their degradation, countries are coming together to making tangible commitments to safeguard global ecosystem services (such as the 2020 targets for the Convention on Biological Diversity) and to assess national and international progress toward those commitments (through, for example, the Group on Earth Observations Biodiversity Observation Network and the Programme on Ecosystem Change and Society; these latter two entities synthesize knowledge for

Farmers on their land, an organic family farm, Chilamate, Costa Rica. © MShieldsPhotos/Alamy.

the Intergovernmental Platform for Biodiversity and Eco-system Services, formally established in 2012). Several new international research efforts aim to feed into these international processes, including the Natural Capital Project, the Resilience Alliance, the Economics of Ecosystems and Biodiversity, and the Stockholm Resilience Centre. Other entities (e.g., the Katoomba Group and the Ecosystem Marketplace, both initiated by Forest Trends) are focused on establishing and tracking ecosystem service markets as one mechanism for bringing greater attention to the benefits of ecosystem services to society.

Future challenges and opportunities

Although scientific and policy communities have made great progress toward envisioning, framing, and applying ecosystem service accounting approaches in decision-making,

the road to mainstreaming these concepts is still long and winding. Some remaining challenges include: (1) the further development of models and tools for modeling ecosystem services that are simple to use, credible, transparent and realistically communicate uncertainty (InVEST, developed by the Natural Capital Project, is one example of a tool that meets some of these needs); (2) better understanding of the conditions under which biodiversity and ecosystem service provision provide opportunities for win-wins and ones in which trade-offs are likely to occur; and (3) additional examples of the ways in which ecosystem services approaches lead to improved outcomes for people and the environments on which they depend.

Ultimately, an ecosystem service perspective can help society perceive the critical services nature provides, appro-priately value natural capital, and help human communities make better choices about the use of these life-sustaining environments. It has the potential to help people see themselves in conservation. The framework of ecosystem services broadens the dialogue of conservation from setting aside parks to preserve nature and its biodiversity to setting aside parks *and* better managing landscapes and seascapes to yield the best outcomes for both partners in coupled human–natural systems. In this way, it is a powerful tool for recovering imperiled species and preventing extinctions in the long run.

Quantifying, mapping, and valuing ecosystem services have the potential to fundamentally change natural-resource deci-sion-making. Making explicit the connections between natural systems and human well-being helps people understand that the *environment* is not *other*. There is no black-and-white checker-board with people in some places and nature in others. Nature is not merely the concern of self-identified environmentalists. It provides for and sustains everyone. Most importantly people are inextricably part of nature, in all its hues.

Resources

Books

Berry, Wendell. *Home Economics*. San Francisco, California: North Point, 1987.

Conte, Marc N. "Valuing Ecosystem Services." In *Encyclopedia of Biodiversity*, 2nd ed, Edited by Simon A. Levin. Oxford: Elsevier, forthcoming.

Cronon, William. *The Trouble with Wilderness; or, Getting Back to the Wrong Nature*. In *Uncommon Ground: Rethinking the Human Place in Nature*, edited by William Cronon, 69–90. New York: Norton, 1995.

Daily, Gretchen C. "Countryside Biogeography and the Provi-sion of Ecosystem Services." In *Nature and Human Society: The Quest for a Sustainable World; Proceedings of the 1997 Forum on Biodiversity*, edited by Peter H. Raven and Tania Williams, 104–113. Board on Biology, National Research Council. Washington, DC: National Academies Press, 1997.

Daily, Gretchen C., ed. *Nature's Services: Societal Dependence on Natural Ecosystems*. Washington, DC: Island Press, 1997.

Daily, Gretchen C., and Katherine Ellison. *The New Economy of Nature: The Quest to Make Conservation Profitable*. Washington, DC: Island Press, 2002.

Freeman, A. Myrick, III. *The Measurement of Environmental and Resource Values: Theory and Methods*. 2nd ed. Washington, DC: Resources for the Future, 2003.

Grabowski, Jonathan H., and Charles H. Peterson. "Restoring Oyster Reefs to Recover Ecosystem Services." In *Ecosystem Engineers: Plants to Protists*, edited by Kim Cuddington, James E. Byers, William G. Wilson, and Alan Hastings, 281–298. Amsterdam: Academic Press, 2007.

Guerry, Anne D., Mark L. Plummer, Mary H. Ruckelshaus, and Daniel S. Holland. "Modeling Marine Ecosystem Services." In *Encyclopedia of Biodiversity*, 2nd ed., edited by Simon A. Levin. Oxford: Elsevier, forthcoming.

Kareiva, Peter, Robert Lalasz, and Michelle Marvier. "Conservation in the Anthropocene." In *Love Your Monsters*, edited by Michael Shellenberger and Ted Nordhaus, Oakland, CA: Breakthrough Institute, 2011. Accessed June 30, 2012.

http://breakthroughjournal.org/content/authors/peter-kareiva-robert-lalasz-an-1/conservation-in-the-anthropoce.shtml.

Kareiva, Peter, Heather Tallis, Taylor H. Ricketts, Gretchen C. Daily, and Stephen Polasky, eds. *Natural Capital: Theory and Practice of Mapping Ecosystem Services*. New York: Oxford University Press, 2011.

Krutilla, John V., and Anthony C. Fisher. *The Economics of Natural Environments: Studies in the Valuation of Commodity and Amenity Resources*. Rev. ed. Washington, DC: Resources for the Future, 1985.

Maris, Emma. *Rambunctious Garden: Saving Nature in a Post-Wild World*. New York: Bloomsbury USA, 2011.

Micheli, Fiorenza, and Anne D. Guerry. "Ecosystem Services." In *Encyclopedia of the Natural World*, edited by Alan Hastings and Louis Gross, 235–240. Berkeley: University of California Press, 2012.

Millennium Ecosystem Assessment. *Ecosystems and Human Well-Being: Synthesis*. Washington, DC: Island Press, 2005. Accessed June 29, 2012. http://www.maweb.org/documents/document.356.aspx.pdf

Naeem, Shahid, Daniel E. Bunker, Andy Hector, Michel Loreau, Charles Perrings, eds. *Biodiversity, Ecosystem Functioning, and Human Wellbeing: An Ecological and Economic Perspective*. New York: Oxford University Press, 2009.

National Research Council. *Approaches for Ecosystem Services Valuation for the Gulf of Mexico after the Deepwater Horizon Oil Spill: Interim Report*. Washington, DC: National Academies Press, 2011.

National Research Council. Committee on Assessing and Valuing the Services of Aquatic and Related Terrestrial Ecosystems. *Valuing Ecosystem Services: Toward Better Environmental Decision-Making*. Washington, DC: National Academies Press, 2005.

Newell, Roger I.E. "Ecological Changes in Chesapeake Bay: Are They the Result of Overharvesting the American Oyster, *Crassostrea virginica*?" In *Understanding the Estuary: Advances in Chesapeake Bay Research*, edited by Maurice P. Lynch and Elizabeth C. Krome, 536–546. CRC Publication 129. Solomons, MD: Chesapeake Research Consortium, 1988. Accessed June 29, 2012. Available from http://www.vims.edu/GreyLit/crc129.pdf

Righter, Robert W. *The Battle over Hetch Hetchy: America's Most Controversial Dam and the Birth of Modern Environmentalism*. New York: Oxford University Press, 2005.

Schulze, Ernst-Detlef, and Harold A. Mooney, eds. *Biodiversity and Ecosystem Function*. Berlin: Springer-Verlag, 1993.

Tallis, Heather, Gretchen C. Daily, and Anne D. Guerry. "Ecosystem Services." In *Encyclopedia of Biodiversity*, 2nd ed., edited by Simon A. Levin. Oxford: Elsevier, forthcoming.

Periodicals

Balvanera, Patricia, Andrea B. Pfisterer, Nina Buchmann, et al. "Quantifying the Evidence for Biodiversity Effects on Ecosystem Functioning and Services." *Ecology Letters* 9, no. 10 (2006): 1146–1156.

Beck, Michael W., Robert D. Brumbaugh, Laura Airoldi, et al. "Oyster Reefs at Risk and Recommendations for Conservation, Restoration, and Management." *BioScience* 61, no. 2 (2011): 107–116.

Bignal, Eric M., and David I. McCracken. "Low-Intensity Farming Systems in the Conservation of the Countryside." *Journal of Applied Ecology* 33, no. 3 (1996): 413–424.

Boyd, James, and Spencer Banzhaf. "What Are Ecosystem Services? The Need for Standardized Environmental Accounting Units." *Ecological Economics* 63, nos. 2–3 (2007): 616–626.

Carpenter, Stephen R., Ruth DeFries, Thomas Dietz, et al. "Millennium Ecosystem Assessment: Research Needs." *Science* 314, no. 5797 (2006): 257–258.

Carpenter, Stephen R., Harold A. Mooney, John Agard, et al. "Science for Managing Ecosystem Services: Beyond the Millennium Ecosystem Assessment." *Proceedings of the National Academy of Sciences of the United States of America* 106, no. 5 (2009): 1305–1312.

Chan, Kai M.A., Anne D. Guerry, Patricia Balvanera, Sarah Klain, Terre Satterfield, Xavier Basurto, Ann Bostrom, et al. "Where are *Cultural* and *Social* in Ecosystem Services? A Framework for Constructive Engagement." *BioScience* 62, no. 8 (2012): 744–756.

Chan, Kai M.A., Robert M. Pringle, Jai Ranganathan, et al. "When Agendas Collide: Human Welfare and Biological Conservation." *Conservation Biology* 21, no. 1 (2007): 59–68.

Chapin, F. Stuart, III, Brian H. Walker, Richard J. Hobbs, et al. "Biotic Control over the Functioning of Ecosystems." *Science* 277, no. 5325 (1997): 500–504.

Chapin, F. Stuart, III, Erika S. Zavaleta, Valerie T. Eviner, et al. "Consequences of Changing Biodiversity." *Nature* 405, no. 6783 (2000): 234–242.

Coen, Loren D., Robert D. Brumbaugh, David Bushek, et al. "Ecosystem Services Related to Oyster Restoration." *Marine Ecology Progress Series* 341 (2007): 303–307.

Cottingham, K.L., B.L. Brown, and J.T. Lennon. "Biodiversity May Regulate the Temporal Variability of Ecological Systems." *Ecology Letters* 4, no. 1 (2001): 72–85.

Daily, Gretchen C., Susan Alexander, and Paul R. Ehrlich. "Ecosystem Services: Benefits Supplied to Human Societies by Natural Ecosystems." *Issues in Ecology*, no. 2 (1997): 1–16.

Daily, Gretchen C., Gerardo Ceballos, Jesús Pacheco, Gerardo Suzán, and Arturo Sánchez-Azofeifa. "Countryside Biogeography of Neotropical Mammals: Conservation Opportunities in Agricultural Landscapes of Costa Rica." *Conservation Biology* 17, no. 6 (2003): 1814–1826.

Daily, Gretchen C., Paul R. Ehrlich, and G. Arturo Sánchez-Azofeifa. "Countryside Biogeography: Use of Human-Dominated Habitats by the Avifauna of Southern Costa Rica." *Ecological Applications* 11, no. 1 (2001): 1–13.

Daily, Gretchen C., Tore Söderqvist, Sara Aniyar, et al. "The Value of Nature and the Nature of Value." *Science* 289, no. 5478 (2000): 395–396.

Ehrlich, Paul R., and Harold A. Mooney. "Extinction, Substitution, and Ecosystem Services." *BioScience* 33, no. 4 (1983): 248–254.

Foley, Jonathan A., Navin Ramankutty, Kate A. Brauman, et al. "Solutions for a Cultivated Planet." *Nature* 478, no. 7369 (2011): 337–342.

Guerry, Anne D., Mary H. Ruckelshaus, Katie K. Arkema, et al. "Modeling Benefits from Nature: Using Ecosystem Services to Inform Coastal and Marine Spatial Planning." *International Journal of Biodiversity Science, Ecosystem Services, and Management.* 8 (2012): 107–121.

Horner-Devine, M. Claire, Gretchen C. Daily, Paul R. Ehrlich, and Carol L. Boggs. "Countryside Biogeography of Tropical Butterflies." *Conservation Biology* 17 (2003):168–177.

Isbell, Forest, Vincent Calcagno, Andy Hector, et al. "High Plant Diversity Is Needed to Maintain Ecosystem Services." *Nature* 477, no. 7363 (2011): 199–202.

Krutilla, John V. "Conservation Reconsidered." *American Economic Review* 57, no. 4 (1967): 777–786.

Liu, Shuang, Robert Costanza, Stephen Farber, and Austin Troy. "Valuing Ecosystem Services: Theory, Practice, and the Need for a Transdisciplinary Synthesis." *Annals of the New York Academy of Sciences* 1185 (2010): 54–78.

Loreau, Michel. "Biodiversity and Ecosystem Functioning: Recent Theoretical Advances." *Oikos* 91, no. 1 (2000): 3–17.

Mayfield, Margaret M., and Gretchen C. Daily. "Countryside Biogeography of Neotropical Herbaceous and Shrubby Plants." *Ecological Applications* 15 (2005): 423–439.

McCann, Kevin Shear. "The Diversity–Stability Debate." *Nature* 405, no. 6783 (2000): 228–233.

McCauley, Douglas J. "Selling Out on Nature." *Nature* 443, no. 7107 (2006): 27–28.

Naidoo, Robin, Andrew Balmford, Robert Costanza, et al. "Global Mapping of Ecosystem Services and Conservation Priorities." *Proceedings of the National Academy of Sciences of the United States of America* 105, no. 28 (2008): 9495–9500.

Nelson, Erik, Guillermo Mendoza, James Regetz, et al. "Modeling Multiple Ecosystem Services, Biodiversity Conservation, Commodity Production, and Tradeoffs at Landscape Scales." *Frontiers in Ecology and the Environment* 7, no. 1 (2009): 4–11.

Peterson, Charles H., Jonathan H. Grabowski, and Sean P. Powers. "Estimated Enhancement of Fish Production Resulting from Restoring Oyster Reef Habitat: Quantitative Valuation." *Marine Ecology Progress Series* 264 (2003): 249–264.

Plummer, Mark L. "Assessing Benefit Transfer for the Valuation of Ecosystem Services." *Frontiers in Ecology and the Environment* 7, no. 1 (2009): 38–45. Accessed June 30, 2012. doi:10.1890/080091.

Ranganathan, Jai, R.J. Ranjit Daniels, M.D. Subash Chandran, et al. "Sustaining Biodiversity in Ancient Tropical Countryside."

Proceedings of the National Academy of Sciences of the United States of America 105, no. 46 (2008): 17852–17854.

Redford, Kent, and M.A. Sanjayan. "Retiring Cassandra." *Conservation Biology* 17, no. 6 (2003): 1473–1474.

Taylor H. Ricketts, Gretchen C. Daily, Paul R. Ehrlich, and John P. Fay. "Countryside Biogeography of Moths in a Fragmented Landscape: Biodiversity in Native and Agricultural Habitats." *Conservation Biology* 15, no. 2 (2001): 378–318.

Sathirathai, S., and E.B. Barbier. "Valuing Mangrove Conservation in Southern Thailand." *Contemporary Economic Policy* 19, no. 2 (2001): 109–122.

Sekercioglu, Cagan H., Paul R. Ehrlich, Gretchen C. Daily, et al. "Disappearance of Insectivorous Birds from Tropical Forest Fragments." *Proceedings of the National Academy of Sciences of the United States of America* 99, no. 1 (2002): 263–267.

Scholes, R.J., G.M. Mace, W. Turner, G.N., et al. "Toward a Global Biodiversity Observing System." *Science* 321 (2008): 1044–1045.

Tallis, Heather, Peter Kareiva, Michelle Marvier, and Amy Chang. "An Ecosystem Services Framework to Support Both Practical Conservation and Economic Development." *Proceedings of the National Academy of Sciences of the United States of America* 105, no. 28 (2008): 9457–9464.

Tallis, Heather, and Stephen Polasky. "Mapping and Valuing Ecosystem Services as an Approach for Conservation and Natural-Resource Management." *Annals of the New York Academy of Sciences* 1162 (2009): 265–283.

Other

United Nations Environment Programme World Conservation Monitoring Centre. *World Database on Protected Areas.* Accessed June 30, 2012. http://www.unep-wcmc.org/world-database-on-protected-areas_164.html.

IUCN. "Global Ocean Protection: Present Status and Future Possibilities". Last modified November 23, 2010. http://www.iucn.org/knowledge/publications_doc/publications/?6500/Global-ocean-protection—present-status-and-future-possibilities.

Gallup.com. "Most Important Problem." Accessed June 30, 2012. http://www.gallup.com/poll/1675/most-important-problem.aspx.

Anne D. Guerry

• • • • •

The relative cost of saving species

Although biodiversity conservation underpins socioeconomic welfare and human well-being, the links between biodiversity and human well-being are often overlooked. In many cases, conservation goals are seen as being different from human well-being goals or even opposed to development goals, promoting a trade-off between investing in nature and investing in economic development or human well-being. Nevertheless, in spite of its invisibility in markets, the benefits obtained from conserving biodiversity usually exceed the opportunity costs of its conservation.

Moving from biodiversity intrinsic values to instrumental values

People's attitudes and preferences are usually the driving force for promoting biodiversity conservation policies. These attitudes and preferences could be related to spiritual, moral, ethical, utilitarian, ecological, or scientific motivations. Therefore, understanding which factors cause changes in people's attitudes and preferences toward biodiversity is essential for designing successful conservation policies. Moreover, understanding people's preferences toward biodiversity, as well as the underlying factors promoting them, helps identify which type of relationship is established between nature and society (e.g., utilitarian, affective). How people perceive biodiversity determines the choices they make regarding how to conserve it (Gatzweiler 2008).

The broad spectrum of attitudes toward species may be grouped under two main conflicting and antagonistic headings: (1) affection, which represents people's affective and emotional responses to species, versus (2) utility, which represents people's self-interest related to the instrumental value of species. This model of two primary motivations for biodiversity conservation has been identified in several scientific investigations during the twentieth century; these studies have used various terms in describing this model, including anthropomorphism versus anthropocentrism, affection versus economic self-interest, empathy/identification versus instrumental self-interest, and affection versus utility (see Serpell 2004 for a review).

This attitudinal structure toward species based on two antagonistic dimensions is rooted in two kinds of human

values—intrinsic value and instrumental value. *Intrinsic value* has been defined as the nonhuman species' rights to exist that cannot be measured in terms of markets; it is strongly related to the human moral obligation to share the planet (Callicott 1986). In the words of Paul R. Ehrlich and Anne Ehrlich (1981, 48), "Our fellow passengers on Spaceship Earth, who are quite possibly our only living companions in the entire universe, have a right to exist." Thus, intrinsic value proposes that biodiversity has a value in itself and is valued as an end in itself, independent of its usefulness (Brondízio et al. 2010). On the contrary, *instrumental value* is a function of utility and assumes that biodiversity is valuable only as a means or tool for people to obtain human welfare, satisfaction, or happiness. Instrumental value lies not in the object itself but in the uses or potential uses it gives or could give.

This dichotomy of values constitutes the core of modern environmentalism in Western culture and has guided the conservation debate in the last decades (Pascual et al. 2010). In fact, the conservation debate in modern Western societies has moved away from biocentric ethics, which refers to promoting conservation based on the idea of any theory that considers all live beings as possessing intrinsic value, to anthropocentric arguments, which consider that biodiversity conservation should be supported because it is a tool to enhance human quality of life, satisfying as many individual preferences, desires, and needs as possible (such as the importance of many plants for medicinal compounds, a plant's ornamental value, or the satisfaction that people feel just knowing that biodiversity will exist in the future). Because the way people relate to nature is reflected not only by their behavior, but also by the rules they apply to articulate values for nature (Brondízio et al. 2010), this conservation debate actually represents how people perceive human–nature relationships and support different biodiversity conservation policies.

As J. Baird Callicott claimed in a 1986 article, a biodiversity conservation movement based on intrinsic values (or on biocentric ethics) is ultimately emotional and derived from feelings and affects toward nonhuman living species and is one in which all of humanity feels responsible. Consequently, the intrinsic value of species could be determined by how they make humans feel, which depends ultimately on certain factors that promote human affections toward species. Several studies have demonstrated that humans devote attention to, and feel affection for, those species that are phylogenetically close to them and to those that are physically or behaviorally similar to them (i.e., vertebrates, mostly mammals). In addition to those factors, Konrad Lorenz showed in a 1971 study that humans have an innate tendency to feel affection for nonhuman species with human neonates' physical features, such as large eyes, large rounded forehead, or short nose.

If intrinsic values (and the underlying factors determining them) have mostly been responsible for the biodiversity conservation policies of the late twentieth and early twenty-first centuries, does this mean that conservation policies have been biased toward those species that evoke positive affections and emotions in humans? Scientific research has demonstrated that, in fact, likability factors, such as different physical traits (e.g., body size, body shape, eye size), as well as whether the species is phylogenetically close to humans, play an important role in defining the conservation policies adopted for a species. As a consequence, the majority of species (i.e., invertebrates, plants, fungi, and microorganisms), which are key to maintaining ecological functions as well as sustaining life, are underrepresented in conservation priority-setting schemes. Biodiversity conservation policies based exclusively on the intrinsic values of nonhuman species trigger people to desire to protect those higher life-forms (i.e., mammals and birds) that produce positive emotions and feelings.

In order to counteract this "vertebrate chauvinism" in biodiversity conservation strategies (so named by Thomas E. Lovejoy; see Kellert 1986), a pragmatic movement has appeared under an anthropocentric utilitarianism lens. Conservation arguments based on the benefits society obtains from nature have exponentially increased in scientific literature under the rubric of the term *ecosystem services* (see Figure 1). In fact, this notion has involved the development of new conceptual and analytical frameworks for understanding human-nature relationships, such as the Millennium Ecosystem

Assessment (MA), The Economics of Ecosystems and Biodiversity (TEEB), and the Intergovernmental Platform on Biodiversity and Ecosystem Services (IPBES).

In this debate, the main issue is whether these two approaches to species conservation are complementary or opposed. The question here focuses again on the society-nature relations model. If one considers humanity's relation with nature to be innate to human evolutionary history—the biophilia hypothesis (Kellert and Wilson 1993)—then one should conclude that all humans have an innate need to connect with nature from both dimensions: ethical or emotional and utilitarian (Brondízio et al. 2010). Under this relations model, biocentric and anthropocentric arguments should coexist in biodiversity conservation strategies. From this point of view, justifying conservation exclusively on the basis of ethical considerations about the right of species to exist (i.e., intrinsic value) ignores an important motivation that people have for preserving species: the importance of biodiversity as a source of human well-being through the delivery of ecosystem services (i.e., instrumental value). In fact, the Millennium Ecosystem Assessment (2005) acknowledges that people can make decisions concerning biodiversity based on their own and others' well-being as well as based on ethical concerns about species, thus recognizing instrumental and intrinsic values.

Conceptualizing biodiversity as a source of ecosystem services

There is a growing awareness of the importance of biodiversity for human well-being through the delivery of ecosystem services to society (see Figure 2). *Ecosystem services* have been defined as the direct and indirect contributions of ecosystems to human well-being (Wittmer and Gundimeda 2010). They have been classified into three main categories: (1) provisioning services, such as food, fiber, raw materials, resources for medicine, and fuel; (2) regulating services, such as climate regulation, air purification, flood control, water purification, erosion control, renewal of soil fertility, pollination, and pest control; and (3) cultural services, such as recreational activities and ecotourism, aesthetic values, spiritual values, and scientific and local ecological knowledge. In addition to the inherent importance of biodiversity for most of the provisioning services, such as food, medicines, and fibers, biodiversity influences recreational activities (e.g., recreational fishing) and the aesthetic experiences of people, promoting psychological well-being; moreover, biodiversity plays a key role in sustaining a number of regulating services, including water purification, flood buffering, global climate and microclimate regulation, air purification, soil formation, and pollination.

Although all components of biodiversity, from genotypes to ecological communities, are essential to the delivery of ecosystem services, there is ample evidence supporting the positive and key effect of functional diversity on delivering many ecosystem services (Díaz et al. 2006), particularly regulating services (see Figure 2). Here, functional groups are understood as those that perform particular ecosystem functions (e.g., nitrogen fixation, seed dispersal, biomass

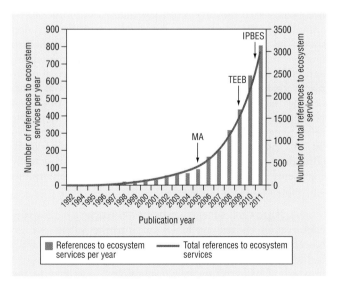

Figure 1. Number of publications using the term of "ecosystem services" between the period 1990–2011 years (search carried out in ISI Web of Science in March 2012). Reproduced by permission of Gale, a part of Cengage Learning.

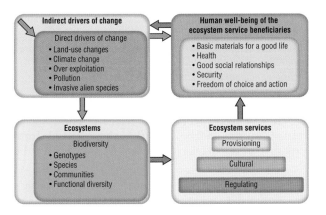

Figure 2. Complex links between biodiversity and human well-being. On one hand, biodiversity is affected by indirect and direct drivers of change but, on the other hand, biodiversity influence on human well-being through the delivery of ecosystem services. Among all the ecosystem services, regulating services are the base for the maintenance of others and provisioning services depend on both regulating, which are strongly related to ecosystem's integrity and those cultural services related to knowledge (local ecological knowledge and scientific knowledge). Reproduced by permission of Gale, a part of Cengage Learning.

decomposition, biomass production) through different physiological, structural, behavioral, or phenological characteristics or traits (e.g., root type, leaf texture, plant height). In addition, the diversity of species and populations within a functional group—that is, its amount of functional redundancy—is the

main biological component that helps to preserve ecosystem services (Elmqvist et al. 2003). In fact, the scientific evidence suggests that the relationship between biodiversity and ecosystem services supply is mostly based on functional diversity and functional redundancy, rather than on species richness.

Reviews probing for evidence of links between biodiversity and the delivery of ecosystem services have shown that the functional role of microorganisms, vegetation, and invertebrates is the component of biodiversity that most influences ecosystem service supply. Table 1 shows, in a simplified way, the links between biodiversity components and the delivery of provisioning and regulating services. Cultural services are not considered in Table 1 because they are generally coproduced by a deeper relationship between people and ecosystems. In fact, most cultural services (e.g., recreational activities, aesthetic enjoyment, spiritual values) require people's experiences in nature. That is, only people who visit nature will benefit from cultural services through experiential enjoyment.

A number of scientific studies have reviewed the role of functional diversity and species in the delivery of ecosystem services (Kremen 2005; Balvanera et al. 2006; Díaz et al. 2006; Luck et al. 2009; Cardinale et al. 2012). As previously mentioned, however, little information exists about the state and tendencies of the key functional groups that provide services to society, despite the essential role they play either in ecological functioning or in the delivery of ecosystem services. Focusing more scientific and political attention on these overlooked functional groups is an undertaking of great importance for the implementation of future biodiversity conservation strategies.

The social and economic importance of biodiversity

Regarding the social dimension, biodiversity conservation has been considered an important tool for achieving poverty reduction and social justice because priority areas of conservation seem to be efficient targets for maintaining human well-being through the services they deliver (Turner et al. 2012). Emerging evidence for this has been found by the World Bank, which estimated that ecosystem services directly support more than one billion people living in poverty (World Bank 2011). Similarly, a 2010 study carried out in Costa Rica and Thailand by Kwaw S. Andam and colleagues has shown that those districts with protected areas experienced between 10 percent and 30 percent less poverty. In addition, WWF International and the International Union for Conservation of Nature offered some different examples in two reports— *Safety Net: Protected Areas and Poverty Reduction* (Dudley et al. 2008) and *Can Protected Areas Contribute to Poverty Reduction? Opportunities and Limitations* (Scherl et al. 2004), respectively—in which they illustrated how protected areas mostly contribute to the wider aspects of human well-being (i.e., promoting basic material resources for living, increasing education rates, enhancing good social relationships in those cases of comanagement) rather than to poverty reduction per se (i.e., increasing per capita income). In fact, biodiversity conservation directly contributes to poverty reduction in at

Ecosystem services	Organizational level at which biodiversity is involved	Main taxonomic groups involved
Provisioning		
Food	Genes, species populations, communities	Mainly vegetation (plant crops, wild fruits, etc.), fish, birds, and mammals. In specific cases, fungi, invertebrates, and other vertebrates
Fibers	Species populations, communities, functional groups	Vegetation and vertebrates
Medicine	Genes, species populations	Microorganisms, fungi, vegetation, and animals
Regulating		
Air purification	Species populations, functional groups	Microorganisms and vegetation
Water purification	Communities, functional groups	Microorganisms, vegetation, and aquatic invertebrates
Hydrological regulation, erosion control and flood mitigation	Species populations, communities, functional groups	Vegetation
Renewal of soil fertility	Communities, functional groups	Soil microorganisms, nitrogen-fixing plants, soil invertebrates, and waste products of animals
Pollination	Species populations, functional groups	Insects, birds and mammals

Table 1. Links between ecosystem services delivery and biodiversity (organizational level of biodiversity and main taxonomic groups involved). Reproduced by permission of Gale, a part of Cengage Learning.

least five ways: food security, health improvements, income generation, reduced vulnerability for human populations, and ecosystem services delivery. Consequently, protecting biodiversity conservation priority areas delivers direct benefits to society and alleviates poverty (Turner et al. 2012).

Regarding the economic dimension, the contributions of biodiversity to human well-being have major economic significance, even though their value is not recognized by conventional markets. The economic value of biodiversity can be divided into *insurance value* and *output value* (Pascual et al. 2010; see Figure 3). Insurance value refers to the ecosystem's capacity to maintain the delivery of ecosystem services to society in the face of disturbance. Output value is obtained from the aggregation of the ecosystem service benefits and usually is recognized as total economic value (TEV).

The components of TEV usually are represented using a value taxonomy (see Figure 4). The main distinction made is between *use* and *nonuse values* (Turner et al. 2008). Use values include *direct use*, *indirect use*, and *option values*. Direct use values are derived from the conscious use and enjoyment of ecosystem services by individuals. They may be extractive, such as agriculture or fishing, or they may be nonextractive, such as recreational activities, nature tourism, and aesthetical

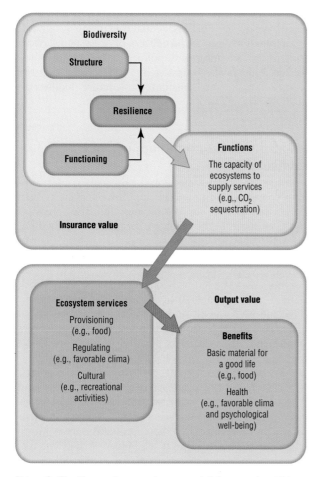

Figure 3. The "ecosystem service cascade" framework, which assess links between the insurance value and the output value measured as Total Economic Value (TEV). Reproduced by permission of Gale, a part of Cengage Learning.

enjoyment of landscapes. Hence, extractive direct use values are strongly related to provisioning services, whereas non-extractive direct use values are strongly related to cultural services. Indirect use values are associated with the regulating services that are provided by biodiversity and ecosystems, without entailing consciousness in their use. Finally, option values are related to the future direct and indirect uses of biodiversity by individuals. These values are related to the individual satisfaction derived from ensuring that an ecosystem service will be available for the future.

Nonuse values reflect the value that arises from people's feelings attributable to the existence of biodiversity and ecosystem services, either related to the satisfaction that individuals derive from the knowledge that an ecosystem service will be available to future generations (bequest value) or related to the individual satisfaction obtained from the knowledge that a species, species population, community, landscape, or ecosystem service continues to exist (existence value). Among nonuse values, some authors also consider altruist values, which are related to the satisfaction derived from ensuring that biodiversity and ecosystem services are available to other people in the current generation. Hence, whereas bequest values are related to intergenerational equity concerns, altruist values are related to intragenerational concerns (Pascual et al. 2010). In addition, because nonuse values are derived from human satisfaction related to moral, ethical, or religious issues, they are connected to particular cultural services, such as spiritual values. Thus, the boundaries of nonuse values are not clear-cut because of their strong relationship with different moral motivations—that is, a person's motivation to support biodiversity conservation is composed of intertwined value types and not just one value type. In fact, the existence and bequest values are sometimes considered a way to capture the intrinsic values that exist under the ecosystem service framework.

In practical terms, the valuation of TEV is estimated by the aggregation of the different ecosystem service values provided by a particular ecosystem that are feasible to quantify. The estimation of the TEV of biodiversity and ecosystem services in a particular area or ecosystem should help to assess the contributions of these services to the economy of the region. The most interesting example illustrating the contributions of biodiversity to economic or financial value is the case of the Catskill watershed, the main source of water for New York City. The citizens of New York avoided spending $6 billion to $8 billion on the construction of a water filtration plant by instead investing $1.5 billion to buy lands in the upper watershed to restore and conserve them (Chichilnisky and Heal 1998). From a scientific point of view, one of the first attempts to estimate the value of ecosystem services on a global basis was a 1997 study by Robert Costanza and colleagues encompassing seventeen ecosystem services delivered by sixteen biomes around the world. These researchers estimated that the value of ecosystem services ranged from $16 trillion to $54 trillion per year, with an average annual value of over $33 trillion, a figure higher than the annual global gross domestic product. In a 2007 study, Will R. Turner and colleagues demonstrated that the areas identified a decade earlier by Costanza and colleagues as priority areas

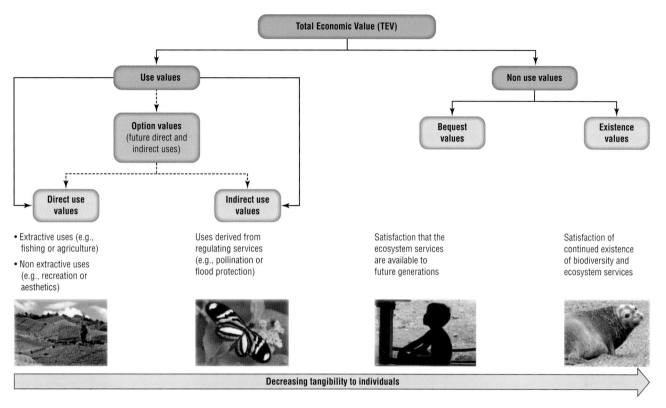

Figure 4. Components of the Total Economic Value (TEV) of biodiversity and ecosystems (Turner et al. 2008). Reproduced by permission of Gale, a part of Cengage Learning.

because of the high economic value of their ecosystem services spatially overlap with high (terrestrial) biodiversity priority areas. This means that a general positive concordance exists between biodiversity conservation priority areas (i.e., areas with a high diversity of species) and areas with high ecosystem services value. Among all terrestrial areas, Turner and colleagues found the following areas of high concordance: Congo, the Amazon, central Chile, the Western Ghats region of India, and parts of Southeast Asia. Consequently, this study shows that tropical forests offer the best opportunities for synergies between conserving biodiversity because of their intrinsic value (high biodiversity) and their instrumental value (high ecosystem service value). In fact, sustainable strategies of tropical forest management show a positive benefit-cost ratio—that is, net marginal benefits—derived from not only timber but also non-timber forest products, water supply, hydrological regulation, flood prevention, carbon sequestration, and the preservation of habitat for endangered species. Unsustainable management, by contrast, was associated with higher private benefits through timber harvesting or conversion to agriculture lands but reduced social benefits because of the degradation of regulating and cultural services. Unfortunately, estimates of land-use changes in forests reveal the highest losses in tropical forests, specifically high economic losses. For example, the economic losses expected in Brazil between the years 2000 and 2050 as a result of the loss of tropical forests range from 2 percent to 6 percent of the country's 2050 gross domestic product ($3.6 billion to $11 billion in 2050 dollars) (Chiabai et al. 2011). These economic losses

are underestimated because they include only the degradation of wood and non-wood forest products (provisioning services), carbon sequestration as regulating service, and the cultural services of ecotourism and existence value. For a review of the economic losses derived from unsustainable management of forests, see the 2011 study by Aline Chiabai and colleagues.

One of the main contributors to the study of the economics of biodiversity in the early twenty-first century has been the international project called the Economics of Ecosystems and Biodiversity (Wittmer and Gundimeda 2010). This project, which quantifies the costs of biodiversity loss and ecosystem services degradation, has analyzed seven different scientific studies that estimate the TEV of sustainable management practices in different ecosystems and compared the figures to those arising from management regimes involving land-use intensification or unsustainable practices. Overall, this comparative analysis demonstrates that the global benefits of biodiversity conservation are about 250 percent greater than the benefits from conversion. Thus, the economic value of managing the ecosystem more sustainably (in all of the cases) is higher than the economic value of converting the ecosystem or managing it unsustainably, even though the private benefits of the provisioning services captured in current markets (e.g., timber, aquaculture, agriculture) would favor conversion or unsustainable practices.

The following three studies, which were discussed in a 2002 article by Andrew Balmford and colleagues, offer a comparison between the net benefits of conserving ecosystems and the net benefits of converting them:

1. A study carried out in a tropical forest of Mount Cameroon estimated that the social benefits obtained from low-impact logging practices, such as non-timber forest products, flood prevention, carbon sequestration, and nonuse values, amount to $1,376 per acre. However, the conversion of forests to small-scale agriculture for food yielded the greatest private benefits, about $809 per acre.

2. Research in Thailand to study which management practice is more profitable—conserving mangroves or converting the land to aquaculture—demonstrates that the net marginal benefits of conserving mangroves exceed at least $324 per acre. The estimated benefits of mangrove ecosystems, which include timber, charcoal, non-wood forest products, offshore fisheries, and storm protection, yield a minimum of $405 and a maximum of $14,569 per acre. In contrast, the conversion to shrimp farming in order to provide food generates benefits of only about $81 per acre.

3. Draining freshwater marshes in agricultural areas in Canada involves a total economic loss of $1,376 per acre. The social benefits of marshes are related to provisioning services, such as hunting; regulating services, such as hydrological regulation; and cultural services, such as recreational activities. The TEV value estimated for these services was on average $2,347 per acre, whereas the value of land converted from wetlands to agriculture totaled $971 per acre.

In these three cases, the average economic losses attributable to conversion or to unsustainable practices amounted to 54.9 percent of the TEV of the sustainable ecosystems. These three studies provide good examples of cases in which the conversion of ecosystems for forestry, aquaculture, or agriculture does not make sense from either an environmental viewpoint or a socioeconomic perspective.

The value of species conservation

As species populations are becoming increasingly more threatened worldwide because of the effects of drivers of change (see Figure 2), it is important to acknowledge an important consideration—the willingness of society to conserve these species. One of the most commonly used measures of people's awareness of species conservation is their willingness to pay (WTP) for the preservation of a particular species. Several studies have found that individuals are willing to pay a small portion of their income toward the aim of conserving endangered and rare species (Richardson and Loomis 2009). The WTP for conserving species could be related to a variety of reasons, such as recreational use (direct use value), pollination or pest control (indirect use values), bequest values, or mostly existence values.

Although this indicator has been frequently used by scientists since the early 1980s, its use is clearly on the rise in the twenty-first century. In a 2008 article, Berta Martín-López and colleagues reviewed 60 studies, published from 1980 to 2005, measuring the WTP for conserving species (see Figure 5a). Overall, scientists have tended to focus on estimating WTP for protecting vertebrates, mainly mammals and birds (see Figure 5b). The economic valuation of species has also focused on studying those species that live in marine and forest ecosystems, whereas species that live in drylands and human-made habitats have been understudied. In addition, the economic studies have focused (among all the components of TEV) on nonuse values, mostly existence values (see Figure 5c). Whereas direct use values, mainly recreational activities, generate the highest economic value for species protection ($54.4 per person in 2005 dollars), indirect use values generate the lowest ($33.9 per person). It seems, however, that people refer to both use and nonuse values in determining their WTP for biodiversity conservation. This indicates that economic valuations of species should explore all components of TEV rather than individual elements. In general, people hold a higher WTP for protecting mammals and birds, in comparison with reptiles and crustacean, which are less valued (see Figure 5d).

It is interesting to note that people around the world are highly willing to pay for big species (both in length and weight) with relative big eyes (see Figure 6). Consequently, the economic value of species conservation obtained through this technique is biased not only toward mammals and birds but also toward those species with neonates' physical characteristics, as Konrad Lorenz showed in a 1971 study. This indicates that people's motivations that underlie existence values (i.e., phylogenetic and physical similarity) also explain society's prioritization regarding species conservation.

In addition, WTP experiments for species conservation show that visitors usually allocate higher values to species protection than local stakeholders, because visitors are likely to have a recreational component to their WTP and are likely to be more knowledgeable about biodiversity. Reaching the opposite conclusion, however, was a 2011 study by Adriana Ressurreição and colleagues that centered on marine biodiversity in southern Europe. When comparing residents and visitors with similar socioeconomic profiles, local residents were found to be more likely to attach higher values to species protection than visitors. This study also demonstrated that people are able to perceive the importance of biodiversity conservation in wider terms than the single-species approach and that they therefore attach greater economic benefits to the conservation of entire ecosystems rather than partial conservation plans of particular species.

To give another illustration of the economic contributions of species to human welfare, Justin G. Boyles and colleagues in a 2011 study found that insectivorous bats provided pest control services worth more than $3.7 billion annually to the US agricultural sector. This figure represents the cost of pesticide applications that the agricultural industry does not have to make because bats eat so many insects.

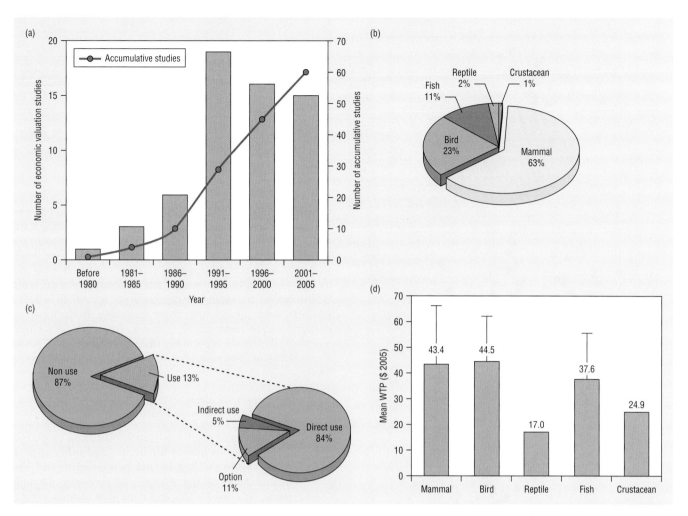

Figure 5. (a) Historical trend of publications studying the individuals' willingness to pay (WTP) for species conservation. (b) Distribution of publications for species conservation among taxonomic groups. (c) Distribution of publications for species conservation among components of the Total Economic Value (TEV). (d) Average economic value (WTP per person) per taxonomic group. Reproduced by permission of Gale, a part of Cengage Learning.

Similarly, John E. Losey and Mace Vaughan, in a 2006 study, estimated the economic value of insects in the United States through an analysis of dung burial, pest control, and pollination regulating services, as well as the role of insects in wildlife nutrition, in order to estimate their contribution to nature tourism and recreational activities. These authors estimated that the value of such ecosystem services provided by insects totaled almost $60 billion per year. The annual contribution of dung beetles with regard to decomposing cattle waste was estimated at $380 million. Regarding pollination, one-third of human food comes from plants pollinated by wild pollinators, and native insect pollinators were found to be responsible for almost $3.7 billion of fruits and vegetables produced in the United States per year. The value of pest control by native insects was estimated at approximately $4.5 billion per year. Finally, the key role of insects as a food source for much of the wildlife involved in hunting, fishing, and bird-watching activities, was estimated to be valued at $49.9 billion annually. Despite the clear benefits they supply, insects are pitifully underrepresented in conservation policies and

conservation programs involving insects are severely underfunded.

Economic benefits of biodiversity versus conservation costs

Based on all the estimates of the economic value of ecosystems, species, or the ecosystem services provided by biodiversity, increased investment in biodiversity conservation is justified. According to a 2002 study by Balmford and colleagues, expanding the biodiversity conservation area network to cover 15 percent of the landscape and 30 percent of the seascape would cost approximately $45 billion per year. In this case, the costs of biodiversity conservation include five main components: acquisition, management, damage, transaction, and opportunity (see Table 2). The ecosystems within that network would supply a diverse flow of ecosystem services with net annual benefits greater than $4.4 trillion per year. Consequently, investing in biodiversity conservation would help to maintain global ecosystem services that are

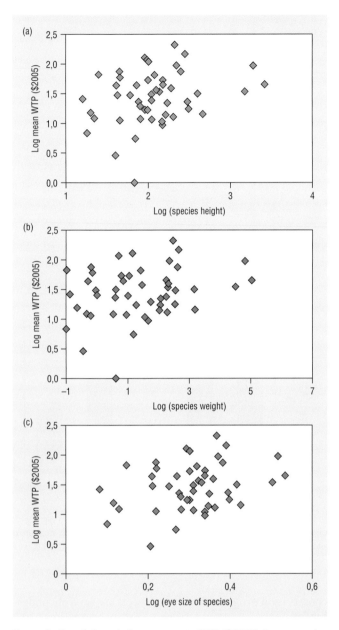

Figure 6. Correlations between average WTP ($2005) for conserving species and species' physical traits: (a) length, (b) weight, and (c) relative eye size. Reproduced by permission of Gale, a part of Cengage Learning.

Type of conservation cost	Description
Acquisition costs	Costs of acquiring property rights to a parcel of land.
Management costs	Costs associated with the management of a conservation strategy, e.g., costs related to establishing and maintaining a protected area.
Transactions costs	Costs associated with negotiating, such as the cost of transferring property rights.
Damage costs	Costs related with damages arising from conservation activities in human welfare, such as the economic losses caused by wild fauna on livestock.
Opportunity costs	Costs of foregone opportunities, such as the economic losses caused when extractive uses are forbidden in protected areas.

Table 2. Different types of conservation costs. Reproduced by permission of Gale, a part of Cengage Learning.

Gundimeda 2010). At the global scale, the economic benefits of conservation are more important than its costs. At the local scale, however, the situation is ambiguous because sometimes benefits outweigh the costs and other times the opposite holds true. Although the management costs of protected areas are usually placed at the national or international levels, the costs of lost access to natural resources, costs related to wildlife conflicts, and opportunity costs stemming from forgoing conversions to nonnatural systems (that produce high commercial value) are often placed at the local scale. Protected areas, however, provide several benefits at the local level:

1. the supply of clean water, particularly from mountains and well-managed natural forests

2. the maintenance of food security because protected areas act as an habitat for insects and other species that provide pollination and pest control, and the maintenance of the subsistence economy through game, fish, wild fruits, and natural medicines

3. a reduction in risks from unpredictable events and natural hazards, such as landslides or floods being mitigated by soil stabilization or hydrological regulation

4. the support of nature tourism, benefiting local people

5. assistance in protecting cultural heritage and spiritual values such as traditional practices and sacred places

For all these services, there is evidence that local economic benefits provided by ecosystems in protected areas exceed the costs of conservation. In summary, the overall magnitude of the global benefits of biodiversity conservation, which was estimated at $33 trillion annually in 1997 by Robert Costanza and colleagues, suggests that the economic benefits clearly exceed the costs of conservation, which Alexander N. James and colleagues estimated at only $6 billion per year in 1999.

Beyond 2020 biodiversity targets

In spite of efforts to improve biodiversity conservation based on higher investments, biodiversity continues to decline and the main target for 2010—that is, to halt the decline of biodiversity—were not reached (Butchart et al. 2010). The economics of biodiversity has been given a key role to play in the strategies outlined to achieve the 2020 targets: (1) to deal with the underlying causes of biodiversity loss through

worth about 100 times more than the costs of designating and managing the conservation network. Therefore, by spending $1 on biodiversity conservation, humanity will reap $100 in benefits because of all the ecosystem services provided by biodiversity. Although this conclusion should be regarded as indicative rather than precise because of the assumptions, generalizations, and various findings of different valuation studies, it seems clear that the benefits of biodiversity conservation do outweigh its conservation costs.

Nevertheless, the economic benefits and costs of biodiversity conservation through the use of protected areas vary significantly depending on the spatial scale (Wittmer and mainstreaming biodiversity across society; (2) to diminish the pressures on biodiversity, (3) to improve the state of biodiversity by protecting ecosystems, species, and genetic diversity, (4) to enhance the benefits obtained from biodiversity and ecosystem services, and (5) to implement conservation strategies based on participatory planning, knowledge management, and capacity building. In this sense, it has been recommended that the economic values of biodiversity should be integrated into national accounts and environmental decision-making. It should be noted, however, that capturing the economic values of biodiversity in terms of markets—sometimes by creating new markets, through such initiatives as wetland banking, endangered species credits, and payments for ecosystem services—could obscure other dimensions of biodiversity value, including intrinsic, ecological, and socio-cultural values. Consequently, while the economic criteria of biodiversity conservation (i.e., that the economic benefits of biodiversity exceed the costs of conservation) is essential to designing conservation policies, it is also important to keep in mind that (1) biodiversity conservation is a combination of intrinsic and instrumental values; (2) within instrumental values, biodiversity conservation should be viewed as a diverse flow of ecosystem services in order to promote sustainable land-use management and avoid social conflicts; and (3) criteria of social equity, of either an intragenerational and intergenerational nature, should be considered.

Resources

Books

Brondízio, Eduardo S., Franz W. Gatzweiler, Christos Zografos, and Manasi Kumar. "The Socio-Cultural Context of Ecosystem and Biodiversity Valuation." In *The Economics of Ecosystems and Biodiversity: Ecological and Economic Foundations*, edited by Pushpam Kumar, 149–181. London: Earthscan, 2010.

Callicott, J. Baird. "On the Intrinsic Value of Nonhuman Species." In *The Preservation of Species: The Value of Biological Diversity*, edited by Bryan G. Norton, 138–172. Princeton, NJ: Princeton University Press, 1986.

Dudley, Nigel, Stephanie Mansourian, Sue Stolton, and Surin Suksuwan. *Safety Net: Protected Areas and Poverty Reduction*. Gland, Switzerland: WWF International, 2008. Accessed July 27, 2012. http://assets.panda.org/downloads/safety_net_final. pdf

Ehrlich, Paul R., and Anne Ehrlich. *Extinction: The Causes and Consequences of the Disappearance of Species*. New York: Random House, 1981.

Kellert, Stephen R. "Social and Perceptual Factors in the Preservation of Animal Species." In *The Preservation of Species: The Value of Biological Diversity*, edited by Bryan G. Norton, 50–73. Princeton, NJ: Princeton University Press, 1986.

Kellert, Stephen R., and Edward O. Wilson, eds. *The Biophilia Hypothesis*. Washington, DC: Island Press, 1993.

Lorenz, Konrad. *Studies in Animal and Human Behaviour*. Vol. 2. Translated by Robert Martin. Cambridge, MA: Harvard University Press, 1971.

Millennium Ecosystem Assessment. *Ecosystems and Human Well-Being: Biodiversity Synthesis*. Washington, DC: World Resources Institute, 2005.

Pascual, Unai, Roldan Muradian, Luke Brander, et al. "The Economics of Valuing Ecosystem Services and Biodiversity." In *The Economics of Ecosystems and Biodiversity: Ecological and Economic Foundations*, edited by Pushpam Kumar, 183–255. London: Earthscan, 2010.

Scherl, Lea M., Alison Wilson, Robert Wild, et al. *Can Protected Areas Contribute to Poverty Reduction? Opportunities and Limitations*. Gland, Switzerland: International Union for Conservation of Nature, 2004.

Turner, R. Kerry, Stavros Georgiou, and Brendan Fisher. *Valuing Ecosystem Services: The Case of Multi-functional Wetlands*. London: Earthscan, 2008.

Wittmer, Heidi, and Haripriya Gundimeda, eds. *The Economics of Ecosystems and Biodiversity for National and International Policy Makers*. London: The Economics of Ecosystems and Biodiversity, 2010.

World Bank. *The Changing Wealth of Nations: Measuring Sustainable Development in the New Millennium*. Washington, DC: World Bank Publications, 2011.

Periodicals

Andam, Kwaw S., Paul J. Ferraro, Katharine R.E. Sims, et al. "Protected Areas Reduced Poverty in Costa Rica and Thailand." *Proceedings of the National Academy of Sciences of the United States of America* 107, no. 22 (2010): 9996–10001.

Balmford, Andrew, Aaron Bruner, Philip Cooper, et al. "Economic Reasons for Conserving Wild Nature." *Science* 297, no. 5583 (2002): 950–953.

Balvanera, Patricia, Andrea B. Pfisterer, Nina Buchmann, et al. "Quantifying the Evidence for Biodiversity Effects on Ecosystem Functioning and Services." *Ecology Letters* 9, no. 10 (2006): 1146–1156.

Boyles, Justin G., Paul M. Cryan, Gary F. McCracken, and Thomas H. Kunz. "Economic Importance of Bats in Agriculture." *Science* 332, no. 6025 (2011): 41–42.

Butchart, Stuart H.M., Matt Walpole, Ben Collen, et al. "Global Biodiversity: Indicators of Recent Declines." *Science* 328, no. 5982 (2010): 1164–1168.

Cardinale, Bradley J., J. Emmett Duffy, Andrew Gonzalez, et al. "Biodiversity Loss and Its Impact on Humanity." *Nature* 486, no. 7401 (2012): 59–67.

Chiabai, Aline, Chiara M. Travisi, Anil Markandya, et al. "Economic Assessment of Forest Ecosystem Services Losses: Cost of Policy Inaction." *Environmental and Resource Economics* 50, no. 3 (2011): 405–445.

Chichilnisky, Graciela, and Geoffrey Heal. "Economic Returns from the Biosphere." *Nature* 391 (1998): 629–630.

Costanza, Robert, Ralph d'Arge, Rudolf de Groot, et al. "The Value of the World's Ecosystem Services and Natural Capital." *Nature* 387, no. 6630 (1997): 253–260.

Díaz, Sandra, Josepth Fargione, F. Stuart Chapin III, and David Tilman. "Biodiversity Loss Threatens Human Well-Being." *PLoS Biology* 4, no. 8 (2006): e277.

Elmqvist, Thomas, Carl Folke, Magnus Nyström, et al. "Response Diversity, Ecosystem Change, and Resilience." *Frontiers in Ecology and the Environment* 1, no. 9 (2003): 488–494.

Gatzweiler, Franz W. "Beyond Economic Efficiency in Biodiversity Conservation." *Journal of Interdisciplinary Economics* 19, nos. 2–3 (2008): 215–238.

James, Alexander N., Kevin J. Gaston, and Andrew Balmford. "Balancing the Earth's Accounts." *Nature* 401, no. 6751 (1999): 323–324.

Kremen, Claire. "Managing Ecosystem Services: What Do We Need to Know about Their Ecology?" *Ecology Letters* 8, no. 5 (2005): 468–479.

Losey, John E., and Mace Vaughan. "The Economic Value of Ecological Services Provided by Insects." *BioScience* 56, no. 4 (2006): 311–323.

Luck, Gary W., Richard Harrington, Paula A. Harrison, et al. "Quantifying the Contribution of Organisms to the Provision of Ecosystem Services." *BioScience* 59, no. 3 (2009): 223–235.

Martín-López, Berta, Carlos Montes, and Javier Benayas. "Economic Valuation of Biodiversity Conservation: The Meaning of Numbers." *Conservation Biology* 22, no. 3 (2008): 624–635.

Naidoo, Robin, Andrew Balmford, Paul J. Ferraro, et al. "Integrating Economic Costs into Conservation Planning." *Trends in Ecology and Evolution* 21, no. 12 (2006): 681–687.

Ressurreição, Adriana, James Gibbons, Tomaz Ponce Dentinho, et al. "Economic Valuation of Species Loss in the Open Sea." *Ecological Economics* 70, no. 4 (2011): 729–739.

Richardson, Leslie, and John Loomis. "The Total Economic Value of Threatened, Endangered, and Rare Species: An Updated Meta-analysis." *Ecological Economics* 68, no. 5 (2009): 1535–1548.

Serpell, James A. "Factors Influencing Human Attitudes to Animals and Their Welfare." *Animal Welfare* 13, supp. 1 (2004): S145–S151.

Turner, Will R., Katrina Brandon, Thomas M. Brooks, et al. "Global Conservation of Biodiversity and Ecosystem Services." *BioScience* 57, no. 10 (2007): 868–873.

Turner, Will R., Katrina Brandon, Thomas M. Brooks, et al. "Global Biodiversity Conservation and the Alleviation of Poverty." *BioScience* 62, no. 1 (2012): 85–92.

Other

"The Economics of Ecosystems and Biodiversity." Accessed July 27, 2012. http://www.teebweb.org.

"Intergovernmental Platform on Biodiversity and Ecosystem Services." Accessed July 27, 2012. http://www.ipbes.net.

"Millennium Ecosystem Assessment." Accessed July 27, 2012. http://www.maweb.org.

Berta Martín-López
Marina García-Llorente

Should extinction be prevented?

Human activities are driving species to extinction and have been doing so since at least the Pleistocene. In the past few centuries, however, the extinction rate has accelerated, and in response several countries have enacted national legislation that seeks to halt or minimally slow the pace at which species are lost, with the United States being the first to do so, in 1973. Unfortunately, national legislation by itself rarely is enough to bring a species back from the brink of extinction. The recovery of species and the prevention of extinction inevitably cost money, require active management, and often necessitate forgoing significant opportunities for commercial gain. For instance, to prevent the extinction of forest-dwelling species, logging may need to be banned, housing developments halted, a job-creating factory closed down, and conversion of lands to farmland blocked (Kareiva and Marvier 2011). In other words, to prevent extinction, short-term human desires and needs may have to be set aside in favor of direct conservation action.

A parable of the potential conflict between saving species and taking care of humans entails a woman with a rifle on top of a ridge, looking down on her daughter playing by the river's edge. She notices a tiger creeping up behind her child. Imagine that the tiger is the last living female tiger, and imagine that this tiger is pregnant with what could be the cubs that will ensure the recovery of tigers from extinction. Should the armed mother shoot and kill the last tiger, thereby causing the extinction of a species to save one child? Most humans would answer this question the same way: The child has precedence over the tiger. But at the same time, if there were no child in imminent danger, most people would not want to be the person to have killed the last tiger in the world. This parable seems to highlight a moral premise: Human life has precedence over species existence. But the parable does so by presenting a contrived choice that probably never occurs—a choice between the life and death of humans and the life and death of a species. In reality, the issue is not about such absolute or simplistic choices but is instead about trade-offs. In particular, what trade-offs in terms of short-term human gain are people willing to accept in order to prevent the loss of species? More specifically, what investment of money and energy should be applied toward saving species from extinction?

Remarkably, most societies and countries have decided that some money and energy should be spent on preventing the extinction of species. For example, the Convention on Biological Diversity requires national strategies for the protection of biological diversity, and it has over 150 nations as signatories. However, the economic cost that can be tolerated in order to prevent species extinction varies from country to country. Developed economies have larger conservation budgets and conservation is a higher national priority than is often the case in underdeveloped countries (Kareiva and Marvier 2011). In all likelihood, not even in the richest countries, such as the United States, is there enough money to prevent the extinction of every species. Resources and political will are limited. As a result, in spite of laws aimed at halting extinction, some species will still be lost. This situation raises the more nuanced ethical questions. One question is whether humans should decide which species to save and which species to let go. Another is whether humans should try to save them all given that doing so means some species will still go extinct. Interested people ask if it be better not to spend energy and money on species that are so-called lost causes so that more money and resources can be expended on species for which conservation dollars are likely to make a difference.

The situation of species nearing extinction and limited resources to sustain them brings to mind the notion of triage, a concept that was first applied to battlefield casualties by Napoléon Bonaparte's chief surgeon during the Russian campaign in 1812 (Iverson and Moskop 2007). Interestingly in the Russian campaign, triage was applied by focusing treatment on those who were most severely injured and leaving the less injured to manage without treatment. The British navy took a different tact later in the nineteenth century, with its surgeon arguing that treatment should be allocated according to whether it was likely to be successful. Hence, those whose wounds were probably fatal should be denied treatment. Medical triage has been applied in wars ever since the nineteenth century, and subsequently in emergency rooms and following disasters, with different rules depending on the situation. The ethics of medical triage is hotly debated, but at least it is debated.

Regarding endangered species, however, rarely does the concept of triage enter the discussion. First, the notion that it might be necessary to give up on some species can feel like a surrender which may sap people's commitment to conservation and diminish their overall conservation energy to the detriment of all species. Second, some find it morally repugnant to give up on any species, in much the same way

The snail darter (*Percina tanasi*) was at the center of debate between the economic benefits of building the Tellico Dam and the risk of species extinction. Jerry A. Payne, USDA Agricultural Research Service, Bugwood.org.

one could find it morally repugnant to give up on a sick or injured human. But there are obviously arguments in favor of species triage, just as there are arguments in favor of medical triage. In the absence of enough money to protect all species, the greatest benefit in terms of averted extinctions is likely to be achieved by targeting species for which there is the greatest likelihood that the investment will significantly improve prospects for survival, while disinvesting in species for which there is little chance that conservation efforts will do much good. This line of reasoning emerges from classical decision theory in which the objective is to minimize loss.

The question of triage in species conservation and the feasibility, as well as the ethics, of sitting by and allowing some species to go extinct, while working hard to save other species, are best explored in a specific context. A good choice for offering that context is the United States, which in 1973 passed into law the Endangered Species Act (ESA) and which is the focus of the remainder of this entry. The ESA reflects the broad public values that favored conservation during what has been called the peak of the wave of environmentalism that swept the United States in the 1960s and 1970s (Barrow 2009). It is worth noting that, although the act itself is comprehensive and extends protection to all vanishing animals and plants, the public's support for the legislation has been rallied primarily by appeals surrounding dwindling charismatic animals such as the bald eagle, the grizzly bear, and whales. The full scope of the act, however, became clearer much later, when conflicts arose between the economic benefits of the Tellico Dam proposed in Tennessee and the extinction risk posed by the dam to the snail darter, a small nondescript fish species (Barrow 2009). The Tellico dam led to the creation of a special committee that asked for the first time whether or not the "cost" of protecting an endangered species was too great in terms of jobs or the economy. Ever since then the ESA has been surrounded by controversy about economics versus species protection, with opponents of conservations signaling out certain species as especially unworthy of the trouble—like a lowly small fish such as the snail darter. This becomes an issue largely because there may be so many species at risk, that for some interest groups the

cost of protecting them all seems prohibitive. Before turning to what is feasible and ethical regarding species triage in the United States, it is important to document the scale of the problem—that is, how many species are indeed at risk, the extent to which resources might be in short supply, the wide range of investment currently being made to rescue species from extinction, and, finally, how a shortage of money for active management can limit species recovery.

Scope of the US extinction problem

As of August 2012, the U.S. Fish and Wildlife Service (USFWS) had listed 1,400 threatened and endangered species (USFWS 2012c). The total amount spent by state and federal agencies on threatened and endangered species in 2010 was reported as about $1.45 billion. Although this might seem like a lot of money, it represents only one-fifth of the funding the USFWS has requested to support its species recovery plans and is far less than is needed for averting extinction of the full list of threatened and endangered species (Miller et al. 2002). What is most striking about USFWS funding for endangered species is how unevenly it is applied (see Figure 1). In 2010 more than $51 million was spent on the chinook salmon in the Snake River, while more than 22 percent of listed animals had received no funding from the USFWS and did not even have an active recovery plan. Overall, half the money spent to protect or recover endangered species in the United States goes to only 36 of the 1,400 species at risk. In other words, only 3 percent of the imperiled species receive over 50 percent of USFWS conservation funding. This uneven distribution leads to the conclusion that the USFWS is, effectively, already practicing triage and allowing some species to go extinct while investing heavily in others. Of course, this is never explicitly stated in any USFWS literature.

Money matters. Analysis of species trends and investments in recovery reveal that the more money is spent, the more likely the species' status is to improve (Miller et al. 2002). Conversely, an absence of money leads to no or negligible

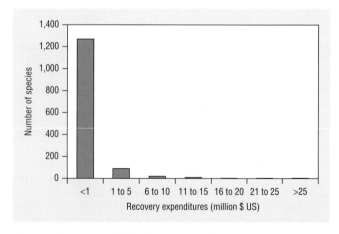

Figure 1. Frequency distribution of expenditures per species showing that the bulk of the species (more than 1,200) receive less than $1 million, while a handful of species receive 20 times that amount (greater than $20 million). Reproduced by permission of Gale, a part of Cengage Learning.

Protection of chinook or King salmon (*Oncorhynchus tshawytscha*) may result in an increase in fishing jobs, spur economic growth, improve recreation and maintain salmon as a food source. © Mark Conlin/Alamy.

improvements in species status. For example, in a 2007 study, Paul J. Ferraro and colleagues analyzed changes in population status for species and found a clear pattern whereby the listing of a species as threatened or endangered accomplished little, but if there was investment in recovery there was evidence of improvement. What this means is that an absence of funding for species recovery is potentially a death sentence for a species. A 2012 plea for saving the world's 100 most threatened species argues similarly, that their fate is squarely in people's hands—these species can be saved if people act (Baillie and Butcher 2012).

Establishing priorities for which species to save

Uneven spending on species recovery makes it clear that there are priorities at play, with some species receiving huge efforts on their behalf and other species being neglected. The question is how these priorities are determined and how they should be determined.

The USFWS has established a priority ranking system for recovery efforts to guide how it should invest in threatened and endangered species protection (see Table 1). In this ranking system a "1" represents the highest priority and an "18" the lowest. In this ranking system the degree of risk is the most important factor, so that the more highly endangered a species is, the higher priority it is given. After risk, "recovery potential" dictates the priority ranking, which can be interpreted as the likelihood that investing in recovery will be effective. The last factor that enters into the USFWS priority ranking system is evolutionary uniqueness as indicated by taxonomic status (e.g., if the species is the only species in a genus it receives the highest priority, and if it is a subspecies that is listed, it receives the lowest priority).

Although the US government has published clear principles for prioritization, actual expenditures by the USFWS do not appear to match its own ranking system. Instead there appear to be taxonomic biases and political factors at play. In general, comprehensive reviews of Endangered Species Act

expenditures suggest that characteristics such as cultural value, historical use, size, and charisma have greater influence on whether a species receives more or less funding than its priority rank (Langpap and Kerkvliet 2010). Charismatic megafauna in particular tend to attract more funds than their strict ranking would suggest they should (Dawson and Shogren 2001).

Birds and mammals are more appealing to most people than snakes or lizards. Thus, it should come as no surprise that reptiles receive less funding per species than mammals and birds, even in cases where priority rankings are similar (Restani and Marzluff 2002). Similarly, recovery plans written for birds and mammals are more likely to be written at the subspecies level (67% and 77%, respectively at the subspecies level), whereas all other animal taxa tend to have recovery plans written at the species level (77% at species level and only 23% at subspecies level; Tear et al. 1995). In a 1995 study, Benjamin M. Simon and colleagues found no correlation between a species' priority ranking and the amount of funding it received, if it received any funding at all. The bald eagle, which had a priority score of 14, which is almost as low as a species can get (the lowest being 18), nonetheless received more than $9.8 million in 2004. That expenditure put the bald eagle in the top 25 species in terms of funding, even though it had an extremely low priority score (Schwartz 2008). This, of course, is not surprising because the bald eagle is the national bird of the United States and an American icon. In general, several surveys of human attitudes toward species have revealed strong biases toward species that are large in size, have neotenic or juvenile features, and have big eyes; examples include bears and otters.

Degree of Threat	Recovery Potential	Taxonomy	Priority
High	High	Monotypic genus	1
	High	Species	2
	High	Subspecies	3
	Low	Monotypic genus	4
	Low	Species	5
	Low	Subspecies	6
Moderate	High	Monotypic genus	7
	High	Species	8
	High	Subspecies	9
	Low	Monotypic genus	10
	Low	Species	11
	Low	Subspecies	12
Low	High	Monotypic genus	13
	High	Species	14
	High	Subspecies	15
	Low	Monotypic genus	16
	Low	Species	17
	Low	Subspecies	18

Table 1. U.S. Fish and Wildlife Service recovery priority ranking system (Federal Register, volume 48, page 51935). Degree of threat: High = extinction is certain. Moderate = postponement of recovery will not result in extinction. Low = temporary population decline may be self-correcting or short term. Recovery Potential: High = Threats are known, minimal management needed and high success potential. Low = Contrary is true. Monotypic genera have greater impact on diversity = higher priority. Reproduced by permission of Gale, a part of Cengage Learning.

The bald eagle (*Haliaeetus leucocephalus*) was taken off the USFWS's endangered species list in 2007. © Shutterstock.com/visceralimage.

How far may people go to prevent extinction?

Many species persist only because of massive efforts on the part of humans to sustain them. An iconic example of an "expensive" recovery effort is the captive breeding and reintroduction program for the California condor (*Gymnogyps californianus*). Between 2000 and 2010 about $19.4 million was spent on the California condor, and in 2010 more than $1.6 million was spent on this species. In June 2012 there were 233 condors living in the wild, which means it has cost on average at least $83,000 for each condor now in the wild (National Park Service 2012). The breeding of condors in captivity has taken years to perfect and involves the use of puppets to feed nestling condors to reduce the chance a bird will imprint on humans. To call this reintroduction a success is something of a stretch, because the condors that do exist in the wild are not really self-sufficient. All birds are monitored using radio or GPS collars; are provided food, vaccinations, and trash removal around active condor sites; and are recaptured semiannually for blood draws to test for lead poisoning and for a physical examination, which can result in long stays in captivity. Lead poisoning is thought to be the main cause of the near extinction of the condor, and nearly 20 percent of free-flying condors in California are treated for lead poisoning annually (Finkelstein et al. 2012). By no means is the condor the only expensive recovery effort. Dozens of other species

might be considered conservation reliant—meaning the species would perish if not for human intervention and continuous management such as food supplementation, predator control, captive breeding to restock wild populations, and removal of invasive species.

Zoos play a major role in maintaining many species that are near extinction. There are more tigers in captivity in the United States than there are tigers in the wild (McKibben 1989). If a species is considered to be on the brink of extinction in the wild, then a captive breeding program is perhaps its only hope of survival. Under such circumstances Association of Zoos and Aquariums (AZA) member institutions draw up a species survival plan (SSP), which establishes a managed breeding program involving all the individuals of a specific species residing in member zoos. As of 2011 all SSP species were categorized into green, yellow, or red programs (listed from least to most dire, with red being the most dire), with the categories determined by the number of individuals in zoos and aquariums as well as genetic diversity. If the captive population totals fewer than fifty individuals, that species is categorized as red. Several species of lemurs, along with fishing cats (*Prionailurus viverrinus*), fishers (*Martes pennanti*), sun bears (*Helarctos malayanus*), and belugas (*Delphinapterus leucas*), had been placed in the red program category by 2012. The overall goal of these survival plans is to increase

The California condor (*Gymnogyps californianus*) is a federally listed endangered species in the United States and subject of a major captive breeding program. © Jeffrey Jackson/Alamy.

the genetic diversity of the captive population within AZA member institutions through the knowledge of individuals' bloodlines and their relation to others within a species (AZA 2011).

In 2012 there were over 300 SSP programs running in AZA member institutions. Many of these plans are focused on the charismatic megafauna that draw crowds, whereas others enrolled in the SSP programs may never have a display or exhibit for visitors (AZA 2012). Captive breeding is difficult and typically results in low success rates. Among the species that do not breed well in captive environments are giant pandas (*Ailuropoda melanoleuca*), western lowland gorillas (*Gorilla gorilla gorilla*), snow leopards (*Uncia uncia*), Andean condors (*Vultur gryphus*), both African elephants (*Loxodonta* spp.) and Asian elephants (*Elephas maximus*), and cheetahs (*Acinonyx jubatus*), to name a few. Even if zoos are successful in captive breeding, species are a long way from being recovered. One reason is many reintroductions to the wild have low success rates (Kareiva and Marvier 2011). The limiting factor here is usually habitat—the habitat a species requires may no

longer exist in the wild or may be so degraded (as is the case with the California condor because of the continued presence of lead in carcasses) that reintroductions fail. For example, the axolotl (*Ambystoma mexicanum*), a unique neotenic salamander, was known from only two lakes in central Mexico, and neither of those lakes exists today. Finally, even if a reintroduction originally appears to succeed, humans may end up eliminating the species a second time from the wild. For example, two decades after reintroducing the Arabian oryx (*Oryx leucoryx*), the wild population suffered a major setback when the population was poached to the point that the wild population was no longer sustainable (Seddon 1999).

What are the ethics of species triage?

Species are unique creations of evolution that may have taken millions of years to develop into their modern forms. Just as it would be immoral to destroy a beautiful piece of art, a religious temple, or a human life, it is immoral to cause the extinction of a species. That said, the issue of triage and allowing some species to go extinct raise difficult questions.

Essentially there are three ethical approaches to species triage.

1. The first is to regard all species as having equal value and then to apply a decision-theoretic approach that minimizes the loss rate of species by allocating resources in a manner such that the net outcome in terms of species survival is maximized. This approach inevitably leads to giving up on some species.

2. The second approach is to assign a priori greater value to some species and lesser value to other species and invest resources for recovery according to those assigned values. The USFWS ranking system appears to mix these two approaches. Ranking species in terms of recovery potential is a result that would come out of a decision-theoretic analysis seeking to minimize overall loss regardless of species identity. Meanwhile, ranking species from a monotypic genus higher than species that do not come from a monotypic genus reflects a value system based on evolutionary uniqueness.

3. A third approach is to essentially use a lottery to decide which species to protect, recognizing there is not enough money to protect all species, but also feeling no one has the right to decide arbitrarily which species should be favored. While it might seem like this approach is never used in the formal way that a lottery might be used (such as the draft lotteries that took place in the United States during the Vietnam War), in some sense this is indeed the current practice for endangered species in the United States. One might draw this conclusion because if a return on investment is not applied (and it is not in the United States), and if no one follows any explicit set of transparent values (they do not in the United States), then the result is an ad hoc triage effort, with ad hoc representing something close to random in the sense that no rational decision process can be articulated.

Between 1990 and 2010, with the ascendance of conservation based on the concept of ecosystem services, arguments were made to prioritize species according to their utility to humans (Kareiva and Marvier 2011). Here the idea is that if a primary purpose of conservation is to protect ecosystem services (the benefits nature provides to people), then perhaps resources should focus on protecting those species that deliver the greatest benefits to humans. No government agency has officially or publicly embraced this logic yet. Under this system salmon might be prioritized because, if they were recovered, they could be a valuable food source. In 2009 California, Oregon, and Washington combined generated $480 million in revenue from commercial fishing, further emphasizing the importance of saving fish species (NMFS 2010). Investing time and money in salmon protection and recovery may increase fishing jobs, spur economic growth, improve recreation, and make salmon more plentiful as a food.

A related but slightly different idea is to rank species according to their total impact on ecosystem structure and function. In other words, one might want to focus on recovering species whose loss fundamentally alters an ecosystem—the so-called keystone species. An example of a keystone species is the North American beaver (*Castor canadensis*). In the seventeenth century the extreme desirability of beaver fur for making felt created an incentive for beaver hunting and trapping throughout North America. Indeed, much of early North American exploration is credited to the high demand for beaver fur. As a result of this intense hunting, by the early 1900s the beaver was nearly extinct. The beaver was spared from extinction largely because fashion changed, easing trapping and hunting pressures. Today, the beaver is increasing in number throughout its range thanks to regulations on hunting and trapping, the relative absence of predators, and habitat availability. Beavers are important to their environment because their act of damming streams creates ponds. These ponds are beneficial to the beaver in terms of creating a site for mating pairs to build their lodges, and, compared to a riffle site, beaver ponds have more than 200 times the biomass of aquatic insects and other species that form the base of the food chain. The soil in flooded sites behind beaver dams also experiences an increase in the retention of sediment and organic matter, creating habitat for countless organisms. Simply by building a dam and flooding the surrounding area, a beaver produces a habitat for numerous other species and essentially engineers an entire ecosystem. Very few if any species can compare to beavers in terms of the impact they have on their environment. A beaver can cut more than a ton of wood within the surrounding 330 feet (100 m) of its pond, and in doing so totally reshaping the local plant community (Naiman et al. 1988). This keystone role, or role as so-called ecosystem engineer, would today make the North American beaver a priority in most people's minds if the species were ever to be threatened again with extinction.

When it comes to providing ecosystem services or shaping the way the world functions, all species are not created equal. In habitats and ecosystems throughout the world there are keystone species and ecosystem engineers—species whose presence fundamentally shapes the entire system in unique

North American beaver (*Castor canadensis*) may be considered a keystone species because of the dramatic impact on the environment they have through the building of their dams. © fotofactory/Shutterstock.com.

ways. If those species were to go extinct, not only are they lost, but the whole ecosystem might change.

Keystone species and ecosystem services use ecological science to establish priorities for protection. While these scientific rationales are important, human desires, choices, and culture are equally critical. In some cases, humans may want species for their direct enjoyment. For example, in 2011, 90.1 million Americans, representing 38 percent of the total US population, participated in hunting, fishing, or other wildlife-associated recreational activities; the $145 billion spent on these activities supported local communities through employment and tax revenue (USFWS 2012d). The salmon mentioned earlier fit into this category, and salmon species made up twenty-three of the top sixty funded species in the 2010 fiscal year. Salmon are also important to many Native American tribes on the Pacific Coast, where they not only provide for traditional livelihoods but also figure prominently in Native American rituals and celebrations (NMFS 2009).

When is a species extinct?

Among human societies, if an individual is in a coma and has limited brain activity, most societies allow family members to make a choice to take the human off life support. The reasoning is that the human is, to all intents and purposes, already dead. An interesting conservation question is whether such a state exists for species. Is there any situation in which humans might be willing to take a species off life support? The evolutionary history of every species is uniquely represented in its DNA, and with carefully constructed genealogies, it is possible to arrange zoo matings so that the loss of genetic variability is minimized. Given adequate resources, some species may be able to persist indefinitely in the zoo community. Such a so-called zoo species would differ from its wild ancestors. Such a species will have lost behaviors and represents a different entity than its members in the wild. But unlike the human in a coma, zoo species would be (and are) very much alive. Are species that can live only in zoos functionally dead and thus not worthy of repeated attempts for re-wilding using zoo-reared animals?

To take this one step further suppose a species has gone extinct, but scientists have samples of its DNA from museum specimens. In this age of biotechnology, genomic techniques are advancing rapidly, and it is becoming feasible to reconstitute the genomes of vanished species, using genetic material from preserved specimens (Long Now Foundation 2012). For example, in 2008 it was reported that Japanese scientists began attempts to clone the DNA taken from woolly mammoth specimens and then insert that DNA into the egg of an African elephant and thereby create a viable newborn mammoth (Nicholls 2008). Such de-extinction is being discussed for a wide variety of species, including the passenger pigeon, the saber-toothed cat, the Carolina parakeet, the dodo, Steller's sea cow, and one species of New Zealand giant moa. The technical hurdles are huge, but not insurmountable. The DNA is typically somewhat degraded so it has to be repaired with the DNA of the closest living relatives. Any taxa resurrected from ancestral DNA would in fact be a new species—albeit it a new species very similar to what had been lost. For extinct taxa such as moas and Steller's sea cows that were driven to extinction by hunting, suitable habitat may not be a hurdle. Technically, so-called de-extinction is a realistic possibility. This does not mean it should be pursued. There are serious ethical questions. Many wonder if people should even be considering this line of action.

The ethics of de-extinction might seem like they are irrelevant to the ethics of triage and extinction. They are not. In both cases the key questions are about the power of humans and what humans do with that power regarding the nature they build and nurture. If it is possible to revive an extinct species via biotechnology, many wonder if people should do so; they wonder if it matters that an extinction was caused by humans. Others ponder whether some species are more desirable to bring back than others. The ethics in improving a revived species—for example, to make a formerly extinct bird resistant to avian malaria—is also pondered. Any species resurrected using biotechnology will be a new species, because it is very difficult to recover all the DNA. However, given that heart transplants are done to save a human life, some people wonder might it not, by analogy, be acceptable to fill in some genetic gaps with the DNA of a closely related living species to resurrect the passenger pigeon. It is widely assumed that any resurrected pigeon would look indistinguishable from the passenger pigeons of the past. But others consider if it would, indeed, be a passenger pigeon and what it ought to be named.

In the end, any discussion of species extinction has to reflect upon the incredible power and responsibility of humans. Humans have been referred to as the "God species" because of the species' impact on the planet (Lynas 2011). Nowhere is this more evident than in the human role in causing the extinction of species, as well as potentially recovering species or even raising species from the dead. What people do with this power is not the arena of science; it is the arena of cultural and social values and ethics. There are no easy answers. The willful and indiscriminate extinction of any species on the planet is unethical. But the question of extinction and the diversity of life on Earth are best addressed for the whole portfolio of life and the birth and death of species. Just as humans are a force of extinction, humans can also be a source of creation and biological innovation. Genetically modified crops with unimaginable combinations of novel genes are now common throughout the world. Salamanders that once lived in small ponds have evolved to make watering tanks for cattle their primary habitat. Humans have domesticated animals and created 240 distinct breeds of dogs that are a source of joy and comfort to many people (AKC 2012). Clearly humans are creating a new "natural" world.

Burning books and destroying art is despicable. So too is destroying species, which are evolution's works of art, albeit works of art that are in a constant state of changing. But instead of thinking of humanity's conservation dilemma one species at a time, which creates false parables such as that of the child and tiger, it is necessary to think of humanity's moral responsibility to nurture biological diversity in total. Biological diversity, which includes human diversity, is a value to be embraced and nurtured, even though it also means a changing biological world in which some species may go extinct, but new species will arise.

Resources

Books

Baillie, Jonathan E.M., and Ellen R. Butcher. *Priceless or Worthless? The World's Most Threatened Species.* London: Zoological Society of London, 2012.

Barrow, Mark V., Jr. *Nature's Ghosts: Confronting Extinction from the Age of Jefferson to the Age of Ecology.* Chicago: University of Chicago Press, 2009.

Kareiva, Peter, and Michelle Marvier. *Conservation Science: Balancing the Needs of People and Nature.* Greenwood Village, CO: Roberts, 2011.

Lynas, Mark. *The God Species: Saving the Planet in the Age of Humans.* Washington, DC: National Geographic Society, 2011.

McKibben, Bill. *The End of Nature.* New York: Random House, 1989.

Periodicals

Barnosky, Anthony D., Nichols Matzke, Susumu Tomiya, et al. "Has the Earth's Sixth Mass Extinction Already Arrived?" *Nature* 471 (2011): 51-57.

Dawson, Deborah, and Jason F. Shogren. "An Update on Priorities and Expenditures under the Endangered Species Act." *Land Economics* 77, no. 4 (2001): 527–532.

D'Elia, Jesse, Michele Zwartjes, and Scott McCarthy. "Considering Legal Viability and Societal Values When Deciding What to Conserve under the U.S. Endangered Species Act." *Conservation Biology* 22, no. 4 (2008): 1072–1074.

Ferraro, Paul J., Craig McIntosh, and Monica Ospina. "The Effectiveness of the US Endangered Species Act: An Econometric Analysis Using Matching Methods." *Journal of*

Environmental Economics and Management 54, no. 3 (2007): 245–261.

Finkelstein, Myra E., Daniel F. Doak, Daniel George, et al. "Lead Poisoning and the Deceptive Recovery of the Critically Endangered California Condor." *Proceedings of the National Academy of Sciences of the United States of America* 109, no. 28 (2012): 11449–11454.

Iverson, Kenneth V., and John C. Moskop. "Triage in Medicine, Part I: Concept, History, and Types." *Annals of Emergency Medicine* 49, no. 3 (2007): 275–287.

Kerkvliet, Joe, and Christian Langpap. "Learning from Endangered and Threatened Species Recovery Programs: A Case Study Using U.S. Endangered Species Act Recovery Scores." *Ecological Economics* 63, nos. 2–3 (2007): 499–510.

Langpap, Christian, and Joe Kerkvliet. "Allocating Conservation Resources under the Endangered Species Act." *American Journal of Agricultural Economics* 92, no. 1 (2010): 110–124.

Miller, Julie K., J. Michael Scott, Craig R. Miller, and Lisette P. Waits. "The Endangered Species Act: Dollars and Sense?" *BioScience* 52, no. 2 (2002): 163–168.

Naiman, Robert J., Carol A. Johnston, and James C. Kelley. "Alteration of North American Streams by Beaver." *BioScience* 38, no. 11 (1988): 753–762.

Nicholls, Henry. "Darwin 200: Let's make a mammoth." *Nature* 456, no. 7220 (2008): 310-314.

Restani, Marco, and John M. Marzluff. "Funding Extinction? Biological Needs and Political Realities in the Allocation of Resources to Endangered Species Recovery." *BioScience* 52, no. 2 (2002): 169–177.

Sarrazin, François, and Robert Barbault. "Reintroduction: Challenges and Lessons for Basic Ecology." *Trends in Ecology and Evolution* 11, no. 11 (1996): 474–478.

Schwartz, Mark W. "The Performance of the Endangered Species Act." *Annual Review of Ecology, Evolution, and Systematics* 39 (2008): 279–299.

Seddon, Philip J. "Persistence without Intervention: Assessing Success in Wildlife Reintroductions." *Trends in Ecology and Evolution* 14, no. 12 (1999): 503.

Simon, Benjamin M., Craig S. Leff, and Harvey Doerksen. "Allocating Scarce Resources for Endangered Species Recovery." *Journal of Policy Analysis and Management* 14, no. 3 (1995): 415–432.

Tear, Timothy H., J. Michael Scott, Patricia H. Hayward, and Brad Griffith. "Recovery Plans and the Endangered Species Act: Are Criticisms Supported by Data?" *Conservation Biology* 9, no. 1 (1995): 182–195.

Other

AKC (American Kennel Club). "AKC Breeds: Complete Breed List." Accessed September 17, 2012. http://www.akc.org/breeds/complete_breed_list.cfm.

AZA (Association of Zoos and Aquariums). "Species Survival Plan (SSP) Program Handbook." Silver Spring, MD: AZA, 2011. Accessed October 29, 2012. http://www.aza.org/uploaded-Files/Animal_Care_and_Management/AZASpeciesSurvival-PlanHandbook_2012.pdf.

AZA (Association of Zoos and Aquariums). "Zoo and Aquarium Statistics." Last modified May 9, 2012. http://www.aza.org/zoo-aquarium-statistics.

Long Now Foundation. "Revive and Restore." Accessed October 29, 2012. http://rare.longnow.org/index.html.

National Park Service. "California Condors—Grand Canyon National Park." Accessed September 15, 2012. http://www.nps.gov/grca/naturescience/california-condors.htm.

NMFS (National Marine Fisheries Service). "Pacific Summary." In *Fishing Communities of the United States, 2006,* 13–20. Silver Spring, MD: National Oceanic and Atmospheric Administration, 2009. Accessed August 31, 2012. http://www.st.nmfs.noaa.gov/st5/publication/communities/Pacific_ALL_Communities.pdf.

NMFS (National Marine Fisheries Service). "Regional Summary: Pacific Region 2009." In *Fisheries Economics of the United States, 2009,* 23–39. Silver Spring, MD: National Oceanic and Atmospheric Administration, 2010. Accessed August 31, 2012. http://www.st.nmfs.noaa.gov/st5/publication/econ/2009/Pacific_ALL_Econ.pdf.

USFWS (U.S. Fish and Wildlife Service). "Federal and State Endangered and Threatened Species Expenditures: Fiscal Year 2004." Accessed September 15, 2012a. http://www.fws.gov/endangered/esa-library/pdf/2004ExpendituresReport.pdf.

USFWS (U.S. Fish and Wildlife Service). "Federal and State Endangered and Threatened Species Expenditures: Fiscal Year 2010." Accessed July 17, 2012b. http://www.fws.gov/endangered/esa-library/pdf/2010.EXP.FINAL.pdf.

USFWS (U.S. Fish and Wildlife Service). "Summary of Listed Species Listed Populations and Recovery Plans." Accessed August 2, 2012c. http://ecos.fws.gov/tess_public/pub/Boxscore.do.

USFWS (U.S. Fish and Wildlife Service). "2011 National Survey of Fishing, Hunting, and Wildlife-Associated Recreation: National Overview." Issued August 2012d. http://library.fws.gov/Pubs/nat_survey2011-national-overview-prelim-findings.pdf.

Whitney P. Crittenden
Peter Kareiva

· · · · ·

The politics of extinction

In November 2011 the International Union for Conservation of Nature (IUCN), arguably the world's most authoritative scientific body on the status of endangered species, declared the African western black rhinoceros (*Diceros bicornis longipes*) extinct. This came as no surprise to observers of the politics of extinction. Sadly, no signs of this species have been spotted since 2006. The IUCN report concluded that "given the wildlife poaching taking place, lack of political will and conservation effort by Cameroon conservation authorities in the past, and increasing illegal demand for rhino horn and associated increased commercial rhino poaching in other range states, it is highly probable that this subspecies is now extinct" (Emslie 2011). This story is indicative of the trajectory of extinction that plagues the present era: increased pressure on mammalian habitat from growing human settlements and agricultural activity; poaching of valuable species for their coveted body parts, often for illegal export and facilitated by corruption along borders; and eventual extinction. All this occurred despite an international campaign to stem the tide that had achieved great success in terms of awareness, but could not bridge the gap between Western expectations and realities on the ground.

The western black rhino's fate is indicative of what scientists often refer to as the sixth great mass extinction in Earth's history, one unfolding at the present time. Post-glacial Pleistocene history can be viewed from the perspective of humans displacing other species and, with the agricultural and industrial revolutions, converting biomass and energy toward human ends. The spread of humanity around the world in the preceding 100,000 years has caused the extinction of megafauna and marine species through overhunting and habitat alteration, as well as the diminishment of biodiverse flora (trees and plants) through agricultural settlement. However, the evidence emerging recently suggests a mass extinction may be underway because of an even more complex confluence of modern anthropogenic (human-influenced) factors.

According to *Global Biodiversity Outlook 3*, the signature publication of the Convention on Biological Diversity (CBD), there are "multiple indications of continuing decline" in all three of the main components of biodiversity: genes, species, and ecosystems.

- Amphibians face the greatest risk and coral species are deteriorating most rapidly in status. Nearly a quarter of plant species are estimated to be threatened with extinction.

- The abundance of vertebrate species, based on assessed populations, fell by nearly a third on average between 1970 and 2006, and continues to fall globally, with especially severe declines in the tropics and among freshwater species.

- Natural habitats in most parts of the world continue to decline . . . although there has been significant progress in slowing the rate of loss for tropical forests and mangroves, in some regions. Freshwater wetlands, sea ice habitats, salt marshes, coral reefs, seagrass beds, and shellfish reefs are all showing serious declines.

- Extensive fragmentation and degradation of forests, rivers and other ecosystems have also led to loss of biodiversity and ecosystem services.

- Crop and livestock genetic diversity continues to decline in agricultural systems. (Secretariat of the CBD 2010, p. 9)

In addition, an extensive report from the conservation group BirdLife International estimates that over 1,250 bird species (or 12.5% of all known extant species) are critically endangered, endangered, or vulnerable (BirdLife International 2012). David Suzuki and David Robert Taylor suggest in their 2009 book *The Big Picture*:

Around the world . . . bird populations are in trouble—largely because human activities are damaging their habitats. Converting prairie grassland to farmland, for example, has resulted in a 60 percent decline in native prairie bird species. Similarly, in Africa, 50 percent of all birds are threatened by agricultural expansion. . . . Longline fishing . . . kills tens of thousands of [endangered] albatrosses every year . . . when they swallow freshly baited hooks. . . . In India, three vulture species face imminent extinction from eating livestock carcasses tainted with diclofenac, an anti-inflammatory drug. (63–64)

Declining vulture populations, according to Suzuki and Taylor, have resulted in increased feral dog populations and the widespread incidence of rabies among humans in India.

The complex web of natural interdependency is being seriously disrupted by human activity.

As this crisis unfolds, there has been a seismic shift in how humanity views the threat of extinction. While the dodo (*Raphus cucullatus*), passenger pigeon (*Ectopistes migratorius*), and many others serve as early examples of hunting species to extinction or endangerment, the principal concern today is with habitat protection. There are numerous cases in which it is clear that the threat to species comes not from direct human exploitation but from changes to the natural living conditions and habitats of species with limited range. Human land use, deforestation, desertification, river diversion, and introduced invasive species are all likely candidates today as causes of extinction. Wildlife corridors and parklands are, in part, efforts to avoid this development, as is *ex situ* (out of the wild) conservation. But these are, in many cases, examples of actions that are proving to be too little, too late, often smacking of desperation or, worse, efforts to manipulate public perception.

As with any discussion of environmental politics today, it is important to look at this issue from a variety of perspectives. Even the process of determining an extinct or endangered species is heavily layered with political interventions at every step: Does adequate funding and the scientific capacity exist to produce accurate assessments of species populations? There is no doubt among biologists, geneticists, and taxonomists that the vast majority of species have not even been identified, let alone monitored (insect species are especially intriguing and could well number in the tens of millions). Even the wealthiest countries lack the necessary scientific infrastructure to adequately gauge the state of species, and though in some of these countries biodiversity has been granted a higher premium than elsewhere, other expenditures, such as those relating to health care, defense, and employment, have been accorded far more public funding. In certain cases, politicians, state bureaucrats, and large corporations have vested interests in minimizing the knowledge released to stakeholders. In other cases, the state of crisis may be deliberately exaggerated to attract funding. Every country has its own extinction-risk classification system, subject in some cases to constant political intervention. There are even disputes among members of small communities regarding how biodiversity is monitored and extinction defined. Overall though, a combination of local knowledge and internationally accepted (peer-reviewed) scientific research expressed through policy networks such as the IUCN can serve to provide general indications regarding the contemporary fate of endangered species. (The IUCN Red List of Threatened Species is probably the most widely used classification system for endangered species, employing nine groups: extinct, extinct in the wild [extirpation], critically endangered, endangered, vulnerable, near threatened, least concern, data deficient, and not evaluated. "Threatened" species include those deemed critically endangered, endangered, or vulnerable. Meanwhile, the Convention on Biological Diversity runs the Global Taxonomy Initiative, a paltry effort to increase taxonomic capacity around the world that has netted little results.)

But any political intervention to change the worst-case scenario (which, according to the existing evidence, is akin to a business-as-usual approach) will need to take place across multiple political levels, as well as within various socioeconomic contexts. The next section outlines this multilevel approach to environmental governance. Subsequent sections discuss specific case studies at the local, regional, and national levels. The final section includes a discussion of the more abstract, but equally important, level of global environmental governance, before offering a brief exploration of possible policy responses and institutional needs. Finally, it is very important to identify what many scientific observers feel is the greatest threat to species today—climate change. This not only complicates the menu of policy choices but also poses further questions about humanity's ability to respond politically to extinction when human suffering and population movements will be exacerbated by changing climatic conditions.

The politics of extinction are, to put it mildly, intense. There are few issues that raise comparable emotions among stakeholders, including those seeking to prevent extinctions and those who feel that other factors, such as property values, human health, agricultural production, and wider species concerns, demand at least equal consideration. The genuine concern for endangered species is often spurred by much broader critiques of modernity, industrial society, and animal rights. The willingness to "let nature take its course" is often wrapped within a broader pattern of denial of anthropogenic climate change or a resolute faith in the inevitably of human ingenuity or divine fatalism. Of course, most natural and social scientists today would concur that one should not think in terms of dichotomies between the maintenance of biodiversity and human progress. Indeed the two are so intertwined it is foolish to separate them. This does not necessarily translate, however, into an easier policy path when the clash of stakeholders with differing perceptions and interests takes place in the real world.

Normative issues and policy dilemmas

Extinction is forever. Of course, viewers of films such as *Jurassic Park*, in which frozen DNA from dinosaurs is used to resurrect various creatures from the deep past (with predictably nasty results) might argue with this preposition. Indeed, there is scientific evidence that suggests that such a process could, eventually, be employed, though there are ethical debates simmering over whether doing so would be appropriate. The various DNA banks being established around the world today, such as the Frozen Zoo in San Diego, are testament to the widespread concern that Earth is on the brink of losing a large swath of its extant life-forms and that humans wish to retain the right to engage in resurrection experiments in the future. Indeed, one of the stated goals of the Global Strategy for Plant Conservation, a program of internationally agreed conservation targets for plant species created in 2002 under the aegis of the Convention on Biological Diversity (CBD), is to save up to 75 percent of all known plant species in *ex situ* conditions.

Ideas about how to protect endangered species differ markedly in various parts of the world. Two central perspectives have emerged at meetings of the Convention on International Trade in Endangered Species of Wild Fauna and Flora (CITES), the main global instrument to regulate trade in endangered species created in Washington, DC, in 1973, and in other venues related to the overexploitation of wildlife. These meetings have become highly polemical battlegrounds between *preservationists*, who argue that certain endangered species and in some cases their habitats should be off limits from any type of human harvesting or disturbance, and *conservationists* (also known as *sustainable use advocates*), who argue that species should be conserved so that they can be used by human populations if and when there is a noticeable recovery in population numbers. While preservationists typically favor "no-go" green zones and the legal use of CITES Appendix I (which bans all commercial trade in listed species), conservationists believe that species will be sustainably used only if they are viewed as valuable and that local human populations should be rewarded for conserving endangered species. Conservationists suggest that the maintenance of commercial trade bans even while formally endangered species are recovering or have reached pre-exploitation populations is tantamount to punishing local communities for successfully implementing conservation programs.

In one of the best-known and most controversial examples, the inclusion of the African elephant in Appendix I of CITES, which bans any commercial trade in any body part, was initially based on the need to control the ivory trade. The establishment of ivory export quotas based on scientifically established management programs at the Fifth Conference of the Parties of CITES and its refinement at the sixth conference was followed by the creation at the tenth conference of two monitoring mechanisms: Monitoring the Illegal Killing of Elephants and the Elephant Trade Information System. Both mechanisms track illegal elephant activities, with the overall objective of obtaining more information to support decision-making and, thus, reduce the polarity of opinions among countries. Local farmers in some areas, notably South Africa, are dealing with rising elephant populations that have expanded beyond protected areas, leaving them to contend with eaten harvests and destroyed crops. Nevertheless, the authority to keep animal populations in check remains at the national level. While most countries continue to ban trade in ivory, some have set quotas in regard to trophy tusks. South Africa, for instance, set a quota of 300 tusks as trophies (from 150 animals) for 2011 but does not allow any other form of ivory exports. Conservationist countries (including Japan, Norway, and Iceland) are content to resume a sustainable trade in these great beasts, and they are determined to drop certain whale species from CITES Appendix I. Preservationists argue that some creatures, including bears, whales, gorillas, and elephants, are too special to kill and trade regardless of how endangered they have become and that accepting their further commodification would be a giant leap backward. Elephant politics are a good example of how the politics of extinction play out from the local to the global level.

Another normative issue relates to the proper point of focus for conservation efforts: Ought it to be individual species, habitats, or ecosystems? Most of the examples discussed below are based primarily on cases in which individual species have attracted great attention. Although it has become customary in scientific circles to bemoan the public's fascination with certain key charismatic megavertebrates, such as the gorilla, the tiger, the panda bear, and great whales, the attention they garner is vital if endangered species are to remain on the political agenda, especially if they are vital keystone species, playing an irreplaceable role in ecosystem and food chain maintenance. The polar bear (*Ursus maritimus*), for example, has been cited as endangered as a result of climate change and disruptions to Arctic ice formation (see below). Its loss would have profound impacts in the fragile Arctic ecosystem, inducing what scientists refer to as a cascading extinction risk:

> Because ringed seals [(*Pusa hispida*)] are the primary prey of polar bears, extinction of this top predator may have positive consequences for ringed seal abundance. Adult ringed seals prey primarily upon Arctic cod and polar cod, and increases in numbers of ringed seals may adversely affect cod abundance and possibly even cod fisheries in the Arctic and subarctic. This, in turn, may have broader consequences for marine food webs, including zooplankton dynamics and nutrient transport. (Post and Brodie 2012, 133–134)

Yet the argument for focusing on habitat is a strong one, especially considering the reality that human populations are dependent on ecosystem services as well as certain key species for survival. According to many established religions, humans have a responsibility to be good stewards of the land. In contrast, religions say relatively little about the morality of driving individual species to extinction. One can also argue that demanding the survival of individual species runs counter to the mechanism of natural selection, though this argument is diminished by the realization that some believe the world has entered a new epoch, the Anthropocene (the climatic age of humankind), whereby human activity has become the most determinative feature of the physical context in which natural selection occurs. This gives rise to other arguments about the validity of *ex situ* conservation methods and animal captivity and whether the goal is to avoid extinction per se or extirpation (the local loss of a species within a part of its natural habitat). Further, some endangered species may indeed survive, but in an artificial environment. It is common to describe such species as being "lost from the wild," but this is a misleading label because it is increasingly difficult to actually identify the "wild" in the first place and especially hard to differentiate it from natural areas managed by or at least strongly affected by human activity.

The advent of contemporary climate change further complicates this picture because many species will find themselves living outside their natural range and even ecosystems as temperatures change, humidity shifts, water levels rise and fall, and other geophysical characteristics are modified. This will increase the range of invasive alien species, further threatening fauna and flora unable to migrate to more comfortable temperature ranges. Indeed, there are two immediate dangers linking climate change and extinction. The first is that species themselves will not be able to adjust, either phenotypically or genetically; the second, generally labeled the "functional extinction" of ecosystems, is that

A thin female polar bear (*Ursus maritimus*) at the edge of the ice in Svalbard, Norway. © J.-L. Klein & M.-L. Hubert/Photo Researchers, Inc.

entire biomes will become uninhabitable for many of the species currently living there. There has been a great deal of work published on this latter phenomenon, which can be observed today (coral reefs and coniferous forests are but two examples) and, according to most predictions, will result in a great extinction rate. Indeed, the phrase "exponential extinctions" has been used in this context. A report published in the highly regarded journal *Nature* in 2004 calculated that over one million species are at risk of extinction as a result of expected ecosystem instability associated with climate change (Thomas et al. 2004; see Hannah 2012 for an extensive discussion). This calculation did not include marine species, though it is of course impossible to deny the widespread threat posed to them by climate change as well, especially in the highly biodiverse and sensitive coral-reef ecosystems; and, again, scientists do not even have a reliable approximation about just how many undiscovered species or unclassified subspecies actually exist.

Ultimately, therefore, the politics of extinction is enmeshed in the science of conservation biodiversity, the political economy of the wildlife trade and agricultural production (including the fisheries), policy debates over pollution control and climate change, and many other, constantly changing, social factors. There are several theoretical models that have been employed to explain humanity's tendency to drive species

toward extinction, but they have limited application. The most popular, for many years, was the "tragedy of the commons" model, advanced in a seminal paper by Garrett Hardin published in 1968. Hardin suggested that common resources would always suffer because individual humans, acting out of self-interest, will cause unintentional aggregate harm. He applied this premise to human propagation in particular, arguing that to "couple the concept of freedom to breed with the belief that everyone born has an equal right to the commons is to lock the world into a tragic course of action" (Hardin 1968, 1246). The crux of the tragedy of the commons is that one individual will benefit from pursuing his or her self-interest, but everyone will suffer from taking a similar path at the aggregate level. Elinor Ostrom (1990) and others have demonstrated this is fallacious in many contexts, but the term sticks, especially as Hardin employed it to refer to the need to overcome the sanctity of the right to procreate. Indeed, many argue that human birth control is an ethical obligation in order to avoid the coming mass extinction of other life-forms.

The tragedy of the commons model was frequently used to explain the diminishment of whale stocks as various countries competed to catch as many whales as possible, even after the advent of the International Convention for the Regulation of Whaling in 1946. Another, related model, labeled the

Coral bleaching occurs when excessively warm sea temperatures force coral to eject the algae that give them their color. This consequence of climate change is seriously threatening many reefs, including the Great Barrier Reef near Queensland, Australia. © Ashley Cooper/Corbis.

"economics of overexploitation" in a landmark 1973 article by Colin W. Clark, suggested that humans overharvest desirable species because of rational investment decisions aimed at profit maximization: It made more sense to kill more whales, even if it meant there would soon be no whales to kill, and invest the proceeds from this economic activity elsewhere, than to conserve them for future killing when their fetching price would be undetermined and quite possibly be lower as other technologies surpassed whale oil and eating habits changed among whale-eating populations. Most theories along these lines assume people act as rational, self-interested decision-making units, in economics as well as politics. Similarly, countries are often assumed to pursue self-interest on the international stage, as are multinational corporations. Of course there is a kernel of truth in this assumption, but there are many other factors that go into decision-making, and the politics of extinction also bring these to light.

For example, the aesthetic value of certain species, such as elephants and whales, is hard to overlook, and the emotions engendered by the extinction debate offer particularly poignant reminders of how people rely on nature for more than just food and other ecosystem services. Different definitions of common welfare help drive the formation of public policy on environmental issues. Marxist explanations focus on the mode of production in society and the use of the state to express and advance the interests of the dominant economic classes, which is often used to explain humanity's thoughtless approach to the conservation of nature: While some argue that large-scale capitalism is inherently bad for natural survival, it is clear that the state-managed economies of the former Soviet empire fared even worse. But this misses the point, which is that the commodification of nature sets in motion a vicious cycle of overexploitation that can be stymied only by changes in the structure of the system or the adoption of new norms and principles by the governing elite.

All of these approaches have some merit, but students of environmental policy and politics today—or, more generally,

ecopolitics—focus on concepts such as authority, legitimacy, regulatory capacity, communications, policy network development, social movements, private–public partnerships, environmental values, risk and justice, and collective action problems. At the heart of the debate is the question of governance: Who should make key decisions, and for whom? At what level should these decisions be made? Efforts to avoid extinction demand a legitimate response to the governance of *collective action problems*. As the name suggests, collective action problems cannot be solved unilaterally, or by any one actor involved, no matter how powerful it may be. Of course, some actors may have more influence and play leading, or debilitating, roles. But it is clear that the successful management of collective ecopolitical dilemmas will not result from reliance on any one level of governance. Indeed, this may be one of the few points most analysts of global ecopolitics can agree on today. While it was previously normal to reflexively argue that either top-down or bottom-up solutions were needed, it is now quite common to realize either group is futile on its own. Hierarchical thinking is limiting, no form of authority has pure sovereignty over others, and sophisticated policies will emerge only when various scales of governance (and not limited to the formal work of governments) are taken into account. National legislation to protect endangered species is only a coercive device bound to incite rebellion if it is not considered legitimate by the local human populations who must live with the species involved.

The local and regional level

Though urbanized readers might be more accustomed to the broader perspective of extinction as a threat to humanity and certain idealized aesthetic values, it is at the local level that extinction is felt the hardest, because people have a closer bond to, and often have a symbiotic relationship with, endangered species. A few brief case studies can illustrate this impact, as well as the complex causality that can lead to extinction or extirpation.

One of the earliest examples of deliberate and ruinous mass extermination of a species was the plight of the Great Plains or American buffalo (also known as the American bison [*Bison*

A sperm whale (*Physeter macrocephalus*) on deck, being cut up for blubber and other products. © Biophoto Associates/Photo Researchers, Inc.

Bison (*Bison bison*) in front of Grand Teton Mountain range. © visceralimage/ShutterStock.com.

bison]), which neared extinction as the European colonization of North America took place in the 1800s. Though historians debate whether this was part of a genocidal government policy, the eradication of millions of buffalo had a tremendous impact on the lives of Native Americans and First Nations peoples in North America. These people had typically used the buffalo in an exceptionally sustainable manner, using every body part, and the animal was central to their nutritional, cultural, and spiritual regimen. The American buffalo was driven to near extinction with the advent of the trained horse, gunpowder, and commercial markets for products, as well as the relentless expansion of European settlements and farms. Its recovery is an equally engrossing story, suggesting that coordinated conservation management efforts can be effective if supported by various stakeholders. Nevertheless, this near extinction stands as an historic example of the local devastation that can result from the widespread slaughter of a once-abundant species.

Another prominent North American example is that of the northern spotted owl (*Strix occidentalis caurina*), which became a cause célèbre for both environmentalists and the so-called "wise use movement" in the United States in the 1990s. This owl makes its home in the old-growth forests of the Pacific Northwest, and the Endangered Species Act (ESA; described in more detail below) demands that the habitat area of

endangered species receive adequate protection in order to avert extinction. When the US Fish and Wildlife Service declared the owl a threatened species in 1990 and the ESA was used to order a stoppage of logging in several forested areas, environmentalists concerned about the impact of clear-cut

Shooting buffalo (*Bison bison*) in front of Grand Teton Mountain range.on the line of the Kansas–Pacific Railroad, Frank Leslies Illustrated Newspaper, June, 1871. © LegendsOfAmerica.com.

logging on delicate old-growth forests celebrated an unfamiliar victory. Loggers, however, were less impressed, and the northern spotted owl quickly became a symbol for the overzealous ambitions of environmentalism gone mad, destroying the livelihoods of entire logging communities with longstanding and proud traditions of forestry. Small sawmills were especially affected. Environmentalists argued that forestry jobs were bound to decline as clear-cutting became unacceptable and/or economically infeasible. The species has continued to decline, largely because of an invasion of the barred owl in spotted owl ranges, but the case remains an iconic example of how divided stakeholders can be, and, according to many representatives of the forestry industry, how legal actions to protect threatened species can threaten local communities as well.

Another striking example took place on the small Pacific island of Guam, where bird extirpations became common following what is now a classic case of bioinvasion. Native to the South Pacific (areas of Indonesia and Australia, as well as Papua New Guinea and the Solomon Islands), the brown tree snake (*Boiga irregularis*) was unintentionally brought to the island of Guam sometime between the end of World War II and 1952, likely as a stowaway released upon the unloading of military equipment. It has, arguably, wreaked greater destruction than the war itself. Introduced into a previously snake-free island habitat with an unusual abundance of prey, the snake proceeded to rapidly propagate, spawning a modern episode of mass extinction. In four decades, the snake has been responsible for extirpating nearly all of the native forest vertebrate species and equally responsible for the decline of the flying fox, a species crucial for the seed dispersal and pollination of tropical trees on the island. Famously, the snake preys heavily on avian species, including insectivores, which in turn allows the insect population to soar unchecked, posing severe problems for agriculture as well as human health. Species of lizards and fruit bats have similarly undergone marked declines since the arrival of the snake. In sum, the cost has been brutally high for Guam's natural wildlife: All species of breeding seabirds have been lost, as have 10 of 13 native species of forest birds, 2 of 3 native mammals, and 6 of 10 to 12 species of native lizards. The impact on human society and culture has been as marked. For example, the decline of bat populations has limited indigenous cultural practices and nutritional intake (see Knights 2008). And the economic impact on Guam has also been significant. Power outages occur on average once every three days as a result of snakes climbing onto electrical lines, and one out of every 1,000 emergency room visits in Guam is due to snakebite. The decline of bird populations harms agricultural output and reduces tourist interest. In his 2006 book titled *Seeking the Sacred Raven*, Mark Jerome Walters tells a similar story about a culturally significant bird extirpation from the raven family in Hawaii.

Another case is that of the Great Lakes of Africa, where several factors led to a mass extinction of unusual breadth, particularly in Lake Victoria, which is shared by Uganda, Kenya, and Tanzania. This was once a vibrant ecosystem with an incredible diversity of fish species, including more than 400 species of cichlids. The commercial introduction of the Nile perch, a large omnivorous fish caught mainly for export, combined with pollution, overfishing, and other invasive species such as the water hyacinth, have cut cichlid numbers in half and created large oxygen-deprived dead zones. Local fishermen have been displaced by commercial fisheries and women in the fish-cleaning sector by fish-processing plants. One might argue this is part and parcel of modern industrial development or, as the Marxist perspective would suggest, the imposition of a new mode of production in a new area, benefiting investors and a new working class but devastating local populations and natural resources. The calamity gave inspiration to a surprisingly popular, if controversially one-sided, film directed by Hubert Sauper, *Darwin's Nightmare* (2004). Restorative efforts are being made, however, and there is some cause for optimism as the regional governments begin to take this problem seriously. But one should not lose sight of the bigger picture: Freshwater lakes and rivers and the species contained within them are in peril around the world. Fresh waters cover less than 1 percent of Earth's surface, but their biodiversity is "unrivaled" (Poff et al. 2012, 309); between 10,000 and 20,000 endemic freshwater species are estimated to be at serious risk of extinction, regardless of the future impact of climate change. As N. LeRoy Poff, Julian D. Olden, and David L. Strayer write in a 2012 article: "The current global freshwater biodiversity crisis...stems from many types of human activity: severe alteration of natural runoff patterns, fragmentation of river corridors by dams, increased addition of sediment and nutrients from poor land use practices, and introduction and spread of harmful nonnative species" (311).

The link between the modern global economy, local hardship, and the threat of extinction is a familiar theme, often punctuated by divisive community politics and layers of corrupt government. This is certainly the main story arc of the imperiled gorillas of Africa, a story that also involves the Great Lakes region. Even in this war-torn region of Central Africa, extinction has played a major role thanks in large part to the endangered mountain gorilla (*Gorilla beringei beringei*) population in the Virunga volcanic mountain ranges bordering Rwanda, Uganda, and the Democratic Republic of the Congo (DRC), as well as the Bwindi Impenetrable National Park in Uganda. Indeed, it is difficult to imagine a region where violence has been as prevalent as the Great Lakes region in Africa, where up to four million people died between 1998 and 2012 in the ongoing conflict in the DRC, and up to 800,000 were slaughtered in the 1994 genocide in Rwanda. Local gorilla populations have been severely reduced; subject to both human population pressure and poaching for bushmeat and body parts (gorilla hands are popular ornaments in some markets and are even used as ashtrays). There are fewer than 700 mountain gorillas remaining (Robbins and Williamson 2008). (A subspecies of the lowland gorilla, meanwhile—the Cross River gorilla (*Gorilla gorilla diehli*)—is also on the verge of extinction, with fewer than 300 remaining.) Logging companies pose threats, as does the transmission of deadly viruses. Yet, the charismatic nature of these magnificent mammals may be their salvation, and the conservation of the gorilla remains a professed steadfast priority of post-conflict governments in the region. This is no doubt linked to the economic value of the species,

The introduction of the Nile Perch (*Lates niloticus*), along with other invasive species and local pollution, has cut cichlid numbers in half and created large oxygen-deprived dead zones. © John Warburton-Lee Photography/Alamy.

because they are a large (some would say too large for the good of their natural habitat) tourist draw. International campaigns to ensure the survival of the mountain gorilla have been quite forceful, led by the visionary work of the American zoologist Dian Fossey—ultimately murdered in 1985 by local thugs protecting the poaching racket—and have resulted in the regional Agreement on the Conservation of Gorillas and Their Habitats, which went into force in 2008. Similar threats face the severely diminished orangutan population in Indonesia, including illegal logging and forest fires and displacement caused by the rapid spread of palm oil plantations. In 2007 the Indonesian president Susilo Bambang Yudhoyono initiated the Orangutan Conservation Strategy and Action

Plan, which many have criticized as a public relations showpiece designed to assure foreign observers that something positive is being done, although it at least represents an official recognition of the problem.

In the case of mountain gorillas, tourism may be part of the fight against extinction. Tourism, however, has an often-deleterious impact on local wildlife, while the money it attracts is an undeniable factor in decision-making in biodiverse travel destinations such as Costa Rica, Madagascar, and Indonesia. Particularly infamous has been the impact of tourism on coral reefs. It should be stressed that many other factors, such as inappropriate fishing techniques

Officials look at four dead mountain gorillas (*Gorilla beringei beringei*) that were illegally killed in the Virunga National Park in the Democratic Republic of Congo. The silverback male and three females were shot in the southern sector of the park, which contains more than a fifth of the world's population of 700 mountain gorillas. © HO/Reuters/Corbis.

(including the use of cyanide and dynamite), sediment disturbance, shipping, and acidification are generally viewed as larger threats. A mass bleaching event occurred off the coast of Australia in 2002, "leading to the death of 5–10 percent of reef-building corals within the Great Barrier Reef Marine Park" (Hoegh-Guldberg 2012, 266); and all species with calcified structures are threatened by acidification. But the presence of ignorant or insensitive tourist traffic— literally, people touching and/or walking on the reefs as they view them—has long been recognized as a threat. In a cruel dilemma, local populations often become dependent on this tourist traffic, and yet they also lose valuable fishing opportunities if the reef ecosystem is damaged beyond repair and becomes functionally extinct. Educating tourists and tourism operators is certainly a noble goal, but if it conflicts with profit-making opportunities this is often an afterthought at best.

The national level

Extinction, as the local examples above demonstrate, means different things to different people, even if there are some common denominators. While traditional lifestyles involving hunting and fishing were sustainable for thousands of years, modern mass commercial activity has driven many species to endangerment or extinction. This has, in many cases, resulted in a shift in norms. For example, the United States has emerged since the 1960s as a champion for the preservationist stance on whale species (cetaceans). But it was American whalers who led the early mass slaughter of many species, and as late as the 1950s orca (killer whales; *Orcinus orca*) were used as target practice by the US Navy. Meanwhile, Japan has adopted a strong international position against the international moratorium on commercial whaling that was passed in 1982, claiming it is an instance of cultural imperialism unfairly targeted at conservationist-oriented countries undertaken by a

hypocritical industrial behemoth that allows its own aboriginals in the states of Alaska and Washington to engage in limited whale hunts. Such national differences do not necessarily reflect marked distinctions between national cultures, however; rather, they reflect the diplomatic stance of the prevailing governments of the day. Nevertheless, once they are ingrained as official policies, they tend to stick with a country's self-image and international reputation.

In many countries, extinction has not been a prominent factor in the national discourse. There is little active discussion in policy circles in Russia, for example, of realized and potential extinctions resulting from the years of environmental neglect under Soviet rule, despite the existence of expansive dead zones resulting from nuclear weapons testing and production (though there were welcome Russian actions to take the lead on the conservation of the Amur [Siberian] tiger [*Panthera tigris altaica*] in 2012). In some countries, such as the United States and the United Kingdom, extinction has become a major national issue, driving innovative (and, as noted above, some would say draconian) legislation and mass popular movements. In some countries it has been a peripheral issue at the national level, whereas in others it has often assumed a prominent spot on the national political agenda. Different discourses have evolved as the threat of extinction has shifted in terms of public perception, scientific awareness, and even national identity. Given the extinction threats posed by invasive species discussed above, some countries, such as New Zealand and Australia, have placed great emphasis on expensive but vital policy regimes designed to prevent alien introductions. Arguably, this has changed the very national character of these countries over time as biosecurity assumes an overt place in the national consciousness.

Another pertinent factor is the domestic political structure and political economy. In federal states such as the United States or Canada, state or provincial levels of government have great control over natural resources and thus are essential players in the process. Democratic countries, while more likely to see the expression of environmentalists' concerns, must engage with a more cumbersome process to instill legislative change than is the case in authoritarian regimes. Governments with limited internal legitimacy, such as Mexico or Somalia, have an even harder time implementing policy decisions. Corruption is often a factor in the implementation of programs, and countries with military governments, such as Burma, are also characterized by the links between military leaders and economic wealth.

In the United States, the risk of extinction became a hot-button political issue in the late 1960s and led to the Endangered Species Act, one of the most far-reaching pieces of environmental legislation in any country, which was an ambitious effort to ensure nationwide protection for endangered species that integrated habitat conservation but one that also placed the onus on private property owners to restrain from altering much of their land, inciting a significant backlash. Indeed, an initially supportive Richard M. Nixon, who was president when the act was ushered through the US Congress in 1973, would shortly come to regret the entire affair; by 1974 Nixon had become "one of environmental law's harshest critics" (Lazarus 2004, 78). Nonetheless, the law has survived countless attacks and even the fierce anti-environmental regulatory stance of the administration of President George W. Bush. Concerns over the near extinction of the iconic bald eagle (*Haliaeetus leucocephalus*) in the 1950s and 1960s helped convince many state regulators to ban the use of DDT and other pesticides; the bald eagle was delisted from the endangered category by the US Department of the Interior in 2007. The listing of many populations of Pacific salmon has had a widespread impact on both inland and marine enterprises, because they are anadromous species, meaning they are born and reproduce in freshwater but live much of their lives in saltwater. The Marine Mammal Protection Act of 1972, equally controversial and resilient, is also regarded as a relatively strong piece of legislation. The recovery of the eastern gray whale population, which migrates from Baja California off Mexico to Alaska each year, is another remarkable success story buttressed by national efforts to outlaw whaling in the early 1970s (although the western Pacific gray whale, found off the Russian coast, remains critically endangered). Of course, one must keep things in healthy perspective. The open concern for extinction in the United States (displayed most visibly in the rise of popular environmental nongovernmental organizations [NGOs] across the country and in the strong preservationist

stance of the US government in diplomatic forums such as CITES and the International Whaling Commission [IWC]) is undeniable. Yet, as the brief description of the American bison above indicates, the United States was forged at least partially by the eradication of many endemic species (as was environmentally conscious Europe), and the US record as the world's largest polluter—arguably, the single-greatest national threat to ecosystems worldwide—is another undeniable fact, despite China's gains in overall greenhouse gas emissions.

In Canada, by contrast, while endangered species elicit quite a bit of attention in many urban centers, traditional ways of life have always centered on the exploitation of natural resources, and governments have been reluctant to take the firm steps necessary to mirror the development of the ESA in the United States. Canada actually withdrew from the IWC when it passed what was in effect a US-driven moratorium on whaling in 1982. It was not until 2003 that the Species at Risk Act was finally put into force in Canada; though it is viewed as a much-needed step toward greater protective efforts for threatened species, critics charge it is weak in both range and ambition, and it remains open to excessive political interference because federal politicians can deny listings, on the basis of economic or social impacts, even in cases in which scientists on the Committee on the Status of Endangered Wildlife in

Adult gray whale (*Eschrichtius robustus*) breaching in the San Ignacio Lagoon, Mexico Sea of Cortez. © Michael S. Nolan/Seapics.com.

Frozen tuna lies awaiting inspection and auction at Tsukiji fish market, the worlds largest daily fish market, Tokyo, Japan. Japan is the world's largest consumer of tuna, and now faces price rises in the cost of the delicacy food, following tuna fishing quota reductions imposed by the International Commission for the Conservation of Atlantic Tunas. © jeremy sutton-hibbert/Alamy.

Canada have recommended them. Environmentalists contend that the weakness in this Canadian legislation was an outcome of the government-organized coalition of stakeholders that in effect stitched the legislation together, a coalition that included the Forest Products Association of Canada and the Mining Association of Canada. To the dismay of many NGOs and marine biologists, Canada sided with Japan in successfully opposing the listing of the commercially valuable bluefin tuna (*Thunnus* sp.; widely acknowledged to be endangered) at the 2010 CITES Conference of the Parties. The Toronto Zoo is well known for its excellent, if financially troubled, *ex situ* conservation efforts, and Canadian scientific expertise and indigenous knowledge are invaluable; but it would be a case of mistaken identity to assume Canada is a preservationist nation by character.

Canada, the United States, and European countries all have committed environmental movements, and it would be unfair to suggest that various extractive industries have not learned that taking wildlife welfare into consideration makes good business sense. In many less democratic and more socially stratified countries, it is even more difficult to launch serious campaigns to avert the extinction of keystone species or functional ecosystems. Indonesia, mentioned above in relation to the orangutan crisis, is an example of a biodiverse hot spot where rapid economic growth, fueled by forestry and palm oil production, has clearly dominated the national agenda. The perilous state of the Sumatran tiger (*Panthera tigris sumatrae*) is indicative of what can happen to large predator populations that need big hunting ranges when agricultural, forestry, and industrial development interests clearly dominate the governance of a country over successive generations. Fewer than 500 Sumatran tigers remain in the wild, and the Asia Pulp and Paper Company, among others, has received international condemnation for its role in the relentless destruction of forest cover. Environmental NGOs such as Greenpeace have used the company's role and the state of the Sumatran tiger as focal points in campaigns that

pushed American corporations toward temporarily black-listing the company and its products. Given the Indonesian government's strong support for the forestry and palm oil industries and its noted history of government corruption in these sectors, one can be excused for withholding some cynicism about the success rate of such campaigns, but it is equally apparent that without this external pressure the Sumatran tiger will be extirpated in short order.

Another cat case that has attracted much attention—and one that involves another charismatic vertebrate—is that of the Bengal tiger (*Panthera tigris tigris*) in India, listed as endangered on the IUCN's Red List of Threatened Species. Globally, the iconic tiger is in great trouble, with more tigers in captivity than in the wild. Three of the nine subspecies of the modern tiger are now extinct, and the others are endangered or critically endangered, dwindling as their ranges shrink as a result of human settlements and their food sources (wild deer, pigs, and other large mammals) disappear with development. This crisis reached national proportions in India when genuine fears of the tiger's complete extirpation surfaced in the 1960s. India's Project Tiger, started in 1972 to protect Bengal tigers, has become a model for some, but it is clearly failing in the fight against the extirpation of this tiger subspecies. More than 40 National Tiger Conservation Authority wildlife reserves cover some 14.6 square miles (37.8 sq km) in India, but despite a spike in population in the 1990s, subsequent surveys indicated that there are fewer than 1,500 Bengal tigers remaining. India attaches great national pride to the tiger (similarly, the jaguar appears on the Brazilian currency—the real—despite the near extirpation of this jungle species in some regions). More robust efforts in the Indian state of Karnataka have established the Tiger Protection Force, designed to tackle poachers and relocate about 200,000 villagers in order to preserve tiger habitat. Does this program have the legitimacy required to succeed? Observers are doubtful: Leaving aside the fact that tigers are known to kill children and livestock, the idea that people should move in order to preserve them is problematic for many. Moreover, the involvement of paramilitary troops, while perhaps necessary given the pernicious nature of poaching and the related corruption in the region, can be seen by many as a further effort to exert state control.

This last point deserves further explication. As noted, there has been much progress toward the establishment of a precautionary approach in many countries. Indeed, the adoption of conservation areas is often viewed as the most effective way to prevent extirpation or extinction of endangered species. However, a critical commentary has emerged on these developments, one that sees them as continuations of historical patterns of colonialism and the enclosure of the commons. Viewing the environment as a security issue may be welcome because it casts much-needed light on environmental issues and demands increased attention from governments. But, as an excellent collection of essays suggests (Peluso and Watts 2001), even conservation areas are part of the quest for resources that can also bolster the ability of authoritative regimes to claim more effective sovereignty over previously ungoverned areas. Indeed the earliest efforts at establishing conservation areas were conducted by the imperial powers occupying land in Africa and Asia (see Warner 2006 for an extensive discussion of the waves of conservationism and its relationship with colonial expansion and imperialism). The impact of modern conservationist regimes on indigenous people can be devastating if as a result they are denied access to the natural resources on which they, and their cultures, have depended for centuries. To understand the political ecology of conservation, it is necessary to look beyond the negative impact of development and also be concerned with the potential negative impact of environmental protection on local human security, as well as the possibility that protective actions might further solidify the power of what are quite often exploitative governance structures. Some critics suggest that actions at the international level (covered in the next section) have a similar effect.

Meanwhile, the advent of climate change has driven some countries to identify closely with the threat of extinction, in a much more human manner than concerns over individual species or even ecosystems. Small island states are the most poignant example. They already face serious extinction threats related to invasive species, but rising sea levels and increasingly intense storms have added a new level of concern. In 2009 Mohamed Nasheed, then president of the Maldives, made international headlines when he took his cabinet underwater for a meeting, a symbolic act reminding other world leaders that the threat of climate change is both real and personal for those living in the archipelago of over 1,000 coral islands off the coast of India. While coastal populations around the world will likely be forced to move inland if the sea-level rises predicted by the United Nations' Intergovernmental Panel on Climate Change are accurate, those on small islands have limited space for migration. Beyond this human dimension, another pertinent factor is that although islands cover only 5 percent of all landmass, they are home to approximately 20 percent of all indigenous plant species. Nasheed spoke openly of the future need to find a new homeland for the 400,000 residents of the Maldives, and the Alliance of Small Island States, which was formed in 1990, is working at the international level to increase whatever small progress has been made at global climate talks. While even in collective numbers these countries have limited political power, and their desperate pleas have had little direct resonance, they are highly vocal reminders of one of the central dilemmas of the twenty-first century and the need for the sustained pursuit of global environmental justice. Many people grew up thinking about national extinction as a possible consequence of full-scale thermonuclear war and nuclear winter, but few people foresaw the possibility that entire countries could literally cease to exist because of pollution and political intransigence on a global scale.

Canadian and US views of the link between extinction and climate change, meanwhile, are clashing on a largely unexpected front: the polar bear. In 2008 the US Department of the Interior and the US Fish and Wildlife Service formally listed the polar bear as a threatened species under the Endangered Species Act, even though they acknowledged that some polar bear populations had actually increased in recent decades. Instead of the past record, the listing was based on projections into the year 2050, specifically on "an extensive review of

modeling studies suggesting that global climate change will have significant negative impacts on key sea ice features used by polar bears for hunting and feeding" (Hannah 2012, 34). At one point it was overhunting that threatened polar bear populations, but that is largely controlled now in Canada, Alaska, and Russia, thanks to the 1976 International Agreement on the Conservation of Polar Bears. The threat now is related to melting ice, on which the species depends for hunting ringed seals. The Canadian government has not followed suit to the US listing, despite a review by scientists suggesting it should do so. The administration of then Alaskan governor Sarah Palin, which had filed suit against the Interior Department's decision on the polar bear, asserted that the listing of "a currently healthy species based entirely on highly speculative and uncertain climate and ice modeling and equally uncertain and speculative modeling of possible impacts on a species would be unprecedented" (State of Alaska 2007). Of course, that was precisely the point of the listing—to set a precedent reflecting current scientific evidence. But it is clear that there will be strong opposition to declaring species threatened with extinction based on long-term projections derived from climate models, no matter how widely accepted those models may become. (There are several other species commonly identified as threatened by climate change in the Arctic, including the walrus [*Odobenus rosmarus*], narwhal [*Monodon monoceros*], spectacled eider [a sea duck; *Somateria fischeri*], ringed seal, and arctic fox [*Alopex lagopus*]. In Antarctica, which is comanaged by members of the Antarctic Treaty System, the Adélie penguin [*Pygoscelis adeliae*] and the chinstrap penguin [*Pygoscelis antarcticus*] are both threatened by changing ice formations.)

The global level

Actors engaged in global environmental governance have not marginalized extinction as an animating theme. There is a wealth of theoretical literature debating just what global governance really is—or even whether it really exists. In general, global environmental governance refers to multilateral efforts (i.e., by many countries) to forge mutually reciprocal and enforceable commitments and arrangements that can manage different aspects of the global political system. There are other definitions, some more supportive, suggesting that a grand architecture of law and custom is being built that will eventually usher in a new age of international cooperation and progress, and some critical, suggesting that the architecture being developed is just another form of domination on a global scale, continuing the ugly tradition of neocolonialism. Most analysts fall between these extremes, contending that efforts to establish international law and concomitant institutions will certainly reflect the interests of the most powerful participants, but they are also potential vehicles for change and progress. Global environmental governance is one of the more evolved areas of global governance, even if it is relatively weak compared to the importance the international diplomatic community places on issues deemed more central to the survival and progress of the nation-state, such as trade and military security. Meanwhile, international environmental law has evolved over time into a complex layer of "soft" governance (see Bodansky 2010).

The international level is vital to understanding the politics of extinction in the twenty-first century, because countries cannot operate in isolation when it comes to policy decisions related to endangered species. The Convention on International Trade in Endangered Species of Wild Fauna and Flora (CITES) is a prominent example of an international convention that has attracted nearly universal membership, in which states routinely participate and publicly display their positions on issues and contribute valuable data. The Convention on Biological Diversity, which operates under the aegis of the United Nations Environment Programme, is another example: It serves as a clearinghouse for information and conservation funding as well as a stimulus for participating states to report on the status of biodiversity within their borders. The Convention on Migratory Species, the International Plant Protection Convention, the Food and Agricultural Organization of the United Nations, the UN Convention on the Law of the Sea, and many other international treaties and secretariats are also important, including financing institutions such as the Global Environment Facility. There are also hundreds of regional and/or species-specific organizations, such as the IWC, the North Atlantic Marine Mammal Commission, and the Convention on the Conservation of European Wildlife and Natural Habitats. Another set of actors at the international level includes scientific and advocacy NGOs, such as the IUCN (and in particular its Species Survival Commission), Conservation International, Greenpeace, and the International Wildlife Federation. These groups play important roles in the policy process, both informing and lobbying decision makers, and they also gather taxonomic data and conduct other research. Finally, multinational corporations are also important, as foreign investment decisions can both threaten and encourage conservation efforts abroad.

The typical meeting at CITES, the IWC, and other conservation bodies still gets mired in the ageless debate between the conservationists (e.g., Japan, Norway, South Africa, and on some issues Canada) and the preservationists (e.g., the United States, much of Europe, and Kenya). It is safe to say, however, that most policy circles at the international level are moving past this debate. Though the international politics of extinction are contentious, they are also increasingly characterized by cross-national policy network developments. Indeed, the overwhelming majority of work now takes place on a transnational level, afforded by educational institutions, NGOs, Internet-based communication, and the media.

As of 2012, much of the international stage is cluttered with the climate change debate, but this may not be a bad focus from the perspective of those concerned about extinction and environmental justice. The IUCN has also gone to great lengths to integrate climate change risks into its listing process. A 2009 report of the results of a global multitaxon analysis suggested that 35 percent of all bird species and 52 percent of all amphibian species (Vié et al. 2009) are threatened with extinction primarily because of anticipated risks associated with climate change. While the international diplomatic debate does not hinge on this alarming projection, it is certainly a contextual element. This debate also encourages open discussions of how climate change adaptation policies and concomitant funding can be applied in sensitive and fair ways. Overall, however, the twisting path toward a truly effective global climate treaty is a long and contentious one, and it would be a monumental strategic error to place faith in the outcome of this process.

Governance issues

There are many formal methods for avoiding extinctions, some of which have been mentioned in the preceding sections. Habitat preservation remains the gold standard, but it is increasingly difficult to pursue given expanded human populations and increased human agricultural needs. *Ex situ* conservation is one of the least desirable responses, but it possibly is the most necessary one. This includes captive breeding as well as efforts at genetic manipulation to ensure survivability, breeding for potentially adaptive traits, and propagule (seed and egg) banking, each of which has its own set of ethical dilemmas. Managed relocations are notoriously difficult, not only because of the adaption needs of relocated and potentially invasive species but also because of the local politics in areas where relocation takes place. Without sustained legitimacy—usually entailing the active participation of and tangible benefits for local communities—such projects are doomed to failure. Moreover, corruption will always be an issue, especially if wildlife protection is not adequately funded or is simply used as an excuse to spread state and police power. Meanwhile, ecosystem restoration, favored by forestry, mining, and oil companies, has become a major industry in its own right, but the reintroduction of species must be done with great care. And of course all of these approaches could be for naught if climate change makes species adaptation difficult because the delicate ecological balance that sustained the extirpated species will be fundamentally altered.

Indeed, in efforts to prevent future extinctions, climate change adaptation will be vital. Biodiversity conservation must be a fundamental aspect of adaptation policies and needs to be reflected in the decision-making processes of local, national, and global institutions. Various in situ approaches to conservation with climate change in mind have been put forward, including strategically managing for resistance for species that live in

> habitat too fragmented to permit successful range extension, or species with narrowly defined habitats that are not available outside their present range … Protection may require that managers not only enhance the ability of target species to resist climate change but also reduce potential competitor or predator species that are naturally extending their ranges into the management area … in some cases, the intensity of such efforts may approach cultivation, essentially creating a wild-animal park or arboretum. (Hellman et al. 2012, 374–375)

This is a tall order, well beyond the capacity of most governments and wildlife agencies where some of the most endangered species live. Indeed, it is almost incomprehensible that such major operations can take place in countries with advanced economies, such as Canada or the United States, let alone in the equatorial zones or countries with seriously challenged economies such as Russia or Indonesia. Efforts to fund climate change adaptation have offered some promise. For example, a climate change adaptation fund has been created under the aegis of the UN Framework Convention on Climate Change, and other instruments, such as the Reduced Emissions from Deforestation and Forest Degradation in Developing Countries initiative, are also promising. However, some critics argue that the anticipated funds, which may or may not materialize in an age of global economic shocks, are miniscule in relation to what is necessary—a Marshall Plan for the environment. Again, these efforts will need to compete with other priorities in both northern and southern countries, where social security, unemployment, health, defense, and education are all competing areas of expenditure. Even within the sciences there is harsh competition for scarce resources. For example, as a field of science, taxonomy has been generally underfunded, and much more research is needed for scientists to even begin to understand the totality of species affected by human activity, as well as to offer good policy advice to politicians and bureaucrats on the cutting edge of conservation efforts.

The normative battle between preservationists and conservationist will continue to rage, especially in public and controversial forums such as the IWC, and the need to link local needs and economic security with wider efforts to save both species and habitat is stronger than ever. While national and international efforts to legalize the protection of endangered species are viewed by some as acts of cultural imperialism, there is a common denominator among both camps: If a species is on the brink of extinction surely it deserves human protection, regardless of the various camps' ethical stances. The even more difficult question that has emerged is whether the protection of large habitat zones, including areas of the commons such as the oceans, deserves such attention, and, if so, how those who will lose income and other opportunities can be compensated.

Institutions engaged in conservation activities will need to perform many functions at several levels if they are to achieve long-term success. They will need to seek democratic legitimacy by involving local stakeholders in their decisions and provide compensatory funding for those who are displaced by conservation efforts. They will need to interact with the local and global legal and policy environment, which demands a subtle understanding of municipal, constitutional, and international law. Institutions at all levels will need to argue their cause to politicians and the general public in order to justify their demands for scarce resources amid universal budgetary constraints. It is vital that they combine sensitivity to local cultures and indigenous knowledge with the most advanced scientific assessments of species status, habitat needs, and ecosystem approaches. Plus they need to constantly explore and enrich links among themselves, in order to avoid needless duplication of work, combine creative energies and physical resources, and reinforce in the public view the idea that extinction is a real and serious issue—and not an easy one to tackle in the short term.

Resources

Books

Bodansky, Daniel. *The Art and Craft of International Environmental Law*. Cambridge, MA: Harvard University Press, 2010.

Hannah, Lee, ed. *Saving a Million Species: Extinction Risk from Climate Change*. Washington, DC: Island Press, 2012.

Hellmann, Jessica J., Vicky J. Meretsky, and Jason S. McLachlan. "Strategies for Reducing Extinction Risk under a Changing Climate." In *Saving a Million Species*, 363–387.

Hoegh-Guldberg, Ove. "Coral Reefs, Climate Change, and Mass Extinction." In *Saving a Million Species*, 261–283.

Lazarus, Richard J. *The Making of Environmental Law*. Chicago: University of Chicago Press, 2004.

Ostrom, Elinor. *Governing the Commons: The Evolution of Institutions for Collective Action*. Cambridge, UK: Cambridge University Press, 1990.

Peluso, Nancy Lee, and Michael Watts, eds. *Violent Environments*. Ithaca, NY: Cornell University Press, 2001.

Poff, N. LeRoy, Julian D. Olden, and David L. Strayer. "Climate Change and Freshwater Fauna Extinction Risk." In *Saving a Million Species*, 309–336.

Post, Eric, and Jedediah Brodie. "Extinction Risk at High Latitudes." In *Saving a Million Species*, 121–138.

Secretariat of the Convention on Biological Diversity (CBD). *Global Biodiversity Outlook 3*. Montreal: Secretariat of the Convention on Biological Diversity, 2010.

Suzuki, David, and David Robert Taylor. *The Big Picture: Reflections on Science, Humanity, and a Quickly Changing Planet*. Vancouver, Canada: Greystone Books, 2009.

Vié, Jean-Christophe, Craig Hilton-Taylor, and Simon N. Stuart, eds. *Wildlife in a Changing World: An Analysis of the 2008 IUCN Red List of Threatened Species*. Gland, Switzerland: International Union for Conservation of Nature, 2009.

Walters, Mark Jerome. *Seeking the Sacred Raven: Politics and Extinction on a Hawaiian Island*. Washington, DC: Island Press, 2006.

Warner, Rosalind. "The Place of History in International Relations and Ecology: Discourses of Environmentalism in the Colonial Era." In *International Ecopolitical Theory: Critical Approaches*, edited by Eric Laferrie`re and Peter J. Stoett, 34–51. Vancouver, Canada: UBC Press, 2006.

Periodicals

Clark, Colin W. "The Economics of Overexploitation." *Science* 181, no. 4100 (1973): 630–634.

Hardin, Garrett. "The Tragedy of the Commons." *Science* 162, no. 3859 (1968): 1243–1248.

Knights, Paul. "Native Species, Human Communities, and Cultural Relationships." *Environmental Values* 17, no. 3 (2008): 353–373.

Thomas, Chris D., Alison Cameron, Rhys E. Green, et al. "Extinction Risk from Climate Change." *Nature* 427, no. 6970 (2004): 145–148.

Other

BirdLife International. "Birds on the IUCN Red List." Accessed March 17, 2012. http://www.birdlife.org/action/science/species/global_species_programme/red_list.html.

Emslie, R. "*Diceros bicornis* ssp. *Longipes*." 2011. IUCN Red List of Threatened Species. Version 2011.2. Accessed April 4, 2012. http://www.iucnredlist.org/apps/redlist/details/39319/0.

Robbins, M., and L. Williamson. "*Gorilla beringei*." 2008. IUCN Red List of Threatened Species. Version 2011.2. Accessed April 4, 2012. http://www.iucnredlist.org/apps/redlist/details/39994/0.

State of Alaska. Office of the Governor. "Response to September 7, 2007, USGS Polar Bear Reports." October 22, 2007. Accessed March 17, 2012. http://www.eswr.com/docs/93008/pb_alaska_comments_usgs_102207.pdf.

Peter Stoett

• • • • •

The future of life on Earth

The study of biological evolution on Earth has increased scientists' understanding of the distant past. One characteristic they have deduced is that evolutionary history has resulted from stochastic, that is, unpredictable events. Another is that the course of evolution, although affected by numerous forces both biological and physical, has been perhaps most importantly affected by—and effected—the course of the physical evolution of Earth, its atmosphere, and its oceans. These two, major characteristics of evolutionary history seem to suggest that making any meaningful predictions about the course and makeup of *future* evolutionary biota a low probability exercise. Scientists cannot predict, for example, when or even if a twin of the 6.2-mile-wide (10-km-wide) asteroid that changed the Mesozoic biota into its Cenozoic successor will again hit Earth sometime during the next 7 billion years—the calculated time that Earth has left before the Sun expands into a red giant, increasing its size to at least the current orbit of Mars and destroying Earth in the process.

Although scientists cannot make accurate predictions of specific kinds and appearances of future evolutionary biota, they can make highly refined estimates about predictable changes in global temperatures, atmospheric and oceanic chemistries, and large-scale geophysical events that will necessarily take place over Earth's remaining lifetime. In this entry, knowable Earth futures are discussed, and the exact appearances, behaviors, and species produced as adaptations to the likely changes are speculated about. The actual course of future evolution remains unknown, but enough is known about Earth's future that generalities can be made.

Earth: A habitable planet

The concept of a habitable planet is based on planetary nurture, with life being the ultimate result of planetary formation and change. Life exists because of a series of systems that recycle important nutrients and maintain near-constant global temperatures.

A *system* can be defined as a group of components that interact. Systems are thus diverse because each system has interrelated parts that function as a complex whole. In vertebrate animals (and in many invertebrates as well), the interacting systems include the circulatory, respiratory,

endocrine, nervous and sensory, lymphatic, reproductive, digestive, and excretory. These systems evolved early in any given animal's history, and they then remained essentially constant for very long periods of time. However, eventually the systems age, that is, become less efficient. That aging process affects different life systems at different times; some systems, such as the circulatory system, age faster than others. This analogy seems to hold for habitable planets, including Earth. In describing Earth's future, the questions are, first, what are Earth's systems, and, second, how will they change through time—and thus influence future biotic evolution on Earth.

Organisms need material and energy to live and grow. They need the matter necessary to build cell walls and organelles, nucleic acids, and polymers, the entire physical superstructure that is life. Thus, organisms are open systems, that is, they require additional material during life for living and growing. Humans—and almost all other organisms—do not live very long without a constant intake of new material. By contrast, Earth continuously recycles material already present. However, it also receives material from space, as in the meteor mentioned above, and more important, it continuously receives an influx of light energy from the Sun. Thus, the distinction between the open system (organisms) and the closed system (Earth) breaks down since Earth receives both material and energy from space. Earth functions mostly as a closed system regarding matter and is generally an open system regarding energy.

For life, the most important of these fluxes are the movement and transformation of the elements carbon, nitrogen, sulfur, phosphorus, and various trace elements. Each of these elements is critical to the existence of life on the planet. These and other elements move in and out of the atmosphere, hydrosphere, and the solid Earth. Because the movement and transformation of matter require energy, Earth system science also examines the energetic underpinnings of the various systems, which largely come from two sources: the Sun and heat generated from the breakdown of radioactive material beneath Earth's surface.

Each of these systems has changed through time and will continue to do so. The presence of life on the planet and the ability of life to evolve and increase in its complexity through time have caused each of the nonbiotic earth systems to be modified and then caused feedbacks affecting life. These couplings linking Earth's organic and inorganic components have evolved in tandem over time as the planet has aged and as life has radically transformed itself, developing increasing diversity and complexity. The study of Earth has yielded accurate information about how these interactions have occurred in the past, enabling scientists to make predictions about how these systems will change in the future.

How the changing Sun will affect future evolution

Life requires energy. Be it a plant directly harvesting the energy from the Sun or an animal ingesting some part of that plant and thus acquiring solar energy secondhand (or third- or fourth-hand), the Sun is the ultimate source of almost all of life's energy on Earth.

The Sun is Earth's primary source of warmth and light in a cold and dark universe. Not only does the Sun's light energy power photosynthesis, its gravity also holds Earth in orbit, and its heat keeps the planet from freezing. It powers the wind, drives the waves, and makes clouds that provide an ocean-covered planet with a nearly endless supply of fresh water. Even at a distance of 93 million miles (150 million km), the light of the midday summer Sun still carries over a kilowatt of power to each 10.8 square feet (1 square meter) of Earth it illuminates.

Unfortunately, the star that Earth orbits is a time bomb, and each tick moves the planet toward a future of drastic change. Although it has nourished life for billions of years, the Sun is always changing and future changes will cause major changes on Earth. The Sun will be *the* major factor that drives Earth's future biotic evolution to its ultimate fate. It is truly ironic that the Sun, a body that has played such a positive role in Earth's history, is also one of the villains, and will eventually be responsible for the ultimate destruction of life on the planet and the physical planet itself.

The Sun is a powerful nuclear reactor, but its stability is a matter of debate. Scientists have no direct long-term record of the Sun's output, and their only direct information comes from observations of similar stars. Stars that have a similar mass and age as the Sun also have nearly the same brightness as the Sun. This correlation suggests that the brightness of these Sun-like stars does not vary greatly. It is expected, however, that long-term changes must occur. Scientists are certain that stars get brighter very slowly. Great energy-generating engines, stars work for billions of years, but near the end of their lives they begin to run out of fuel, and they undergo extraordinary and complex changes. Unlike most other engines, they do not fade as they age but rather become ever more powerful and energetic.

The Sun, like all similar stars, becomes brighter because the number of atoms in its center decreases. The pressure on the inside of the Sun has to exactly support the weight of its overlying mass. The size of the Sun does not change over long periods of time, so the pressure in its center remains constant.

The pressure is produced by the cumulative impacts of vast numbers of particles. As each atom bounces off the surface of core, it exerts a small outward force. The total pressure is the net effect of all of the particles. Based on what is known as the gas law, the pressure in core depends on just two factors: the number of particles and the temperature of the gas. Given that the number of particles is constantly decreasing, the temperature constantly rises if the pressure is to remain constant.

As the Sun evolves, the number of particles in the core decreases. The chain of nuclear reactions effectively takes four protons and turns them into one helium nucleus. If all the hydrogen were converted to helium, the number of particles left would be only one-fourth of the number that the Sun had initially. As the number of particles gradually decreases, the temperature of the Sun's core rises. As the temperature rises, hydrogen travels at higher speed, collisions are more energetic, and the production of helium and the total amount of energy generation rises. This slow ramp-up of energy generation occurs for the full ten billion years that the Sun generates all of its energy by fusion of hydrogen to helium.

Increase in the Sun's brightness is slow but continuous and inevitable. All stars like the Sun share this same characteristic. The Sun has increased in brightness by about 30 percent in the last 4.5 billion years of its life. The rise in brightness increases the intensity of the sunlight that illuminates its planets. This change would cause the oceans to be lost to space and create hellish conditions, similar to those that exist on Venus. The brightness growth of the Sun is accelerating, and in 4 billion years the Sun will be twice as bright as it was 4 billion years ago. At that time the Sun will be in its middle age. Ultimately, at about 10 billion years, the nuclear burning process will move outward, as the outer hydrogen surrounding a nearly pure helium core. At this point the Sun will enter what is known as the red giant stage, whereupon its surface will become cooler but its diameter will expand so much that its overall brightness will increase thousands of times. The slow brightening of the Sun during its middle age is much less dramatic than the events that will occur when it becomes a red giant at approximately 10 billion years old. During its middle age, the Sun will brighten by only a factor of two, but nonetheless this change will produce significant stress and change on the inner planets, Mercury, Venus, Earth, and Mars. On Earth, theoretically, the rise of solar heating by a factor of two would increase the surface temperature by about 180°F (100°C).

For all its history, Earth has been within the temperate zone of the solar system. That is, Earth is the "right" range of distance from the Sun to have surface temperatures that allow oceans and animals to exist without freezing or frying. This *habitable zone* extends from a well-known limit just inside Earth's orbit to a less understood outer limit near Mars or possibly beyond. The habitable zone moves outward as the Sun becomes brighter, and in the future the zone will pass Earth and leave it behind. Earth will in essence become the Venus of today. The inner edge of the habitable zone is only about 9.3 million miles (15 million km) away, and it will effectively reach Earth in a half a billion or a billion years from now (or possibly less). After this time, the Sun will be too bright for organisms to survive on Earth.

The final stage of stars is reasonably well understood from a century of research. The final stage of stellar lifetimes is relatively short. For the Sun, its normal state (when it is similar to what is observed today) lasts around 10 billion years, while the advanced state—the red giant phase—will last less than a billion years.

The planetary thermostat and its end

The steadily rising amount of energy hitting Earth from the Sun would have long ago ended life on Earth—as it did on Venus (assuming that Venus ever had life)—except for one of the most important of all of the planetary life support systems, which is sometimes called the *planetary thermostat*. For more than 3 billion years (and perhaps 4 billion years) this system has kept the global average temperature of Earth between the freezing and boiling points of water—thus allowing the most important requirement for life, liquid water, to continually exist on the surface of the planet for that immense amount of time. Just as important, life, which evolved within tight temperature limits, has been able to maintain essentially similar physiologies and internal chemical reactions that are temperature dependent.

The planetary thermostat is composed of three important subsystems: plate tectonics, the carbon cycle, and the carbonate–silicate cycle. The planetary thermostat is greatly affected by plate tectonics, or continental drift. On Earth, the plate tectonics system is essential in maintaining surface temperatures at levels that allow the existence of liquid water. The upward and downward movement of earth material buries some material and liberates others, and it causes chemical changes, through new mineral formation, heating, and the liberation of gases. All these aspects play a role in maintaining the planet's relatively constant temperature. Rising temperature and a dearth of atmospheric carbon dioxide are the two processes that in combination will have the greatest effect on future biotic evolution.

While many circulating elements, such as sulfur, phosphorus, oxygen, and nitrogen, are essential for life, perhaps the most important element dictating the future of evolution is carbon. The carbon cycle is the main process for regulating long-term temperatures as well as atmospheric composition, and it is especially important in controlling the future climate as the Sun bombards Earth with ever more energy.

Give that carbon is the crucial atom of terrestrial life, its rapid exchange between inorganic compounds (the most important being carbon dioxide [CO_2]) and organic compounds is crucial to life. In addition to being required for life to exist and for organisms to allow new cell growth and repair, carbon is also of overriding importance to Earth's temperature when it occurs as CO_2 in the atmosphere. Carbon dioxide is a greenhouse gas, a gas that warms a planet's surface by absorbing radiant heat, also known as infrared radiation, and sending some of it back toward Earth's surface, rather than allowing it to escape out into space. Methane (CH_4) and even water vapor are also highly effective greenhouse gases.

The long-term climate—and the maintenance of a global thermostat setting between the freezing and boiling points of water over billions of years of time—is largely controlled by what has come to be called the silicate–carbonate geochemical cycle, an integral part of the carbon cycle. The silicate–carbonate geochemical cycle involves the movement (transfer) of carbon to and from the crust and mantle of the planet, a movement accomplished by the plate tectonic system described earlier. This cycle balances inorganic reactions taking place deep in the planet with interactions between the atmosphere and the surface of the planet, and it involves living organisms. This balance keeps atmospheric CO_2 levels essentially constant—and hence keeps Earth's surface temperature relatively constant—for the long time scale of geological time.

Two quite different processes are also essential. The first of these is carbonate precipitation. If calcium is combined with carbonic acid under correct temperature and pressure conditions, it can combine to form calcium carbonate—the common rock type also known as limestone, and the most common mineral used by skeletonized organisms. The rate at which limestone forms on Earth's surface has important consequences for the long-term climate.

The second major reaction involves the weathering of a class of rocks known as silicates. Weathering is the chemical or physical breakdown of rocks and minerals. When silicate rocks weather, the by-products can combine with other compounds to produce calcium, silicon, water, and carbonic acid. The weathering of the silicate rocks eventually removes CO_2 from the atmosphere, an action that is of enormous importance to life. Slight perturbations to the rates of carbon will spell ultimate doom for plant life and eventually all life on Earth.

As the rate of weathering increases, more silicate material is made available to react with the atmosphere, and more CO_2 is removed—thus causing *cooling*. Yet as the planet cools, the rate of weathering decreases, and the CO_2 content of the atmosphere begins to rise, causing warming to occur. In this fashion, Earth's temperature oscillates between warmer and cooler as a result of the carbonate–silicate weathering and precipitation cycles.

The planetary thermostat requires a balance between the amount of CO_2 being pumped into the atmosphere through volcanic action and the amount being taken out, largely by organisms, through the formation of limestone. The entire system is driven by heat emanating from Earth's interior, which causes plate tectonics. But as noted, there is more to this cycle than simply heating from the interior. Weathering on Earth's surface is crucial as well, and the rate of weathering is highly sensitive to temperature, because reaction rates involved in weathering tend to increase as temperature increases. This process causes silicate rocks to break down faster and thus creates more calcium, the building block of limestone. With more calcium available, more limestone can form. But the rate of limestone formation affects the CO_2 content of the atmosphere, and when more limestone forms, there is less and less CO_2 in the atmosphere, causing the climate to cool.

The rises and falls of CO_2 are now fairly well documented for the last 500 million years—the time of animals. Oxygen, a

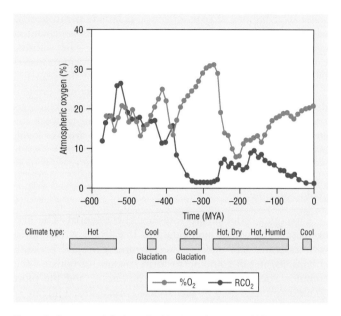

Figure 1. Oxygen and Carbon dioxide through the last 600 million years. Reproduced by permission of Gale, a part of Cengage Learning.

requirement of all animals, is obviously important too. The model levels of these two gases—oxygen and carbon dioxide—during the time of animals is shown in Figure 1.

One of the curious aspects that research from the early twenty-first-century has shown is what appears to be a significant correlation between oxygen levels and diversity. Diversity is higher during times of higher oxygen, and the changes in both can be seen in Figure 2, which compares oxygen levels with changes in diversity modeled by a group led by John Alroy.

Although the fluctuations in oxygen do not seem to have a predictable trend, those of CO_2 do. As can be seen in Figure 2, there has been a long-term reduction. Because CO_2 is a greenhouse gas, that long-term trend should have been accompanied by long-term cooling—and it has been. From this graph one could predict that cooling might continue into the far future. However, that is not the case.

It is not cooling from a lowering of the concentration of CO_2 in the atmosphere that will be a hallmark of the aging Earth. It will be heating. The increasing heat from the Sun will utterly dwarf the cooling effects of diminishing CO_2. When the average global temperature temperatures rises to perhaps 120 to 140°F (50 to 60°C), Earth will begin to lose its oceans to space.

The ever-increasing energy output of the Sun, a phenomenon of all so-called main-sequence stars, will ultimately cause the loss of Earth's oceans (sometime in the next 2 to 3 billion years). When the oceans are lost to space, planetary temperatures will rise to uninhabitable levels. But long before that, life will have died out on Earth's surface because photosynthetic organisms, from microbes to higher plants, will no longer be able to survive in the low CO_2 atmosphere. This dwindling carbon resource will then cause a further

reduction of planetary habitability, because the CO_2 drop will trigger a drop in atmospheric oxygen to a level too low to support animal life.

This process is already observable. The amount of CO_2 in Earth's atmosphere has been steadily dropping over the last 200 million years. It is life that makes most calcium carbonate deposits, such as coral skeletons, and thus life that ultimately causes the drop in CO_2, because CO_2 has to be taken out of the atmosphere to build this kind of skeleton. Life will continue to do this, until a lethal lower limit CO_2 is reached.

Future biodiversity

The most important questions about any future evolution concern the future of biodiversity—the number of species on Earth. Two questions that arise are will there be more than now and if so for how long. But to begin to answer these questions—as is so often the case—one needs to look into the past.

The history of biodiversity—the assembly and measurement of diversity and biomass through time—was first considered in

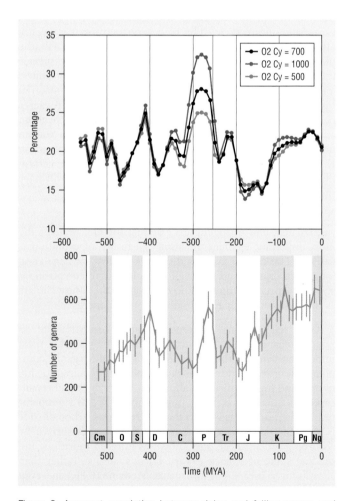

Figure 2. Apparent correlation between rising and falling oxygen and global diversity. Reproduced by permission of Gale, a part of Cengage Learning.

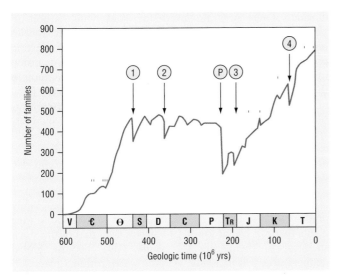

Figure 3. Twentieth century estimate of diversity through time. Reproduced by permission of Gale, a part of Cengage Learning.

the work of John Phillips, who is credited with subdividing the geological time scale through the introduction of the concepts of the Paleozoic, Mesozoic, and Cenozoic eras (see Figure 3). Phillips, who published his monumental work in 1860, recognized that major mass extinctions in the past could be used to subdivide geological time, because the aftermath of each such event resulted in the appearance of a new fauna as recognized in the fossil record. But Phillips did far more than recognize the importance of past mass extinctions and define new geological time terms: He proposed that diversity in the past was far lower than in the modern day and that the rise of biodiversity has been one of wholesale increases in the number of species, except during and immediately after the mass extinctions. His scheme recognized that mass extinctions—short intervals of time that witnessed disproportionably high amounts of biotic extinction at the species level—greatly slowed down diversity. But, as is now known, they did so only temporarily.

Phillips's view of the history of diversity was completely new. Yet a century passed before the topic was again given scientific attention. In the late 1960s, the paleontologists Norman Newell and James Valentine again considered the problem of exactly when, and at what rate, the world became populated with species of animals and plants. Both wondered if the real pattern of diversification was a rapid increase in species following the so-called Cambrian explosion of about 520 to 540 million years ago, followed by an approximate steady state.

The debate on whether diversity has shown a rapid increase through time or achieved a high level early on and has stayed approximately steady ever since dominated paleontological research for much of the latter part of the twentieth century. In the 1970s massive data sets derived from published records of fossil appearances and disappearances began to be assembled by J. John Sepkoski Jr. of the University of Chicago (and his colleagues and students). These data, compiling the record of marine invertebrates in the sea, as well as other data sets for both terrestrial plants and

for vertebrate animals, seemed to vindicate Phillips's early view. In particular, the curves discovered by Sepkoski showed a quite striking record, with three main pulses of diversification carried out by different assemblages of organisms (see Figure 4). The first was seen in the Cambrian (with the so-called Cambrian fauna composed of trilobites, brachiopods, and other archaic invertebrates) and was followed by a second in the Ordovician. The Ordovician led to an approximate steady state throughout the rest of the Paleozoic era (the Paleozoic fauna were composed of reef-building corals, articulate brachiopods, cephalopods, and archaic echinoderms) but culminated in a rapid increase beginning in the Mesozoic. Diversification then quickly accelerated in the Cenozoic to produce the highest levels of diversity seen in the world today; the evolution of the modern fauna happened during this time—gastropods and bivalve mollusks, most vertebrates, and echinoids, among other groups.

The net view of biodiversity over the last 500 million years was the same as that of Phillips in 1860: There are more species on the planet than any time in the past. Even more comforting, the trajectory of biodiversity seemed to show that the engine of diversification—the processes producing new species—was in high gear, suggesting that in the future the planet would continue to have ever more species. These findings certainly do not suggest that Earth is in any sort of planetary old age. All in all, the 130-year belief, from the time and work of Phillips to that of Sepkoski—that there are more species now than at any time in the past—remained a comforting view. This long-held scientific belief suggested to many that the current times are the best of biological times (at least in terms of global biodiversity) and that there is every reason to believe that better times, an even more diverse and productive world, still lie ahead.

It may be, however, that diversity peaked early in the history of animals and, in contrast to most views since the time of Phillips, that it has remained in an approximate steady state since or perhaps may already be in decline. Although many new innovations, such as the adaptation that enabled the

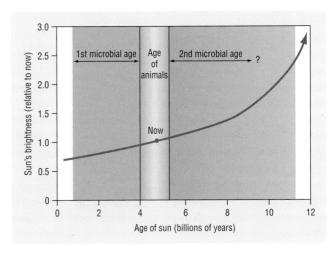

Figure 4. Projected estimates of major events affecting future evolution. Reproduced by permission of Gale, a part of Cengage Learning.

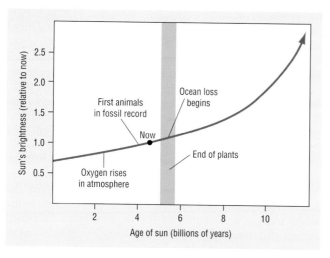

Figure 5. Projected estimates of major events affecting future evolution. Reproduced by permission of Gale, a part of Cengage Learning.

dioxide. There was no need for conserving carbon in physiological processes. Even today, many plant species require a minimum of 150 ppm of CO_2, and James F. Kasting pointed out in a 1997 article that there is a second large group of plants, including many of the grassy species so common in the midlatitudes of the planet, that use a quite different form of photosynthesis and can exist at lower CO_2 concentrations—sometimes as low as 10 ppm. These plants would last far longer than their more CO_2-addicted cousins and would considerably extend the life of the biosphere even in a world in which CO_2 levels had fallen far, far below present-day values.

These are the so-called C_4 plants, and one can safely predict that the future evolution of plant life will be toward plants that can live at lower CO_2 levels than that of the stock ancestral to them, the C_3 plants. Also, because global temperatures will be rising, keeping water within a plant will be an increasing problem. Plants will have two conflicting needs—ever larger holes in their exterior to let the small amount of carbon dioxide in the atmosphere to get into the interior, where photosynthesis can take place, at the same time trying to reduce the loss of water molecules through these same pores. At a minimum, one can expect a future flora of tough, waxy plants that would completely close down all portals to the outside world when there is no sunlight for photosynthesis.

With new plants with tougher exteriors, leaves, at least in their present form, might be expected to disappear. The same will happen to grass; the loss of water from plants with relatively high surface-area-to-volume ratios will doom grass blades and thin leaves alike. All of this, of course, will require a marked change in animal life.

As early as 500 million years from now, or perhaps as much as 1 billion years or so into the future, the level of CO_2 in the atmosphere will reach a point at which familiar plant life will no longer be able to exist. The changeover, at first, will be in no way dramatic. All over the plants will slowly die. But the planet will not immediately become brown. For as one suite of plants dies, their places will be taken immediately by another cohort of plant life that may look nearly identical to those dying. Deep inside the tissues of these two groups of plants, however, fundamental processes of photosynthesis will be radically different. After this changeover, life on Earth will continue in ways probably not too dissimilar from that which came before—at least for a time.

There is also the possibility that plants will continue to evolve other photosynthetic pathways to compensate for lower CO_2 levels. In this case, some sort of plant life may survive at minimal CO_2 levels. Eventually, however, even these last holdouts will die out. All models suggest that CO_2 will continue to drop in volume, ultimately arriving at the critical level of 10 ppm.

When this is projected to happen is debated. Early models projected that this lethal blow to life on Earth—the loss of plant life—would take place in as little as 100 million years from now. More sophisticated models pushed that date forward—to perhaps more that 500 million years from

evolution of land plants and animals, certainly caused there to be many new species added to the planet's biodiversity total, it may be that since late in Paleozoic time the number of species on the planet has been approximately constant. The implication of this speculation is important: Perhaps the planet, rather than still growing in biodiversity totals, has already peaked and is already sliding back into lesser numbers. This possibility is tied to the realization by planetary scientists and biologists that not just diversity but the very mass of life on Earth will predictably drop as global temperatures rise and the pool of available carbon decreases.

This is the second way to look at the relative abundance of life on Earth, past, present, and, through modeling, future. Instead of looking at the number of species, a model of the overall biomass of the planet can be constructed, allowing informed estimates. A highly pertinent model was published in 2006 by the German atmospheric scientist Siegfried Franck and his colleagues. Their graph, shown in Figure 5, models the future carbon dioxide path through time. The striking aspect of this graph is the long-term trend of decreasing CO_2, with a drop of as much as five orders of magnitude. Current levels are at 380 parts per million (ppm), and thus the earliest Earth may have had 10,000 times more CO_2. That would mean that a significant percentage—perhaps one-third—of the ancient atmosphere would have been CO_2, not the trace it makes up today.

Predictions based on changing solar radiation and CO_2 levels

General predictions about future evolution can be based on the deadly twins of dropping carbon dioxide and increasing temperature. The first of these predictions deals with plant life.

When plants first colonized Earth's surface some 450 million years ago, they did so in an atmosphere rich in carbon now—although one group suggested that, as a result of the biotic enhancement of weathering, there will be sufficient CO_2 for plants until about 1 billion years from now. Some other researchers even suggest that CO_2 will hover at the critical level and never dip below it, thus allowing some minimal amount of vegetation to continue to exist on the planet. Yet even this best-case scenario produces a world vastly different from that of today, and one in which there is little advanced life. Whatever the time frame, the loss of plants will be dramatic and world changing.

It seems ironic: Plants will begin to die for no apparent reason. The world will not be a hothouse—although it will certainly be hotter than now, it may be no hotter than during the Cretaceous period, some 100 million years ago. All other aspects of the planet will seem normal. Yet the plants will indeed begin to die.

The first to go will be those plants with the C_3 pathway. If plants remain in the present configurations for C_3 and C_4 species, the world would undergo a radical episode of deforestation, leaving behind mainly grasslands and species adapted for high heat and low moisture—the cactus and succulent floras and their ilk. Because the drop in CO_2 will have been occurring for hundreds of millions of years, it can be expected that evolutionary adaptations to this new environmental reality will have spurred evolutionary processes to evolve whole new types of plants in response to the lower CO_2 levels. But this may not be the case, as there will still be C_3 plants right up until they are no longer viable, resulting in a first wave that will remove the forests from the planet. By the time the first wave of lower CO_2 levels begins to kill off plants, there may already be a global flora using the C_4 pathway.

It will not just be the plant flora that is traumatized by the lower CO_2 levels. Larger marine plants and perhaps plankton as well will be affected similarly. Marine communities thus will be strongly affected, because the base of most marine communities is phytoplankton, a single-celled plant that floats in the seas. A reduction in CO_2 will directly affect these as well as land plants. Yet the disappearance of land plants will also cause a drastic reduction in the biomass of marine plankton, even without accounting for CO_2 effects on plant volumes in the seas. Marine phytoplankton are severely nutrient limited in most ocean settings. The influx of nitrates and phosphates into the oceans each season causes phytoplankton to bloom. But the source of this phosphate and nitrate is rotting terrestrial vegetation, brought into the oceans through river runoff from the land. As land plants diminish in volume, so too will the volume of nutrients be diminished. The seas will be starved for nutrients, and the volume of plankton will catastrophically decline. This decline will never be reversed, for even if land plants rebound at low levels, as outlined above, they will never again reach the enormous mass of material that is present in a world (such as that of today) where CO_2 starvation does not exist.

On land and sea the base of the food chains as they are constructed today will disappear. The effect of this changeover from a planet coated with a veneer of plants to one without will be dramatic. Earth will no longer be recognizable

to those living during this current time of plants. The changes brought about by the loss of plants will affect and alter all four of Earth's systems: the biosphere, the hydrosphere, atmosphere, and the solid Earth systems. One small link in the various systems of Earth will be damaged, and, as a result, all of the systems will be shaken and, in some cases—such as the biosphere—ultimately destroyed by this perturbation.

The loss of plants will suddenly cause global productivity—a measure of the amount of life on the planet—to plummet, but scientists wonder by how much. As catastrophic as the loss of multicellular plants will be, there will still be life, and lots of it. For while terrestrial plants will die off, organisms capable of photosynthesis will not. There will still be great masses of bacteria, such as cyanobacteria, also known as blue-green algae, that will continue to thrive, because these hardy single-celled organisms can live at CO_2 levels that are below those necessary to keep multicellular plants alive.

The question remains how much of the world's productivity is tied up in green plants. While a glance at most habitats on Earth, with their abundance of green plants ranging from grass and moss to giant trees, would suggest that most productivity would end, a more balanced view is that there would still be a great deal of productively taking place—because of bacteria.

Multicellular green plants on land make up the majority of land productivity, whereas single-celled green algae in the sea provide the majority of the oceans' productivity. But there are photosynthetic bacteria in both places, as well as an unknown but probably gigantic amount of bacteria in the soil and even solid rock that also fixes carbon. Estimates of productivity from bacterial and Archean microbes alone might account for half of all of the productivity of the planet.

Cutting world productivity so drastically will affect all other life on the planet, from bacteria to animals, and undoubtedly life on planet Earth will become far rarer. No longer will falling leaves create giant volumes of reduced carbon that make their way into the soil, the sea, and the sedimentary rock record. No longer will coal and oil begin its process of formation. The carbon, nitrogen, and phosphorus cycles will be radically changed. There will no longer be spring plankton blooms. As land plants disappear, the soil will erode, leaving beyond bare rock. This will, in turn, perturb the hydrological cycle and even the pathway of liberalization on the planet. Giant transfers of carbon between the various land, ocean, and sedimentary record reservoirs will occur.

The disappearance of plants will drastically affect landforms and the nature of the planet's surface. As roots disappear, and surface layers become less stable, the very nature of rivers will change. The large, meandering rivers of the modern era date back, at most, to the Silurian period of some 400 million years ago, when land plants first colonized the surface of the planet, for it takes root stability to maintain the banks of meandering rivers. When plants die out, or are not present because of slope, soil, or other inopportune environmental conditions, a different kind of river exists—braided rivers or streams, the kinds of flows found on desert alluvial fans or in front of glaciers, two types of environments not conducive to rooted plant life. This was the nature of rivers before the advent of land plants, and it will again be the way that rivers flow when CO_2 drops to the plant die-off threshold.

The loss of soils will be no less dramatic. As soils are blown away, they will leave behind bare rock surfaces. As this condition begins to occur over the surface of the planet, it will change the albedo—the reflectivity of Earth. Far more light will reflect back into space, thereby affecting Earth's temperature balance. The atmosphere and its heat transfer and precipitation patterns will be radically changed. Blowing wind will begin to carry the grains of sand created by the action of heat, cold, and running water on the bare rock surfaces. While chemical weathering will lessen as a result of the loss of soil, this mechanical weathering will build up enormous volume of blowing sand. The surface of the planet will become a giant series of dune fields.

Although this event could signal the final extinction of all plant life on land (and perhaps in the sea as well), it is more likely that a long period of time (perhaps in the hundreds of millions of years) will ensue in which CO_2 levels hover at the level causing plant death. As the levels drop to lethal limits, plants will die off, reducing weathering and allowing CO_2 to again accumulate in the atmosphere, once again allowing any small surviving seeds or rootstocks to germinate and, at least for some millennia, to flourish at least at low population numbers. As plant life again spreads across land surfaces, weathering rates will again increase, CO_2 will be again reduced, and plants will again die off.

Green plants and oxygen

Animal life is dependent on an oxygen atmosphere. There are no animals capable of living in zero- or even low-oxygen conditions. With the loss of plants, one wonders what happens to atmospheric oxygen. While some scientists have thought that the loss of plants will have little effect on atmospheric oxygen values, more recent studies suggest just the opposite. The loss of plants will shut off the major oxygen-producing pathway on the planet: photosynthesis. But the loss of plants will have no effect on the most important oxygen sink—the oxidation of dead matter on the surface and volcanic gases emanating from Earth's interior. It is the latter that will most rapidly deplete the oxygen supply. A study by David C. Catling has suggested that by about 15 million years after the death of plants, less than 1 percent of the atmosphere will be oxygen in contrast to the 21 percent volume that Earth's atmosphere contains today.

The next mass extinction

Stephen Jay Gould listed three tiers of evolution—the levels at which it takes place. These are at the level of the individual (microevolution), the level of the species (macro-evolution), and the level of the entire biota at any given time—by mass extinction. Perhaps there can be no larger effects on the future evolution of life on Earth than from the mass extinctions to come. One wonders what might these be.

The brief outline of events given above suggests that most of the mass extinctions seem to have been caused by the forces generated by the planet itself. Some still remain mysterious as to their cause, however, and there are other potential catastrophic events that might have affected Earth.

In one way or another, all mass extinctions appear to have been immediately caused by changes in the global atmosphere inventory. The specific killing agents arise through changes in the makeup and behavior of the atmosphere or through factors such as temperature and circulation patterns that are dictated by properties of the atmosphere. Those in turn can be caused by many factors: an asteroid or comet impact, the loss of carbon dioxide or other gases into the oceans and atmosphere during flood basalt extrusion (when great volumes of lava flow out onto Earth's surface), the loss of gases caused by the liberation of organic-rich oceanic sediments during sea-level change, or changes in ocean circulation patterns.

Earth's geological record suggests, however, that more than a single cause is usually associated with any mass extinction. Sometimes multiple events occur at the same time; sometimes they are separated by hundreds of thousands of years. Perhaps one perturbation stresses the planet, making it more susceptible to the next perturbation.

One such perturbation, difficult to detect, would have involved changes in the amount of energy coming from the Sun, or periods of intense solar flares and storms of greater magnitude than anything humans have observed in the very short time that scientific observation of the Sun has taken place at all. It is known that the Sun has been increasing in its energy output over time and that since the formation of Earth the amount of energy hitting the planet has increased by one-third. If that energy were greater still, over short intervals of time, Earth might experience the fate of Venus—something called a *runaway greenhouse event* that would cause the loss of Earth's oceans. This will be the ultimate fate of the oceans, but that is thought to still be at least 2 billion years in the future. But that could change if the Sun turns out to be less stable than scientists now think.

One might ask what about the opposite—not a hot greenhouse, but a deep freeze. This has already happened several times. Scientists wonder if it could happen again, or, if not on a planetary scale, could the planet be heading for a new interval of glaciation, similar to the times and temperatures of maximum ice cover of the Plio-Pleistocene ice ages of the last 2.5 million years. The input of massive volumes of human-made greenhouse gases such as CO_2 makes this unlikely in the near term. But if civilization stumbles or runs out of coal or oil, this is a distinct possibility. A new glacial interval would be even more catastrophic (especially to agriculture) than the globally tropical times that the planet seems to be heading toward because of human effects on the atmosphere.

Humans are not the only producers of greenhouse gases. There is increasing evidence that the occurrences of flood basalts in Earth's past were coincident with mass extinctions. At the end of the Permian, for instance, the Siberian Traps—the largest known flood basalt on Earth—were extruded, while at the end of the Triassic the Central Atlantic Magmatic

Province was extruded. These events involve the short-term extrusion of large volumes of basalt onto Earth's surface, as well as the outgassing of large volumes of carbon dioxide, a greenhouse gas, and sulfur dioxide, a poisonous gas. In the early 2000s, however, it is not known how the extrusion of large amounts of lava onto Earth's surface translates to a killing mechanism on land or in the sea. One idea is that short-term global heating is caused by huge amounts of greenhouse gases, including methane (previously bound in clathrates), being liberated at these times. Modeling of temperatures at the end of the Permian confirms a short-term heat spike. It would be naive to think that such magmatic episodes have come to an end, and if such a gigantic geological event were to occur, no amount of human engineering could stop it.

Dangers from space

Thanks to Hollywood, the most famous of the potential future catastrophes is the impact of a large asteroid or comet on Earth. But many specialists now think that it would not even take the hit of a big rock from space, such as the 6.2-mile-wide (10-km-wide) body that led to the extinction of the dinosaurs. Even a small asteroid strike would surely perturb agricultural production by throwing up great volumes of dust into the atmosphere to the point of creating global famine. While most of the Earth-crossing asteroids in the solar system have been found and are being tracked, there are many bodies with a diameter of less than 0.6 miles (1 km) that would be too small to detect until it is too late for any sort of intervention. Any hit in the ocean would also cause huge tsunamis that would make the 2004 Indian Ocean disaster seem petty by comparison.

Scientists wonder what the odds are of such a strike. Every century Earth is struck by something large enough to perturb the climate, and thousand-year events could conceivably wipe out appreciable numbers of humans because of the resulting famine.

Outer space is clearly a dangerous place, and not just because it holds asteroids and comets that can and will continue to hit Earth. There are other potential killers out there as well. The two most dangerous are surely supernovas and gamma-ray bursts.

A supernova is a star that explodes, violently, at the end of its lifetime. In 1995 it was calculated that a star going supernova within 30 light-years of the Sun would release fluxes of energetic electromagnetic radiation and charged particles (cosmic rays) sufficient to destroy Earth's ozone layer in 300 years or less. The removal of the ozone layer would then prove calamitous to the biosphere, as it would expose both marine and land organisms to potentially lethal solar ultraviolet radiation. Photosynthesizing organisms such as phytoplankton and reef communities would be particularly affected. Based on the average number of stars near the Sun and the rates of supernova explosions (only very rare stars, those with masses at least eight times that of the Sun, go supernova), these researchers concluded that supernova explosions within 30 light-years of Earth—and therefore presumably dangerous—

occur on average every 200 to 300 million years. Thus, a supernova might have caused several of the still mysterious mass extinctions of the past.

A subsequent proposed agent of mass extinction are gamma-ray bursts, one type of which lasts a second or less, peaks at about 200 kiloelectron volts of energy, and is thought to arise from the merging of two neutron stars into a black hole (most observed bursts originate in very distant galaxies) or is related somehow to supernova explosions. Such a burst of gamma rays hitting Earth could have many deleterious effects. The following are a few of those:

1. delivering enough ionizing energy to Earth's surface (after the original gamma rays are transformed into ultraviolet spectral lines) to cause significant genetic damage to organisms, despite the relatively thick atmosphere of Earth;

2. stripping the ozone layer and thus allowing solar ultraviolet radiation to damage life;

3. causing changes to atmospheric chemistry that could lead to significant changes in surface temperature or biogeochemical cycles; or

4. producing harmful radioactive species on Earth.

A critical issue is the frequency of these postulated events in any galaxy. A detailed analysis in 2002, indicated that the Milky Way should have one gamma-ray burst every 2 million years on average, and it may be that a gamma-ray burst happening *anywhere* in the galaxy may well seriously affect Earth's life (as well as any other life in the galaxy, no matter where it was located) with a photon jolt. The long-term biological effects, however, are uncertain regarding what levels of radiation would be necessary to cause significant mutations and other evolutionary effects and how these would vary for different species.

The most probable dangers

Setting aside the obvious horrors of human warfare, famine from overpopulation, and global epidemic of new and lethal diseases, humanity is faced with some serious threats. Two seem paramount. The first is the impact of a small asteroid, as described above. The second, paradoxically, is a change in oceanic water currents.

The Day After Tomorrow (2004), a very silly climate change movie, posits a sudden freezing of many parts of the planet as a result of a change in oceanic water currents. Although the narrative and science are so far from reality as to be laughable, the basic premise, that global warming could cause catastrophic climate change, is not. One of the greatest threats to humanity is a change in Earth's oceans from their present-day mixed and oxygenated conditions to a new kind of ocean, with no oxygen at its bottom and with current systems different from those of the present day. Evidence is beginning to suggest that such changes could occur relatively quickly.

The presence of oxygen and salinity similarities from the top to the bottom of the oceans is not a characteristic of only the modern oceans but stretches back some 30 to 40 million years. Oceanographers call this type of ocean a *mixed* one because the great temperature differences between the warm tropics and the frigid arctic cause the transport of heat by surface currents, as well as because the heat differences result in a variance in ocean salinity. Thus, while the mixed ocean is oxygenated throughout, density differences produced by the warming or cooling of water—with greater salinity making it denser, and less salinity making it cooler—provide the energy for the three-dimensional mixing that spans the entire globe.

Humans have scrutinized the ocean since the dawn of history. Yet, in spite of the ever more sophisticated monitoring, sampling and modeling, many aspects of the processes leading to mixing are still poorly understood. In addition to their surface currents, oceans have a deeper current system that has enormous influence on the global climate. Because agricultural success depends on climate stability, anything that suddenly cause massive climate change is a potential threat to human civilization. In certain polar regions (both arctic and antarctic), seawater that has been subjected to extreme cooling sinks and flows toward the equator along the ocean bottom. The obvious result of all of this movement is seen in weather patterns. These weather patterns are occurring today. Indeed, Europe would be much cooler than it is today if this so-called thermohaline system were not in place. If the positions where water rises and sinks change, weather will change as a consequence, and this change could negatively affect agriculture.

A second danger stemming from the warming of the planet and its oceans comes from the potential for the formation of poisonous gases in the ocean that could make their way into the atmosphere. This could happen if the oceans change from mixed to *stratified* (that is, arranged in layers such that the bottom of the sea has no oxygen, as is the current case in the Black Sea). If the mixed oceans of today change to stratified, disaster could ensue.

The fossil record offers abundant testimony that the current mixed oceans are relatively new. Black shales extending back some 600 million years ago, and possibly even before that, tell a tale of unmixed oceans, ones that are stratified. The stratification involved temperature and salinity, as well as two more factors that are far more important for life: dissolved oxygen and organic (reduced) carbon. Anoxic oceans are stratified. Moreover, there is an important—even phenomenal—consequence to this type of ocean. Organic material reaching the anoxic sea bottom can stay on the bottom any amount of time and remain as reduced carbon, with potentially dangerous results.

There are plenty of examples of stratified oceans during the time of animals. Yet some may have been more dangerous than others. Some not only were low in oxygen but also became filled with giant populations of noxious bacteria that converted sulfur compounds into energy—and created the poisonous gas hydrogen sulfide in the process. So toxic were these oceans that they may have reduced animal life or even inhibited its first evolution for millions of years during the long-ago Precambrian time intervals tracking back from 600 million years ago to the time of life's origin.

There seems to be two reasons behind this thinking. First is the obvious toxicity of the hydrogen sulfide, but just as important may have been the microbe's inhibition of nitrogen formation in compounds useful for plant life. While many kinds of microbes can fix nitrogen, an essential element for life, into compounds that are biologically useful, the eukaryotes cannot do so, and they depend on microbes to do the job for them. When there is a preponderance of sulfur bacteria, however, little nitrogen becomes available because this kind of bacteria actually inhibits other microbes from supplying nitrogen. A nitrogen-poor ocean would have been an ocean literally needing fertilizer and not getting it. It would have been just like a soil in which all the nitrogen has somehow been leached out—only a small amount of plant life will grow in such a medium. This type of environment would have been nasty, and the nastiest aspect is that the poisonous hydrogen sulfide might come out of the ocean to poison land life. Many humans have died of hydrogen sulfide poisoning in natural gas–producing regions.

Scientists are just beginning to learn about the conditions that led to these kinds of oceans, and they are also coming to understand that these events are dangerous to life. The mass extinctions at the end of the Permian and the end of the Triassic seem to have been caused by a change from an ocean like that of today to an ocean with large amounts of methane and toxic hydrogen sulfide. Scientists ponder what may have caused this change. The answer, one relevant to humanity in the twenty-first century, is global warming. In the past, global warming was produced by massive amounts of carbon dioxide and methane entering the atmosphere from volcanic sources. But a greenhouse gas is a greenhouse gas, regardless of its source, be it a volcano—or a car.

The future evolution of humanity

Scientists wonder if humanity's future as an evolving species is one of bigger-brained humans with high foreheads and higher intellect. Bigger brains are probably not in humanity's future. The fossil record shows that the days of rapid brain increase, at least based on skull sizes over the past several thousand generations, seem to be over. But if not giant brains, then scientists wonder what *might* evolution hold for the human species. They also wonder if humans will continue to evolve. Another intriguing question is whether the human species has undergone *any* significant evolution since its formation some 200,000 years ago.

Until the late twentieth century, tracking human evolution in the far to near past was the province of paleoanthropologists, for it was only through the study of ancient bone morphology that science could track evolutionary change. But the revolution in studying evolution made possible by the many DNA techniques for studying present and past genomes have opened a whole new world of for understanding recent human evolution. The surprising revelation is that not only has the human genome undergone some major reshuffling since the species' formation, but it appears that the rates of human evolution, if anything, have been increasing over the past thirty millennia. The following short overview looks at these studies and uses them to make some modest predictions about the future.

Evolution in the far and recent past

Paleoanthropologists have deciphered the location and timing of the speciation event that produced the human species. The human family, called the Hominidae, seems to have begun as much as 5 to 6 million years ago with the appearance of a small protohuman called *Australopithecus afarensis*. Since then, the human family has had as many as nine species, although there is ongoing debate about the number, which changes as both new discoveries and new interpretations of past bones make their way into print. Nonetheless, the most important descendent of the early, pre-Pleistocene hominids is the first member of the human genus, *Homo*, a species named *Homo habilis* (meaning "handyman," so named for its ability to use tools), which dates to about 2.5 million years ago. This creature gave rise to *Homo erectus* about 1.5 million years ago, and *H. erectus* either gave rise to the human species, *Homo sapiens*, either directly about 200,000 years ago or through an evolutionary intermediate known as *Homo heidelbergensis*. The human species has been further subdivided into a number of separate varieties. Some scientists consider the Neanderthals to be a variety, whereas others interpret it as a separate species—*Homo neanderthalensis*.

Each formation of new human species occurred when a small group of hominids somehow became separated from a larger population for many generations. In the 1960s and 1970s there was a view that modern humans came about from what has been called a *candelabra* pattern of evolution—that all over the planet, separate stocks of archaic hominids such as *Homo erectus* all evolved into *Homo sapiens* at different times and places; this notion now seems unlikely. The fossil record reveals that the oldest member of the human species, variably called a *modern* to distinguish it from more archaic forms of *Homo sapiens*, lived 195,000 years ago in what is now Ethiopia. It is unknown, and not terribly important, whether this fossil represents the oldest human tribe, or was from a group that wandered in from a previous place and was fortuitously fossilized in Ethiopia. But very soon after, members of this band set out walking, to the farthest southern regions of the African continent and then turned northward, finding a way out of Africa through Eurasia—and in so doing they spread out across the globe, effectively isolating themselves from others of the species, and thus adapting to the very different environmental conditions in which these wanderers found themselves. Survival in the Sun-starved, ice-covered north depended on quite different adaptations, from morphological to physiological, than did those who remained on the plains of Africa, as well as all areas in between. As the species' numbers grew, so too did its variation—and its various evolutionary changes. But all of these changes occurred within the same species.

By 10,000 years ago humans had successfully colonized each of the continents except Antarctica, and adaptations to the many locales led to what were later called the various human races. While it was long thought that such obvious features as skin color were purely adaptations to varying amounts of exposure to sunlight, subsequent work suggested that much of what are called *racial* characteristics might simply be adaptations brought about by sexual selection, rather than increases to fitness in various environments. But many other adaptations, most invisible to morphologists, were happening as well.

With the globe fairly well covered (with a few later invasions of such larger islands and island groupings as Madagascar, New Zealand, Polynesia, and Hawaii), one might expect that the time for evolving was mostly finished. But that turns out not to be the case, as shown by multiple studies.

Studies showing recent evolution

Scientists wonder about the rate of recent human evolution A study by Henry C. Harpending and John Hawks has given a dramatic answer. The findings of these researchers suggest that over the past 5,000 years, humans have evolved as much as 100 times more quickly than at any time since the split of the earliest hominid from the ancestors of modern chimpanzees some 6 million years ago. Moreover, rather than seeing a reduction of evolution of those characteristics that in combination are used to distinguish human races, until very recently the human races in various parts of the world have become *more*, not less distinct. Only in the past century, through the revolution in human travel and the more open behavioral attitudes of most humans to those of other races, has this pattern seemed to have slowed.

To arrive at this conclusion, the researchers analyzed data from the international haplotype map of the human genome, as well as genetic markers in 270 people from four groups: Han Chinese, Japanese, Africa's Yoruba, and northern Europeans. They found that at least 7 percent of human genes have undergone recent evolution. Some of the changes were tracked back to just 5,000 years ago.

Regarding the evolutionary changes affecting people in different parts of the globe, researchers noted that in China and most of Africa, few people can digest fresh milk into adulthood. Yet in Sweden and Denmark, the gene that makes the milk-digesting enzyme lactase remains active, so almost everyone can drink fresh milk. This may explain why dairy farming is more common in Europe than in the Mediterranean and Africa.

Other studies have discovered evidence for recent change owing to natural selection, rather than random mutation—in other words, evolutionary change to improve the fitness of various geographic populations of humans. The kinds of evolutionary change discovered by the study included resistance to one of the great scourges of humanity in Africa, the virus causing Lassa fever. Other gene changes were associated with changes in skin pigmentation and a particular development of hair follicles among Asians, the evolution of lighter skin and blue eyes in northern Europe, and partial resistance to diseases other than Lassa, such as malaria, among some African populations.

Although these studies seem to reaffirm that humans are not yet finished being first-class evolvers, others take quite a different tack. It is clear that modern medicine is successful at keeping alive individuals who would otherwise die before reaching sexual maturity. The large numbers of premature babies are but one example, and while such early births may be unrelated to genetics, they are certainly evidence that technology is impacting survivorship—which is itself the driver of evolution. Some evolutionists point to this so-called *survival of the unfit*, as well as the near absence of human predators (another common driver of evolutionary change in natural prey species), among many other aspects of fitness, to suggest that natural selection no longer applies to humanity.

Future evolution?

Humans thus seem to be first-class evolvers, or at least they were until very recently. With that known, it is possible to speculate about what the future might hold for the human species in terms of further evolutionary change—assuming that the species gets its few million years that appears to be the average longevity of any mammal species. Because much of the observed evolutionary changes of the past 5,000 years involved adaptation to particular environments, it is fair to ask how the future world, with the expectation of larger populations than now, and with larger cities and agricultural fields amid the other offshoots of technology, might affect the species' evolutionary outcome—or will it be affected at all. There are many questions, such as the following:

1. Will humans become larger or smaller, gain or lose intelligence, be it intellectual or emotional?

2. Will humans become more or less tolerant of oncoming environmental problems, such as a dearth of freshwater, an abundance of ultraviolet radiation, and an increase in global temperatures?

3. Will humans produce a new species, or is the species now evolutionarily sterile; might the future evolution of humanity be not within the species' genes, but through the addition of silicon expression and memory augmentation to human brains through neural connections with inorganic machines?

4. Is humanity but the builders of the next dominant intelligence on Earth—the machines?

Fossils, including those from humanity's own ancestors, reveal that evolutionary change is not a continuous process but rather occurs in fits and starts. Moreover, it is not progressive or directional. Organisms get smaller as well as larger, more simple as well as more complex. Furthermore, while most lineages evolve through time in some manner, the most dramatic evolutionary changes often take place when a new species first appears. If this is the case, then future *morphological* evolution in *Homo sapiens* may be minimal. At the same time, however, humans may show radical change in their behavior and perhaps physiology. Perhaps—and this is the biggest perhaps—a new species of human will evolve in the not-so-distant (or distant) future. But such an evolutionary change will almost surely require some sort of geographic isolation of a population of humans. As long as humans are restricted to Earth's surface, such an event seems unlikely.

Since the time of Charles Darwin (1809–1882), it has been accepted that the forces that produce new species usually occur when small populations of an already existing species get cut off from the larger population and can no longer interbreed with the larger number of individuals. Gene flow, the interchange of genetic material that maintains the integrity and identity of any species, thus gets cut off. Of course, genetic isolation generally means geographic separation, which means new environmental conditions, different from those experienced before. Adding these complications into the mix creates a recipe for making a new species—given enough time and the continued separation of the two groups.

Although major structural changes in *Homo sapiens* might now be over, the studies showing substantial evolutionary change promoting adaptation among the widespread human masses also showed that over the past 5,000 years, there has been ever more variability in humanity. But such may not be the case for the past century with its unprecedented population increase among technologically mobile human populations. Never has the gene pool had such common mixing of what were heretofore entirely separated local populations of the species. In fact, the new mobility of humanity might instead be bringing about the homogenization of the species: Humanity's former geographic isolation has been broached by the ease of transportation and the dismantling of social barriers once keeping minor genetic differences of the various human racial groups intact. Now, according to some evolutionists, with the species' new population size, humans are more like inbred laboratory mice than any wild-type animal at equilibrium.

Forces opposing human evolution

Natural selection in the human species is being thwarted on many fronts by humanity's technology, medicines, and rapidly changing behavior and moral values. Babies no longer die in large numbers in many parts of the globe, and babies with the gravest types of genetic damage that were once certainly fatal in prereproductive stages are permitted to live; predators too no longer affect the rules of survival. Tools, clothes, technology, and medicine—all have increased the species' fitness for survival, but at the same time they have thwarted the very mechanisms that brought about the species' creation through natural selection.

In stature most human populations seem to be getting bigger, yet this is surely not a genetic feature: With improved nutrition humans are simply maximizing the height potentials carried by their genes. Some of this change is cultural.

The role of culture in human evolution is a new field. The physicist Freeman Dyson has noted that what might be called the basic "Darwinian interlude"—the ritualized sexual exchange of information—is now reverting to a biologically "original" form, in which the fairly recently discovered microbial process known as horizontal gene transfer (where large segments of DNA [information] are swapped between microbial species) will prevail through the intervention of human culture. Microbes used gene transfer long before the advent of sex, and human culture surely does something similar. "We are moving rapidly into the post-Darwinian era, when species other than our own will no longer exist, the rules of Open Source sharing will be extended from the exchange of software to the exchange of genes," Dyson wrote in a 2007 article.

Human behavior and directed evolution

Because humans have directed the evolution of so many animal and plant species, it might be natural for humans to direct their own, which is a relevant factor when discussing the topic of potential human evolution. Some may ask why wait for natural selection to do the job when humans can do it faster and in ways beneficial to themselves. This is precisely the tack taken by many behavioral geneticists who are searching for ways to manipulate human genes. Behavioral genetics is a branch of science that asks what in the genes of humans makes individual people different from each other (versus what makes humans different from other species or what makes people human). Scientists working in this field try to track down the genetic components of behavior—not just problems and disorders, such as those profiled above, but everyday behaviors that may very well be heritable traits: overall disposition, the predilection for addiction or criminality, many aspects of sexuality, aggressiveness, and competitiveness. These are traits that are known intuitively to be at least partially heritable.

The implications for the future of the species are incalculable. It seems likely that ever-greater segments of society—or at least of wealthy societies—will eventually accede to the idea that DNA samples are given to genetic specialists. When this happens, elaborate screenings for an individual's genetic makeup will become commonplace, and specific genes for depression and other behavioral abnormalities will be detected. The second step will be the application of behavioral drugs using newly discovered chemical pathways. But the third path will be to actually change people's genes. This can be done in two ways: either somatically, by changing the genes in a relevant organ only, or through what is known as germ-line therapy, a process by which the entire genome of an individual can be changed. Because germ-line therapy involves changes in the genetic code of a person's egg and sperm, it will not help the individual in question, but it will help that person's children.

The major obstacle to genetic engineering in humans is a property known as *pleiotropy*: Usually genes perform more than one function, and many functions are coded on far more than a single gene. All genes involved in behavior are probably pleiotropic. Such is surely the case, for example, with the many genes involved in human intelligence. (In fact, neuroanatomists and behavioral geneticists believe that the genes involved in intelligence are probably involved in many basic brain functions as well.) Therefore, far more will have to be known about the human genome before wholesale tinkering can begin, because slight changes in genome frequency can potentially lead to drastic changes in the species-level genome. As has often been noted, a mere 1 percent difference in the genome is all that is necessary to produce the vast gulf between chimps and humans, and a 10 percent change can bring humans back to the level of yeast.

Many may wonder then why scientists consider changing genes. In all probability, the pressure will come from parents wanting to improve their children to guarantee certain valued traits and to eliminate the possibility of negative ones. The motives are there, and they are strong. The Human Genome Project, now completed, had for much of its motivation (whatever it supporters argue) the desire to find bad genes. Once these are found, it will require new herculean efforts to eliminate them. So assuming that it does become practical to change the nature of human genes, it remains to be seen how that affect the future evolution of humanity. Probably a great deal if the practice continues for millennia.

If natural selection is unlikely to produce a new human species—an event foreseen by H.G. Wells in his novel *The Time Machine* (1895)—the same end result could certainly be completed by directed human effort. As easily as breeding into existence new varieties of agricultural animals, humans have it in their power to bring a new human race, variety, or even a separate species into the world. Whether humanity chooses to follow such a path will be for future generations to decide.

Just as the push by parents to genetically enhance their children will be societally irresistible, so too will the assault on human aging be an area in which the human species will unnaturally evolve. Much recent research shows that aging is not so much a simple wearing down of body parts as it is a combination of programmed decay, much of it genetically controlled. It is highly probable that the next century of genetic research will unlock numerous genes controlling many aspects of aging and that these genes can be manipulated. An individual human lasting between one and two centuries is an obtainable goal. Whether it should be obtained in light of human population growth is another question.

Here is a scenario already posited by several scientists (and science fiction writers) that could potentially lead to a new human species or at least a new variety. Parents allow their unborn children to be genetically altered, with enhanced intelligence, looks, and longevity. They might be made as smart as they are long lived—IQs of 150 and a maximum age of 150 as well. Unlike current humans, these new humans can breed for eighty years or more. Thus, they have more children—and because they are both smart and live a long time, they accumulate wealth in ways different from current humans. Very quickly there will be pressure on these new humans to breed with others of their kind. Just as quickly, they may become rulers of everyone else. With some sort of presumably self-imposed geographic or social segregation, genetic drift might occur, given enough time, allowing a differentiation of these forms as a new human species.

The cyborg route

Humans are no longer simply toolmakers. Now humans are machine makers as well, and not all of the machines can be considered tools. In ways perhaps even less predictable than humanity's use of genetic manipulation, it may be humanity's manipulation of machines—or they of humanity—that creates the most profound evolutionary change in the human species. This would not be a change of simply morphology, or even of behavior (although that might happen too)—but of something far more consequential. Some wonder if the ultimate evolution of the human species will be one of symbiosis with machines, to produce a human–machine synthesis.

This view was starkly enunciated by George Dyson in his book *Darwin among the Machines*. According to Dyson, future intellectual progress will not be a product of Darwinian evolution among populations of *Homo sapiens*, but rather of an ongoing symbiosis with the machines humanity builds:

> Everything that human beings are doing to make it easier to operate computer networks is at the same time, but for different reasons, making it easier for computer networks to operate human beings. (10)

New human species

The human lineage has produced new species in the past, but many wonder what the future holds. Speciation requires an isolating mechanism of some sort. The most common means is *allopatric* speciation, where a small population gets cut off from the larger gene-pool reservoir and then transforms its own set of genes sufficiently so that it can no longer successfully reproduce with the parent population. Most species have undergone this process through geographic isolation, yet the very size and efficiency of transport of humanity make this possibility low—at least on Earth. If, however, human colonies are set up on distant worlds and then cut off from common gene flow, new human species could indeed arise.

Perhaps humans will lose (or voluntarily discard) the technology sufficient to allow the global interchange of human genes from continent to continent. A large asteroid hitting Earth could certainly halt the rate of gene flow now on the planet, if not extinguish the human species itself. If separation lasts long enough, and if conditions on separated continents are sufficiently different, it is possible that a new human species could arise as a result of geographic separation, as apparently occurred with *Homo floresiensis*, a hominin species dating back 12,000 years discovered in 2003 on the Indonesian island of Flores.

In regard to alternative scenarios of humanity in the future, putting aside the notion that the human species will soon go extinct somehow leaves the species with what Rod Taylor (in the 1960 film version of H.G. Wells's *The Time Machine*) described as "all the time in the world." With hundreds of thousands to millions to perhaps billions of years to yet play with before the Sun becomes a red giant, and perhaps more billions if humanity develops interstellar transportation, what might the human species evolve into? Here are three scenarios:

1. *Stasis*: In this scenario the species largely stays as it is now: isolated individuals. Minor changes may occur, mainly through the merging of the various races.

2. *Speciation*: Through some type of isolating mechanism a new human species evolves, either on Earth or on another world following space travel and colonization.

3. *Symbiosis with machines*: The evolution of a collective intelligence through the integration of machines and human brains.

Resources

Books

Bostrom, Nick. "The Future of Human Evolution." In *Death and Anti-death*. Vol. 2, *Two Hundred Years after Kant, Fifty Years after Turing*, edited by Charles Tandy, 339–371. Palo Alto, CA: Ria University Press, 2004.

Comings, David E. *The Gene Bomb: Does Higher Education and Advanced Technology Accelerate the Selection of Genes for Learning Disorders, ADHD, Addictive, and Disruptive Behaviors?* Duarte, CA: Hope Press, 1996.

Dyson, George B. *Darwin Among the Machines: The Evolution of Global Intelligence*. Addison-Wesley Pub. Co., 1997.

Hallam, Anthony, and Paul B. Wignall. *Mass Extinctions and Their Aftermath*. Oxford: Oxford University Press, 1997.

Lasaga, Antonio C., Robert A. Berner, and Robert M. Garrels. "An Improved Geochemical Model of Atmospheric CO_2 Fluctuations over the Past 100 Million Years." In *The Carbon Cycle and Atmospheric CO_2: Natural Variations, Archean to Present*, edited by E.T. Sundquist and W.S. Broecker, 397–411. Washington, DC: American Geophysical Union, 1985.

Phillips, John. *Life on Earth: Its Origin and Succession*. Cambridge, London: Macmillan and Co., 1860.

Schneider, Stephen H., and Penelope J. Boston, eds. *Scientists on Gaia*. Cambridge, MA: MIT Press, 1991.

Schwartzman, David. *Life, Temperature, and the Earth: The Self-Organizing Biosphere*. New York: Columbia University Press, 1999.

Stanley, Steven M. *Extinction*. New York: Scientific American Library, 1987.

Ward, Peter. *The End of Evolution: On Mass Extinctions and the Preservation of Biodiversity*. New York: Bantam Books, 1994.

Ward, Peter. *Future Evolution*. New York: Times Books, 2001.

Ward, Peter. *Out of Thin Air: Birds, Dinosaurs, and Earth's Ancient Atmosphere*. Washington, DC: Joseph Henry Press, 2006.

Ward, Peter, and Donald Brownlee. *The Life and Death of Planet Earth: How the New Science of Astrobiology Charts the Ultimate Fate of Our World*. New York: Times Books, 2003.

Periodicals

Barreiro, Luis B., Guillaume Laval, Hélène Quach, et al. "Natural Selection Has Driven Population Differentiation in Modern Humans." *Nature Genetics* 40, no. 3 (2008): 340–345.

Berner, Robert A. "A Model for Atmospheric CO_2 over Phanerozoic Time." *American Journal of Science* 291, no. 4 (1991): 339–376.

Berner, Robert A. "Weathering, Plants, and the Long-Term Carbon Cycle." *Geochimica et Cosmochimica Acta* 56, no. 8 (1992): 3225–3231.

Berner, Robert A. "Paleozoic Atmospheric CO_2: Importance of Solar Radiation and Plant Evolution." *Science* 261, no. 5117 (1993): 68–70.

Berner, Robert A. "GEOCARB II: A Revised Model of Atmospheric CO_2 over Phanerozoic Time." *American Journal of Science* 294, no. 1 (1994): 56–91.

Berner, Robert A. "The Rise of Plants and Their Effect on Weathering and Atmospheric CO_2." *Science* 276, no. 5312 (1997): 544–546.

Berner, Robert A., Antonio C. Lasaga, and Robert M. Garrels. "The Carbonate–Silicate Geochemical Cycle and Its Effect on Atmospheric Carbon Dioxide over the Past 100 Million Years." *American Journal of Science* 283, no. 7 (1983): 641–683.

Berner, Robert A., and Danny M. Rye. "Calculation of the Phanerozoic Strontium Isotope Record of the Oceans from a Carbon Cycle Model." *American Journal of Science* 292, no. 2 (1992): 136–148.

Caldeira, Ken, and James F. Kasting. "The Life Span of the Biosphere Revisited." *Nature* 360, no. 6406 (1992): 721–723.

Caldeira, Ken, and James F. Kasting. "Susceptibility of the Early Earth to Irreversible Glaciation Caused by Carbon Dioxide Clouds." *Nature* 359, no. 6392 (1992): 226–228.

Cerling, Thure E., James R. Ehleringer, and John M. Harris. "Carbon Dioxide Starvation, the Development of C4 Ecosystems, and Mammalian Evolution." *Philosophical Transactions of the Royal Society* B 353, no. 1365 (1998): 159–171.

Cerling, Thure E., John M. Harris, Bruce J. MacFadden, et al. "Global Vegetation Change through the Miocene/Pliocene Boundary." *Nature* 389, no. 6647 (1997): 153–158.

Dyson, Freeman. "Our Biotech Future." *New York Review of Books*, July 19, 2007. Accessed August 12, 2012. http://www.nybooks.com/articles/archives/2007/jul/19/our-biotech-future/

Ehleringer, James R., and Thure E. Cerling. "Atmospheric CO_2 and the Ratio of Intercellular to Ambient CO_2 Concentrations in Plants." *Tree Physiology* 15, no. 2 (1995): 105–111.

Ehleringer, James R., Thure E. Cerling, and Brent R. Helliker. "C_4 Photosynthesis, Atmospheric CO_2, and Climate." *Oecologia* 112, no. 3 (1997): 285–299.

Franck, Siegfried, A. Block, Werner von Bloh, et al. "Reduction of Biosphere Life Span as a Consequence of Geodynamics." *Tellus* B 52, no. 1 (2000): 94–107.

Franck, Siegfried, and Christine Bounama. "Continental Growth and Volatile Exchange during Earth's Evolution." *Physics of the Earth and Planetary Interiors* 100, nos. 1–4 (1997): 189–196.

Kasting, James F. "Warming Early Earth and Mars." *Science* 276, no. 5316 (1997): 1213.

Kuhn, William R., James C.G. Walker, and Hal G. Marshall. "The Effect on Earth's Surface Temperature from Variations in Rotation Rate, Continent Formation, Solar Luminosity, and Carbon Dioxide." *Journal of Geophysical Research* 94, no. D8 (1989): 11129–11136.

Lenton, Thomas M., and Werner von Bloh. "Biotic Feedback Extends the Life Span of the Biosphere." *Geophysical Research Letters* 28, no. 9 (2001): 1715–1718.

Lovelock, J.E., and M. Whitfield. "Life Span of the Biosphere." *Nature* 296, no. 5857 (1982): 561–563.

Pauls, D.L., D.J. Cohen, K.K. Kidd, and J.F. Leckman. "Tourette Syndrome and Neuropsychiatric Disorders: Is There a Genetic Relationship?" *American Journal of Human Genetics* 43, no. 2 (1988): 206–217.

Raup, David M., and J. John Sepkoski Jr. "Mass Extinctions in the Marine Fossil Record." *Science* 215, no. 4539 (1982): 1501–1503.

Sabeti, Pardis C., Patrick Varilly, Ben Fry, et al. "Genome-Wide Detection and Characterization of Positive Selection in Human Populations." *Nature* 449, no. 7164 (2007): 913–918.

Sagan, Carl, and Christopher Chyba. "The Early Faint Sun Paradox: Organic Shielding of Ultraviolet-Labile Greenhouse Gases." *Science* 276, no. 5316 (1997): 1217–1221.

Sagan, Carl, and George Mullen. "Earth and Mars: Evolution of Atmospheres and Surface Temperatures." *Science* 177, no. 4043 (1972): 52–56.

Schwartzman, David W., and Tyler Volk. "Biotic Enhancement of Weathering and the Habitability of Earth." *Nature* 340, no. 6233 (1989): 457–460.

Voight, Benjamin F., Sridhar Kudaravalli, Xiaoquan Wen, and Jonathan K. Pritchard. "A Map of Recent Positive Selection in the Human Genome." *PLoS Biology* 4, no. 3 (2006): e72.

Walker, James C.G., P.B. Hays, and James F. Kasting. "A Negative Feedback Mechanism for the Long-Term Stabilization of Earth's Surface Temperature." *Journal of Geophysical Research* 86, no. C10 (1981): 9776–9782.

Peter D. Ward

Further reading

Adams, Lowell W. *Urban Wildlife Habitats: A Landscape Perspective*. Minneapolis: University of Minnesota Press, 1994.

Adams, Robert McC. *The Evolution of Urban Society: Early Mesopotamia and Prehispanic Mexico*. New Brunswick, NJ: AldineTransaction, 2005.

Adds, John, Erica Larkcom, and Ruth Miller. *Genetics, Evolution, and Biodiversity*. Rev. ed. Cheltenham, UK: Nelson Thornes, 2004.

Alexander, Edward P., and Mary Alexander. *Museums in Motion: An Introduction to the History and Functions of Museums*. 2nd ed. Lanham, MD: AltaMira Press, 2008.

Alroy, John, and Gene Hunt, eds. *Quantitative Methods in Paleobiology*. Paleontology Society Papers 16. Boulder, CO: Paleontological Society, 2010.

Archibald, J. David. *Extinction and Radiation: How the Fall of Dinosaurs Led to the Rise of Mammals*. Baltimore: Johns Hopkins University Press, 2011.

Armstrong, Susan J., and Richard G. Botzler, eds. *Environmental Ethics: Divergence and Convergence*. 3rd ed. New York: McGraw-Hill, 2004.

Aubry, Marie-Pierre, Spencer G. Lucas, and William A. Berggren, eds. *Late Paleocene–Early Eocene Climatic and Biotic Events in the Marine and Terrestrial Records*. New York: Columbia University Press, 1998.

Baillie, Jonathan E.M., and Ellen R. Butcher. *Priceless or Worthless? The World's Most Threatened Species*. London: Zoological Society of London, 2012.

Barlow, Connie. *The Ghosts of Evolution: Nonsensical Fruit, Missing Partners, and Other Ecological Anachronisms*. New York: Basic Books, 2000.

Barnosky, Anthony D. *Heatstroke: Nature in an Age of Global Warming*. Washington, DC: Island Press/Shearwater Books, 2009.

Barrow, Mark V., Jr. *Nature's Ghosts: Confronting Extinction from the Age of Jefferson to the Age of Ecology*. Chicago: University of Chicago Press, 2009.

Beissinger, Steven R., and Dale R. McCullough, eds. *Population Viability Analysis*. Chicago: University of Chicago Press, 2002.

Bennett, Elizabeth L., and John G. Robinson. *Hunting of Wildlife in Tropical Forests: Implications for Biodiversity and Forest Peoples.* Washington, DC: World Bank, Environment Department, 2000.

Benton, Michael J., ed. *The Fossil Record 2*. London: Chapman and Hall, 1993.

Benton, Michael J. *When Life Nearly Died: The Greatest Mass Extinction of All Time*. New York: Thames and Hudson, 2003.

Benton, Michael J. *Vertebrate Palaeontology*. 3rd ed. Malden, MA: Blackwell, 2005.

Berra, Tim M. *Freshwater Fish Distribution*. Chicago: University of Chicago Press, 2007.

Bissonette, John A., and Paul R. Krausman, eds. *Integrating People and Wildlife for a Sustainable Future*. Bethesda, MD: Wildlife Society, 1995.

Bodansky, Daniel. *The Art and Craft of International Environmental Law*. Cambridge, MA: Harvard University Press, 2010.

Bowler, Peter J. *Evolution: The History of an Idea*. 4th ed. Berkeley, CA: University of California Press, 2009.

Burney, David A. *Back to the Future in the Caves of Kaua'i: A Scientist's Adventures in the Dark*. New Haven, CT: Yale University Press, 2010.

Carson, Rachel. *Silent Spring*. Boston: Houghton Mifflin, 1962.

Case, Ted J. *An Illustrated Guide to Theoretical Ecology*. New York: Oxford University Press, 2000.

Challenger, Melanie. *On Extinction: How We Became Estranged from Nature*. London: Granta, 2011.

Chaloner, William G. and Anthony Hallam, eds. *Evolution and Extinction*. Cambridge, UK: Cambridge University Press, 1994.

Cheke, Anthony, and Julian Hume. *Lost Land of the Dodo: An Ecological History of Mauritius, Réunion, and Rodrigues*. New Haven, CT: Yale University Press, 2008.

Cochrane, Mark A. *Tropical Fire Ecology: Climate Change, Land Use, and Ecosystem Dynamics*. Berlin: Springer, 2009.

Coe, Angela L., ed. *The Sedimentary Record of Sea-Level Change*. Milton Keynes, UK: Open University; Cambridge, UK: Cambridge University Press, 2003.

Colborn, Theo, Dianne Dumanoski, and John Peterson Myers. *Our Stolen Future: Are We Threatening Our Fertility,*

Intelligence, and Survival? A Scientific Detective Story. New York: Penguin, 1997.

Collins, James P., and Martha L. Crump. *Extinction in Our Times: Global Amphibian Decline*. Oxford: Oxford University Press, 2009.

Conn, Steven. *Museums and American Intellectual Life, 1876–1926*. Chicago: University of Chicago Press, 1998.

Cooney, Gabriel. *Landscapes of Neolithic Ireland*. Milton Park, UK: Routledge, 2000.

Corlett, Richard T., and Richard B. Primack. *Tropical Rain Forests: An Ecological and Biogeographical Comparison*. 2nd ed. Chichester, UK: Wiley-Blackwell, 2011.

Courchamp, Franck, Luděk Berec, and Joanna Gascoigne. *Allee Effects in Ecology and Conservation*. Oxford: Oxford University Press, 2008.

Courtillot, Vincent. *Evolutionary Catastrophes: The Science of Mass Extinction*. Translated by Joe McClinton. New York: Cambridge University Press, 1999.

Cox, C. Barry, and Peter D. Moore. *Biogeography: An Ecological and Evolutionary Approach*. 8th ed. Hoboken, NJ: Wiley, 2010.

Cracraft, Joel, and Niles Eldredge, eds. *Phylogenetic Analysis and Paleontology*. New York: Columbia University Press, 1979.

Crump, Marty. *In Search of the Golden Frog*. Chicago: Chicago University Press, 2000.

Cushman, Samuel A., and Falk Huettmann, eds. *Spatial Complexity, Informatics, and Wildlife Conservation*. Tokyo: Springer, 2010.

Daily, Gretchen C., and Katherine Ellison. *The New Economy of Nature: The Quest to Make Conservation Profitable*. Washington, DC: Island Press/Shearwater Books, 2002.

Daniels, Tom. *When City and Country Collide: Managing Growth in the Metropolitan Fringe*. Washington, DC: Island Press, 1999.

Darvill, Timothy. *Prehistoric Britain*. 2nd ed. Milton Park, UK: Routledge, 2010.

Davis, Mark A. *Invasion Biology*. Oxford: Oxford University Press, 2009.

Dayton, Paul K., Simon Thrush, and Felicia C. Coleman. *Ecological Effects of Fishing in Marine Ecosystems of the United States*. Arlington, VA: Pew Oceans Commission, 2002.

Dobzhansky, Theodosius. *Genetics of the Evolutionary Process*. New York: Columbia University Press, 1970.

Donald, Paul F., Nigel J. Collar, Stuart J. Marsden, and Deborah J. Pain. *Facing Extinction: The World's Rarest Birds and the Race to Save Them*. London: T&AD Poyser, 2010.

Doub, J. Peyton. *The Endangered Species Act: History, Implementation, Successes, and Controversies*. Boca Raton, FL: Taylor and Francis, 2013.

du Toit, Johan T., Kevin H. Rogers, and Harry C. Biggs, eds. *The Kruger Experience: Ecology and Management of Savanna Heterogeneity*. Washington, DC: Island Press, 2003.

Dypvik, Henning, Filippos Tsikalas, and Morten Smelror, eds. *The Mjølnir Impact Event and Its Consequences: Geology and Geophysics of a Late Jurassic/Early Cretaceous Marine Impact Event*. Berlin: Springer, 2010.

Eldredge, Niles. *Life in the Balance: Humanity and the Biodiversity Crisis*. Princeton, NJ: Princeton University Press, 1998.

Eldredge, Niles, and Joel Cracraft. *Phylogenetic Patterns and the Evolutionary Process: Method and Theory in Comparative Biology*. New York: Columbia University Press, 1980.

Erwin, Douglas H. *The Great Paleozoic Crisis: Life and Death in the Permian*. New York: Columbia University Press, 1993.

Erwin, Douglas H. *Extinction: How Life on Earth Nearly Ended 250 Million Years Ago*. Princeton, NJ: Princeton University Press, 2006.

Estes, James A., Douglas P. Demaster, Daniel F. Doak, et al., eds. *Whales, Whaling, and Ocean Ecosystems*. Berkeley: University of California Press, 2006.

Fa, John E., Stephan M. Funk, and Donnamarie O'Connell. *Zoo Conservation Biology*. Cambridge, UK: Cambridge University Press, 2011.

FAO Fisheries and Aquaculture Department. *The State of World Fisheries and Aquaculture, 2012*. Rome: Food and Agriculture Organization of the United Nations, 2012.

Fitter, Richard, and Maisie Fitter, eds. *The Road to Extinction: Problems of Categorizing the Status of Taxa Threatened with Extinction; Proceedings of a Symposium Held by the Species Survival Commission, Madrid, 7 and 9 November 1984*. Gland, Switzerland: International Union for Conservation of Nature, 1987.

Fleagle, John G. *Primate Adaptation and Evolution*. 2nd ed. San Diego, CA: Academic Press, 1999.

Forman, Richard T.T. *Land Mosaics: The Ecology of Landscapes and Regions*. Cambridge, UK: Cambridge University Press, 1995.

Fortey, Richard. *Survivors: The Animals and Plants That Time Has Left Behind*. London: Harper Press, 2011.

Fowler, Sarah L., Rachel D. Cavanagh, Merry Camhi, et al., comps. and eds. *Sharks, Rays, and Chimaeras: The Status of the Chondrichthyan Fishes*. Gland, Switzerland: International Union for Conservation of Nature, 2005.

Frankham, Richard, Jonathan D. Ballou, and David A. Briscoe. *Introduction to Conservation Genetics*. 2nd ed. Cambridge, UK: Cambridge University Press, 2010.

Freeman, A. Myrick, III. *The Measurement of Environmental and Resource Values: Theory and Methods*. 2nd ed. Washington, DC: Resources for the Future, 2003.

Fuller, Errol. *Extinct Birds*. Rev. ed. Ithaca, NY: Comstock, 2001.

Gardner, Alfred L., ed. *Mammals of South America*. Vol. 1, *Marsupials, Xenarthrans, Shrews, and Bats*. Chicago: University of Chicago Press, 2007.

Gaston, Kevin J., ed. *Biodiversity: A Biology of Numbers and Difference*. Oxford: Blackwell Science, 1996.

Ghazoul, Jaboury, and Douglas Sheil. *Tropical Rain Forest Ecology, Diversity, and Conservation*. Oxford: Oxford University Press, 2010.

Gibbs, George. *Ghosts of Gondwana: The History of Life in New Zealand*. Nelson, New Zealand: Craig Potton, 2006.

Goodman, Steven M., and Bruce D. Patterson, eds. *Natural Change and Human Impact in Madagascar*. Washington, DC: Smithsonian Institution Press, 1997.

Gotelli, Nicholas J., and Aaron M. Ellison. *A Primer of Ecological Statistics*. 2nd ed. Sunderland, MA: Sinauer, 2013.

Gould, Stephen Jay. *Time's Arrow, Time's Cycle: Myth and Metaphor in the Discovery of Geological Time*. Cambridge, MA: Harvard University Press, 1987.

Gould, Stephen Jay, ed. *The Book of Life: An Illustrated History of the Evolution of Life on Earth*. New York: Norton, 2001.

Gradstein, Felix M., James G. Ogg, Mark D. Schmitz, and Gabi Ogg, eds. *The Geologic Time Scale, 2012*. 2 vols. Boston: Elsevier, 2012.

Green, Roger H. *Sampling Design and Statistical Methods for Environmental Biologists*. New York: Wiley, 1979.

Grill, Andrea, Paolo Casula, Roberta Lecis, and Steph Menken. "Endemism in Sardinia." In *Phylogeography of Southern European Refugia: Evolutionary Perspectives on the Origins and Conservation of European Biodiversity*, edited by Stephen Weiss and Nuno Ferrand, 273–296. Dordrecht, Netherlands: Springer, 2007.

Groom, Martha J., Gary K. Meffe, and C. Ronald Carroll. *Principles of Conservation Biology*. 3rd ed. Sunderland, MA: Sinauer, 2006.

Guerrant, Edward O., Jr., Kayri Havens, and Mike Maunder, eds. *Ex Situ Plant Conservation: Supporting Species Survival in the Wild*. Washington, DC: Island Press, 2004.

Hallam, Anthony. *Phanerozoic Sea-Level Changes*. New York: Columbia University Press, 1992.

Hallam, Anthony. *Catastrophes and Lesser Calamities: The Causes of Mass Extinctions*. Oxford: Oxford University Press, 2004.

Hallam, Anthony, and Paul B. Wignall. *Mass Extinctions and Their Aftermath*. Oxford: Oxford University Press, 1997.

Hannah, Lee, ed. *Saving a Million Species: Extinction Risk from Climate Change*. Washington, DC: Island Press, 2012.

Haynes, Gary, ed. *American Megafaunal Extinctions at the End of the Pleistocene*. New York: Springer, 2009.

Hedeen, Stanley. *Big Bone Lick: The Cradle of American Paleontology*. Lexington: University Press of Kentucky, 2008.

Helfman, Gene S. *Fish Conservation: A Guide to Understanding and Restoring Global Aquatic Biodiversity and Fishery Resources*. Washington, DC: Island Press, 2007.

Helfman, Gene S., Bruce B. Collette, Douglas E. Facey, and Brian W. Bowen. *The Diversity of Fishes: Biology, Evolution, and Ecology*. 2nd ed. Chichester, UK: Wiley-Blackwell, 2009.

Hetherington, Renée, and Robert G.B. Reid. *The Climate Connection: Climate Change and Modern Human Evolution*. Cambridge, UK: Cambridge University Press, 2010.

Hiscock, Peter. *Archaeology of Ancient Australia*. London: Routledge, 2008.

Hofmeijer, Gerard Klein. *Late Pleistocene Deer Fossils from Corbeddu Cave: Implications for Human Colonization of the Island of Sardinia*. BAR International Series 663. Oxford: British Archaeological Reports/John and Erica Hedges, 1997.

Hunter, Luke, and Priscilla Barrett. *Carnivores of the World*. Princeton Field Guides. Princeton, NJ: Princeton University Press, 2011.

Hunter, Malcolm L., Jr., and James P. Gibbs. *Fundamentals of Conservation Biology*. 3rd ed. Malden, MA: Blackwell Publishing, 2007.

Hunter, Michael, Alison Walker, and Arthur MacGregor, eds. *From Books to Bezoars: Sir Hans Sloane and His Collections*. London: British Library Publishing, 2012.

Jackson, Dana L., and Laura L. Jackson, eds. *The Farm as Natural Habitat: Reconnecting Food Systems with Ecosystems*. Washington, DC: Island Press, 2002.

Johnson, Chris. *Australia's Mammal Extinctions: A 50,000 Year History*. Cambridge, UK: Cambridge University Press, 2006.

Johnson, Elizabeth A., and Michael W. Klemens, eds. *Nature in Fragments: The Legacy of Sprawl*. New York: Columbia University Press, 2005.

Kareiva, Peter, and Michelle Marvier. *Conservation Science: Balancing the Needs of People and Nature*. Greenwood Village, CO: Roberts, 2011.

Kareiva, Peter, Heather Tallis, Taylor H. Ricketts, et al., eds. *Natural Capital: Theory and Practice of Mapping Ecosystem Services*. New York: Oxford University Press, 2011.

Koeberl, Christian, and Kenneth G. MacLeod, eds. *Catastrophic Events and Mass Extinctions: Impacts and Beyond*. Geological Society of America Special Paper 356. Boulder, CO: Geological Society of America, 2002.

Koike, Fumito, Mick N. Clout, Mieko Kawamichi, et al., eds. *Assessment and Control of Biological Invasion Risks*. Kyoto, Japan: Shoukadoh Book Sellers; Gland, Switzerland: IUCN—the World Conservation Union, 2006.

Kull, Christian A. *Isle of Fire: The Political Ecology of Landscape Burning in Madagascar*. Chicago: University of Chicago Press, 2004.

Kumar, Pushpam, ed. *The Economics of Ecosystems and Biodiversity: Ecological and Economic Foundations*. London: Earthscan, 2010.

Kurlansky, Mark. *Cod: A Biography of the Fish That Changed the World*. New York: Walker, 1997.

LaBastille, Anne. *Mama Poc: An Ecologist's Account of the Extinction of a Species*. New York: Norton, 1990.

Ladle, Richard J., and Robert J. Whittaker, eds. *Conservation Biogeography*. Chichester, UK: Wiley-Blackwell, 2011.

Laferrière, Eric, and Peter J. Stoett, eds. *International Ecopolitical Theory: Critical Approaches*. Vancouver, Canada: UBC Press, 2006.

Lannoo, Michael, ed. *Amphibian Declines: The Conservation Status of United States Species*. Berkeley: University of California Press, 2005.

Laurance, William F., and Carlos A. Peres, eds. *Emerging Threats to Tropical Forests*. Chicago: University of Chicago Press, 2006.

Lawton, John H., and Robert M. May, eds. *Extinction Rates*. Oxford: Oxford University Press, 1995.

Lazarus, Richard J. *The Making of Environmental Law*. Chicago: University of Chicago Press, 2004.

Leakey, Richard, and Roger Lewin. *The Sixth Extinction: Patterns of Life and the Future of Humankind*. New York: Doubleday, 1995.

Leopold, Aldo. *A Sand County Almanac, and Sketches Here and There*. New York: Oxford University Press, 1949.

Leppäkoski, Erkki, Stephan Gollasch, and Sergej Olenin, eds. *Invasive Aquatic Species of Europe: Distribution, Impacts, and Management*. Dordrecht, Netherlands: Kluwer Academic Publishers, 2002.

Levy, Paul S., and Stanley Lemeshow. *Sampling of Populations: Methods and Applications*. 4th ed. Hoboken, NJ: Wiley, 2008.

Levy, Sharon. *Once and Future Giants: What Ice Age Extinctions Tell Us about the Fate of Earth's Largest Animals*. Oxford: Oxford University Press, 2011.

Lister, Adrian, and Paul Bahn. *Mammoths: Giants of the Ice Age*. Rev. ed. Berkeley: University of California Press, 2007.

Lockwood, Charles. *The Human Story: Where We Come from and How We Evolved*. New York: Sterling, 2008.

Long, John A., Michael Archer, Timothy Flannery, and Suzanne Hand. *Prehistoric Mammals of Australia and New Guinea: One Hundred Million Years of Evolution*. Sydney: University of New South Wales Press, 2002.

Losos, Jonathan B., and Robert E. Ricklefs, eds. *The Theory of Island Biogeography Revisited*. Princeton, NJ: Princeton University Press, 2010.

Lynas, Mark. *The God Species: Saving the Planet in the Age of Humans*. Washington, DC: National Geographic Society, 2011.

MacArthur, Robert H., and Edward O. Wilson. *The Theory of Island Biogeography*. Princeton, NJ: Princeton University Press, 1967.

Macdonald, David W., ed. *The New Encyclopedia of Mammals*. Oxford: Oxford University Press, 2001.

Macdonald, David W., and Andrew J. Loveridge, eds. *Biology and Conservation of Wild Felids*. Oxford: Oxford University Press, 2010.

MacLeod, Norman. *The Great Extinctions: What Causes Them and How They Shape Life*. London: Natural History Museum Publications, 2013.

MacPhee, Ross D.E. *Extinctions in Near Time: Causes, Contexts, and Consequences*. New York: Kluwer Academic/Plenum, 1999.

Mahoney, John J., and Millard F. Coffin, eds. *Large Igneous Provinces: Continental, Oceanic, and Planetary Flood Volcanism*. Geophysical Monograph Series 100. Washington, DC: American Geophysical Union, 1997.

Maris, Emma. *Rambunctious Garden: Saving Nature in a Post-wild World*. New York: Bloomsbury USA, 2011.

Marsh, Helene, Thomas J. O'Shea, and John E. Reynolds III. *Ecology and Conservation of the Sirenia: Dugongs and Manatees*. Cambridge, UK: Cambridge University Press, 2012.

Martin, Paul S. *Twilight of the Mammoths: Ice Age Extinctions and the Rewilding of America*. Berkeley: University of California Press, 2005.

Martin, Paul S., and Richard G. Klein, eds. *Quaternary Extinctions: A Prehistoric Revolution*. Tucson: University of Arizona Press, 1984.

Marzluff, John M., Reed Bowman, and Roarke Donnelly, eds. *Avian Ecology and Conservation in an Urbanizing World*. Boston: Kluwer Academic, 2001.

Matthiessen, Peter. *Wildlife in America*. Rev. ed. New York: Viking, 1987.

McGhee, George R., Jr. *The Late Devonian Mass Extinction: The Frasnian/Famennian Crisis*. New York: Columbia University Press, 1996.

McGowan, Alistair J., and Andrew B. Smith, eds. *Comparing the Rock and Fossil Records: Implications for Biodiversity Studies*. Geological Society Special Publication 358. London: Geological Society, 2011.

McIntyre, Alasdair D., ed. *Life in the World's Oceans: Diversity, Distribution, and Abundance*. Chichester, UK: Wiley-Blackwell, 2010.

Merrick, J.R., M. Archer, G.M. Hickey, and M.S.Y. Lee, eds. *Evolution and Biogeography of Australasian Vertebrates*. Oatlands, Australia: Auscipub, 2006.

Millennium Ecosystem Assessment. *Ecosystems and Human Well-Being: Synthesis*. Washington, DC: Island Press, 2005.

Mills, L. Scott. *Conservation of Wildlife Populations: Demography, Genetics, and Management*. 2nd ed. Hoboken, NJ: Wiley-Blackwell, 2013.

Mittermeier, Russell A., Patricio Robles Gil, Michael Hoffman, et al. *Hotspots Revisited: Earth's Biologically Richest and Most Threatened Terrestrial Ecoregions*. Mexico City: Cemex, Conservation International, and Agrupación Sierra Madre, 2005.

Mittermeier, Russell A., Janette Wallis, Anthony B. Rylands, et al., eds. *Primates in Peril: The World's 25 Most Endangered Primates, 2008–2010*. Arlington, VA: IUCN/SSC Primate Specialist Group, International Primatological Society, and Conservation International, 2009.

Montgomery, David R. *Dirt: The Erosion of Civilizations*. Berkeley: University of California Press, 2007.

Morris, William F., and Daniel F. Doak. *Quantitative Conservation Biology: Theory and Practice of Population Viability Analysis*. Sunderland, MA: Sinauer, 2002.

Morris, William F., Daniel F. Doak, Martha Groom, et al. *A Practical Handbook for Population Viability Analysis*. Arlington, VA: Nature Conservancy, 1999.

Musick, John A., ed. *Life in the Slow Lane: Ecology and Conservation of Long-Lived Marine Animals*. American Fisheries Society Symposium 23. Bethesda, MD: American Fisheries Society, 1999.

Mussi, Margherita. *Earliest Italy: An Overview of the Italian Paleolithic and Mesolithic*. New York: Kluwer Academic/Plenum, 2001.

Myers, Norman. *The Sinking Ark: A New Look at the Problem of Disappearing Species*. Oxford: Pergamon Press, 1979.

Naeem, Shahid, Daniel E. Bunker, Andy Hector, et al., eds. *Biodiversity, Ecosystem Functioning, and Human Wellbeing: An Ecological and Economic Perspective*. New York: Oxford University Press, 2009.

National Geographic. "Mass extinctions. What causes animal die-offs?" Accessed February 26, 2013. http://science.nationalgeographic.co.uk/science/prehistoric-world/mass-extinction.

National Research Council. *Science and the Endangered Species Act*. Washington, DC: National Academy Press, 1995.

National Research Council. *Approaches for Ecosystem Services Valuation for the Gulf of Mexico after the Deepwater Horizon Oil Spill: Interim Report*. Washington, DC: National Academies Press, 2011.

National Research Council. Committee on Assessing and Valuing the Services of Aquatic and Related Terrestrial Ecosystems. *Valuing Ecosystem Services: Toward Better Environmental Decision-Making*. Washington, DC: National Academies Press, 2005.

National Research Council. Committee on Ecosystem Effects of Fishing. *Dynamic Changes in Marine Ecosystems: Fishing, Food Webs, and Future Options*. Washington, DC: National Academies Press, 2006.

Nelson, Joseph S. *Fishes of the World*. 4th ed. Hoboken, NJ: Wiley, 2006.

Nichols, Douglas J., and Kirk R. Johnson. *Plants and the K–T Boundary*. Cambridge, UK: Cambridge University Press, 2008.

NOAA Fisheries. *Endangered Species Act*. Accessed February 26, 2013. http://www.nmfs.noaa.gov/pr/laws/esa.

Novacek, Michael J., and Quentin D. Wheeler, eds. *Extinction and Phylogeny*. New York: Columbia University Press, 1992.

Nowak, Ronald M. *Walker's Mammals of the World*. 6th ed. 2 vols. Baltimore: Johns Hopkins University Press, 1999.

Oldfield, Sara, Charlotte Lusty, and Amy MacKinven. *The World List of Threatened Trees*. Cambridge, UK: World Conservation Press, 1998.

Over, D. Jeffrey, Jared R. Morrow, and Paul B. Wignall, eds. *Understanding Late Devonian and Permian–Triassic Biotic and Climatic Events: Towards an Integrated Approach*. Amsterdam: Elsevier, 2005.

Paris, Scott G., ed. *Perspectives on Object-Centered Learning in Museums*. Mahwah, NJ: Lawrence Erlbaum, 2002.

Pauly, Daniel, and Jay Maclean. *In a Perfect Ocean: The State of Fisheries and Ecosystems in the North Atlantic Ocean*. Washington, DC: Island Press, 2003.

Perfecto, Ivette, John Vandermeer, and Angus Wright. *Nature's Matrix: Linking Agriculture, Conservation, and Food Sovereignty*. London: Earthscan, 2009.

Perrin, William F., Bernd Würsig, and J.G.M. Thewissen, eds. *Encyclopedia of Marine Mammals*. San Diego, CA: Academic Press, 2002.

Pianka, Eric R., and Laurie J. Vitt. *Lizards: Windows to the Evolution of Diversity*. Berkeley: University of California Press, 2003.

Prothero, Donald R. *The Eocene–Oligocene Transition: Paradise Lost*. New York: Columbia University Press, 1994.

Prothero, Donald R. *After the Dinosaurs: The Age of Mammals*. Bloomington: Indiana University Press, 2006.

Prothero, Donald R. *Greenhouse of the Dinosaurs: Evolution, Extinction, and the Future of Our Planet*. New York: Columbia University Press, 2009.

Prothero, Donald R., Linda C. Ivany, and Elizabeth A. Nesbitt, eds. *From Greenhouse to Icehouse: The Marine Eocene–Oligocene Transition*. New York: Columbia University Press, 2003.

Quammen, David. *The Song of the Dodo: Island Biogeography in an Age of Extinctions*. New York: Scribner, 1996.

Raup, David M. *Extinction: Bad Genes or Bad Luck?* New York: Norton, 1991.

Raven, Peter H., and Tania Williams, eds. *Nature and Human Society: The Quest for a Sustainable World; Proceedings of the 1997 Forum on Biodiversity*. Washington, DC: National Academy Press, 2000.

Righter, Robert W. *The Battle over Hetch Hetchy: America's Most Controversial Dam and the Birth of Modern Environmentalism*. New York: Oxford University Press, 2005.

Roberts, Callum. *The Unnatural History of the Sea*. Washington, DC: Island Press/Shearwater Books, 2007.

Rong, Jiaju, and Zongjie Fang, eds. *Mass Extinction and Recovery: Evidences from the Palaeozoic and Triassic of South China*. [In Chinese.] Hefei: University of Science and Technology of China Press, 2004.

Rowley, Trevor. *The English Landscape in the Twentieth Century*. London: Hambledon Continuum, 2006.

Rudwick, Martin J.S. *The Meaning of Fossils: Episodes in the History of Palaeontology*. 2nd ed. New York: Science History Publications, 1976.

Rudwick, Martin J.S. *Bursting the Limits of Time: The Reconstruction of Geohistory in the Age of Revolution*. Chicago: University of Chicago Press, 2005.

Rudwick, Martin J.S. *Worlds before Adam: The Reconstruction of Geohistory in the Age of Reform*. Chicago: University of Chicago Press, 2008.

Rudwick, Martin J.S., and Georges Cuvier. *Georges Cuvier, Fossil Bones, and Geological Catastrophes: New Translations and Interpretations of the Primary Texts*. Chicago: University of Chicago Press, 1997.

Ryder, Graham, David Fastovsky, and Stefan Gartner, eds. *The Cretaceous–Tertiary Event and Other Catastrophes in Earth History*. Geological Society of America Special Paper 307. Boulder, CO: Geological Society of America, 1996.

Sale, Peter F., ed. *Coral Reef Fishes: Dynamics and Diversity in a Complex Ecosystem*. San Diego, CA: Academic Press, 2002.

Sawyer, G.J., Viktor Deak, Esteban Sarmiento, et al. *The Last Human: A Guide to Twenty-Two Species of Extinct Humans*. New Haven, CT: Yale University Press, 2007.

Sax, Dov F., John J. Stachowicz, and Steven D. Gaines, eds. *Species Invasions: Insights into Ecology, Evolution, and Biogeography*. Sunderland, MA: Sinauer, 2005.

Scheiner, Samuel M., and Michael R. Willig, eds. *The Theory of Ecology*. Chicago: University of Chicago Press, 2011.

Schuh, Randall T., and Andrew V.Z. Brower. *Biological Systematics: Principles and Applications*. 2nd ed. Ithaca, NY: Cornell University Press, 2009.

Schwartzman, David. *Life, Temperature, and the Earth: The Self-Organizing Biosphere*. New York: Columbia University Press, 1999.

Semlitsch, Raymond D., ed. *Amphibian Conservation*. Washington, DC: Smithsonian Books, 2003.

Sepkoski, J. John, Jr. *A Compendium of Fossil Marine Animal Genera*. Bulletins of American Paleontology, no. 363. Ithaca, NY: Paleontological Research Institution, 2002.

Shellenberger, Michael, and Ted Nordhaus, eds. *Love Your Monsters: Postenvironmentalism and the Anthropocene*. Oakland, CA: Breakthrough Institute, 2011.

Shogren, Jason F., and John Tschirhart, eds. *Protecting Endangered Species in the United States: Biological Needs, Political Realities, Economic Choices*. Cambridge, UK: Cambridge University Press, 2001.

Shoshani, Jeheskel, and Pascal Tassy, eds. *The Proboscidea: Evolution and Palaeoecology of Elephants and Their Relatives*. Oxford: Oxford University Press, 1996.

Shukla, O.P., Omkar, and A.K. Kulshretha. *Pesticides, Man, and Biosphere*. New Delhi: APH Publishing, 1999.

Simmons, Alan H., and associates. *Faunal Extinction in an Island Society: Pygmy Hippopotamus Hunters of Cyprus*. New York: Kluwer Academic/Plenum, 1999.

Simpson, George Gaylord. *Tempo and Mode in Evolution*. New York: Columbia University Press, 1944.

Sodhi, Navjot S., and Paul R. Ehrlich, eds. *Conservation Biology for All*. Oxford: Oxford University Press, 2010.

Souder, William. *A Plague of Frogs: The Horrifying True Story*. New York: Hyperion, 2000.

Soulé, Michael E., ed. *Conservation Biology: The Science of Scarcity and Diversity*. Sunderland, MA: Sinauer, 1986.

Stanley, Steven M. *Macroevolution: Pattern and Process*. San Francisco: W.H. Freeman, 1979.

Stanley, Steven M. *Extinction*. New York: Scientific American Library, 1987.

Steadman, David W. *Extinction and Biogeography of Tropical Pacific Birds*. Chicago: University of Chicago Press, 2006.

Stein, Bruce A., Lynn S. Kutner, and Jonathan S. Adams, eds. *Precious Heritage: The Status of Biodiversity in the United States*. Oxford: Oxford University Press, 2000.

Stewart, Kathlyn M., and Kevin L. Seymour, eds. *Palaeoecology and Palaeoenvironments of Late Cenozoic Mammals*. Toronto: University of Toronto Press, 1996.

Stringer, Chris, and Peter Andrews. *The Complete World of Human Evolution*. London: Thames and Hudson, 2005.

Suzuki, David, and David Robert Taylor. *The Big Picture: Reflections on Science, Humanity, and a Quickly Changing Planet*. Vancouver, Canada: Greystone Books, 2009.

Talent, John A., ed. *Earth and Life: Global Diversity, Extinction Intervals, and Biogeographic Perturbations through Time*. Dordrecht, Netherlands: Springer, 2012.

Taylor, Paul D., ed. *Extinctions in the History of Life*. Cambridge, UK: Cambridge University Press, 2004.

Tennyson, Alan J.D., and Paul Martinson. *Extinct Birds of New Zealand*. Wellington, New Zealand: Te Papa Press, 2006.

Thompson, Steven K. *Sampling*. 3rd ed. Hoboken, NJ: Wiley, 2012.

Thomson, Keith. *The Legacy of the Mastodon: The Golden Age of Fossils in America*. New Haven, CT: Yale University Press, 2008.

Tilson, Ronald and Philip J. Nyhus, eds. *Tigers of the World: The Science, Politics and Conservation of* Panthera tigris. 2nd ed. London: Academic Press, 2010.

Turner, Alan, and Mauricio Antón. *The Big Cats and Their Fossil Relatives: An Illustrated Guide to Their Evolution and Natural History*. New York: Columbia University Press, 1997.

Turner, Monica G., Robert H. Gardner, and Robert V. O'Neill. *Landscape Ecology in Theory and Practice: Pattern and Process*. New York: Springer, 2001.

Turner, R. Kerry, Stavros Georgiou, and Brendan Fisher. *Valuing Ecosystem Services: The Case of Multi-functional Wetlands*. London: Earthscan, 2008.

Turvey, Samuel T., ed. *Holocene Extinctions*. Oxford: Oxford University Press, 2009.

Valentine, James W., ed. *Phanerozoic Diversity Patterns: Profiles in Macroevolution*. Princeton, NJ: Princeton University Press; San Francisco: American Association for the Advancement of Science, Pacific Division, 1985.

van der Geer, Alexandra, George Lyras, John de Vos, and Michael Dermitzakis. *Evolution of Island Mammals: Adaptation and Extinction of Placental Mammals on Islands*. Chichester, UK: Wiley-Blackwell, 2010.

Vermeij, Geerat J. *Evolution and Escalation: An Ecological History of Life*. Princeton, NJ: Princeton University Press, 1987.

Vié, Jean-Christophe, Craig Hilton-Taylor, and Simon N. Stuart, eds. *Wildlife in a Changing World: An Analysis of the 2008 IUCN Red List of Threatened Species*. Gland, Switzerland: International Union for Conservation of Nature, 2009.

Wagner, Warren L., and V.A. Funk, eds. *Hawaiian Biogeography: Evolution on a Hot Spot Archipelago*. Washington, DC: Smithsonian Institution Press, 1995.

Walters, Mark Jerome. *Seeking the Sacred Raven: Politics and Extinction on a Hawaiian Island*. Washington, DC: Island Press/Shearwater Books, 2006.

Ward, Peter D. *The End of Evolution: On Mass Extinctions and the Preservation of Biodiversity*. New York: Bantam Books, 1994.

Ward, Peter D. *The Call of Distant Mammoths: Why the Ice Age Mammals Disappeared*. New York: Springer-Verlag, 1998.

Ward, Peter D. *Future Evolution*. New York: Times Books, 2001.

Ward, Peter D. *Out of Thin Air: Dinosaurs, Birds, and Earth's Ancient Atmosphere*. Washington, DC: Joseph Henry Press, 2006.

Ward, Peter D., and Donald Brownlee. *The Life and Death of Planet Earth: How the New Science of Astrobiology Charts the Ultimate Fate of Our World*. New York: Times Books, 2003.

Weishampel, David B., Peter Dodson, and Halszka Osmólska, eds. *The Dinosauria*. 2nd ed. Berkeley: University of California Press, 2004.

Whitmore, T.C., Timothy Charles, and J.A. Sayer, eds. *Tropical Deforestation and Species Extinction*. London: Chapman and Hall, 1992.

Whitney-Smith, Elin. *The Second-Order Predation Hypothesis of Pleistocene Extinctions: A System Dynamics Model*. Saarbrüken, Germany: VDM Verlag Dr. Müller, 2009.

Whittaker, Robert J., and José Maria Fernández-Palacios. *Island Biogeography: Ecology, Evolution, and Conservation*. 2nd ed. Oxford: Oxford University Press, 2007.

Wilkinson, Clive, ed. *Status of Coral Reefs of the World: 2008*. Townsville, Australia: Global Coral Reef Monitoring Network, Reef and Rainforest Research Centre, 2008.

Wilson, Edward O. *The Diversity of Life*. Rev. ed. New York: Norton, 1999.

Wilson, Edward O. *The Future of Life*. New York: Knopf, 2002.

Wing, Scott L., Philip D. Gingerich, Birger Schmitz, and Ellen Thomas, eds. *Causes and Consequences of Globally Warm Climates in the Early Paleogene*. Geological Society of America Special Paper 369. Boulder, CO: Geological Society of America, 2003.

Wittmer, Heidi, and Haripriya Gundimeda, eds. *The Economics of Ecosystems and Biodiversity for National and International Policy Makers*. London: The Economics of Ecosystems and Biodiversity, 2010.

Wood, Bernard. *Human Evolution: A Very Short Introduction*. Oxford: Oxford University Press, 2005.

World Bank. *The Changing Wealth of Nations: Measuring Sustainable Development in the New Millennium*. Washington, DC: World Bank, 2011.

Wormworth, Janice, and Cagan H. Sekercioglu. *Winged Sentinels: Birds and Climate Change*. Port Melbourne, Australia: Cambridge University Press, 2011.

Worthy, Trevor H., and Richard N. Holdaway. *The Lost World of the Moa: Prehistoric Life of New Zealand*. Christchurch, New Zealand: Canterbury University Press, 2002.

Organizations

American Museum of Natural History
Central Park West at 79th Street
New York, NY 10024-5192 USA
Phone: (212) 769-5100
http://www.amnh.org

Association of Zoos and Aquariums
8403 Colesville Road, Suite 710
Silver Spring, MD 20910-3314 USA
Phone: (301) 562-0777
Fax: (301) 562-0888
http://www.aza.org

Barcode of Life
National Museum of Natural History
Smithsonian Institution
P.O. Box 37012, MRC 105
Washington, DC 20013-7012 USA
Phone: (202) 633-0808
Fax: (202) 633-2938
Email: cbol@si.edu
http://www.barcodeoflife.org

Center for Biological Diversity
P.O. Box 710
Tucson, AZ 85702-0710 USA
Phone: (866) 357-3349
Fax: (520) 623-9797
Email: center@biologicaldiversity.org
http://biologicaldiversity.org

Committee on Recently Extinct Organisms
Department of Ichthyology
American Museum of Natural History
Central Park West at 79th Street
New York, NY 10024 USA
Phone: (212) 769-5797
Fax: (212) 769-5642
Email: http://creo.amnh.org/index.html

Earth Impact Database
Planetary and Space Science Centre
Room 221, 2 Bailey Drive
Department of Earth Sciences
P.O. Box 4400
University of New Brunswick
Fredericton, NB, E3B 5A3
Phone: (506) 453-3560

Fax: (506) 447-3004
http://www.passc.net/AboutUs/index.html

Encyclopedia of Life
http://www.eol.org

The Field Museum
1400 S. Lake Shore Drive
Chicago, IL 60605-2496 USA
Phone: (312) 922-9410
http://www.fieldmuseum.org

GenBank
http://www.ncbi.nlm.nih.gov/genbank

Global Biodiversity Information Facility
GBIF Secretariat
Universitetsparken 15
DK-2100 Copenhagen Ø Denmark
Phone: +45 35 32 14 70
Fax: +45 35 32 14 80
Email: info@gbif.org
http://www.gbif.org

Impact Earth!
Purdue University and Imperial College London
http://impact.ese.ic.ac.uk

Intergovernmental Panel on Climate Change
World Meteorological Organization
7bis Avenue de la Paix
C.P. 2300
CH-1211 Geneva 2 Switzerland
Phone: +41-22-730-8208/54/84
Fax: +41-22-730-8025/13
Email: IPCC-Sec@wmo.int
http://www.ipcc.ch

International Barcode of Life
http://ibol.org

International Commission on Zoological Nomenclature
Natural History Museum
Cromwell Road
London SW7 5BD, UK
Phone: +44-20-7942-5653
Email: iczn-em@nhm.ac.uk
http://iczn.org

International Union for Conservation of Nature
    Rue Mauverney 28
    1196 Gland Switzerland
    Phone: +41 22 999-0000
    Fax: +41 22 999-0002
    Email: mail@iucn.org
    http://www.iucn.org

International Union for Conservation of Nature Red List of
    Threatened Species
    IUCN Global Species Programme Red List Unit
    IUCN UK Office
    219c Huntingdon Road
    Cambridge CB3 0DL, UK
    Phone: +44-1223-277966
    Fax: +44-1223-277845
    Email: redlist@iucn.org
    http://www.iucnredlist.org

International Whaling Commission
    The Red House
    135 Station Road
    Impington
    Cambridge
    Cambridgeshire CB24 9NP, UK
    Phone: +44 1223 233 971
    Fax: +44 1223 232 876
    Email: secretariat@iwc.int
    http://iwc.int

Large Igneous Provinces Commission
    http://www.largeigneousprovinces.org

Museum of Vertebrate Zoology
    University of California, Berkeley
    3101 Valley Life Sciences Building
    Berkeley, CA 94720-3160 USA
    Phone: (510) 642-3567
    Fax: (510) 643-8238
    http://mvz.berkeley.edu

National Aeronautics and Space Administration
    Near Earth Object Program
    http://science.nationalgeographic.co.uk/science/prehistoric-
        world/mass-extinction

National Center for Ecological Analysis and Synthesis
    735 State Street, Suite 300
    Santa Barbara, CA 93101 USA
    Phone: (805) 892-2500
    Fax: (805) 892-2510
    Email: nceas@nceas.ucsb.edu
    http://www.nceas.ucsb.edu/

National Evolutionary Synthesis Center
    2024 W. Main Street, Suite A200
    Durham, NC 27705-4667 USA
    Phone: (919) 668-4551

Fax: (919) 668-9198
    Email: info@nescent.org
    http://www.nescent.org

National Oceanic and Atmospheric Administration
    1401 Constitution Avenue, NW
    Room 5128
    Washington, DC 20230
    http://www.noaa.gov

National Science Foundation
    4201 Wilson Boulevard
    Arlington, VA 22230 USA
    Phone: (800) 877-8339
    Email: info@nsf.gov
    http://www.nsf.gov

National Wildlife Refuge System
    U.S. Fish and Wildlife Service
    U.S. Department of the Interior
    1849 C Street NW
    Washington, DC 20240 USA
    Phone: (800) 344-9453
    http://www.fws.gov/refuges

Natural History Museum
    Cromwell Road
    London SW7 5BD, UK
    Phone: +44 20-7942-5000
    http://www.nhm.ac.uk

Sea Level Rise: Understanding the Past—Improving Predictions
    for the Future
    http://www.cmar.csiro.au/sealevel

The Sixth Extinction Website
    Email: extinct-animals@petermaas.nl
    http://extinct.petermaas.nl

Smithsonian Institution
    P.O. Box 37012
    SI Building, Room 153, MRC-010
    Washington, DC 20013-7012 USA
    Phone: (202) 633-1000
    Email: info@si.edu
    http://www.si.edu

Tree of Life Web Project
    http://tolweb.org/tree/phylogeny.html

World Wildlife Fund
    1250 24th Street, N.W.
    Washington, DC 20037 USA
    P.O. Box 97180
    Washington, DC 20090-7180 USA
    Phone: (202) 293-4800
    http://worldwildlife.org

Glossary

Acid rain—Rain (or any other form of precipitation) that is unusually acidic; caused in the modern environment by the uptake of anthropogenic carbon dioxide (CO_2) from the atmosphere.

Acidification (oceanic)—Process by which marine waters become progressively more acidic; caused largely in modern oceans by the uptake of anthropogenic carbon dioxide (CO_2) from the atmosphere. This process increases the extinction susceptibility of organisms whose external skeletons are composed of carbonate (e.g., aragonite, calcite).

Acritarchs—Small, organic fossils that cannot be assigned to any more precise taxonomic group.

Actualism—In geology, the doctrine that all natural phenomena are the result of processes that are operating at the present time. Actualism logically opposes supernaturalism and is related to—though not synonymous with—uniformitarianism.

Adaptation—Process by which organisms modify their body forms, physiology, life history, and so on in order to survive to reproduce in their environment(s), or the state of being able to live and reproduce in a set of environments. Individuals and species become extinct when their environments change at a rate, or to an extent, that no longer allows them to survive to reproductive age and/or reproduce successfully.

Agassiz, Louis (1807–1873)—Swiss-born American paleontologist and geologist prominent for his rejection of both evolution and Darwinian natural selection and most famous for his proposal that Earth had been subject to an "ice age" in which large glaciers had covered much of its surface. Agassiz proposed that the Pleistocene megafaunal extinctions resulted from climate changes associated with this ice age and that the mammoths, mastodons, cave bears, and large cats that once roamed throughout Europe and North America were all animals adapted to tropical climate conditions.

Agglutination—The clumping of small particles together; most commonly used to describe those animals who create protective coverings for themselves by cementing small detrital particles together (e.g., agglutinated or agglutinating foraminifera).

Allee effect—The commonly observed positive correlation between population size and/or population density and mean individual fitness. This term refers generically to the propensity for small, geographically localized populations to have a higher likelihood of becoming extinct relative to large, widely distributed populations.

Allen's rule—See Rule (Allen's).

Ambergris—A solid, waxy, substance produced by the digestive systems of sperm whales (*Physeter macrocephalus*) and found either floating on the surface of marine waters (having been expelled with the feces and regurgitated through the mouth) or in the animal's intestines. With time ambergris acquires a sweet odor, and it has traditionally been used as a fixative for perfumes. Ambergris has also been regarded as having aphrodisiac properties.

Ambrein—A triterpene alcohol that is the primary active ingredient in ambergris.

American incognitum—A common name given to fossils of the American mastodon (*Mammut americanum*) prior to this species being identified as an ancient, extinct elephant species.

Ammonites—A large group of extinct cephalopod mollusks distinguished by their unique possession of conical shells, often formed into a spiral, subdivided by partitions (septa) whose joints with the shell's surface can range from gently undulating (goniatitic) to complexly convoluted (ammonitic). Ammonites existed from the Devonian through the Cretaceous periods of Earth history.

Amnion—Name given to the membranous sac that surrounds and protects the embryos of some tetrapod species, chiefly reptiles, mammals, and birds. Common possession of this morphological feature, along with the physiological processes that create it, is regarded as evidence of common ancestry for those groups.

Amniote—The group of tetrapod animal species that exhibit an amnion. These species are regarded as constituting a natural evolutionary unit based on common ancestry.

Anadromous—The type of fish species that spend most of their lives in the sea but return to freshwater habitats to breed (e.g., salmon).

Anagenesis—Mode of evolutionary change that involves an entire population acquiring an adaptive trait as a group, through the spread of a series of mutations throughout the population via mating, over time. Also known as phyletic evolution. Anagenesis is distinguished from cladogenesis.

Anoxia—The condition of total absence of oxygen in an environment or medium (e.g., anoxic waters).

Apomorphy—A unique characteristic or feature of an organism that provides evidence of common ancestry and can be used to identify or diagnose members of a unified evolutionary group or clade. For example, shared possession of the amnion is an apomorphy of the Amniota.

Appalachia—The island continent composed of the land currently located along the Eastern Seaboard of North America that existed during the Upper Cretaceous period as a result of the Late Cretaceous sea-level rise that flooded the North American Craton and created the Western Interior Seaway.

Aragonite—One of the two common and naturally occurring forms of calcium carbonate ($CaCO_3$), aragonite is characterized by an orthorhombic crystal lattice with common acicular crystals. Aragonite can be formed by physical (e.g., evaporite) or biological processes but lacks the broad stability of the other crystal form, calcite. Owing to this narrow stability range, organisms that form their skeletons from aragonite (e.g., scleractinian corals) may be restricted to a narrow range of environmental conditions.

Aridification—Process whereby landscapes and habitats become progressively drier, as a result of lower levels of annual precipitation and/or the drying up of surface-water bodies.

Asteroid—Orbiting the Sun inside the orbit of Jupiter, asteroids are small, naturally occurring objects distinct from comets. Asteroids are thought to be the remains of bodies (planetesimals) that formed within the solar nebula and that failed to coalesce or aggregate into planet-sized bodies.

Aurochs—An extinct ungulate species (*Bos primigenius*) that is regarded as the ancestor of domestic cattle. Aurochs occurred throughout Europe, Asia, and North Africa but went extinct in the early 1600s.

Autocorrelation—The property of repeated patterning in a set of time-series data as a function of time.

Autotroph—Any species that has the ability fix carbon completely as a result of internal physiochemical processes and so does not require the input of organic carbon from the environment to support basic physiological processes and growth. Green plants are examples of autotrophs.

Background extinction—See Extinction (background).

Baleen—A series of long hairy plates formed from keratin and found within the mouths of mysticete whales. This system of plates form a key component of the filter-feeding system these species use to extract small planktonic and nektonic organisms from seawater.

Bathymetry—The measurement and/or summarization of data pertaining to the depth of water in an ocean, lake, or river system.

Benthic species—See Species (benthic).

Bergmann's rule—See Rule (Bergmann's).

Biodiversity—The degree of variation exhibited by a group of organisms at a particular locality or over time.

Biodiversity deficit—The difference between the extinction rate and the speciation rate for a set of species at a particular locality or over time.

Biodiversity hot spots—A localized area, region, or biome that contains an unusually large number of species, including many endemic species. Between 25 and 35 biodiversity hotspots have been identified. Together, these contain as much as 60 percent of Earth's plant and tetrapod species, along with (presumably) a large number of invertebrate and protist species.

Biofacies—Stratigraphic unit differing in its biotic content from other time-equivalent or space-equivalent rock units.

Biogeography—The study of the distribution of species and ecosystems in space and time.

Biomass—The sum total of living biological material occurring within an area or region at a specific time or over a specified time interval.

Biomere—A biostratigraphical unit bounded by the abrupt non-phyletic extinction of the dominant elements of a single organismal group (e.g., trilobites).

Blitzkrieg extinction—See Extinction (blitzkrieg).

Blubber—A thick layer of fat that occurs under the skin of marine mammals and serves to insulate the body from heat loss resulting from the cold temperature of the surrounding seawater.

Bolide—There are two definitions of this term. Astronomers use the term *bolide* to refer to a fireball that reaches the luminosity magnitude of 14 or higher. Geologists use the term *bolide* to refer to a large (0.6 to 6.2 miles [1–10 km] in diameter) Earth impactor of nonspecific origin (e.g., asteroid, comet).

Boreal realm—The subarctic climate zone typically characterized by taiga and conifer-dominated forests.

Bycatch—The set of fish caught unintentionally as a result of fishing operations designed to target a specific species or set of species.

Cabinets of curiosity—Encyclopedic collections of objects amassed by well-to-do European gentlemen from the time of the Renaissance onward designed to convey the assembler's control over the world via its reproduction in microcosm. Cabinets of curiosity are regarded as the forerunners of natural history museums.

Calcite—One of the two common and naturally occurring forms of calcium carbonate ($CaCO_3$), calcite is characterized by a trigonal-rhombohedral crystal lattice. Calcite can be formed by physical or biological processes and is broadly stable under a wide range of environmental conditions.

Cambrian explosion—Term used to describe the relatively sudden appearance of species representing most of the known animal phyla around 530 million years ago. Biologically this diversification event may be more apparent than real and involve the acquisition of hard parts that would preserve as fossils by a number of species that had long, though paleontologically invisible, prehistories.

Cannonball Sea—See Western Interior Seaway.

Carnivore—Any organism that obtains all or most of its nutrient requirements from a diet of animal tissue, whether through active predation or scavenging.

Carrying capacity—The maximum number of organisms an environment can sustain indefinitely.

Carson, Rachel (1907–1964)—North American marine biologist, conservationist, and author whose books (e.g., *The Sea Around Us* [1951], *Silent Spring* [1962]), along with her other writings, are widely credited with bringing the conservation movement—in particular the dangers of the uncontrolled use of pesticides such as DDT—to the attention of the public worldwide.

Catastrophism—In geology the doctrine that Earth has been repeatedly subjected to unusual, sudden, violent events of short duration that affected both marine and terrestrial environments simultaneously. Catastrophism logically opposes actualism and uniformitarianism.

Cenozoic—The current era of the geological time scale, encompassing the interval of Earth history from 66 million years ago to the present and including the Paleogene, Neogene, and Quaternary periods.

Charismatic species—See Species (charismatic).

Chemocline—A marked change in the concentration of a chemical with depth in a body of water.

Chronology—The science of arranging, or inferring the arrangement of, a series of events in time.

Citizen science—Scientific research that uses, in whole or in part, amateur and/or nonprofessional workers usually to collect and/or participate in the analysis of data. This component of scientific investigation is also known as crowdsourcing.

Clade—A monophyletic group of organisms or species consisting of the ancestral species and all of its descendants.

Cladogenesis—Mode of evolutionary change that involves one or a small number of local populations of a species acquiring an adaptive trait through the spread of a series of mutations throughout the population via mating, over time, and then diverging reproductively until backcrossing with the parent species is either impossible or produces sterile offspring. Cladogenesis increases biodiversity because the original (parent) species does not become extinct. Also known as branching evolution. Cladogenesis is distinguished from anagenesis.

Coextinction—The obligate extinction of one species as a direct consequence of the extinction of another (e.g., the extinction of a host-specific parasite species as a result of the extinction of the host).

Committee on Recently Extinct Organisms (CREO)—A research program established at the American Museum of Natural History to improve the quality of information available on recently extinct organisms to support conservation biology research.

Conservation biology—The scientific study of the conservation status of Earth's species and the processes by which species can avoid extinction.

Continent—A large, continuous, discrete landmass that stands above sea level, is typically composed of relatively light siliceous rocks and sediments, and is usually (but not always) separated from other continents by expanses of water.

Continental drift—Term used for the inferred historical movement of the continental landmasses from a condition when they were all joined together, as evidenced by the geographical correspondence between the shape of their coastlines and/or continental margins, to their present positions. The idea of continental drift superseded the modern theory of plate tectonics—the physical mechanism that drives continental drift.

Continental shelf—The submerged perimeter of a continental landmass that consists of the portion of the coastal plane that lies below sea level at any given place and time.

Continental slope—The seaward border of the continental margin that represents the transition zone between the continental shelf and the deep-ocean basin.

Convention on Biological Diversity (CBD)—A legally binding treaty on the conservation of biological diversity, sustainable use of biodiversity, and sharing of the benefits from the discovery and commercialization of genetic resources signed by 193 countries in 1992, as part of the Earth Summit in Rio de Janeiro, Brazil.

Convention on International Trade in Endangered Species of Wild Fauna and Flora (CITES)—A multilateral treaty enacted in 1963 whose purpose is to ensure that the international trade in plants and animals does not put species at risk of extinction. The CITES accord has been signed by all but 17 member states of the United Nations.

Convergent evolution—The process by which characteristics or traits evolve independently in species from unrelated lineages to solve similar ecological problems.

Cope's rule—See Rule (Cope's)

Cosmopolitanism—In species, the exhibition of broad environmental tolerances that allow them to inhabit many different biomes successfully. Species with an ecologically cosmopolitan distribution are usually able to tolerate a wide range of environmental conditions.

Cryptofauna—Term referring to species that are rarely seen. In some cases it is uncertain whether members of the cryptofauna are extinct, while in others the uncertainty centers on whether they existed in the first place (e.g., are mythological).

Cut mark—A gouge or scratch in natural materials (e.g., bone) created by the action of another animal cutting, biting, or gnawing the material. Genuine cut marks constitute evidence of predation. It is often difficult, however, to distinguish cut marks from marks made by non-food-processing-related activities (e.g., trampling).

Cuvier, Baron Georges (1769–1832)—French anatomist, zoologist, paleontologist, and naturalist best known for his anatomical research, establishment of the field of vertebrate paleontology, establishment of the fact of extinction, and advocacy of the doctrine of catastrophism.

Darwin, Charles (1809–1882)—English naturalist, explorer, and author credited with establishing the fact of organic evolution via the presentation of copious examples drawn his observations of nature and best know for proposing the theory of natural selection as the natural process that results in the formation of new species over time.

Dating (radiocarbon)—A isotopic dating technique based on the radiogenic decay of carbon-14 (^{14}C) used to estimate the age of wood and leather up to about 60,000 years old.

Dating (radioisotopic)—A set of isotopic dating methods that estimate the ages of material by measuring the proportion of radioactive isotopes and their decay products in natural materials; also termed radiometric dating.

Deccan Traps—See Traps (Deccan).

Deforestation—The removal of a stand or trees or an entire forest after which the land is converted to some other use (e.g., agriculture, cattle grazing, development).

Demography—The statistical study of the processes that affect birth and death rates, age structure, and population sizes, as well as the effect of these on population dynamics.

Density-dependence—Any factor that influences a population, species, higher taxonomic group, ecological guild, functional group, and so on, to a degree that varies in response to how many individuals assignable to the group exist.

Dinosaurs—The group of organisms consisting of the ornithischian tetrapods, sauropodomorph tetrapods, and theropod tetrapods along with their common ancestor and all descendants. As one of the direct descendants of the theropod dinosaurs are modern birds, the traditional class Aves—which includes all ancient and modern birds—are considered to be members of the Dinosauria.

Dinosaurs (non-avian)—The group of dinosaurs excluding the birds (traditional class Aves). While this is a paraphyletic group, it is a useful collective term often used to describe the set of dinosaurs whose last representatives disappeared from the fossil record during the end-Cretaceous extinction event.

Disconformity—A stratigraphic unconformity, representing an interval of missing time, between parallel layers of sedimentary rock. Because disconformities can have only very subtle representations physically, they are often difficult to recognize in stratigraphic successions and/or cores. Apparent coordinated or simultaneous extinction patterns can be produced artificially by disconformities.

Diversity—A measure of the number of biological groups (e.g., species, higher taxonomic groups, clades, ecological guilds, functional groups) and the apportionment of individuals to these groups existing at a particular time and/or place. The former parameter is termed richness and the latter evenness or equitability.

Dysaeroby—Any environment characterized by less than 0.1 ml–1.0 ml of oxygen per liter of medium (e.g., water).

Dysoxic conditions—Environmental conditions that conform to state of dysaeroby.

Ecology—The study of the patterns and factors influencing the abundance and distribution of species.

Ecophenotypy—Nonheritable changes in the expressing of an organism's phenotype induced by local differences in environmental conditions. This phenomenon is commonly observed in plants but also occurs in animals (e.g., dwarfing of fish species living in restricted environments such as aquaria) and humans (e.g., obesity resulting from a lack of the need to exercise).

Ecospace—The full and complete range of resources used and environmental conditions tolerated by a set of organisms at a given locality and time. A single species usually occupies only a portion of the ecospace available at any given locality. This concept is related to the concept of the ecological niche.

Ecosystem—The network of interactions among and between organisms and both the living and nonliving components of their environment.

Ecosystem services—The set of interactions and processes provided by ecosystems to a group of organisms that are necessary for the group to remain alive and viable.

Effect (Allee)—See Allee effect.

Elvis species—See Species (Elvis).

Embryo—The earliest stage of development in the life cycle of a multicellular, genetically diploid organism.

Emigration—The state or act of moving away from a locality or region to take up residence in another region or locality.

Endangered species—See Species (endangered).

Endangered Species Act of 1969 (ESA)—A US law, enacted in 1969, that expended the provisions of the Endangered Species Preservation Act of 1966 to cover species and subspecies threatened with worldwide extinction and prohibited the importation into the United States of any species or subspecies or any product thereof (e.g., skins, eggs) on that list. Limited exceptions to the provisions of this act were made for scientific, education, conservation, and economic purposes.

Endangered Species Act of 1973 (ESA)—A US law, enacted in 1973, designed to protect imperiled or endangered species from extinction as a consequence of unregulated economic growth and/or development. This legislation grew out of the Endangered Species Preservation Act of 1966.

Endangered Species Preservation Act (ESPA)—A US law, enacted in 1966, that permitted the listing of native US animals as endangered and allowed limited protection to be placed on those animals. This legislation was the forerunner of the Endangered Species Act and also gave the US secretary of the interior the authority to administer and manage the U.S. National Wildlife Refuge System.

Endemism—The state of being unique or confined to a particular geographic location (e.g., island, nation).

Epicontinental sea—The portion of an ocean or sea lying on or extending onto a continental shelf. Epicontinental seas tend to occur and be extensive during times of high eustatic sea level and to be restricted or rare during times of low eustatic sea level. Both terrestrial and marine populations can be affected by changes to the size and character of epicontinental seas. Also termed an epeiric sea.

Eustatic sea-level change—See Sea-level change (eustatic).

Eutherian mammals—See Mammals (placental).

Eutrophy—The condition of being rich or well-endowed with mineral and organic nutrients.

Euxinic conditions—See Anoxia.

Evaporite—Term used to describe any sedimentary deposit formed from water-soluble minerals and created as a consequence of the evaporation of a large body of water. The evaporation of both marine and fresh waters can result in the creation of evaporite deposits.

Event (Hangenberg)—The second phase of the extended Late Devonian extinction event that took place in the temporal vicinity of the boundary between the Devonian and Carboniferous periods.

Event (Kellwasser)—The first phase of the extended Late Devonian extinction event that took place in the temporal vicinity of the boundary between the Frasnian and Famennian stratigraphic stages.

Exotic species—See Species (exotic).

Extinct—The state of a population, species, or other organismal group that once existed but suffered the death of all individuals.

Extinction—The death of the last individual of a species or other organismal group. By convention, extinction refers to the death of the last individual worldwide, whereas extirpation refers to the extinction of a local population.

Extinction (background)—The set of stratigraphic stage-level extinction events that fall within the range of typical variation for these stage-level data as a whole. Background extinction events are distinguished from mass extinction events. It is generally thought that the normal Darwinian (microevolutionary) processes of competition and selection predominate during background extinction intervals.

Extinction (blitzkrieg)—A variant of the overkill hypothesis positing that the Pleistocene megafaunal extinctions in North America (and perhaps elsewhere) were caused by human populations actively seeking out and hunting animals to the point of local extinction over a short period of time and then moving on to new hunting grounds where the process was repeated.

Extinction (mass)—Any anomalously large extinction event. Traditionally five to six mass extinctions have been recognized in the Phanerozoic (Late Cambrian, end-Ordovician, Late Devonian, end-Permian, end-Triassic, and end-Cretaceous), but more recent research has shown that this group of events is not a discrete category of extinction events; rather, it is the upper region of a continuous extinction-intensity spectrum.

Extinction (quasi-)—The state of a population or species that exists at a density or size below the capacity necessary to preserve it in the face of random environmental variation. Quasi-extinct species still exist, but their long-term viability in the absence of targeted conservation support is regarded as doubtful. It is presently unknown how many populations or species are quasi-extinct.

Extinction debt—The delay between the occurrence of an event or change that is ultimately responsible for a species' extinction and the occurrence of the species' actual extinction. It is suspected that many modern species are in, or close to, a state of extinction debt.

Extirpation—The localized extinction of a population species or other group.

Fecundity—A measure of the potential reproductive capacity of an individual.

Financial capital—The money used by businesspeople to purchase the resources, goods, and services they need in order to manufacture the products and/or provide the services they sell.

Food web—The pattern of feeding connections that exist within an ecological community.

Foraminifera—A group of amoeboid marine zooplankton characterized by a network of thin, anastomosing pseudopodia that radiate out from the central body mass and are used to catch prey items and effect limited movement in benthic species. Many species secrete or assemble a shell composed of individual chambers and separated by walls or septa that are pierced by one or more openings. Foraminiferal species may be planktonic or benthic.

Forams—A colloquial terms used to refer to foraminifera.

Fossil—The preserved remains—including marks and molecules—of life-forms that existed in Earth's ancient past.

Fossil record—The set of fossils of various types and ages that have been preserved in sedimentary rocks over the course of Earth history.

Gamma-ray burst—Intense concentrations of gamma rays that emanate from extremely energetic explosions that have been observed to take place in distant galaxies by astronomers. This mechanism has been proposed by some researchers as a potential cause of mass extinction events in Earth history.

Genetics—The science of genes, heredity, and heritable variation in living organisms.

Genus—The category of organization in biological taxonomy that lies between the species and the family. A genus is composed of a set of similar species. A family is composed of a set of similar genera.

Geological time scale—A system of stratigraphic categories, usually associated with measured or inferred absolute dates, that relates positions and distances in stratigraphical succession to times and time intervals.

Global boundary stratotype section and point (GSSP)—Designated horizon in a particular stratigraphic succession at a particular locality that serves as the ultimate reference for a particular stratigraphic boundary. As of 2012, 64 of the 101 stratigraphic stages have been standardized via internationally agreed-on GSSPs.

Gloger's rule—See Rule (Gloger's).

Gondwana—Term used to describe the southern super-continent, composed of a amalgamation of the present-day continents of Africa, South America, Antarctica, and Australia, that existed in the Paleozoic prior to the assembly of Pangaea and in the Mesozoic following the rifting of Pangaea and prior to the internal rifting that, over the course of the Mesozoic and Cenozoic, resulted in the modern continental identities and distributions.

Grade—A group of organisms whose similarity results from the attainment of a common level of biological organization rather than common ancestry. For example, the traditional taxonomic category of Reptilia (a class) is a grade because the characteristics that separates it from the class Mammalia and the class Aves (possession of skin covered by scales) are also present in the latter two groups but in an altered (i.e., derived) form (hair and feathers).

Great Ordovician Biodiversification Event (GOBE)—The Paleozoic evolutionary transition that marked the change from the marine fauna that characterized the Cambrian—one dominated by the trilobite, polychaete, archaeocyathid, monoplacophoran, and inarticulate brachiopod groups—to the marine fauna that characterized the remainder of the Paleozoic, which centered on the articulate brachiopod, cephalopod, coral, crinoid, and bryozoan groups.

Greenhouse climate—The global climate regime characterized by the absence of continental ice sheets, warm sea-surface temperatures, and high concentrations of greenhouse gases in the atmosphere. Earth was last in a greenhouse state about 40 million years ago.

Greenhouse gas—Any gas that absorbs radiation and reemits it in the infrared thermal range (wavelengths 0.74 m to 0.3 mm). Common greenhouse gases includes water vapor (H_2O), carbon dioxide (CO_2), methane (CH_4), nitrous oxide (N_2O), and ozone (O_3).

Habitat—The range of ecological and environmental conditions that are inhabited by a particular species at a particular time.

Habitat fragmentation—The development of spatial discontinuities in a species habitat that have the effect of subdividing a previously large, contiguous population into a set of smaller subpopulations, each of which is more susceptible to extinction via density-dependent and Allee effects.

Habitat tracking—The geographic movement of a local population and/or species undertaken in an effort to maintain its habitat within the tolerance limits set by the species' adaptations. Different species differ in their ability to track the spatial migration of their preferred habitat. Habitat tracking is especially common during times of global climate change.

Hangenberg event—See Event (Hangenberg).

Herbivore—Any animal anatomically and physiologically adapted to obtain all or most of its nutrient requirements from a diet of plant tissue.

Hermatypy—Term referring to corals that form reef-sized aggregations or colonies.

Heterotroph—Any species that lacks the ability fix carbon completely as a result of internal physiochemical processes and so requires the input of organic carbon from the environment to support basic physiological processes and growth. Most animals are heterotrophs.

Holocene—The geological epoch that includes the time interval from the end of the Pleistocene (about 11,700 years ago) to the present day.

Holotype—An individual specimen designated as the formal representative of a species by the species' author. Also referred to as the "type specimen" or "type."

Homoplasy—The state of similarity between two or more species that is the result of convergent evolution rather than common ancestry.

Homotroph—See Autotroph.

Hyperdisease hypothesis—A variant of the overkill hypothesis positing that the Pleistocene megafaunal extinctions in North America (and perhaps elsewhere) were caused by human populations introducing one or more virulent diseases into mammal populations that, over a comparatively short time, drove the native populations to extinction.

Hypothetical extinct species—See Species (hypothetical extinct)

IBC—See International Botanical Congress.

Icehouse climate—The global climate regime characterized by the presence of continental ice sheets, cool sea-surface temperatures, and low concentrations of greenhouse gases in the atmosphere. Earth's climate is currently in an icehouse state.

ICZN—See International Commission on Zoological Nomenclature.

Immigration—The state or act of moving into a locality or region to take up residence.

Impact winter—A prolonged interval of unseasonably cold weather brought about from the injection of dust, ash, and other materials into Earth's atmosphere as a result of an asteroid or comet impact.

Instrumental value—See Value (instrumental).

Insular environment—The environment of a physical or ecological island or island region. For species, insular environments may constitute refugia from the processes that lead to extinction (see Refugium).

Intergovernmental Panel on Climate Change (IPCC)—An intergovernmental body established by the United Nations in 1988 to provide scientific, technical, and socioeconomic information about the risk posed by climate change as a result of human activity and to develop options that governments and/or businesses may enact to mitigate this risk.

International Botanical Congress (IBC)—An international meeting of botanists that since 1900 has been held every six years under the auspices of the International Association of Botanical and Mycological Societies. Proposed changes and/or amendments to the International Code of Botanical Nomenclature are approved in the plenary session of this meeting.

International Commission on Zoological Nomenclature (ICZN)—Charitable organization that serves as an adviser and arbiter to the zoological community at large on issues relating to the correct use and validity of scientific names of animals. ICZN members write and the organization publishes the International Code of Zoological Nomenclature.

International Union for Conservation of Nature (IUCN)—An international organization established for the purpose of finding pragmatic solutions to pressing environmental and developmental challenges. The IUCN supports research into and collates information about the conservation status of species and summarizes this information in the form of its Red List of Threatened Species.

International Whaling Commission (IWC)—International organization established in 1946 to conserve and manage whale stocks in a sustainable manner on behalf of the whaling industry.

Intertropical convergence zone—Area near Earth's equator where the northeastern and southeastern trade winds come together. This zone is usually marked by a band of low-surface wind speeds and thunderstorms that circles the globe.

Intrinsic value—See Value (intrinsic).

Introduced predator—See Predator (introduced).

Invasive species—See Species (invasive).

Invertebrate—A paraphyletic group consisting of all animals other than those assigned to the subphylum Vertebrata, all of which lack the ability to develop a vertebral column as part of their body's axial structure.

IPCC—See Intergovernmental Panel on Climate Change.

Island—Any region where a unique habitat or microclimate exists that serves to isolate it from the surrounding area or region. Species that inhabit islands often exhibit a high susceptibility to extinction as a result of their small population sizes and dependence on the ability of the island habitat to maintain itself.

Isotope—A variant of a chemical element that differs from other such variants in the number of neutrons it possess and thus its atomic weight as well. Different isotopes of the same element may exhibit slightly different chemical properties or differentially participate in some types of chemical reactions.

IUCN—See International Union for Conservation of Nature.

IWC—See International Whaling Commission

Kellwasser event—See Event (Kellwasser).

Keystone species—Any species whose presence (or absence) has an effect on its local ecosystem disproportionate to its relative abundance.

Laramidia—The island continent composed of the land currently located along the western seaboard of North America that existed during the Upper Cretaceous period as a result of the Late Cretaceous sea-level rise that flooded the North American Craton and created the Western Interior Seaway.

Large igneous province (LIP)—Any large accumulation (greater than 38,600 square miles [100,000 sq km]) of intrusive or extrusive igneous rocks in Earth's continental or oceanic crust. These accumulations are the remnants of significant magmatic activity in Earth history. Extrusive LIP events are often associated with high levels of species extinctions worldwide.

Larvae (lecithotrophic)—Larvae characterized by an internal source of nutrition (usually a yolk sac) that sustains or augments their physiological needs, usually during migration to a new locality when they undergo the transition to the adult stage of their life cycle. Lecithotrophic larvae are usually limited in the distances over which they can disperse, but they exhibit a lower larval mortality.

Larvae (planktotrophic)—Larvae of marine species characterized by the ability to capture and eat food items during this phase of their life cycle. Possession of this ability often allows planktotrophic larvae to disperse over large distances during this phase of their life cycle but to suffer correspondingly higher larval mortalities.

Larval stage—A distinct postembryonic but preadult developmental stage or phase many animals undergo as part of their life cycles. A caterpillar is the larval stage of a butterfly's or moth's life cycle. Possession of larval developmental stages are typical of many insects, amphibians, and cnidarians.

Lava—Molten or formerly molten rock extruded on the Earth surface during a volcanic eruption.

Lava flow—Bodies of actual or once extrusive molten rock emplaced as a result of a volcanic eruption.

Lazarus species—See Species (Lazarus).

Lecithotrophic larvae—See Larvae (lecithotrophic).

Leopold, Aldo (1886–1948)—American professor (University of Wisconsin, Madison), author, ecologist, and environmentalist and the widely recognized founder of the science of wildlife management. Leopold is most famous for his book *A Sand County Almanac* (1949) and his land ethic, along with his recognition of the conservation significance of keystone species and trophic cascades.

Linnaeus, Carolus (1707–1778)—Swedish physician and natural historian wildly recognized as the founder of the modern science of taxonomy and famous for his system of binomial nomenclature in the naming of species.

Logistic growth—Either organism or population growth that conforms to a logistic or sigmoid mathematical function in which the increase begins slowly, followed by a phase of exponential increase that then slows and diminishes progressively to an asymptotic upper limit. A simple mathematical description of this type of relationship can be obtained by graphing the function $P(t) = 1/1+e^{-t}$ for t and $P(t)$.

Lyell, Charles (1797–1875)—British lawyer and geologist famous for his textbook *Principles of Geology* (1830–1833) in which he opposed catastrophism and popularized James Hutton's principle of uniformitarianism.

Macroevolution—Evolution involving patterns and processes that operate above the level of the taxonomic species, which is to say on a scale of separate gene pools. Although the processes that might give rise to macroevolutionary patterns remain controversial, the existence of long-term, directional trends in taxonomic and evolutionary data sets that appear to have a low correlation with patterns of environmental change is a commonly observed feature of the fossil record.

Magma—A mixture of viscous, fluid rock, mineral crystals, and dissolved gas that exists below the earth's surface in areas of volcanic activity.

Mammals—A clade within the Tetrapoda consisting of all tetrapods possessing hair, three middle-ear bones,

mammary glands (in females), and a neocortex region in the brain.

Mammals (eutherian)—See Mammals (placental).

Mammals (marsupial)—A group of mammals characterized by the common possession of a style of development in which an underdeveloped larva-like offspring is born and migrates to a pouch of skin—the marsupium—containing one or more nipples from which the offspring feeds and, over time, completes its development to the juvenile stage within this pouch. Marsupial mammals are currently placed in the infraclass Metatheria.

Mammals (metatherian)—See Mammals (marsupial).

Mammals (placental)—A large group of mammals that give birth to live young at a juvenile grade of development, which, prior to birth, are nourished through a specialized embryonic organ—the placenta—attached to the uterine wall. Placental mammals are currently placed in the infraclass Eutheria.

Marine—Pertaining to or living either in or on the ocean.

Marine protected area—Any marine area in which human activity is restricted in order to conserve the natural environment and the species living therein.

Marine regression—A large-scale, quasi-permanent decline in the level of the ocean usually resulting in the subareal exposure of former seafloor.

Marine transgression—A large-scale, quasi-permanent increase in the level of the ocean usually resulting in the submergence of former land surface.

Marsupial mammals—See Mammals (marsupial).

Mass extinction—See Extinction (mass).

MEA—See Millennium Ecosystem Assessment.

Megafauna—Any large animal greater than about 100 pounds (45 kg) in weight.

Mesozoic—The time interval in Earth history that begins at the base of the Triassic Period (c. 252 MYA) and ends at the top of the Cretaceous Period (c. 66 MYA).

Metapopulation—A group of spatially separated and distinct populations that interact in a variety of ways and exhibit a global stability in the face of environmental change greater than that of any of the component populations by itself.

Metatherian mammals—See Mammals (marsupial).

Meteor—A visible streak of light that can be seen in the sky (usually at night) caused by the heating of a meteoroid (a particle of debris, ranging in size from a grain of sand to a boulder, left over form the formation of the solar system) as it passes through the atmosphere.

Methane hydrate—A form of methane in which an amount of this gas residing beneath marine sediments becomes trapped within the crystal structure of water, forming a solid similar to ice. Also known as methane clathrate or natural gas hydrate. It is thought that increasing global temperatures melting methane hydrate deposits residing on the shallow continental shelves, releasing large amounts of methane into the atmosphere and setting off a runaway greenhouse effect, is a mechanism for causing some episodes of mass extinction.

Microcontinents—Relatively small fragments of continents that have, through plate tectonic processes, become separated from the main continental landmasses and posses a unique physiographic history.

Microevolution—Any heritable, naturally occurring change in allele frequencies resulting from mutation and/or selection. Neosynthetic evolutionary theory holds that microevolutionary processes are sufficient to explain all observed evolutionary phenomena—that a unique set of macroevolutionary process do not exist.

Millennium Ecosystem Assessment (MEA)—A synthesis of the data, interpretations, and opinions or over 1,000 biologists, published in 2005, that documents the state of Earth's ecosystems and provides guidelines for political, commercial, and academic decision makers.

Monomorphy—The state of having only one form (e.g., genotype, body form, ecology, developmental sequence) through a series of developmental or environmental changes.

Muktuk—English word for a traditional Inuit-Yupik and Chukchi meal of whale skin and whale blubber.

National Oceanic and Atmospheric Administration (NOAA)—Agency within the US Departmet of Commerce charges with monitoring and and providing advice on the conditions of the Earth's oceans and atmosphere.

Natural capital—The set of economic goods and services relating to and resulting from the natural environment.

Natural selection—The natural, nonrandom process by which biological traits and characteristics become more (or less) common in a population as a result of the differential reproductive success (or failure) of the individuals possessing those traits in competition with other individuals and species in their local environment.

Near-Earth asteroid (NEA)—Any asteroid whose orbit brings it into the proximity of Earth (c. 0.983–1.3 astronomical units from the Sun). NEAs may conceivable be captured by Earth's gravitational force and impact on its surface.

Near-Earth object (NEO)—Any object whose orbit brings it into proximity of the Earth (c. 0.983–1.3 astronomical units from the Sun). NEOs may conceivable be captured by Earth's gravitational force and impact on its surface.

Nekton—The group of aquatic organisms capable of moving independently of local water currents.

Neolithic—The interval of human history characterized by the widespread use of stone tools (c. 10,200–2000 BC).

Neontology—That part of biology dealing with currently living organisms. Neontology is usually contrasted with paleontology, which deals primary with extinct organisms.

New (evolutionary) synthesis—The union of several previously separate theories of evolution and fields of evolutionary research, including natural selection, Mendelian genetic, population genetics, and taxonomy; achieved for the most part between 1936 and 1947.

Newell, Norman D. (1909–2005)—Influential professor of geology (Columbia University), curator (American Museum of Natural History), and author who emphasized the importance of extinction studies and, along with Otto H. Schindewolf, developed the concept of mass extinction.

Nongovernmental organization (NGO)—A legally constituted organization that operates independently from any government and is distinct from a commercial business. The United Nations is an example of an NGO.

Oceanic acidification—See Acidification (oceanic).

Oligotroph—Any species capable of living in an environment that supplies few organic nutrients.

Overkill hypothesis—Generic term for the idea that the Pleistocene megafaunal extinctions in North America (and perhaps elsewhere) took place over a relatively short period of time (c. 1,000 years) and were caused by the activity of human populations entering the continent.

Paleocene–Eocene thermal maximum—A relatively sudden and extreme warming of Earth (an increase of mean annual surface temperature of 11°F [6°C]) that occurred close to the boundary between the Paleocene and Eocene stages of Earth history and is associated with both major biological changes in ecosystems and physical changes in the carbon cycle globally.

Paleontology—The scientific study of all aspects of pre-historic life.

Paleosol—An ancient soil layer usually preserved beneath sediments or lithified into rock.

Paleothermometer—Any method or technique used for determining past temperatures using proxy observations derived from nature (e.g., isotopic abundances, sediments, ice layers, tree rings, leaf shapes).

Paleozoic—The earliest of the three eras of the geological time scale; encompasses the Cambrian through Permian periods (c. 541–252 MYA).

Pangaea—Late Paleozoic supercontinent formed during the Late Carboniferous through Permian by the collision of Euramerica (formed from the Early Paleozoic continents of Laurentia, Baltica, and Avalonia) and Gondwana.

Pathogen—An infectious agent—usually a microorganism such as a virus, bacterium, prion, or fungus—that causes disease in its host.

Pelagic species—See Species (pelagic).

Phanerozoic—The time interval over which abundant, multicellular life has been present on Earth; roughly, the past 541 million years.

Phillips, John (1800–1874)—English geologist who published the first geological time scale and coined the term *Mesozoic*.

Photic zone—Depth to which a sufficient amount a light will penetrate a water column to allow photosynthesis to take place; the aquatic zone within which virtually all primary productivity occurs.

Phylogenetic diversity—Any measure of species richness or diversity that takes into consideration an estimate of the phylogenetic differences between species.

Phylogenetic supertree—A composite cladogram estimated from more restricted cladograms sharing some, but not necessarily all, terminal taxa.

Phylogeny—The pattern of genealogical relations expressing ancestry and descent among species through time. A phylogeny is distinct from a cladogram in that the former requires specific ancestral species to be identified whereas the latter does not.

Phytoplankton—Any autotrophic member of the plankton.

Placental mammals—See Mammals (placental).

Plankton—Any aquatic organism incapable of independent lateral movement through the water column.

Planktotrophic larvae—See Larvae (planktotrophic).

Plate tectonics—The scientific theory that both describes and explains the disposition and historical movement of ocean basins and continental landmasses over the earth's surface; formed from a synthesis of continental drift and seafloor spreading theory.

Polymorphy—The state of being able to exist in many different forms (e.g., genotype, body form, ecology, developmental sequence) through a series of developmental or environmental changes.

Population—A monospecific group of organisms living in the same locality at the same time, all of whom are potentially capable of interbreeding.

Population growth rate—Average percentage increase in numbers of a population, represented by the total proportion of recruits (births and immigrants) minus the total proportion of animals lost (by death and emigration).

Population viability analysis—A species-specific method of extinction risk assessment used by conservation biologists to determine the likelihood a species will become extinct in an a priori–defined time interval.

Potlatch—A gift-giving ceremony practiced by certain northwestern North American Indian tribes whose main purpose was the confirmation of friendly relations between family units or tribes by wealth redistribution and gifting reciprocity.

Predator (introduced)—Any predator species not native to an area or region that has been introduced as a result of human activity. This introduction may be harmful, damaging, or neutral to populations resident in the area.

Primary type—See Type (primary).

Probability—A measure of how likely an event is or how likely a statement is to be true.

Productivity—A measure of the rate of biomass generation in a natural system.

Pseudoextinction—term used to refer to the extinction of a taxonomic name that occurs when a species over time develops a unqiue characteristic via anagenesis such that it can be distinguished from the older form of the species and, in recognition of this acquisaition, is given a new taxonomic name. Also referred to as phyletic extinction.

Pycnocline—A marked change in the density with depth in a body of water.

Quasi-extinction—See Extinction (quasi-).

Quaternary—The most recent of the periods of the geological time scale, encompassing the last 2.59 million years of Earth history and including the Pleistocene and Holocene epochs.

Radiocarbon dating—See Dating (radiocarbon).

Radioisotopic dating—See Dating (radioisotopic).

Rain forest—Any forest characterized by a high amount of rainfall annually (e.g., 68 inches [1,750 mm] or higher). Rain forests are largely, though not exclusively, confined to tropical and subtropical regions.

Random walk—A mathematical representation of a sequence that consists of a series of random steps. Random walks are often used as mathematically precise formulations of null hypotheses in biological statistics.

Rate—The ratio of two measurements, quantities, or observations often used to quantify the amount of change that takes place in one variable relative to another. Time is typically used as the referent (denominator) in the calculation of rates.

Recruitment rate—Absolute number of individuals added to a population over a defined time period, often per year.

Red Queen hypothesis—The evolutionary-ecological hypothesis that regards species as having to constantly improve their state of adaptation to local environmental conditions not just to survive but to maintain their state of adaptation relative to other species.

Reef—An organosedimentary structure that stands up above the surrounding floor of submerged aquatic environments and is constructed by the interaction of organisms with their environment.

Refugium—A locality or region, usually characterized by a distinct microenvironment, providing shelter for a relict population of a formerly widespread species exists.

Richness (species)—The number of species that exist in a locality or region at a specified time.

Ruderal species—See Species (ruderal)

Rule (Allen's)—The commonly observed tendency for endothermic animals and species that inhabit colder climates to possess shorter or smaller appendages and/or limbs relative to those from warmer climates.

Rule (Bergmann's)—The commonly observed tendency for endothermic animals and species that inhabit colder climates to exhibit larger adult body sizes relative to those that live in warmer climates.

Rule (Cope's)—The commonly observed tendency for lineages of successive fossil species to increase in body size over time.

Rule (Gloger's)—The commonly observed tendency for endothermic animals and species that inhabit more humid climates to exhibit darker pigmentation relative to those that live in drier climates.

Sample—A small, but representative quantity of, or collection from, a larger entity (e.g., a population) acquired for the purpose of describing the characteristics of, or making inferences about, the larger entity.

Schindewolf, Otto H. (1896–1971)—German professor of geology (University of Tübingen) and director of the Geological Survey of Berlin who is famous for his writings on the tempo and mode of evolutionary change, the "hopeful monster" theory of saltational evolution, and the recognition of the reality of mass extinction in Earth history.

Scrimshaw—Carvings and/or engravings done on bone or ivory, especially with reference to those that use the teeth of sperm whales (*Physeter macrocephalus*) as their medium.

Sea-level change—Any secular and long-term change in the average height of the ocean surface.

Sea-level change (eustatic)—Any change in the global level of the ocean surface relative to a common and global reference point resulting from changes to volume of water in Earth's ocean basins, the volume of the ocean basins themselves, and/or global temperature variations.

Sea-level change (local)—Any change in the local or regional level of the ocean surface as a result of isostatic adjustment, local atmospheric pressure changes, ocean currents, local temperature changes, and/or changes in the volume of local ocean basins.

Secondary type—See Type (secondary)

Shocked quartz—A form of the mineral quartz whose microscopic structure has been altered as a result of being subjected to high pressures but relatively low temperatures. Shocked quartz is identified by the observation of multiple sets of planar deformation features in the crystal. The occurrence of shocked quartz is usually interpreted as direct evidence of an asteroid or comet impact.

Siberian Traps—See Traps (Siberian).

Signor–Lipps effect—A collective term referring to the assumption that the first or last observed occurrence of a fossil species will constitute an underestimate of the true first or last appearance of the fossil species both locally and globally. The Signor–Lipps effect suggests that few inferences can be drawn with confidence regarding the timing of either species or extinction events in Earth's past from the observed patterns of first and last occurrences of fossils in local stratigraphic sections and/or cores.

Southern Oscillation event—A quasiperiodic pattern of sea-surface temperature variation in the eastern Pacific Ocean and surface air pressure variation in the tropical western Pacific Ocean that affects Earth's climate worldwide.

Species—A group of individuals capable of surviving unaided in a natural environment, interbreeding, and producing fertile offspring.

Species (benthic)—Species that live on or within the sediment found at the bottom of a body of water (e.g., ocean, lake).

Species (charismatic)—Species with widespread popular appeal often used by those seeking to call pubic attention to environmental and/or conservation concerns.

Species (Elvis)—Species morphologically similar to species thought to be extinct that have been mistaken for the extinct species on cursory examination by lay observers and/or taxonomists.

Species (endangered)—Species judged to be at a high risk of local or global extinction.

Species (hypothetical extinct)—Species that have been recorded from a specific locality or region but whose existence have not been demonstrated recently by empirical evidence or observation by a qualified taxonomist.

Species (invasive)—Collective term referring either to a species nonnative to an area or region or to a native species that has achieved an artificially large population size as a result of human activity. This introduction/stimulation may be harmful, damaging, or neutral to (other) populations resident in the area.

Species (Lazarus)—Species long thought to be extinct for which viable populations have been rediscovered.

Species (pelagic)—Planktonic or nektonic species that live near the surface of a body of water.

Species (ruderal)—A species that is the first of its type to appear and survive in a recently disturbed habitat. This term is more typically applied to plant species than animal species.

Species (zombie)—This term has been used in two senses. The first usage refers to a fossil specimen (e.g., a dinosaur tooth) that has been washed out of (i.e., reworked from) older sediments and redeposited in sediments that were deposited after the species' extinction (referred to as the "resurrected dead"). The second usage refers to a living species that lacks the ability to reproduce and will become extinct when the last of its current members die (referred to as the "living dead").

Species–area relationship—A descriptive mathematical relation between the size of a habitat and the number of species (richness) living in the habitat. The investigation of the range of natural process that determine the character of this relation has become a biological research program in its own right.

Species Survival Plan (SSP)—A program developed by the Association of Zoos and Aquariums, a US nonprofit organization, designed to maintain the quality of, and coordinate efforts to, implement captive breeding programs to ensure the survival of selected species in zoos and aquaria, especially those species that are endangered in the wild. This program has been running since 1981.

Spermaceti—A waxy substance often found in the head cavities of sperm whales (*Physeter macrocephalus*). There is no consensus as to the function of this substance in sperm whale biology. Before alternatives were discovered, however, spermaceti was used to make candles of a constant photometric value, to dress fabrics, and an excipient in the manufacture of certain pharmaceutical remedies. The economic value of spermaceti, along with other products coming from this whale species, was a reason for its differential targeting by whaling operations.

Stratigraphic unconformity—See unconformity stratigraphic.

Stochasticity—A pattern of results of observations produced by any nondeterministic system or process.

Subtropics—Latitudinally delimited regions immediately north and south of the tropics, usually defined as extending between 23.44° and 40° north and south latitude. This region typically has a mean annual temperature of 50°F (10°C) or above, which supports relatively high plant productivity.

Symbiosis—A collective term for any physiologically close and long-term interaction between two species. There are several different types of symbiosis (e.g., mutualism, commensalism, parasitism).

Tabua—The polished tooth of a sperm whale (*Physeter macrocephalus*) traditionally exchanged by members of Fijian societies as gifts of esteem and/or atonement.

Taxonomic compendia—A concise but comprehensive summary of information about living and/or fossil species. In extinction studies, taxonomic compendia have been instrumental in assessing the dynamics of extinction history over geological time. Their use in this context was pioneered by Otto H. Schindewolf and Normal D. Newell.

Taxonomy—The study of the membership and structure of groups of organisms on the basis of shared characteristics.

Terrestrial—Pertaining to or living either in or on the land.

Tertiary—An older term for the Cenozoic.

Tetrapod—A monophyletic group (superclass Tetrapoda) consisting of all vertebrates that possess four limbs at any stage of their life cycle, including both living and extinct amphibians, reptiles, birds, and mammals.

Thermocline—A marked change in the temperature with depth in a body of water.

Total economic value (TEV)—The value to human populations derived from a natural resource, including both use value (the value of a collectible resource) and nonuse value (the value of a noncollectible resource).

Traps (Deccan)—A large accumulation of multiple layers of solidified flood basalt that covers the west-central portion of India and extends out into the western Indian Ocean. This volcanic activity took place from 65 to 66 million years ago, spanning the time interval within which the end-Cretaceous extinction event occurred. The Deccan Traps is one of the largest volcanic features on Earth.

Traps (Siberian)—A large accumulation of multiple layers of solidified flood basalt that covers the north-central portion of Siberia. This volcanic activity took place from 251 to 250 million years ago, spanning the time interval within which the end-Permian extinction event occurred. The Siberian Traps is one of the largest volcanic features on Earth.

Tree of life—This term is used in two related senses: as a general metaphor illustrating the fact that all species are interconnected and share a common ancestry and as a specific goal of the phylogenetic research program that seeks to discover the specific pattern of genealogical relationships between all species.

Triage—Any process that seeks to establish the priority of actions based on an estimation of the severity of the consequences of not acting. Triage was developed in a medical context and based treatment priorities on the severity of soldier's wounds. The concept has since been generalized, however, to a form of cost–benefit analysis.

Trophic cascade—Any change in the overall trophic structure of a food web that results from the reduction or elimination of a component part of the food web. For example, the removal of an apex predator can result in an increase in the abundance of some herbivore species, a corresponding decrease in the abundance of primary producers, and a consequent decrease in the abundances of herbivores and predators throughout the system. Trophic cascades can occur with either a top-down or bottom-up initial polarity.

Tropical climate—Any area typified by warm to hot conditions and often with large amounts of precipitation that may be confined to distinct intervals during the yearly cycle (e.g., the monsoon season).

Tropics—A latitude-delimited zone encompassing 23.44° north and south of the equator that marks the region within which the Sun appears directly overhead at some point during the year. Given the effects of plate tectonics, continents (and portions thereof) that are outside the tropical zone today may have been within this zone at previous times in Earth's past.

Type (primary)—A specimen on which the description of a species is based; usually the holotype.

Type (secondary)—Any specimen other than the one on which the description of a species is based.

Unconformity (stratigraphic)—An erosional of non-depositional surface that separates two rock bodies of different ages. Presence of an unconformity in a stratigraphic succession means that an interval of time has gone unrecorded in the sequence.

Uniformitarianism—A variant of the principle of actualism positing that all observed phenomena are the result of processes that are operating at present and that Earth history is unendingly cyclical, with, in the words of James Hutton, "no vestige of a beginning, no prospect of an end." Charles Lyell equated uniformitarianism with actualism, and this usage has been maintained by most geologists ever since.

United Nations Environment Programme (UNEP)—Program within the United Nations (UN) that coordinates UN environmental activities and provides advice to member countries related to the implementation of environmentally sound practices along with the development of national and international environmental policy.

U.S. Fish and Wildlife Service (USFWS)—An agency of the US federal government administratively located within the U.S. Department of the Interior and responsible for the conservation and management of fish and wildlife populations along with their natural habitats for the benefit of US citizens.

Urban sprawl—The development of low-density residential communities on rural land around major urban areas as a consequence of the proliferation of automobiles and systems of public transportation that enable workers to commute long distances to and from the urban center.

Vagility—A measure of the extent to which a species can move about freely within its environment and/or migrate to different environments.

Value (instrumental)—A value assigned to any object (either physical or hypothetical) that reflects the ability of the object to aid the user in achieving some other end (e.g., the manufacture of another product). The value of an object as a means to an end. Instrumental value is the opposite of intrinsic value.

Value (intrinsic)—A value assigned to any object (either physical or hypothetical) for its own sake, as an intrinsic property of the object. The value of an object as an end in itself. Intrinsic value is the opposite of instrumental value.

Vertebrate—A monophyletic group consisting of all animals that possess the ability to develop a vertebral column as part of their body's axial structure.

Volcanic eruption—See Large igneous province.

Western Interior (Pierre) Seaway—A large epicontinental sea that occupied the middle of the North American continent at times during the Late Cretaceous; also referred to as the Cannonball Sea.

Whale oil—A naturally occurring oil sourced from the blubber of whales. Prior to the development of alternatives such as kerosene and vegetable oils, whale oil was used in oil lamps and to make soap and margarine. Demand for whale oil in Europe and North America drove development of the whaling industry in the 1800s and 1900s and reduced whale populations to dangerous levels globally.

Zombie species—See Species (zombie).

Zooplankton—Any heterotrophic member of the plankton.

Zooxanthellae—An older, colloquial name for any golden-brown endosymbiont residing in the tissues of foraminifera, sponges, cnidarians, flatworms, and mollusks. Examples include diatoms and dinoflagellates. The dependence of the host organism on its endosymbionts limits or restricts the range of environments it can inhabit.

Index

• • • • •

A

Abramovitz, Janet N., 1:160–161
Abrupt extinctions
  Australia, 2:580
  end-Cretaceous extinction theories, 2:508
  stratigraphic confidence intervals, 1:9–10
  trilobites, 1:236
Abundance
  adaptive cluster sampling, 1:154
  Cambrian trilobites, 1:235
  dinosaurs, 1:281
  fishing and species abundance, 2:801
  life on Earth, 2:896
  rare species monitoring, 1:153–154
  rarity mechanisms, 1:198
  terrestrial *vs.* marine fossil record,
    1:199–200
  time issues in species monitoring, 1:155
  See also Population dynamics
Acadian flycatchers, 2:772
Accidental extinction, 1:120
Acclimatization societies, 2:601
Acid rain, 1:323
Acidification, ocean. *See* Ocean acidification
Acinonyx jubatus, 1:370, 380, 2:760
Acipenseriformes, 2:793
Acroporid corals, 2:686
Actualism
  extinction patterns, 1:198–199
  geological gradualism, 2:694
  Lyell, Charles, 1:3, 36, 41
Adams, Michael Friedrich, 1:38
Adams, Robert McCormick, 2:761
Adansonia grandidieri, 2:591
Adaptation
  background extinctions, 1:6
  climate change, 2:739–741
  Darwin, Charles, theories of, 1:41
  dinosaur extinction theories, 1:283
  escalation hypothesis, 1:122
  geographic range, 1:8
  mammal-like traits in non-mammalian
    synapsids, 1:260–261
  marine reptiles, 1:269, 269–270
  Ordovician faunas, 1:438–439
  Ordovician trilobites, 1:236–239
  Red Queen hypothesis, 1:120–122
Adaptive cluster sampling, 1:154, 154f
Aeromonas, 1:340
Aetokremnos, 1:182–183, 183

Aetosaurs, 1:482
Afghanistan, 2:755
Africa
  bushmeat, 2:813–820
  hominin species, 1:295
  late Quaternary, 1:366
  politics and policies, 2:881–882
  primates, 1:409, 412–413
  wildlife underpass, 2:775
African megaherbivores, 1:394, 394f
  body size and extinction vulnerability, 1:389–391
  causes of extinctions, 1:391–392
  elephants, 1:66, 389
  late Pleistocene, 1:387(f2), 388t
  Pleistocene, 1:385–387, 389
  Red List, 1:393t
  relative biomass, 1:382f
  reproductive rates, 1:390f
  taxonomic diversity, 1:386t
  timing of extinctions, 1:387(f1)
  See also Megafauna
African pygmy hippopotamuses, 1:180, 180–181
Agasiz's diagram, 1:77f
Agassiz, J. Louis R., 1:76
Age of Armored Fishes, 1:452
Age structure
  fishing, effect of, 2:797
  hunting, effect of, 2:817
  population models, 1:136–137
Agreement on the Conservation of Gorillas and
  Their Habitats, 2:882
Agriculture, 2:761
  Afghanistan, 2:755
  chemicals, 1:323
  conservation efforts, 1:32
  coral reefs, pollution of, 2:684
  Holocene oceanic island extinctions, 2:663
  industrial era, 2:761–765
  island faunas extinctions, 1:181
  Madagascar, 2:592
  modern biodiversity crisis, 2:620, 625, 628, 630
  prehistoric, 2:760–761
  transition from hunting to, 2:759–760
  tree extinctions, 1:421
Aichi Biodiversity Targets, 1:88
Aigialosaurs. *See* Mosasauroids
Akotiri *Aetokremnos*, 1:182–183, 183
Alaska, 2:538, 886
Albatross, 2:630
Alberta, Canada, 1:287

Alcover, Josep Antoni, 2:606–607
Algae, 2:525–526
Algeo, Thomas J., 1:205
Algific talus slope, 1:146–147, 147(f2)
Aliasing in species monitoring, 1:155
Alievium, 1:12f
Allee effect, 1:135–136, 136f, 2:782
Allen's rule, 2:740
Alligators, 1:348, 352
Allocation of resources, 2:833–834, 837–839
Allopatric speciation, 2:904
Alroy, John, 1:106
Altai Mountains, 1:145
Alvarez, Luis W., 1:247, 2:508, 693–694
Alvarez, Walter, 2:508, 693
Amazon region
  bushmeat, 2:817–818, 818f
  deforestation, 1:421, 421, 2:675
  tropical rain forests, 2:674
Amazon River dolphins, 1:403, 403
Ambiguous gaps, 1:79
American alligators, 1:348, 352
American bison, 1:163–164, 164, 2:879–880, 880
American crocodiles, 1:348
American martens, 2:752, 752, 753
American mastodons, 2:551
American oystercatchers, 1:115–116, 116
American West, 1:37
Ammonites, 1:39, 247
  Brahmaites brahma, 1:248
  Cambrian extinctions, 1:427
  clots, 1:54, 54f
  Diplomoceras cylindraceium, 1:251
  diversity, changes in, 1:246
  early ideas about fossils, 1:36
  end-Cretaceous extinctions, 1:56f, 2:506–507
  end-Jurassic diversity, 2:488
  end-Jurassic extinctions, 2:489f, 490–493,
    490f, 491
  extrapolation techniques, 1:59
  habitat, 1:246
  Hoploscaphites constricturs, 1:250
  Pachydiscus neubergicus, 1:247
  Phylloceras surya, 1:251
  Pseudophyllites indra, 1:248
  Sphenodiscus binckhorsti, 1:249
  Triassic–Jurassic boundary, 1:47
  unbinned range-end sequences, 1:56–57,
    57(f6)
  Zumaya ammonite data, 1:9–11

extinction rates based on, 1:106
   geologic time scale 2012, 1:44–48
   sea-level change, 2:722–723
   terrestrial vertebrates, 1:188
Stratigraphic *vs.* phylogenetic analysis, 1:14
Streams and rivers. *See* Freshwater fish
Strix occidentalis caurina, 2:880–881
Structured populations, 1:115–117
Sturgeons, 1:*319*, 2:793, 798
Sturnella neglecta, 1:*113*
Sub-sampling, 1:57–58, 58*f*
Subfossil lemurs, 1:*410*, 410–411
Subrecent oceanic island extinctions, 2:659–665
Subsidies, 2:736, 764, 766, 838
Subsistence fishing, 1:325
Subspecies
   bird extinctions, 1:357
   endangered African megaherbivores, 1:394
   endangered marine mammals, 1:399–404
   large cats, 1:375–376
   leopards, 1:378–379, 381
   marine mammal extinctions, 1:397
   pumas, 1:379–380
   taxonomic compendia, 1:74
   tigers, 1:377
Suddenness of extraterrestrial impacts, 2:697
Suez Canal, 2:*784*, 784–785, 786
Sulfur
   large igneous province volcanism, 2:708–710
   sea-level change, effects of, 2:719
Sumatran orangutans, 1:*369*, *414*, 415
Sumatran tigers, 1:382, *382*
Sumbar, 1:250
Sun, 2:892–893, 896–898
Supernovas, 2:899
Surfaces, hardening of, 1:99
Surlyk, Finn, 2:505
Survival in unexplored regions theories, 1:7
"Survival of the unfit," 2:902
Survivorship
   age-structured population models, 1:136
   Cambrian extinctions, 1:428
   current thriving species, 2:628
   cynodonts, 1:264
   end-Ordovician, 1:435, 438–440
   end-Triassic extinction survivors, 1:482–483
   extinction rates, 1:104, 107–109
   extraterrestrial impacts, 2:697
   graph for general carnivorous placental
      mammals, 1:119*f*
   human evolution, 2:901–902
   late Devonian extinction, 1:451
   late Quaternary extinctions, 2:536, 540, 541
   nannofossil species, 2:503
   planktonic foraminifera, 2:504
   stratigraphic completeness, 1:58
   survivorship curves, 1:80, 80*f*, 119–123
Susceptibility to extinction. *See* Extinction
   vulnerability
Sustainable development, 2:774
Sustainable economics. *See* Ecological economics
Sustainable fishing, 2:649, 792, 796–797
Sustainable hunting, 2:817–818, 818*f*, 820
Sustainable scale, 2:829–832, 835–836, 839–841
Susus, 1:403
Suzuki, David, 2:875
Svensen, Henrik, 2:517
Swamp flora survivors, 1:303
Sweden, 2:775–776
Sweet, Arthur R., 1:303

Swenson, Jon E., 2:783
Swimming
   ammonite extinctions, 1:251–252
   island faunas, 1:178
   marine reptile adaptations, 1:269–270
Symbols, 1:165
Synergies, 1:101, 2:745
Systema Naturae (Linnaeus), 1:73
Systematic random sampling, 1:151*f*
Systematics, natural history museums,
   1:230–232

T

Tabarelli, Marcelo, 1:422
Taghanic onlap, 1:240
Takahes, 2:602, 603
Taking, 1:166–168
Tamarins, 1:*416*, 2:574
Tammar wallabies, 2:783, *783*
Tampa Bay, Florida, 1:315
Tana River red colobus monkey, 1:412
Tanabe, Kazushige, 1:252
Tanner, Lawrence H., 1:475
Tapinocephalus Assemblage Zone, 1:262, 263
Tapirs, 2:569
Tarsiers, 1:413
Tarsium tumpara, 1:413
Tasmanian devils, 2:582
Tasmanian tigers, 2:581, *581*
Taxa
   invasive species, susceptibility to, 2:785
   last appearances, documenting, 1:187
   missing, 1:53, 53*f*
   phylogenetic analysis, 1:14
   plants, dominance exchange in, 1:201–202, 202*f*
   survivorship and within-lineage extinction
      rates, 1:107–109
   taxonomic ranks in compendia, 1:78–79
Taxation, 2:770–771, 836, 838
Taxonomic compendia
   Books of Life, 1:73–74
   data quality, 1:76, 78–79
   early use of in the earth sciences, 1:76, 78*f*
   extant species estimations, 1:79–80
   marine invertebrates, 1:187–188
   neontological *vs.* paleontological use, 1:75–76
   Paleobiology Database, 1:191
   terrestrial vertebrates, 1:189
   uses of, 1:74–75
Taxonomic diversity, 1:439
Taxonomic extinction, 1:12–13, 78–79
Taxonomic hierarchy, 1:75
Taxonomic keys, 1:74
Taxonomic opinions, 1:74
Taxonomic partitioning, 1:8
Taxonomic patterning, 1:11–14
Taxonomy
   African ungulate diversity, 1:386*t*
   amphibians, 1:337
   Books of Life, 1:73–74
   character complexity, 1:22
   fishes, 1:317
   large cats, 1:375–376
   natural history museums, 1:230–232
   Red List bias, 1:89
   trait extinction, 1:17–19
Taylor, David Robert, 2:875

Taylor, Martin F. J., 1:169
TDWG. *See* Biodiversity Information
   Standards group
Technical Guidelines on the Management of Ex situ
   Populations for Conservation (IUCN), 1:209
Technology
   human evolution, 2:901–904
   natural history museums, 1:231–232
Tectonic chance, 1:59
Tectonic movement, 1:441–443
Teeth
   hominin species, 1:292
   ichthyosaurs, 1:272
   non-mammalian synapsids, 1:256
   shark tooth spiral, 1:*330*
Teilhardina brandti, 2:519, *519*
Tellico Dam, 2:868
Temnospondyl amphibians, 1:467, 482, *482*
Tempo and Mode in Evolution (Simpson), 1:76
Temporal bins, 1:103
Temporal patterning, 1:8–11
Tennessee Valley Authority v. Hill, 1:167
Terminology. *See* Definitions and terminology
Terocephalians, 1:263
Terrestrial fauna
   end-Permian extinctions, 1:467, 470
   end-Triassic extinctions, 1:481–482
   Eocene–Oligocene extinctions, 2:526–527, *527*
   late Paleocene, 2:517
Terrestrial mammals
   bushmeat availability, 2:815
   Red List threat patterns, 1:89*f*
Terrestrial species
   diversification, by major organism groups,
      1:201(*f*3)
   diversity models, 1:203–204
   extinctions since 1850, 2:652–654
   Red List bias, 1:89
Terrestrial to marine adaptation, 1:*269*,
   269–270
Terrestrial vertebrates
   birds, 2:*620*
   diversity curve, 1:189*f*
   end-Cretaceous evolution, 2:509
   end-Cretaceous extinctions, 2:497–500
   extinction, patterns of, 1:191–192, 194
   fossil record, 1:188–189
   New Zealand extinctions, 2:*596*, 599*t*–600*t*
   patterns of extinction, 1:191–192, 194
   Red List, 1:84(*f*2)
   in zoos, 1:210*f*
   See also Non-mammalian synapsids
Terrestrial *vs.* marine extinction patterns
   actualism, 1:198–199
   data quality, 1:201
   ecological effects of taxonomic loss,
      1:200–201
   ecological succession, 1:197–198
   external factors, 1:205
   extinction cascades and collateral damage,
      1:199
   extinction comparisons, 1:201–203
   extinction realms, 1:205
   fossil record, 1:199–200
   intrinsic factors, 1:203–205
   Permian–Triassic boundary extinctions,
      1:205*f*
Tethyan ammonites, 2:490
Tetraceratops, 1:257
Tetrapodomorphs, 1:449–451, *451*